Nutrition and Gene Expression

Edited by

Carolyn D. Berdanier, Ph.D.
James L. Hargrove, Ph.D.

Department of Foods and Nutrition
University of Georgia
Athens, Georgia

CRC Press
Boca Raton   Ann Arbor   London   Tokyo

Library of Congress Cataloging-in-Publication Data

Nutrition and gene expression / editors, Carolyn D. Berdanier, James
  L. Hargrove.
      p.   cm.
    Includes bibliographical references and index.
    ISBN 0-8493-6961-4
    1. Nutrition.   2. Genetic regulation.   3. Gene expression.
  I. Berdanier, Carolyn D.   II. Hargrove, James L.
  QP144.G45N87 1992
  612.3′9—dc20   92-16664
                                                                      CIP

© 1993 by CRC Press, Inc.

International Standard Book Number 0-8493-6961-4

Library of Congress Card Number 92-16664

Printed in the United States of America   1 2 3 4 5 6 7 8 9 0

Printed on acid-free paper

PREFACE

The science of nutrition has changed radically since its inception. It has evolved from the discovery of nutrients which cured well-described diseases to a recognition that appropriate food choices can delay the development of a variety of degenerative diseases. In the past, nutritionists worked hard to devise recommendations to the public for nutrient intakes that would ensure the absence of deficiency diseases. In the future, nutritionists will make nutrient intake recommendations not on such characteristics as sex, height, weight or age, but on the basis of the individual's phenotypic expression for health while suppressing that individual's phenotypic expression for disease. We are a long way from being able to make these recommendations. However, as a first step we are beginning to recognize the interaction of specific nutrients with the genetic code possessed by all nucleated cells. This book and its sequels will provide the reader with some of the insights gained by foremost scientists in the field of nutrient-gene interactions. Some of these interactions have been researched in depth whereas others are in their infancy. As the reader proceeds through the book, one hopes that the imagination and dedication of the scientists who have contributed to this volume will be appreciated. To these authors who struggled to meet our stringent deadline, and who wrote of their life's work, we give our thanks.

University of Georgia
Athens, Georgia

Carolyn D. Berdanier
James L. Hargrove

EDITORS

Carolyn D. Berdanier, Ph.D., is a Professor of Nutrition at The University of Georgia in Athens, Georgia. She received a B.S. degree from the Pennsylvania State University and the M.S. and Ph.D. degrees from Rutgers University in Nutrition in 1966. After a post-doctoral fellowship year with Dr. Paul Griminger at Rutgers, she served as a Research Nutritionist with the Human Nutrition Institute which is part of ARS, a unit of the U.S. Department of Agriculture. In 1975 she moved to the University of Nebraska College of Medicine where she continued her research in nutrient gene interactions. In 1977 she moved to the University of Georgia where she served as Head of the Department of Foods and Nutrition. She stepped down from this post ten years later and devoted her full time efforts to research and teaching in her research area. Her research on the diet and genetic components of diabetes and vascular disease has been supported by NIH, USDA, U.S. Department of Commerce, The National Livestock and Meat Board and the Egg Board. She is a member of the American Institute of Nutrition, the American Society for Clinical Nutrition, The Society for Experimental Biology and Medicine and several honorary societies in science. She has served on the Editorial Boards of the *FASEB Journal, The Journal of Nutrition, Nutrition Research* and *Biochemistry Archives*. She is also a Contributing Editor for *Nutrition Reviews* and Editor of the AIN News Notes. Current research interests include studies on aging, the role of diet in damage to mitochondrial DNA, and the role of specific dietary ingredients in the secondary complications of diabetes.

James L. Hargrove, Ph.D., is an Associate Professor of Nutrition at The University of Georgia in Athens, Georgia. He received a B.S. in Zoology from the University of Washington in Seattle, Washington in 1969, and earned a doctorate in Physiology from Utah State University in 1975. His interest in nutritional and hormonal effects on synthesis of mRNA and enzymes was stimulated by post-doctoral training with Daryl K. Granner, M.D., in the Diabetes and Endocrinology Research Center at the University of Iowa. Dr. Hargrove joined the Department of Anatomy and Cell Biology at Emory University School of Medicine in 1982, and was promoted to Associate Professor. He joined the Department of Foods and Nutrition at The University of Georgia in 1989.

Dr. Hargrove's studies in enzyme regulation and the kinetics of gene expression have been supported by grants from the N.I.H., the American Diabetes Association, the American Cancer Society, and the National Kidney Foundation. He is a member of the American Institute of Nutrition and the American Society for Biochemistry and Molecular Biology. Current research interests include computer-based modeling of gene expression and metabolic control, and effects of dietary protein on organ growth and gene expression.

CONTRIBUTORS

Victoria E. Allgood, Ph.D.
Department of Cell Biology
Baylor College of Medicine
Houston, Texas

John B. Allred, Ph.D.
Department of Food Science and
  Technology
The Ohio State University
Columbus, Ohio

Roderick A. Barke, Ph.D.
Department of Food Science and
  Nutrition and Surgery
University of Minnesota
St. Paul, Minnesota

Carolyn D. Berdanier, Ph.D.
Department of Foods and
  Nutrition
The University of Georgia
Athens, Georgia

Diana F. Bowers, Ph.D.
Department of Food Science and
  Technology
The Ohio State University
Columbus, Ohio

Linda J. Brady, Ph.D.
Department of Food Science and
  Nutrition
University of Minnesota
St. Paul, Minnesota

Paul S. Brady, Ph.D.
Department of Food Science and
  Nutrition
University of Minnesota
St. Paul, Minnesota

Charles O. Brostrom, Ph.D.
Department of Pharmacology
Robert Wood Johnson
  Medical School
Piscataway, New Jersey

Margaret A. Brostrom, Ph.D.
Department of Pharmacology
Robert Wood Johnson
  Medical School
Piscataway, New Jersey

Hans A. Büller, M.D., Ph.D.
Department of Pediatrics
Academic Medical Center
Amsterdam, The Netherlands

David K. C. Chan, Ph.D.
Department of Foods and
  Nutrition
The University of Georgia
Athens, Georgia

Sylvia Christakos, Ph.D.
Department of Biochemistry and
  Molecular Biology
University of Medicine and
  Dentistry of New Jersey
New Jersey Medical School
Newark, New Jersey

John A. Cidlowski, Ph.D.
Professor
Departments of Physiology,
  Biochemistry and Biophysics
University of North Carolina
Chapel Hill, North Carolina

Steven D. Clarke, Ph.D.
Professor
Department of Food Science and
  Human Nutrition
Colorado State Univerity
Fort Collins, Colorado

Robert J. Cousins, Ph.D.
Center for Nutritional Sciences
University of Florida
Gainesville, Florida

Nicholas O. Davidson, M.D.
Associate Professor
Department of Medicine
University of Chicago
Chicago, Illinois

Jan Dekker, Ph.D.
Department of Pediatrics
Academic Medical Center
Amsterdam, The Netherlands

R. S. Eisenstein, Ph.D.
Department of Nutritional
  Sciences
College of Agriculture and Life
  Science
University of Wisconsin
Madison, Wisconsin

Nick Gekakis, B.S.
Student
Department of Nutrition
Harvard School of Public Health
Boston, Massachusetts

Rajbir K. Gill, Ph.D.
Department of Biochemistry and
  Molecular Biology
University of Medicine and
  Dentistry of New Jersey
New Jersey Medical School
Newark, New Jersey

Howard P. Glauert, Ph.D.
Associate Professor
Department of Nutrition and Food
  Science
University of Kentucky
Lexington, Kentucky

Richard J. Grand, Ph.D.
The Floating Hospital
New England Medical Center
Tufts University
School of Medicine
Boston, Massachusetts

James L. Hargrove, Ph.D.
Associate Professor
Department of Foods and
  Nutrition
The University of Georgia
Athens, Georgia

Ann Jerkins, Ph.D.
Research Fellow
Department of Nutrition
Harvard School of Public Health
Boston, Massachusetts

Thomas L. Jetton, Ph.D.
Research Associate
Department of Molecular
  Physiology and Biophysics
Vanderbilt, Tennessee

Donald B. Jump, Ph.D.
Department of Physiology
Michigan State University
East Lansing, Michigan

Z. Kikinis, Ph.D.
Human Nutrition Research Center
  on Aging
United States Department of
  Agriculture
Boston, Massachusetts

Rolf F. Kletzien, Ph.D.
Metabolic Diseases Research
The Upjohn Company
Kalamazoo, Michigan

Gerald J. Lepar, Ph.D.
Department of Physiology
Michigan State University
East Lansing, Michigan

Jane B. Lian, Ph.D.
Professor
Department of Cell Biology
University of Massachusetts
Medical Center
Worcester, Massachusetts

Pauline Kay Lund, Ph.D.
Associate Professor
Department of Physiology and
  Pediatrics
University of North Carolina
Chapel Hill, North Carolina

Ormond A. MacDougald, Ph.D.
Department of Biological
  Chemistry
Johns Hopkins University
Baltimore, Maryland

Mark A. Magnuson, M.D.
Associate Professor
Department of Molecular
  Physiology and Biophysics
Vanderbilt University
Nashville, Tennessee

Robert K. Montgomery, Ph.D.
The Floating Hospital
New England Medical Center
Tufts University School of
  Medicine
Boston, Massachusetts

Naima Moustaid, Ph.D.
Department of Nutrition
Harvard School of Public Health
Boston, Massachusetts

H. M. Munro, Ph.D.
Human Nutrition Research Center
  on Aging
United States Department of
  Agriculture
Boston, Massachusetts

Tamio Noguchi, Ph.D.
Assistant Professor
Department of Nutrition and
  Physiological Chemistry
Suita, Osaka, Japan

Antoine Puigserver
Center de Biochemie et de
  Biologie Moleculaire
CNRS
Marseille, Cedex, France

Edmond H. H. M. Rings, M.D.
Department of Pediatrics
Academic Medical Center
Amsterdam, The Netherlands

Kenji Sakamoto, Ph.D.
Department of Nutrition
Harvard School of Public Health
Boston, Massachusetts

Richard I. Schwartz, Ph.D.
Division of Cell and Molecular
  Biology
Lawrence Berkeley Laboratory
Berkeley, California

Neil F. Shay, Ph.D.
Center for Nutritional Sciences
University of Florida
Gainesville, Florida

Cynthia Smas, M.S.
Graduate Student
Department of Nutrition
Harvard School of Public Health
Boston, Massachusetts

Gary S. Stein, Ph.D.
Professor
Department of Cell Biology
University of Massachusetts
Worcester, Massachusetts

Hei Sook Sul, Ph.D.
Associate Professor
Department of Nutrition
Harvard School of Public Health
Boston, Massachusetts

J. W. Suttie, Ph.D.
Departments of Biochemistry and
  Nutritional Sciences
College of Agricultural and Life
  Sciences
University of Wisconsin-Madison
Madison, Wisconsin

Takehiko Tanaka, M.D., Ph.D.
Professor
Department of Nutrition and
  Physiological Chemistry
Osaka University Medical School
Suita, Osaka, Japan

Douglas B. Tully, Ph.D.
Department of Physiology
University of North Carolina
Chapel Hill, North Carolina

Malcolm Watford, Ph.D.
Department of Nutritional
  Sciences
Cook College
Rutgers
New Brunswick, New Jersey

Catherine Wicker-Planquart
Centre de Biochimie et de
  Biologie Moleculaire
CNRS
Marseille, Cedex, France

TABLE OF CONTENTS

Chapter 1

NUTRIENT RECEPTORS AND GENE EXPRESSION

James L. Hargrove and Carolyn D. Berdanier

TABLE OF CONTENTS

0-8493-6961-4/93/$0.00 + $.50

I. INTRODUCTION

The scope of potential nutrient-gene interactions is enormous. More than 40 nutrients are known to be essential to mammals, and health may be impaired by a deficit in any one of them.[1] Normal growth, reproduction, immune function, and lifespan require adequate intakes of foods containing these essential dietary constituents. Just as deficient intakes can be deleterious, excessive intakes can also be harmful, as is true for vitamins A and D, selenium, iron, and total caloric intake. Genes are thought to play a fundamental role in adaptation to the day-to-day variability in nutrient supply, such that the nutritional status and health of the individual can be relatively constant despite the variable supply. That compensatory mechanisms exist is not in doubt.[2-5] However, the means by which individual nutrients are detected by cells, and the details of how altered nutrient supply leads to a reprogramming of the genetic apparatus, are still lacking in many important cases.

Not only is the current explosion in genetic technology increasing our understanding of how genetic mechanisms participate in monitoring nutrient supply and utilization, it is also clear that genes participate in higher order processes such as food-seeking behavior. Ingestion of nutrients quickly activates complex neural and endocrine circuits that affect gene expression not only in the gut and peripheral tissues, but also in the brain. Genes are activated as a means of feedback control to deal with undersupply and oversupply of particular nutrients, which requires the central nervous system to integrate internal, metabolic cues and environmental conditions.

The purpose of this chapter is to provide a context for the topic of nutrient-gene interactions, starting with the question of how the transducing systems that mediate such effects may have arisen during evolution. Even though few mammalian systems have been characterized that show direct effects of nutrients on genes, numerous examples are under study in microorganisms, and some of these are described to illustrate some of the means by which nutrients may affect genetic activity in the absence of hormonal mediators. Next, the presence of nutrient-sensitive chemoreceptors in mammals is discussed, including neural and endocrine mechanisms. To complete this general introduction, temporal and quantitative aspects of gene expression in mammals are reviewed.

II. THE EVOLUTION OF GENETIC CONTROL MECHANISMS

Evolutionary theory postulates that present-day genetic mechanisms are inherited from ancestral patterns that have been modified by mutation, variation, and natural selection during competition for limited resources. Foremost among those limited resources are the nutrients, which are defined as any chemical substances that are necessary to life processes such as cellular metabolism, reproduction, motility, and sensation. Because the availability of nutrients has always been sporadic, organisms that are best able to detect, take up, and assimilate nutrients have the best chance of contributing their genotype to the next generation so that it is maintained in the population.

Implicit in the theory of evolution is the idea of feedback control; living organisms can not function optimally unless they are able to detect and use information concerning the environment. The nutrients found in food can not be detected unless they interact specifically with appropriate components, such as ligand-binding proteins that may be linked to processes for their uptake or utilization. The need to detect concentrations indicates that nutrients serve two distinct functions: one which is a *metabolic* function and one which is *informational.* These two functions may be related; for example, glucose both provides energy to the pancreatic β-cells and generates a signal that promotes insulin secretion. The functions may also be separate, as in the use of glutamic acid for neuronal protein synthesis or for neurotransmission.

Appropriate regulation of genes in response to diet requires two types of controlling elements. The first is a means of sensing the level of nutrient that is present; this is the function of *sensors* or *receptors,* which may detect one particular nutrient, members of one class of nutrients, or hormones secreted in response to the nutrients. We are most familiar with this aspect of nutrients through the senses of taste and smell, by which neuronal chemoreceptors detect the presence of sugars, amino acids, salts, and other substances. Second, there must be a *transducing mechanism* by which the level of nutrient affects cellular activities, including the activity of genes that regulate the processes by which nutrients are used.

How might organisms have acquired the capacity to bind nutrients with specificity and to use them in feedback control circuits? The exon-shuffling hypothesis suggests that discrete segments of DNA encode units of structure and function in mRNAs and proteins. These unitary elements may be recombined to produce new molecules with two or more structural or functional elements. This theory is now being applied to analyze functional elements in genes that encode nutrient receptors and enzymes that participate in nutrient metabolism. For instance, plasma membranes of hepatocytes and Kupffer cells contain lectins, or receptors that bind carbohydrates.[6] The carbohydrate-binding portion of the proteins is located at the carboxylic terminals and is encoded by three exons at the 3′ end of the gene. However, in the related mannose-binding protein, a single exon encodes the binding function. Separate exons encode a transmembrane domain and other structural elements.

Current theory based on the exon-shuffling hypothesis suggests that most mammalian proteins belong to superfamilies whose members originated by recombination of exons that encoded short functional domains. Subsequent mutations and recombination then caused divergence among family members. Another example of this mechanism is the gene for the vitamin D-binding protein; the gene contains 13 exons.[7] This protein contains a single binding site for sterols such as 25-hydroxy D_3 that is thought to be encoded at the 5' end of the gene on exons 1 and 2. The protein is multifunctional, and a region that binds actin may be encoded on exon 10.[7] Vitamin D-binding protein is thought to be related to albumin and α-fetoprotein. Ray et al. have proposed that the latter proteins may have been generated from an ancestral gene that resembled the gene for vitamin D-binding protein.[7]

The rise of multicellular organisms permitted functionally distinct groups of cells to be formed, including some that participated in storing or mobilizing nutrient reserves. Functional specialization required that means for communication among groups of cells be developed, so that a signaling system would permit nutrients to be withdrawn as needed by metabolically active cells. Although secretion of metabolites and other small molecules may have served for communication at first, finer controls eventually developed that gave rise to paracrine and endocrine signals, intracellular signal transduction, and the specialization of cells to coordinate the transfer of information.

Even though endocrine and neural mechanisms have supplanted nutrients as primary signals in mammalian signaling networks, nutrient availability still represents the primary signal for activating or inactivating most genetic networks in unicellular organisms. It is also likely that neural and endocrine signals in multicellular organisms have not entirely replaced the early genetic mechanisms, because neurotransmitters and hormones cooperate with nutrients in directing nutrient utilization. For these reasons, it is pertinent to review the components of genetic feedback systems and the means by which nutrients regulate genetic activity in unicellular organisms.

III. ELEMENTS OF GENETIC CONTROL

All information that affects gene transcription converges on regulatory elements in chromosomes. The *promoter* is the segment of DNA that is responsible for binding the enzyme, RNA polymerase II, plus general transcription factors IIA-IIF, and other DNA-binding proteins that are necessary to initiate transcription in mammals[8] (Figure 1A). Because promoters are located on the same strand of DNA as the structural genes that encode messenger RNAs, they are referred to as *cis-acting elements*. The structural genes are regulated by *trans-acting factors* that are products of separate genes, which may be located on different chromosomes. *Trans*-acting factors are usually, but not exclusively, proteins. Promoters for different genes compete for these factors, and it has been customary to assign genes to two classes according to promoter strength, an index of the ability to utilize transcription factors. Genes that are active at low levels in most cells are referred to as "housekeeping genes"; with few exceptions, such genes contain no TATA box (the site at which RNA polymerase binds), but contain multiple binding sites for factor Sp1.[8-10] Although housekeeping genes were once considered to be expressed constitutively (at relatively fixed rates), it is now apparent that

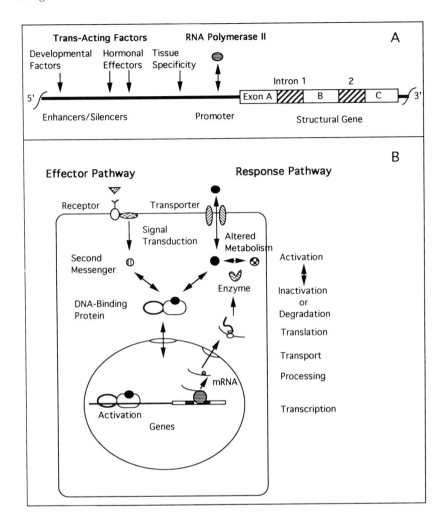

FIGURE 1. (A) Components of a typical, mammalian gene include the promoter element, which binds RNA polymerase II and related transcription factors; enhancer and silencer elements; and the structural gene, which is divided into exons (the mRNA coding segments) and introns (segments that are removed during processing of the primary transcript as mature mRNA is formed). Trans-acting factors include proteins (and possibly RNA molecules) that modify the rate of transcription by binding to regulatory elements in the gene. (B) Responses of genes to nutrients require an effector pathway (left side of figure) that confers specificity of response according to cell type and kind of nutrient, and a response pathway (right side of figure) that may be affected at any of several levels by the nutrient or its metabolites.

most genes are subject to multiple controls. However, genes that are fully active in only one or a few tissues tend to be transcribed at a higher basal rate, contain a TATA element, and respond actively to signal molecules such as hormones and second messengers. Such genes are said to be ''inducible'', and the processes by which enzymes and mRNAs increase to new levels are called *induction*. This process occurs in response to proteins that bind to regulatory portions of genes that may be adjacent to the promoter, but can also occur at long distances away in either direction. These regulatory segments are termed *enhancer elements* if they increase transcription, and *silencer elements* if they decrease transcription. By affecting proteins that bind to these parts of genes, nutrients and hormones may provide specificity of genetic responses.

IV. THE IMPORTANCE OF RECEPTORS IN NUTRIENT-GENE INTERACTIONS

In order to produce specific responses of any kind, nutrients or their metabolites must interact with cellular components termed *receptors*. A receptor should be capable of binding to a nutrient or metabolite with high specificity and with an affinity that is appropriate to physiological concentrations of the ligand. Binding should then initiate a change in cellular function, possibly by altering conformation of the receptor protein or by activating a linked enzymatic activity. For instance, receptors for amino acids such as glutamic acid and glycine in mammalian nerves are coupled to ion channels and transducing mechanisms that alter neuronal membrane potential, calcium metabolism, and phosphoinositide turnover.[11,12] Stimulation of the glutamatergic system not only alters electrical potential, but also activates transcription from cellular oncogenes and can overstimulate the neuron and bring about its death.

The simplest example of a receptor affecting gene expression would be a protein whose binding to specific segments of DNA is altered in the presence of the nutrient, its metabolites, or second messengers produced in response to the nutrient (Figure 1B). Numerous examples of this type of response have been characterized in bacteria and yeast, but few examples have been demonstrated in mammals. Metal ions represent an exception; the ability of zinc and other ions to alter transcription provides a well-known example.[13] Similarly, the ability of iron to alter the stability or translation of the mRNAs encoding the transferrin receptor and ferritin depends upon a protein that binds to the 3' and 5' ends of the mRNAs.[14] Although less direct than effects on transcription, the result in each case is an altered level of a protein that affects iron accumulation.

Receptor systems for many nutrients may be much more complex than those for metal ions, as is true of the system in the β cell of the pancreas that regulates output of insulin.[15] Evidence suggests that glucose and its metabolites initially alter insulin synthesis at the translational level and trigger insulin secretion through a complex system that includes effects on energy metabolism, synthesis of cyclic adenosine 3',5'-monophosphate (cAMP), ion transport, and numerous other cellular processes. One potential component of the transducing system for glucose has been suggested to be the enzyme, glucokinase.[15] In addition, binding of glucose or its metabolites to plasma membranes prepared from an islet-cell tumor decreases the activity of calcium channels over physiological concentrations.[16] Whatever intervening mechanisms may be involved, the result of elevated glucose is increased secretion of insulin from the β cell and decreased secretion of glucagon from the α cell. For cells elsewhere in the body, changes in glucose levels are accompanied by appropriate changes in the ratio of insulin to glucagon. Therefore, many genetic consequences of ingesting nutrients are mediated by the endocrine system, for which receptors and transducing systems are well defined. Even so, the presence of nutrients often modifies the response to hormones.

V. NUTRITIONAL REGULATION OF GENES IN UNICELLULAR ORGANISMS

A. NUTRIENT-GENE INTERACTIONS IN BACTERIA

The first genetic responses to nutrients were identified in bacteria and fungi in the late 19th century (refer to Monod's review[17] of early studies). In testing the ability of the mold, *Aspergillus,* to grow on different substances, Duclaux found that cultures at first grew slowly when protein-containing media were substituted for carbohydrate-containing media, but then grew rapidly after a few days of acclimation. He showed that this mold secreted proteases at an elevated rate after the treatment, and that the type of enzyme produced was appropriate to the nutritional substrate.

In microorganisms, sugars, amino acids, and inorganic compounds, such as phosphate ion, initiate or terminate transcription from specific genes by affecting appropriate DNA-binding proteins. Unicellular organisms monitor available nutrients, activate those genes that will provide enzymes capable of synthesizing all needed components from the available nutrient sources, and repress transcription from unneeded genes. Unlike mammals, bacteria and yeast have the capacity to synthesize all of the amino acids and enzyme cofactors from simple sugars, ammonia, and ions present in the substrate.

The idea that a regulatory segment of DNA can control genetic responses to nutrients by affecting the rate at which gene products are synthesized derived from studies of microbial metabolism. The classic series of papers that culminated in this idea focused on the ability of the sugar, lactose, to increase the activity of a permease system that transports lactose into the bacterial cell, and also to increase the quantity of the enzyme, β-galactosidase, which cleaves lactose to release glucose and galactose.[18] Binding proteins for nutrients or their metabolites serve vital roles in this process. Within the periplasmic space that separates the bacterial cell wall from the plasma membrane are located at least two dozen distinct proteins that specifically bind sugar isomers, amino acids, or ions such as sulfate and phosphate.[19,20] The binding proteins that complex with sugars and some amino acids have been studied by X-ray crystallography, and many of the functional groups that interact with parts of the ligands have been identified. For instance, the arabinose-binding protein will complex with arabinose, glucose, or fucose by forming hydrogen bonds with sugar hydroxyl groups; the protein binds the sugars through functional groups of amino acids in its primary structure, including amides, hydroxyls, and carboxyl groups. The ligand is further stabilized by stacking between hydrophobic residues that hold the sugar ring in place. Interaction with ligands alters conformation of the binding proteins, which are thought to interact with other proteins in the plasma membrane that initiate active transport of the nutrient or affect flagellar motility or chemotaxis. At the atomic level, similar kinds of interactions might be expected to occur in mammalian proteins that bind particular nutrients.

Intracellular binding proteins mediate the effects of sugars on bacterial gene expression with the same high degree of specificity that is typical of other protein-ligand interactions such as enzymatic catalysis. In the case of the lactose operon, a protein called the *lac* repressor normally prevents transcription from this promoter. The *lac* repressor is a tetrameric protein that binds the inducing sugar with high affinity, and also has two binding sites that recognize a specific DNA sequence present in the *lac* promoter.[21] When the inducing metabolite, allolactose, binds to the repressor, the repressor dislodges from promoter DNA, and transcription of the gene by RNA polymerase may begin. Because the repressor mediates the genetic effect, it could be considered a *nutrient receptor* for activation of the *lac* operon.

It might be argued that genes in most mammalian cells differ from those in bacteria because they do not respond directly to nutrients, but instead are controlled by hormones or second messengers that are generated in response to nutrients. However, bacteria have similar mechanisms by which nutritional status is encoded by a signal molecule. For example, dissociation of the *lac* repressor by itself is ineffective in increasing transcription from the *lac* operon. Also needed is a positive regulator called the catabolite activator protein (CAP). Metabolism of glucose inactivates adenylyl cyclase and causes levels of cAMP to decline, and the *lac* operon can only be activated if the complex of cAMP and the activator protein is bound to the promoter DNA. Low levels of cAMP indicate that glucose is being utilized, and that there is no need to make use of lactose. The production of cAMP in mammalian liver may indicate low glucose status in the whole organism as reflected by output of insulin and glucagon by the pancreas. However, it should be expected that many genetic responses of mammals might not involve unmodified nutrients, but a signal related to nutrient utilization. For instance, glycolysis is regulated by availability of several compounds of low molecular weight, including glucose-6-phosphate, fructose-2, 6-(bis)phosphate, citrate, NAD^+, and phosphorylation state of the cytosol (ratio of ATP to ADP $+$ AMP $+$ P_i).[22] It is feasible that one or more of these compounds could serve as a metabolic signal that activates transcription of genes whose products are needed for integrating glycolysis with other metabolic pathways.

B. NUTRITIONAL REGULATION OF GENES IN YEAST

Because yeast are eukaryotic organisms, many aspects of their structure and metabolism are similar to features of mammalian cells.[23] Yeast DNA is invested with histones and other proteins typical of chromatin, which is separated from the cytoplasm by a nuclear envelope. Like mammalian cells, yeast contain mitochondria and separate mechanisms for protein synthesis in mitochondria and endoplasmic reticulum. Glycolysis takes place in the soluble fraction of the cell, whereas enzymes of the tricarboxylic acid cycle are mitochondrial. Despite numerous similarities to mammalian cells, yeast differ in being capable of growth under anaerobic conditions, in having a cell wall in addition to a plasma membrane, and in their ability to synthesize complex organic cofactors and amino acids from simple precursors. Yeast resemble bacteria in that nutrients and their metabolites directly activate or inactivate sets of genes, with hormonal regulation being of lesser importance. The genes for individual biosynthetic pathways are not organized in operons as in bacteria, but are distributed on different chromosomes and activated by trans-acting elements.

Glucose is the preferred carbon source for yeast, and it is taken up by a membrane carrier and quickly phosphorylated by a hexokinase or glucokinase.[23,24] The level of enzymes needed for utilization of glucose and other substrates depends on glucose availability and oxygen tension. Glucose repression refers to a long-term, adaptive system that allows yeast to utilize hexoses through the glycolytic pathway and produce ethanol.[24] When glucose, mannose, or fructose is available, the synthesis of enzymes that metabolize maltose, sucrose, or galactose is repressed, as are enzymes of the gluconeogenic pathway and tricarboxylic acid cycle. When glucose is not available, yeast acquire the capacity to grow on galactose by inducing enzymes that convert it to glucose-6-phosphate. These include galactose permease, galactokinase, galactose-1-phosphate uridyltransferase, and uridine diphosphoglucose 4-epimerase, whose genes are linked on chromosome II in baker's yeast *(Saccharomyces cerevisiae)*. For example, products of the *GAL*1 and *GAL*10 genes are needed for utilization of galactose. Transcription from these genes is prevented by two types of negative control.[25,26] If glucose is available, the genes are repressed because of inhibition exerted through elements called *operators* that are present in multiple copies in the upstream activation sequence of these two genes (Figure 2). The mechanism by which the operators repress transcription is not understood.

Absence of glucose is necessary but not sufficient to allow transcription from *GAL*1 and *GAL*10. Transcription from these genes is stimulated by an inducing protein, gal4p, which is a product of the *GAL*4 locus. Unless galactose is present, however, transcription is prevented by a trans-acting repressor produced from the *GAL*80 locus that binds to the inducer, gal4p (Figure 2). Induction is thought to occur when a metabolite that is generated when galactose is present binds to the *GAL*80 gene product and relieves inhibition. The inducer is neither galactose itself nor galactose-1-phosphate.[27] During

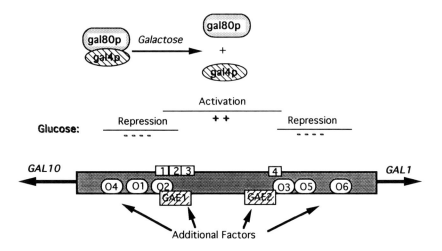

FIGURE 2. Elements of a genetic switch for regulation of galactose utilization by the yeast, *Saccharomyces cerevisiae,* are shown. The switch contains an upstream activation sequence (shaded bar) that controls transcription from two genes, *GAL*1 and *GAL*10 (indicated by right and left arrows, respectively). Transcription is inhibited by two major inhibitory elements: (1) Multiple copies of the operators O1 to O6 are present, which prevent transcription when glucose is available; (2) The inhibitory protein, gal80p, binds to a transcriptional activator unless galactose is present. An inducer is then thought to be produced that causes gal80p to dissociate from gal4p. When gal4p binds to multiple activating elements (boxes 1 to 4), transcription is stimulated by several orders of magnitude compared to the repressed rate. Other activating elements, such as GAE1 and 2, are present in this complex promoter; the figure was drawn based on data presented by Finley et al.[25] and Yun et al.[26]

the induction process, the protein, gal4p, becomes phosphorylated, and evidence suggests that a protein kinase is necessary for derepression of glucose-repressible genes. It is of considerable interest that the product of the *TUP*1 locus, which is needed for glucose repression, is structurally similar to mammalian G proteins that participate in signal transduction.[28]

Although yeast do possess adenyl cyclase, the role of cAMP in glucose repression is uncertain. However, addition of glucose to derepressed cells results in production of cAMP and a cascade of protein phosphorylation, which causes a transition to the repressed state. Transport of glucose is not sufficient to generate cAMP; the cells must also contain a hexokinase or glucokinase. It is not certain whether the sugar must be phosphorylated, or if another function is served by the hexose kinases and/or transporters. At least two genes encode hexose transport systems, which are homologous to the mammalian glucose-transport proteins and contain leucine zipper motifs that might have potential for binding to DNA. It has been postulated that they may function as glucose sensors.[29] In addition, the glucose-induced signal requires a RAS protein that is a member of a proto-oncogene family, and a CDC protein that forms a complex with the RAS protein and adenyl cyclase.[30]

It will be extremely interesting to learn whether or not mammalian glucose-transport proteins and hexokinase isoenzymes may play similar roles in mediating effects of glucose on genetic activity in mammalian cells.

Pathways that control the biosynthesis and catabolism of amino acids make use of similar trans-acting factors, and are subject to both positive and negative controls that result in induction and repression. The system that mediates nitrogen catabolite repression is analogous to the glucose-catabolite system. When urea, glutamine, or other high-quality source of nitrogen is present, enzymes for amino acid biosynthesis are induced while proteases and transport systems for poorer nitrogen sources are repressed. Products that are generated by metabolism of these compounds act as negative and positive regulators of mRNA transcription for specific pathways. For instance, the *LEU*3 gene product, leu3p, binds to upstream activating DNA sequences found in several genes for enzymes that participate in synthesis of branched-chain amino acids.[31] Leu3p contains functional domains for nuclear targeting, for binding of α-isopropylmalate (the coinducer, an intermediate in leucine biosynthesis), for binding to DNA sequences, and a separate area for activation of transcription. In the absence of a coinducer, the inducer molecule binds to DNA but does not activate transcription. When isopropylmalate accumulates, it binds to leu3p and activates transcription from the genes that participate in synthesis of branched-chain amino acids. These examples are but a few of the numerous systems involved in regulating the uptake and utilization of nutrients by microorganisms. Do any related systems occur in mammals?

VI. NUTRIENT SENSORS IN MAMMALS

Although few nutrient receptors have been characterized in vertebrates, there is no question that transducing mechanisms exist for water, salts, sugars, and amino acids. Receptors for these molecules are present on neurons that mediate the senses of taste and smell. It is true that many responses to nutrients result secondarily from release of hormones and transmitters from endocrine cells or neurons. Nevertheless, numerous cells in the digestive tract, vagus nerve, and central nervous system respond to nutrients either by secreting hormones into the bloodstream or generating action potentials, and these cells must have receptors for nutrients or their metabolites.[32-36] The hormones and action potentials convey information to the brain and other organs concerning the levels of specific nutrients in the gut or the bloodstream.

At least three structural elements in the gastrointestinal tract respond to nutrients. These elements include the afferent, sensory fibers of the vagus nerves, the enteroendocrine cells of the gastric and intestinal mucosa, and probably sensory fibers from the enteric nervous system. Evidence for glucoreceptors has come from recordings made from nerve bundles and single nerve fibers. Sensory fibers are ends of unmyelinated nerves that possess swellings, mitochondria, and vesicles.[32] In some cases, they are found in proximity to endocrine cells, and it is uncertain whether responses are direct or mediated by locally secreted hormones. Both pathways are possible, because receptors for peptides such as cholecystokinin are found on vagal afferent neurons. Vagal sensory fibers in the gut are often inactive in the absence of substrate, but an infusion of glucose leads to production of action potentials within a few seconds. This very short latency is evidence that at least some effects may be due to direct interactions of nutrients with neurons. Similar results have been obtained after infusion of the intestine with solutions containing amino acids.[32] In both cases, the responses were relatively specific and did not occur after stimulation with KCl or by mechanical means. It is interesting that nerve fibers in the hepatic vagus and area postrema are inhibited rather than stimulated by glucose.[33-36] These fibers are not silent, but generate action potentials spontaneously. It appears that the brain receives distinct patterns of information concerning nutrients from different parts of the viscera and brainstem.

The cells of the diffuse endocrine system in the intestinal mucosa are also thought to be sensitive to individual nutrients. For instance, secretion of cholecystokinin occurs in the presence of amino acids and some fatty acids but not glucose.[37] In contrast, glucose is a strong stimulus to secretion of gastric inhibitory peptide, which stimulates insulin secretion.[38] The plasma concentration of hormones from the enteroendocrine system increases severalfold after a mixed meal, and the hormones are then carried to target sites elsewhere in the body.[37-40] Receptors for the hormones have been characterized in liver, gall bladder, pancreas, and brain (including the area postrema). Thus, neural and endocrine signals initiated by nutrients in the gut are relayed to other parts of the body, where appropriate responses occur that may include altered genetic activity. For instance, ingestion of a protein-containing meal results in increased transcription from the gene for cholecystokinin in endocrine cells of the duodenum.[41]

Regulatory circuits for detection of nutrients can be completed starting with the sensory system in the gut, which conveys information to the brain by neuronal and endocrine means, and modulates behavior, metabolism, and function of the autonomic nerves (Figure 3). Sensory information from the vagus nerves impinges on the receptive field in the brainstem, including the nucleus of the tractus solitarius (NTS).[42,43] Information is relayed to the parabrachial nucleus and then to hypothalamic nuclei and the median eminence, with potential effects on secretion of hormones from the pituitary.[44,45] Humoral circuits follow similar paths, because chemosensitive neurons and binding sites for hormones are found in the area postrema.[43] This region is supplied by fenestrated capillaries through which nutrients and peptides readily pass. The area postrema is specialized to sample the chemical content of the bloodstream and is probably important for feeding behavior as well as cardiovascular function. Neurons in the area postrema project to the NTS and also to the parabrachial nucleus, both of which project to hypothalamic nuclei. Although the median eminence is also supplied by fenestrated capillaries, it

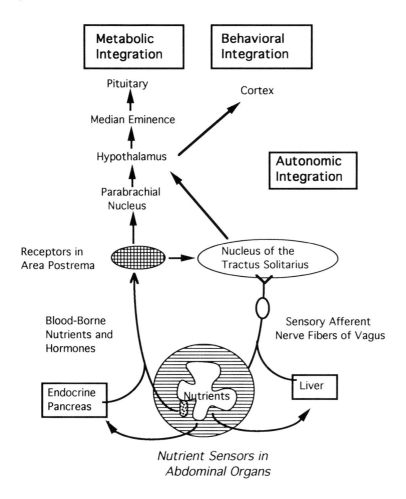

FIGURE 3. Neural and endocrine systems for sensing nutrients communicate between the gut, brain, and peripheral organs. Receptors for glucose and amino acids are thought to be present on sensory endings of afferent fibers in the vagus nerve (right side of figure). These fibers pass to the nucleus of the tractus solitarius, which is a major relay for chemical sensation and baroreflexes; it sends information to the hypothalamus and also controls efferent activity of the sympathetic and parasympathetic nervous system. Nutrient receptors are also present in endocrine cells of the gut and glands such as the pancreas, which secrete numerous peptide hormones in response to dietary constituents (left side of figure). Blood passing to the liver, peripheral organs, and brain contains nutrients and hormones that may alter gene expression. The blood-brain barrier prevents much of this information from reaching the brain directly, but circumventricular organs, such as the area postrema, lack this barrier. They contain receptors for nutrients and hormones and relay information to other parts of the brain. Small molecules such as glucose pass through the blood-brain barrier and may directly stimulate neurons in the hypothalamus. These interactions serve to integrate autonomic function, metabolic activity, and behavior.

contains secretory nerve endings that are not thought to convey information to other parts of the brain.

Some neurons in the brain are sensitive to nutrients, particularly glucose and amino acids. Glucose inhibits the rate at which action potentials are generated by neurons in the area postrema and in the ventromedial hypothalamus, indicating a possible chemosensory function.[34] Amino acids in the bloodstream are capable of affecting nerve function in the circumventricular organs; for instance, large doses of glutamate cause cell death in the area postrema.[46] In contrast, neurons that are protected by the blood-brain barrier are spared, but utilize amino acids as transmitters. Receptors for excitatory amino acids in the circumventricular organs could function as nutrient sensors, but the gating function of the blood-brain barrier makes this less likely for neurons that are sited deeper in the brain.

No matter how indirect the route, information concerning nutrient status alters genetic activity in the brain. Messenger RNAs corresponding to neuropeptide Y, cholecystokinin, and pituitary hormones are affected by fasting and dietary composition.[44,45] Recent work by Oomura and colleagues[47] has identified a striking effect of feeding or glucose administration on release of a protein from ependymal cells that line the third ventricle of the lateral hypothalamus. Owing to its growth-promoting activity, the protein is named fibroblast growth factor (FGF), but this may be misleading. The concentration of the protein in the cerebrospinal fluid (CSF) increases as much as 10,000-fold when glucose concentrations in the CSF increase 2- to 4-fold after a meal. FGF suppressed the activity of glucose-sensitive neurons in the lateral hypothalamic area, but not in the ventromedial hypothalamus, and suppressed feeding. When antibodies to FGF were injected bilaterally into the third ventricle, nocturnal food intake increased. This evidence suggests that ependymal cells may relay information concerning nutrient status to neurons involved in ingestive behavior, and possibly to neurons that produce releasing factors involved in output of pituitary hormones that respond to nutrient status.

VII. TEMPORAL AND QUANTITATIVE ASPECTS OF GENE EXPRESSION

Regulatory pathways that require activation of transcription are excellent devices for relatively long-term control (hours to days), but are relatively sluggish and not suited to short-term controls. To provide a specific example, it is interesting to compare genetic and nongenetic responses in the liver to hormones that increase glucose production. The classic studies of Sutherland and colleagues established that glucagon and epinephrine activated glycogen phosphorylase[48] and glucose production[49] by increasing the level of the second messenger, cAMP, within minutes of being added to liver slices, and that the response persisted for at least an hour in the continued presence of hormone (Figure 4A and B). However, glucagon and cAMP also activate glucose

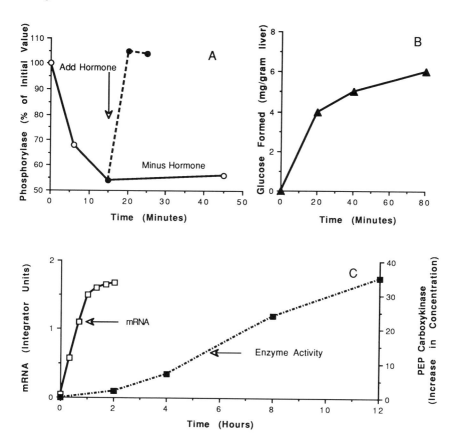

FIGURE 4. The time scale of genetic responses is much slower than metabolic control exerted by phosphorylation/dephosphorylation in response to the same initiating stimulus. (A) Response of hepatic glycogen phosphorylase to addition of epinephrine (arrow) or glucagon;[48] (B) output of glucose from liver slices after addition of glucagon;[49] (C) response of the gene for PEPCK in glucose-fed rats to addition of cAMP or starvation gives rise to elevated mRNA and enzyme (only the increase in enzyme level, and not the total concentration, is plotted).[50,51] Note that phosphorylase is maximally activated within 5 min after addition of hormone, compared to several hours needed for transcription to fully alter enzyme levels. Panels A through C were drawn from data in the references cited.

production for a longer period by increasing synthesis of phosphoenolpyruvate carboxykinase and its mRNA. Normally, this occurs in response to fasting. The time course over which this enzyme responds is relatively long because the half-life of the enzyme is about 5 to 7 h, and a full day is required for the enzyme to respond to onset of starvation (Figure 4C).[50] In contrast, the half-life of the mRNA is less than 1 h, and levels of the RNA shift to new values within 2 to 3 h after injection of a cAMP analog that permeates cell membranes.[51]

The slow time course of enzyme induction can be understood through a principle of kinetics that applies to gene products as well as nutrients and drugs. Namely, elimination constants (or biological half-lives) are very important determinants of the time course of biological effects.[52] The theory suggests that if a stimulus elevates the rate of transcription for a gene, the time required to obtain a proportional elevation in mRNA concentration in the cytoplasm is proportional to the half-life of the mRNA. Figure 5 depicts the basis for the simplest kinetic model that can describe this result.[53] In this model, the gene serves as a template for catalyzing conversion of nucleotides into mRNA, and the mRNA serves as a template for converting amino acids into specific proteins. Factors that promote transcription, processing, and transport of mRNA from the nucleus to the cytoplasm contribute to a rate of mRNA formation, and this is considered to be constant during specific intervals unless a new stimulus increases or decreases the rate. Therefore, synthesis of mRNA would follow zero-order kinetics. In contrast, degradation of most mRNAs, proteins, and other biological molecules is proportional to the amount present in the cells of interest. By definition, this is first-order decay, and the amount of decay equals the product of the first-order decay constant and the quantity of mRNA present (Figure 5). When this is true, product will accumulate according to a simple, exponential function that can be expressed as follows:

$$C = \frac{k_s}{k_d} (1 - e^{-k_d t}) \tag{1}$$

This equation predicts that, at equilibrium, the concentration of a product (C) equals the ratio between its rate constant for synthesis (k_s) and the rate constant for decay (k_d). If one of the kinetic parameters changes, the concentration will approach a new steady-state with a period defined by the rate constant for decay. The exponential term, $1 - e^{-k_d t}$, is a fraction that varies between the values of 0 and 1 as a function of time and the decay constant. What is the message buried in this equation? In any system to which these assumptions apply, the time course for change is governed more by the rate of product elimination than by the rate at which it is synthesized or introduced!

The half-lives of different mRNAs and intracellular proteins range from abut half an hour to several days,[54,55] and the time required to fully alter gene expression equals at least five half-lives. The maximum period required may be much longer, particularly if amino acid availability limits the rate of protein synthesis. This kinetic model is fully valid only when applied to transitions between steady states, and assumes that all mRNA enters the translated pool, and that the supply of energy and activated amino acids does not limit protein synthesis. Despite its limitations, this model makes an important prediction that is supported by experimental data. Half-lives of mRNAs and proteins are unrelated, and a delay that is proportional to protein half-life should occur

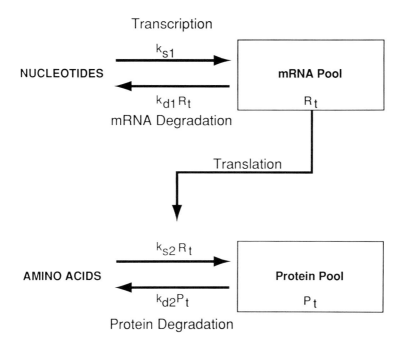

FIGURE 5. Basis for a quantitative model that shows why gene products accumulate slowly in response to altered rates of transcription. Specific mRNAs are synthesized from nucleotides at relatively constant rates unless regulatory molecules alter their transcription; the rate is not proportional to mRNA concentration. In contrast, the rate of mRNA decay is usually proportional to mRNA concentration. This gives rise to a model based on zero-order synthesis and first-order decay, which predicts that a period equal to the half-life of an mRNA or protein must pass before induction can be 50% complete. Supporting data are discussed elsewhere.[53]

before a change in mRNA concentration can produce an equal change in the concentration of the protein it encodes.[53] This results from the fact that different decay constants apply to the separate pools of mRNA and protein (Figure 5). This is a physical principle that produces a biological delay between altered transcription and altered product concentration.

How much time is required to complete each step involved in generating mRNAs and proteins that are encoded by mammalian genes? The periods required for the mammalian enzyme, phosphoenolpyruvate carboxykinase (PEPCK), will be used as an example of a gene product that is regulated relatively rapidly. Transcription from the gene for PEPCK attains a maximum within 30 min after treatment of hepatoma cells with cAMP analogues and declines at a similar rate after addition of insulin.[56,57] The rate of transcription of this gene by RNA polymerase II has been estimated to be about 1000 nt/min in the presence of insulin and 2500 nt/min in the presence of cAMP.[58] A period of about 2 to 5 min is required to complete transcription for this 6 kb gene, which generates a mature transcript of about 2.6 kb. Transcription of other class II genes has been estimated to proceed at about 3000 nt/min.[59] Processing and transport of mRNA from the nucleus delay the accumulation of product in the cytoplasm. In the case of the mRNA for PEPCK, the first mature mRNA appeared within 10 to 20 min after addition of cAMP,[60] indicating that processing of the primary transcript required no more than 10 min. The concentration of this mRNA was maximal in nuclei by 45 min after treatment, but it continued to increase in the cytoplasm for about 3 h. The period required to translate mRNAs is similar to the time needed to complete transcription; transit time of ribosomes on typical mRNAs is about 2 to 3 min.[61] A minimum of 12 h is needed to attain a new, steady-state concentration of PEPCK.[50] This implies that decay constants are the major determinants of the period required to attain new, steady-state levels of gene products, a conclusion that is supported by data derived from studies of numerous mRNAs and proteins.[53] This explains why genetic processes are so well suited for long-term control of metabolism, and why short-term regulation requires activation or inactivation of preexisting enzymes. For instance, many structural proteins, such as components of the cytoskeleton, the contractile apparatus, the nuclear matrix, or the extracellular matrix, are very stable once they have become part of mature structures. Short-term regulation of these stable proteins is usually achieved through a variety of covalent modifications. Even when the final product is stable, however, newly synthesized proteins may be unstable, and this pool may be subject to rapid regulation.

VIII. CONCLUSIONS AND PROSPECTS

The energy-yielding macronutrients are the substrates for numerous metabolic reactions, and metal ions and vitamins are cofactors for the enzymes that catalyze the interconversion of nutrients. Examples of nutrient-gene interactions that occur in microorganisms demonstrate that transcription of genes for biosynthetic reactions tends to be stimulated by precursors but inhibited by products of the enzyme reactions, whereas transcription of genes for catabolic reactions is often stimulated by presence of substrate. Similar principles apply to the regulation of genes in mammals, with the added complexities of hormonal and other, higher-order regulation. Even though the majority of genetic controls are positive in mammals, negative controls are important in preventing futile cycles, such as may occur in glycolysis and gluconeogenesis. Negative controls also terminate production of cholesterol when adequate levels are available. In order for feedback and feed-forward controls to be effective, the level of substrate or product must be sensed by means of receptors for nutrients or signals related to nutrient use. This chapter has provided examples of some of these systems, and explained the utility of genes in controlling processes that take place over several hours or days, as in adaptation to novel diets or periods of fasting. A clearer picture of the relationship between structural elements in genes and functional elements in proteins that participate in nutrient binding and feedback mechanisms is now emerging, and should suggest unifying principles of great importance to the science of nutrition.

Much current work dealing with nutrient-gene interactions emphasizes effects that are regulated by nutrients but secondary to hormonal secretion. At the time of this writing, there is no example that shows the pathway by which a macronutrient controls activity of a mammalian gene independently of hormonal controls. A few approaches to this challenging problem are discussed in this volume, and results of current studies using DNA transfection and gene transfer suggest that examples will soon be available for glucose, amino acids, and fatty acids.[61,62] One problem is that cells that monitor nutrient availability are often scarce and scattered among other cells, making them difficult to study, and few established cell lines exhibit these properties. It would be very desirable to understand the complete transduction pathway by which carbohydrates, lipids, and amino acids affect synthesis of mammalian gene products and to identify the receptors and metabolites involved. Equally worthwhile would be to learn how nutrients interact with hormones in affecting gene expression, as exemplified by the phenomenon of glucose-dependent repression of synthesis of gluconeogenic enzymes and stimulation of enzymes involved in synthesis of fatty acids and triglycerides. In the following chapters, expert investigators will describe current knowledge in some of these areas and studies that are seeking answers to these important questions.

REFERENCES

1. **Shils, M. E. and Young, V. R.,** *Modern Nutrition in Health and Disease,* 7th edition, Lea and Febiger, Philadelphia, 1988.
2. **Young, V. R. and Marchini, J. S.,** Mechanisms and nutritional significance of metabolic responses to altered intakes of protein and amino acids, with reference to nutritional adaptation in humans, *Am. J. Clin. Nutr.,* 51, 270, 1990.
3. **Eisenstein, R. S. and Harper, A. E.,** Relationship between protein intake and hepatic protein synthesis in rats, *J. Nutr.,* 121, 1581, 1990.
4. **Shay, N. F., Nick, H. R., and Kilberg, M. S.,** Molecular cloning of an amino acid-regulated mRNA (amino acid starvation-induced) in rat hepatoma cells, *J. Biol. Chem.,* 265, 17844, 1990.
5. **Munro, S. and Pelham, H. R. B.,** An Hsp70-like protein in the ER: identity with 78 kd glucose-regulated protein and immunoglobulin heavy chain binding protein, *Cell,* 46, 291, 1986.
6. **Hoyle, G. W. and Hill, R. L.,** Structure of the gene for a carbohydrate-binding receptor unique to rat Kupffer cells, *J. Biol. Chem.,* 266, 1850, 1991.
7. **Ray, K., Wang, X., Zhao, M., and Cooke, N. E.,** The rat vitamin D binding protein (Gc-globulin) gene. Structural analysis, functional and evolutionary correlations, *J. Biol. Chem.,* 266, 6221, 1991.

8. **Wingender, E.,** Transcription regulating proteins and their recognition sequences, *Crit. Rev. Eukaryotic Gene Expr.,* 1, 11, 1990.
9. **Boyer, T. G. and Maquat, L.,** Minimal sequence requirements for the initiation of transcription from an atypical, TATATAA box-containing housekeeping promoter, *J. Biol. Chem.,* 265, 20524, 1990.
10. **Dynan, W. S. and Tjian, R.,** Control of eukaryotic messenger RNA synthesis by sequence-specific DNA-binding proteins, *Nature (London),* 316, 774, 1985.
11. **Betz, H.,** Ligand-gated ion channels in the brain: the amino acid receptor superfamily, *Neuron,* 5, 383, 1990.
12. **Verdoorn, T. A., Burnashev, N., Monyer, H., Seeburg, P. H., and Sakmann, B.,** Structural determinants of ion flow through recombinant glutamate receptor channels, *Science,* 252, 1715, 1991.
13. **Durnam, D. M. and Palmiter, R. D.,** Transcriptional regulation of the mouse metallothionein-I gene by heavy metals, *J. Biol. Chem.,* 256, 5712, 1981.
14. **Thiel, E. C.,** Regulation of ferritin and transferrin receptor mRNAs, *J. Biol. Chem.,* 265, 4771, 1990.
15. **Matschinsky, F.,** Glucokinase as glucose sensor and metabolic signal generator in pancreatic β-cells and hepatocytes, *Diabetes,* 39, 647, 1990.
16. **Hoenig, M., Lee, R. J., and Ferguson, D. C.,** Glucose inhibits the high-affinity (Ca^{+2}-Mg^{+2})-ATPase in the plasma membrane of a glucose-responsive insulinoma, *Biochim. Biophys. Acta,* 1022, 333, 1990.
17. **Monod, J.,** The phenomenon of enzymatic adaptation and its bearing on problems of genetics and cellular differentiation, *Growth,* 11, 223, 1947.
18. **Jacob, F. and Monod, J.,** On the regulation of gene activity, *Cold Spring Harbor Symp. Quant. Biol.,* 26, 193, 1960.
19. **Quiocho, F. A., Vyas, N. K., Sack, J. S., and Vyas, M. N.,** Atomic protein structures reveal basic features of binding of sugars and ionic substrates, and calcium cation, *Cold Spring Harbor Symp. Quant. Biol.,* 52, 453, 1987.
20. **Vyas, N. K., Vyas, M. N., and Quiocho, F. A.,** Sugar and signal-transducer binding sites of the *Escherichia coli* galactose chemoreceptor protein, *Science,* 242, 1290, 1988.
21. **Pace, H. C., Lu, P., and Lewis, M.,** *lac* repressor: crystallization of intact tetramer and its complexes with inducer and operator DNA, *Proc. Natl. Acad. Sci. U.S.A.,* 87, 1870, 1990.
22. **Pilkis, S. J., El-Maghrabi, M. R., and Claus, T. H.,** Hormonal regulation of hepatic gluconeogenesis and glycogenolysis, *Annu. Rev. Biochem.,* 57, 755, 1988.
23. **Strathern, J. N., Jones, E. W., and Broach, J. R., Eds.,** *The Molecular Biology of the Yeast Saccharomyces. Metabolism and Gene Expression,* Cold Spring Harbor Laboratory, New York, 1982.
24. **Entian, K.-D.,** Glucose repression: a complex regulatory system in yeast, *Microbiol. Sci.,* 3, 366, 1986.
25. **Finley, R. L., Jr., Chen, S., Ma, J., Byrne, P., and West, R. W., Jr.,** Opposing regulatory functions of positive and negative elements in UAS_G control transcription of the yeast *GAL* genes, *Mol. Cell. Biol.,* 10, 5663, 1990.
26. **Yun, S.-J., Hiraoka, Y., Nishizawa, M., Takio, K., Titani, K., Nogi, Y., and Fukusawa, T.,** Purification and characterization of the yeast negative regulatory protein *GAL*80, *J. Biol. Chem.,* 266, 693, 1991.
27. **Bhat, P. J. and Hopper, J. E.,** The mechanism of inducer formation in *gal*3 mutants of the yeat galactose system is independent of normal galactose metabolism and mitochondrial respiratory function, *Genetics,* 128, 233, 1991.
28. **Williams, F. E. and Trumbly, R. J.,** Characterization of *TUP*1, a mediator of glucose repression in *Saccharomyces cerevisiae, Mol. Cell. Biol.,* 10, 6500, 1990.
29. **Lewis, D. A. and Bisson, L. F.,** The *HXT*1 gene product of *Saccharomyces cerevisiae* is a new member of the family of hexose transporters, *Mol. Cell. Biol.,* 11, 3804, 1991.

30. **Van Aelst, L., Jans, A. W. H., and Thevelein, J. M.,** Involvement of the *CDC*25 gene product in the signal transmission pathway of the glucose-induced RAS-mediated cAMP signal in the yeast, *Saccharomyces cerevisiae, J. Gen. Microbiol.,* 137, 341, 1991.

31. **Zhou, K. and Kohlhaw, G.,** Transcriptional activator *LEU*3 of yeast. Mapping of the transcriptional activation function and significance of activation domain tryptophans, *J. Biol. Chem.,* 265, 17409, 1990.

32. **Mei, N.,** Intestinal chemosensitivity, *Physiol. Rev.,* 65, 211, 1985.

33. **Niijima, A.,** Glucose-sensitive afferent nerve fibers in the liver and their role in food intake and blood glucose regulation, *J. Autonomic Nervous Syst.,* 9, 207, 1983.

34. **Oomura, Y. and Yoshimatsu, H.,** Neural networks of glucose monitoring system, *J. Autonomic Nervous Syst.,* 10, 359, 1984.

35. **Vigas, M., Tatar, P., Jurcovicova, J., and Jezova, D.,** Glucoreceptors located in different areas mediate the hypoglycemia-induced release of growth hormone, prolactin, and adrenocorticotropin in man, *Neuroendocrinology,* 51, 365, 1990.

36. **Ewart, W. R.,** Medullary integration of afferent information from the gastrointestinal tract, in *Brain-Gut Interactions,* Tache, Y. and Wingate, D., Eds., CRC Press, Boca Raton, FL, 1991, 109.

37. **Rehfeld, J. F.,** Cholecystokinin, in *Handbook of Physiology, Secton 6: The Gastrointestinal System,* Vol. II, Schultz, S., Makhlouf, G. M., and Rauner, B. B., Eds., 1989, 337.

38. **Brown, J. C., Buchan, A. M. J., McIntosh, C. H. S., and Pederson, R. A.,** Gastric inhibitory peptide, *Brain-Cut Interactions,* Tache, Y. and Wingate, D., Eds., CRC Press, Boca Raton, FL, 1991, 109.

39. **Taylor, I. L.,** Pancreatic polypeptide family: pancreatic polypeptide, neuropeptide Y, and peptide YY, *Brain-Gut Interactions,* Tache, Y. and Wingate, D., Eds., CRC Press, Boca Raton, FL, 1991, 109.

40. **Miselis, R. R., Shapiro, R. E., and Hyde, T. M.,** The area postrema, in *Circumventricular organs and body fluids,* Vol. II, Gross, P. M., Ed., CRC Press, Boca Raton, FL, 1987, 185.

41. **Liddle, R. A., Carter, J. D., and McDonald, A. R.,** Dietary regulation of rat intestinal cholecystokinin gene expression, *J. Clin. Invest.,* 81, 2015, 1988.

42. **Czyzyk-Krzeska, M. F., Bayliss, D. A., Seroogy, K. B., and Millhorn, D. E.,** Gene expression for peptides in neurons of the petrosal and nodose ganglia of the rat, *Exp. Brain Res.,* 83, 411, 1991.

43. **Adachi, A., Kobashi, M., Miyoshi, N., and Tsukamoto, G.,** Chemosensitive neurons in the area postrema of the rat and their possible functions, *Brain Res. Bull.,* 26, 137, 1991.

44. **Brady, L. S., Smith, M. A., Gold, P. W., and Herkenham, M.,** Altered expression of hypothalamic neuropeptide mRNAs in food-restricted asnd food-deprived rats, *Neuroendocrinology,* 52, 441, 1990.

45. **Chua, S. C., Jr., Liebel, R. L., and Hirsch, J.,** Food deprivation and age modulate neuropeptide gene expression in the murine hypothalamus and adrenal gland, *Mol. Brain Res.,* 9, 95, 1991.

46. **Phelix, C. F. and Hartle, D. K.,** Systemic glutamate induces degeneration of a subpopulation of serotonin-immunoreactive neurons in the area postrema of rats, *Neurosci. Lett.,* 117, 31, 1990.

47. **Oomura, Y., Sasaki, K., Suzuki, K., Muto, T., Li, A., Ogita, A.-I., Hanai, K., Tooyama, I., Kimura, H., and Yanaihara, N.,** A new brain glucosensor and its physiological significance, *Am. J. Clin. Nutr.,* 55, 278S, 1992.

48. **Sutherland, E. W.,** The effect of the hyperglycemic factor and epinephrine on enzyme systems of liver and muscle, *Ann. N.Y. Acad. Sci.,* 54, 693, 1951.

49. **Rall, T. W., Sutherland, E. W., and Wosilait, W. D.,** The relationship of epinephrine and glucagon to liver phophorylase, *J. Biol. Chem.,* 218, 483, 1956.

50. **Hopgood, M. F., Ballard, F. J., Reshef, L., and Hanson, R. W.,** Synthesis and degradation of phosphoenolpyruvate carboxylase in rat liver and adipose tissue. Changes during a starvation-re-feeding cycle, *Biochem. J.,* 134, 445, 1975.

51. **Beale, E. G., Hartley, J. L., and Granner, D. K.** N^6, $O^{2'}$-dibutyryl cAMP and glucose regulate the amount of messenger RNA coding for hepatic phosphoenolpyruvate carboxykinase (GTP), *J. Biol. Chem.,* 257, 2022, 1982.

52. **Goldstein, A., Aronow, L., and Kalman, S. M.,** *Principles of Drug Action,* 2nd ed., Harper and Row, New York, 1976, 280.

53. **Hargrove, J. L., Hulsey, M. G., and Beale, E. G.,** The kinetics of mammalian gene expression, *BioEssays,* 13, 667, 1991.

54. **Moore, R. E., Goldsworthy, T. L., and Pitot, H. C.,** Turnover of 3′-polyadenylate-containing RNA in livers from aged, partially hepatectomized, neonatal, and Morris 5123C hepatoma-bearing rats, *Cancer Res.,* 40, 1449, 1980.

55. **Shapiro, D. J., Blume, J. E., and Nielsen, D. A.,** Regulation of messenger RNA stability in eukaryotic cells, *BioEssays,* 6, 221, 1987.

56. **Granner, D., Andreone, T., Sasaki, K., and Beale, E.,** Inhibition of transcription of the phosphoenolpyruvate carboxykinase gene by insulin, *Nature,* 305, 549, 1983.

57. **Sasaki, K., Cripe, R., Koch, S., Andreone, T. L., Petersen, D. D., Beale, E. G., and Granner, D. K.,** *J. Biol. Chem.,* 259, 15242, 1984.

58. **Sasaki, K. and Granner, D. K.,** Regulation of phosphoenol-pyruvate carboxykinase gene transcription by insulin and cAMP: reciprocal actions on initiation and elongation, *Proc. Natl. Acad. Sci. U.S.A.,* 85, 2954, 1988.

59. **Sehgal, P., Derman, E., Molloy, G. R., Tamm, I., and Darnell, J. E.,** 5,6-Dichloro-1-β-D-ribofuranosylbenzimidazole inhibits initiation of nuclear heterogeneous RNA chains in HeLa cells, *Science,* 194, 431, 1976.

60. **Chrapkiewicz, N. B., Beale, E. G., and Granner, D. K.,** Induction of the messenger ribonucleic acid coding for phosphoenolpyruvate carboxykinase in H4IIE cells. Evidence for a nuclear effect of cAMP, *J. Biol. Chem.,* 257, 14428, 1982.

61. **Palmiter, R. D.,** Quantitation of parameters that determine the rate of ovalbumin synthesis, *Cell,* 4, 189, 1975.

62. **Thompson, K. S. and Towle, H. C.,** Localization of the carbohydrate response element of the rat L-type pyruvate kinase gene, *J. Biol. Chem.,* 266, 8679, 1991.

63. **McGrane, M. M., Yun, J. S., Patel, Y. M., and Hanson, R. W.,** Metabolic control of gene expression: *in vivo* studies with transgenic mice, *Trends Biochem. Sci.,* 17, 40, 1992.

Chapter 2

LACTOSE INTOLERANCE AND REGULATION OF SMALL INTESTINAL LACTASE ACTIVITY

Robert K. Montgomery, Hans A. Büller, Edmond H. H. M. Rings, Jan Dekker, and Richard J. Grand

TABLE OF CONTENTS

0-8493-6961-4/93/$0.00 + $.50
© 1993 by CRC Press, Inc.

23

I. INTRODUCTION: MILK DRINKING AND LACTOSE INTOLERANCE

The use of milk as food for adults has its origins in prehistory. The average North American or Northern European has been taught to think of milk (usually cow's milk) as the perfect food. This attitude is so pervasive that an inability to drink milk is considered a serious abnormality. Although the terminology of milk or lactose intolerance and lactase deficiency is well established in the literature, they are misnomers and misleading, as will be discussed below. Milk intolerance can be due to allergies to milk proteins or to the inability to digest milk sugar or lactose; the remainder of this review will be confined to the latter: lactose intolerance. Lactose, the major carbohydrate in milk, is hydrolyzed to glucose and galactose by the small intestinal enzyme, lactase-phlorizin hydrolase (E.C. 3.2.1.23; 3.2.1.62), subsequently referred to as lactase. Lactose intolerance is due to reduced levels of lactase in the small intestine. The clinical aspects of lactose intolerance and nutrition,[1] and the population distribution of high and low lactase levels, have been published in detail[2] and will be reviewed only briefly. The major part of the discussion will focus on recent progress in understanding the structure and function of lactase and its regulation.

It is now clear that although milk intolerance is a frequent problem in human adults, it is not a disease, nor an abnormality, but the normal condition for most humans.[3,4] Although milk intolerance was first described in 1901,[5] a clear understanding of the digestion and absorption of lactose depended on the availability of reliable tests. The development of the glucose oxidase method by Dahlqvist[6] provided a rapid, accurate method for assaying lactase in tissue samples, including intestinal biopsies. In combination with measurements of blood glucose levels following ingestion of a standardized lactose dose, this method enabled investigators to link intestinal lactase activity levels with lactose absorption and lactose intolerance.[7] Numerous studies using these methods have delineated the distribution of lactase along the small intestine, during human development, and in different human populations throughout the world (summarized in References 2 and 8 through 10).

Milk intolerance was first described by Jacobi et al.,[5] who suggested a relationship between carbohydrate ingestion and infantile diarrhea. Mendel and Mitchell[11] observed a decline in lactase activities from high levels in neonatal calves, through a slow decrease, to low adult levels. The same pattern was confirmed in other animals, which led to the hypothesis that lactase levels were the result of milk consumption, and that continuation of milk ingestion would maintain lactase enzyme at newborn levels. However, as early as 1907, Plimmer[12] examined this hypothesis in adult rabbits and rats, and reported no adaptation of lactase activity to the presence of milk or lactose in the diet. Neither prolonged ingestion of lactose in humans[13] nor exclusion of lactose from the diet[14] influences lactase activity, strongly suggesting that the enzyme is not directly regulated by availability of substrate. However, because of the frequency and variable course of lactose intolerance, this subject has remained controversial, and a great deal of effort has been devoted to determining if lactose induction of lactase activity occurs and to examining the mechanism of lactase decline.

Lactose is a disaccharide unique to milk. In the lactating mammary gland, membrane-bound galactosyl transferase interacts with the soluble milk protein, alpha-lactalbumin, to form lactose synthetase. Lactose synthetase catalyzes the reaction of UDP-galactose with glucose to form lactose. Although the lactose group also occurs as part of oligosaccharides, it is found as a free disaccharide only in milk. It is not clear, however, what advantage lactose confers, that it should be the sugar in milk. In the milks which have been analyzed, the percentage of lactose remains remarkably constant; it varies by approximately 50%, from a high of 7% in human milk to a low of about 4% in the dog and others.[15] In contrast, the fat content of milk can vary by 30-fold, with human milk having a low fat content and that of aquatic mammals having a very high fat content, presumably because of the high energy requirements necessitated by their environment. Although suggestions (discussed below) of possible selective advantages conferred by lactose have been advanced, there are few data to support the hypotheses. Furthermore, there are significant examples of milks which do not contain lactose: the pinnipeds which have been examined (and probably other aquatic mammals) neither synthesize lactose in their milk nor have lactase activity,[16] most likely a recent evolutionary adaptation to their aquatic habitat. A number of Australian marsupials that have been examined do not have neutral beta-galactosidase (lactase) in their intestine[17] and secrete little or no lactose in their milk,[18] probably an evolutionarily primitive stage of mammalian development. Unhydrolyzed lactose is not absorbed in any significant quantity, but is passed through the small intestine unchanged. The symptoms grouped as ''lactose intolerance'' are due to the osmotic effects of lactose on the intestine and the formation of gas when lactose is digested by the colonic flora.

II. DEFINITION OF CLINICAL SYMPTOMS OF LACTOSE INTOLERANCE

A variety of terms has been used to describe clinical symptoms induced by the ingestion of milk and/or milk products. Typical complaints include abdominal pain, cramps or distention, nausea, flatulence, and diarrhea; in children and adolescents, vomiting may predominate. While patients commonly identify these symptoms as a consequence of milk intolerance, they can be based either on the inability to digest lactose or sensitivity to milk proteins.[4] Lactose intolerance is characterized by symptoms, as described above, after the ingestion of lactose-containing foods or a test dose either of lactose in water or of milk. The term lactose malabsorption is reserved for those patients in whom the intestinal malabsorption of lactose has been investigated using an appropriate test of lactose absorption (lactose absorption test) or malabsorption (lactose breath hydrogen test).[19] Lactase deficiency is defined only when low (<2 SD below the mean) or, very rarely, no level of lactase activity has been found in a small intestinal biopsy sample appropriately assayed.[6,7,20] Lactase deficiency is either a primary or secondary event. Primary lactase deficiency occurs as a developmental process in premature infants or as a rare clinical syndrome.[10,21] It also appears as ''late onset lactase deficiency'' (a misnomer, as discussed below) in the majority of the world's population around the age of 5 years. Secondary lactase deficiency is found following mucosal injury.[21]

Several factors account for the variability of symptoms produced by lactose ingestion in people with lactose intolerance accompanying low lactase activity; these include the osmolarity and fat content of the food in which lactose is ingested, the rate of gastric emptying, sensitivity to intestinal distension produced by the osmotic load of unhydrolyzed lactose in the upper small bowel, the rate of intestinal transit, and the response of the colon to the lactose load. In general, the higher the osmolarity of gastric contents and the higher the fat content of the diet containing lactose, the slower the gastric emptying, and the lesser the symptoms induced by lactose. Consumption of a meal simultaneously with ingestion of lactose-containing beverages was demonstrated to significantly reduce symptoms in low lactase individuals, probably due to delayed gastric emptying.[22] Different individuals appear to have more or less sensitivity to abdominal distension, and complain differently when ingested lactose stimulates an influx of water into the lumen of the small intestine. Thus, those with greater tolerance will report fewer symptoms. These subjective responses are difficult to quantify. Intestinal transit also is influenced by the quality of the diet and individual motility patterns. Accordingly, some lactose intolerant people experience very rapid movement of lactose to the cecum, while others have slower motility. If responses to lactulose (a nonabsorbable disaccharide) simulate those to lactose, transit time from the stomach to the cecum may be as rapid as 10 min.[23] Finally, fecal flora are known to adapt to ingested carbohydrate, and it is clear that non-absorbed lactose can be salvaged by the colonic flora by fermentation to short chain fatty acids, hydrogen, methane, water, carbon dioxide, and the production of energy.[24] If lactose is provided slowly over a long period of time in many ''intolerant'' people, the flora may adapt to the load and symptoms produced by gas and acid in the colon may be reduced or eliminated.[1] This mechanism of lactose tolerance in people with low lactase levels accounts for the discrepancy in some studies in which ''lactase deficient patients'' are not lactose intolerant.[25] Colonic distension may also vary among lactose intolerant people, providing another variable in the expression of symptoms after lactose ingestion.

A. DIAGNOSIS

The diagnosis of lactose malabsorption and its pathogenesis is based on a combination of clinical findings and the results of appropriate tests.[19,26-29] The use of screening tests for lactose malabsorption in the diagnosis of lactose intolerance has received wide attention.[19,26]

B. FECAL pH AND REDUCING SUBSTANCES

The presence of low fecal pH and reducing substances indicates lactose malabsorption, but these tests are only valid when lactose has been ingested, intestinal transit time is rapid, stools are collected fresh and assayed immediately, and when bacterial metabolism of colonic carbohydrate is incomplete.[26] In general, confirmation of lactose malabsorption is best accomplished using more specific tests.

C. LACTOSE ABSORPTION TEST AND LACTOSE BREATH HYDROGEN TEST

The capacity for lactose absorption can be measured using the lactose absorption test.[19] In adults, it has a sensitivity and specificity of 75 and 96%, respectively. However, in children, it is cumbersome, invasive, and time-consuming, and has largely been replaced by the lactose breath hydrogen test.[28,30] Although this test really measures lactose nonabsorption rather than lactose hydrolysis and monosaccharide uptake, its sensitivity (100%) and specificity (100%)[19] in adults with ''late onset lactase deficiency'' is actually superior to those for the lactose absorption test, and it is simple and noninvasive.[19,25-30] Breath hydrogen testing is discussed in detail elsewhere.[31]

D. LACTASE ACTIVITY IN INTESTINAL BIOPSY

The assay of lactase activity in small bowel biopsy samples establishes the presence of lactase deficiency and has been used to define populations with ''late onset lactase deficiency''.[20] However, when lactase deficiency accompanies intestinal injury, the lesion may be focal or patchy; consequently, intestinal biopsy samples may not yield an abnormal result. Furthermore, this test is invasive, time-consuming, and assays are available only in a few centers. Normal values have been published.[20]

E. COMPARISON OF TESTS

Studies are available which have compared intestinal histology, lactase activity, and breath hydrogen test results in children with chronic diarrhea or abdominal pain.[25,27] In patients with abnormal intestinal histology, approximately 75% had abnormal lactose breath hydrogen values (sensitivity 75%). However, in patients with normal histology, the lactose breath hydrogen test was only 54% specific, reflecting the presence of a patchy villus lesion. Very similar values were found when lactase activities on biopsy were compared with results of lactose breath hydrogen tests.[25,29] Accordingly, an abnormal lactose breath hydrogen test is of value in identifying those patients who will be symptomatic after lactose ingestion. In children less than 5 years of age, an abnormal lactose breath hydrogen test always signifies abnormal intestinal mucosa, which usually needs further definition with a small intestinal biopsy. A normal lactose breath hydrogen test does not rule out an intestinal mucosal abnormality and should not be used to avoid an intestinal biopsy in the diagnosis of suspected mucosal disease (e.g., gluten sensitive enteropathy).[25]

F. TREATMENT OF LACTOSE INTOLERANCE

The treatment of lactose intolerance includes four general principles: (1) reduction or restriction of dietary lactose, (2) substitution of alternative nutrient sources to avoid reduction in energy and protein intake, (3) regulation of calcium intake, and (4) use of a commercially available enzyme substitute. When lactose restriction is necessary, the patient must be instructed to read labels of commercially prepared foods, as hidden lactose may be difficult to identify. Calcium is supplemented in the form of calcium carbonate (''Tums'' and ''OsCal'' are popular and effective). In infants, liquid calcium gluconate is readily tolerated and available.[32]

Commercially available ''lactase'' preparations are actually bacterial or yeast β-galactosidases. When added to lactose-containing food or ingested with meals containing lactose, these are effective in reducing symptoms and breath hydrogen values in many lactose intolerant subjects.[33] However, they are not capable of completely hydrolyzing all dietary lactose, and the results achieved in individual patients are variable. Live-culture yogurt, which contains endogenous β-galactosidase, is a useful alternative source of both calcium and calories and may be well tolerated by a number of lactose intolerant patients.[34] However, yogurts which contain milk products added back after fermentation may produce symptoms. While consumption of yogurt alone by low lactase individuals reduced symptoms, consumption of yogurt together with additional lactose did not reduce symptoms.[22]

III. INFLUENCE OF LACTOSE INTOLERANCE ON NUTRITION

The discrepancy between measured intestinal lactase levels by either breath test or biopsy has given rise to a wide variety of interpretations of the capacity of populations to tolerate milk intake. A study by Bayless et al.[35] reported that 59% of low lactase subjects were intolerant of 240 ml of milk and had a minor milk intake. Other studies[36] have claimed that the great majority of low lactase subjects can tolerate one glass of milk and have a normal milk intake. A study by Lisker et al.[37] correlated milk drinking habits with intestinal lactase activity as judged by a lactose tolerance test. These authors concluded that intestinal lactase activity was important in determining the extremes of milk consumption: either four or more glasses of milk in a day or none, but that it had little influence on the intermediate pattern. A number of individuals with measured low lactase could tolerate ingestion of amounts of milk less than 250 ml at one time. Possible explanations for varying tolerance are discussed above.

Low lactase activity may influence the absorption of calcium, an important nutrient, in two ways. Milk is a major source of dietary calcium, and no or reduced milk intake due to low lactase activity and lactose intolerance may result in calcium deficiency. On the other hand, several studies suggest that lactose in the intestine may enhance the uptake of calcium. Studies by Kocian et al.[38] indicated that while calcium absorption in lactose-intolerant subjects was reduced, calcium retention and the total absorbed were normal, suggesting that compensatory slowing of intestinal transit or other compensatory mechanisms allowed normal levels of calcium absorption. A more recent study by Griessen et al.[39] also indicated that low lactase individuals did not have decreased calcium absorption when ingesting milk containing either lactose or glucose.

IV. LACTASE "DEFICIENCY" IN HUMAN POPULATIONS

Since lactose is a key nutrient, lactase plays a critical role in the nutrition of mammalian neonates. As mentioned above, human congenital lactase deficiency, present from birth, is extremely rare and inherited as an autosomal recessive gene.[10] In adulthood, milk drinking is considered normal among populations of Northern European extraction and a few other groups. However, for most humans and all other adult mammals, significant milk ingestion results in mild to severe gastrointestinal symptoms, caused by the inability to digest lactose,[2] due to low levels of intestinal lactase. The initial human studies were performed in adult patients of Northern European origin who had elevated lactase. The data obtained led to the assumption that high lactase

TABLE 1 Distribution of Lactase Phenotypes in Selected Populations in the United States	
Population	Low lactase (%)
Northern European	7
Whites	22
Blacks	65
American Indians	95
Vietnamese	100

Data from Reference 4.

TABLE 2 Distribution of Lactase Phenotypes in Selected European Populations		
Country	Population	Low lactase (%)
Sweden	Swedes	1
Netherlands	Dutch	0
Austria	Austrians	20
France	French	32
	Southern French	44
Italy	Northern Italians	50
	Southern Italians	72
	Sicilians	71

Data from Reference 4.

activity throughout life was normal. Subsequent observations by Dahlqvist et al. in 1963[7] of the almost total absence of lactase activity in small intestinal biopsies of healthy adults with normal mucosal histology, and the findings of ethnic distribution of lactose malabsorption in studies by Cuatrecasas et al.[8] and Bayless and Rosensweig[9] led to the designation of these patients as having lactase "deficiency". Numerous studies have delineated the detail of the population patterns of lactase "deficiency". There is a definite geographic pattern, which has been worked out in considerable detail.[2] In Europe, there is a general north to south gradient of increasing frequency of low lactase activity, although there are specific areas where this generalization does not hold, presumably because of migration patterns.[40] (Selected population examples are shown in Tables 1 and 2.)

The general pattern of lactase distribution was established early in the analysis of digestive enzymes. The first investigators studied adult and juvenile mammals, as well as birds, reptiles, and amphibians. Their findings were summarized by Plimmer:[12] "Neither frog nor fowl have lactase in their intestine, and we may conclude that animals lower than mammals do not possess this ferment." These early studies also indicated that lactase was distributed differently in different regions of the intestine and that juvenile mammals exhibited lactase activity, while mature mammals did not.[11] Subsequent workers have confirmed these findings and provided a more detailed picture of the comparative biology of lactase and the regional distribution of lactase in the small intestine.

That a similar pattern prevails in the human population has been known for some time, yet the common terminology still implies that low lactase levels in adult humans represent an abnormality (e.g., see discussion in Reference 4). Extensive population studies in the late 1960s and early 1970s (summarized in Reference 4) showed that lactase is "deficient" in the majority of the world's adult population. Unfortunately, the term "adult lactase deficiency" has become established in the literature, even though it is now clear that low lactase activity is not a deficiency, but the normal adult human condition (as for all mammals). Although population genetic analysis indicates that elevated lactase activity is inherited as a single autosomal dominant gene,[4] a generally accepted mechanism of regulation has not yet been described.

The extensive population studies of the distribution of elevated and low lactase levels in adults have demonstrated that while the majority of human adults have low lactase activities, there are specific groups with elevated lactase levels. Today, the populations with elevated lactase activity also use milk from cattle as a major food source. Since the primitive human pattern was clearly one of low lactase activity in adulthood, there must have been some selective pressure to maintain and increase the number of people with elevated lactase levels in adulthood. It has been suggested that in the case of nomadic herdsman with little else available as a food source, there was strong selection in favor of the ability to utilize milk.[4] This argument is less persuasive for Northern European populations, which have other resources, but which express elevated lactase activity in nearly 100% of those studied. It is also somewhat puzzling that lactose should be the carbohydrate in milk instead of some other, suggesting that there may be some selective advantage to lactose. Flatz[4] combined these arguments and suggested that lactose might enhance the absorption of calcium and that this conferred an advantage in Northern Europe where reduced sunshine decreased vitamin D levels and thereby calcium absorption. There are some data to support this argument, but they do not make a wholly convincing case. It also necessitates an explanation of how the sea lion and other marine mammals maintain calcium balance without ingesting lactose at any time in their lives.[41,42]

V. LACTASE GENETICS AND DEVELOPMENT

A. ADAPTATION VS. INHERITANCE

In order to explain the regional and ethnic distribution of lactose malabsorption in adults, two conflicting basic explanations have been suggested. The first hypothesis assumed lactase induction by lactose and argued that in areas where milk and milk products are not constituents of the traditional diet of young children and adults, such as in Africa and Asia, the decline in lactase activity was due to the absence of lactose.[43] The second hypothesis suggested that lactase "deficiency" is genetic in origin and due to homozygosity for an autosomal recessive gene.[44]

Although early studies showed that rats and rabbits did not adapt their levels of lactase enzyme after ingestion of lactose-rich feedings, the adaptive theory remained widely accepted until the 1970s. It was frequently observed that patients, who on biopsy were lactase deficient, did not report symptoms of lactose intolerance.[45,46] Another major source of confusion was the disappearance of symptoms of lactose malabsorption after prolonged periods of lactose ingestion in lactose intolerant patients. In addition to clinical observations, the induction hypothesis was supported by studies which showed that lactase activity increased in adult rats after very high lactose intake.[47,48] However, the changes in lactase activity were relatively small compared to the magnitude of decrease that occurs around the weaning period, and high glucose intake produced comparable effects on lactase activity. In the studies by Lebenthal et al.,[48] lactase activity was higher during prolonged nursing than in controls, but the decline of lactase activity around weaning was not prevented by these experimental conditions.

Kretchmer[41] described Nigerian medical students with clinically proven lactose intolerance who remained unable to digest lactose, when tested by lactose tolerance tests, after 6 months of daily lactose intake (50 g). Nevertheless, the students gradually adjusted to this large lactose intake and did not show any symptoms of intolerance. It is now clear that the explanation for this adaptive response was the emergence of fecal flora capable of fermenting the nonabsorbed lactose to short chain fatty acids.[49] Thus, there is no compelling evidence for adaptive changes in small intestinal lactase activity in humans.

The genetic hypothesis was first put forward in 1966 when Bayless and co-workers showed that 70% of a group of black Americans examined in Baltimore were lactose malabsorbers, while only 8% of the white subjects were malabsorbers.[9] Many subsequent studies correlated racial and ethnic distribution of lactase enzyme levels and lactose intolerance in adults (summarized in References 2, 4, 50). The necessary genetic analysis was provided by Sahi, who showed Mendelian inheritance of lactose malabsorption as an autosomal recessive trait in a large family study from Finland.[51] Similar studies performed subsequently support the genetic hypothesis.[52] The available data indicate that persistence of high levels of lactase enzyme is probably an autosomal dominant trait.[52] The persistence in adulthood of the capacity for lactose digestion in a few human groups is probably a recent evolutionary development, while the majority of the world's population, and all placental mammals studied, show a reduction of lactase activity in adulthood. It has been hypothesized that elevated lactase levels in adults emerged over a span estimated to be a minimum of 10,000 years in several loci around the world, probably coincident with the development of dairying.[50]

B. LACTASE DEVELOPMENT

In rats and rabbits, the two mammals which so far have been studied in greatest detail, the specific activity of lactase (expressed in units per mg protein) exhibits a similar developmental pattern.[53] Lactase activity is undetectable until a few days before birth; then there is a late gestational rise, with a peak in specific activity shortly after birth, and a fall during weaning to the low levels which are seen in adulthood.[53] This pattern is partially a reflection of the short gestation and immaturity of these mammals. The guinea pig, which has a gestation of 68 days and is born in a more mature state, shows detectable lactase activity at 30 days of gestation and a steady increase until shortly before birth when levels fall slightly.[54] Guinea pigs are also reported to show very little decrease in lactase levels from the time of birth to adulthood.[55] In rats, the initial appearance of enzyme activity coincides with the initial morphogenesis of the enterocyte, which occurs about three days prior to birth.[53,56]

In contrast to these patterns, the human intestine begins to synthesize lactase at approximately the 10th week of gestation, but levels remain low until 27 to 32 weeks, when enzyme levels increase rapidly.[57] In view of the relatively low lactase levels in infants born prematurely (28 to 32 weeks), it is of interest that few develop clinical signs of carbohydrate intolerance. It has been demonstrated[58-60] that lactose is malabsorbed in these infants and reaches the colon where it is salvaged by colonic flora.

In the majority of the world's population who develop "late onset lactase deficiency" in mid-childhood, the pattern is similar to that in other mammals; the late gestational rise is followed by persistence of lactase specific activity until approximately 5 to 7 years with a fall thereafter to low adult levels.[20] In the Caucasian population, especially those peoples from or derived from Northern Europe, and certain localized clusters of other racial origins, lactase-specific activity rises late in gestation and remains at, or slightly below, this level throughout adult life—in other words, the maintenance of a juvenile trait (summarized in References 2, 4).

C. LACTASE DISTRIBUTION AND HORMONAL EFFECTS

Intestinal cell fractionation methods, pioneered by Miller and Crane,[61] coupled with the Dahlqvist assay,[6] demonstrated that lactase, as well as other intestinal enzymes, was associated with the microvillus membrane of the absorptive epithelial cells of the small intestine. Nordström et al.[62] first described the characteristic crypt to villus increase in lactase activity, which is typical of a number of digestive enzymes and has been termed the "vertical" activity gradient.[63] Studies from Kretchmer's laboratory first described the rapid increase in lactase-specific activity prior to birth, peak around birth, and decline at weaning, which represents the developmental pattern of rat lactase.[53] Shortly after, studies from the laboratory of Auricchio[64] described the pattern in human intestine. These studies also delineated the pattern of a peak of activity in the mid intestine, with decreased activity proximally and distally, the characteristic distribution of lactase in the intestine, which has been termed the "horizontal" gradient of enzyme activity by Gordon.[63] Newcomer and McGill[45] showed that both normal and deficient human adults displayed this pattern, with the "normal" subjects having markedly elevated activities along the intestine, except at the most proximal and most distal points It should be emphasized that in adult humans as well as adult rats and rabbits, even though lactase activity levels are decreased significantly from levels found in sucklings, there are still measurable levels of activity. At the time of weaning, lactase activity is not decreased to zero nor "switched off".

Animal studies indicate that thyroxine is a major regulator of the developmental pattern of lactase at the time of weaning in rats.[65] Hypophysectomy and thyroidectomy retard the decrease in lactase activity during the third postnatal week, while thyroxine replacement restores the normal pattern.[65] Glucocorticoids enhance rat lactase activity in the first weeks of life.[66] Recent detailed studies in rats demonstrated cooperative effects of thyroxine and glucocorticoid hormones in modulating the postnatal development of lactase.[67] Malo and Menard[68] observed an increase in lactase-specific activity following injection of epidermal growth factor (EGF); Foltzer-Jourdainne and Raul[69] have recently reported on the effects of EGF on intestinal enzymes. In hypophysectomized animals, the decrease is retarded but still occurs.[65] Studies of transplanted fetal intestine have shown that the decrease in lactase activity occures according to the age of the tissue, not of the host.[70] Thus, the underlying pattern appears to be programmed or intrinsic to the tissue, not dependent on hormonal changes.

There are few data available on hormonal regulation of lactase in human intestine. Experiments with explants of 12- to 14-week gestation human intestine showed that hydrocortisone induced an increase in lactase levels.[71] However, the major increase in human lactase occurs in the last several weeks of fetal life. A late gestational upsurge in fetal serum cortisol levels has been described in humans,[72] but whether or not this has a regualtory role in humans is currently unknown.

VI. CELLULAR AND MOLECULAR BIOLOGY OF LACTASE

One hypothesis to explain the decrease in lactase-specific activity, which occurs at weaning, that has received considerable attention is structural modification of the enzyme such that its activity is decreaased. This idea has prompted numerous investigations of the structure of lactase.

Despite a sizeable literature devoted to the measurements of the activity of lactase in various animal and human populations, and many studies of lactose intolerance in humans, there remain important gaps in our knowledge of the synthesis, glycosylation, and processing of lactase, although an overall pattern has emerged in the last several years. A number of mechanisms have been proposed to explain the decrease in lactase-specific activity at weaning, although no consensus has yet emerged. In addition, there still remain a number of important, unresolved questions, such as the number of active sites on the enzyme and the structure of the active enzyme on the microvillus membrane.

A. STRUCTURE AND FUNCTION

Lactase exhibits at least three characteristic enzyme activities: lactase (β-D-galactoside galactohydrolase, EC 3.2.1.23), phlorizin hydrolase (phlorizin glucohydrolase, EC 3.2.1.62), and glycosylceramidase (EC 3.2.1.45-46).[73-78] In addition, microvillus membrane lactase is one of three intestinal epithelial cell enzymes with β-galactosidase activity. The enterocyte also contains a lysosomal acid β-galactosidase, which in addition to its main function also hydrolyzes lactose, and a cytosolic β-galactosidase, which has no specificity for lactose.[74,75]

Lactase has been studied primarily in relation to its hydrolysis of lactose, although the enzyme also hydrolyzes a number of other less well known substrates. Earlier publications[76-78] indicated that lactase also exhibited glycosylceramidase activity. However, in those studies, the enzyme was isolated using column chromatography, and the possibility that more than one protein had been copurified was never excluded. Recent work from our laboratory clearly demonstrates the presence of multiple enzyme specificities (not only lactase, but also glycosylceramidase activities) on the single protein immunoprecipitated with a monoclonal antibody from intestinal epithelium in young and adult rats.[73] Additional studies of immunoprecipitated human lactase have also demonstrated enzymatic activities against lactose and phlorizin, as well as several glycolipids. The strong affinity of this enzyme for glycosylceramides, in combination with the finding of appreciable amounts of active lactase enzyme in the adult intestine, suggests a possible role for lactase in adulthood in the digestion of glycolipids.[73]

Although multiple substrate specificity for lactase has been well established, the number and location of the active sites is still unclear. Heat inactivation data suggested that there were two sites.[79] The active sites of lactase have been studied using conduritol-β-epoxide (CBE) as an affinity label.[80] CBE is an inhibitor that binds covalently to amino acids in the groups that are essential for the hydrolysis of glucosides. CBE, first introduced by Legler,[81] has been used in the study of a variety of β-glucosidases[81,82] as well as the β-glucosidase, sucrase-isomaltase.[82,83] Preliminary results showed that CBE inactivates lactase, and all other known galactosidase and glucosidase activities, in human and rat lactase immunoprecipitated with a monoclonal antibody. These findings are the first to identify the inactivation of a galactosidase by CBE. Stoichiometric studies showed that CBE bound to lactase in a molar ratio of two moles of CBE to one mole of lactase from both rat and human intestine, strongly indicating the existence of two distinct active sites.[80]

B. GLYCOSYLATION, PROCESSING, AND TRANSPORT

As with other membrane glycoproteins,[84] glycosylation influences intracellular transport and degradation of lactase,[85] but whether or not glycosylation plays a role in intracellular targeting of lactase to the MVM or in the developmental changes in expression of lactase activity on the MVM remains unclear.

In a compositional analysis of chromatographically purified rat lactase, Birkenmeier and Alpers[79] found that the enzyme was a glycoprotein containing 17% carbohydrate by weight. The carbohydrate composition suggested the presence of both N-linked and O-linked oligosaccharides. Recent studies[86] have shown that the initial intracellular form of lactase could readily be labeled with [2-^{3}H] mannose, suggesting that this form represents the high mannose-type glycosylated precursor. In follow-up studies,[87] confirmation was obtained by the susceptibility of this 205-kDa intracellular precursor peptide to both endo-H and *N*-glycanase. These high mannose oligosaccharides are then modified by a series of glycosidases and glycosyltransferases located in the ER and Golgi complex, as described for many glycoproteins.[88] Further O-linked glycosylation may take place in the trans-Golgi complex.

Following final modification of the carbohydrate composition in the Golgi, rat lactase is transported by an as yet unknown mechanism, and inserted into the MVM as a complex glycosylated, high molecular mass precursor of 220 kDa.[86] Studies of human biopsies in organ culture and Caco-2 cells have documented a similar pattern, although the human protein is somewhat larger.[89-91] Danielsen et al.[92] presented evidence that human lactase was cleaved intracellularly to the mature 160-kDa form prior to its arrival on the cell surface. Based on these data, Mantei et al.[93] hypothesized that the intracellular cleavage step produced lactase and another undiscovered enzyme from a large synthetic precursor, as discussed below. In the rat, the 220-kDa form is proteolytically cleaved on the microvillus membrane in two separate steps: first, from 220 to 180 (a transient intermediate observed by metabolic labeling) and then from 180 to the final 130-kDa form.[86] A similar analysis in humans did not identify an intermediate step, but direct processing from a 240-kDa precursor to the 160-kDa cell surface form.[90,92] The proteolytic processing, most likely by integral membrane proteases, is independent of the presence of luminal (pancreatic) proteases,[86] in contrast to sucrase-isomaltase where the role of pancreatic proteases has been clearly established.[94,95] Recent studies in rat intestine by Yeh et al.[96] support the model of final cell surface processing of lactase by integral membrane proteases. Additional support for this model is provided by the data of Naim et al.,[97] who showed that transfection of COS cells with a full length human lactase cDNA resulted in the synthesis and transport to the cell surface of both a high mannose and a complex glycosylated high molecular weight form of lactase, with full enzymatic activity. Addition of exogenous trypsin cleaved

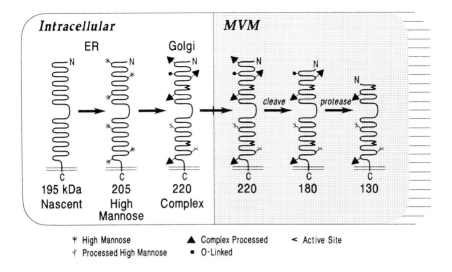

FIGURE 1. Representational model of the synthesis and processing of rat lactase-phlorizin hydrolase. See text for details. (Data from Reference 86.)

the cell surface enzyme to a 160-kDa form very similar to the mature form found in normal intestine. The authors feel, however, that because the COS cells are not epithelial cells, the processing of lactase may be aberrant. A schematic diagram summarizing the data from studies in the rat on lactase synthesis and processing is shown in Figure 1.

Studies of intestinal microvillus membrane composition have demonstrated a change at the time of weaning from high sialic acid to high fucose content.[98] Rat intestinal MVM glycoproteins undergo sialylation during the suckling period, while fucosylation begins at weaning, indicating developmental alterations in glycosylation. Concurrent with this change, the enzymes mediating fucosylation and sialylation also change in activity.[99] Based on lectin binding studies, Nsi-Emvo et al.[100] proposed that the postweaning decline of lastase activity is due to changes in glycosylation with age, leading to formation of an unprocessed, inactive precursor. These data remain to be confirmed.

So far, no other studies have reported an inactive, high molecular weight precursor form of lactase in adult intestine, although a recent study from Gray's laboratory describes the accumulation of an inactive 100-kDa processed form of the enzyme.[101] On the contrary, lactase immunoprecipitated with a monoclonal antibody shows comparable protein forms in suckling and adult animals.[102] When the immunoprecipitated enzyme is analyzed, suckling and adult forms show identical specific activities as well as K_m values.[73] Recent studies of the age-specific changes in glycosylation of rat lactase indicate that the core structure, consisting of both N-linked and O-linked oligosaccharides, remains constant during development, although terminal sugars shift from predominantly sialic acid during the suckling period to fucose in adulthood. This alteration in glycosylation of the protein occurs in a different pattern from the postweaning decline in lactase-specific activity. Consequently, age-dependent changes in glycosylation are unlikely to result in an inactive precursor nor account for the decrease in lactase-specific activity observed during development.[103]

C. MOLECULAR STRUCTURE OF LACTASE

Recently, the primary amino acid structures of human and rabbit lactase have been deduced from the sequence of the cloned cDNAs.[93] The two showed almost 80% homology at both the nucleotide and amino acid level. Human lactase is encoded by an mRNA 6274 nucleotides in length. An unexpected finding was that the mRNA is much longer than required to code for the microvillus form of the mature active protein. The amino terminus of human lactase starts at amino acid 869, beginning the sequence AFTFPS . . . .[127] The subsequent sequence represents the luminal portion of the enzyme, including the active site(s), with the 3′ portion coding for the hydrophobic, presumptive transmembrane anchor sequence. It is presently unknown what function, if any, is represented by the nearly 2 kb of mRNA which do not code for lactase. Mantei et al.[93] suggested that it directed the synthesis of a separate, active enzyme which still remains to be discovered. Because of the difficulty in estimating sizes of such large glycoproteins on SDS-PAGE, it is not clear if the large precursor forms of lactase reported in several studies may be of sufficient size to encompass the complete initial synthetic product. If the size is estimated from the 1927 amino acids which code for the complete human sequence from initiation site to stop site, and taking 110 daltons as an average amino acid, the estimated size of the nonglycosylated precursor is approximately 210 kDa. This is reasonably close to the size of 215 kDa estimated from immunoprecipitated nascent human lactase analyzed by SDS-PAGE.[90] Whether the proteolytic cleavage of the amino terminal fragment of the lactase molecule occurs intracellularly or at the cell surface, if it is no longer attached when lactase reaches the cell surface, this fragment may rapidly disappear into the intestinal lumen and be lost. The complete nucleotide sequence for rat lactase has recently been published.[104] Rat lactase mRNA is of a similar size to that reported for human and rabbit.[104,105] The three cDNA sequences reported to date show a high degree of homology and structural similarity (Figure 2), including the location of possible glycosylation sites.

Analysis of the derived amino acid sequence reveals a fourfold symmetry, including a high degree of homology among all four regions, suggesting the possibility that the large enzyme molecule arose through gene duplication.[93]

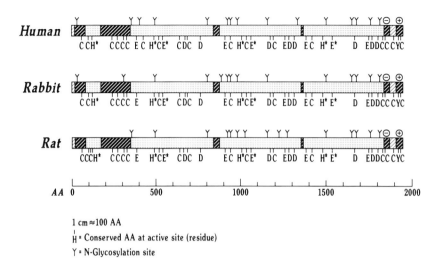

FIGURE 2. Comparison of derived amino acid sequences from human, rabbit, and rat lactase-phlorizin hydrolase. Region II begins at approximately residue 360; region III at approximately 882; and region IV at approximately 1370. (Prepared by Dr. J. Dekker; data from References 93 and 104.)

Studies of chromosome loss in cell hybrids indicate that the gene for human lactase is located on chromosome 2;[106] there are no data available on the chromosomal location of rat or rabbit lactase. Hybridization of the human lactase probe to human DNA digested with a panel of restriction enzymes in a search for restriction fragment length polymorphisms (RFLP) revealed a common MspI polymorphism. This RFLP may provide a marker for the lactase gene in genetic analysis.[107]

It has recently been shown that intestinal isomaltase exhibits sequence homology with lysosomal alpha-glucosidase, suggesting a common evolutionary origin.[108] A similar relationship might be postulated for lactase and lysosomal enzymes with similar specificities. However, Mantei et al.[93] compared their human lactase sequence with that of lysosomal glucocerebrosidase and found no homology. Although the cytosolic and lysosomal beta-galactosidases from the enterocyte are not well characterized, the complete nucleotide sequence of human lysosomal beta-galactosidase from placenta has recently been published.[109] Comparison of this sequence with that of human lactase by dot plot and sequence alignment (our unpublished data) showed no homology at either the nucleotide or amino acid level. A provocative recent study found that a newly cloned β-glucosidase (cellobiase) from *Clostridium thermocellum* showed extensive homology to human lactase, particularly at the putative active sites.[110] The *Clostridium* enzyme is homologous to one of the duplicated regions of human lactase, providing support for the idea that the enzyme structure is the result of duplication of the gene. This provides a link between a bacterial enzyme and the three mammalian enzymes which have been analyzed, but does not address the issue of intermediate steps or how the two may be related.

D. COMPARISON OF LACTASE AMINO ACID SEQUENCES FROM THREE SPECIES

Although functional data are currently lacking, alignment of the derived amino acid sequences from the three species so far available, human, rabbit, and rat, demonstrates some intriguing similarities, as shown in Figure 2. A number of important amino acids are conserved in all three species, suggesting that they play important structural or functional roles in the protein. All of the sequences show the same fourfold symmetry, suggesting that the postulated gene duplication was a relatively ancient event in mammals. Based on the *Clostridium* data, the active site is the segment containing histidine and glutamic acid which is in the same relative position in the *Clostridium* enzyme and the human segment IV. These positions are conserved in all three species. In addition, aspartic acid residues, which may also be important components of the active sites, are highly conserved, as shown by their consistent position in all three lactases. The potential N-glycosylation sites are also largely conserved, although there are some exceptions, such as position 1033 in the rat. All three species also show very similar cytoplasmic tails, with an inner positive segment, a membrane spanning hydrophobic region, and an outer negative region. Each has a tyrosine residue, potentially a phosphorylation site, on the cytoplasmic portion, and a cysteine residue, potentially a palmitoylation site, in the membrane spanning region. In region I, there are two areas which are not homologous to the other three regions, but which do show homology to the corresponding regions I in the other lactases. These areas contain conserved cysteine residues in the same positions in each species, suggesting that they may be important in folding and processing of the protein. The conservation of these regions may have important implications for understanding the fate of the large molecular mass synthetic precursor or its function. If this region is included in the precursor form which is inserted into the microvillus membrane,[86,89,97] it may contain targeting information. On the other hand, there appear to be amino acids homologous to the active site of the *Clostridium* enzyme in both regions I and II; so, there may be a second, as yet unidentified, enzyme produced, as suggested by Mantei et al.[93] The other three regions also contain cysteine residues, which are conserved in all three species, and, thus, may be important in protein folding.

VII. REGULATION OF LACTASE DECREASE

The major focus of interest in lactase has been the mechanism which regulates the decrease in lactase activity at weaning. As discussed above, the most obvious explanation would be that lactase levels depend on the level of the substrate lactose, so that when lactose decreased in the diet, lactase decreased also. However, it is now clear that enzyme activity levels do not depend upon ingested lactose. The available data on lactase structure, function, size, and glycosylation provide no compelling evidence for changes in any of these attributes which occur with weaning and account for the decrease in enzyme-specific activity. On the contrary, the available evidence indicates that the same protein is present before and after weaning, or in lactase sufficient and deficient humans. Those individuals with low levels of activity have less lactase protein. Although a number of other mechanisms have been suggested and investigated, no consensus has been reached and the subject is still controversial.

Tsuboi et al.[110] suggested that the decline in the specific activity of lactase was dependent upon a change in the rate of enterocyte turnover, and that its synthesis was constant, resulting in a decrease in specific activity. At the time of weaning, as shown by others and confirmed by Tsuboi et al.,[110] cell migration rates do increase while lactase-specific activity is decreased. However, a recent study by Yeh showed that corticoid injection induced both an increase in cell migration and an increase in lactase levels, not a decrease, in suckling rats,[111] suggesting that a change in enterocyte migration rate is not sufficient to account for the decrease in lactase activity. In contrast, Jonas et al.[112] found that lactase synthesis fell in parallel with the decline seen in lactase-specific activity during development, indicating a decrease in lactase synthesis as a mechanism. Smith and James[113] suggested that a two-stage process was involved: early in the postnatal period, the fall in the specific activity of lactase was due to increased cell turnover, and later was dependent upon a decreased rate of synthesis. Another mechanism which has received little attention is increased degradation of lactase. Studies of dietary effects on adult lactase levels suggest that these may occur at least partially through modifications of lactase degradation.[114] It is known that the level of luminal proteases, which modulate lactase turnover, increases with weaning.[115] However, there is currently no direct evidence of increased turnover in the decline in lactase-specific activity.

The studies described all measured lactase-specific activity, which may lead to deceptive conclusions. The studies in our laboratory[112] indicated substantial synthesis of lactase even in adult rats, suggesting that the whole concept of lactase decline at weaning needed reexamination. In fact, our recent work[73] shows that *total* lactase activity remains nearly constant from day 12 to adulthood during development of rat intestine, although the specific activity decreases over time. This is due to a substantial increase in total cellular protein and a probable reduction in the number of lactase enzyme molecules per cell. These findings are in agreement with studies by Ekstrom et al.,[116] demonstrating significant total lactase activity in pigs past the time of weaning, as well as observations by Moog of significant postweaning levels of lactase in mice.[117] This persistence of lactase activity into adulthood challenges previous concepts of the regulation of the enzyme and alters the interpretation of previous data in the literature suggesting a marked loss of enzyme activity during maturation. It also implies that this enzyme might maintain important function after the weaning period.

Lactase protein levels reflect the synthetic rate of the protein, which in turn is dependent upon the steady-state level of mRNA. There is presently no convincing evidence of a significant role for posttranslational mechanisms, such as modifications of glycosylation, as an overall mechanism for the developmental pattern of lactase activity. Increased degradation may play a major role, but there is currently little supportive evidence for such a step. As described below, the primary regulation of lactase levels in the small intestine is probably transcriptional, as has been found for most other mammalian proteins.[118]

A. MOLECULAR BIOLOGY OF LACTASE DECLINE

The cloning of human, rabbit, and rat lactase probes has made it possible to examine lactase mRNA levels and extend studies of regulatory mechanisms to the level of gene transcription.

The first such study was reported by Sebastio et al. who presented an examination of the molecular basis of lactase regulation in a group of surgical patients in Naples.[119] They found high enzyme activity with high mRNA, low enzyme activity with high mRNA, and high enzyme activity with low mRNA in different individuals. Based on this lack of correlation between enzyme and mRNA levels, the authors concluded that lactase enzyme levels must be regulated posttranscriptionally. Freund et al.[120] have recently presented similar data from a rat model in which they found discordant changes in lactase mRNA levels and enzyme activities. They suggest that this was due to posttranscriptional regulation of lactase enzyme levels and that regulation of lactase levels in jejunum and ileum occurs by different mechanisms.

In contrast, a comparison of the level of rat lactase mRNA and total lactase enzyme activity during development indicated that the two changed coordinately,[105] as shown in Figure 3. The pattern of regional distribution along the intestine was not completely coordinated with enzyme activity, but appeared to be affected by the level of pancreatic proteases. These findings are consistent with a fundamental transcriptional regulation, modified by specific regional factors.

Two recent studies in which lactase synthesis was examined suggest that human intestinal lactase levels are determined by the rate of synthesis of the enzyme. Sterchi et al.[121] recently reported a careful examination of lactase synthesis in 18 lactose tolerant and 14 intolerant patients. Their data indicate that the lactose intolerant subjects had reduced levels of lactase enzyme, due

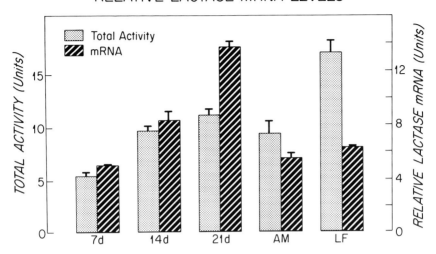

FIGURE 3. Coordinate changes in total lactase mRNA and total lactase enzyme activity during development of rat intestine. The time shown along the horizontal axis refers to postnatal age in days. AM refers to adult males; LF refers to lactating females. The increased lactase activity seen in LF is probably a posttranscriptional event. (Data from Reference 105.)

to a greatly decreased rate of lactase synthesis compared to the tolerant subjects. The structure of lactase was identical in both groups. They also identified some specific defects in intracellular transport and processing in some of the individuals with reduced lactase activity. Witte et al.,[122] identified markedly reduced synthesis in three deficient patients with low lactase enzyme levels and decreased conversion of precursor to mature lactase in a fourth deficient patient. Together, these studies suggest that (1) the major determinant of lactose ''intolerance'' is reduced synthesis of lactase enzyme, and (2) within racial groups which commonly display high lactase activity, there are individuals with specific defects in synthesis and processing. The overall pattern would be consistent with a decrease in levels of lactase mRNA in those individuals with reduced levels of lactase enzyme synthesis.

A further possible mechanism is suggested by the recent report by Maiuri[123] of both complete absence and mosaic expression of lactase immunoreactivity in the intestinal epithelium in a group of adult patients with hypolactasia. This pattern is very similar to that demonstrated in young mouse intestine by Schmidt et al.,[124] when the crypt cells are still polyclonal in origin. In mice, the crypts become monoclonal, that is, derived from a single lineage stem cell type, by two weeks of age. Such data are not available on human crypt cell lineages.

TABLE 3
Distribution of Sucrase/Lactase Activity
Ratios and Presence of Lactase mRNA
(+) in Human Intestinal Biopsy Samples
in Three Racial Groups

	(Number of subjects)		
Data	Orientals	Blacks	White
Ratio > 2	12	4	6
Ratio < 2	0	2	7
Total	12	6	13
Lactase mRNA − and ratio > 2	12	4	6
Lactase mRNA + and ratio < 2	0	2	7

Note: A ratio less than 2 signifies elevated or "normal" lactase enzyme activity. The patients with ratios below two and positive for lactase mRNA are the same individuals.

B. MOLECULAR BIOLOGY OF LACTASE PERSISTENCE IN HUMANS

Data from our laboratory indicate that human lactase expression is a direct consequence of the corresponding concentration of lactase mRNA.[125] We correlated lactase-specific activity levels and expression of lactase mRNA in biopsies from adult human orientals, blacks, and whites. Because of the variability in enzyme levels obtained by biopsy, the ratio of sucrase to lactase activity has been shown to be a useful criterion of lactase deficiency.[46] All of the Orientals examined had low levels of lactase activity (and high ratios) and no detectable lactase mRNA. Two groups of black and white patients could be distinguished. One had high lactase levels and a ratio of sucrase to lactase specific activities less than two, and readily detectable lactase mRNA. The second group had ratios greater than two and did not have readily detectable lactase mRNA. There was an exact correlation between high levels of lactase mRNA and high levels of lactase enzyme expression (Table 3 and Figure 4). Similar studies by Lloyd et al. reached the same conclusions.[126] Therefore lactase in humans also appears to be regulated at the level of transcription.[125]

VIII. CONCLUSIONS

Although the literature still frequently refers to "lactose intolerance" and "lactase deficiency" as if these were abnormalities, it is clear that low levels

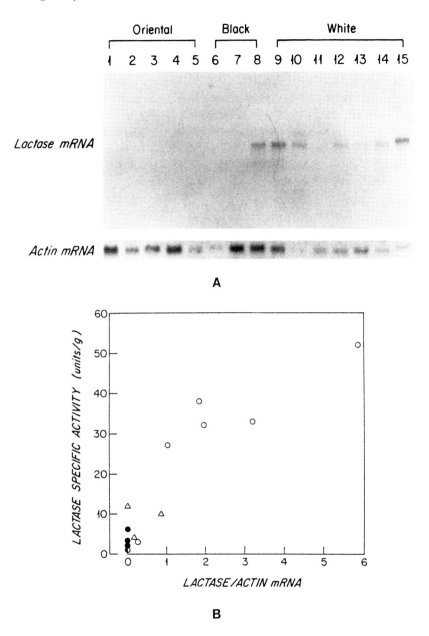

A

B

FIGURE 4. Comparison of lactase activity and mRNA levels in biopsies from human intestine. (A) The Northern blot shows results from 5 Oriental, 3 black, and 7 white patients. Four μg of total human intestinal RNA were prepared and hybridized to an appropriately labeled human cDNA probe (h-LPH-cDNA-1).[125] The blots at the bottom of the figure represent hybridization to a human β-actin probe. (B) Lactase mRNA content/actin mRNA content were obtained by densitometry and a ratio calculated; this was compared to lactase specific activity for each patient: (●) Oriental patients, (△) black patients, (○) white patients. (Data from Reference 125; reprinted with permission of the Rockefeller University Press, New York.)

of lactase, and the associated intolerance for significant intake of lactose, are the normal condition for most of the world's human population. An extensive literature has correlated the occurrence of lactose intolerance with decreased levels of lactase in the small intestine. However, it is worth reiterating that, with the exception of frank disease states, humans (as well as most mammals examined) have measurable levels of lactase in their small intestine. As discussed above, "lactase" may actually have other functions in the adult intestine, explaining why it is still present. In this sense, rats and rabbits are useful models for studying the decline in lactase levels in humans. The currently available evidence suggests that the decreased expression of lactase protein is a reflection of decreased expression of lactase mRNA.

The overall pattern in the human population thus appears similar to that described in rats, where enzyme levels and mRNA levels are coordinate in development.[105] Elevated lactase levels in specific population groups, which have been correlated with the presence of an autosomal dominant gene, appear due to an elevated transcription of lactase mRNA. The molecular control of this elevated expression remains to be elucidated.

The regulation of the decline in lactase levels is probably a complex combination of factors, including decreased mRNA transcription, the resulting decreased enzyme synthesis, and increased enzyme degradation. The unique and more interesting question is what mechanism controls the persistence in some human groups of high juvenile levels of lactase into adulthood.

IX. SUMMARY

The small intestinal enzyme lactase-phlorizin hydrolase, which hydrolyzes lactose, the sole carbohydrate in milk, plays a critical role in the nutrition of the mammalian neonate. Milk is also an important source of nutrients, especially calcium, for many adults, particularly those of Northern Europe and North America. Lactose intolerance, which can preclude significant milk intake, is common in adult humans, usually due to low levels of small intestinal lactase. Low lactase levels result from either intestinal injury, or, in the majority of the world's adult population, alterations in the genetic expression of lactase. Although the mechanism of decreased lactase levels has been the subject of intensive investigation, no concensus has yet emerged. Recent studies have begun to define the cellular and molecular biology of this enzyme. In animals and humans, a glycosylated precursor is proteolytically cleaved to yield the mature enzyme on the microvillus membrane of the enterocyte, bound to the lipid bilayer only by a hydrophobic anchor sequence. The enzyme hydrolyzes lactose, phlorizin, and glycosylceramides. A decline in lactase-specific activity occurs at the time of weaning in most mammalian species; in the majority of humans, who have low lactase activity as adults, the decline occurs at approximately 5 to 7 years of age. In a few human groups, the elevated juvenile level of lactase-specific activity persists throughout adulthood. These developmental patterns of lactase expression are influenced by a number of factors, but most likely the fundamental regulation is at the level of gene transcription.

ACKNOWLEDGMENTS

Portions of this chapter are reprinted from the *FASEB Journal* with permission of the Federation of American Societies for Experimental Biology, Rockville, MD 20852.

Work in the authors' laboratories has been supported by National Institutes of Health research grant DK-32658; a grant from the Whitaker Health Sciences Fund; grant P30 DK 34928 to the Center for Gastroenterology Research on Absorptive and Secretory Processes, New England Medical Center Hospitals; a grant from Nutricia, Zoetermeer, The Netherlands; and a NATO collaborative research grant.

REFERENCES

1. **Paige, D. M. and Bayless, T. M., Eds.,** *Lactose Digestion,* The Johns Hopkins University Press, Baltimore, 1981.
2. **Scrimshaw, N. S. and Murray, E. B.,** The acceptability of milk and milk products in populations with a high prevalence of lactose intolerance, *Am. J. Clin. Nutr.,* 48, 1083, 1988.
3. **Johnson, J. D., Kretchmer, N., and Simoons, F. J.,** Lactose malabsorption: its biology and history, in *Advances in Pediatrics,* Vol 21, Schulman, I., Ed., Year Book Medical Publishers, Chicago, 197, 1974.
4. **Flatz, G.,** Genetics of lactose digestion in humans, in *Advances in Human Genetics,* Harris, H. and Hirschhorn, K., Eds., Plenum Press, New York, 1, 1987.
5. **Jacobi, A.,** Milk sugar in infant feeding, *Trans. Am. Pediatr. Soc.,* 13, 150, 1901.
6. **Dahlqvist, A.,** Method for assay of intestinal disaccharidases, *Anal. Biochem.,* 7, 18, 1964.
7. **Dahlqvist, A., Hammond, J. B., Crane, R. K., Dunphy, J. V., and Littman, A.,** Intestinal lactase deficiency and lactose intolerance in adults, *Gastroenterology,* 45, 488, 1963.
8. **Cuatrecasas, P., Lockwood, D. H., and Caldwell, J. R.,** Lactase deficiency in the adult, a common occurrence, *Lancet,* 1, 14, 1965.
9. **Bayless, T. M. and Rosensweig, N. S.,** A racial difference in the incidence of lactase deficiency: a survey of milk intolerance and lactase deficiency in healthy males, *JAMA,* 197, 968, 1966.
10. **Savilahti, E., Launiala, K., and Kuitunen, P.,** Congenital lactase deficiency: a clinical study on 16 patients, *Arch. Dis. Child,* 58, 246, 1983.
11. **Mendel, L. B. and Mitchell, P. H.,** Chemical studies on growth: the inverting enzymes of the alimentary tract, especially in the embryo, *Am. J. Physiol.,* 20, 81, 1907.

12. **Plimmer, R. H. A.,** On the presence of lactase in the intestine of animals and on the adaptation of the intestine to lactose, *J. Physiol. (London),* 35, 81, 1906.
13. **Gilat, T., Russo, S., Gelman-Malachi, E., and Aldor, T. A. M.,** Lactase in man, a non-adaptable enzyme, *Gastroenterology,* 62, 1125, 1972.
14. **Kogut, M. D., Donnell, G. N., and Shaw, K. N. F.,** Studies of lactose absorption in patients with galactosemia, *J. Pediatr.,* 71, 75, 1967.
15. **Palmiter, R. D.,** What regulates lactose content in milk, *Nature (London),* 221, 912, 1969.
16. **Kretchmer, N. and Sunshine, P.,** Intestinal disaccharidase deficiency in the sea lion, *Gastroenterology,* 53, 123, 1967.
17. **Crisp, E. A., Czolij, R., and Messer, M.,** Absence of beta-galactosidase (lactase) activity from intestinal brush borders of suckling macropods: implications for mechanism of lactose absorption, *Comp. Biochem. Physiol.,* 88B, 923, 1987.
18. **Jenness, R., Regehr, E. A., and Sloan, R. E.,** Comparative biochemical studies of milks. II. Dialyzable carbohydrates, *Comp. Biochem. Physiol.,* 13, 339, 1964.
19. **Newcomer, A. D., McGill, D. B., Thomas, P. J., and Hofmann, A. F.,** Prospective comparison of indirect methods for detecting lactase deficiency, *N. Engl. J. Med.,* 293, 1232, 1975.
20. **Welsh, J. D., Poley, J. R., Bhatia, M., and Stevenson, D. E.,** Intestinal disaccharidase activities in relation to age, race, and mucosal damage, *Gastroenterology,* 75, 847, 1978.
21. **Mobassaleh, M., Montgomery, R. K., Biller, J. A., and Grand, R. J.,** Development of carbohydrate absorption in the fetus and neonate, *Pediatrics,* 75, 160, 1985.
22. **Martini, M. C. and Savaiano, D. A.,** Reduced intolerance symptoms from lactose consumed during a meal, *Am. J. Clin. Nutr.,* 47, 57, 1988.
23. **Bond, J. H. and Levitt, M. D.,** Investigation of small bowel transit time in man utilizing pulmonary hydrogen (H_2) measurements, *J. Lab. Clin. Med.,* 85, 546, 1975.
24. **Levitt, M. D.,** Production and excretion of hydrogen gas in man, *N. Engl. J. Med.,* 281, 122, 1969.
25. **Hyams, J. S., Stafford, R. J., Grand, R. J., and Watkins, J. B.,** Correlation of lactose breath hydrogen test, intestinal morphology, and lactase activity in young children, *J. Pediatr.,* 97, 609, 1980.
26. **Newcomer, A. D.,** Screening tests for carbohydrate malabsorption, *J. Pediatr. Gastroenterol. Nutr.,* 3, 6, 1984.
27. **Barr, R. G., Perman, J. A., Schoeller, D. A., and Watkins, J. B.,** Breath test in pediatric gastrointestinal disorders: new diagnostic opportunities, *Pediatrics,* 62, 393, 1978.
28. **Douwes, A. C., Fernandes, J., and Degenhardt, H. J.,** Improved accuracy of lactose tolerance test in children, using expired H_2 measurement, *Arch. Dis. Child.,* 53, 939, 1978.
29. **Barr, R. G., Levine, M. D., and Watkins, J. B.,** Recurrent abdominal pain of childhood due to lactose intolerance: a prospective study, *N. Engl. J. Med.,* 300, 1449, 1979.
30. **Ostrander, C. R., Cohen, R. S., Hopper, A. O., and Stevenson, D. K.,** Breath hydrogen analysis: a review of the methodologies and clinical applications, *J. Pediatr. Gastroenterol. Nutr.,* 2, 525, 1983.
31. **Perman, J. A.,** Clinical application of breath hydrogen measurements, *Can. J. Physiol. Pharmacol.,* 69, 111, 1991.
32. **Queen, P. M. and Henry, R. R.,** Growth and nutrient requirements of children, in *Pediatric Nutrition: Theory and Practice,* Grand, R. J., Sutphen, J. L., and Dietz, W. H., Eds., Butterworths, Boston, 1987, 341.
33. **Rosado, J. L., Solomons, N. W., Lisker, R., and Bourges, H.,** Enzyme replacement therapy for primary adult lactase deficiency, *Gastroenterology,* 87, 1072, 1984.
34. **Newcomer, A. D. and McGill, D. B.,** Clinical importance of lactase deficiency, *N. Engl. J. Med.,* 310, 42, 1984.

35. **Bayless, T. M., Rothfeld, B., Massa, C., Wise, L., Paige, D. M., and Bedine, M. S.,** Lactose and milk intolerance: clinical implications, *N. Engl. J. Med.,* 292, 1156, 1975.
36. **Stephenson, L. S. and Latham, M. C.,** Lactose intolerance and milk consumption: the relation of tolerance to symptoms, *Am. J. Clin. Nutr.,* 27, 296, 1974.
37. **Lisker, R., Aguilar, L., and Zavala, C.,** Intestinal lactase deficiency and milk drinking capacity in the adult, *Am. J. Clin. Nutr.,* 31, 1499, 1978.
38. **Kocian, J., Skala, I., and Bakos, K.,** Calcium absorption from milk and lactose-free milk in healthy subjects and patients with lactose intolerance, *Digestion,* 9, 317, 1973.
39. **Griessen, M., Cochet, B., Infante, F., Jung, A., Bartholdi, P., Donath, A., Loizeau, E., and Courvoisier, B.,** Calcium absorption from milk in lactase-deficient subjects, *Am. J. Clin. Nutr.,* 49, 377, 1989.
40. **Cavalli-Sforza, L. T., Strata, A., Barone, A., and Cucurachi, L.,** Primary adult lactose malabsorption in Italy: regional differences in prevalence and relationship to lactose intolerance and milk consumption, *Am. J. Clin. Nutr.,* 45, 748, 1987.
41. **Kretchmer, N.,** Memorial lecture: lactose and lactase—a historical perspective, *Gastroenterology,* 61, 805, 1971.
42. **Büller, H. A. and Grand, R. J.,** Lactose intolerance, *Annu. Rev. Med.,* 41, 141, 1990.
43. **Bolin, T. D. and Davis, A. E.,** Primary lactase deficiency: genetic or acquired?, *Am. J. Dig. Dis.,* 15, 679, 1970.
44. **Rosensweig, N. S.,** Adult lactase deficiency: genetic control or adaptive response?, *Gastroenterology,* 60, 464, 1971.
45. **Newcomer A. D. and McGill, D. B.,** Distribution of disaccharidase activity in the small bowel of normal and lactase-deficient subjects, *Gastroenterology,* 51, 481, 1966.
46. **Welsh, J. D.,** Isolated lactase deficiency in humans: report on 100 patients, *Medicine,* 49, 257, 1970.
47. **Bolin, T. D., Pirola, R. C., and Davis, A. E.,** Adaptation of intestinal lactase in the rat, *Gastroenterology,* 60, 432, 1969.
48. **Lebenthal, E., Sunshine, P., and Kretchmer, N.,** Effect of prolonged nursing on the activity of intestinal lactase in rats, *Gastroenterology,* 64, 1136, 1973.
49. **Bond, J. H. and Levitt, M. D.,** Quantitative measurement of lactose absorption, *Gastroenterology,* 70, 1058, 1976.
50. **Simoons, F. J.,** The geographic hypothesis and lactose malabsorption, *Dig. Dis. Sci.,* 23, 963, 1978.
51. **Sahi, T., Isokoski, M., Jussila, J., Launiala, K., and Pyörälä, K.,** Recessive inheritance and adult type-lactose malabsorption, *Lancet,* 2, 823, 1973.
52. **Sahi, T. and Launiala, K.,** More evidence for the recessive inheritance of selective adult type lactose malabsorption, *Gastroenterology,* 73, 231, 1977.
53. **Doell, R. G. and Kretchmer, N.,** Studies of small intestine during development. I. Distribution and activity of β-galactosidase, *Biochim. Biophys. Acta.,* 62, 353, 1962.
54. **Bailey, D. S., Cook, A., McAllister, G., Moss, M., and Mian, N.,** Structural and biochemical differentiation of the mammalian small intestine during fetal development, *J. Cell Sci.,* 72, 195, 1984.
55. **Blaxter, K. L.,** Lactation and the growth of the young, in *Milk: The Mammary Gland and Its Secretion,* Vol. II, Kon, S. K. and Cowie, A. T., Eds., Academic Press, New York, 1961, 305.
56. **Montgomery, R. K., Kothe, M. J. C., Büller, H. A., and Grand, R. J.,** Rat lactase mRNA appears concurrently with development of columnar epithelial cells in fetal intestine, *Gastroenterology,* 98, A423(Abstr.), 1990.
57. **Antonowicz, I., Chang, S. K., and Grand, R. J.,** Development and distribution of lysosomal enzymes and disaccharidases in human fetal intestine, *Gastroenterology,* 67, 51, 1974.

58. **MacLean, W. C. and Fink, B. B.,** Lactose malabsorption by premature infants: magnitude and clinical significance, *J. Pediatr.,* 97, 383, 1980.

59. **Kien, C. L., Liechty, E. A., Myerberg, D. Z., and Mullett, M. D.,** Dietary carbohydrate assimilation in the premature infant: evidence for a nutritionally significant bacterial ecosystem in the colon, *Am. J. Clin. Nutr.,* 46, 456, 1987.

60. **Murray, R. D., Boutton, T. W., Klein, P. D., Gilbert, M., Paule, C. L., and MacLean, W. C.,** Comparative absorption of [13C] glucose and [13C] lactose by premature infants, *Am. J. Clin. Nutr.,* 51, 59, 1990.

61. **Miller, D. and Crane, R. K.,** Digestive function of the epithelium of the small intestine. II. Localization of disaccharide hydrolysis in the isolated brush border portion of intestinal epithelial cells, *Biochim. Biophys. Acta,* 52, 293, 1961.

62. **Nordström, C., Dahlqvist, A., and Josefsson, L.,** Quantitative determination of enzymes in different parts of the villi and crypts of rat small intestine, *J. Histochem. Cytochem.,* 15, 713, 1968.

63. **Gordon, J. I.,** Intestinal epithelial differentiation: new insights from chimeric and transgenic mice, *J. Cell Biol.,* 108, 1187, 1989.

64. **Auricchio, S., Rubino, A., and Murset, G.,** Intestinal glycosidase activites in the human embryo, fetus and newborn, *Pediatrics,* 35, 944, 1965.

65. **Yeh, K. Y. and Moog, F.,** Intestinal lactase activity in the suckling rat: influences of hypophysectomy and thyroidectomy, *Science,* 183, 77, 1974.

66. **Koldovskỳ, O. and Sunshine, P.,** Effect of cortisone on developmental pattern of the neutral and acid beta-galactosidase of the small intestine of the rat, *Biochem. J.,* 117, 467, 1970.

67. **Yeh, K. Y., Yeh, M., and Holt, P. R.,** Thyroxine and cortisone cooperate to modulate postnatal intestinal enzyme differentiation in the rat, *Am. J. Physiol.,* 260, G371, 1991.

68. **Malo, C. and Menard, D.,** Influence of epidermal growth factor on the development of suckling mouse intestinal mucosa, *Gastroenterology,* 83, 28, 1982.

69. **Foltzer-Jourdainne, C. and Raul, F.,** Effect of epidermal growth factor on the expression of digestive hydrolases in the jejunum and colon of newborn rats, *Endocrinology,* 127, 1763, 1990.

70. **Montgomery, R. K., Sybicki, M. A., and Grand, R. J.,** Autonomous biochemical and morphological differentiation in fetal rat intestine transplanted at 17 and 20 days of gestation, *Dev. Biol.,* 87, 76, 1981.

71. **Arsenault, P. and Menard, D.,** Influence of hydrocortisone on human fetal small intestine in organ culture, *J. Pediatr. Gastroenterol. Nutr.,* 4, 893, 1985.

72. **Murphy, B. E. P.,** Human fetal serum cortisol levels related to gestational age: evidence of midgestational fall and a steep late gestational rise, independent of sex or mode of delivery, *Am. J. Obstet. Gynecol.,* 144, 276, 1982.

73. **Büller, H. A., Wassenaer, A. G. van, Raghavan, S., Montgomery, R. K., Sybicki, M. A., and Grand, R. J.,** New insights into the lactase and glycosylceramidase activities of rat microvillus membrane lactase-phlorizin hydrolase, *Am. J. Physiol.,* 257, G616, 1989.

74. **Asp, N. G. and Dahlqvist, A.,** Human small intestine β-galactosidases: specific assay of three different enzymes, *Anal. Biochem.,* 47, 527, 1972.

75. **Gray, G. M. and Santiago, N. A.,** Intestinal β-galactosidases. I. Separation and characterization of 3 enzymes in normal human intestine, *J. Clin. Invest.,* 48, 716, 1969.

76. **Brady, R. O., Gal, A. E., Kanfer, J. N., and Bradley, R. M.,** The metabolism of glucocerebrosides, III, Purification and properties of glucosyl- and galactosylceramide cleaving enzyme from rat intestinal tissue, *J. Biol. Chem.,* 240, 3766, 1965.

77. **Kobayashi, T. and Suzuki, K.,** The glycosylceramidase in the murine intestine, *J. Biol. Chem.,* 256, 7768, 1981.

78. **Leese, H. J. and Semenza, G.,** On the identity between the small intestinal enzymes phlorizin hydrolase and glycosylceramidase, *J. Biol.Chem.,* 248, 8170, 1973.
79. **Birkenmeier, E. and Alpers, D. H.,** Enzymatic properties of rat lactase-phlorizin hydrolase, *Biochim. Biophys. Acta,* 350, 100, 1974.
80. **Haringsma, J., Büller, H. A., Matsudaira, P. T., Montgomery, R. K., and Grand, R. J.,** Human lactase-phlorizin hydrolase, function and active sites, *Eur. J. Clin. Invest.,* 20 A26 (Abstr.), 1990.
81. **Legler, G.,** Untersuchungen zum wirkungsmechanismus glycosidspaltender enzyme. III. Markierung des aktiven zentrums einer beta-glucosidases aus aspergillus wentii mit conduritol-b-epoxid, *Z. Physiol. Chem.,* 349, 767, 1968.
82. **Dinur, T., Osiecki, K. M., Legler, G., Gatt, S., Desnick, R. J., and Grabowski, G. A.,** Human acid β-glucosidase: isolation and amino acid sequence of a peptide containing the catalytic site, *Proc. Natl. Acad. Sci. U.S.A.,* 83, 1660, 1986.
83. **Quaroni, A. and Semenza, G.,** Partial amino acid sequences around the essential carboxylate in the active sites of the sucrase-isomaltase complex, *J. Biol.Chem.,* 251, 3250, 1976.
84. **Rousset, M., Trugnan, G., Brun, J. L., and Zweibaum, A.,** Inhibition of the posttranslational processing of microvillar hydrolases is associated with a specific decrease of sucrase-isomaltase and an increased turnover of glucose in CaCo-2 cells treated with monensin, *FEBS Lett.,* 208, 34, 1986.
85. **Danielsen, E. M. and Cowell, G. M.,** Biosynthesis of intestinal microvillar proteins: processing of N-linked carbohydrate is not required for surface expression, *Biochem. J.,* 240, 777, 1986.
86. **Büller, H. A., Montgomery, R. K., Sasak, W. V., and Grand, R. J.,** Biosynthesis, glycosylation, and intracellular transport of intestinal lactase-phlorizin hydrolase in rat, *J. Biol. Chem.,* 262, 17206, 1987.
87. **Büller, H. A., Rings, E. H. H. M., Montgomery, R. K., Sasak, W. V., and Grand, R. J.,** Further studies of glycosylation and intracellular transport of lactase-phlorizin hydrolase in rat small intestine, *Biochem. J.,* 263, 249, 1989.
88. **Hubbard, C. S. and Ivatt, R. J.,** Asparagine-linked oligosaccharides, *Annu. Rev. Biochem.,* 50, 555, 1981.
89. **Hauri, H. P., Sterchi, E. E., Bienz, D., Fransen, J. A. M., and Marxer, A.,** Expression and intracellular transport of microvillus membrane hydrolases in human intestine, *J. Cell Biol.,* 101, 838, 1985.
90. **Naim, H. Y., Sterchi, E. E., and Lentze, M. J.,** Biosynthesis and maturation of lactasephlorizin hydrolase in the human small intestinal epithelial cells, *Biochem. J.,* 241, 427, 1987.
91. **Skovbjerg, H., Danielsen, E. M., Noren, O., and Sjöström, H.,** Evidence for biosynthesis of lactase-phlorizin hydrolase as a single-chain high molecular weight precursor, *Biochim. Biophys. Acta,* 798, 247, 1984.
92. **Danielsen, E. M., Skovbjerg, H., Noren, O., and Sjöström, H.,** Biosynthesis of intestinal microvillar proteins, intracellular processing of lactase-phlorizin hydrolase, *Biochim. Biophys. Res. Comm.,* 122, 82, 1984.
93. **Mantei, N., Villa, M., Enzler, T., Wacker, H., Boll, W., James, P., Hunziker, W., and Semenza, G.,** Complete primary structure of human and rabbit lactase-phlorizin hydrolase: implications for biosynthesis, membrane anchoring and evolution of the enzyme, *EMBO J.,* 7, 2705, 1988.
94. **Hauri, H. P., Quaroni, A., and Isselbacher, K. J.,** Biogenesis of intestinal plasma membrane: posttranslational route and cleavage of sucrase-isomaltase, *Proc. Natl. Acad. Sci. U.S.A.,* 76, 5183, 1979.

95. **Montgomery, R. K., Sybicki, M. A., Forcier, A. G., and Grand, R. J.,** Rat intestinal microvillus membrane sucrase-isomaltase is a single high molecular weight protein and fully active enzyme in the absence of luminal factors, *Biochim. Biophys. Acta,* 661, 346, 1981.

96. **Yeh, K., Yeh, M., Pan, P., and Holt, P. R.,** Post-translational cleavage of rat intestinal lactase occurs at the luminal side of the brush border membrane, *Gastroenterology,* 101, 312, 1991.

97. **Naim, H. Y., Lacey, S. W., Sambrook, J. F., and Gething, M.-J.,** Expression of a full-length cDNA coding for human intestinal lactase-phlorizin hydrolase reveals an uncleaved, enzymatically active, and transport-competent protein, *J. Biol. Chem.,* 266, 12313, 1991.

98. **Srivastava, O. P., Steele, M. I., and Torres-Pinedo, R.,** Maturational changes in terminal glycosylation of small intestinal microvillar proteins in the rat, *Biochim. Biophys. Acta,* 914, 143, 1987.

99. **Chu, S. W. and Walker, W. A.,** Developmental changes in the activities of sialyl- and fucosyltransferases in rat small intestine, *Biochim. Biophys. Acta,* 883, 496, 1986.

100. **Nsi-emvo, E., Launay, J. F., and Raul, F.,** Is adult-type hypolactasia in the intestine of mammals related to changes in the intracellular processing of lactase, *Cell Mol. Biol.,* 33, 335, 1987.

101. **Quan, R., Santiago, N. A., Tsuboi, K. K., and Gray, G. M.,** Intestinal lactase, shift in intracellular processing to altered, inactive species in the adult rat, *J. Biol. Chem.,* 265, 15882, 1990.

102. **Quaroni, A.,** Pre- and postnatal development of differentiated functions in rat intestinal epithelial cells, *Dev. Biol.,* 111, 280, 1985.

103. **Büller, H. A., Rings, H. H. M., Pajkrt, D., Montgomery, R. K., and Grand, R. J.,** Glycosylation of lactase-phlorizin hydrolase in rat small intestine during development, *Gastroenterology,* 98, 667, 1990.

104. **Duluc, I., Boukamel, R., Mantei, N., Semenza, G., Raul, F., and Freund, J.,** Sequence of the precursor of intestinal lactase-phlorizin hydrolase from fetal rat, *Gene,* 103, 275, 1991.

105. **Büller, H. A., Kothe, M. J. C., Goldman, D. A., Grubman, S. A., Sasak, W. V., Matsudaira, P. T., Montgomery, R. K., and Grand, R. J.,** Coordinate expression of lactase-phlorizin hydrolase mRNA and enzyme levels during development, *J. Biol. Chem.,* 265, 6978, 1990.

106. **Kruse, T. A., Bolund, L., Grzeschik, K. H., Ropers, H. H., Sjöström, H., Noren, O., Mantei, N., and Semenza, G.,** The human lactase-phlorizin hydrolase gene is located on chromosome 2, *FEBS Lett.,* 240, 123, 1988.

107. **Kruse, T. A., Bolund, L., Byskov, A., Sjöström, H., Noren, O., Mantei, N., and Semenza, G.,** Mapping of the human lactase-phlorizin hydrolase gene to chromosome 2, *Cytogenet. Cell Genet.,* 51, 1026, (Abstr.), 1989.

108. **Hoefsloot, L. H., Hoogeveen-Westerveld, M., Kroos, M. A., Beeumen, J. van, Reuser, A. J. J., and Oostra, B. A.,** Primary structure and processing of lysosomal alpha-glucosidase; homology with the intestinal sucrase-isomaltase complex, *EMBO J.,* 7, 1697, 1988.

109. **Oshima, A., Tsuji, A., Nagao, Y., Sakuraba, H., and Suzuki, Y.,** Cloning, sequencing, and expression of cDNA for human beta-galactosidase, *Biochem. Biophys. Res. Comm.,* 157, 238, 1988.

110. **Tsuboi, K. K., Kwong, L. K., D'Harlingue, A. E., Stevenson, D. K., Kerner, J. A., and Sunshine, P.,** The nature of maturational decline of intestinal lactase activity, *Biochim Biophys. Acta,* 840, 69, 1985.

111. **Yeh, K., Yeh, M., and Holt, P. M.,** Intestinal lactase expression and epithelial cell transit in hormone-treated suckling rats, *Am. J. Physiol.,* 260, G379, 1991.

112. **Jonas, M. M., Montgomery, R. K., and Grand, R. J.,** Intestinal lactase synthesis during postnatal development in the rat, *Pediatr. Res.,* 19, 956, 1985.

113. **Smith, M. W. and James, P. S.,** Cellular origin of lactase decline in post weaned rats, *Biochim. Biophys. Acta,* 905, 503, 1987.

114. **Goda, T., Bustamante, S., and Koldovský, O.,** Dietary regulation of intestinal lactase and sucrase in adult rats: quantitative comparison of effect of lactose and sucrose, *J. Pediatr. Gastroenterol. Nutr.,* 4, 998, 1985.

115. **Seetharam, B., Yeh, K. Y., and Alpers, D. H.,** Turnover of intestinal brush border proteins during postnatal development in rat, *Am. J. Physiol.,* 239, G524, 1980.

116. **Ekstrom, K. E., Benevenga, N. J., and Grummer, R. H.,** Changes in the intestinal lactase activity in the small intestine of two breeds of swine from birth to 6 weeks of age, *J. Nutr.,* 105, 1032, 1975.

117. **Moog, F., Denes, A. E., and Powell, P. M.,** Disaccharidases in the small intestine of the mouse: normal development and influence of cortisone, actinomycin D, and cycloheximide, *Dev. Biol.,* 35, 143, 1973.

118. **Johnson, P. F. and McKnight, S. L.,** Eukaryotic transcriptional regulatory proteins, *Annu. Rev. Biochem.,* 58, 799, 1989.

119. **Sebastio, G., Villa, M., Sartorio, R., Guzzetta, V., Poggi, V., Auricchio, S., Boll, W., Mantei, N., and Semenza, G.,** Control of lactase in human adult-type hypolactasia and in weaning rabbits and rats, *Am. J. Hum. Genet.,* 45, 489, 1989.

120. **Freund, J. N., Duluc, I., and Raul, F.,** Lactase expression is controlled differently in the jejunum and ileum during development in rats, *Gastroenterology,* 100, 388, 1991.

121. **Sterchi, E. E., Mills, P. R., Fransen, J. A. M., Hauri, H. P., Lentze, M. J., Naim, H. Y., Ginsel, L., and Bond, J.,** Biogenesis of intestinal lactase-phlorizin hydrolase in adults with lactose intolerance, *J. Clin. Invest.,* 86, 1329, 1990.

122. **Witte, J., Lloyd, M., Lorenzsonn, V., Korsmo, H., and Olsen, W.,** The biosynthetic basis of adult lactase deficiency, *J. Clin. Invest.,* 86, 1338, 1990.

123. **Maiuri, L., Raia, V., Potter, J., Swallow, D., Ho, M. W., Fiocca, R., Finzi, G., Cornaggia, M., Capella, C., Quaroni, A., and Auricchio, S.,** Mosaic pattern of lactase expression by villous enterocytes in human adult-type hypolactasia, *Gastroenterology,* 100, 359,1991.

124. **Schmidt, G. H., Wilkinson, M. M., and Ponder, B. A. J.,** Cell migration pathway in the intestinal epithelium: an *in situ* marker system using mouse aggregation chimeras, *Cell,* 40, 425, 1985.

125. **Escher, J. C., de Koning, N. D., Engen, C. G. J. van, Arora, S., Büller, H. A., Montgomery, R. K., and Grand, R. J.,** Molecular basis of lactase levels in adult humans, *J. Clin. Invest.,* 89, 480, 1992.

126. **Lloyd, M., Mevissen, G., Fischer, M., Olsen, W., Goodspeed, D., Genini, M., Boll, W., Semenza, G., and Mantei, N.,** Regulation of intestinal lactase in adult hypolactasia, *J. Clin. Invest.,* 89, 524, 1992.

127. **Büller, H. A.,** unpublished data.

Chapter 3

REGULATION OF GASTROINTESTINAL LIPASE GENE EXPRESSION BY DIETARY LIPIDS

Catherine Wicker-Planquart and Antoine Puigserver

TABLE OF CONTENTS

0-8493-6961-4/93/$0.00 + $.50

I. INTRODUCTION

Triacylglycerol is the main component of dietary lipids which are consumed by most mammalian species. It represents 95 to 99% of lipid intake. This corresponds to a daily intake in human of 100 to 150 g of triglyceride, vs. 2 to 10 g phospholipids, 0.2 to 0.8 g cholesterol, and about 16 mg liposoluble vitamins.[1] Intestinal absorption of triglycerides requires prior hydrolysis into monoglycerides and fatty acids. This is ensured by the combined action of prepyloric and pancreatic lipases. The contribution made by each enzyme to the digestion of triglycerides is influenced by diet and physiological state of the organism. Other enzymes, phospholipase A2 and carboxyl ester hydrolase, contribute to lipid digestion[2] (see Table 1). Pancreatic phospholipase A2 (MW = 14 kDa) hydrolyzes alimentary and biliary phospholipids, producing lysophospholipids and fatty acids while pancreatic carboxyl ester hydrolase (MW = 64 to 100 kDa) hydrolyzes steryl and lipovitamin esters, which are solubilized in the aqueous phase in the intestinal content, producing cholesterol, free vitamins, and fatty acids.

We will focus this chapter on the enzymes responsible for triglyceride lipolysis (preduodenal and pancreatic lipases) and the effects of dietary triacylglycerol on lipolytic activity, with emphasis on the adaptation of pancreatic lipase to dietary lipid intake. Effects of other dietary components, as well as neural or hormonal control of lipases, will not be developed. Finally, the mechanisms of pancreatic lipase and colipase adaptation to dietary lipids (transcriptional, pretranslational, translational, posttranslational) will be discussed.

II. LIPID DIGESTION WITHIN THE GASTROINTESTINAL TRACT

A. PREDUODENAL LIPASES

As early as in 1946, Schonheyder and colleagues demonstrated that the human stomach was able to digest dietary lipids and reported the presence of a lipase enzyme in gastric juice.[3] Preduodenal lipases have been detected in most animal species and are widely distributed. Their synthesis and secretion differ from one species to another. Depending on the species considered, preduodenal lipase is almost entirely produced by the tongue as in the rat and mouse,[4,5] pharynx as in the calf and the sheep,[6,7] or stomach as in the man, rabbit, dog, hog, horse, cat, guinea pig, baboon, and macaque.[3,8-11]

In the rat, lingual lipase is located in the acinar cells of serous von Ebner glands and in the lingual mucous gland.[12,13] By means of specific polyclonal antibodies to human and rabbit gastric lipase, human gastric lipase has been demonstrated to be exclusively located in the chief cells of the fundic mucosa of the human stomach,[14-16] whereas the cells producing rabbit gastric lipase are restricted in the upper base of the gastric fundic gland.[10,17] Lipase activity in human gastric content is 15 to 75 U/ml[18] or 4700 U/g of fresh mucosa in men under 50 years.[15] Gastric lipolytic activity decreases in aged persons, to reach 700 U/g of fresh mucosa in subjects over 60 years of age.[15]

Most of the biochemical studies have been performed with human and rabbit gastric lipases. Preduodenal lipases are acid enzymes (the pH optima for preduodenal lipases generally fall in the range of 2 to 6[2]), which may work in extreme pH conditions. They can act in the stomach, where the pH is 4.5 to 5.5 after a meal intake, to initiate the hydrolysis and digestion of dietary fat. Human gastric lipase does not hydrolyze a tributyrin emulsion in

TABLE 1
Classification of Digestive (Phospho)lipases

Main site of action			
Stomach (prepyloric lipases)	**Small intestine (postpyloric lipases)**		
Proposed Generic Names			
Acid lipases	Colipase-dependent lipases	Bile salt-dependent lipases	Phospholipases A2
Tissue and Cellular Origin			
Tongue (Von Ebner glands)	Exocrine pancreas (acinar cells)	Human milk	Exocrine pancreas (acinar cells)
Pharynx		Exocrine pancreas (acinar cells)	Intestine (Paneth cells)
Stomach (chief cells)			
Other Names			
Lingual lipase	Pancreatic lipase (EC 3.1.1.3)	Bile salt stimulated lipase (BSSL)	Phospholipase A2 (EC 3.1.1.4)
Salivary lipase		Carboxylic ester hydrolase (CEH)	
Pharyngeal lipase		Cholesterol esterase (CE)	
Gastric lipase		Carboxyl ester lipase (CEL)	
Pregastric lipase		Monoglyceride lipase	
Nonpancreatic lipase		Nonspecific lipase	
Preduodenal lipase		Lysophospholipase	
Substrate Specificity			
sn-3 Position of glycerides (also sn-1 and sn-2 positions of glycerides)	sn-1 And sn-3 positions of glycerides	Cholesterol esters	sn-2 Position of L-α-glycerophospholipids
		Fat-soluble vitamin esters	
		Glycerides	
		Glycerophospholipids	
		Primary, secondary, and tertiary carboxylic esters	

Adapted from Reference 2.

the absence of bile salts or alimentary proteins. The presence of most of the bile salts (taurocholate, glycocholate, taurodeoxycholate, glycodeoxycholate, or bile salts mixture), which must be added prior to the enzyme in order to prevent irreversible lipase denaturation, activates human gastric lipase on the tributyrin substrate.[2] This can be explained by the fact that human gastric lipase activity is restricted to a triacylglycerol/water interfacial tension range of 8 to 13 dyn/cm. Interfacial tension at tributyrin/water interface is 15 dyn/cm, which rapidly inactivates human gastric lipase. The fact that the enzyme has a single disulfide bridge[19] can explain the rapid interfacial denaturation of human gastric lipase. Naturally occurring bile salts like taurodeoxycholate in the concentrations prevailing in the upper small intestine reduce the interfacial tension to about 9 dyn/cm. Other amphiphiles, like most alimentary proteins, which decrease the tributyrin/water interfacial tension to values between 8 to 13 dyn/cm, also allow the expression of lipase activity.

Prepyloric lipases can hydrolyze a variety of lipid substrates including milk fat globule triacylglycerols. They are essential for the digestion of milk fat in the newborn. It was thought until recently that intragastric digestion was specific for short and medium chain triglycerides and had very little effect on long chain triglycerides, the major constituents of dietary fat. However, using purified human gastric lipase, Gargouri et al.[19] have shown that human gastric lipase catalyzes the hydrolysis of both long and short chain triacylglycerols at comparable rates. Sn-3 positions of ester bonds in triglycerides are preferentially hydrolyzed by the enzyme. An important role for gastric lipolysis is to prepare triacylglycerols for further intraduodenal digestion.

B. PANCREATIC LIPASES

Pancreatic lipase (triacylglycerol acyl hydrolase, EC 3.1.1.3) is, along with phospholipase A_2, the predominant pancreatic lipolytic enzyme. It amounts to 4.5% of the total proteins in pancreatic juice.[20] It is synthesized and secreted by pancreatic acinar cells and is present in duodenum aspirate. The optimum pH for pancreatic lipase is at around pH 8 to 9.[21] Pancreatic lipase irreversibly loses its activity below pH 5.0.[21]

Pancreatic function is immature in neonates of most species (man, dog, rat). However, a pancreatic lipase activity, although weak, is detectable during the last week of fetal development in the rat.[22] Lipase level is still reduced in the rat during the period from birth to weaning, lipase concentration reaching 40% of that in adult at the end of weaning.[22] However, in healthy adults, pancreatic lipase is considered to be of prime importance in the digestion of dietary lipids.

Pancreatic lipase (52 kDa for the pig enzyme) hydrolyzes long chain triacylglycerols (preferentially by attacking the external chains at positions *sn*-1 and *sn*-3 of glycerides), producing first diacylglycerol, then 2-mono-acylglycerol and free fatty acids. Pancreatic lipase, similarly to the gastric enzyme, displays almost no activity when substrate is in the monomeric state.[23]

Formation of substrate aggregates or emulsified particles which occurs by increasing the concentration of substrate above saturation results in a marked lipolysis by pancreatic lipase. Activation of the enzyme by adsorption to the interfaces is about 10^3-fold.[24] Presence of various amphiphiles (bile salts, phospholipids) in the duodenum inhibits *in vivo* pancreatic lipase action due to a desorption of the enzyme from the amphiphile coated interface. Pancreatic lipase requires the presence of a cofactor, the colipase, to restore its activity.[25] This cofactor is a small protein (about 10 kDa), cosecreted with lipase in pancreatic juice. Its primary role is to adsorb to bile-salt covered interface and to act as an anchor for lipase.[26] Colipase binds lipase in a 1:1 complex either in the presence or in the absence of bile salts. Lipase/colipase binding, as well as the binding of colipase to bile salts, is enhanced by free fatty acids.[27,28] Another function for colipase is to stabilize lipase, by protecting it from proteolytic digestion. Concentrations of lipase and colipase in the gut are closely equimolar. They are about $2 \times 10^{-6}\ M$ in the small intestine of man after the ingestion of a liquid test meal[29] and $2 \times 10^{-7}\ M$ in the pig under basal conditions.[30]

C. SEQUENTIAL EVENTS IN FAT DIGESTION AND ABSORPTION

With the major mechanical steps of physical-chemical behavior of fat during duodenal digestion now understood,[31,32] it is possible to describe more accurately the sequential levels of dietary lipid digestion.

1. Intragastric Lipolysis

Lipids are mixed and dispersed with the other nutrients in the stomach. Hydrolysis of alimentary lipids, catalyzed by one of the preduodenal lipases secreted from the serous glands of the tongue or from the gastric mucosa,[2,3-11] takes place in the stomach. It has been shown that these lipases are particularly stable under the acidic conditions prevailing in the stomach (pH 1 to 3) and are activated by various dietary proteins.[33] During *in vivo* gastric lipolysis, triacylglycerol is hydrolyzed to primarily α,β-diacylglycerol and free fatty acid. Production of free fatty acids facilitates emulsification of triglycerides and diglycerides.[34] The amount of hydrolysis products (diglycerides and free fatty acids) increases over time following ingestion to reach a maximum corresponding to 10 to 30%[17] to 65%[35] of total hydrolysis. The lack of complete hydrolysis is most likely due to preduodenal lipolysis inhibition by long chain free fatty acids.[33,36]

2. Intraduodenal Lipolysis

Pancreatic lipase and colipase are discharged into the duodenum and associate first to bile lipid micelles and vesicles prior to adsorption to emulsified particles from the stomach.[37] Triglyceride-rich emulsion particles containing 2 to 3% triglycerides are thus covered by a mixed lipid monolayer composed of phospholipids, free cholesterol, fatty acids, and vitamins.[38] When hydrolysis of diglycerides and triglycerides by pancreatic lipase occurs, monoglycerides and fatty acids are produced, which accumulate and form multilamellar bilayers (lamellar liquid crystal) at the emulsion surface.[32,39-42] Increased surface pressure due to the accumulation of hydrolysis products leads to a phase separation of core and surface lipids, i.e., coexistence of large liquid-crystalline structures and small unilamellar vesicles at the emulsion-water interface generating from the multilamellar liquid-crystal. Unilamellar vesicles represent primary dispersed product of fat digestion.[32] The vesicles are heterogenous in size (hydrodynamic radius, $\overline{Rh}$ = 200 to 600 Å by quasielastic light scattering) and in lipid composition.[32] In the presence of bile salts, a two-phase unilamellar vesicle-plus-saturated mixed micelle system will form.[40]

3. Intestinal Absorption of Dietary and Biliary Lipids

According to Hernell et al.,[32] saturated mixed micelles of bile salt and lipolytic products provide the ideal thermodynamic conditions for highest efficiency in absorption rates of lipids. However, a vesicle-mediated absorption could also occur in particular physiological conditions. The maximal absorption capacity of the rat small intestine is about 500 μmol fatty acid per hour.[43] Lipid absorption and transport is a rapid process, occurring in 14 min in the rat.[44]

D. SIGNIFICANCE OF GASTRIC AND PANCREATIC DIGESTION

In physiological conditions, pancreatic levels of lipase are significantly higher (about 10 times) than those required to ensure complete digestion of dietary triglycerides (100 to 150 g daily in human).[1] However, digestion of triacylglycerols in the stomach is not negligible, since gastric lipase hydrolyzes between 10 to 65% of dietary lipids.[15,35] In adults rats, in which gastric lipolysis depends exclusively on oral lipases, when saliva is diverted from the stomach by ligation of the esophagus about 8.6% bovine milk triglycerides are hydrolyzed within 1 h, the reaction products being diglycerides (7.9%), monoglycerides (0.7%), and free fatty acids (3.2%) compared to 18.7% hydrolysis in control animals (the reaction products being diglycerides [16.4%], monoglycerides [2.3%] and free fatty acids [7.3%]).[45] Absence of gastric lipolysis leads to a marked decrease in fat hydrolysis in the duodenum and ileum since free fatty acids amount to 77.3% of total fatty acid in control animals vs. 54.3% in experimental animals. There is a synergy of hydrolysis between gastric and pancreatic lipases: prehydrolysis of triglycerides by pre-duodenal lipase enhances pancreatic lipase action.[45] Partially digested triglycerides are hydrolyzed at much higher rates than triglycerides by pancreatic lipase. Furthermore, the binding of colipase to lipase and to bile salts is enhanced by the presence of free fatty acids.[27,28]

The gastric phase of fat digestion is also thought to play an essential role in milk digestion in preterm or fullterm neonates, where bile salt levels[46] and pancreatic lipase secretions[47-49] are quite low. Functional immaturity of exocrine pancreatic function, although associated with excretion of up to 25% of ingested fat, does not appear to have a major clinical impact.[50,51] It has been shown that 14 to 60% of milk lipid is hydrolyzed to diglyceride, monoglyceride, and free fatty acid in the first 30 to 60 min following ingestion.[35] Subsequent hydrolysis is achieved by a milk lipase (optimum pH = 8.5), which acts in the intestine and requires bile salts to be active. This bile salt stimulated lipase (BSSL) is identical to carboxyl ester hydrolase found in pancreatic juice.[52] *In vivo,* the enzyme hydrolyzes monoglyceride to free glycerol and fatty acid. Prior intragastric digestion greatly enhances BSSL activity where 30% hydrolysis by BSSL is achieved, compared to 5% hydrolysis without prior exposure.[35] Thus, concerted action of the two enzymes (preduodenal lipase and BSSL) explains the efficiency of fat digestion in the suckling neonate.

Isolated lipase deficiency,[53-55] colipase deficiency,[56] and combined lipase-colipase deficiencies[57,58] have been reported. These genetic diseases and pancreatic lipase deficiency caused by cystic fibrosis or chronic alcoholism are commonly associated with fat excretion in the stools[54,58] but are not related to growth deficit or malnutrition. Dietary fat of 50 to 70% in severe pancreatic insufficiency[59] or in isolated lipase and colipase deficiency[58] is indeed absorbed. In these situations, gastric enzyme plays an essential role in the digestion, acting first in the stomach and then in the upper small intestine where the postprandial pH is still acidic.[60]

III. ADAPTIVE RESPONSE OF GASTROINTESTINAL LIPASES TO DIETARY FAT

A. PREDUODENAL ADAPTATION
1. Gastrointestinal Transit

Inhibitory effect of fat in the small intestine on the rate of gastric emptying has been reported.[61] However, the importance of the recent dietary history in small bowel transit time has been emphasized.[62] Both gastric emptying and mouth to cecum transit time are significantly faster after consumption of a high fat diet for 14 days than after consumption of a low fat diet in human[62] (half-time for gastric emptying = 98 vs. 147 min, mouth to cecum transit time = 240 vs. 360 min, respectively). According to the authors, the changes observed in gastrointestinal transit time could be related to the release or sensitivity to humoral transmitters such as cholecystokinin (CCK) and peptide YY (PYY). These peptides have been shown to be released by the presence of lipid in the small intestine and to delay gastric and small intestinal transit.[63-65]

2. Prepyloric Lipases

In contrast with pancreatic lipase, which level is very low in the newborn,[47-49] the concentration of human gastric lipase at birth is quite comparable to that found in the adult.[18] In the rat, lingual lipase activity was found to raise exponentially immediately after birth.[66] This suggests that ingestion of milk might be of major importance in eliciting preduodenal lipase activities during the first days of postnatal life.

Only a few studies have clearly established that preduodenal lipases were able to undertake an adaptive response to dietary fat. The first preliminary observation was made in 1978 by Hamosh,[4] who showed that rat lingual lipase content in animals fed a high fat diet (22% corn oil) for 2 weeks was 1.4-fold higher than that of animals fed a similar purified diet containing 4% corn oil. A more detailed examination of the effect of dietary fat on lingual lipase activity established that even 10% fat in the diet (composed of a mixture of half sunflower oil and half lard), when given to rats for 3 weeks, was able to induce the specific activity of lingual lipase by a 1.23-fold factor, the maximum increase (1.30-fold) being reached when rats were fed a 20% fat diet.[67] Adding more fat to the diet (up to 30%) did not further enhance lingual lipase activity. Maximum values of enzyme-specific activities were measured after 1 week of feeding a 20% fat diet, and no additional change in lingual lipase specific activity occurred when rats were fed the 20% fat diet for 4 weeks. Whatever the type of fat added in the diet (lard, sunflower oil, olive oil, peanut oil, soybean oil, corn oil, salmon oil, or butter), no significant differences in lipase induction factor could be measured, i.e., the adaptive response of lingual lipase to fat was not affected either by the chain length, by the number, or the position of the double bonds in the fatty acid molecule, even when dietary fat was provided in a moderate amount (10% by weight).[67] Evidence of an adaptive response to dietary fat has also been described for the gastric enzyme in the rabbit.[68] Similar findings concerning the amount of fat inducing a maximum increase in gastric lipase activity (6% fat in the diet for fundus lipase and 12% for cardia lipase) or the lack of a specific effect of the fatty acid composition of dietary triglyceride molecules on lipase adaptation were reported.[68] However, feeding rabbits a 12% dietary fat diet for 1 week did not change the specific activity of lipase in the cardia or fundus mucosae, as compared with feeding a 2.7% dietary fat diet. A 2-week period was necessary to achieve adaptation of gastric enzyme.

B. PANCREATIC LIPASE (AND COLIPASE) ADAPTATION

Adaptation of pancreatic lipase to the lipid content of the diet is clearly established in a number of species, including the rat,[69-75] dog,[76] chicken,[77] and hog.[78,79] The response of pancreatic lipase may be influenced by a number of factors such as the mode of meal intake (intestinal perfusate, liquid, or solid meal), the type of dietary triglycerides, the degree of saturation of fatty acid chains, the other components of the diet, the species studied, and the age of animal. This can explain the differences measured in the extent of induction of pancreatic lipase.

In the rat, pancreatic lipase concentration is high at birth but falls after 1 d.[48] Emergence of lipase is quite slow since it represents 40% of the value in adult rats just before weaning.[22] This suggests that an increase in pancreatic lipase before weaning is not a direct consequence of milk fat intake during suckling. Low pancreatic lipase secretions have also been reported in the human infant.[49-51] and in the neonatal dog.[35] According to Snook,[48] lipase accumulation before weaning occurs mainly in response to a developmental clock that may be affected only indirectly by the type of food eaten. In fact, weaning rats on a high lipid diet (50% corn oil) does not increase the rate of lipase accumulation beyond that observed in rats weaned on to a high starch diet during the first 25 d after birth.[70] By contrast, 30-d-old rats on the high fat diet have higher lipase content than 30-d-old rats on the high starch diet.[70] Thus, adaptation of pancreatic lipase to nutritional components only occurs after the usual weaning period (about 3 weeks in the rat).

1. Pancreatic Secretion after Intraduodenal Perfusate or Meal Intake
a. *Intraduodenal Perfusates*

Intestinal perfused fat digestive products have various effects on pancreatic secretion.[80-82] Undigested triglycerides (corn oil) do not stimulate more pancreas output than a control perfusion with 0.15 M NaCl. However, when the same amount of triglyceride is incubated with lipase-active pancreatic juice, bile, and saline, prior to perfusion, pancreatic bicarbonate and protein secretion are significantly enhanced. Triglyceride hydrolysis is thus required to stimulate pancreatic secretion.[80] Monoglyceride itself is a luminal stimulus of pancreatic secretion. Ability of fatty acids to stimulate pancreatic secretion has been related to their chain length.[80,82] Fatty acids of longer than eight carbons in chain length increase output of pancreatic bicarbonate and protein. The pancreatic response to fatty acids is dose-related over fatty acids concentrations ranging from 5 to 160 mM. The mode of dispersion of fatty acid can also affect pancreatic response since oleate dispersed as a micellar solution of fatty acid salt (soap) or in polysorbate-80 is a more efficient stimulant than when dispersed in monolein-bile stabilized micelles, or than nondispersed oleate.[80]

b. *Response to Meal Intake*

The secretory response of the pancreas may be quite different, depending on the mode of lipid administration (intraduodenal perfusions or meals). In the healthy human, increased amounts of fat in the diet (10 to 40%) are associated with enhanced rates of interdigestive and postprandial exocrine pancreatic enzyme secretion and lipase output.[83] In long-term adapted dogs, consumption of a diet rich in olive oil (= rich in monounsaturated fatty acids) induces no change of the flow of pancreatic secretion throughout the entire postprandial period, whereas a diet rich in sunflower oil (= high polyunsaturated fatty acid content) significantly enhances the volume of pancreatic juice secretion during the first 3 h after the meal, as well as bicarbonate levels and protein output during the first 2 h after eating.[84] Lipase activity and specific activity did not significantly differ between the two high fat-fed groups. However, due to the enhanced volume of secreted pancreatic juice throughout the experimental period, lipase production was significantly higher in the group fed the sunflower diet than in the group fed the olive oil diet. This suggests that the degree of unsaturation of dietary fat can affect lipase production.

Pancreatic enzyme response to perfusates and to meals is thought to occur via a hormone-mediated pathway. Oleic acid is known to be a potent releaser of CCK.[63,64] Some other hormones, such as secretin, also mediate pancreatic output stimulation.[85,86] According to Ballestra et al.,[84] the lack of response of bicarbonate and protein content in pancreatic juice to the olive oil meal might be due to the fact that oleic acid also releases other peptides from the gut such as pancreatic peptide (PP)[87] and PYY,[65] which inhibit CCK- and secretin-stimulated pancreatic secretion. However, a direct effect of lipid on pancreatic secretion cannot be excluded since a fat-rich diet (34% fat = 17% soybean oil + 17% coconut fat) is able to stimulate pancreatic secretion as well as lipase secretion of an isolated perfused rat pancreas,[88] indicating an action independent of altered hormonal stimulation and neural reflexes.

2. Effect of Type of Fat, Chain Length, and Degree of Saturation
a. Type of Fat

Comparison of the response of pancreatic lipase to type of fat, changes in chain length, and degree of saturation of fatty acids reported from different studies is difficult, because of varied experimental procedures, controls, and dietary compositions. We have tentatively compared three studies,[70,73,74] performed in the rat, where lipase assays were measured in tissue homogenates and under conditions where the presence of colipase and bile salts was unnecessary for full lipase activity,[70,74] or in the presence of an excess of crude exogenous colipase.[73] Table 2 reports induction factors for lipase (and for colipase, when measured) in response to high triglyceride fat diets. Diets rich in phospholipids (20%) significantly increase lipase-specific activity in pancreatic tissue and juice, whereas a diet enriched with cholesterol does not induce any change in pancreatic lipase content in the rat.[89]

b. Chain Length

According to Sabb et al.,[74] no influence of chain length (from C_{12} to C_{18}) was seen on lipase activity in weanling rats (40 to 60 g) fed high fat diets (29% lipid, i.e., 67% kcal as fat) for 1 week since coconut oil, the dietary

TABLE 2
Fatty Acid Composition of Diets, Induction Factors for Lipase and Colipase in Response to the Ingestion of Fat Diets

Diet	Main FA[a] (%)	Unsaturated FA (%)	Induction Factors			
			Lipase			Colipase
			Deschodt-Lanckman et al.[70]	Sabb et al.[74]	Saraux et al.[73]	Saraux et al.[73]
Tricaprylin	98% $C_{8:0}$	<1	1.8	—	—	—
Medium chain triglycerides	99% $C_{8:0}$–$C_{10:0}$	<1	—	—	1.5	1.3
Tristearin	2% $C_{16:0}$ 96% $C_{18:0}$	<2	1.8	—	—	—
Coconut oil	20% $C_{8:0}$–$C_{10:0}$ 45% $C_{12:0}$	<15	—	1.8	2.0	1.8
Lard	29% $C_{16:0}$ 23% $C_{18:0}$ 33% $C_{18:1}$	45	1.7	2.1	2.0	2.0
Olive oil	14% $C_{16:0}$ 65% $C_{18:1}$	80	3.6	1.5	—	—
Corn oil	14% $C_{16:0}$ 33% $C_{18:1}$ 49% $C_{18:2}$	82	4.0	1.9	—	—
Sunflower oil	7% $C_{16:0}$ 20% $C_{18:1}$ 69% $C_{18:2}$	91	3.4	—	2.4	2.0

a   FA stands for fatty acid.

fat with the greatest percentage (45%) of C_{12} chain length used in their study, stimulated lipase similarly to the other fats (corn oil, lard, safflower oil, butter, olive oil) (Table 2). Regarding the effects of lard (99% fatty chain superior to $C_{12:0}$: 2% $C_{14:0}$, 29% $C_{16:0}$, 23% $C_{18:0}$, 2% $C_{16:1}$, 33% $C_{18:1}$, 10% $C_{18:2}$) and coconut oil (20% $C_{8:0}$ to $C_{10:0}$, 45% $C_{12:0}$, 18% $C_{14:0}$, 10.9% $C_{16:0}$, 9.8% $C_{18:0}$) in adult rats (180 g) fed 40% fat diet for 3 weeks, Saraux et al.[73] observed that lipase was stimulated to the same extent (about twofold). However, lipase- and colipase-specific activities were less efficiently increased (only 1.5- and 1.3-fold, respectively) by diets containing medium chain triglycerides (Table 2). The fact that absorption of the medium chain fatty acids resulting from hydrolysis of medium chain triglycerides occurs via the portal route, the fatty acids being rapidly transported to the liver and oxidized,[70] might explain their relative inaction with respect to inducing maximal pancreatic lipase content. Unfortunately, in Sabb's and Saraux's studies, all experimental diets, except the medium chain triglyceride one, contained varying amounts of unsaturated lipids, which, in turn, may influence induction factor values for pancreatic lipase. Indeed, Deschodt-Lanckman et al.[70] reported that a 50% (w:w) fat diet with tricaprylin as the lipid source, given to adult rats (180 to 220 g) for 5 d, poorly increased lipase content, when compared to other lipid diets. This effect seemed to be related to the degree of saturation rather than to the chain length of the dietary fat since tricaprylin (98% $C_{8:0}$) and tristearin (96% $C_{18:0}$) diets led to the same induction of lipase-specific activity in pancreas homogenates (1.8-fold)[70] (Table 2).

c. Effect of Degree of Saturation

Only a few studies have dealt with the adaptative response of the pancreas as a function of the degree of unsaturation of triglycerides,[70,73,74,79] reaching opposite conclusions. In the rat, as far as high fat diets are concerned, unsaturated fats were better inducers than saturated according to Deschodt-Lanckman et al. (using 50% fat diet[70]), whereas for Saraux et al. (using 40% fat diets[73]) and for Sabb et al. (using 29% fat diets, i.e., 67% kcal as fat[74]), the degree of saturation was found to have no impact on lipase[73,74] and colipase.[73] On the other hand, in Sabb's study, when rats were fed diets containing 17.4% lipid (i.e., 40% kcal fat), only the highly unsaturated safflower oil (polyunsaturated/saturated ratio = 7.9) increased pancreatic lipase activity 1.63-fold compared to the low fat control diet, 5% lipid (i.e., 11% kcal as corn oil). By contrast, 17.4% corn oil, lard, butter, olive oil, and coconut oil diets did not significantly change lipase levels.[74] In the pig, only one study concerning the degree of saturation of dietary lipids is available.[79] This study reported that pancreatic lipase was highly affected by the nature of dietary lipids since lipase activity was 1.6- and 2.75-fold enhanced in pancreata of pigs fed 21% lard and 21% sunflower oil, respectively, as compared to the control diet (3.5% lard + 3.5% sunflower oil). Together, these results suggest that pancreatic lipase is sensitive to the degree of unsaturation of triacylglycerol fatty acids and is affected only below the level of maximal enzyme response.

3. Response of Pancreatic Lipase and Colipase to Amount of Dietary Fat

a. Lipase-Specific Activity

The effect of varying the proportion of fat to carbohydrate in the diet while keeping the diet isonitrogenous and isocaloric is reported in Table 3 (from Reference 75). Indeed, colipase, but not lipase, is particularly responsive to protein intake.[90] It is therefore necessary to investigate adaptation of lipase and colipase to dietary fat keeping the protein intake at a constant level. Reduction in sunflower oil (0 to 1%, by weight), as compared to the control diets (3 or 5%), resulted in nonsignificant changes in lipase activity. Lipase rose significantly between 5 and 10% fat in the diet (1.4-fold) and even more between 10 and 25% dietary lipid (2.5-fold, as compared to the control diet). Between 25 and 30% dietary fat, lipase activity was at its maximum level. Thus, the amount of sunflower oil (polyunsaturated/saturated ratio = 5.19) necessary to induce an enzyme response is around 10% (by weight), corresponding to 25% dietary calories as fat. This amount depends on the type of fat ingested (Table 4). When highly saturated butter is given to animals (polyunsaturated/saturated ratio = 0.03), more than 34% of the fat content (55% of total energy) must be provided to enhance lipase activity in the tissue.[71] Examination of previous investigations on specific tissue content in pancreatic lipase in response to changes in dietary lipids at the expense of carbohydrate[67,71,73-75] (Table 4) seems to indicate that the threshold of fat content above which there is a significant induction of pancreatic lipase is lower for unsaturated fat than for saturated fat. However, whatever the type of lipid in the diet, the lipase activity in the tissue rises as the level of dietary fat increases, and maximal lipase induction factors vary from 1.8[71] to 2.8.[74] The four-[70] and sixfold[91] increases in lipase activity reported in other studies performed in the rat may be due to the fact that rats were not fed isocaloric diets, which could have altered the amounts of nutrients ingested. In the pig, inductibility of pancreatic lipase varies from 3 when specific activity was measured in pancreatic homogenates[79] to 7 when the amount was determined in pancreatic juice.[78] Once more, differences in experimental procedures may explain these different inductions.

b. Colipase-Specific Activity

As far as the existence of an adaptation of pancreatic colipase to nutritional lipids is concerned, contradictory findings have been reported. No adaptation of colipase to dietary lipids was observed in either rats or mice by Vandermeers-Piret et al.,[92] whereas several studies have described colipase adaptation

TABLE 3
Specific Tissue Content in Lipase and Colipase Activity and Relative Synthesis of Lipase in Pancreatic Lobules in Response to Changes in Lipid in the Diet[75]

Sunflower oil (%, by weight)	% Energy provided as fat	Enzyme-specific activity in pancreatic tissue (U/mg)		Lipase/Colipase	Lipase relative synthesis (% of total synthesis)
		Lipase	Colipase		
0	0	207 ± 27	230 ± 29	0.90 ± 0.05	6.9 ± 0.9
1	2.5	190 ± 30	193 ± 29	0.99 ± 0.10	7.8 ± 0.4
3	7.4	186 ± 25	175 ± 30	1.08 ± 0.20	7.6 ± 0.6
5	12	219 ± 39	224 ± 42	0.99 ± 0.20	7.4 ± 0.3
10	25	259 ± 51	253 ± 44	1.01 ± 0.04	9.8 ± 0.8
20	49	372 ± 38	280 ± 36	1.33 ± 0.10	15.3 ± 1.7
25	62	462 ± 130	396 ± 90	1.16 ± 0.10	13.6 ± 0.9
30	74	447 ± 64	332 ± 65	1.37 ± 0.17	15.6 ± 2.0

TABLE 4
Experimental Procedures and Type of Dietary Fat Related to the Amount of Lipids Necessary to Induce Minimal and Maximal Lipase Response in Rat Pancreatic Tissue

		Experimental conditions and resulting lipase activity adaptation				
	Authors	Snook[71]	Saraux et al.[73]	Armand et al.[67]	Sabb et al.[74]	Wicker and Puigserver[75]
Experimental procedures	Dietary fat	Butter	Lard	Sunflower oil/lard	Corn oil	Sunflower oil
	Polyunsaturated/saturated (from Ref. 67)	0.03	0.13	2.6	5.0	5.19
	adaptation period	2 weeks	3 weeks	3 weeks	1 week	10 d
Minimum amount of fat-inducing lipase adaptation	% By weight of dietary fat	34.6	23	20	23.2	10
	% Dietary calories as fat	55	45.4	39.3	54	25
	Resulting induction factor for lipase	1.2	1.57	1.89	2.4	1.38
Maximum induction of lipase activity	% Dietary calories as fat	77	73.2	52.7	67	62
	Maximum inducation factor	1.83	2.15	2.15	2.85	2.48

to dietary fat in the rat[73,75,93,94] as well as in the pig.[78] These conflicting findings may be explained by quantitative and qualitative differences in dietary protein ingestion.[73]

The effect on rat colipase activity of changing relative proportions of dietary fat and carbohydrate while keeping the diet isonitrogenous (22% protein, the daily protein intake being around 4 g per day) and isocaloric[79] is described in Table 3. In contrast with lipase, colipase activity was found to exhibit a significant enhancement (1.3-fold) in the pancreas of rats fed the lipid-free diet, as compared to rats fed the standard diet. Between 3 and 25% fat in the diet, a 2.3-fold increase in colipase-specific activity was measured in the pancreas, which was lower than that of pancreatic lipase. Colipase activity then slightly declined between 25 and 30% dietary fat but was still significantly higher (1.9-fold) than that in control rats. Due to the differential adaptation of lipase and its cofactor, the colipase/lipase ratio progressively decreased from 0.93 in control animals to a minimal value of 0.73 in 30% lipid-fed rats. Nonparallel adaptations of lipase and colipase, where colipase/lipase ratio dropped from 0.78 to 0.51, have also been reported in the rat[94] when dietary fat increased from 3 to 19.5% and in the pig,[78] where the ratio decreased from 2.35 in 25% fat-fed pigs to 1.77 in 5% fat-fed animals. In hyperlipidic diet-fed rats, intrinsic colipase level is therefore not high enough to allow full activation of pancreatic lipase when bile salts are present.

Saraux et al.[73] showed that lipid intake modulates colipase in the rat only when the protein intake is between 3.5 and 6.0 g per day. When protein ingestion is below 3.5 g, colipase does not adapt to dietary lipids (from 2 to 40%, by weight), while above 6.0 g protein intake, colipase level is maximal, even with 2% lipid in the diet. Between 3.5 and 6.0 g, lipid intake modulates colipase and lipase in a parallel way, colipase/lipase ratio being always around 0.7. In their conditions, a correlation coefficient of colipase with fat intake of 0.64 was reported. The correlation of colipase with protein intake was even stronger (0.85).[73] The particular sensitivity of colipase to nutritional proteins might explain that their lipidic diets did increase colipase content to the same extent as lipase since, unfortunately, their experimental diets were not isonitrogenous, an increase in the protein content of the diet (from 17 to 37%) accompanying the rise in fat (from 0.7 to 45%, by weight, corresponding to 1.7 to 73.2% of total energy). Nevertheless, these findings suggest that there is a cooperative effect on colipase activity between lipid and protein in the rat in the range of 3.5 to 6.0 g daily protein intake.[73].

c. *Lipase Biosynthesis*

There is little information on the role of dietary lipids in the regulation of the earlier steps in the secretory pathway of acinar cells, especially their effect on protein synthesis. Data concerning lipase biosynthesis, as influenced by dietary lipids, are quite scarce.[71,75,95] Snook,[71] when directly measuring protein synthesis *in vitro* by the rate of incorporation of ^{14}C-valine into pancreas slices from rats fed a high fat diet (providing more than 55% of the dietary calories as fat) or a control diet, reported that slices from rats fed the high fat diet incorporated more label into pancreatic proteins. We found that the increase in the synthesis of total pancreatic secretory proteins was about twofold in magnitude[75] by incubating pancreatic lobules from rats fed high fat (25 and 30% sunflower oil) and standard (3% sunflower oil) diets in the presence of ^{35}S-methionine for 30 min and measuring incorporation of labeled methionine into the pancreatic secretory proteins.

The increase in pancreatic lipase content by the high fat diets was directly caused by an enhanced synthesis of the enzyme, as indicated by the good correlation between the increases in lipase-specific activity and relative synthesis. The specific enhancement in lipase fractional synthetic rate (expressed as the percentage of radioactivity incorporated into pancreatic lipase, compared to radioactivity incorporated into the total mixture of exocrine proteins) was about twofold in the high fat diet fed animals (Table 3), a plateau in lipase biosynthesis being reached when rats were given the 20% fat diet.

4. Time Course of the Variations in Lipase and Colipase Activities and Biosynthesis in Pancreatic Tissue

a. *Lipase Activity and Biosynthesis*

Time studies over a 5-d period of high lipid diet ingestion[70,95] have shown that lipase content was significantly enhanced (by 2-fold[70] and 1.4-fold[95]) in the pancreas of fat-fed rats as early as the second day of dietary induction (Figure 1). Further consumption of the high lipid diet resulted in a further increase in lipase-specific activity (3-fold[70] and 1.9-fold[95]). Adaptive response of lipase activity is rapid, although several days of induction are necessary for lipase to reach a maximal steady-state level. Long-term (2 to 4 weeks) dietary fat feedings lead to increases in lipase activity similar to those measured in 1-week fed rats.[67] Lipase dietary adaptation is reversible since shifting lipid-adapted rats (on a 50% corn oil diet) to a standard diet (4% corn oil) results in a decrease in lipase activity, the kinetics of lipase decay being almost exactly opposite to those previously observed in rats shifted from a low fat to high fat diet.[70]

Ingestion of a 25% sunflower oil diet has an even more rapid effect on total protein and lipase synthesis, as measured by incorporation rates of ^{35}S-labeled methionine into pancreatic proteins.[95] Increase in the rate of total pancreatic protein synthesis is almost immediate (1.6-fold on the first day of adaptation) and persists throughout the ingestion period of the lipid-rich diet.[95] When measuring rates of synthesis of individual exocrine proteins, the relative synthetic rate of lipase increases by a 1.4-fold factor, as early as on the first day of diet intake. This corresponds to a calculated induction of 2.2-fold of lipase absolute synthesis. This rise in lipase synthesis can be easily visualized on following the separation of radioactive labeled pancreatic proteins by using PAGE electrophoresis and gel fluorography (Figure 2). Enhancement in lipase biosynthesis is progressive, and a 2.2-fold increase in fractional rate of lipase synthesis is reached on the 5th day of induction.

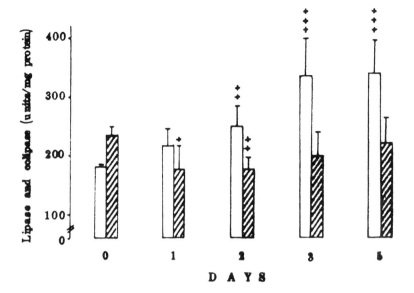

FIGURE 1. Short-term dietary adaptation of lipase (empty blocks) and colipase (shaded blocks) activities in rat pancreatic tissue, as a function of the ingestion period (0 to 5 d) of a high fat diet (25% sunflower oil).[95] Each point represents the mean value of five determinations. The vertical bars indicate the SE. The P values, calculated from Student's t-test comparing 25% lipidic diet to the control diet (3% lipid), are $<0.05(+)$, $<0.01(++)$, $<0.001(+++)$.

b. Colipase Activity

Short-term adaptation of colipase does not resemble that of lipase. Colipase responds more slowly than lipase; its content in the pancreas decreases initially and then increases gradually in response to dietary fat (Figure 1). Coenzyme level on day 5 is quite comparable to that of rats fed the standard diet.[95] Increasing the adaptation period to 10 d leads to a twofold induction in colipase activity.[75]

Although the alterations in food composition change the lipase and colipase content, the mechanisms for the changes are probably distinct, as suggested by the differential short-term adaptation responses of both pancreatic proteins.

C. POSSIBLE REGULATORS OF PANCREATIC LIPASE ADAPTATION TO DIETARY FAT

1. Nutritional Regulators

The role of triglyceride hydrolysis products on pancreatic secretion is now well documented.[80-82] We have seen that intraluminal free fatty acids and monoglycerides (but not undigested triglycerides) stimulated pancreatic secretion in man, dog, and rat.[80-82] However, contradictory findings have been

D A Y S

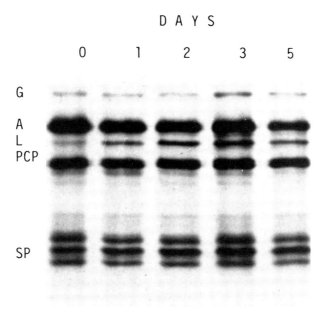

FIGURE 2.   Fluorographic pattern of [³⁵S]-methionine-labeled pancreatic secretory proteins as influenced by a 0 to 5 day ingestion period of fat diet (25% sunflower oil).[95] Pancreatic lobules prepared from rats fed the lipid diet for 0 to 5 d were incubated in a Krebs-Ringer buffer in the presence of [³⁵S]-methionine (3.7 MBq/ml) for 30 min and homogenized. Radiolabeled proteins (200,000 cpm per lane) were separated by PAGE electrophoresis and the gel was impregnated with a fluorographic reagent, prior to autoradiography. Radioactive glycoprotein, amylase, lipase, procarboxypeptidases A and B, and serine proteases are indicated as G, A, L, PCP, and SP, respectively.

reported, as far as the effect of intravenous administration of lipid on pancreatic secretion is concerned: intravenous fat has been reported either to exert no stimulation of pancreatic secretion in dog[96,97] and human[98] or to increase the lipase content in dogs[99] and rat.[100] The first findings (i.e., no effect of intravenous lipid on pancreatic secretion) suggest an indirect action of triglyceride hydrolysis products, through the release of a second messenger from the intestinal mucosa. By contrast, the latter results (i.e., increase of lipase consecutive to intravenous fat) indicate that the hydrolysis products have to pass through the intestinal wall to induce pancreatic response. Other intermediates in triglyceride metabolism, such as ketones,[101] have also been thought to play a role in the regulation of lipase. The circulating levels of ketones in the blood are increased during adaptation to a high fat diet. When infused into rats for 5 days, they lead to a 40% raise in lipase activity.[101] However, their role in lipase induction seems to be relatively limited. Furthermore, the direct effects of β-hydroxybutyrate (β-HBA) on a primary culture of pancreatic acinar cell is unclear. β-HBA affects lipase activity only in cells from low fat fed rats (11% energy as fat corresponding to 5% fat) cultured in low glucose medium for 48 h, when the antecedent diet is a semipurified diet, prepared with well-defined components.[102] The magnitude of lipase induction in these conditions is relatively small (33%). On the other hand, when the antecedent diet, providing 10% energy as fat, is prepared from crude components, no increase in cellular lipase activity was observed.[102] According to the authors,[102] their results support a role for the ketones in the regulation of pancreatic lipase; however, ketones are not the sole mediator and might act in conjunction with other food components.

2. Hormonal Regulators

Hormonal factors are thought to be involved in lipase response to dietary fat. Fat hydrolysis products are assumed to act on the intestinal mucosa (more likely on the proximal part of the small intestine[103]), leading to a secretion of peptide hormones. In fact, the presence of lipid in the small intestine leads to a release of CCK and secretin in man,[63] dog,[64,85,86] and rat[104] but not in the pig.[105] CCK does not seem to be the mediator of lipase adaptation since intravenous administration of caerulein, a synthetic analog of CCK, does not modify lipase synthesis.[106] On the other hand, secretin is a possible intermediate between ingested fat and pancreatic lipase regulation since the hormone exerts stimulatory effects on lipase concentration in pancreatic tissue[107] and on total protein[108] and lipase[109] synthesis. On the other hand, secretin does not alter the lipase content of cultured pancreatic cells.[110] Injection of secretin for 15 days increases the pancreatic weight, RNA amount, and lipase content in the rat pancreas.[107] These findings have been confirmed and studied more extensively by others.[108,109] Infusion of synthetic secretin in conscious rats (16 clinical units per kg and h) elicits total protein synthesis after a 2- to 3-h lag period, a maximal twofold enhancement being obtained after a 12-h infusion.[108] When measuring rates of synthesis of individual exocrine proteins by following the incorporation of a mixture of 15 ^{14}C-labeled amino acids into proteins, a pronounced increase (by a 2.5 factor) in the relative synthetic rate of lipase was observed.[109] This leads to a calculated 4.9- and 4.1-fold change with time, after 12- and 24-h stimulation, respectively, in lipase absolute synthetic rate. These values are very close to those obtained from high fat fed rats,[75,95] as far as relative and absolute synthesis rates of lipase are concerned. Therefore it can be concluded that secretin remains a very possible mediator in lipase adaptation. Whether other hormones in the secretin family such as glucagon, vasoactive intestinal polypeptide (VIP), and gastric inhibitory polypeptide (GIP) also have a stimulatory effect on lipase synthesis has not yet been demonstrated.

Finally, other nonpeptidic hormones, such as steroid ones, are not responsible for nutritional changes of lipase synthesis.

D. MECHANISMS INVOLVED IN THE REGULATION OF
PANCREATIC LIPASE AND COLIPASE BY DIETARY LIPIDS
1. Outline of the Expression of Pancreatic Enzymes
a. *General Scheme of Eukaryotic Gene Expression*

Numerous studies in many laboratories have contributed to our present understanding of gene organization and the control of their expression (for review, see References 111 through 113). All eukaryotic cells follow the same sequence of operations to express one gene which codes for a protein. The essential steps of pancreatic gene expression are summarized in Figure 3. Control of gene expression can be exerted at any one of these essential steps.

The first level of control is that of gene transcription. Protein-coding genes are transcribed by RNA polymerase II (RNA pol II or B). The sequences required for accurate and efficient initiation of transcription (= promoters) are dispersed upstream from the start site of transcription, generally locating between about -40 and -200 bp. Some promoter elements, such as TATA, GC, and CCAAT boxes, are common to many genes transcribed by pol II. These sequences interact with general transcription factors and RNA polymerase II leading to the formation of stable preinitiation complexes. Less common regulatory signals may confer inducible transcription, for example in response to a specific hormone, and are also binding sites for specific transcription factors. There are multiple DNA binding factors which can act on gene transcription as positive or negative regulatory elements, vary in abundance or in binding ability, and compete with other factors for DNA binding sites. The frequency of initiation of mRNA synthesis depends on all these interactions of transcription factors with gene promoters in a sequence- or conformation-specific manner.

Another class of transcriptional controlling elements is that of enhancers which, in contrast with that of promoters, are not located in the vicinity of regulated genes but can function over large distances of more than 20 kilobases. They can enhance the transcription from a gene more than 100-fold. Stimulation is possible regardless of their orientation and their position from the gene they are turning on. Enhancer-like sequences have been found for the genes of chymotrypsinogen,[114] elastase,[115] and other pancreatic exocrine enzymes.[116] Enhancers are often responsible for tissue- or development-specific control of gene expression. These regulations also occur via the interaction of DNA binding proteins with consensus sequences.

The eukaryotic gene is constituted of discrete regions on the DNA molecule (exons) which contain all coding information for the final gene product, separated by extra DNA stretches (introns) which do not carry any coding information. The DNA gene is integrally transcribed into heterogenous nuclear RNA (hn RNA) in the cell nucleus.

Although transcriptional control in eukaryotes is fundamentally important, several posttranscriptional steps must be correctly realized in order to maintain the steady state levels of functional mRNAs in the cytoplasm. The primary

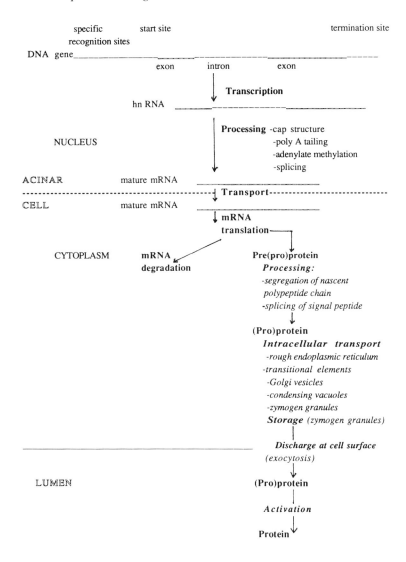

FIGURE 3.   Essential steps of gene expression of pancreatic secretory proteins.

RNA transcript is converted to mRNA in the nucleus through several modifications: a cap of 7-methylguanosine and a tail of polyadenosine tract are added to the 5′ and 3′ end of RNA, respectively. Internal adenine residues (1 out of 400 Ap) are methylated and introns are eliminated through RNA splicing at consensus sequences. Mature mRNA is then transported to the cytoplasm (about 5% of the mass of mRNA) where it will be translated.

The events of pancreatic protein synthesis are carried out on the ribosomes which are attached on the rough endoplasmic reticulum (RER). Protein synthesis is a complex, multistep process, which can be affected by a variety of physiological and pharmacological stimuli (for review, see Reference 117). Translational regulation occurs mainly at the initiation step of protein synthesis.

The rate of mRNA turnover, or degradation, can also significantly affect the rate of protein synthesis.[118] The half-lives of mRNAs in eukaryotes vary widely, from seconds to several hours, or even days. Poly(A) tract in conjunction with poly(A)-binding protein has been shown to protect mRNA chains against exonucleolytic attack. The nucleotide sequences in the 5' noncoding segment of mRNA may also be involved with the decay process. The influence of the coding region on mRNA turnover has also been demonstrated. The stabilization of effect appears to involve ribosome interaction.

b. Synthesis and Subsequent Processing of Pancreatic Secretory Proteins

Pancreatic enzyme level can also be regulated at a posttranslational stage. The general pathway and the time course of intracellular transport of secretory proteins are essentially due to the work of Jamieson and Palade.[119] Pancreatic proteins are synthesized as large precursor proteins (presecretory protein) on membrane-bound ribosomes. During synthesis, proteins are vectorially transported across the microsomal membrane where the signal peptide is spliced.[120] Newly synthesized secretory proteins, once segregated in the cisternal space of the RER, pass successively from the vesicular space of the RER into small vesicles to be conveyed to the Golgi complex and to condensing vacuoles. Glycosylation of some proteins on asparagine residues is initiated in the RER and completed in the Golgi apparatus. Condensing vacuoles progressively concentrate to form mature zymogen granules. The pathway for all pancreatic enzymes is essentially the same. Secretory proteins are stored within mature zymogen granules until their discharge from the apical portion of the acinar cell upon hormonal or nutritional stimulation. A mean transit time of 30 min has been estimated in the guinea pig. Activation of pancreatic proteins occurs in the gut, by cleavage of trypsinogen activation peptide by enterokinase, and subsequent removal of activation peptides of other pancreatic proteins by trypsin.

2. Induction of Lipase and Colipase mRNAs in Response to High Fat Diet

We have shown that lipase accumulation in pancreatic tissue, consecutive to fat absorption, resulted from a selective stimulation of enzyme synthesis.[75,95] The question arises then as to the molecular level of the mechanisms of enzyme induction.

A major problem encountered in the purification of mRNA from the pancreatic tissue is the presence of high levels of pancreatic ribonuclease, especially in the rat where the mean content is about 260 μg ribonuclease per gram wet weight of pancreas.[121] Methods of isolation of undegraded mRNA have been reported,[122] which allow, with due precaution, satisfactory preparation of pure, biologically active material. Quantitation of specific mRNA expression has been then rendered possible thanks to the establishment of lipase and colipase cDNA clones. The cloning and characterization of pancreatic colipase cDNA has been performed in man,[123,124] rat,[125] and dog,[126] those of pancreatic lipase cDNA in man,[127] dog,[128] and rat recently.[129] We have chosen, in the assay of lipase and colipase mRNAs, to hybridize mRNA immobilized on filters to the corresponding synthetic labeled cDNA. This is a more sensitive method than the translational one in a cell-free system.

a. Lipase mRNA

The effects of a 10-d ingestion of 10% or 25% sunflower oil on lipase mRNA level result in a 2.2- and 3.9-fold increase in the lipase messenger, respectively.[130] These data support a pretranslational mechanism by which triglycerides regulate pancreatic adaptation.

When we examined the kinetics of the cellular mRNA concentration to dietary stimulation, two distinct responses were obtained.[95,131] Lipase mRNA increased in a biphasic manner in adapted rats.[95] This may have been the result of the presence of a rather large amount of sodium bicarbonate (18.5% by weight) in the experimental diet (25% sunflower oil). Indeed, we had emphasized in another study[130] the possible interference of sodium bicarbonate with the lipase response, since sodium bicarbonate had been shown to be responsible for a variety of side effects, such as sodium overload, impaired absorption of calcium, and an increase in plasma CCK level. On the other hand, lipase mRNA was found to be maximally induced as soon as the second day of lipid diet intake (by twofold) when no sodium bicarbonate was present in the 25% sunflower diet and did not further increase up to 10 d.[131] Nevertheless, all these data suggest that the mechanism of lipase adaptation by dietary lipids is mostly pretranslational. Changes in lipase mRNA levels consecutive to lipid diet ingestion probably result from an increased transcriptional rate of lipase gene. At present, we cannot exclude yet other mechanisms, such as RNA processing, regulation of lipase messenger transport, and changes in mRNA stability.

The rise in lipase mRNA was consistent with the increase in the measured values for lipase relative synthesis.[130,131] However, when the total increase in pancreatic protein synthesis was taken into account, lipase absolute synthesis rates increased more than the mRNA level. It is therefore possible that a control mechanism acting at the translational level favors the translation of lipase mRNA among the different pancreatic mRNAs. There is now a great deal of evidence that such translational discriminations between different mRNAs exist.[117,132]

The effect of a high fat diet on lipase mRNA levels in pancreatic tissue allows us to eliminate the steroid hormones as possible regulators in diet induced lipase activity. In fact, adrenalectomy plus castration[133] or adrenal-ectomy alone,[134] although increasing lipase content in rat pancreas, do not alter lipase mRNA levels.[135] This suggests a translational control mechanism by these hormones.

b. Colipase mRNA

The level of colipase mRNA, as measured by dot-blot hybridization with the corresponding cDNA probe, moderately increases, by a 1.5-fold factor, in the pancreas of high fat adapted rats (25% sunflower oil), a plateau being reached on the 3rd day of diet ingestion.[125] This finding suggests that ad-aptation of pancreatic colipase to dietary lipid is genetically regulated. It is nevertheless difficult to correlate colipase mRNA level and cofactor activity in pancreatic tissue, since alterations in cofactor synthesis and secretion may occur. The magnitude of the rise in colipase cytoplasmic mRNA concentra-tion, although significant, is less than that in lipase mRNA under the same experimental conditions.[131] This can explain the differences in induced activity levels of both proteins in adapted rats.[75,95]

3. Transcriptional Regulation of Pancreatic Lipase Gene Expression

Even if there is good evidence that the control of pancreatic gene expres-sion in the rat is exerted in some cases at the translational level[136] and that the stability of some mRNAs may vary according to nutritional situations,[137] transcription is most likely a key step in this regulation.

Measurements of specific gene transcription can be directly realized with isolated nuclei from control and adapted animals. In the run on transcription assay, nuclei are incubated with radioactive UTP, and ^{32}P-labeled RNAs in nuclei are purified and measured after hybridization to nitrocellulose filters to which an excess of plasmid DNA containing the gene sequences had been bound. This assay predominantly estimates the elongation of *in vivo* initiated RNA chains and directly indicates the relative number of polymerases on the specific gene.

The time course response of synthesis of pancreatic lipase mRNA to a fat diet ingestion (25% sunflower oil) has been determined by run-on transcription[131] (Table 5). A progressive induction in lipase mRNA synthesis occurs as a function of time of adaptation to the fat diet. This increase in lipase gene transcription is rapid, beginning on the first day of diet intake. The stimulation of lipase transcription can entirely explain messenger accu-mulation in the cytoplasm during dietary adapation.

IV. PERSPECTIVES AND GENERAL CONCLUSION

In the present chapter, we have seen that both prepyloric and pancreatic lipases are of prime importance in the hydrolysis of dietary triglycerides. These enzymes act sequencially, and prior hydrolysis of triglycerides by

TABLE 5
Changes in Lipase Relative Synthesis, mRNA Levels, and Transcriptional Rates, Induced by a Lipid-Rich Diet[131]

Changes	Days on diet				
	0	1	2	3	5
Biosynthesis[a]	6.1 ± 0.5	6.7 ± 0.6	7.8 ± 0.8	8.9 ± 0.5	9.6 ± 0.7
mRNA[b]	1	1.4 ± 0.08	1.79 ± 0.28	1.83 ± 0.16	1.99 ± 0.33
Transcriptional rate[c]	3.3	4.7	ND	ND	5.7

[a] Relative synthesis is expressed as a percentage of radioactivity incorporated into lipase compared to that incorporated into total exocrine proteins.
[b] mRNA content is normalized to that in the pancreas of rats fed the control diet (day 0).
[c] Transcriptional rates are normalized to that of actin, included as an internal standard in the hybridization experiments.

preduodenal lipase enhances pancreatic lipase action. Both lipases show rapid adaptation to the amount of fat ingested. Such nutritional regulations are thought to occur via the release of gastrointestinal peptide(s). It is believed that secretin largely controls the pancreatic adaptation to dietary lipids. Lipase accumulation in pancreatic tissue has been shown to result from a selective stimulation of enzyme synthesis. The increases in lipase synthesis and cofactor activity in the pancreas of rats adapted to a lipid-rich diet are due to rises in both messengers. As far as colipase is concerned, maximal change in mRNA level occurs on the third day after alteration in dietary composition, thus preceding changes in tissue levels of the cofactor. Adaptation of lipase mRNA to a lipid-rich diet is faster than that of colipase, since changes in messenger concentration are observed as early as 1 d after dietary manipulation. This suggests that different effectors might be involved in lipid adaptation of each pancreatic protein. The magnitude of lipase mRNA response is also greater than that of its cofactor, explaining the nonparallel changes in lipase and colipase activity following fat ingestion. Finally, the mechanism by which lipids alter the expression of pancreatic lipase is essentially transcriptional, as directly demonstrated by nuclear run-on assays.

Investigations into the structure-function relationships of pancreatic lipase gene, as well as DNA-mediated transfection techniques to introduce the functional lipase gene into cells, will help in studying lipase gene regulation by nutrients. The only available structure of a pancreatic lipase gene and DNA sequence of the 5' flanking region has been performed in the dog.[138] Unfortunately, analysis of canine lipase gene by DNA transfer methods in pancreatic acinar cells, such as AR4–2J or 266–6 for the presence of regulatory signals in the vicinity of the canine structural gene, failed to reveal these regulating stretches of DNA.[138] It is possible that these cell lines, derived from pancreatic acinar cells, one from rat (AR4–2J) and the other from mouse (266–6), are not suitable for the study of the canine gene. Another possibility is that regulatory elements are positioned in another region of the gene. It would be of interest to test the 5' flanking region of the rat gene, when rat lipase gene cloning will be performed, and analyze the upstream region to help to define genetic elements that mediate the induction of the lipase gene by nutritional agents and secretin.

In the case of other pancreatic genes, it has been possible, when constructing expression recombinant plasmids containing the 5' flanking region of rat chymotrypsin gene linked to the coding sequence of a reporter gene such as chloramphenicol acetyltransferase (CAT), to obtain specific expression of chymotrypsin gene recombinant in the rat exocrine pancreas tumor line cells AR4–2J, but not in HIT cells, a transformed line from hamster pancreatic endocrine cells.[114] Careful examination of the DNA sequences upstream for nine rat pancreatic genes (amylase, carboxypeptidases A1 and A2, chymotrypsin B, elastases I and II, ribonuclease, trypsins I and II) coupled with deletion experiments, have led to the identification of a consensus sequence: GTCACCTGTGCTTTTCCCTG located − 89 to − 213 bp from the cap site; this confers preferential expression of these genes in pancreatic exocrine cells.[116] These regulatory sequences are similar to other enhancers discovered so far in that they work whether they are located before or after the genes and whether they are in the normal orientation or are inverted.[116] Enhancers are believed to work through the binding of pancreatic nuclear protein, which leads to a change in the structure of the chromosome in the vicinity of the gene. Indeed, two teams independently have demonstrated binding of pancreatic nuclear protein with the enhancer consensus sequence of mouse amylase gene[139] and rat chymotrypsin gene.[140] The specific factor for amylase gene consensus element isolated from rat pancreatic nuclei interacts also with elastase I gene consensus sequence, as indicated by competition experiments for protein binding to the amylase sequences.[139] Site-directed mutagenesis of the amylase sequence results in concomitant loss of protein binding and enhancer activity, which indicates that the nuclear protein is a positive regulator, controlling the activity of the amylase gene enhancer.[139] Another factor, isolated from 266–6 and HeLa cells, having the DNA-binding specificity of MLTF (major late transcription factor of adenovirus), has been shown to bind to the chymotrypsin sequence *in vitro*.[140] Site-directed mutagenesis of chymotrypsin sequence into a perfect MLTF site results in the loss of enhancer activity in acinar cells, but not of the binding of MLTF-like factor.[140] This finding seems to indicate that MLTF-like factor acts as a negative regulator of chymotrypsin enhancer function. Such transcriptional repressions have also been described in other systems.[141,142]

Finally, another exciting new area of active research is the generation of transgenic mice which can be analyzed and tested under a variety of physiological circumstances. Such studies have already led to pancreas-specific expression of amylase/CAT construct in transgenic lines and to the study of insulin dependence of the hybrid amylase gene[143] or to the study of islet-specific expression of human insulin and its regulation by glucose and amino acids.[144]

All these techniques constitute powerful tools which will be exploited in the future and will open up other sets of experimental possibilities for the study of lipase and colipase gene regulation.

REFERENCES

1. **Lairon, D. and Borel, P.,** La régulation nutritionnelle des enzymes de la digestion des lipides, *Cah. Nutr. Diét.,* 24, 413, 1989.
2. **Gargouri, Y., Moreau, H., and Verger, R.,** Gastric lipases: biochemical and physiological studies, *Biochem. Biophys. Acta,* 1006, 255, 1989.
3. **Schonheyder, F. and Volquartz, K.,** The gastric lipase in man, *Acta Physiol. Scand.,* 11, 349, 1946.
4. **Hamosh, M.,** Rat lingual lipase: factors affecting enzyme activity and secretion, *Am. J. Physiol.,* 235, E416, 1978.
5. **Field, R. B. and Scow, R. O.,** Purification and characterization of rat lingual lipase, *J. Biol. Chem.,* 258, 14563, 1983.
6. **Sweet, B. J., Matthews, L. C., and Richardson, T.,** Purification and characterization of pregastric esterase from calf, *Arch. Biochem. Biophys.,* 234, 144, 1984.
7. **Bernback, S., Hernell, O., and Bläckberg, L.,** Purification and molecular characterization of bovine pregastric lipase, *Eur. J. Biochem.,* 148, 233, 1985.
8. **Cohen, M., Morgan, R. G. H., and Hofmann, A. F.,** Lipolytic activity of human gastric and duodenal juice against medium and long-chain triglycerides, *Gastroenterology,* 60, 1, 1971.
9. **Perret, J. P.,** Lipolyse gastrique chez le lapereau. Origine et importance physiologique de la lipase, *J. Physiol. (Paris),* 78, 221, 1982.
10. **DeNigris, S. J., Hamosh, M., Kasbekar, D. K., Lee, T. C., and Hamosh, P.,** Lingual and gastric lipases: species differences in the origin of prepancreatic digestive lipases and in the localization of gastric lipase, *Biochim. Biophys. Acta,* 959, 38, 1988.
11. **Moreau, H., Gargouri, Y., Lecat, D., Junien, J. L., and Verger, R.,** Screening of preduodenal lipases in several mammals, *Biochem. Biophys. Acta,* 959, 247, 1988.
12. **Roberts, I. M. and Jaffe, R.,** Lingual lipase: immunocytochemical localization in the rat von Ebner gland, *Gastroenterology,* 90, 1170, 1986.
13. **Roberts, I. M., Nochomovitz, L. E., Jaffe, R., Hanel, S. I., Rojas, M., and Agostini, R. A.,** Immunocytochemical localization of lingual lipase in serous cells of the developing rat tongue, *Lipids,* 22, 764, 1987.
14. **Abrams, C. K., Hamosh, M., Lee, T. C., Ansher, A. F., Collen, M. J., Lewis, J. H., Benjamin, S. B., and Hamosh, P.,** Gastric lipase: localization in the human stomach, *Gastroenterology,* 95, 1460, 1988.
15. **Moreau, H., Laugier, R., Gargouri, Y., and Verger, R.,** Human gastric lipase is entirely of gastric fundic origin, *Gastroenterology,* 95, 1221, 1988.
16. **Moreau, H., Bernadac, A., Gargouri, Y., Benkouka, F., Laugier, R., and Verger, R.,** Immunocytolocalization of human gastric lipase in chief cells of the fundic mucosa, *Histochemistry,* 91, 419, 1989.

17. **Moreau, H., Bernadac, A., Trétout, N., Gargouri, Y., Ferrato, F., and Verger, R.,** Immunocytochemical localization of rabbit gastric lipase and pepsinogen, *Eur. J. Cell Biol.,* 51, 165, 1990.

18. **Fredrickzon, B. and Hernell, O.,** Role of feeding on lipase activity in gastric contents, *Acta Paediatr. Scand.,* 66, 479, 1977.

19. **Gargouri, Y., Piéroni, G., Rivière, C., Saunière, J. F., Lowe, P. A., Sarda, L., and Verger, R.,** Kinetic assay of human gastric lipase on short- and long-chain triacylglycerol emulsions, *Gastroenterology,* 91, 919, 1986.

20. **Schick, J., Verspohl, R., Kern, H., and Scheele, G.,** Two distinct adaptative responses in the synthesis of exocrine pancreatic enzymes to inverse changes in protein and carbohydrate in the diet, *Am. J. Physiol.,* 247, G611, 1984.

21. **Verger, R.,** Pancreatic lipases, in *Lipases,* Borgstrom, B. and Brockman, H. L., Eds., Elsevier, Amsterdam, 1984, 84.

22. **Saraux, B. and Girard-Globa, A.,** Development of pancreatic enzymes in fetal and suckling rats with emphasis on lipase and colipase, *J. Dev. Physiol.,* 4, 121, 1982.

23. **Sarda, L. and Desnuelle, P.,** Action de la lipase pancréatique sur les esters en émulsion, *Biochem. Biophys. Acta,* 30, 513, 1958.

24. **Sémériva, M. and Desnuelle, P.,** Pancreatic lipase and colipase. An example of heterogeneous biocatalysis, *Adv. Enzymol.,* 48, 319, 1979.

25. **Maylié, M.-F., Charles, M., Gache, C., and Desnuelle, P.,** Isolation and partial identification of a pancreatic colipase, *Biochem. Biophys. Acta,* 229, 286, 1971.

26. **Chapus, C., Sari, H., Sémériva, M., and Desnuelle, P.,** Role of colipase in the interfacial adsorption of pancreatic lipase at hydrophobic interfaces, *FEBS Lett.,* 58, 155, 1975.

27. **Patton, J. S., Donner, J., and Borgström, B.,** Lipase-colipase interactions during gel filtration high and low affinity binding situations, *Biochem. Biophys. Acta,* 529, 67, 1978.

28. **Sauve, P. and Desnuelle, P.,** Interactions of pancreatic colipase with taurodeoxycholate-oleate mixtures above the critical micellar concentrations, *FEBS Lett.,* 122, 91, 1980.

29. **Borgström, B. and Hildebrand, H.,** Lipase and colipase activities of human small intestinal contents after a liquid test meal, *Scand. J. Gastroenterol.,* 10, 585, 1975.

30. **Patton, J. S., Albertsson, P.-A., Erlanson, C., and Borgström, B.,** Binding of porcine pancreatic lipase and colipase in the absence of substrate studied by two-phase partition and affinity chromatography, *J. Biol. Chem.,* 253, 4195, 1978.

31. **Staggers, J. E., Hernell, O., Stafford, R. J., and Carey, M. C.,** Physical-chemical behavior of dietary and biliary lipids during intestinal digestion and absorption. I. Phase behavior and aggregation states of model lipid systems patterned after aqueous duodenal contents of healthy adult human beings, *Biochemistry,* 29, 2028, 1990.

32. **Hernell, O., Staggers, J. E., and Carey, M. C.,** Physical-chemical behavior of dietary and biliary lipids during intestinal digestion and absorption. II. Phase analysis and aggregation states of luminal lipids during duodenal fat digestion in healthy adult human beings, *Biochemistry,* 29, 2041, 1990.

33. **Gargouri, Y., Pieroni, G., Lowe, P. A., Sarda, L., and Verger, R.,** Human gastric lipase. The effect of amphiphiles, *Eur. J. Biochem.,* 156, 305, 1986.

34. **Borgstrom, B.,** Importance of phospholipids, pancreatic phospholipase A2 and fatty acid for the digestion of dietary fat, *Gastroenterology,* 78, 954, 1980.

35. **Iverson, S. J., Kirk, C. L., Hamosh, M., and Newsome, J.,** Milk lipid digestion in the neonatal dog: the combined actions of gastric and bile salt stimulated lipases, *Biochim. Biophys. Acta,* 1083, 109, 1991.

36. **Hamosh, M.,** Fat digestion in the newborn: role of lingual lipase and preduodenal digestion, *Pediatr. Res.,* 13, 615, 1979.

37. **Lairon, D., Nalbone, G., Lafont, H., Leonardi, J., Domingo, N., Hauton, J. C., and Verger, R.,** Possible roles of bile lipids and colipase in lipase adsorption, *Biochemistry,* 17, 5263, 1978.

38. **Miller, K. W. and Small, D. M.,** The phase behavior of triolein, cholesterol and emulsions, *J. Colloid Int. Sci.,* 89, 466, 1982.
39. **Carey, M. C., Small, P. M., and Bliss, C. M.,** Lipid digestion and absorption, *Annu. Rev. Physiol.,* 45, 651, 1983.
40. **Hofman, A. and Borgström, B.,** Physical-chemical state of lipids in intestinal content during their digestion and absorption, *Fed. Proc.,* 21, 43, 1962.
41. **Patton, J. S. and Carey, M. C.,** Watching fat digestion, *Science,* 204, 145, 1979.
42. **Lindström, M., Ljusberg-Wahren, H., and Larsson, K.,** Aqueous lipid phases of relevance to intestinal fat digestion and absorption, *Lipids,* 16, 749, 1981.
43. **Clark, B. S., Lawergren, B., and Martin, J. V.,** Regional intestinal absorptive capacities for triolein: an alternative to markers, *Am. J. Physiol.,* 225, 574, 1973.
44. **Tso, P., Pitts, V., and Granger, D. N.,** Role of lymph flow in intestinal chylomicron transport, *Am. J. Physiol.,* 249, G21, 1985.
45. **Plucinski, T. M., Hamosh, M., and Hamosh, P.,** Fat digestion in rat: role of lingual lipase, *Am. J. Physiol.,* 237, E541, 1979.
46. **Watkins, J. B., Ingall, D., and Szczepanik, P.,** Bile salt metabolism in the newborn, *New Engl. J. Med.,* 288, 431, 1973.
47. **Robberecht, P., Deschodt-Lanckman, M., Camus, J., Bruylands, J., and Christophe, J.,** Rat pancreatic hydrolases from birth to weaning and dietary adaptation after weaning, *Am. J. Physiol.,* 221, 376, 1971.
48. **Snook, J. T.,** Effect of diet on development of exocrine pancreas of the neonatal rat, *Am. J. Physiol.,* 221, 1388, 1971.
49. **Zoppi, G., Andreotti, G., Pajno-Ferrara, F., Njai, D. M., and Gaburro, D.,** Exocrine pancreas function in premature and full term neonates, *Pediatr. Res.,* 6, 880, 1972.
50. **Fomon, S. J., Ziegler, E. R., Thomas, L. N., Jensen, R. L., and Filer, L. J.,** Excretion of fat by normal full term infants fed various milks and formulae, *Am. J. Clin. Nutr.,* 23, 1299, 1970.
51. **Katz, L. and Hamilton, J. R.,** Fat absorption in infants of birth weight less than 1300 g, *J. Pediatr.,* 85, 608, 1974.
52. **Blackberg, L., Lombardo, D., Hernell, O., Guy, O., and Olivecrona, T.,** Bile-salt-stimulated lipase in human milk and carboxyl ester hydrolase in pancreatic juice. Are they identical enzymes?, *FEBS Lett.,* 136, 284, 1981.
53. **Sheldon, W.,** Congenital pancreatic lipase deficiency, *Arch. Dis. Child.,* 39, 268, 1964.
54. **Larbre, F., Hartman, E., Cotton, J. B., Mathieu, M., Charrat, A., and Moreau, P.,** Diarrhée chronique par absence de lipase pancréatique, *Pediatric,* 24, 807, 1969.
55. **Figarella, C., DeCaro, A., Leupold, D., and Poley, J.,** Congenital pancreatic lipase deficiency, *J. Pediatr.,* 96, 412, 1980.
56. **Hilderbrand, H., Borgström, B., Bokassy, A., Erlanson-Albertsson, C., and Helin, I.,** Isolated colipase deficiency in two brothers, *Gut,* 23, 243, 1982.
57. **Fayez, K., Ghishan, J., Moran, R., Durie, P. R., and Greene, H. L.,** Isolated congenital lipase-colipase deficiency, *Gastroenterology,* 86, 1580, 1984.
58. **Ligumsky, M., Granot, E., Branski, D., Stankiewicz, H., and Goldstein, R.,** Isolated lipase and colipase deficiency in two brothers, *Gut,* 31, 1416, 1990.
59. **Ross, C. A. C. and Sammons, H. G.,** Nonpancreatic lipase in children with pancreatic fibrosis, *Arch. Dis. Child.,* 30, 428, 1955.
60. **Abrams, C. K., Hamosh, M., VanHubbard, S., Dutta, S. K., and Hamosh, P.,** Lingual lipase in cystic fibrosis: quantitation of enzyme activity in the upper small intestine of patients with exocrine pancreatic insufficiency, *J. Clin. Invest.,* 73, 374, 1984.
61. **Cooke, A. R.,** Control of gastric emptying and motility, *Gastroenterology,* 68, 804, 1975.
62. **Cunningham, K. M., Daly, J., Horowitz, M., and Read, N. W.,** Gastrointestinal adaptation to diets of differing fat composition in human volunteers, *Gut,* 32, 483, 1991.

63. **Hopman, W. P. N., Jansen, J. B. M. J., and Lamers, C. B. H. W.,** Comparative study of the effects of equal amounts of fat, protein, and starch on plasma CCK in man, *Scand. J. Gastroenterol.,* 20, 843, 1985.

64. **Konturek, S. J., Tasler, J., Bilski, J., Jong, A. J., Jansen, J. B. M. J., and Lamers, C. B.,** Physiological role and localization of CCK release in dogs, *Am. J. Physiol.,* 250, G391, 1986.

65. **Aponte, G. W., Fink, A. S., Meyer, J. H., Takemoto, K., and Taylor, I. L.,** Regional distribution and release of peptide YY (PYY) with fatty acids of different chain length, *Am. J. Physiol.,* 249, G745, 1985.

66. **Hamosh, M. and Hand, A. R.,** Development of secretory activity in serous cells of the rat tongue. Cytological differentiation and accumulation of lingual lipase, *Develop. Biol.,* 65, 100, 1978.

67. **Armand, M., Borel, P., Cara, L., Senft, M., Chautan, M., Lafont, H., and Lairon, D.,** Adaptation of lingual lipase to dietary fat in rats, *J. Nutr.,* 120, 1148, 1990.

68. **Borel, P., Armand, M., Senft, M., Andrey, M., Lafont, H., and Lairon, D.,** Gastric lipase: Evidence of an adaptive response to dietary fat in the rabbit, *Gastroenterology,* 100, 1582, 1991.

69. **Bucko, A. and Kopec, Z.,** Adaptation of enzyme activity of the rat pancreas on altered food intake, *Nutr. Dieta,* 10, 276, 1968.

70. **Deschodt-Lanckman, M., Robberecht, P., Camus, J., and Christophe, J.,** Short-term adaptation of pancreatic hydrolases to nutritional and physiological stimuli in adult rats, *Biochimie,* 53, 789, 1971.

71. **Snook, J. T.,** Dietary regulation of pancreatic enzymes in the rat with emphasis on carbohydrate, *Am. J. Physiol.,* 221, 1383, 1971.

72. **Gidez, L. I.,** Effect of dietary fat on pancreatic lipase levels in the rat, *J. Lipid. Res.,* 14, 169, 1973.

73. **Saraux, B., Girard-Globa, A., Ouagued, M., and Vacher, D.,** Response of the exocrine pancreas to quantitative and qualitative variations in dietary lipids, *Am. J. Physiol.,* 243, G10, 1982.

74. **Sabb, J. E., Godfrey, P. M., and Brannon, P. M.,** Adaptive response of rat pancreatic lipase to dietary fats: effects of amount and type of fat, *J. Nutr.,* 116, 892, 1986.

75. **Wicker, C. and Puigserver, A.,** Effects of inverse changes in dietary lipid and carbohydrate on the synthesis of some pancreatic secretory proteins, *Eur. J. Biochem.,* 162, 25, 1987.

76. **Behrman, H. R. and Kare, M. R.,** Adaptation of canine pancreatic enzymes to diet composition, *J. Physiol.,* 205, 667, 1969.

77. **Wills, J. R. and Hinners, S. W.,** Effect of dietary lipids upon the lipase, pancreatic cholesterol esterase, plasma triglyceride and total plasma lipids level in the chick, *Poult. Sci.,* 47, 1732, 1968.

78. **Mourot, J. and Corring, T.,** Adaptation of the lipase-colipase system to dietary lipid content in pig pancreatic tissue, *Ann. Biol. Anim. Biochim. Biophys.,* 19, 119, 1979.

79. **Simoes Nunes, C.,** Adaptation of pancreatic lipase to the amount and nature of dietary lipids in the growing pig, *Reprod. Nutr. Dévelop.,* 26, 1273, 1986.

80. **Meyer, J. H. and Jones, R. S.,** Canine pancreatic responses to intestinally perfused fat and products of fat digestion, *Am. J. Physiol.,* 226, 1178, 1974.

81. **Malagelada, J.-R., DiMagno, E. P., Summerskill, W. H. J., and Go, V. L. W.,** Regulation of pancreatic and gall bladder function by intraluminal fatty acids and bile acids in man, *J. Clin. Invest.,* 58, 493, 1976.

82. **Singer, M. V.,** Pancreatic secretory response to intestinal stimulants: a review, *Scand. J. Gastroenterol.,* 22, 1, 1987.

83. **Boivin, M., Lanspa, S. J., Zinsmeister, A. R., Go, V. L. W., and DiMagno, E. P.,** Are diets associated with different rates of human interdigestive and postprandial pancreatic enzyme secretion? *Gastroenterology,* 99, 1763, 1990.

84. **Ballestra, M. C., Manas, M., Mataix, F. J., Martinez-Victoria, E., and Seiquer, I.,** Long-term adaptation of pancreatic response by dogs to dietary fats of different degrees of saturation: olive and sunflower oil, *Br. J. Nutr.,* 64, 487, 1990.

85. **Kim, Y. C., Faichney, A., and Ky, L.,** Endogenous release of secretin by sodium oleate in dog, *Gastroenterology,* 78, 1195, 1980.

86. **Faichney, A., Chey, W. Y., Kim, Y. C., Lee, K. Y., Kim, M. S., and Chang, T. M.,** Effect of sodium oleate on plasma secretin concentration and pancreatic secretion in dogs, *Gastroenterology,* 81, 458, 1981.

87. **Fink, A. S., Taylor, I. L., Luxemburg, M., and Meyer, J. H.,** Pancreatic polypeptide release by intraluminal fatty acids, *Metabolism,* 32, 1063, 1983.

88. **Sommer, H. and Kasper, H.,** Dietary effects on pancreatic exocrine function. Experiments on the isolated perfused rat pancreas, *Ann. Nutr. Metab.,* 25, 381, 1981.

89. **Belleville, J., Prost, J., and Gillet, M.,** Effets des régimes riches en triglycérides et en phospholipides sur les activites de la lipase et de la phospholipase A du suc pancréatique et du pancréas de rat, *Arch. Int. Physiol. Biochim.,* 86, 631, 1978.

90. **Ouagued, M., Saraux, B., Girard-Globa, A., and Bourdel, G.,** Differential regulation of lipase and colipase in the rat pancreas by dietary fat and proteins, *J. Nutr.,* 110, 2302, 1980.

91. **Robberecht, P., Deschodt-Lanckman, M., Camus, J., Bruylands, J., and Christophe, J.,** Rat pancreatic hydrolases from birth to weaning and dietary adaptation after weaning, *Am. J. Physiol.,* 221, 376, 1971.

92. **Vandermeers-Piret, M. C., Vandermeers, A., Wijns, W., Rathe, J., and Christophe, J.,** Lack of adaptation of pancreatic colipase in rats and mice, *Am. J. Physiol.,* 232, E131, 1977.

93. **Girard-Globa, A. and Simond-Cote, E.,** Nutritional and circadian variations in lipase activity and colipase saturation in rat pancreas, *Ann. Biol. Anim. Biochim. Biophys.,* 17, 539, 1977.

94. **Lairon, D., Lacombe, C., Borel, P., Corraze, G., Nibbelink, M., Chautan, M., Chanussot, F., and Lafont, H.,** Beneficial effects of wheat germ on circulating lipoproteins and tissue lipids in rats fed a high fat, cholesterol-containing diet, *J. Nutr.,* 117, 838, 1987.

95. **Wicker, C. and Puigserver, A.,** Changes in mRNA levels of rat pancreatic lipase in the early days of consumption of a high-lipid diet, *Eur. J. Biochem.,* 180, 563, 1989.

96. **Stabile, B. E., Borzatta, M., Stubbs, R. S., and Debas, H. T.,** Intravenous mixed aminoacids and fats do not stimulate exocrine pancreatic secretion, *Am. J. Physiol.,* 246, G274, 1984.

97. **Burns, G. P. and Stein, T. A.,** Pancreatic enzyme secretion during intravenous fat infusion, *J. Parenteral Enteral Nutr.,* 11, 60, 1987.

98. **Edelman, K. and Valenzuela, J. E.,** Effect of intravenous lipid on human pancreatic secretion, *Gastroenterology,* 85, 1063, 1983.

99. **Konturek, S. J., Tasler, J., Cieszkowski, M., Jaworek, J.,and Konturek, J.,** Intravenous amino acids and fat stimulate pancreatic secretion, *Am. J. Physiol.,* 236, E678, 1979.

100. **Lavau, M., Bazin, R., and Herzog, J.,** Comparative effects of oral and parenteral feeding on pancreatic enzymes in the rat, *J. Nutr.,* 104, 1432, 1974.

101. **Bazin, R., Lavau, M., and Herzog, J.,** Pancreatic lipase and ketogenic conditions, *Biomedicine,* 28, 160, 1978.

102. **Hirschi, K. K., Sabb, J. E., and Brannon, P. M.,** Effects of diet and ketones on rat pancreatic lipase in cultured acinar cells, *J. Nutr.,* 121, 1129, 1991.

103. **Simoes Nunes, C.,** Lack of pancreatic enzyme adaptation to diet carbohydrates and lipids after proximal small intestine bypass in the pig, *Digestion,* 25, 108, 1982.

104. **Douglas, B. R., Woutersen, R. A., Jansen, J. B. M. J., De Jong, A. J. L., and Lamers, C. B. H. W.,** The influence of different nutrients on plasma cholecystokinin levels in the rat, *Experimentia,* 44, 21, 1988.

105. **Corring, T. and Chayvialle, J. A.,** Diet composition and the plasma levels of some peptides regulating pancreatic secretion in the pig, *Reprod. Nutr. Develop.,* 27, 967, 1987.
106. **Schick, J., Kern, H., and Scheele, G.,** Hormonal stimulation in the exocrine pancreas results in coordinate and anticoordinate regulation of protein synthesis, *J. Cell Biol.,* 99, 1569, 1984.
107. **Solomon, T. E., Petersen, H., Elashoff, I., and Grossman, M. J.,** Interaction of caerulein and secretin on pancreatic size and composition in the rat, *Am. J. Physiol.,* 235, E714, 1978.
108. **Rausch, U., Vasiloudes, P., Rüdiger, K., and Kern, H. F.,** In-vivo stimulation of rat pancreatic acinar cells by infusion of secretin. I. Changes in enzyme content, pancreatic fine structure and total rate of protein synthesis, *Cell Tissue Res.,* 242, 633, 1985.
109. **Rausch, U., Vasiloudes, P., Rüdiger, K., and Kern, H. F.,** In-vivo stimulation of rat pancreatic acinar cells by infusion of secretin. II. Changes in individual rates of enzyme and isoenzyme biosynthesis, *Cell Tissue Res.,* 242, 641, 1985.
110. **Kemper, K. A. and Brannon, P. M.,** Effects of secretin on enzymes of cultured pancreatic acinar cells, *FASEB J.,* 2, A731, 1988.
111. **Brown, D. D.,** Gene expression in eukaryotes, *Science,* 211, 667, 1981.
112. **Darnell, J. E.,** Variety in the level of gene control in eukaryotic cells, *Nature,* 297, 365, 1982.
113. **Mitchell, P. J. and Tjian, R.,** Transcriptional regulation in mammalian cells by sequence-specific DNA binding proteins, *Science,* 245, 371, 1989.
114. **Walker, M., Edlund, T., Boulet, A., and Rutter, W.,** Cell specific expression controlled by the 5′ flanking region of insulin and chymotrypsin genes, *Nature,* 306, 557, 1983.
115. **Swift, G., Hammer, R., MacDonald, R., and Brinster, R.,** Tissue specific expression of the pancreatic elastase I gene in transgenic mice, *Cell,* 38, 639, 1984.
116. **Boulet, A. M., Erwin, C. R., and Rutter, W. J.,** Cell-specific enhancers in the rat exocrine pancreas, *Proc. Natl. Acad. Sci. U.S.A.,* 83, 3599, 1986.
117. **Austin, S. A. and Kay, J. E.,** Translational regulation of protein synthesis in eukaryotes, *Essays Biochem.,* 18, 79, 1982.
118. **Ross, R. J.,** Messenger RNA turnover in eukaryotic cells, *Mol. Biol. Med.,* 5, 1, 1988.
119. **Jamieson, J. D. and Palade, G. E.,** Synthesis, intracellular transport and discharge of secretory proteins in stimulated pancreatic exocrine cells, *J. Cell Biol.,* 50, 135, 1971.
120. **Blobel, G. and Dobberstein, B.,** Transfer of proteins across membranes. Presence of proteolytically processed and unprocessed nascent immunoglobulin light chains on membrane-bound ribosomes of murine myeloma, *J. Cell Biol.,* 67, 835, 1975.
121. **Barnard, E. A.,** Biological function of pancreatic ribonuclease, *Nature,* 221, 340, 1969.
122. **Chirgwin, J. M., Przybyla, A. E., MacDonald, R. J., and Rutter, W. J.,** Isolation of biologically active ribonucleic acid from sources enriched in ribonuclease, *Biochemistry,* 18, 5294, 1979.
123. **Lowe, M. E., Rosenblum, J. L., McEven, P., and Strauss, A. W.,** Cloning and characterization of the human colipase cDNA, *Biochemistry,* 29, 823, 1990.
124. **Renaud, W. and Dagorn, J. C.,** cDNA sequence and deduced amino acid sequence of human preprocolipase, *Pancreas,* 6, 157, 1991.
125. **Wicker, C. and Puigserver, A.,** Rat pancreatic colipase mRNA: nucleotide sequence of a cDNA clone and nutritional regulation by a lipidic diet, *Biochem. Biophys. Res. Commun.,* 167, 130, 1990.
126. **Fukuoka, S.-I., Taniguchi, Y., Kitagawa, Y., and Scheele, G.,** Full length cDNA sequence encoding canine pancreatic colipase, *Nucleic Acids Res.,* 18, 5549, 1990.
127. **Lowe, M. E., Rosenblum, J. L., and Strauss, A. W.,** Cloning and characterization of human pancreatic lipase cDNA, *J. Biol. Chem.,* 264, 20042, 1989.

128. **Kerfelec, B., LaForge, K. S., Puigserver, A., and Scheele, G.,** Primary structures of canine pancreatic lipase and phospholipase A2 messenger RNAs, *Pancreas,* 1, 430, 1986.

129. **Wicker-Planquart, C. and Puigserver, A.,** Primary structure of rat pancreatic lipase mRNA, FEBS Lett., 296, 61, 1992.

130. **Wicker, C., Scheele, G. A., and Puigserver, A.,** Pancreatic adaptation to dietary lipids is mediated by changes in lipase mRNA, *Biochimie,* 70, 1277, 1988.

131. **Wicker, C. and Puigserver, A.,** Expression of rat pancreatic lipase gene is modulated by a lipid-rich diet at a transcriptional level, *Biochem. Biophys. Res. Commun.,* 166, 358, 1990.

132. **Kern, H. F., Rausch, U., and Scheele, G.,** Regulation of gene expression in pancreatic adaptation to nutritional substrates or hormones, *Gut,* 28, 89, 1987.

133. **Beaudoin, A. R., Grondin, G., St-Jean, P., Vacherear, A., Cabana, C., and Grossman, A.,** Steroids and secretory function of the exocrine pancreas, *Endocrinology,* 119, 2106, 1986.

134. **Duan, R.-D. and Erlanson-Albertsson, C.,** The anticoordinate changes of pancreatic lipase and colipase activity to amylase activity induced by adrenalectomy in normal and diabetes rats, *Int. J. Pancreatol.,* 6, 271, 1990.

135. **Duan, R.-D., Poensgen, J., and Erlanson-Albertsson, C.,** Adrenalectomy anticoordinately changes the synthesis of rat pancreatic enzymes, *Digestion,* 43, 138, 1989.

136. **Wicker, C., Puigserver, A., Rausch, U., Scheele, G., and Kern, H.,** Multiple-level caerulein control of the gene expression of secretory proteins in the rat pancreas, *Eur. J. Biochem.,* 151, 461, 1985.

137. **Puigserver, A., Wicker, C., and Gaucher, C.,** Aspects moléculaires de l'adaptation des enzymes pancréatiques et intestinales au régime alimentaire, *Reprod. Nutr. Dév.,* 25, 787, 1985.

138. **Mickel, F. S., Weidenbach, F., Swarovsky, B., LaForge, K. S., and Scheele, G. A.,** Structure of the canine pancreatic lipase gene, *J. Biol. Chem.,* 264, 12895, 1989.

139. **Howard, G., Keller, P. R., Johnson, T. M., and Meisler, M. H.,** Binding of a pancreatic nuclear protein is correlated with amylase enhancer activity, *Nucleic Acids Res.,* 17, 8185, 1989.

140. **Meister, A., Weinrich, S. L., Nelson, C., and Rutter, W. J.,** The chymotrypsin enhancer core. Specific factor binding and biological activity, *J. Biol. Chem.,* 264, 20744, 1989.

141. **Nir, U., Walker, M. D., and Rutter, W. J.,** Regulation of rat insulin 1 gene expression: evidence for negative regulation in nonpancreatic cells, *Proc. Natl. Acad. Sci. U.S.A.,* 83, 3180, 1986.

142. **Zhang, Z. X., Kumar, V., Rivera, R. T., Chisholm, J., and Biswas, D. K.,** Cis-acting negative regulatory element of prolactin gene, *J. Biol. Chem.,* 265, 4785, 1990.

143. **Osborn, L., Rosenberg, M. P., Keller, S. C., Ting, C.-N., and Meisler, M. H.,** Insulin response of a hybrid amylase/CAT gene in transgenic mice, *J. Biol. Chem.,* 263, 16519, 1988.

144. **Selden, R. F., Skoskiewicz, M. J., Burke Howie, K., Russell, P. S., and Goodman, H. M.,** Regulation of human insulin gene expression in transgenic mice, *Nature (London),* 321, 525, 1986.

Chapter 4

NUTRITIONAL CONTROL OF GASTROINTESTINAL HORMONE GENE EXPRESSION

Pauline Kay Lund

TABLE OF CONTENTS

0-8493-6961-4/93/$0.00 + $.50

91

I. INTRODUCTION

The gastrointestinal (GI) hormones are a large and complex group of peptides that are synthesized by endocrine cells in the mucosal lining of the gut.[1] They regulate the functioning of the GI tract and the secretion of electrolytes and digestive enzymes that are necessary for optimal digestion and absorption of ingested nutrients. Radioimmunoassays of plasma and extracts of gastrointestinal mucosa have established that the secretion of different GI hormones is regulated by specific nutrients.[1-3] Until recently there has been little emphasis on whether these effects are linked to, or mediated by, effects on GI hormone gene expression. An understanding of nutrient regulation of GI hormone gene expression is relevant to understanding the factors that underlie normal development and functioning of the GI tract and for defining the role of specific nutrients and GI hormones in the etiology, pathophysiology, and treatment of GI disease. The cloning and characterization of the genes and complementary DNAs (cDNAs) encoding many of the GI hormones have provided the potential for investigations of nutritional control of GI hormone gene expression. Progress in this area will be reviewed with particular emphasis on gastrin, cholecystokinin (CCK), somatostatin, and enteroglucagon. These hormones represent only a subset of the large array of hormones expressed in the GI tract. However, each of these hormones has a relatively well-defined physiological role, and collectively there is more information available about their regulation by nutrients than is available for other GI hormones.

II. GENERAL CONSIDERATIONS

To date, most studies of nutritional regulation of GI hormone gene expression have focused on quantification of the steady-state concentrations of the particular mRNA in total RNA or total messenger RNA (mRNA) extracted from the entire wall of a given region of the gut or from gut mucosa. The extracted RNA is then analyzed by Northern blot hybridization, dot or slot blot hybridization, or solution hybridization using a radiolabeled DNA or RNA probe that is complementary to the GI hormone mRNA of interest. There are some general features of the GI endocrine system which pose particular challenges in understanding nutritional control of gene expression using these or other methods:

Heterogeneity—In contrast to ''classical endocrine glands'' that are comprised of primarily endocrine cells, the GI endocrine system is a diffuse and heterogeneous system. Multiple endocrine cell types are dispersed within the mucosal layer of the gut wall. The majority of cell types in the mucosa are ''non-endocrine'' epithelial cells and supporting mesenchymal cells. In addition to mucosa, the gut wall contains underlying submucosa, muscle layers, and serosa. From a practical standpoint this means that the mRNAs encoding an individual GI hormone represent a relatively small fraction of the total mRNA expressed in the entire gut wall or mucosa. Highly sensitive hybridization methods must therefore be employed to analyze GI hormone mRNAs. The heterogeneity of cell types within the gut wall also complicates the interpretation of nutrient dependent changes in steady-state concentrations of GI hormone mRNAs *in vivo*. Such changes may reflect a true alteration in the level of expression of the gene encoding the GI hormone or a change in the stability of the mRNA. Alternatively, nutrient-dependent changes in steady-state levels of GI hormone mRNAs may be secondary to changes in relative abundance of other, more predominant mRNAs derived from the ''non-endocrine'' cell types. One way to distinguish these possibilities is by use of quantitative or semiquantitative *in situ* hybridization histochemistry in parallel with the hybridization analyses on extracted RNA. Quantitative *in situ* hybridization could establish if altered levels of steady-state mRNA reflect altered levels of mRNA in individual GI endocrine cells. To date, this has been done infrequently, probably because of the inherent difficulties in developing quantitative *in situ* hybridization methods.

Turnover and renewal—Cells within the mucosal lining of the GI tract have a high rate of turnover and renewal and these processes are highly susceptible to nutritional status.[4,5] Both the amount and the type of ingested nutrient exert profound effects on the mass of mucosal cells within a given gut region by altering the rate of stem cell division and the rate of renewal of the mucosal lining. Such alterations in mucosal mass are termed adaptive growth of the GI tract. For example, fasting results in hypoplasia and atrophy of the gut mucosa while hyperphagia results in mucosal hyperplasia.[4,5] It is not clear whether alterations in nutrient load or type affect the rate of turnover of GI endocrine cells to the same degree as other mucosal cell types. Indeed, there is evidence that some GI endocrine cells turn over much more slowly than other epithelial cells.[6] Changes in the concentrations of GI hormone mRNAs may therefore be secondary to differential effects of nutrient status on relative mass of endocrine and ''non-endocrine'' cell types and may not reflect the effect of nutrient on gene expression per se. Histological analyses of the mucosa performed in parallel with quantification of GI hormone mRNAs could help discern if apparent alterations in the levels of GI hormone mRNAs are secondary to alterations in relative mass of endocrine cells. To date, such combined studies have been performed very infrequently.

Chemical and mechanical stimuli—Ingested nutrients present a complex stimulus to the GI tract. Mechanical stimulation results from distention of the gut by food. Chemical stimulation occurs via nutrients within the lumen of the GI tract or via altered concentrations of absorbed nutrients in the circulation. Gut responses to these nutritional stimuli can involve long neural reflexes mediated by the central nervous system and short neural reflexes mediated by the intrinsic neuronal plexuses in the gut wall. Mechanical and chemical stimuli may stimulate secretion of GI hormones into the circulation or local secretion into the extracellular fluid. Nutrient effects on GI hormone gene expression therefore may be mediated by direct effects of nutrients on the GI endocrine cells, by indirect effects of neuronal reflexes, or by indirect effects of other GI hormones acting in an endocrine and paracrine manner. Thus, careful experimental design is required to elucidate precisely which of these multiple nutrient-mediated events underlies a nutrient-dependent change in GI hormone gene expression.

Hormone expression—A given GI hormone may be expressed in multiple regions of the gut or accessory tissues. A nutrient-mediated effect observed in one gut region or cell type may not necessarily reflect effects in other regions or cells.

***Cis-* and *trans*-acting factors**—Cell-specific and regulated expression of most genes appear to be determined by *cis*-acting regulatory elements comprised of DNA sequences that flank the transcribed portions of the gene or lie within the structural gene.[7] *Trans*-acting factors, usually proteins, bind to these DNA regulatory elements and alter the rate of gene transcription.[7] It is increasingly clear that steady-state levels of many mRNAs are regulated by posttranscriptional mechanisms, probably via effects on RNA binding factors that influence mRNA stability and turnover.[8] Ultimately, an understanding of nutrient regulation of GI hormone-gene expression and synthesis will require characterization of the *cis*-elements, *trans*-acting factors, and posttranscriptional events that mediate the nutrient effects. Even though many of the genes for gastrointestinal hormones were cloned in the early 1980s, remarkably little progress has been made in characterizing the promoters and regulatory elements that determine basal, cell-specific, and regulated expression in enteroendocrine cells or GI tract *in vivo*. This likely relates to the limited availability of useful model culture systems for analyses of GI hormone-gene expression. It is difficult to maintain gut regions or slices in culture. The diffuse nature of the GI endocrine system creates problems in isolation and culture of homogeneous GI endocrine cell populations. Relatively few cell lines express GI hormones and those that do are often phenotypically very different from enteroendocrine cells. Progress towards development and use of model systems for the analyses of GI hormone-gene expression will be described.

III. GASTRIN GENE EXPRESSION

A. BACKGROUND

Gastrin is synthesized primarily in G cells of the antrum of the stomach and exists as two major forms of 34 (G34) and 17 (G17) amino acids.[1] The two forms derive from differential processing of a single precursor encoded by a single gene and mRNA.[1] Gastrin has two major physiological roles, the stimulation of gastric acid secretion and stimulation of mucosal growth in the acid-secreting portion of the stomach.[1] Since gastric acid secretion is a major determinant, if not causative factor in peptic ulcer disease, understanding the molecular basis of nutritional effects on gastrin synthesis and release has implications for understanding the etiology and pathophysiology of peptic ulcer.

Fasting results in decreases in plasma gastrin, decreases in antral gastrin concentrations, and a decrease in antral G cell density.[1,9,10] These effects are reversed by refeeding.[1,9] Gastrin is released into the circulation in response to food ingestion.[1] The chemical composition of food influences gastrin release. Proteins, digestion products of proteins, and individual amino acids are particularly strong stimulants of gastrin secretion.[1,11,12] Neural reflexes also influence gastrin release.[1] Recent studies, conducted primarily in rat and dog, suggest that the effects of fasting and nutrient-induced changes in circulating and antral gastrin concentrations are mediated at the level of gastrin mRNA (Figure 1). These findings therefore indicate that nutrient status influences gastrin gene transcription of mRNA stability.

B. FASTING REDUCES GASTRIN mRNA ABUNDANCE

Fasting results in parallel decreases in antral gastrin concentrations and antral gastrin mRNA.[13] Effects are rapid and are observed within hours of

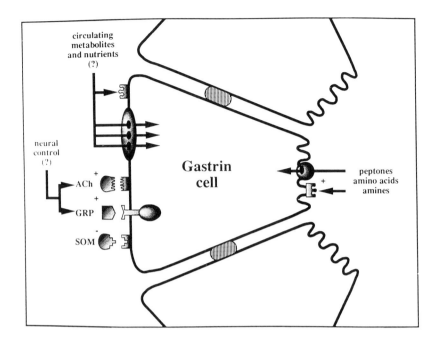

FIGURE 1.   Schematic to summarize nutrient effects on the gastrin cell that are implicated in regulation of gastrin gene expression: + = positive effect; − = negative effect; ? = indirect evidence for an effect. Peptones, amino acids, and amines appear to act via the lumen either by cell surface receptors or after uptake into the G cell as indicated.

fasting. Effects are also progressive since a continued decrease in gastrin mRNA occurs with prolonged fasting.[13] It is possible that decreases in relative numbers of G cells contribute to the progressive decrease in antral gastrin mRNA observed after a prolonged fast.[9] However, alterations in relative mass of G cells are not likely to be rapid enough to account for a decline in antral gastrin mRNA within 12 hours of fasting.[13] It seems likely, therefore, that fasting alters the rate of gastrin gene transcription or gastrin mRNA stability although this has not been tested directly. The mechanism underlying these changes is not well defined. Fasting leads to a decline in gastric pH, in part due to the loss of buffering capacity of food. Low gastric pH is known to inhibit gastrin secretion.[1] Achlorhydria induced by pharmacological inhibition of gastric acid secretion or fundectomy leads to an increase in antral gastrin mRNA.[14,15] Direct effects of hydrogen ions on the G cell to decrease gastrin mRNA during a fast cannot be excluded. It seems likely, however, that neuronal reflexes or endocrine or paracrine consequences of reduced gastric pH contribute to the fasting-induced decrease in antral gastrin mRNA.

Somatostatin is a primary candidate as mediator of fasting-induced decreases in gastrin mRNA. *In vivo,* reduced gastric pH stimulates somatostatin release from gastric D cells, and somatostatin decreases gastrin release.[1,16-18] Infusion of a somatostatin analogue can prevent the increases in antral gastrin mRNA induced by achlorhydria.[15] Recent studies in isolated dog antral mucosa provide direct evidence that somatostatin inhibits gastrin gene expression. In this system, somatostatin has no effect on basal gastrin gene transcription but does decrease gastrin gene transcription in response to a number of stimuli and also increases gastrin mRNA turnover.[19,20] Definitive evidence that somatostatin mediates fasting-induced decreases in antral gastrin mRNA will require additional studies, such as immunoneutralization of somatostatin in fasted rats *in vivo* and analyses of effects on antral gastrin gene transcription and mRNA.

C. SPECIFIC NUTRIENTS INCREASE GASTRIN mRNA ABUNDANCE

Refeeding of fasted rats results in rapid increases in gastrin mRNA.[21] Studies where different nutrients were directly instilled into the stomach indicate that the effects of refeeding on gastrin mRNA depend on the type of nutrient. Gastric instillation of peptones or amino acids, such as phenylalanine, elicits the greatest increases in gastrin mRNA and effects are apparent within 15 min.[13] More modest increases in antral gastrin mRNA occur in response to gastric instillation of fat but not in response to glucose or saline.[13]

The rapid induction of antral gastrin mRNA by gastric instillation of peptones, amino acids, and fat suggests effects on gastrin gene transcription or mRNA stability (Figure 1). As with fasting, the precise mediators of these effects are not well defined. Decreased release of somatostatin and thereby a reduced negative influence of somatostatin on gastrin synthesis may mediate the nutrient dependent increases in antral gastrin mRNA. Compared with effects on gastrin mRNA, gastric instillation of nutrients has reciprocal effects on gastric somatostatin mRNA and peptide.[13] The effects of nutrient on gastric somatostatin or somatostatin mRNA are, however, slower than effects on gastrin mRNA.[13] Thus, if reduced somatostatin does mediate nutrient effects on gastrin mRNA, a decreased release of somatostatin, locally in the stomach, may precede nutrient dependent changes in somatostatin synthesis.

As well as somatostatin, direct effects of nutrients on gastrin gene expression cannot be excluded (Figure 1). Peptone, amino acids, and amines that derive from decarboxylation of amino acids can all stimulate gastrin release from isolated G cells in culture suggesting direct effects of these nutrients on G cell function.[12] To date, however, the direct effect of nutrients on gastrin gene expression in isolated G cells has not been evaluated.

Gastrin release *in vivo* and *in vitro* is stimulated by acetylcholine released from vagal efferents and by gastrin releasing peptide (GRP; the mammalian counterpart of the amphibian peptide bombesin) released from efferent nerve endings in proximity to the G cell.[1] To date, little is known about the role of neural influences in nutrient effects on gastrin gene expression (Figure 1).

D. MODEL SYSTEMS TO ANALYZE THE MOLECULAR BASIS FOR NUTRIENT EFFECTS ON GASTRIN GENE EXPRESSION

It it clear that additional studies are required to establish definitively that the effects of nutrients on antral gastrin mRNA are mediated at the level of gene transcription and to elucidate the precise stimuli and molecular mechanisms that underlie these effects. Progress has been made in the development of a number of model systems to address these issues.

Isolated G cells represent one useful model system to study nutrient effects on gastrin gene expression.[22] Isolated G cells offer the advantage that they are not transformed cells but are limited by the small numbers of cells that can be isolated and by their limited viability. The development of highly sensitive techniques to measure gene expression, such as quantitative polymerase chain reaction (PCR), should facilitate analyses of gastrin gene expression in isolated G cells.

Isolated antral mucosa maintained in short-term tissue culture also represents a useful model system to analyze gastrin gene expression. This system has the advantage that it has been used successfully to measure gastrin gene transcription in response to a number of stimuli and retains some of the paracrine and neural influences that may influence nutrient effects.[19,20] The major limitation of this system is its short-term viability.

Recently, several cell lines have been used to analyze gastrin gene expression and have led to the characterization of the gastrin promoter and *cis*-elements that mediate changes in gastrin gene transcription in response to hormones or neuropeptides.[23-27] In GH_4 pituitary cell lines and a number of insulinoma cell lines, somatostatin inhibits gastrin gene expression, and GRP may have modest stimulatory effects on gastrin gene expression.[24,27] These findings provide indirect evidence that somatostatin or GRP may be involved in fasting or nutrient effects on gastrin gene expression. The applicability of such model cell lines to nutrient effects on gastrin gene transcription may be limited since most cell lines are neoplastic as well as divorced from the normal G cell environment. The progress made using these cell lines to define putative gastrin *cis*-regulatory elements should facilitate the development of other systems to assess nutrient effects on gastrin gene expression. One potentially promising approach is to generate transgenic animals with putative gastrin *cis*-elements linked to a readily assayable reporter gene. Such transgenic models should help elucidate the *cis*-regulatory elements that mediate nutrient effects on gastrin gene expression *in vivo*.

IV. CHOLECYSTOKININ (CCK) GENE EXPRESSION

A. BACKGROUND

CCK is structurally related to gastrin but derives from a different gene.[28-30] CCK is synthesized by endocrine cells in the mucosa of the small intestine as well as expression in neurons within the central and peripheral nervous system.[1] It has long been recognized that CCK is a stimulant of pancreatic enzyme secretion and gallbladder contraction.[31,32] More recent evidence suggests that CCK exerts trophic effects on the exocrine pancreas.[33] Understanding the nutritional effects on CCK gene expression has relevance for understanding the role of nutrients in normal development and functioning of the exocrine pancreas as well as the pathophysiology of pancreatic disease such as pancreatitis.

CCK is secreted into the circulation after ingestion of nutrients although there are species differences in the responses. In man, fat, fatty acids, protein, and amino acids are potent stimulants of CCK secretion and glucose is a less potent stimulant.[34] In dogs, intraduodenal fatty acids and amino acids stimulate CCK secretion whereas intact proteins are poor stimulants.[35,36] In the rat, ingested protein appears to be the major stimulant of CCK secretion whereas amino acids, carbohydrates, and fats are poor stimulants.[37,38]

CCK secretion is controlled by negative feedback. Active proteases in the intestinal lumen inhibit CCK secretion. CCK secretion is stimulated by inactivation of intestinal proteases or by binding of proteases to substrates or trypsin inhibitor.[38] In other species there is emerging evidence for feedback control of CCK secretion via intestinal proteases.[39] Recent studies have elucidated one mechanism underlying the feedback control of CCK secretion in the rat. In this species, the rate of CCK secretion is tightly linked to the amount of protein within the duodenal lumen via a novel CCK releasing peptide, termed monitor peptide.[39-41] Rat monitor peptide has been sequenced and cloned.[4] Monitor peptide possesses trypsin inhibitor activity and is co-secreted from exocrine pancreas into the intestinal lumen together with other digestive enzymes.[39-41] Monitor peptide appears to compete with ingested protein for binding to trypsin. In the presence of ingested protein, monitor peptide remains free from trypsin to stimulate CCK secretion from duodenal endocrine cells. In the absence of ingested protein, monitor peptide forms a tight complex with trypsin so that the CCK releasing activity of monitor peptide is inhibited.[39-41] These findings raise interesting possibilities that as well as CCK secretion, CCK gene expression may be influenced by monitor peptide or other luminal protease inhibitors (Figure 2).

Recent studies in rat indicate that some of the nutrient effects on CCK secretion are mediated at the level of CCK mRNA and CCK gene transcription (summarized in Figure 2). It is probably important to note that findings in the rat may not be directly extrapolated to other species including man since there are species differences in nutrient effects on CCK secretion.[33-38]

B. FASTING REDUCES CCK mRNA ABUNDANCE IN DUODENUM

Prolonged fasting produces a gradual and progressive decline in CCK mRNA.[42,43] It is not clear whether the decrease in CCK mRNA represents

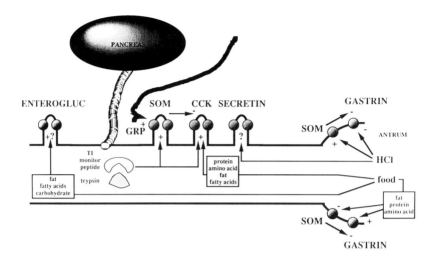

FIGURE 2. Summary of evidence for nutrient mediated effects on expression of gastrin, somatostatin (SOM), secretin, enteroglucagon (enterogluc), and CCK genes: + = positive effect on gene expression; − = negative effect; ? = indirect evidence for an effect; TI = monitor peptide.

effects on CCK gene transcription, mRNA stability, or abundance of CCK cells relative to other mucosal cell types. The mechanism underlying the fasting-induced decrease in CCK mRNA is not known but is not paralleled by effects on duodenal somatostatin mRNA or peptide.[43] If monitor peptide influences CCK gene transcription, fasting-induced decreases in CCK mRNA may relate to reduced availability of free monitor peptide for stimulation of CCK producing cells (Figure 2). This possibility has not been tested.

C. REFEEDING AND SPECIFIC NUTRIENTS INCREASE CCK mRNA ABUNDANCE IN DUODENUM

After only one day of refeeding of fasted rats, CCK mRNA levels are restored to normal.[43] When elemental diet is given to rats by perfusion into the intestinal lumen, there is no effect on CCK secretion or CCK mRNA.[44] In contrast, when a soybean trypsin inhibitor is perfused into the intestinal lumen concomitantly with the elemental diet, there is a rapid increase in CCK secretion and in CCK mRNA.[44] The enhanced CCK secretion is presumably because the trypsin inhibitor competes with the endogenous monitor peptide for binding to trypsin and thereby ''allows'' the monitor peptide to stimulate CCK producing cells. The increase in CCK mRNA in response to soybean trypsin inhibitor is mediated at the level of CCK gene transcription.[31] The precise stimuli and mechanisms underlying this effect are unknown. Clearly, monitor peptide is a primary candidate, but effects of monitor peptide on CCK gene transcription have not been reported. Direct stimulatory effects of protein or other nutrients on the CCK producing cell also cannot be excluded (Figure 2).

Another candidate mediator of nutrient-mediated effects on CCK gene transcription is GRP which is known to stimulate CCK secretion and is localized in efferent nerve endings throughout the gastrointestinal tract.[1] Evidence againt GRP as a mediator of increased CCK gene transcription in response to luminal nutrient stems from studies demonstrating that intravenous bombesin (the amphibian counterpart of mammalian GRP) transiently increases plasma levels of CCK but does not influence CCK mRNA levels in duodenum.[45] This finding is of interest in that it indicates that increases in CCK secretion are not obligatorily linked to increases in CCK mRNA.[45] On the other hand, a role for GRP in nutrient effects to increase CCK mRNA cannot entirely be excluded since intravenous bombesin may not necessarily elicit the same effects as activation of GRP efferents by nutrients within the intestinal lumen.

The soybean trypsin inhibitor model is a well defined model to study effects of luminal protein on CCK mRNA since overall nutrient status of the animals is not significantly affected.[44] Using this model Kanayama and Liddle have demonstrated recently that somatostatin can inhibit the induction of CCK mRNA in response to soybean trypsin inhibitor.[46] This raises the possibility that, as with gastrin, a component of nutrient effects to stimulate CCK gene transcription may be mediated by removal of a negative influence of somatostatin (Figure 2). To date, however, there are no data to directly support this possibility. Thus, the precise mediators of nutrient effects on CCK gene transcription are not defined (Figure 2).

D. MODEL SYSTEMS TO ANALYZE NUTRIENT EFFECTS ON CCK GENE EXPRESSION

In contrast to gastrin there has been little progress to date in developing model cell culture systems to study regulation of the CCK gene. Isolation and culture of enriched populations of enteric CCK endocrine cells has not yet been reported, and CCK gene expression has not been studied in short-term culture of intestinal mucosa. Recent studies have indicated that chimeric genes containing the rat CCK 5′ flanking region and a chloramphenicol acetyltransferase gene are active when transfected into a number of cell lines.[47,48] These cell lines have been used as model systems to define *cis*-elements that determine basal expression of the CCK gene and increased CCK expression induced in response to elevated cyclic AMP or phorbol esters.[47,48] Specific nuclear proteins appear to bind to these *cis*-elements and therefore are candidate *trans*-acting factors for regulation of CCK gene transcription.[47] It is not clear that findings in these systems will relate to control of the CCK gene in enteroendocrine cells. Progress in understanding nutritional control of the CCK gene in endocrine cells of the small intestine thus awaits development of useful model systems and progress in characterizing the CCK *cis*-regulatory elements.

V. SOMATOSTATIN GENE EXPRESSION

A. BACKGROUND

Somatostatin is a tetradecapeptide originally isolated from the hypothalamus.[49] It is now established that the GI tract and pancreas are major sites of somatostatin synthesis.[1,50,51] In stomach and pancreas, like brain, the predominant form of somatostatin is a tetradecapeptide whereas a 28 amino acid form predominates in the intestine.[51-53] These two forms of somatostatin derive from differential processing of a precursor encoded by a single somatostatin gene and mRNA.[54-56] Somatostatin is synthesized in D cells of the pancreatic islets and D cells within the mucosa of the stomach, small intestine, and large intestine.[1,50-53,57,58] In some species, somatostatin is also synthesized by neurons within the intrinsic neuronal plexuses of the gut.[1,58]

Somatostatin has widespread actions that have been reviewed elsewhere and are beyond the scope of this chapter.[1,50] In general, somatostatin is an inhibitory peptide. In the GI tract, somatostatin acts to inhibit gastric acid secretion, pancreatic exocrine secretion, and the secretion of hormones from pancreatic islet and GI endocrine cells.[1] Available evidence suggests that rather than acting in an endocrine manner, gastrointestinal somatostatin may act in a paracrine manner to modulate target cell function, possibly via release of somatostatin from cytoplasmic processes projecting from the D cells onto target cells.[59]

A number of studies suggest effects of nutrients on somatostatin concentrations in plasma and somatostatin content of different regions of the GI tract, especially stomach and pancreas.[1,3,50] The interpretation of such findings is complicated by the multiple cellular sites of somatostatin biosynthesis. In addition, since somatostatin acts in a paracrine manner, tissue content of somatostatin may remain unchanged even in the face of local alterations in somatostatin synthesis or secretion by specific cell populations. As well as providing information about nutrient regulation of somatostatin gene expression, analyses of nutrient effects on somatostatin mRNA content in different regions of GI tract may therefore aid in the interpretation of nutrient effects on D cell function.

B. FASTING ELICITS REGION-SPECIFIC EFFECTS ON GASTROINTESTINAL SOMATOSTATIN mRNA

Within 12 h of fasting there are increases in somatostatin mRNA in the gastric antrum that are maintained with prolonged fasting.[13,14] Since short duration fasting does not alter the relative number of gastric D cells,[10] these observations suggest that fasting increases somatostatin gene transcription or mRNA stability. In contrast to somatostatin mRNA, fasting does not signif-icantly alter the concentrations of immunoreactive somatostatin in the an-trum.[14] Measurement of somatostatin mRNA may therefore provide an in-dication of alterations in gastric D cell function that are not readily apparent by the measurement of peptide. Fasting induces no significant change in somatostatin mRNA or peptide within the duodenum, suggesting that fasting induces region-specific effects on somatostatin gene transcription/mRNA sta-bility within the GI tract.[43]

While the mechanism underlying fasting-induced increases in antral so-matostatin mRNA is not defined, it likely relates to increased acidity and reduced pH of antral contents. Effects of acidification of antral contents on antral somatostatin mRNA have not been tested directly but reduced gastric pH stimulates somatostatin secretion from gastric D cells.[1,16] Antral soma-tostatin mRNA is decreased in response to elevation of gastric pH by fundectomy or inhibition of acid secretion from parietal cells using omeprazole, a blocker of the parietal cell hydrogen/potassium ATPase.[14,15] Thus, a number of lines of evidence suggests that reduced gastric pH may mediate fasting-induced increases in antral somatostatin gene expression (Figure 2). If so, this could provide an explanation for the region-specific effects of fasting on somato-statin mRNA. Pancreatic juice and bile serve to neutralize gastric acid as it enters the duodenum so that duodenal D cells would not be subject to the same low pH as antral D cells.

C. NUTRIENT EFFECTS ON GASTRIC AND INTESTINAL SOMATOSTATIN GENE EXPRESSION

To date, few studies have addressed the role of specific nutrients in regulation of somatostatin gene expression. Refeeding of fasted rats with solid food, or gastric instillation of peptone, amino acids, or oil, results in a rapid decline in antral somatostatin mRNA.[13,14] Specific nutrients within the stom-ach therefore appear to inhibit somatostatin gene expression or reduce mRNA stability.[13,14] As with fasting, effects of nutrients on antral somatostatin mRNA are more pronounced than effects on peptide.[14] An increase in gastric pH as a result of the buffering capacity of nutrients may account for their effects on antral somatostatin mRNA. Indirect effects via gastrin release or other events that may be modulated by the nutrients cannot be excluded.[14]

Duodenal somatostatin mRNA is not altered by refeeding of fasted rats.[43] Perfusion of the intestinal lumen with soybean trypsin inhibitor does, however, increase duodenal somatostatin mRNA (Figure 2).[44] It is not clear whether this is a direct effect on duodenal D cells or an indirect effect due to stimulation of CCK secretion, secretion of other hormones, or secretion of pancreatic enzymes. Somatostatin appears able to exert "autoregulation" of its own synthesis in the duodenum since intravenous infusion of somatostatin inhibits the trypsin inhibitor induced increase in duodenal somatostatin mRNA.[46] This may be a direct effect of somatostatin on D cells since similar autoregulation of somatostatin synthesis and secretion is observed in isolated canine D cells.[60,61] Alterations in duodenal somatostatin mRNA induced by trypsin inhibitor or somatostatin are not accompanied by effects on peptide, again pointing to the potential advantages of analyses of somatostatin mRNA as a measure of D cell function.

Intravenous bombesin produces rapid, quite dramatic yet transient increases in duodenal somatostatin mRNA, implicating the neuropeptide GRP as one potential mediator of nutrient effects on somatostatin gene expression (Figure 2).[45] It is not clear whether the effect of bombesin is specific for duodenal D cells or also influences gastric D cells.

D. MODEL SYSTEMS TO ANALYZE NUTRIENT EFFECTS ON SOMATOSTATIN GENE EXPRESSION

Accumulating evidence, therefore, suggests a role of nutrients in modulating somatostatin gene expression and suggests differential effects of nutrient status in different regions of the GI tract (Figure 2). The complexity of the somatostatin system in the GI tract suggests that additional approaches to nutrient control of somatostatin gene expression *in vivo* are warranted. The paracrine mode of action of somatostatin suggests that highly localized changes in somatostatin gene expression may occur in subsets of D cells within a given region of gut. Such changes may be missed or underestimated by analyses of gene expression in a whole organ or gut region. The multiplicity of cell types that express the somatostatin gene in the gastrointestinal tract also raises the possibility for cell-type specific effects of nutrients on somatostatin gene expression, but few or no *in vivo* studies have addressed this possibility. Interesting information on cell specific regulation of somatostatin gene expression may emerge from a combination of histological methods, such as *in situ* hybridization, with other approaches used to date.

The *cis*-acting elements that dictate somatostatin gene expression have been studied extensively in cultured cell lines. A cyclic AMP responsive octanucleotide TGACGTCA (CRE) is important for basal and cyclic AMP regulated expression of the somatostatin gene in a number of cell lines.[62] This element is bound by a CRE binding protein CREB which belongs to a growing family of structurally related transcription factors. The CRE together with other sequences that lie within 65 base pairs of the transcription initiation site is sufficient for specific expression in pancreatic islet cell lines.[63] Islet cell-specific expression of the somatostatin gene also appears to depend on other DNA binding proteins in addition to the more ubiquitously expressed CREB.[63] As with gastrin and CCK, progress in defining the *cis*-elements and *trans*-acting factors that determine somatostatin gene expression in gastric and enteroendocrine D cells is hampered by lack of suitable systems. Isolated gastric D cells in or short-term cultures of antral mucosa represent promising systems but suffer from the same limitations as described for gastrin.[60,61]

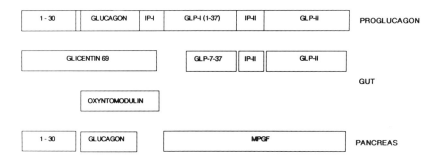

FIGURE 3. Schematic of mammalian proglucagon and products of posttranslational processing in gut and pancreas.

Isolated fetal rat intestinal cells in culture (FRIC cells) are another promising system which has already been used to demonstrate positive regulation of somatostatin gene expression by cyclic AMP and phorbol esters.[64] Nonetheless, progress in elucidating nutrient effects on gastrointestinal somatostatin gene expression will require a combination of information derived from these *in vitro* systems and *in vivo* approaches such as transgenic animals.

VI. GLUCAGON/ENTEROGLUCAGON

A. BACKGROUND

The proglucagon gene is a particularly interesting and complex system with regard to nutrient control of expression. Mammals possess a single glucagon gene and mRNA, but the encoded precursor, proglucagon, is a polyprotein precursor that contains three glucagon related peptides as well as other peptides (Figure 3).[1,65-68] A common proglucagon precursor is processed very differently in the pancreas and intestine to result in different biologically active peptides with different biological activities.[69-73] The structure of the common pancreatic and intestinal proglucagon is shown in Figure 3. Proglucagon contains the 29 amino acid pancreatic glucagon in the midregion with amino acid extensions at amino and carboxyl termini. A long carboxyl terminal extension contains two glucagon-like peptides (GLPs) with structural homology to glucagon, and these have been termed GLP-I and GLP-II. Spacer or intervening peptides (IPs) separate glucagon from GLP-I and GLP-I from GLP-II.

In the pancreas, proglucagon is synthesized in pancreatic A cells.[1,74] Products of posttranslational processing of proglucagon in the pancreas are a 29 amino acid pancreatic glucagon, a 30 amino acid peptide comprised of the amino terminus of proglucagon, and a ''Major proglucagon fragment'' (Figure 3).[69-73] The only product of pancreatic proglucagon processing with defined biological activity is glucagon.[74] The primary biological role of pancreatic glucagon is well established. Glucagon regulates metabolic pathways in liver and adipose tissue and maintains euglycemia during fasting or nutrient perturbation.[74] Pancreatic glucagon is secreted into the circulation in response to fasting, decreases in plasma glucose, and increases in plasma amino acids and neural influences.[71,74] Pancreatic glucagon secretion is suppressed by high plasma glucose.[71,74]

In the intestine, proglucagon is processed very differently than in the pancreas. The products of intestinal proglucagon are enteroglucagons (alternately named glucagon-like immunoreactants [GLIs]), GLP-I (73–77), GLP-II, and IP-II (Figure 3).[69-73] Enteroglucagons or GLIs include glicentin and oxyntomodulin (Figure 3). Glicentin is composed of the first 69 amino acids of proglucagon and includes the entire sequence of glucagon in the midregion flanked by amino acid extensions at both amino and carboxyl termini (Figure 3). Oxyntomodulin consists of glucagon with a hexapeptide extension at the carboxyl terminus (Figure 3). At physiological concentrations, glicentin and oxyntomodulin appear not to share the biological actions of pancreatic glucagon but appear to act primarily to inhibit gastric acid secretion and thus are enterogastrones.[1,71,75] Of the other intestinal proglucagon derived peptides, only GLP-I has been extensively characterized in terms of biological activity. GLP-I 7–37 is the primary form of GLP-I in intestine and is a potent stimulant of insulin secretion and insulin gene expression.[71,76,77] The biological roles of GLP-II and IP-II are not defined. Based on observations that plasma concentrations of enteroglucagons are elevated in most situations of increased cell proliferation of the intestinal mucosa, the enteroglucagons (glicentin/oxyntomodulin) or other products of intestinal proglucagon processing are candidate enterotrophic hormones.[68,78]

Products of intestinal proglucagon-derived peptides are secreted in response to different stimuli than pancreatic products. Available evidence suggests that all of the intestinal products of proglucagon are secreted in parallel.[69-72] Plasma enteroglucagon levels are reduced by fasting and elevated after food ingestion.[71] Intrajejunal glucose or carbohydrate and fatty acids are especially strong stimulants of enteroglucagon secretion.[1,79] The secretion of enteroglucagons in response to luminal glucose but suppression of pancreatic glucagon by high plasma glucose indicates that the pancreatic A cell and intestinal L cell have very different responses to glucose.

To date, the regulation of glucagon gene expression by nutrients has not been studied extensively. The molecular basis for nutrient control of proglucagon gene expression is likely to be both complex and interesting given the different secretory responses of two different proglucagon expressing cell types, the A cell and the L cell, in response to nutritional stimuli.

B. NUTRIENT CONTROL OF GLUCAGON GENE EXPRESSION IN PANCREAS

In rats after a 3-d fast we find no significant change in pancreatic proglucagon mRNA abundance as assessed by Northern blot hybridization of

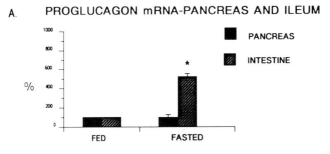

A.   PROGLUCAGON mRNA-PANCREAS AND ILEUM

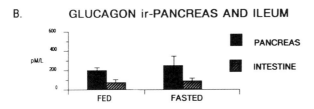

B.   GLUCAGON ir-PANCREAS AND ILEUM

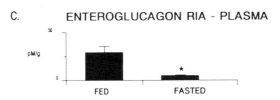

C.   ENTEROGLUCAGON RIA - PLASMA

FIGURE 4. Proglucagon mRNA and peptide in *ad libitum* fed and fasted rats. (A) Proglucagon mRNA in pancreas and small intestine. Proglucagon mRNA was analyzed in poly A + RNA by Northern hybridization and abundance assessed as concentration per µg polyA + RNA. Shown is the abundance in pancreas or intestine of fasted animals expressed as % of abundance of fed animals; (B) Glucagon immunoreactivity in pancreas and intestine of fed and fasted rats; (C) Plasma enteroglucagon immunoreactivity in fed and fasted rats. * = significant difference in fasted animals vs. fed animals ($p < 0.05$).

total mRNA extracted from pancreas (Figure 4). In contrast, Chen et al., using *in situ* hybridization histochemistry, observed a significant increase in pancreatic proglucagon mRNA in response to a 4-d fast that induced signif-icant glucopenia.[80] It is possible that the different durations of fasting account for these different results. However, it also seems possible that the different results reflect the different methodologies utilized. Our inability to detect a change in pancreatic proglucagon mRNA by Northern hybridization may be due to the fact that pancreatic A cells represent a very small proportion of the total pancreatic mass. Alterations in proglucagon mRNA in such small cell populations may be masked by changes in other more abundant mRNAs. This illustrates the potential advantages of a combined approach using both hybridization analyses of extracted mRNA and histological *in situ* hybridization analyses. The increase in A cell proglucagon mRNA in response to fasting indicates that low plasma glucose or neural and hormonal consequences of fasting result in increases in proglucagon gene transcription or mRNA stability.[80] Chen et al. also found that hyperglycemia reduces the abundance of proglucagon mRNA in pancreatic A cells suggesting negative effects of hyperglycemia on pancreatic proglucagon gene transcription or mRNA stability.[80] Of interest, however, are other observations that pancreatic proglucagon mRNA is elevated in rats with streptozotocin-induced, insulin-dependent diabetes, despite hyperglycemia.[81] Insulin administration and restoration of euglycemia rapidly reduces pancreatic proglucagon mRNA in streptozotocin diabetic rats.[81] Together, these findings suggest that insulin is involved in the control of pancreatic proglucagon gene expression by glucose.[81]

C. NUTRIENT CONTROL OF GLUCAGON GENE EXPRESSION IN THE INTESTINE

To date, there is virtually no information on nutrient control of glucagon gene expression in the intestine. This is surprising given the potential relevance of intestinal proglucagon-derived peptides in regulating normal, adaptive, or aberrant growth of intestinal mucosa.[78] In recent studies we observed an increase in proglucagon mRNA abundance in intestine or rats after a 3-d fast (Figure 4) even though plasma enteroglucagon is reduced by fasting (Figure 4) and there is no parallel decline in enteroglucagon immunoreactivity in intestine (Figure 4). It seems likely that the fasting-induced increase in intestinal proglucagon mRNA derives from a reduction in mass of nonendocrine cells in intestinal mucosa, since fasting induces significant mucosal hypoplasia.[4,15] We are currently addressing this possibility using *in situ* hybridization histochemistry. These difficulties in data interpretation attest to the need for caution in the interpretation of nutrient-dependent changes in steady-state GI hormone mRNAs in the intestine as discussed at the beginning of this chapter.

Effects of specific nutrients on intestinal proglucagon gene expression have not been reported, but luminal glucose and fats represent likely candidates given their effects on enteroglucagon secretion (Figure 2).

D. NUTRIENT-DEPENDENT AND INDEPENDENT CHANGES IN INTESTINAL PROGLUCAGON GENE EXPRESSION AFTER PROXIMAL SMALL BOWEL RESECTION

In recent studies, we and others have focused attention on intestinal proglucagon gene expression in a model of altered growth of intestinal mucosa, the rat after proximal small bowel resection. In the resection model there is compensatory hyperplasia of the mucosa of the remnant ileum to restore digestive and absorptive capacity.[4,78] This model is of relevance to man since resection is used as treatment for morbid obesity, and some studies indicate that resection is associated with enhanced carcinogenesis in the gut. Many studies have established that the adaptive growth of ileum after resection is correlated with elevated plasma levels of enteroglucagon,[4,78] supporting the possibility that enteroglucagons or other proglucagon-derived peptides are enterotrophic hormones. We and Taylor et al. have established that adaptive hyperplasia is associated with increases in intestinal proglucagon mRNA.[68,82,83] Such increases do not appear to reflect increased L cell number and therefore likely reflect an increase in intestinal proglucagon gene transcription or mRNA stability.[84] It is well established that resection-induced adaptive growth after resection requires luminal nutrient,[4,5] as does the resection-induced increases in plasma enteroglucagon.[85] Thus, we analyzed the role of luminal nutrient in resection-induced changes in intestinal proglucagon mRNA. Of interest are our observations that there are nutrient-independent increases in intestinal proglucagon mRNA within hours of resection and preceding the adaptive growth response.[82] Such observations are consistent with one or more intestinal proglucagon derived peptides as initiators of or early markers of adaptation. However, sustained increases in intestinal proglucagon mRNA after resection require luminal nutrient, suggesting an interplay between nutrient-independent and nutrient-dependent factors in regulating intestinal proglucagon gene expression and adaptive growth.[82,83] In follow-up studies, we are investigating the role of specific nutrients in resection-induced changes in expression of the proglucagon gene expression and other intestinal hormones and growth factors.[81,88] Such studies should have relevance for understanding the role of nutrient effects on GI hormone gene expression in regulating normal and aberrant growth of intestinal mucosa.

E. MODEL SYSTEMS TO ANALYZE NUTRIENT CONTROL OF PROGLUCAGON GENE EXPRESSION

A number of useful model systems are available for analyses of nutrient control of pancreatic glucagon gene expression. The *in vitro* perfused pancreas represents one useful model.[86] Pancreatic islets can be isolated from exocrine pancreas and maintained in culture for analyses of islet hormone biosynthesis.[70] These systems have been used to analyze insulin gene expression and represent promising systems for future studies of nutrient control of glucagon gene expression. They offer the advantage that many of the paracrine mechanisms that may mediate nutrient effects remain intact. A number of established insulinoma/islet cell lines express the proglucagon gene and have been used to analyze the *cis*-elements that determine islet-specific expression of the proglucagon gene.[87-89] Using such systems three DNA elements in the 5' flanking region of the glucagon gene have been shown to underlie islet-specific expression.[89] Using these islet cell lines, it also has been established that pancreatic proglucagon gene expression is upregulated by protein kinase C-dependent pathways and not by cyclic AMP-dependent pathways, even though the glucagon 5′ flanking region contains a cyclic AMP response element.[87] Such findings raise the possibility that nutrient control of pancreatic proglucagon gene expression is mediated by effects on protein kinase C. This possibility can be tested in transgenic animals with the kinase C responsive *cis*-elements linked to a reporter gene. Progress has been made towards development of such models since transgenic mice have been developed that show pancreas specific expression of a reporter gene driven by 900 base pairs of the proglucagon 5′ flanking region.[90]

As with other GI hormones, the availability of model systems to analyze nutrient control of intestinal proglucagon expression is more limited. Drucker and Brubaker have used fetal rat intestinal cells in culture (FRIC cells) as a model system.[90] Using this system they have demonstrated that control of proglucagon gene expression differs in fetal intestine and islet cells. In FRIC cells, protein kinase C-dependent pathways stimulate enteroglucagon secretion but not gene expression.[91] In contrast to islet cells, proglucagon gene expression is upregulated by cyclic AMP-dependent pathways in FRIC cells.[90] These findings suggest that nutrient effects on intestinal proglucagon gene expression may involve cyclic AMP-dependent pathways. Furthermore, differences in pancreatic and intestinal proglucagon gene expression may be mediated by protein kinase C and cyclic AMP-dependent pathways, respectively. Additional studies in the FRIC cell system and studies of transgenic mice with the proglucagon *cis*-elements linked to a reporter gene should help elucidate mechanisms that underlie the nutrient control of intestinal proglucagon gene expression. It is noteworthy, however, that 900 base pairs of proglucagon 5′ flanking region are not sufficient to drive expression of a reporter gene in intestinal L cells of transgenic mice, even though this construct was active in A cells and in brain.[90] Thus, additional studies are required to define the cell-specific elements that underlie proglucagon gene expression in L cells before nutrient regulation can be addressed in transgenic mice.

VII. CONCLUSIONS AND FUTURE DIRECTIONS

In the past few years, progress has been made towards understanding nutrient control of expression of the genes encoding a number of GI hormones. Nonetheless it is probably clear that progress has been hampered by the complexity of the GI endocrine system and the paucity of useful enteroendocrine cell lines. The utility of the transgenic approach has been alluded to above. With the development of some transgenic animals, and the increased use of these models, it seems likely that substantial and rapid progress will be forthcoming. In addition, potentially useful model enteroendocrine cell lines have been developed recently.[91] STC-1 cell lines derive from enteroendocrine tumors that spontaneously developed in mice generated by crossbreeding of two transgenic lines.[91] The transgenic animals used for cross-breeding were ones with an insulin promoter used to drive the SV40 large T antigen gene and ones with the insulin promotor linked to mouse polyoma small T antigen. The SV40 transgenics develop pancreatic insulinoma but not enteroendocrine tumors.[91] The polyoma transgenics do not develop tumors in pancreas or gut.[91] On crossbreeding, however, a large proportion of animals develop enteroendocrine tumors that express a spectrum of GI hormones. STC-1 cell lines derived from the enteroendocrine tumors express a wide range of GI hormone. The molecular basis for the development of enteroendocrine tumors in these crossbred mice is of itself very interesting since enteroendocrine tumors are relatively rare. For those working in the field of GI hormone expression, the cell lines derived from these tumors may represent a very important and unique tool for future studies.

ACKNOWLEDGMENTS

This work was supported by NIH grant DK 40247. Thanks are expressed to Dr. Rodger Liddle for extremely useful input and discussion. Blanche Jones provided invaluable editorial and secretarial assistance.

REFERENCES

1. **Walsh, J. H.,** Gastrointestinal hormones, in *Physiology of the Gastrointestinal Tract,* Johnson, L. R., Ed., Raven Press, New York, 1987, 181.
2. **Brown, J. C., McIntosh, C. H. S., and Pederson, R. A.,** The gastrointestinal peptides and nutrition, *Can. J. Physiol. Pharmacol.,* 61, 282, 1982.
3. **Yamada, T.,** Gut hormone release induced by food ingestion, *Am. J. Clin. Nutr.,* 42, 1033, 1985. (DB 44)
4. **Dowling, R. H.,** Small bowel adaptation and its regulation, *Scand. J. Gastroenterol.,* 74, 53, 1982.
5. **Lipscomb, H. L. and Sharp, J. G.,** Effects of reduced food intake on morphometry and cell production in the small intestine of the rat, *Virchows Arch.,* 41, 285, 1982.
6. **Fujimoto, S., Kimoto, K., Yamashita, S., Kawai, K., and Hattori, T., and Fugita, S.,** Tritiated thymidine autoradiographic study on origin and renewal of gastrin cells in antral area of hamsters, *Gastroenterology,* 79, 785, 1985.
7. **Mitchell, P. J. and Tjian, R.,** Transcriptional regulation in mammalian cells by sequence-specific DNA binding proteins, *Science,* 245, 371, 1989.
8. **Raghowr, R.,** Regulation of messenger RNA turnover in eukaryotes, *Trends Biochem. Sci.,* 12, 358, 1987.
9. **Lichtenberger, L. M., Lechago, J., and Johnson, L. R.,** Depression of antral and serum gastrin concentration by food deprivation in the rat, *Gastroenterology,* 68, 1473, 1975.

10. **Schwarting, H., Koop, H., Gellert, G., and Arnold, R.,** Effect of starvation on endocrine cells in the rat stomach, *Regul. Pept.,* 14, 33, 1986.

11. **Taylor, I. L., Byrne, W. J., Christie, D. L., Ament, M. E., and Walsh, J. H.,** Effect of individual L-amino acids on gastric acid secretion and serum gastrin and pancreatic polypeptide release in humans, *Gastroenterology,* 83, 272, 1982.

12. **Lichtenberger, L. M.,** Importance of food in the regulation of gastrin release and formation, *Am. J. Physiol.,* 243, G429, 1982.

13. **Wu, V., Sumii, K., Tari, A., Sumii, M., and Walsh, J. H.,** Regulation of rat antral gastrin and somatostatin gene expression during starvation and after refeeding, *Gastroenterology,* 101, 1552, 1991.

14. **Wu, S. V., Sumii, K., Tari, A., Mogard, M., and Walsh, J. H.,** Regulation of gastric somatostatin gene expression, *Metabolism,* 39, 125, 1990. (DB41)

15. **Brand, S. J. and Stone, D.,** Reciprocal regulation of antral gastrin and somatostatin gene expression by omeprazole-induced achlorhydria, *J. Clin. Invest.,* 82, 1059, 1988.

16. **Schusdziarra, V., Harris, V., Conlon, J. M., Arimura, A., and Unger, R.,** Pancreatic and gastric somatostatin release in response to intragastric and intraduodenal nutrients and HCl in the dog, *J. Clin. Invest.,* 62, 509, 1978.

17. **Bloom, S. R., Mortimer, C. H., and Thorner, M. O.,** Inhibition of gastrin and gastric acid secretion by growth hormone release inhibitory hormone, *Lancet,* 1, 1106, 1974.

18. **Harty, R. F., Maico, D. G., and McGuigan, J. E.,** Somatostatin inhibition of basal and carbachol stimulated gastrin release in rat antral organ culture, *Gastroenterology,* 81, 707, 1981.

19. **Karnik, P. S., Monahan, S. J., and Wolfe, M. M.,** Inhibition of gastrin gene expression by somatostatin, *J. Clin. Invest.,* 83, 367, 1989.

20. **Karnik, P. S. and Wolfe, M. M.,** Somatostatin stimulates gastrin mRNA turnover in dog antral mucosa, *J. Biol. Chem.,* 265, 2550, 1990.

21. **Wu, S. V., Sumii, K., and Walsh, J. H.,** Studies of regulation of gastrin synthesis and post-translational processing by molecular biology approaches, *Ann. N.Y. Acad. Sci.,* 597, 17, 1990. (DB18)

22. **Lichtenberger, L. M., Forssman, W. G., and Ito, S.,** Functional responsiveness of an isolated and enriched fraction of rodent gastrin cells, *Gastroenterology,* 79, 447, 1980.

23. **Theill, L. E., Wiborg, O., and Vuust, J.,** Cell-specific expression of the human gastrin gene: evidence for a control element located downstream of the TATA box, *Mol. Cell. Biol.,* 7, 4329, 1987.

24. **Brand, S. J. and Wang, T. C.,** Gastrin gene expression and regulation in rat islet cell lines, *J. Biol. Chem.,* 263, 16597, 1988. (DB 106)

25. **Wang, T. C. and Brand, S. J.,** Islet cell-specific regulatory domain in the gastrin promoter contains adjacent positive and negative DNA elements, *J. Biol. Chem.,* 265, 8908, 1990. (DB63)

26. **Merchant, J. L., Demediuk, B., and Brand, S. J.,** A GC-rich element confers epidermal growth factor responsiveness to transcription from the gastrin promoter, *Mol. Cell. Biol.,* 11, 2686, 1991. (DB61)

27. **Godley, J. M. and Brand, S. J.,** Regulation of the gastrin promoter by epidermal growth factor and neuropeptides, *Proc. Natl. Acad. Sci. U.S.A.,* 86, 3036, 1989. (DB62)

28. **Takahashi, Y., Fukushige, S., Murotsu, T., and Matsubara, K.,** Structure of human cholecystokinin gene and its chromosomal location, *Gene,* 50, 353, 1986.

29. **Deschenes, R. J., Lorenz, L. J., Haun, R. S., Roos, B. A., Collier, K. J., and Dixon, J.,** Cloning and sequence analysis of a cDNA encoding rat preprocholecystokinin, *Proc. Natl. Acad. Sci. U.S.A.,* 81, 726, 1984.

30. **Gubler, U., Chua, A. O., Hoffman, B. J., and Collier, K. J.,** Cloned cDNA to cholecystokinin mRNA predicts an identical preprocholecystokinin in pig brain and gut, *Proc. Natl. Acad. Sci. U.S.A.,* 81, 4307, 1984.

31. **Ivy, A. C. and Oldberg, E.,** A hormone mechanism for gallbladder contraction and evacuation, *Am. J. Physiol.,* 85, 381, 1928.

32. **Harper, A. A. and Raper, H. S.,** Pancreozymin, a stimulant of secretion of pancreatic enzymes in extracts of small intestine, *J. Physiol.,* 102, 115, 1943.

33. **Niederau, C., Luthen, R., Niederau, M., Strohmeyer, G., Ferrell, L. D., and Grendell, J. H.,** Effects of long term CCK stimulation and CCK blockade on pancreatic and intestinal growth, morphology and function, *Digestion,* 46, 217, 1990.

34. **Liddle, R. A., Goldfine, I. D., Rosen, M. S., Taplitz, R. A., and Williams, J. A.,** Cholecystokinin bioactivity in human plasma, *J. Clin. Invest.,* 75, 1144, 1985.

35. **Konturek, S. J., Radecki, T., Thor, P., and Dembinski, A.,** Release of cholecystokinin by amino acids, *Proc. Soc. Exp. Biol. Med.,* 143, 305, 1973.

36. **Meyer, J. H. and Jones, R. S.,** Canine pancreatic responses to intestinally perfused fat and products of fat digestion, *Am. J. Physiol.,* 226, 1178, 1974.

37. **Liddle, R. A., Green, G. M., Conrad, C. K., and Williams, J. A.,** Proteins but not amino acids, carbohydrates, or fats stimulate cholecystokinin secretion in the rat, *Am. J. Physiol.,* 251, G243, 1986.

38. **Liddle, R. A., Goldfine, I. D., and Williams, J. A.,** Bioassay of plasma cholecystokinin in rats: effects of food, trypsin inhibitor, and alcohol, *Gastroenterology,* 87, 542, 1984.

39. **Iwai, K., Fushiki, T., and Fukuoka, S.-I.,** Pancreatic enzyme secretion mediated by a novel peptide: monitor peptide hypothesis, *Pancreas,* 3, 720, 1988.

40. **Iwai, K., Fukuoka, S.-I., Fushiki, T., Tsujikawa, M., Hirose, M., Tsunasawa, S., and Sakiyama, F.,** Purification and sequencing of a trypsin-sensitive cholecystokinin releasing peptide from rat pancreatic juice. Its homology with pancreatic secretory trypsin inhibitor, *J. Biol. Chem.,* 262, 8956, 1987.

41. **Fukuoka, S. and Scheele, G. A.,** Rapid and selective cloning of monitor peptide, a novel cholecystokinin-releasing peptide, using minimal amino acid sequence and the polymerase chain reaction, *Pancreas,* 5, 1, 1990.

42. **Greenstein, R. J., Isola, L., and Gordon, J.,** Differential cholecystokinin gene expression in brain and gut of the fasted rat, *Am. J. Med. Sci.,* 299, 32, 1990.

43. **Kanayama, S. and Liddle, R. A.,** Influence of food deprivation on intestinal cholecystokinin and somatostatin, *Gastroenterology,* 100, 909, 1991. (DB58)

44. **Liddle, R. A., Carter, J. D., and McDonald, A. R.,** Dietary regulation of rat intestinal cholecystokinin gene expression, *J. Clin. Invest.,* 81, 2015, 1988. (DB40)

45. **Kanayama, S. and Liddle, R. A.,** Regulation of intestinal cholecystokinin and somatostatin mRNA by bombesin in rats, *Am. J. Physiol.,* 261, G71, 1991.

46. **Kanayama, S. and Liddle, R. A.,** Somatostatin regulates duodenal cholecystokinin and somatostatin messenger RNA, *Am. J. Physiol.,* 258, 1990, G358. (DB22)

47. **Haun, R. S. and Dixon, J. E.,** A transcriptional enhancer essential for the expression of the rat cholecystokinin gene contains a sequence identical to the -296 element of the human c-fos gene, *J. Biol. Chem.,* 265, 15455, 1990.

48. **Monstein, H.-J. and Folkesson, R.,** Phorbol 12-myristate-13-acetate (PMA) stimulates a differential expression of cholecystokinin (CCK) and c-fos mRNA in a human neuroblastoma cell line, *FEBS,* 293, 145, 1991.

49. **Brazeau, P., Vale, W., Burgus, R., Ling, N., Butcher, M., Riveir, J., and Guillemin, R.,** Hypothalamic polypeptide that inhibits the secretion of immunoreactive pituitary growth hormone, *Science,* 179, 77, 1973.

50. **Newman, J. B., Lluis, F., and Townsend, C. M, Jr.,** Somatostatin, in *Gastrointestinal Endocrinology,* Thompson, J. C., Greeley, G. H., Jr., Rayford, P. L., and Townsend, C. M., Jr., Eds., McGraw-Hill, Toronto, 1987, 286.

51. **Patel, Y. C. and Reichlin, S.,** Somatostatin in hypothalamus, extrahypothalamic brain, and peripheral tissues of the rat, *Endocrinology,* 102, 523, 1978.

52. **Pradayrol, L., Jornvall, H., Mutt, V., and Ribet, A.,** N-terminally extended somatostatin: the primary structure of somatostatin-28, *FEBS,* 109, 55, 1980.

53. **Trent, D. F. and Weir, G. C.,** Heterogeneity of somatostatin-like peptides in rat brain, pancreas, and gastrointestinal tract, *Endocrinology,* 108, 2033, 1981.

54. **Shen, L.-P. and Rutter, W.,** Human somatostatin I: sequence of the cDNA, *Proc. Natl. Acad. Sci. U.S.A.,* 224, 168, 1984.

55. **Funckes, C. L., Minth, C. D., Deschenes, R., Magazin, M., Tavianini, M. A., Sheets, M., Collier, K., Weith, H. L., Aron, D. C., Roos, B. A., and Dixon, J. E.,** Cloning and characterization of a mRNA-encoding rat preprosomatostatin, *J. Biol. Chem.,* 258, 8781, 1983.

56. **Goodman, R. H., Aron, D. C., and Roos, B. A.,** Rat pre-prosomatostatin: structure and processing by microsomal membranes, *J. Biol. Chem.,* 258, 5570, 1983.

57. **Hofler, H., Childers, H., Montminy, M. R., Goodman, R. H., Leehan, R. M., DeLellis, R. A., Tischler, A. S., and Wolfe, H. J.,** Localization of somatostatin mRNA in the gut, pancreas and thyroid gland of the rat using antisense RNA probes for in situ hybridization, *Acta Histochem.,* 34, 101 (Suppl.), 1987.

58. **Vinik, A. I., Gaginella, T. S., O'Dorisio, T. M., Shapiro, B., and Wagner, L.,** The distribution and characterization of somatostatin-like immunoreactivity in epithelial cells, submucosa, and muscle of the rat stomach and intestine, *Endocrinology,* 109, 1921, 1981.

59. **Larrson, L.-I., Goltermann, N., De Magistris, L., Rehfeld, J. F., and Schwartz, T. W.,** Somatostatin cell processes as pathways for paracrine secretion, *Science,* 205, 1393, 1979.

60. **Chiba, T., Park, J., and Yamada, T.,** Biosynthesis of somatostatin in canine fundic D cells, *J. Clin. Invest.,* 81, 282, 1988.

61. **Park, J., Chiba, T., Yokotani, K., DelValle, J., and Yamada, T.,** Somatostatin receptors on canine fundic D-cells: evidence for autocrine regulation of gastric somato-statin, *Am. J. Physiol.,* 257, G235, 1989.

62. **Montminy, M. R., Sevarino, K. A., Wagner, J. A., Mandel, G., and Goodman, R. H.,** Identification of a cyclic-AMP-responsive element within the rat somatostatin gene, *Proc. Natl. Acad. Sci. U.S.A.,* 83, 6682, 1986.

63. **Powers, A. C., Tedeschi, F., Wright, K. E., Chan, J. S., and Habener, J. F.,** Somatostatin gene expression in pancreatic islet cells is directed by cell-specific DNA control elements and DNA-binding proteins, *J. Biol. Chem.,* 264, 10048, 1989.

64. **Brubaker, P. L., Drucker, D. J., and Greenberg, G. R.,** Synthesis and secretion of somatostatin-28 and -14 by fetal rat intestinal cells in culture, *Am. J. Physiol.,* 258, G974, 1990.

65. **Habener, J. F., Goodman, R. H., and Lund, P. K.,** Complementary DNAs encoding precursors of glucagon and somatostatin, in *Biogenetics of Neurohormonal Peptides,* Hakanson, R. and Thorell, J., Eds., Academic Press, San Diego, 1985, 47.

66. **Drucker, D. J. and Brubaker, P. L.,** Glucagon biosynthesis in fetal rat intestine, *Biomed. Res.,* 9 (Suppl. 3), 29, 1988.

67. **Novak, U., Wilks, A., Buell, G., and McEwen, S.,** Identical mRNA for preproglucagon in pancreas and gut, *Eur. J. Biochem.,* 164, 553, 1987.

68. **Hynes, M. A., Shiow-Lian, J., Ulshen, M. H., Simmons, J. G., and Lund, P. K.,** Characterization, localization and regulation of extra pancreatic proglucagon mRNAs, *Biomed. Res.,* 9 (Suppl. 3), 147, 1988.

69. **Moody, A. J., Holst, J., Thim, L., and Jensen, S.,** Relationship of glicentin to proglucagon and glucagon in the porcine pancreas, *Nature (London),* 289, 514, 1981.

70. **Patzelt, C. and Chiltz, E.,** Conversion of proglucagon in pancreatic alpha cells: the major endproducts are glucagon and a single peptide, the major proglucagon fragment, that contains two glucagon-like sequences, *Proc. Natl. Acad. Sci. U.S.A.,* 81, 5007, 1984.

71. **Holst, J. J., Orskov, C., Olsen, J., Buhl, T., Schjoldager, B., and Kofod, H.,** Secretion and effects of the naturally occurring products of proglucagon, *Biomed. Res.,* 9 (Suppl. 3), 181, 1988.

72. **Mojsov, S., Heinrich, G., Wilson, I. B., Ravazzola, M., Orci, L., and Habener, J. F.,** Preproglucagon gene expression in pancreas and intestine diversifies at the level of post-translational processing, *J. Biol. Chem.,* 261, 11880, 1986.

73. **Orskov, C., Holst, J., Knuhtsen, S., Baldissera, F. G. A., Poulsen, S. S., and Nielsen, O. V.,** Glucagon-like peptides GLP-1 and GLP-2, predicted products of the glucagon gene are secreted separately from pig small intestine but not pancreas, *Endocrinology,* 119, 1467, 1986.

74. **Unger, R. and Oric, L.,** Physiology and pathophysiology of glucagon, *Physiol. Rev.,* 56, 778, 1976.

75. **Jarrousse, C., Audousset-Puech, M.-P., Dubrasquet, M., Niel, H., Martinez, J., and Bataille, D.,** Oxyntomodulin (glucagon-37) and its C-terminal octapeptide inhibit gastric acid secretion,. *FEBS Lett.,* 188, 81, 1985.

76. **Holst, J. J., Orskov, C., Vagn Hielsen, O. V., and Schwartz, T. W.,** Truncated glucagon-like peptide I, an insulin-releasing hormone from the distal gut, *FEBS Lett.,* 211, 169, 1987.

77. **Drucker, D. J., Philippe, J., Mojsov, S., Chick, W. L., and Habener, J. F.,** Glucagon-like peptide I stimulates insulin gene expression and increases cyclic AMP levels in a rat islet cell line, *Proc. Natl. Acad. Sci. U.S.A.,* 84, 3434, 1987.

78. **Bloom, S. R. and Polak, J. M.,** The hormonal pattern of intestinal adaptation: a major role for enteroglucagon, *Scand. J. Gastroenterol.,* 74 (Suppl.), 93, 1982.

79. **Holst, J. J., Christiansen, J., and Kuhl, C.,** The enteroglucagon response to intrajejunal infusion of glucose, triglycerides, and sodium chloride, and its relation to jejunal inhibition of gastric acid secretion in man, *Scand. J. Gastroenterol.,* 11, 297, 1976.

80. **Chen, L., Komiya, I., Inman, L., O'Neill, J., Appel, M., Alam, T., and Unger, R. H.,** Effects of hypoglycemia and prolonged fasting on insulin and glucagon gene expression, *J. Clin. Invest.,* 84, 711, 1989.

81. **Chen, L., Komiya, I., Inman, L., McCorkle, K., Alam, T., and Unger, R. H.,** Molecular and cellular responses of islets during perturbations of glucose homeostasis determined by *in situ* hybridization histochemistry, *Proc. Natl. Acad. Sci. U.S.A.,* 86, 1367, 1989.

82. **Rountree, D. B., Ulshen, M. H., Selub, S., Fuller, C. R., Bloom, S. R., Ghatei, M. A., and Lund, P. K.,** Rapid, nutrient independent increases in proglucagon and ODC mRNAs after jejuno-ileal resection, *Gastroenterology,* in press.

83. **Taylor, R. G., Verity, K., and Fuller, P. J.,** Ileal glucagon gene expression: ontogeny and response to massive small bowel resection, *Gastroenterology,* 99, 724, 1990.

84. **Buchan, A. M. J., Griffiths, C. J., Morris, J. F., and Polak, J. M.,** Enteroglucagon cell hyperfunction in rat small intestine after gut resection, *Gastroenterology,* 88, 8, 1985.

85. **Sagor, G. R., Al-Mukntar, M. Y. T., Ghatei, M. A., Wright, N. A., and Bloom, S. R.,** The effect of altered luminal nutrition on cellular proliferation and plasma concentrations of enteroglucagon and gastrin after small bowel resection in rat, *Br. J. Surg.,* 69, 14, 1982.

86. **Grodsky, G. M. and Fanska, R. E.,** The in vitro perfused pancreas, *Methods Enzymol.,* 39, 364, 1975.

87. **Philippe, J., Drucker, D. J., Chick, W. L., and Habener, J. F.,** Transcriptional regulation of genes encoding insulin, glucagon, and angiotensinogen by sodium butyrate in a rat islet cell line, *Mol. Cell Biol.,* 7, 560, 1987.

88. **Philippe, J., Drucker, D. J., and Habener, J. F.,** Glucagon gene transcription in an islet cell line is regulated via a protein kinase C-activated pathway, *J. Biol. Chem.,* 262, 1823, 1987.

89. **Philippe, J., Drucker, D. J., Knepel, W., Jepeal, L., Misulovin, Z., and Habener, J. F.,** Alpha-cell-specific expression of the glucagon gene is conferred to the glucagon promoter element by the interactions of DNA-binding proteins, *Mol. Cell. Biol.,* 8, 4877, 1988.

90. **Efrat, S., Teitelman, G., Anwar, M., Ruggiero, D., and Hanahan, D.,** Glucagon gene regulatory region directs oncoprotein expression to neurons and pancreatic A cells, *Neuron,* 1, 605, 1988.

91. **Drucker, D. J. and Brubaker, P. J.,** Proglucagon gene expression is regulated by a cyclic AMP-dependent pathway in rat intestine, *Proc. Natl. Acad. Sci. U.S.A.,* 86, 3953, 1989.

92. **Rindi, G., Grant, S. G. N., Yiangou, Y., Ghatei, M. A., Bloom, S. R., Bautch, V. L., Solcia, E, and Polak, J. M.,** Development of neuroendocrine tumors in the gastrointestinal tract of transgenic mice, *Am. J. Pathol.,* 136, 1349, 1990.

Chapter 5

CALCIUM HOMEOSTASIS, ENDOPLASMIC RETICULAR FUNCTION, AND THE REGULATION OF mRNA TRANSLATION IN MAMMALIAN CELLS

Margaret A. Brostrom and Charles O. Brostrom

TABLE OF CONTENTS

0-8493-6961-4/93/$0.00 + $.50

I. OVERVIEW

Cells have an extraordinary ability to impose order upon complex arrays of substituents and reactions. It is fascinating to reflect upon potential mechanisms whereby intracellular processes are coordinated and prioritized in the maintenance of this organization. Ordered relationships must exist among housekeeping processes, such as energy production, protein synthesis, and membrane transport systems, that maintain the health of the cell, and cell-specific functions involving stimulus-response coupling. For example, provision of the correct types and amounts of cellular enzymes and structural proteins would appear, *a priori,* to mandate that protein synthesis be rigorously coupled to and subordinated to the overall nutritional and functional status of the cell. It is inconceivable that this energy-intensive process would smoothly service the cell unless it were interdigitated with other cellular processes through coordinating regulation. Such regulation would presumably involve not only longer term adaptive alterations in the relative amounts of each protein to be synthesized requiring hours to days but also more immediate adjustments in the overall rate of peptide chain assembly. This brief review is concerned with highlighting a developing literature relating the control of protein synthesis at translation to the Ca^{2+} signal transduction system basic to the regulation of a wide variety of specialized processes. It is designed to update and extend an earlier review pertaining to Ca^{2+}-dependent regulation of protein synthesis in intact mammalian cells.[1]

II. THE CENTRAL ROLE OF THE ENDOPLASMIC RETICULUM IN Ca^{2+} HOMEOSTASIS AND PROTEIN SYNTHESIS

The endoplasmic reticulum (ER), as commonly described, is comprised of a convoluted, bilayer membrane sheathing a continuous luminal or cisternal space that occupies up to 10% or more of the total cell volume. Conventionally the ER is viewed as consisting primarily of the "rough" or ribosomally decorated region and the "smooth" or ribosomal-free region. Morphologically, the rough ER penetrates to the deepest recesses of the cell in being continuous with the outer membrane of the nucleus. The rough ER tends to occur in plate-life folds extending throughout much of the cytoplasm and possessing an intraluminal space of 20 to 30 nm. At various loci these folds taper into elements of the smooth ER, a dynamic network of anastomosing tubules 30 to 60 nm in diameter.[2] In addition to these traditional divisions, various subregions of the ER have been described possessing cisternae associated with cytoplasmic glycogen, mitochondria, cytoskeletal components, and the plasmalemma. Vesicles that transport protein for processing to the Golgi are derived from and represent another ER compartment. These generalities, however, fail to communicate a sense of the remarkable diversity in ER content and structure existing from one cell type to another. While both the rough and smooth ER possess considerable overlap in their protein content, a number of proteins concerned with ribosomal docking and protein processing are enriched in the rough ER.[3,4] Prominent among lumenal resident proteins, or reticuloplasmins, are endoplasmin or GRP94, GRP78 or BiP, and protein disulfide isomerase. An oxidizing environment is thought to exist within the lumen supporting protein disulfide bond formation and protein hydroxylation. Functions commonly associated with the ER include early protein processing, phospholipid biosynthesis, oxidative metabolism of hydrophobic molecules including many drugs, and storage of Ca^{2+} releasable in response to extracellular stimuli. It is apparent, therefore, that the ER is a highly complex organelle that infiltrates the cytoplasmic space of the cell and interfaces with and produces much of the structural material comprising other organelles. Overall, the ER possesses the structural and functional properties that would be expected for an organelle functioning in the integration and coordination of major cellular processes.

A wealth of literature supports the central role of Ca^{2+} in intracellular signaling related to stimulus-response coupling.[4-12] Prominent processes regulated by the cation include secretion, membrane transport and permeability, glycogen metabolism, and muscle contraction. As intracellular-free Ca^{2+} in response to a stimulus rises severalfold from resting values near $0.1 \mu M$, the cation binds to Ca^{2+} receptor proteins, such as calmodulin and troponin C, which then become capable of activating various enzymatic processes. Contributions of Ca^{2+} to the free pool are derived from the extracellular fluid and/or from intracellular sites of storage; release of cation from either source is driven by concentration gradients on the order of 10^4. Ca^{2+} entry across the plasmalemma involves voltage and/or ligand gated Ca^{2+} channels;[6,8] Ca^{2+} efflux is generally achieved by a Na^+/Ca^{2+} antiport and an active Ca^{2+} pump. Despite much investigative effort, controversy continues regarding the relative contribution of intracellular sequestered Ca^{2+} from one cell type to another in Ca^{2+} signaling, the localization and characteristics of Ca^{2+} stores, and their function in maintaining Ca^{2+} homeostasis.[8-11] At least two stores of Ca^{2+} are frequently proposed to exist, one of which is sensitive to release by inositol 1,4,5-trisphosphate (IP_3) generated in response to hormonal action and another which may support spontaneous oscillations in intracellular free Ca^{2+}.[7-12]

With the discrediting of the mitochondrion as a prominent storage site for Ca^{2+} in normal cells[5] and the increasing appreciation of IP_3 as a second messenger triggering release of intracellular sequestered Ca^{2+} from the microsomal fraction,[7] the ER gained acceptance as a Ca^{2+} storing organelle. Unfortunately the ER is sufficiently heterogeneous in terms of its biochemical, functional, and morphological properties, that subsequent characterization of its role in Ca^{2+} homeostasis proved to be elusive and controversial.[9-11] Cellular contents of ER and IP_3-releasable Ca^{2+} varied widely from one cell type to another. Discrepancies appeared to exist in the distribution of various enzyme markers for the ER and Ca^{2+} uptake, IP_3 binding and IP_3-induced Ca^{2+} release for some tissues, and the distribution of various membrane-associated Ca^{2+} binding proteins. Ambiguities of this nature ultimately led to the controversial proposal that IP_3-sensitive Ca^{2+} storage occurred in a separate organelle, termed the "calciosome".[13] After much additional study and considerable debate, the earlier conclusion that elements of the ER support Ca^{2+} storage subject to release by IP_3 has been largely reaffirmed and strengthened.[9-11,14] Oxalate augments Ca^{2+} uptake into the IP_3 sensitive pool, a property of the sarcoplasmic reticulum (SR) that would also be expected of the ER. Electron probe microanalysis of saponin permeabilized, oxalate loaded liver, and photoreceptor cells have provided strong evidence for an IP_3-sensitive Ca^{2+} storage site associated with the ER.[9,14] The structure of the ER and its ability to store and release Ca^{2+} have been particularly well characterized in the ventral nerve photoreceptor of Limulus.[9] Accumulation of Ca^{2+} by the ER occurs in both the photosensitive rhabdomeral lobe and the insensitive arhabdomeral lobe. Only the rhabdomeral lobe releases Ca^{2+} in response to the injection of IP_3. Thapsigargin, an inhibitor of active accumulation of Ca^{2+}, that acts preferentially, if not specifically, on the ER influences an IP_3-sensitive pool in saponin permeabilized cells.[15] Antibodies against purified IP_3 receptor have revealed immunoreactivity in Purkinje cells associated with both the rough and smooth ER as well as with the *cis* cisternae of the Golgi and subplasmalemmal cisternae but not the plasmalemma or mitochondria. More quantitative analysis of gold-immunolabeled ultrathin cryosections revealed that the IP_3 receptor was much more concentrated on smooth cisternae than on the perinuclear membrane or rough cisternae.[16] As described later, sequestered Ca^{2+} is required for early protein processing, which is established to occur in the ER.[3,4] Depletion of cellular Ca^{2+} accelerates protein degradation in the ER[17] and induces the synthesis of GRP78, an ER resident protein.[18] Calreticulin, a high capacity, low affinity Ca^{2+} binding protein, is reported to exist in both the SR of muscle cells and the ER of a number of cell types.[19]

The emerging picture is that the ER has the ability to accumulate Ca^{2+} throughout most of its structure including the outer nuclear membrane. In contrast, the IP_3 receptor may localize in particularly high densities at cisternae of the smooth ER, especially those in close approximation to the plasmalemma. The distribution of calreticulin appears to largely parallel that of the IP_3 receptor. Assuming that Ca^{2+} has the ability to migrate once inside the ER, it would naturally accumulate in regions containing the highest concentrations of Ca^{2+} binding sites. Should such migration prevail, the ER would function to recover Ca^{2+} from the deeper areas of the cell for subsequent transport to cisternae located near the cell surface. While this function at this point is entirely hypothetical, Ca^{2+} uptake by the SR, a closely related structure, reportedly is separated from the Ca^{2+} release sites of the junctional region of the terminal cisternae. A number of extremely interesting questions remain unresolved at this point. Are the interior components of the ER organized into a matrix that is influenced by Ca^{2+}? What is the role of Ca^{2+} in early protein processing? If the calciosome exists in isolatable form, is it a distinct organelle or a specialized region of the ER that may detach within some cells but not others?

III. REGULATION OF TRANSLATION BY Ca^{2+}

A. EFFECT OF Ca^{2+} ON TRANSLATIONAL INITIATION

Although the primary regulation of specific gene expression in mammalian cells occurs at the transcriptional step of protein synthesis, mRNA translation is currently thought to represent a second important site of control. The translational process has traditionally been divided into the stages of initiation, elongation, and termination, each of which involves complex interactions of proteins and RNA. Cytoplasmic proteins are often perceived, perhaps questionably, as being synthesized on "free" ribosomes, whereas proteins destined for insertion into membranes or for secretion are synthesized on membrane-bound ribosomes and inserted into the ER. Insertion may occur cotranslationally or following completion of polypeptide. Translational control in mammalian cells is viewed as occurring largely, although not exclusively, at the level of initiation.[20-22] Translational initiation is thought to be influenced by the intrinsic translatability of a given mRNA, by the ease of movement of mRNAs from mRNPs into polysomes, and by the phosphorylation states of initiation factors such as eIF-2 and eIF-4F. The following discussion is concerned with a growing body of evidence supporting the proposal that intracellular Ca^{2+} functions in the maintenance of high rates of translational initiation in intact cells. As discussed in later sections of this review and outlined in Table 1, the Ca^{2+} requirement is indirect; the rate of translational initiation appears to be closely controlled by the functional status of the ER within which Ca^{2+} serves to support glycoprotein processing.

Cells that are exposed to EGTA buffered medium normally exhibit an extremely rapid fall (seconds) of cytosolic-free Ca^{2+} concentration as monitored with Ca^{2+} fluorescent dyes.[23] With continuing exposure to EGTA, intracellular-sequestered Ca^{2+} is gradually depleted by spontaneous release to the cytosolic pool. Some cell types, such as GH_3 pituitary cells, release their Ca^{2+} stores relatively rapidly and become largely depleted within 15 to 30 min, whereas other cell types such as HeLa and CHO cells are relatively resistant to depletion by this procedure. Amino acid incorporation into nearly all protein populations in normal rat hepatocytes,[24] C6 glial tumor cells,[25] and GH_3 pituitary cells[26-29] was found to be inhibited 80 to 90% upon depletion of intracellular Ca^{2+} pools with EGTA buffered media within 30 min. Addition of 1 mM Ca^{2+} in excess of chelator restored the rate of protein synthesis

TABLE 1
The Functional Status of the Endoplasmic Reticulum and Regulation
of Translational Initiation

Metabolic/nutritional alteration	Functional status of the ER	Translational initiation	Cytosolic free Ca^{2+}
None	Ca^{2+} Sequestration unimpaired Oxidizing environment present Protein processing optimal	Optimal	Resting level (~0.1 μM)
Ca^{2+}-Mobilizing agonist (minutes)	Ca^{2+}-Depleted Glycoprotein processing retarded	Suppressed	Transiently elevated
Reducing environment (minutes)	Hypoxic Disulfide bond formation impaired	Suppressed	At or near resting level
Ca^{2+}-Mobilizing agonist or reducing environment (hours)	GRP78 Content increased Malprocessed proteins removed Protein translocation facilitated	Partly restored	At or near resting level

Note: Relationships observed to exist among ER sequestered Ca^{2+} and GRP78 concentrations, protein processing, and mechanisms of translational control are highlighted above. Inhibition of protein processing by the ER is seen to interdict translational initiation, such that the synthesis of almost all proteins, including those not subject to processing, is curtailed. The ER is proposed to coordinate mRNA translation with protein processing in a manner coupled to its functions: (1) as reservoir for hormonally releasable Ca^{2+} and (2) in providing an environment suited for protein disulfide bonding. In such an arrangement, rates of protein synthesis would be subordinated to the overall metabolic or nutritional status of the cell via monitoring of sequestered Ca^{2+} or changes in oxidation/reduction status. The accommodation of translation observed following extended incubations in low Ca^{2+} medium or under reducing conditions is consistent with a cellular adaption to assure sufficient rates of protein synthesis for survival during chronic stress. Acquisition of translational tolerance is viewed to arise as the consequence of increased concentrations of newly synthesized GRP78 within the ER lumen. This adaptive response is observed to function prominently against a variety of chronic nutritional, metabolic, physical, and chemical stresses in partially restoring protein synthesis. Although translational elongation is thought to be inhibited when cytosolic free Ca^{2+} is elevated, translational initiation appears to be unaffected by the transient changes in cytosolic free Ca^{2+} that may occur in response to Ca^{2+}-mobilizing agonists.

within 7 to 10 min to that of nondepleted control preparations. Ca^{2+} specifically among physiologically occurring cations restored amino acid incorporation over a broad range of Mg^{2+}, Na^+, and K^+ concentrations, pH, and osmolarity in both minimal and enriched media either with or without sera. The effects of Ca^{2+} depletion were not traceable to changes in amino acid uptake, aminoacylation of transfer RNA, RNA synthesis, protein catabolism, removal of cells from growth surfaces, or changes in viability as measured by dye exclusion, replating, and determinations of ATP and GTP contents.

Ca^{2+} depletion with 1 mM EGTA resulted in the disappearance of polysomes and an accumulation of monosomes and ribosomal subunits typical of slowed rates of translational initiation, as well as a large decrease in 43S preinitiation complex (40S·eIF-2·Met-tRNA$_f$·GTP).[30] Reintroduction of Ca^{2+} rapidly (min) restored cellular contents of 43S ribosomal preinitiation complex and polysomes with corresponding decreases in monosomal and ribosomal subunits. Comparable polysomal profiles were found for Ca^{2+}-depleted and restored cells exposed to cycloheximide which slowed polypeptide chain elongation to rate-limiting values in the overall translation process. Average ribosomal transit times for both Ca^{2+}-depleted and restored cells were identical and were extended in parallel as a function of increasing cycloheximide concentration, indicating that neither elongation nor termination is directly affected by depletion/repletion of the cation. Lysates of GH$_3$ cells exhibited incorporation that was proportional to the polysomal contents derived from the original intact cell preparations, lacked the ability to initiate new peptide synthesis, and were not directly affected by Ca^{2+} or EGTA. Lysates from either Ca^{2+}-depleted or restored cells exposed to cycloheximide provided lysates with identical elongation activity and polysomal contents to lysates prepared from Ca^{2+}-restored controls. Ca^{2+} was therefore proposed to support translation by maintaining the rate of initiation rather than polypeptide chain elongation or termination.

B. REQUIREMENT FOR SEQUESTERED Ca^{2+}

Hormones such as angiotensin II, vasopressin, and α-adrenergic agonists act on hepatocytes to generate IP$_3$ which in turn mobilizes approximately 50% of sequestered Ca^{2+} stores from the ER to the cytoplasm over a few min.[31] Most of this Ca^{2+} is immediately pumped into the extracellular fluid in a manner retarded by high concentrations of the cation. Over several min of exposure, these hormones were found to inhibit protein synthesis in isolated hepatocytes incubated with Ca^{2+} concentrations within the physiologic range and at physiologic pH.[24] This inhibition was largely reversible by the addition of supraphysiologic concentrations of Ca^{2+} to the extracellular medium. Corresponding reductions were observed in the polysomal contents of excised portions of perfused rat liver in response to those hormones and manipulations of the Ca^{2+} content of the perfusing medium.[32] Since protein synthesis was inhibited both by EGTA, which extracts cellular Ca^{2+}, and by hormones, which mobilize sequestered Ca^{2+} with some elevation of cytosolic-free Ca^{2+}, it was proposed that Ca^{2+} sequestered by the ER, rather than cytosolic-free Ca^{2+}, was required for maintenance of adequate rates of translational initiation. The divalent cation ionophores, A23187 and ionomycin, are widely used as Ca^{2+} ionophores to perturb Ca^{2+} homeostasis. While conventionally viewed as irreversible promoters of Ca^{2+} influx across the plasmalemma, these agents also act to release internally sequestered Ca^{2+} as analyzed by

Ca^{2+} kinetics in fluorescent dye and ^{45}Ca loaded cells. At low concentrations these ionophores have been found to promote nearly quantitative efflux of Ca^{2+} from GH_3 and P3X63Ag8 myeloma cells, whereas influx of the cation occurred at higher concentrations.[33] The net direction of Ca^{2+} movement for a given cell type and ionophore concentration appears to be determined by whether ionophore fosters entry from the extracellular fluid at rates that overwhelm the plasmalemmal transport systems pumping the cation from the cell. Low concentrations of Ca^{2+} ionophores inhibited translation comparably to EGTA and Ca^{2+}-mobilizing hormones.[34] Amino acid incorporation was reduced in a graded fashion in Ca^{2+}-restored cells upon addition of nM to μM concentrations of ionophore. Inhibitions were rapid and extensive, with 90 to 95% reduction in protein synthesis observed after 10 min exposure to 1 μM drug. Incorporation in cells depleted of Ca^{2+} with EGTA, however, was largely unaffected by ionophore. Rapid disruption of polysomes and reduced formation of 43S preinitiation complexes were observed upon treatment of Ca^{2+}-restored preparations with A23187. Average ribosomal transit times were unchanged. Ca^{2+} ionophores also provided a means of revealing the strong underlying Ca^{2+} dependence of initiation in cell types, such as HeLa and CHO, that possess only a modest overt Ca^{2+}-dependent component in the presence of EGTA. The agents were proposed to function as initiation inhibitors by provoking mobilization of sequestered intracellular Ca^{2+}. The inhibitory effects of ionomycin and A23187 on amino acid incorporation and on Ca^{2+} accumulation have been recently found to be reversible upon exposure of the cells to fatty acid-free bovine serum albumin.[33]

Free arachidonic acid, which has been associated with the actions of various secretagogues[35] and with ischemic stress,[36] directly releases sequestered Ca^{2+} from cells[35,37] and isolated organelles.[38] The potential role of arachidonate and related fatty acids in the regulation of protein synthesis via Ca^{2+} mobilization was therefore examined.[39] Unsaturated fatty acids at μM concentrations inhibited protein synthesis in a variety of mammalian cell types in a manner dependent on degree of unsaturation and cell number. Arachidonate was generally the most, and the fully saturated arachidic acid the least, potent of the fatty acids tested. At 2×10^6 GH_3 cells/ml, amino acid incorporation into a broad spectrum of polypeptides was inhibited 80 to 90% by 10 to 20 μM fatty acid. Inhibition was maximal at 4 to 8 min and was attenuated by 1 to 2 h and more pronounced at lower pH. Indomethacin did not relieve the inhibition indicating that cyclooxygenase pathway metabolites were not involved. Protein synthesis was maximally inhibited when arachidonate mobilized approximately 40% of cell-associated Ca^{2+}. At lower concentrations (10 μM), arachidonate suppressed translational initiation with the inhibition being reversed as extracellular Ca^{2+} concentrations were increased to supraphysiologic values. These properties of arachidonate were consistent with those expected for a physiologic regulator that inhibits protein synthesis through mobilizing Ca^{2+} sequestered in the ER. At higher concentrations (20

μM), however, arachidonate inhibited peptide chain elongation in a Ca^{2+}-independent manner. While the mechanism of elongation blockade was not determined, low concentrations (0.0007%) of the nonionic detergent Lubrol PX were noted to confer comparable elongation block without significant mobilization of Ca^{2+} or cell lysis. Both arachidonate and Lubrol PX blocked elongation, but not initiation, in reticulocyte lysates. The effects of arachidonate in intact cells were reversible with time via its metabolism. Concentrations of arachidonate synthesized during ischemic stress appear adequate for inhibition of translation by either mechanism.[36]

Synthetic oligopeptides specific for inhibition of metalloendoprotease activity suppress a variety of events involving membrane contact and fusion,[40-42] supporting the concept that proteolysis functions directly in such events. These inhibitors also alter transmembranal Ca^{2+} movements[43] and, at concentrations that block the constitutive secretory pathway in HepG2 hepatoma cells, inhibit amino acid incorporation without affecting ATP concentration or viability.[41] The possibility that these agents inhibit protein synthesis by perturbing sequestered Ca^{2+} stores was therefore explored.[44] Cbz-Gly-Phe-NH_2 provided rapid inhibition of amino acid incorporation into a broad spectrum of proteins in various cultured cells possessing ER but not in reticulocytes. Polysome accumulation and incorporation in GH_3 cells were reduced concurrently, indicating that the dipeptide acted to slow translational initiation. Inhibitions were largest at low extracellular Ca^{2+}, were reversed by increasing extracellular Ca^{2+}, were comparable to those achieved in the presence of EGTA or Ca^{2+} ionophores, and were observed with assorted metalloendoprotease antagonists but not with leupeptin. At concentrations inhibitory to protein synthesis, Cbz-Gly-Phe-NH_2 mobilized cell-associated ^{45}Ca, lowered cytosolic-free Ca^{2+}, and did not generate inositol phosphates. It was concluded that metalloendoprotease antagonists suppress translational initiation as a consequence of their capacity to mobilize sequestered Ca^{2+} stores. Two hypotheses were advanced to explain the effects of the antagonists on Ca^{2+} homeostasis. The first assumes that vesicular fusion events, requiring the action of metalloendoprotease(s), function in maintenance of Ca^{2+} sequestration by the ER. The return from Golgi to ER of transport vesicles carrying the Ca^{2+}-binding proteins of the ER lumen has been proposed to function in the restoration of the Ca^{2+} lost initially during movement of such transport vesicles from ER to Golgi.[45] This process for recovery of ER luminal components, referred to as the "salvage" pathway,[18] may be inactivated in the presence of membrane fusion inhibitors, such as Cbz-Gly-Phe-NH_2, resulting in the draining of ER and cytosolic cation. Alternatively Cbz-Gly-Phe-NH_2 may directly activate a Ca^{2+} extrusion system that lowers both cytosolic-free and sequestered cation.

Several conclusions emerge from the preceding body of information. Amino acid incorporation in a number of cell types is suppressed by extraction of cellular Ca^{2+} with EGTA or by Ca^{2+} mobilizing agents including hormones that generate IP_3, Ca^{2+} ionophores, arachidonate, and various dipeptides. All of these agents are reversed to various degrees by high extracellular Ca^{2+}. It is also clear that thapsigargin, an agent that blocks Ca^{2+} accumulation by the ER, inhibits amino acid incorporation.[46] Not all reports, however, are harmonious with the preceding body of data supporting a role for sequestered Ca^{2+} in initiation. Based on results with lysates, Kumar et al.[47] have concluded that a Ca^{2+} requirement for protein synthesis in Ehrlich ascites tumor cells relates to direct effects of the free cation on the translational apparatus. The Ca^{2+}-selective chelator BAPTA at concentrations of 1 to 2 mM depressed amino acid incorporation by lysates obtained from these cells; the inhibition was reversed by Ca^{2+} but not by several other cations. The BAPTA-treated preparations exhibited decreased levels of 43S preinitiation complexes and decreased binding of factors eIF-2 and eIF-3 to 40S subunits. Similar effects of high concentrations of chelator have been seen, however, in reticulocyte lysates,[25] a system that does not appear to be supported by Ca^{2+}.[48]

In further studies of Ehrlich cells,[49] calmodulin antagonists were found to inhibit protein synthesis in both intact and cell-free preparations. Initiation was preferentially inhibited as indicated by an increase in 80S monomers and 43S complexes accompanied by disaggregation of polysomes. It was suggested that Ca^{2+} maintains translation by activation of a mechanism involving calmodulin or a related Ca^{2+}-binding protein. The relatively high concentrations of calmodulin antagonists required to suppress translation, however, were significantly in excess of those required for inhibition of most calmodulin-dependent enzymes[50] and were well within the range reported to block Ca^{2+} entry through Ca^{2+} channels.[28] A supporting role for calmodulin at initiation would not fit well with the extensive body of information reporting inhibition of peptide chain elongation via a calmodulin dependent mechanism (see below).

The behavior of intact reticulocytes with respect to Ca^{2+} and A23187 was found to be at variance with that of nucleated cells, perhaps because reticulocytes lack a well-defined, functional ER.[51] Rates of amino acid incorporation in intact rabbit reticulocytes were unaffected by depletion of Ca^{2+} with EGTA. The Ca^{2+} ionophore A23187 strongly inhibited incorporation in reticulocytes incubated with 1 mM Ca^{2+} but not with EGTA. Polysomal profiles and the extension of average ribosomal transit times of cells treated with Ca^{2+} ionophore at 1 mM Ca^{2+} were characteristic of translational elongation block. Reticulocyte lysates are a popular system for investigating the properties of translational initiation because the activity survives cell lysis and will support the synthesis of protein from exogenously added mRNA. This property is remarkable since initiation in most mammalian cells is nearly eliminated by cell damage or disruption. It is interesting to speculate that reticulocyte lysates retain high degrees of initiation because the process has been uncoupled from control by vesicular membranes prevalent in most other cell types.

C. EFFECT OF Ca^{2+} ON THE PHOSPHORYLATION/ DEPHOSPHORYLATION OF INITIATION FACTORS

The activities of a number of initiation factors are potentially subject to regulation via protein phosphorylation.[20-22] In particular the phosphorylation of the α subunit and accompanying inhibition of eIF-2 has been extensively studied. This factor catalyzes the formation of the ternary complex eIF-2·-GTP·Met-tRNA, which functions in the binding of initiator Met-tRNA to the 40S ribosomal subunit to form the 43S preinitiation complex essential to the initiation of all mRNAs. As noted earlier, this complex is depressed in Ca^{2+}-depleted cells. Reactivation of eIF-2 after utilization of the ternary complex depends on an exchange of GDP bound to eIF-2 for GTP, catalyzed by the factor eIF-2B. Phosphorylation of eIF-2α by any one of a number of kinases results in an increased affinity of eIF-2B such that eIF-2B is sequestered into an inactive eIF-2B·-eIF-2α(P) complex. This complex is incapable of supporting G nucleotide exchange in the recycling of eIF-2 and translational initiation therefore becomes suppressed. Kimball and Jefferson[52] observed that the activity of eIF-2B in extracts of perfused livers was reduced to 53% of the control value in response to vasopressin, a Ca^{2+} mobilizing hormone. An increase in the proportion of the α subunit of eIF-2 in the phosphorylated form was also observed. These authors proposed that inhibition of translational initiation in vasopressin-treated livers resulted from eIF-2α phosphorylation. Alternatively, phosphorylation of eIF-2α could result as a consequence of translational initiation block.

Another frequently advanced mechanism for translational regulation involves phosphorylation of factor eIF-4E, a subunit of eIF-4F which serves to recognize the 5' terminal mRNA cap in early initiation. Phosphorylation of eIF-4E correlates with enhanced rates of protein synthesis and growth in mammalian cells.[20] The phosphorylation state of the subunit in Ca^{2+}-depleted GH_3 cells, however, did not differ from that in Ca^{2+}-restored preparations.[53] Phosphorylation of ribosomal protein S6, which is observed to correlate with increased formation of ribosome initiation complexes, with ribosomal entry into elongation, and with increased growth,[54] similarly was not altered as a function of cellular Ca^{2+} depletion.

D. CONDITIONS THAT DISSOCIATE INITIATION FROM REGULATION BY Ca^{2+}

Certain types of major stress, including thermal or chemical stress, result in an initial shutdown of normal protein synthesis in eukaryotic cells accompanied by loss of polysomal contents. Over several hours, a series of stress proteins are induced and overall amino acid incorporation is largely restored, but with minimal Ca^{2+} dependence. For example, after thermal stress at 46° or exposure to sodium arsenite, amino acid incorporation into Ca^{2+}-replete GH_3 cells was reduced to that of unstressed, Ca^{2+}-depleted cells,[55] and preparations originally stressed in the Ca^{2+}-depleted condition were subsequently unable to accumulate polysomes in response to Ca^{2+}. Ca^{2+} did not influence either the recovery of amino acid incorporation or the induction of the heat shock proteins (hsp) that occurred during the next several hours. Restoration of the Ca^{2+} requirement occurred by 24 h. More modest degrees of thermal stress (41°) resulted in the induction of hsp 68 without loss of Ca^{2+}-stimulated initiation. Under this condition hsp 68 was synthesized in a Ca^{2+}-dependent manner.

Viral infections interrupt host cell protein synthesis by various mechanisms while reprogramming the translational apparatus to synthesize viral coded proteins. HEp-2 cells infected with HSV-1 were employed to examine the Ca^{2+} requirement for translation of viral mRNAs.[56] Early in viral infection, the synthesis of nearly all proteins, including viral (α) proteins, was sensitive to inhibition by Ca^{2+} ionophore or EGTA. By 4 to 6 h after infection, overall polypeptide synthesis in infected cells had become resistant to depletion of Ca^{2+} stores. Specific viral mRNAs were readily detected in polysomes, and the synthesis of viral polypeptides of early (β) and late (γ) kinetic classes was found to be insensitive to the effects of Ca^{2+} depletion. Production of mature forms of viral glycoproteins, however, was reduced by treatment with EGTA and ionophore. Thus, productive infection of cells by HSV-1, like heat shock or chemical stress, results in a modification that overcomes the requirement of translational initiation for stored Ca^{2+}. Production of most viral proteins may then proceed in a Ca^{2+}-independent fashion. A second, Ca^{2+}-sensitive step affecting the maturation (glycosylation and/or transport) of viral glycoproteins was also apparent which was not bypassed during HSV-1 infection.

E. EFFECT OF Ca^{2+} ON PEPTIDE CHAIN ELONGATION

Peptide chain elongation in healthy cells is readily seen to be rate-limiting over initiation in that polysomal content accumulates to degrees that are not increased by cycloheximide, an inhibitor that reduces incorporation by slowing elongation and extending average ribosomal transit times. An inhibitory regulatory input at this site would be expected to slow protein synthesis quite rapidly.[22] In recent years, evidence has accrued for the existence of a nonselective and transitory mechanism for elongational slowing coupled to increased concentrations of intracellular free Ca^{2+}. The physiologic relevance of this mechanism, which involves the rapid and transient phosphorylation of elongation factor 2 (eEF-2) has not as yet been clearly established. Elongation of polypeptide chains in mammalian cells is catalyzed by two factors.[22] Elongation factor 1 (eEF-1) is responsible for the codon-dependent binding of aminoacyl tRNA to the acceptor (A) site on the ribosome, whereas eEF-2 promotes translocation of the charged tRNA from the A site to the P (peptide bond formation) site. The activity of eEF-2 is associated with a monomeric 100 kDa, GTP-binding protein found in all eukaryotic cells. Immediately prior to translocation, the anticodon end of the peptidyl-tRNA is bound in the A site and the peptidyl end is located in the P site. Although details of the mechanism have not been elucidated, it is likely that binding of the eEF-2·GTP complex promotes movement of the mRNA and the anticodon portion of the peptidyl-tRNA into the ribosomal P site accompanied by the hydrolysis of the nucleotide to GDP.

Cytoplasmic extracts of mammalian cells are routinely found to incorporate phosphate from ATP into a major 100-kDa species.[57] In 1983 Palfrey[58] reported that such incorporation was enhanced by Ca^{2+} and the ubiquitous Ca^{2+} mediator protein calmodulin. Independently, Palfrey and Nairn[58,59] and Guroff and colleagues[60,61] isolated the 100-kDa substrate and employed it in procedures for partial purification of the protein kinase responsible for the *in vitro* phosphorylation. The kinase, which possessed a molecular mass of 130 to 140 kDa, was identified in a broad variety of tissues, required Ca^{2+} and calmodulin for optimal activity, and was distinct in substrate specificity from previously characterized forms of Ca^{2+}/calmodulin-dependent protein kinase.[59,62] The enzyme, named Ca^{2+}/calmodulin-dependent protein kinase III by Nairn et al.,[59] was unable to phosphorylate a variety of substrates utilized by other protein kinases. Its 100-kDa substrate was not phosphorylated by other protein kinases.

Identification of the 100-kDa substrate as eEF-2 was reported in early 1987 by Ryazanov[63] and later by Nairn and Palfrey.[64] Since the factor constituted the only known substrate for the enzyme, Ryazanov proposed the enzyme be named eEF-2 kinase.[65] Partially purified eEF-2 kinase was observed to phosphorylate three threonine residues of eEF-2 from reticulocytes[66] with threonine 56 advanced as the primary phosphorylation site.[67] The comparable factor from chick embryo exhibited additional phosphorylation sites on tyrosine and serine residues.[68] The threonine phosphorylation sites of reticulocyte eEF-2 were localized to a region near the N terminus that may constitute the ribosome binding site.[66] Phosphorylation was found to stabilize the factor against the action of trypsin.[69] Dephosphorylation of phosphorylated factor in cell-free systems could be accomplished by incubation with various phosphatases, with type 2A phosphatase being the most effective.[70-72] Recent functional studies support the hypothesis that phosphorylated eEF-2 cannot sustain elongation. Okadaic acid, a tumor promoter that potently inhibits protein phosphatases, was found by Redpath and Proud[71] to increase the net phosphorylation of eEF-2 in reticulocyte lysates while provoking inhibition of peptide chain elongation. In contrast the stimulation of elongation rate observed when reticulocyte lysates are incubated with high concentrations of cAMP correlated with dephosphorylation of eEF-2.[70] The dephosphorylated form of eEF-2 was found to support poly(U)-directed polyphenylalanine synthesis in a reconstituted elongation system when combined with eEF-1, whereas phosphorylated eEF-2 was ineffective.[64,73] Phosphorylated eEF-2 is reported to bind to 80S ribosomes but to be unable to catalyze translocation of peptidyl tRNA from the A site to the P site for ribosomes carrying poly(U) and phenylalanyl tRNA.[74]

Transient phosphorylation of eEF-2 has been reported for a wide variety of cell types following exposure to Ca^{2+}-mobilizing substances.[57] For example, treatment of fibroblasts with serum, bradykinin, vasopressin, epidermal growth factor (EGF), or Ca^{2+} ionophores, each of which provoked transient increases in $[Ca^{2+}]_i$, resulted in two- to tenfold increases in eEF-2 phosphorylation.[75] Phosphorylation of the factor was maximal at 0.5 to 1 min and attenuated at 5 min. Thrombin and histamine, which elevate $[Ca^{2+}]_i$ in umiblical vein endothelial cells, also provided a rapid and transient phosphorylation of eEF-2, whereas phorbol esters or cAMP-elevating agents were ineffective.[76] Phosphorylation of eEF-2 in cells exposed to bradykinin, serum, vasopressin, histamine, or thrombin was not reduced when conducted in Ca^{2+}-depleted medium, whereas phosphorylation involving EGF or ionophore required the presence of extracellular cation. Phosphorylation of the factor following depolarization of PC-12 pheochromocytoma cells was observed by Nairn and colleagues[77] to be overturned by nerve growth factor (NGF) or cAMP-elevating agents. Significant decreases in eEF-2 kinase activity were noted within 1 h of exposure of the cells to NGF or cAMP-elevating agent. Data supporting the hypothesis that downregulation of eEF-2 kinase is attributable to the action of cAMP-dependent protein kinase have recently been reported.[78] Of additional interest is recent evidence from the laboratory of Nygard et al.[79] that the phosphorylated form of eEF-2 kinase is the active species.

Given the transitory nature of eEF-2 phosphorylation in most cells, the consequent inhibition of translation would be expected to be comparably short-lived and therefore difficult to measure experimentally by standard techniques. It is not clear that eEF-2 becomes sufficiently phosphorylated in intact non-erythroid cells to inhibit the rate of peptide bond formation significantly. If, however, phosphorylated eEF-2 were to bind to ribosomes with higher affinity than nonphosphorylated factor, then only a fraction of the pool of factor would need to be phosphorylated for elongation to be inhibited. The physiologic role played by a brief slowing of translational elongation is also not immediately obvious. Conceivably inhibition through this mechanism would rapidly divert energy toward supporting hormonally activated cell-specific functions. This inhibition would attenuate with depletion of sequestered stores of Ca^{2+} and the concomitant development of inhibition of translational initiation. Such model building would be precluded by an alternative proposal by Ryazanov[57] that eEF-2 kinase may be active only at mitosis. Evidence cited in support of this proposal includes a report by Celis et al.[80] that the proportion of phosphorylated eEF-2 variants present in amnion cells increased considerably during mitosis, a period during which the rate of translation is believed to decrease and $[Ca^{2+}]_i$ to rise briefly. Additionally Severinov et al.[81] found that eEF-2 kinase activity in *Xenopus* oocytes decreased substantially during the final stages of oogenesis and was absent in fully grown oocytes. The mechanism was hypothesized to enhance the degradation of short lived repressors of genes, such as proto-oncogenes, that code for proliferative factors, thereby recruiting new mRNAs to translation. Activation of such gene expression has been reported following exposure to mitogenes and nonspecific protein synthesis inhibitors such as cycloheximide.[82,83] In summary, several major questions relating to the effects of Ca^{2+} on translation remain to be clarified. What is the overall mechanism through which mobilization of sequestered Ca^{2+} from the ER inhibits translational initiation? What events occur during its reversal? What is the physiologic function of either this inhibition or the inhibition of translational elongation directed by cytosolic Ca^{2+}?

IV. Ca^{2+} AND THE INDUCTION OF GRP78/BiP

Deprivation of sequestered Ca^{2+} is currently viewed as a form of metabolic stress to which mammalian cells respond with the transcription-dependent induction of various resident ER proteins.[84] The most prominently induced of these is a 78-kDa protein that has been localized to the lumen. Commonly termed GRP78 or BiP,[18] this reticuloplasmin is induced by agents or conditions that alter metabolic or nutritional status, such as inhibitors of protein glycosylation, glucose deprivation severe enough to deplete glycosyl units, sulfhydryl reducing agents, amino acid analogs, viral infection, and overexpression of secretory proteins.[84-88] GRP78 has been variously hypothesized to function in the correct folding of proteins during early protein processing,[89] in the retention of improperly folded proteins that accumulate within the ER lumen when processing is distressed,[18] or in the translocation of proteins from the cytosol to the ER for processing.[90,91] Transcription of the GRP78 gene is thought to be activated consequent to altered rates of early protein processing via mechanisms not as yet defined. The GRP78 gene, which possess a highly conserved promoter region that confers Ca^{2+} ionophore inducibility and binds *trans*-acting transcription factors,[92,93] is required for the continued viability of yeast cells.[94] The following discussion describes additional findings that implicate GRP78 in a mechanism whereby mammalian cells rapidly develop translational tolerance to depletion of sequestered Ca^{2+} and other metabolic stresses such as a reducing environment.

GH_3 cells depleted of sequestered Ca^{2+} maintain their viability for several hours with stable, albeit reduced, rates of protein synthesis. When exposed for 2 to 3 h to a phorbol ester and/or a cAMP-elevating agent, however, such preparations exhibited increased incorporation of amino acids into a broad spectrum of polypeptides.[29,95] Phorbol esters and cAMP fostered the accumulation of small polysomes without affecting average ribosomal transit times, indicative of increased rates of translational initiation. In EGTA-buffered or ionophore-containing medium, a six- to eightfold stimulation of protein synthesis occurred over 3 h, as compared to relatively small increments in Ca^{2+}-repleted controls. In contrast to the behavior of GH_3 cells, Ca^{2+}-depleted C6 glial tumor cells displayed spontaneous increases in incorporation rates with time that were not affected by cAMP or phorbol ester.

Cells first accommodated to low Ca^{2+} medium and then restored with optimal Ca^{2+} exhibited rates of incorporation identical to those of nontreated controls, but remained resistant to inhibition on subsequent challenge with EGTA or ionophore. Acquisition of translational tolerance to these agents did not involve alterations of cellular capacity or affinity for Ca^{2+}, but was invariably preceded specifically by transcriptional-dependent induction of GRP78.[95] In GH_3 and immunoglobulin-secreting myeloma cells deprived of Ca^{2+}, the synthesis of GRP78 was promoted by phorbol ester and cAMP. Interestingly, the development of translational tolerance to ionophore correlated directly with new synthesis of GRP78 as measured by pulse labeling of the protein with [^{35}S]-methionine, rather than with total GRP78 measurable by staining. Cells treated with the reducing agent dithiothreitol (DTT), an efficacious inducer of GRP78 that does not mobilize sequestered Ca^{2+}, exhibited acute shutdown of translational initiation followed by the subsequent development of translational tolerance not only to DTT but also to Ca^{2+} ionophore and EGTA. Chronic tolerance to these agents was found for NS-1 myeloma cells, which are defective in immunoglobulin processing and constitutively express high contents of GRP78 and its corresponding mRNA.[96] Antisense oligonucleotides directed against GRP78 mRNA reduced amino acid incorporation by tolerant GH_3 cells, but not in nontolerant controls. GH_3 cells treated several h with the metalloendoprotease antagonist Cbz-Gly-Phe-NH_2 also acquired translational tolerance occurring in tandem with GRP78 induction.

In investigation of the roles of cAMP and phorbol ester in the induction of GRP78 and the coordinate development of translational tolerance to Ca^{2+} ionophore in GH_3 cells,[97] induction of the reticuloplasmin was found to depend on two factors. First, cAMP and phorbol ester enhanced GRP78 gene transcription. GRP78 mRNA was induced 3- to 6-fold with ionophore alone as compared to 12- to 20-fold with ionophore plus cAMP-elevating agent and phorbol ester. GRP78 gene transcription in nuclei isolated from ionophore-treated cells was also increased by cAMP and phorbol ester. Induction was not apparent in the absence of ionophore regardless of the presence of cAMP and phorbol esters, and these activators did not affect GRP78 mRNA stability despite the increased polysomal content of cells undergoing translational recovery. This cAMP/phorbol ester requirement for optimal GRP78 mRNA induction was initially surprising in that induction in other cell types was not known to require the activators. However, examination of the sequence of the rat GRP78 promoter leads us to speculate that a cAMP response element (CRE), which differs from the phorbol ester response sequence (TRE) by only one nucleotide,[98] potentially exists in this region. The sequence from -189 to -182 is 5'-TGACGTGA-3 and conforms to the CRE-consensus.[99] The human GRP78 gene has a similar sequence from -153 to -147. These sequences correspond to protected domains and are footprinted on the noncoding strand.[100] With development of promoter deletion mutants, it should be possible to identify sequences that mediate regulation of GRP78 gene expression by cAMP and phorbol ester. The second factor favoring GRP78 induction in stressed GH_3 preparations was the preferential translation of GRP78 mRNA. This mRNA associated almost exclusively with polysomes, whereas other mRNAs were subject to initiation block.[97] This efficient initiation of GRP78 may be explained by the recent observation[101] that the 5' leader of GRP78 mRNA, unlike that of most cellular mRNAs, can confer initiation by an internal ribosome binding mechanism.

On the basis of these and other findings, alterations in translational activity following Ca^{2+} deprivation or exposure to a reducing environment appeared to emanate from perturbation of ER function. As noted earlier, the ER is thought to function prominently in maintenance of Ca^{2+} homeostasis and early protein processing reactions requiring an oxidizing environment. Agents that deplete sequestered Ca^{2+} stores or that generate a hypoxic environment acutely suppress translation at initiation. On chronic exposure to either class of agent, transcriptionally dependent translational accommodation is seen to develop that appears to depend on the rapid induction of GRP78/BiP, an ER resident protein proposed to serve in protein translocation and folding. Major open questions regarding GRP78 include: (1) How does the ER signal the nucleus to transcribe new mRNA for the protein? (2) How does GRP78 promote the recovery of amino acid incorporation in Ca^{2+} depleted cells? Does GRP78 coordinate the relative rates of translational initiation and protein processing? Evidence that both sequestered Ca^{2+} and adequate oxidizing eqivalents are needed to sustain early protein processing events is summarized in the following section.

V. A Ca^{2+} REQUIREMENT FOR SECRETORY PROTEIN PROCESSING AND EXPORT

Although cytosolic free Ca^{2+}, rather than sequestered Ca^{2+}, is conventionally considered to support regulated metabolic processes, certain findings favor the possibility that sequestered Ca^{2+} maintains posttranslational processing. For example, the activity *in vitro* of $\alpha 1,2$ mannosidase, a vesicular enzyme involved in processing of N-linked oligosaccharides, is reportedly stimulated by Ca^{2+}.[102] Depletion of cellular Ca^{2+} stores has been proposed to accelerate degradation of certain heterologous proteins in the ER.[17] Of particular relevance are the findings that A23187 and ionomycin are efficacious inhibitors of the processing and secretion of α_1-antitrypsin, and, to a lesser extent other proteins, by HepG2 cells.[103] Following ionophore treatment secretory glycoproteins with a high mannose configuration accumulated in vesicles with the density of the rough ER. It was unclear from these studies, however, that Ca^{2+} mobilization mediated the observed secretory arrest and that resumption of normal glycoprotein processing and export was possible upon removal of the stress.

HepG2 human hepatoma cells were employed as a model system to further investigate relationships between early protein processing and Ca^{2+} storage by the ER.[104] Three lines of direct evidence were provided that Ca^{2+} per se is required for processing and secretion. First, secretion of α_1-antitrypsin and to a lesser degree other newly synthesized polypeptides was reduced following treatments in EGTA-containing medium. Second, inhibition of glycoprotein processing by Cbz-Gly-Phe-NH_2 was found to depend on Ca^{2+}. Inhibitions of α_1-antitrypsin processing and of complement factor 3 (C3) secretion by dipeptide were largest in Ca^{2+}-depleted medium and were either markedly reduced or prevented at elevated extracellular Ca^{2+}. Third, it was demonstrated that arrest of α_1-antitrypsin processing and export in the presence of ionomycin was fully reversed by Ca^{2+} following extraction of the drug with BSA. In addition to establishing a Ca^{2+} involvement in the mechanism of ionophore inhibition, accomplishment of the reversal illustrated (1) that ionomycin-treated preparations retained the normal precursor form of α_1-antitrypsin rather than a malprocessed intermediate destined for degradation, and (2) that both vesicular processing and protein transport were not permanently disrupted by ionophore treatment. In contrast to agents depleting Ca^{2+} stores, exposure to DTT reduced albumin export while affecting α_1-antitrypsin export minimally. Although the inhibition of albumin export was not accompanied by significant intracellular accumulation of native polypeptide, considerable amounts of several cell-associated, immunoreactive polypeptides of differing molecular mass were invariably seen to develop in DTT-treated cells. These peptides appeared to be aggregative/degradative forms of albumin collectively more than doubling cell associated retention of the protein.

Ca^{2+} has been proposed to be required for the transport of secretory proteins from ER to Golgi.[105] That albumin secretion is affected minimally by ionomycin and not affected by EGTA argues against a prominent role for Ca^{2+} in transport of all proteins from the ER. Alternatively, Ca^{2+} stored in the ER may be required for accurate folding which, in turn, is necessary for transport of secretory glycoproteins to other vesicular compartments. This mechanism assumes more stringent folding requirements for transport of glycoproteins, such as α_1-antitrypsin and C3, than for nonglycosylated secretory species such as albumin. An additional possibility consistent with available data is that depletion of sequestered Ca^{2+} specifically restricts the enzymatic trimming of high mannose forms of glycoproteins, resulting in the accumulation of species not recognized by transport systems. By contrast, these data did not support a reciprocal relationship between sequestered Ca^{2+} and the rate of degradation of secretory proteins. Furthermore, sequestered Ca^{2+} did not appear to function critically in protein disulfide bond formation. Processing and export of albumin, a protein with 17 intrachain disulfide linkages,[106] was largely unimpeded by EGTA or ionomycin.

VI. SUMMARY

Inhibition of mRNA translation consequent to interruption of ER protein processing appears to be of general significance in that suppression of amino acid incorporation into total cellular proteins of HepG2 cells accompanied inhibitions of protein processing by agents depleting sequestered Ca^{2+} stores or by DTT.[104] Tunicamycin, which interferes with transfer of the core oligosaccharide structure to newly synthesized polypeptides and retards secretion of the nonglycosylated species, was found to inhibit translational initiation in GH_3 cells.[95] It is most appealing, therefore, to propose that the ER effectively coordinates rates of mRNA translation with protein processing. While translation obviously provides the substrates for the processing apparatus of the ER and Golgi, there is no evidence of any other influence of translation on processing.[107] In contrast it is quite clear that some inhibitors of protein processing interdict translational initiation, such that the synthesis of almost all proteins, including those not subject to processing, is curtailed. It will be of interest to ascertain whether cells expressing translational tolerance subsequent to induction of GRP78 are capable of efficient protein processing when challenged by stresses that perturb ER function. Other questions of interest include: Where is the Ca^{2+} requirement in glycoprotein processing exerted? Does sequestered Ca^{2+} regulate the rate of glycoprotein processing?

REFERENCES

1. **Brostrom, C. O. and Brostrom, M. A.,** Calcium dependent regulation of protein synthesis in intact mammalian cells, *Annu. Rev. Physiol.,* 52, 577, 1990.
2. **Lee, C. and Chen, L. B.,** Dynamic behavior of endoplasmic reticulum in living cells, *Cell,* 54, 37, 1988.
3. **Green, M. and Mazzarella, R. A.,** Biosynthesis and sorting of proteins of the endoplasmic reticulum, in *Protein Transfer and Organelle Biogenesis,* Das, R. C. and Robbins, P. W., Eds., Academic Press, Orlando, 1988, 243.
4. **Koch, G. L. E.,** The endoplasmic reticulum and calcium storage, *Bioessays,* 12, 527, 1990.
5. **Carafoli, E.,** Intracellular calcium homeostasis, *Annu. Rev. Biochem.,* 56, 395, 1987.
6. **Meldolsi, J. and Pozzan, T.,** Pathways of Ca^{2+} entry at the plasma membrane: voltage-, receptor-, and second messenger-operated channels, *Exp. Cell Res.,* 171, 271, 1987.

7. **Berridge, M. J.,** Inositol phosphates and cell signalling, *Nature (London)*, 341, 197, 1989.

8. **Tsien, R. Y.,** Calcium channels, stores, and oscillations, *Annu. Rev. Cell Biol.*, 6, 715, 1990.

9. **Walz, B. and Baumann, B.,** Calcium-sequestering cell organelles: *in situ* localization, morphological and functional characterization, *Prog. Histochem. Cytochem.*, 20, 1, 1989.

10. **Meldolesi, J., Madeddu, L., and Pozzan, T.,** Intracellular Ca^{2+} storage organelles in non-muscle cells: heterogeneity and functional assignment, *Biochim. Biophys. Acta*, 1055, 130, 1990.

11. **Rossier, M. F. and Putney, J. W., Jr.,** The identity of the calcium-storing, inositol 1,4,5-trisphosphate-sensitive organelle in non-muscle cells: calciosome, endoplasmic reticulum . . . or both?, *Trends Neurosci.*, 14, 310, 1991.

12. **Berridge, M. J.,** Cytoplasmic calcium oscillations: a two pool model, *Cell Calcium*, 12, 63, 1991.

13. **Volpe, P., Krause, K.-H., Hashimoto, S., Zorzato, F., Pozzan, T., Meldolesi, J., and Lew, D. P.,** "Calciosome," a cytoplasmic organelle: The inositol 1,4,5-trisphosphate-sensitive Ca^{2+} store of nonmuscle cells?, *Proc. Natl. Acad. Sci. U.S.A.*, 85, 1091, 1989.

14. **Baumann, O., Walz, B., Somlyo, A. V., and Somlyo, A. P.,** Electron probe microanalysis of calcium release and magnesium uptake by endoplasmic reticulum in bee photoreceptors, *Proc. Natl. Acad. Sci. U.S.A.*, 88, 741, 1991.

15. **Bian, J., Ghosh, T. K., Wang, J.-C., and Gill, D. L.,** Identification of intracellular calcium pools. Selective modification by thapsigargin, *J. Biol. Chem.*, 266, 8801, 1991.

16. **Satoh, T., Ross, C. A., Villa, A., Supattapone, S., Pozzan, T., Snyder, S. H., and Meldolesi, J.,** The inositol 1,4,5-trisphosphate receptor in cerebellar purkinje cells: quantitative immunogold labeling reveals concentration in an ER subcompartment, *J. Cell Biol.*, 111, 615, 1990.

17. **Wileman, T., Kane, L. P., Carson, G. R., and Terhorst, C.,** Depletion of cellular calcium accelerates protein degradation in the endoplasmic reticulum, *J. Biol. Chem.*, 266, 4500, 1991.

18. **Pelham, H. R. B.,** Control of protein exit from the endoplasmic reticulum, *Annu. Rev. Cell Biol.*, 5, 1, 1989.

19. **Milner, R. E., Baksh, S., Shemanko, C., Carpenter, M. R., Smillie, L., Vance, J. E., Opas, M., and Michalak, M.,** Calreticulin, and not calsequestrin, is the major calcium binding protein of smooth muscle sarcoplasmic reticulum and liver endoplasmic reticulum, *J. Biol. Chem.*, 26, 7155, 1991.

20. **Sonenberg, N.,** Cap-binding proteins of eukaryotic messenger RNA: functions in initiation and control of translation, *Prog. Nucleic Acid Res.*, 35, 173, 1988.

21. **Hershey, J. W. B.,** Protein phosphorylation controls translation rates, *J. Biol. Chem.*, 264, 20823, 1989.

22. **Hershey, J. W. B.,** Translational control in mammalian cells, *Annu. Rev. Biochem.*, 60, 717, 1991.

23. **Albert, P. R. and Tashjian, A. H., Jr.,** Relationship of thyrotropin-releasing hormone-induced spike and plateau phases in cytosolic free Ca^{2+} concentrations to hormone secretion, *J. Biol. Chem.*, 259, 15350, 1984.

24. **Brostrom, C. O., Bocckino, S. B., Brostrom, M. A., and Galuska, E. M.,** Regulation of protein synthesis in isolated hepatocytes by calcium-mobilizing hormones, *Mol. Pharmacol.*, 29, 104, 1986.

25. **Brostrom, C. O., Bocckino, S. B., and Brostrom, M. A.,** Identification of a Ca^{2+} requirement for protein synthesis in eukaryotic cells, *J. Biol. Chem.*, 258, 14390, 1983.

26. **Brostrom, M. A., Brostrom, C. O., Bocckino, S. B., and Green, S. S.,** Ca^{2+} and hormones interact synergistically to stimulate rapidly both prolactin production and overall protein synthesis in pituitary tumor cells, *J. Cell. Physiol.*, 121, 291, 1984.

27. **Wolfe, S. E., Brostrom, C. O., and Brostrom, M. A.,** Mechanisms of action of inhibitors of prolactin secretion in GH_3 pituitary cells. I. Ca^{2+}-dependent inhibition of amino acid incorporation, *Mol. Pharmacol.,* 29, 411, 1986.

28. **Wolfe, S. E. and Brostrom, M. A.,** Mechanisms of action of inhibitors of prolactin secretion in GH_3 pituitary cells. II. Blockade of voltage dependent Ca^{2+} channels, *Mol. Pharmacol.,* 29, 420, 1986.

29. **Brostrom, M. A., Chin, K.-V., Cade, C., Gmitter, D., and Brostrom, C. O.,** Stimulation of protein synthesis in pituitary cells by phorbol esters and cyclic AMP, *J. Biol. Chem.,* 262, 16515, 1987.

30. **Chin, K.-V., Cade, C., Brostrom, M. A., Galuska, E. M., and Brostrom, M. A.,** Calcium-dependent regulation of protein synthesis at translational initiation in eukaryotic cells, *J. Biol. Chem.,* 262, 16509, 1987.

31. **Abdel-Latif, A. A.,** Calcium-mobilizing receptors, polyphosphoinositides, and the generation of second messengers, *Pharmac. Rev.,* 38, 227, 1986.

32. **Chin, K.-V., Cade, C., Brostrom, M. A., and Brostrom, C. O.,** Regulation of protein synthesis in intact rat liver by calcium mobilizing agents, *Int. J. Biochem.,* 20, 1313, 1988.

33. **Gmitter-Yellen, D., Brostrom, C. O., Kuznetsov, G., and Brostrom, M. A.,** unpublished data, 1991.

34. **Brostrom, C. O., Chin, K.-V., Wong, W. L., Cade, C., and Brostrom, M. A.,** Inhibition of translational initiation in eukaryotic cells by calcium ionophore, *J. Biol. Chem.,* 264, 1644, 1989.

35. **Wolf, B. A., Turk, J., Sherman, W. R., and McDaniel, M. L.,** Intracellular Ca^{2+} mobilization by arachidonic acid, *J. Biol. Chem.,* 261, 3501, 1986.

36. **Yasuda, H., Kishiro, K., Izumi, N., and Nakanishi, M.,** Biphasic liberation of arachidonic and stearic acids during cerebral ischemia, *J. Neurochem.,* 45, 168, 1985.

37. **Kolesnick, R. N., Musacchio, I., Thaw, C., and Gershengorn, M. C.,** Arachidonic acid mobilizes calcium and stimulates prolactin secretion from GH_3 cells, *Am. J. Physiol.,* 246, E458, 1984.

38. **Chan, K.-M. and Turk, J.,** Mechanism of arachidonic acid-induced Ca^{2+} mobilization from rat liver microsomes, *Biochim. Biophys. Acta,* 928, 186, 1987.

39. **Rotman, E. I., Brostrom, M. A., and Brostrom, C. O.,** Inhibition of protein synthesis in intact mammalian cells by arachidonic acid, *Biochem. J.,* 282, 487, 1992.

40. **Strittmatter, W. J., Couch, C. B., and Mundy, D. I.,** Role of metalloendoprotease in the fusion of biological membranes, in *Cell Fusion,* Sowers, A. E., Ed., Plenum Press, New York, 1987, 99.

41. **Strous, G. J., van Kerkhof, P., Dekker, J., and Schwartz, A. L.,** Metalloendoprotease inhibitors block protein synthesis, intracellular transport, and endocytosis in hepatoma cells, *J. Biol. Chem.,* 263, 18197, 1988.

42. **Hammerschlag, R., Bolen, F. A., and Stone, G. C.,** Metalloendoprotease inhibitors block fast axonal transport, *J. Neurochem.* 52, 268, 1989.

43. **Lelkes, P. I. and Pollard, H. B.,** Oligopeptide inhibitors of metalloendoprotease inhibit catecholamine secretion from bovine adrenal chromaffin cells by modulating intracellular calcium homeostasis, *J. Biol. Chem.,* 262, 15496, 1987.

44. **Brostrom, M. A., Prostko, C. R., Gmitter-Yellen, D., Grandison, L. J., Kuznetsov, G., Wong, W. L., and Brostrom, C. O.,** Inhibition of translational initiation by metalloendoprotease antagonists, *J. Biol. Chem.,* 266, 7037, 1991.

45. **Kelly, R. B.,** Tracking an elusive receptor, *Nature (London),* 345, 480, 1990.

46. **Wong, W. L., Brostrom, M. A., and Brostrom, C. O.,** unpublished data, 1991.

47. **Kumar, R. V., Wolfman, A., Panniers, R., and Henshaw, E. C.,** Mechanism of inhibition of polypeptide chain initiation in calcium-depleted Erhlich ascites tumor cells, *J. Cell Biol.,* 108, 2107, 1989.

48. **Wong, W. L., Brostrom, M. A., and Brostrom, C. O.,** Effects of Ca²⁺ and ionophore A23187 on protein synthesis in intact rabbit reticulocytes, *Int. J. Biochem.,* 23, 605, 1991.

49. **Kumar, R. V., Panniers, R., Wolfman, A., and Henshaw, E. C.,** Inhibition of protein synthesis by antagonists of calmodulin in Ehrlich ascites tumor cells, *Eur. J. Biochem.,* 195, 313, 1991.

50. **Asano, M. and Hidaka, H.,** Biopharmacological properties of naphthalenesul fonamides as potent calmodulin antagonists, in *Calcium and Cell Function,* Vol. 5, Cheung, W. Y., Ed., Academic Press, Orlando, 1984, 123.

51. **West, J. B.,** *Physiological Basis of Medical Practice,* 11th ed., Williams and Wilkins, Baltimore, 1985, 346.

52. **Kimball, S. R. and Jefferson, L. S.,** Mechanism of inhibition of protein synthesis by vasopressin in rat liver, *J. Biol. Chem.,* 265, 16794, 1990.

53. **Fawell, E. H., Boyer, I. J., Brostrom, M. A., and Brostrom, C. O.,** A novel calcium-dependent phosphorylation of a ribosome-associated protein, *J. Biol. Chem.,* 264, 1650, 1989.

54. **Traugh, J. A. and Pendergast, A. M.,** Regulation of protein synthesis by phosphorylation of ribosomal protein S6 and aminoacyl tRNA synthetases, *Prog. Nuclei Acid Res. Mol. Biol.,* 33, 195, 1986.

55. **Brostrom, M. A., Lin, X., Cade, C., Gmitter, D., and Brostrom, C. O.,** Loss of a calcium requirement for protein synthesis in pituitary cells following thermal or chemical stress, *J. Biol. Chem.,* 264, 1638, 1989.

56. **Pancake, B., Prostko, C. R., Brostrom, M. A., and Brostrom, C. O.,** unpublished data, 1989.

57. **Ryazanov, A. G. and Spirin, A. S.,** Phosphorylation of elongation factor 2: a key mechanism in regulating gene expression in vertebrates, *The New Biologist,* 2, 843, 1990.

58. **Palfrey, H. C.,** Presence in many mammalian tissues of an identical major substrate (M_r 100,000) for calmodulin-dependent protein kinase, *FEBS Lett.,* 157, 183, 1983.

59. **Nairn, A. C., Bhagat, B., and Palfrey, H. C.,** Identification of calmodulin-dependent protein kinase III and its major M_r 100,000 substrate in mammalian tissues, *Proc. Natl. Acad. Sci. U.S.A.,* 82, 7939, 1985.

60. **End, D., Tolson, N., Hashimoto, S., and Guroff, G.,** Nerve growth factor-induced decrease in the cell free phosphorylation of a soluble protein in PC12 cells, *J. Biol. Chem.,* 258, 6549, 1983.

61. **Hama, T. and Guroff, G.,** Distribution of Nsp100 and Nsp100 kinase, a nerve growth factor-sensitive phosphorylation system, in rat tissues, *J. Neurochem.,* 45, 1279, 1985.

62. **Togari, A. and Guroff, G.,** Partial purification and characterization of a nerve growth factor-sensitive kinase and its substrate from PC12 cells, *J. Biol. Chem.,* 260, 3804, 1985.

63. **Ryazanov, A. G.,** Ca²⁺/calmodulin-dependent phosphorylation of elongation factor 2, *FEBS Lett.,* 214, 331, 1987.

64. **Nairn, A. C. and Palfrey, H. C.,** Identification of the major M_r 100,000 substrate for calmodulin-dependent protein kinase III in mammalian cells as elongation factor 2, *J. Biol. Chem.,* 262, 17299, 1987.

65. **Ryazanov, A. G., Natapov, P. G., Shestakova, E. A., Severin, F. F., and Spirin, A. S.,** Phosphorylation of the elongation factor 2: the fifth Ca²⁺/calmodulin-dependent system of protein phosphorylation, *Biochimie,* 70, 619, 1988.

66. **Ovchinnikov, L. P., Motuz, L. P., Natapov, P. G., Averbuch, L. J., Wettenhall, R. E. H., Szyzka, R., Kramer, G., and Hardesty, B.,** Three phosphorylation sites in elongation factor 2, *FEBS Lett.,* 275, 209, 1990.

67. **Price, N. T., Redpath, N. T., Severinov, K. V., Campbell, D.G., Russell, J. M., and Proud, C. G.,** Identification of the phosphorylation sites in elongation factor-2 from rabbit reticulocytes, *FEBS Lett.,* 282, 253, 1991.

68. **Kim, Y. W., Kim, C. W., Kang, K. R., Byun, S. M., and Kang, Y.-S.,** Elongation factor-2 in chick embryo is phosphorylated on tyrosine residues as well as serine and threonine, *Biochem. Biophys. Res. Commun.,* 175, 400, 1991.
69. **Nilsson, L. and Nygard, O.,** Altered sensitivity of eukaryotic elongation factor 2 for trypsin after phosphorylation and ribosomal binding, *J. Biol. Chem.,* 266, 10578, 1991.
70. **Sitikov, A. S., Simonenko, P. N., Shestakova, E. A., Ryazanov, A. G., and Ovchinnikov, L. P.,** cAMP-dependent activation of protein synthesis correlates with dephosphorylation of elongation factor 2, *FEBS Lett.,* 228, 327, 1988.
71. **Redpath, N. T. and Proud, C. G.,** The tumor promoter okadaic acid inhibits reticulocyte-lysate protein synthesis by increasing the net phosphorylation of elongation factor 2, *Biochem. J.,* 262, 69, 1989.
72. **Gschwendt, M., Kittstein, W., Mieskes, G., and Marks, F.,** A type 2A protein phosphatase dephosphorylates the elongation factor 2 and is stimulated by the phorbol ester TPA in mouse epidermis *in vivo, FEBS Lett.,* 257, 357, 1989.
73. **Ryazanov, A. G., Shestakova, E. A., and Natapov, P. G.,** Phosphorylation of elongation factor 2 by EF-2 kinase affects rate of translation, *Nature (London),* 334, 170, 1988.
74. **Ryazanov, A. G. and Davydova, E. K.,** Mechanism of elongation factor 2 (EF-2) inactivation upon phosphorylation, *FEBS Lett.,* 251, 187, 1989.
75. **Palfrey, H. C., Nairn, A. C., Muldoon, L. L., and Villereal, M. L.,** Rapid activation of calmodulin-dependent protein kinase III in mitogen-stimulated human fibroblasts, *J. Biol. Chem.,* 262, 9875, 1987.
76. **Mackie, K. P., Nairn, A. C., Hampel, G., Lam, G., and Jaffe, E. A.,** Thrombin and histamine stimulate the phosphorylation of elongation factor 2 in human umbilical vein endothelial cells, *J. Biol. Chem.,* 264, 1748, 1989.
77. **Nairn, A. C., Nichols, R. A., Brady, M. J., and Palfrey, H. C.,** Nerve growth factor treatment or cAMP elevation reduces Ca^{2+}/calmodulin-dependent protein kinase III activity in PC12 cells, *J. Biol. Chem.,* 262, 14265, 1987.
78. **Brady, M. J., Nairn, A. C., Wagner, J. A., and Palfrey, H. C.,** Nerve growth factor-induced down-regulation of calmodulin-dependent protein kinase-III in PC-12 cells involves cyclic AMP-dependent protein kinase, *J. Neurochem.,* 54, 1034, 1990.
79. **Nygard, O., Nilsson, A., Carlberg, U., Nilsson, L., and Amons, R.,** Phosphorylation regulates the activity of the eEF-2-specific Ca^{2+}- and calmodulin-specific protein kinase III, *J. Biol. Chem.,* 266, 16425, 1991.
80. **Celis, J. E., Madsen, P., and Ryazanov, A. G.,** Increased phosphorylation of elongation factor 2 during mitosis in transformed human amnion cells correlates with a decreased rate of protein synthesis, *Proc. Natl. Acad. Sci. U.S.A.,* 87, 4231, 1990.
81. **Severinov, K. V., Melnikova, E. G., and Ryazanov, A. G.,** Downregulation of the translation elongation factor 2 kinase in *Xenopus laevis* occytes at the final stages of oogenesis, *The New Biologist,* 2, 887, 1990.
82. **Reed, J. C., Alpers, J. D., Nowell, P. C., and Hoover, R. G.,** Sequential expression of protooncogenes during lectin-stimulated mitogenesis of normal human lymphocytes, *Proc. Natl. Acad. Sci. U.S.A.,* 83, 3982, 1986.
83. **Greenberg, M. E., Hermanowski, A. L., and Ziff, E. B.,** Effect of protein synthesis inhibitors on growth factor activation of c-fos, c-myc, and actin gene transcription, *Mol. Cell. Biol.,* 6, 1050, 1986.
84. **Lee, A. S.,** Coordinated regulation of a set of genes by glucose and calcium-ionophore in mammalian cells, *Trends Biochem. Sci.,* 12, 20, 1987.
85. **Kim, Y. K., Kim, K. S., and Lee, A. S.,** Regulation of the glucose-regulated genes by β-mercaptoethanol requires *de novo* protein synthesis and correlates with inhibition of protein glycosylation, *J. Cell. Physiol.,* 133, 553, 1987.
86. **Watowich, S. S. and Morimoto, R. I.,** Complex regulation of heat shock- and glucose-responsive genes in human cells, *Mol. Cell. Biol.,* 8, 393, 1988.

87. **Sarnow, P.,** Translation of glucose-regulated protein 78/immunoglobulin heavy-chain binding protein mRNA is increased in poliovirus-infected cells at a time when cap-dependent translation of cellular mRNAs is inhibited, *Proc. Natl. Acad. Sci. U.S.A.,* 86, 5795, 1989.

88. **Dorner, A. J., Wasley, L. C., and Kaufman, R. J.,** Increased synthesis of secreted proteins induced expression of glucose-regulated proteins in butyrate-treated Chinese hamster ovary cells, *J. Biol. Chem.,* 264, 10602, 1989.

89. **Hendershot, L. M.,** Immunoglobulin heavy chain binding protein complexes are dissociated *in vivo* by light chain addition, *J. Cell Biol.,* 111, 829, 1990.

90. **Vogel, J. P., Misra, L. M., and Rose, M. D.,** Loss of BiP/GRP78 function blocks translocation of secretory proteins in yeast, *J. Cell Biol.,* 110, 1885, 1990.

91. **Nguyen, T. H., Law, D. T. S., and Williams, D. B.,** Binding protein BiP is required for translocation of secretory proteins into the endoplasmic reticulum in *Saccharomyces cerevisiae, Proc. Natl. Acad. Sci. U.S.A.,* 88, 1565, 1991.

92. **Ting, J. and Lee, A. S.,** Human gene encoding the 78,000-dalton glucose-regulated protein and its pseudogene: structure, conservation and regulation, *DNA,* 7, 275, 1988.

93. **Li, X. and Lee, A. S.,** Competitive inhibition of a set of endoplasmic reticulum protein genes (GRP78, GRP94, and ERp72) retards cell growth and lowers viability after ionophore treatment, *Mol. Cell. Biol.,* 11, 3446, 1991.

94. **Normington, K., Kohno, K., Kozutsumi, Y., Gething, M.-J., and Sambrook, J.,** *S. cerevisiae* encodes an essential protein homologous in sequence and function to mammalian BiP, *Cell,* 57, 1223, 1989.

95. **Brostrom, M. A., Cade, C., Prostko, C. R., Gmitter-Yellen, D., and Brostrom, C. O.,** Accommodation of protein synthesis to chronic deprivation of intracellular sequestered calcium, *J. Biol. Chem.,* 265, 20539, 1990.

96. **Nakaki, T., Deans, R. J., and Lee, A. S.,** Enhanced transcription of the 78,000-dalton glucose-regulated protein (GRP78) gene and association of GRP78 with immunoglobulin light chains in a nonsecreting B-cell myeloma line (NS-1), *Mol. Cell. Biol.,* 9, 2233, 1989.

97. **Prostko, C. R., Brostrom, M. A., Galuska-Malara, E. M., and Brostrom, C. O.,** Stimulation of GRP78 gene transcription by phorbol ester and cAMP in GH₃ pituitary cells, *J. Biol. Chem.,* 266, 19790, 1991.

98. **Fink, J. S., Verhave, M., Walton, K., Mandel, G., and Goodman, R. H.,** Cyclic AMP and phorbol ester-induced transcriptional activation are mediated by the same enhancer element in the human vasoactive intestinal polypeptide gene, *J. Biol. Chem.,* 266, 3882, 1991.

99. **Montminy, M. R., Gonzalez, G. A., and Yamamoto, K. R.,** Regulation of cAMP-inducible genes by CREB, *Trends Neurosci.,* 13, 184, 1990.

100. **Resendez, E., Wooden, S. K., and Lee, A. S.,** Identification of highly conserved regulatory domains and protein-binding sites in the promoters of the rat and human genes encoding the stress inducible 78-kilodalton glucose-regulated protein, *Mol. Cell. Biol.,* 8, 4579, 1988.

101. **Macejak, D. G. and Sarnow, P.,** Internal initiation of translation mediated by the 5' leader of a cellular mRNA, *Nature (London),* 353, 90, 1991.

102. **Schutzbach, J. S. and Forsee, W. T.,** Calcium ion activation of rabbit liver α1,2-mannosidase, *J. Biol. Chem.,* 265, 2546, 1990.

103. **Lodish, H. F. and Kong, N.,** Perturbation of cellular calcium blocks exit of secretory proteins from the rough endoplasmic reticulum, *J. Biol. Chem.,* 265, 10893, 1990.

104. **Kuznetsov, G., Brostrom, M. A., and Brostrom, C. O.,** Identification of a calcium requirement for secretory protein processing and export. Differential effects of calcium and dithiothreitol, *J. Biol. Chem.,* 267, 3932, 1992.

105. **Balch, W. E.,** Biochemistry of interorganelle transport, *J. Biol. Chem.,* 264, 16965, 1989.
106. **White, A., Handler, P., Smith, E. L., Hill, R. L., and Lehman, I. R.,** *Principles of Biochemistry,* McGraw-Hill, New York, 1978, 136.
107. **Wieland, F. T., Gleason, M. L., Serafini, T. A., and Rothman, J. E.,** The rate of bulk flow from the endoplasmic reticulum to the cell surface, *Cell,* 50, 289, 1987.

Chapter 6

TISSUE-SPECIFIC REGULATION OF GLUCOKINASE

Mark A. Magnuson and Thomas L. Jetton

TABLE OF CONTENTS

I. INTRODUCTION

Euglycemia is characterized by blood glucose concentrations in the 4 to 7 mM range and must be maintained in the face of constantly changing dietary intake and metabolic needs. Deviations in the blood glucose concentration on either side of this narrow range can lead to a myriad of acute and chronic pathological states. While several tissues play roles that are involved in maintaining euglycemia, the liver and pancreatic β cell are probably the most important. The liver is a major site of glucose uptake during periods of hyperglycemia and removes glucose from the circulation for conversion into glycogen or other energy forms. However, during periods of hypoglycemia, the liver produces glucose and releases it into the bloodstream. The pancreatic β cell, on the other hand, contributes to glucose homeostasis by sensing changes in the plasma glucose concentration and adjusting the rate at which insulin is synthesized and released. While the physiologic roles of the hepatocyte and pancreatic β cell in the maintenance of euglycemia are clearly different, both of these cell types share the ability to alter their utilization of glucose as blood glucose concentrations either rise or fall. In the liver, the ability to utilize glucose is primarily dependent upon the relative amounts of circulating insulin and glucagon with contributory effects from other hormones (e.g., glucocorticoids and thyroid hormones). As insulin levels rise, the liver responds by a coordinate induction of enzymes mediating glycolysis and glycogenesis, while at the same time those enzymes that favor glycogenolysis and gluconeogenesis are inhibited.[1] Conversely, when insulin levels fall and glucagon level rise, a coordinate induction of enzymes mediating gluconeogenesis and glycogenolysis occurs, thus allowing the liver to produce glucose. In contrast, glycolysis in the β cell is regulated by glucose and not by insulin. A common feature of both the liver and β cell that contributes to their respective roles in maintaining glucose homeostasis is the expression of hexokinase type IV, more commonly known as glucokinase (ATP: D-hexose 6-phosphotransferase; E.C. 2.7.1.1).[2]

In this chapter we discuss the findings of recent molecular genetic studies that have helped to clarify aspects of the tissue-specific structure and regulation of glucokinase. The physiologic importance of this enzyme in glucose homeostasis suggests the need for a variety of control points for modulating glucokinase gene expression. Indeed, studies in this field have identified several mechanisms by which the enzyme can be expressed in both the hepatocyte and β cell, yet can also be regulated differently in these cells. These observations have provided a basis for understanding the tissue-specific function of the enzyme glucokinase and have provided a basis for future studies in this field. The tissue-specific expression of an active glucokinase molecule is achieved by regulation occurring at several steps in the transcription and processing of the glucokinase mRNA. These include the use of tissue-specific promoter regions in the glucokinase gene, the modulation of the rate of

TABLE 1
Structural and Functional Features of Mammalian Hexokinases

Hexokinase type	Mass (kDa)	K_m for glucose	G-6-P inhibition	Major sites of expression
I	~100	0.04[a]	Yes	Brain, kidney
II	~100	0.13[a]	Yes	Muscle, fat
III	~100	0.02[b]	Yes	Liver, kidney
IV (Glucokinase)	~50	5.0[c]	No	Liver, islet

Note: The mass, K_m for glucose, inhibition by glucose-6-P, and major tissue sites of expression are shown for hexokinase types I through IV.

[a] Hyperbolic kinetics.
[b] Inhibited by excess glucose.
[c] Sigmoidal kinetics.

transcription by hormones in the liver, and by tissue-specific patterns of RNA splicing. We will discuss what is known about each of these control mechanisms and how, together, they determine the tissue-specific regulation of this enzyme.

II. HEXOKINASE GENE FAMILY

The phosphorylation of glucose and other simple sugars at the sixth position is the first step in their metabolism. Since carbohydrates are nearly a universal energy source, hexokinases may also be ubiquitous in their phylogenetic distribution as most organisms depend on glycolysis as the primary pathway for glucose catabolism. The most thoroughly studied hexokinases are those in mammals and yeast. Among mammals, four distinct hexokinases have been identified, termed types I through IV (also A through D), that have been defined by their order of elution from ion-exchange resins.[3] These isozymes share certain structural and functional features and, thus, are thought to be evolutionarily related. The distribution, kinetic, and structural features of each is shown in Table 1.

Glucokinase (hexokinase type IV) differs in several ways from hexokinases types I through III. It is about half the size (~50 vs. ~100 kDa), has a much higher K_m for glucose (~10 mM vs. ~20 to 130 mM), is not subject to end product inhibition by physiologic concentrations of G-6-P, and has sigmoidal kinetics. The enzyme has been thought to be more specific for glucose than are the other hexokinases, thus leading it to be termed "glucokinase"; however, this distinction may be incorrect from an enzymologic viewpoint.[4] Physiologically, glucose may be the most important substrate for the enzyme, so the unique name remains appropriate. Nevertheless, when taken together, these various characteristics of glucokinase suggest that the enzyme is more similar to yeast hexokinase than it is to the mammalian hexokinase types I through III.

The similar sizes of glucokinase and the yeast hexokinases led to the suggestion that the hexokinases types I through III evolved by a gene duplication and tandem ligation event of a common ancestral gene similar to that of the yeast hexokinases or glucokinase.[3,5,6] Comparison of the sequence of glucokinase,[7] yeast hexokinase,[8,9] and mammalian hexokinases[10-12] has allowed this notion to gain further acceptance, although a comparison of the intron/exon structures of the glucokinase gene with a hexokinase gene has not yet been reported. All of these hexokinases have now been shown to share a high degree of amino acid sequence homology. Glucokinase is 33% identical with yeast hexokinase and 53% identical with the carboxy terminal domain of rat brain hexokinase I.[7] When conservative substitutions are considered, these similarities are even more striking. Conservation of primary structure is especially evident in the putative glucose- and ATP-binding domains of these molecules.[7] Indeed, similarity of a region of these hexokinases with the ATP-binding sites in protein kinases led to the identification of a similar domain in glucokinase.[7] In addition, several residues that were identified as playing a role in glucose binding, as originally determined by X-ray crystallographic analysis of yeast hexokinase,[13] are also conserved among the different hexokinases.[7,14] Mutagenesis of one of these residues in hepatic glucokinase (aspartate-205) decreases activity of the enzyme to 1/500th of that of the wild type enzyme thus establishing its importance in catalysis.[14] Lastly, conservation of the carboxy- and amino-terminal halves of mammalian hexokinase type I has further reinforced the notion of gene evolution by duplication and tandem ligation. When the two halves of the protein were compared, they were found to share over 50% primary sequence homology.[10,11]

The distribution of the different mammalian hexokinases is tissue-specific, although all four hexokinase isoforms are present in rat liver.[3,4] Glucokinase is the most restricted in distribution and has been found only in the liver and in the pancreatic islet.[15] Interestingly, the tissue-specific expression of the four different hexokinases roughly mirrors that of the different glucose transporters, although there can be significant overlapping of expression in certain tissues. For instance, both hepatocytes and pancreatic β-cells express a high K_m glucose transporter termed GLUT-2.[16] This glucose transporter has a K_m for glucose of ~ 30 mM, thus allowing the cytoplasm of hepatocytes and β-cells to rapidly equilibrate to plasma concentrations of glucose. Brain and kidney, which express hexokinase type I, also express GLUT 1,[17] a glucose transporter with a very low K_m for glucose, whereas striated muscle and adipose tissue coexpress hexokinase II and the insulin regulatable GLUT-4,[18,19] a transporter with an intermediate K_m for glucose. The coexpression of certain hexokinases with specific glucose transporters suggests they are functionally coupled, although direct interactions have not been demonstrated.

III. DUAL FUNCTIONAL ROLES OF GLUCOKINASE

While glucokinase catalyzes the rate-limiting step in glycolysis in both hepatocytes and β cells, thus showing the biochemical similarity of these different cells at this regulatory control point, the enzyme is regulated differentially in the hepatocyte and β cell. The tissue-specific regulation of glucokinase is thought to allow the hepatocyte and pancreatic β cell to perform their different physiologic functons in order to maintain glucose homeostasis.

A. ROLE IN THE β CELL

The pancreatic β cell responds to changes in glucose concentration and constantly adjusts the rate of insulin secretion. Physiologic glucose concentrations monitored by the β cell are in the mM range, while the concentration of most ligands that interact with cell surface receptors are in the nM or pM range. In the absence of a surface receptor for glucose, the β cell senses changes in the circulating concentration of glucose by a mechanism that involves changes in the rate of glucose metabolism,[20] the so-called "fuel hypothesis" of glucose sensing.[21,22] After glucose enters the pancreatic β cell via GLUT 2, the high K_m glucose transporter,[23,24] its metabolism is initiated by phosphorylation at the sixth position, a reaction catalyzed by both glucokinase and at least one other hexokinase whose isotype is not known.[25-29] At low glucose concentrations, the lower K_m hexokinase isotype contributes substantially to glycolysis. However, as glucose concentrations rise, glucokinase becomes the predominant glucose phosphorylating enzyme in the β cell.[30]

The unique kinetic characteristics of glucokinase allows the rate of glucose metabolism to vary in proportion to glucose concentration. By recognizing and modulating the rate of glucose phosphorylation, glucokinase acts as the rate-limiting determinant for glucose usage by the β cell.[30] The high K_m and the lack of feedback inhibition by glucose-6-P are important kinetic parameters of glucokinase that enable β cell glucose responsiveness. Additionally, with the half-maximal saturation of glucokinase near the middle of the physiologic range of glucose fluctuations and its sigmoidal kinetics, the enzyme is ideally suited for the role of controlling the metabolism of glucose by the β cell. Since the K_m of GLUT 2 is ∼30 mM, glucose transport into the β cell is not thought to be rate-limiting except, perhaps, in certain pathologic states.[31] The role of glucokinase as the so-called glucose sensor for the β cell, first proposed in 1968 by Matshinsky and Ellerman,[20] is now widely accepted.

The metabolism of glucose by the β cell is, in turn, tightly linked to the rate of insulin secretion by a series of intermediate steps as shown in Figure 1. Glycolysis is thought to result in an increase in the ATP/ADP ratio, thereby inhibiting an ATP-sensitive K$^+$ channel which results in a decrease of K$^+$ efflux, the effect of which is to depolarize the membrane of the β cell.[32,33] The change in membrane potential is then thought to activate one or more

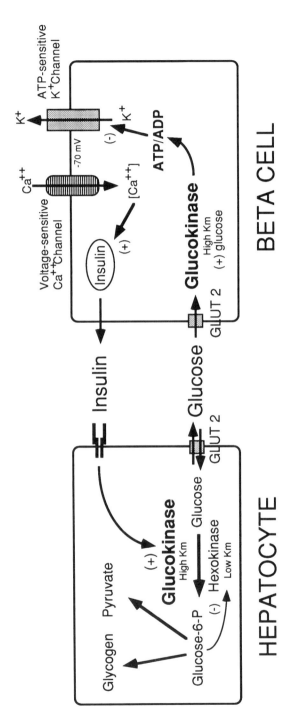

FIGURE 1. Different regulation and functional roles of glucokinase in the hepatocyte and β cell. The different roles of hepatic and β cell glucokinase in maintaining glucose homeostasis are diagramatically illustrated. Both of these cell types express GLUT-2, a high K_m glucose transporter, and glucokinase, a high K_m hexokinase. In the liver, glucokinase serves as an insulin-inducible determinant of glucose usage, whereas in the β cell, the enzyme functions as the so-called "glucose sensor". Glucose metabolism in the β cell is thought to be linked to insulin secretion through several intermediate steps, as illustrated.

voltage-dependent Ca^{2+} channels leading to a sharp increase in the intracellular Ca^{2+} concentration and stimulating the mechanisms involved in the exocytosis of insulin from storage granules.[34]

Glucokinase levels in the β cell are inducible by glucose, thus permitting glucose to serve as a positive mediator of this process.[29] However, for glucokinase to determine the rate of glucose usage by the β cell, and act as the so-called "glucose sensor",[30] it must not be affected by differing concentrations of insulin, particularly at the source where the concentrations are highest. Instead, the enzyme must function as a stable, insulin-independent determinate for glucose usage by the β cell. Indeed, in cultured islets, insulin does not affect glucokinase activity.[29] The mechanism by which glucose modulates glucokinase activity in the β cell is not known; however, glucokinase activity in cultured islets can vary as much as fivefold in 7-day cultured islets in either 3- or 30-mM glucose.[29]

B. ROLE IN THE LIVER

In liver, glucokinase is regulated differently than in the islet, thus reflecting the different physiologic role of the enzyme in the hepatocyte. The entry and utilization of glucose by the liver, in so far as the most proximal steps are concerned, is similar to that in the β cell as shown in Figure 1. Hepatocytes also express GLUT-2, the same high K_m glucose transporter as the β cell, so the entry of glucose into the hepatocyte is not rate-limiting. During periods of hyperglycemia and rising insulin levels, hepatic glucokinase activity is induced.[35] The distinct kinetic features of glucokinase permit the phosphorylation of glucose to proceed at a rate proportional to the concentration of glucose, as in the β cell. Hexokinases I through III, because they are inhibited by glucose-6-P and have lower Michaelis' constants, do not contribute to glucose phosphorylation at the higher (mM) concentrations encountered in the portal vein after a meal. Glucokinase in the liver, therefore, serves as an insulin-sensitive determinant for glucose entry into the hepatocyte, essentially trapping glucose and thus maintaining a gradient for its entry during periods of hyperglycemia. Glucose transport can proceed in either direction depending upon its concentration gradient, so GLUT 2 gene expression need not be regulated.[1] During periods of fasting or starvation, plasma insulin levels decline and glucagon secretion by pancreatic α cells increases. Concomitantly, hepatic glucokinase activity falls while glucose-6-phosphatase levels rise, thus allowing glucose to diffuse out of the liver into the bloodstream.[1]

Thus, in the liver, glucokinase acts as an insulin-dependent modulator of glucose utilization, whereas in the β cell, glucokinase acts as a glucose-dependent glucose sensor. Indeed, the differential regulation of glucokinase activity in the liver and β cell by insulin and glucose, respectively, may permit feedback regulation of glucokinase activity.[36-38]

IV. CLONING OF GLUCOKINASE

In order to understand the mechanisms underlying the tissue-specific regulation of glucokinase, the cloning of a cDNA for the enzyme was required. Since glucokinase is expressed at a higher level in liver than in the islet, particularly after refeeding or insulin treatment, and liver was a more available source of tissue, the cloning of hepatic glucokinase was accomplished first.

A. HEPATIC GLUCOKINASE cDNAs

Hepatic glucokinase cDNAs were cloned initially by two different groups. Iynedjian et al.[39] utilized an antibody to hepatic glucokinase to screen a rat liver cDNA expression library made from 21S hepatic RNA and was the first to report a cDNA for the enzyme. Andreone et al.[7] obtained a cDNA by a route that involved purifying glucokinase to homogeneity, obtaining amino acid sequence information, then developing synthetic oligonucleotides that were used to screen a rat liver cDNA library. Both groups used the glucokinase cDNAs obtained to examine the regulation hepatic glucokinase mRNA and showed that it was markedly induced by both insulin administration to diabetic animals and during the fasted-to-fed transition. Iynedjian et al.[39] also examined the developmental expression of glucokinase mRNA and found it to be induced at weaning, thus showing that expression of glucokinase mRNA coincides with that of hepatic glucokinase activity.

The sequence of hepatic glucokinase was deduced by Andreone et al.[7] who showed that the enzyme was 465 amino acids in length and was homologous to both yeast hexokinases and rat hexokinase I. A map of a full-length hepatic glucokinase cDNA (GK.Z2) is shown in Figure 2, and the amino acid sequence of the enzyme is shown in Figure 3. Within the coding sequence of the enzyme, the putative glucose- and ATP-binding domains were identified based on their similarity to sequences in other hexokinases and protein kinases. While the identity of the cDNA cloned by Andreone et al.[7] as coding for glucokinase was established by finding the sequence of multiple peptides within the reading frame deduced from the cDNA, additional proof was obtained by the expression of a hepatic cDNA in bacteria and measurement of the kinetic properties of the product that positively identified the cDNA as, in fact, coding for glucokinase.[40]

B. ISLET GLUCOKINASE cDNAs

The cloning of hepatic glucokinase permitted studies of islet glucokinase gene expression to be initiated. Previous biochemical and immunological studies that compared hepatic and islet glucokinase found them to be indistinguishable by several criteria, e.g., kinetic characteristics, mass, charge, and immunologic cross-reactivity. Thus, it was thought that the islet isoform was very similar, if not identical, to the hepatic isoform.[15] Further support for a single form of glucokinase was provided by Southern blot analysis of

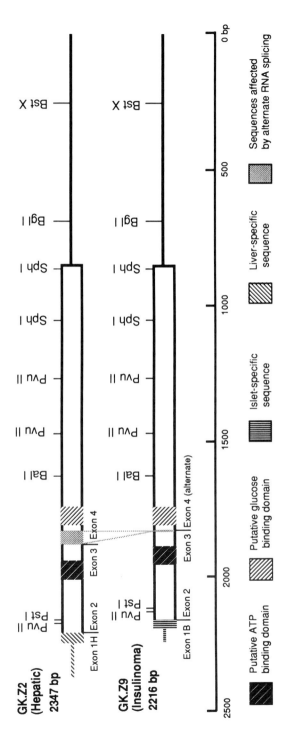

FIGURE 2. Comparison of glucokinase cDNAs from rat liver and insulinoma tissues. The structural differences of glucokinase cDNAs isolated from different tissue sources is illustrated. The GK.Z2 cDNA clone was isolated from a rat liver cDNA library and the GK.Z9 cDNA clone was isolated from a rat insulinoma cDNA library. Differences in the cDNAs at the 5′ end and within the coding sequence of the enzyme are indicated. The differences coincide with the location of splice junctions indicating that alternate RNA splicing is involved in generating glucokinase isoform diversity.

Rat islet glucokinase:
465 amino acids
52,087 kDa
pI 5.08

MLDDRARMEATKKEKVEQILAEFQLQEEDLKKVMSRMQKEMDRGLRLETH
EEASVKMLPTYVRSTPEGSEVGDFLSLDLGGTNFRVMLVKVGEGEAGQWS
VKTKHQMYSIPEDAMTGTAEMLFDYISECISDLLDKHQMKHKKLPLGFTF
SFPVRHEDLDKGILLNWTKGFKASGAEGNNIVGLLRDAIKRRGDFEMDVV
AMVNDTVATMISCYYEDRQCEVGMIVGTGCNACYMEEMQNVELVEGDEGR
MCVNTEWGAFGDSGELDEFLLEYDRMVDESSANPGQQLYEKIIGGKYMGE
LVRLVLLKLVDENLLFHGEASEQLRTRGAFETRFVSQVESDSGDRKQIHN
ILSTLGLRPSVTDCDIVRRACESVSTRAAHMCSAGLAGVINRMRESRSED
VMRITVGVDGSVYKLHPSFKERFHASVRRLTPNCEITFIESEEGSGRGAA
LVSAVACKKACMLAQ

Rat hepatic glucokinase:
465 amino acids
51, 924 kDa
pI 5.01

MAMDTTRCGAQLLTL . . .

FIGURE 3. Cell type-specific sequences of the islet and liver isoforms of glucokinase. The different amino acids sequences of the islet (top) and hepatic (bottom) glucokinase isoforms are shown (underlined).

rat genomic DNA that showed there was only a single gene for glucokinase in the rat.[35,41] However, surprisingly, when the liver-derived cDNA was used as a probe in RNA blot experiments of islet RNA, both Magnuson and Shelton[41] and Iynedjian et al.[35] found that the pancreatic islet form of rat glucokinase mRNA was longer by 200 to 400 bases than that in liver. Moreover, Iynedjian et al.[35] found that the effects of a fasting/refeeding regimen on hepatic and islet glucokinase mRNA and protein levels were also different. While the amount of hepatic mRNA and immunoreactive protein varied with fasting and refeeding, islet glucokinase mRNA and immunodetectable protein did not change under these conditions.[35] To determine the structural basis for the different sizes of the mRNAs, and to explore the basis for the differential regulation, cDNAs for islet glucokinase mRNA were cloned by both groups.

Initial studies identifying structural differences in glucokinase isoforms yielded intriguing results. The glucokinase cDNAs cloned by Magnuson and Shelton[41] were obtained from a rat insulinoma cDNA library, whereas the cDNAs obtained by Hughes et al.[42] were cloned from a rat islet cDNA library. Both groups determined that glucokinase transcripts from islet tissue and hepatocytes were not equivalent and that this difference was located at the

5' end of the mRNAs. The nature of these differences suggested that additional mechanisms, such as alternate RNA splicing and/or the utilization of distinct promoter regions, were involved in producing tissue-specific glucokinase mRNAs. Indeed, when the insulinoma and hepatic cDNAs for glucokinase were compared by Magnuson and Shelton,[41] two differences were found that were due to both the use of alternate promoter regions and to alternate RNA processing, as shown in Figure 2. The sequences at the 5' ends of the cDNAs were distinct and the site at which this difference occurred coincided with the splice junction between the first and second exons, thus indicating that the hepatic and islet glucokinase mRNAs utilized different first exon sequences in the liver and islet, respectively. This finding implicated the use of different promoter regions in the production of glucokinase mRNAs in the different cell types. Moreover, since the translation initiation codons for liver and islet glucokinase were located in different first exons, the amino acid sequence at the amino terminus of the enzyme from liver and pancreatic islets was affected, as previously shown in Figure 3. In addition to sequence differences related to the use of alternate promoter regions, there were also sequence differences within the reading frame of the enzyme in the GK.Z9 cDNA clone, thus implicating alternate RNA processing in second region. Together, the differences in the hepatic and islet glucokinase cDNAs indicated that variant glucokinase isoforms were being produced as a result of alternate RNA splicing of the glucokinase gene product.

V. ALTERNATE PROMOTERS IN THE GLUCOKINASE GENE

The transcription unit that produces the hepatic glucokinase mRNA was characterized by Magnuson et al.[43] and found to contain 10 exons that span about 15 kb. The location of the β cell first exon, however, was not present in the genomic DNA fragments characterized in the cloning of the hepatic gene despite having 7.5 kb of 5' flanking DNA sequence in the lambda phage clone studied (λGK5). To determine the location of the genomic DNA sequences that gave rise to the 5' noncoding sequence in the islet glucokinase mRNA, another genomic DNA fragment was isolated.[41] This new DNA fragment (λGK7) did not overlap with the 5' end of λGK5, the clone containing the liver-specific first exon and promoter region, thus indicating that the upstream promoter region had to be at least 12 kb upstream from the downstream promoter region. Although the exact distance separating the different promoter regions has not yet been determined, an upper limit for this distance has been suggested by the cloning of cosmid DNAs for the gene.[44] Both promoter regions are contained in several cosmid clones that have an insert size of approximately 40 kb, thus placing an upper limit on the distance. However, since these clones have not been fully mapped, the actual distance

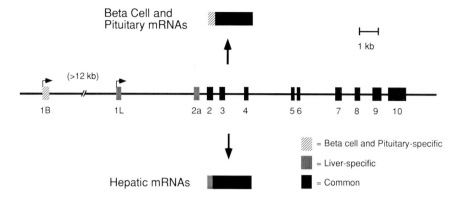

FIGURE 4. Glucokinase gene structure. The intron/exon structure of the glucokinase gene is diagramatically illustrated. The gene has 2 different promoter regions that are separated by at least 12 kb and which are linked to cell type-specific first exon sequences. The upstream promoter region is active in the pancreatic β cell and pituitary, whereas the downstream promoter region is active in the liver. In addition, the location of a cassette exon (2A) is shown which is utilized in the production of about 5% of the glucokinase mRNA present in liver.[59] (From Magnuson, M. A., *J. Cell. Biochem.*, 48, 115, 1992. With permission.)

is not known and may, in fact, be considerably less. A map of the glucokinase gene, as currently understood, is shown in Figure 4. The gene spans over 27 kb and contains 12 different exons that are used in a tissue-specific manner. Among the glucokinase isoforms that are functional, only 10 exons are utilized with the complete sequence of exons 2 through 10 being necessary for catalytic activity.

The tissue-specific expression of the upstream and downstream promoter regions has been tested by RNA-PCR amplification experiments. In these studies, the glucokinase mRNAs containing exon 1β sequences have been detected only in the islet and pituitary, while sequences containing the 1H sequences have only been identified in the liver.[41,45]

A. DOWNSTREAM PROMOTER REGULATION

The location of the downstream (hepatic) promoter region was established by using primer extension and S1 nuclease assays to map the start sites of transcription. In liver, transcription begins at a site 127 bases upstream from the translation initiation codon[43] and occurs over a 4 to 5 bp region, perhaps related to nonoptimal TATA homology (TATTT at -29 to -25 relative to the $+1$ transcription start site). Mapping of the start site of transcription of the hepatic glucokinase mRNA has also been reported by Noguchi et al.[46] with placement of the strongest start site one base upstream of that reported by Magnuson et al.[43] The sequence of approximately 1448 bp of DNA upstream from the transcription start site has been determined. An analysis of this sequence for potential regulatory elements has revealed several sequences

TABLE 2
Differential Regulation of Hepatic and
β Cell Glucokinase

	Stimulatory effect	Inhibitory effect
Hepatic glucokinase:	Insulin[a]	cAMP[a]
	Biotin[a]	
	Triiodothyronine[a]	
β Cell glucokinase:	Glucose	Exercise training

[a] Regulation at the level of gene transcription has been established.

From Magnuson, M. A., *J. Cell. Biochem.*, 48, 115, 1992. With permission.

that might be important in liver-specific transcriptional activity.[43] A possible binding site for the liver-specific transcription factors HNF-1 is present at −171 to −166.[43] In addition, there are 3 tandem repeats of the binding site for liver factor A1 at −69 to −53 and 2 single sites further upstream.[43] Whether these sites play a role in directing expression of glucokinase to the liver has not been determined but awaits development of a system by which the function of these elements can be tested.

1. Effects of Hormones on Hepatic Glucokinase
Gene Transcription

The effects of hormones and other agents on hepatic glucokinase appears to be controlled primarily at the level of gene transcription. Insulin,[43,47] triiodothyronine,[48,49] and biotin[50] stimulate gene transcription while glucagon (acting via cAMP) is inhibitory,[51] as shown in Table 2. In diabetic rats, insulin increases transcription at least 20-fold within 45 min.[47] Similarly, an inductive effect of insulin on glucokinase mRNA can be seen in freshly cultured primary hepatocytes. Comparison of the potency of insulin, proinsulin, and insulin-like growth factor I indicates that the induction occurs through the insulin receptor.[51] Moreover, the rapid inductive effect of insulin on hepatic glucokinase mRNA suggests that the effect is direct and involves preexisting cellular proteins. In addition, glucose does not alter the effect of insulin on glucokinase gene transcription.[51] The addition of cAMP, even in the presence of insulin, has an inhibitory action on transcription, completely eliminating RNA extension within 30 min of addition to cultured hepatocytes.[51] In addition, cAMP appears to accelerate decay of glucokinase mRNA, decreasing the half-life of the mRNA from ~300 min to only 35 min.[51]

Biotin, a water soluble vitamin, increases glucokinase gene transcription but the effect is short lived. In starved rats, biotin causes a 20-fold increase within 1 h, but this peak decays rapidly.[50] Triiodothyronine (T_3) also stimulates transcription of the gene in liver, but the effect is thought to be permissive since it is only observed during refeeding and not during starvation.[48] The effect of T_3 is more pronounced in hepatocytes from neonatal rats where it causes a dose-dependent increase of glucokinase mRNA of a magnitude similar to that seen with insulin.[49]

The *cis*-regulatory elements that mediate the hormonal effects on transcription from the hepatic glucokinase promoter region have not been identified. The absence of hepatoma cell lines that express the gene has hampered studies in this direction. However, even when fusion genes have been transfected into primary hepatocytes, no hormonal effects have been observed. For instance, fusion gene experiments using DNA fragments up to -5.5 kb from the start site in liver, fused to the chloramphenicol acetyltransferase gene, have failed to show evidence for regulation by insulin when transfected into primary hepatocytes.[46] In addition, the signal obtained in the primary hepatocytes was small, suggesting that these constructs also lack liver-specific enhancer elements.[46] Thus it is not yet known where the hormone regulatory elements reside in the glucokinase gene. However, since the gene is positively regulated by insulin, T_3, and biotin, response elements for these agents should be located somewhere within the regulatory region of this gene. In addition, the mechanism by which cAMP inhibits gene transcription is unexplained but presumably is also acting through a response element in the hepatic promoter region of the gene.

B. UPSTREAM PROMOTER REGULATION

The upstream promoter region of the glucokinase gene is both spatially separate and functionally distinct from the downstream promoter region. Various structural features also set the two promoter regions apart. First, the pattern of transcription initiation from the upstream promoter region is distinct. While transcription initiation from the downstream region occurs over a 4 to 5 bp region, transcription initiation from the upstream region is even more diffuse. Using RNA isolated from solid rat insulinoma tumors in both RNAse protection and primer extension assays, transcription initiation was mapped to a 60 to 65 bp region that included multiple start sites. This wide region of transcripton initiation may be explained by the absence of any TATA homology in the proximal promoter region. In this regard, the transcription initiation pattern is similar to that of other genes that lack a TATA box. Second, the sequences of the two different DNA regions are distinct. DNA sequence information up to 2.3 kilobase pairs 5' of the transcription initiation sites has been obtained. Inspection of this DNA sequence shows no detectable homologies with the downstream promoter region. However, examination of the upstream promoter sequence revealed a region that was similar at 7 of 8 bp to an element that occurs in the promoters of the insulin genes.[41] Subsequent studies have shown that the regulation of the glucokinase promoter is distinct from that of the insulin genes.[52] Indeed, since the glucokinase gene is also expressed in the pituitary, whereas the insulin gene is not, a different set of regulatory factors must be utilized to direct this different pattern of tissue expression. With the exception of the pituitary, where the upstream promoter region is utilized for mRNA expression, no other tissue has been found to express this gene.

Fusion gene experiments have revealed that a small fragment of the upstream promoter region is sufficient for expression in cultured β cell lines. Indeed, a 294-bp fragment of DNA containing this region appears to be all that is necessary for specific expression in β cells.[52] Within this DNA fragment are several novel elements that appear to contribute to transcription in β cells.[52] Moreover, in the islets of euglycemic rats, glucokinase appears to be expressed specifically in the β cell as studied by immunohistochemical analysis.[53] Interestingly, glucokinase immunoreactivity among different pancreatic β cells is not uniform. Some β cells display striking amounts of glucokinase immunoreactivity while others have much less.[53] Evidence is accumulating in support of the functional heterogeneity of pancreatic β cells although the basis for this heterogeneity is not known.[54,55] However, the observation of variable glucokinase immunoreactivity itself is important since different amounts of glucokinase activity in different β cells may cause the glucose sensitivity of individual cells to be different.

Hormonal effects on the activity of the upstream glucokinase promoter region have not been observed; however, this has been more difficult to study given the lower abundance of islet glucokinase and the general difficulties encountered in isolating pancreatic islets. However, there is evidence that exercise training has a pronounced effect on islet glucokinase mRNA levels. Koranyi et al.[56] determined that rats which had undergone an exercise training regimen have decreased expression of both proinsulin and glucokinase mRNAs. Whether the change in the amount of glucokinase mRNA in response to exercise is due to a change in the rate of gene transcription or due to an alteration in the stability of the mRNA has not been determined.

Regulation of glucokinase expression in the islet may not necessarily be occurring at the level of gene transcription. For instance, the different 5′ noncoding sequences in the hepatic and islet glucokinase mRNAs might allow for regulation at the translational level. The length of the leader sequence is different in the liver and islet, which might affect the translational efficiencies of the mRNAs. Moreover, the islet mRNA has two ATGs upstream of the actual initiator ATG. The upstream ATGs have poor Kozac consensus sequences but may act to diminish translational efficiency of the RNA.[41]

VI. MULTIPLE GLUCOKINASE ISOFORMS

The existence of different glucokinase isoforms, established by the cloning and comparison of hepatic and islet glucokinase cDNAs, raised questions about whether the structural differences had any functional consequences.

The cell type-specific isoforms of glucokinase, produced by the different translation initiation codons encoded within the different first exons in the glucokinase gene, have different amino acid sequences at their amino terminal ends. Interestingly, though, both the islet and hepatic isoforms of glucokinase are still exactly the same length despite having a portion of their sequence encoded by different first exons. Only 3 of the 15 amino terminal amino acids are conserved in both the liver and islet glucokinase isoforms, as shown in Figure 3. While it has not been established whether the amino acids at the N terminus of the different glucokinase isoforms are of any functional importance, such differences could conceivably affect either the kinetics of the enzyme or its partitioning into distinct subcellular domains.[37,57] For instance, the amino terminal sequences of hexokinase I enable it to bind to the outer mitochondrial membrane protein by interactions with a voltage-dependent anion channel called porin.[58] However, despite differences at the amino terminus in the liver and islet glucokinase isoforms, direct immunolocalization studies using different antibodies to the enzyme indicate that the enzyme is located in the cytoplasm of both the hepatocyte and pancreatic β cell,[53] as has been previously thought.

In addition to the tissue-specific glucokinase isoforms related to the alternate first exons in the glucokinase gene, other isoforms are produced by additional alternate RNA splicing events involving the glucokinase pre-mRNA. One such example of this is the glucokinase cDNA cloned from insulinoma tissue.[41] The cDNA isolated was missing 51 bp in the portion coded by exon 4 due to the use of an alternate splice acceptor site. The cognate RNA encodes a glucokinase isoform that has an in-frame deletion of 17 amino acids. This finding generated two questions that needed to be answered. First, does this isoform have different kinetic features, and secondly, if it does, was this a potential means of regulating the activity of glucokinase? The impact of the alternate splicing on glucokinase activity became apparent in experiments that expressed the different isoforms.

A. ACTIVITY DIFFERENCES OF GLUCOKINASE ISOFORMS

Starting with glucokinase cDNAs isolated from rat liver (encoding the isoform termed L1) and a cDNA isolated from insulinoma (encoding the isoform termed B2) it was possible to make chimeric cDNAs that encoded two additional isoforms (termed L2 and B1) by digesting and ligating the different fragments obtained via a *Pst I* site in the cDNA. The use of the four different cDNAs made it possible to test the effect of the different amino terminal sequences (L1 compared with B1) and test the effect of the deletion due to the use of an alternate splice acceptor site in exon 4 (e.g., L1 vs. L2 and B1 vs. B2). The coding regions of these glucokinase isoforms were inserted into a eukaryotic expression vector that utilized the strong cytomegalovirus intermediate promoter to drive expression of these cDNAs in NIH-3T3 cells. The activity and kinetics of the expressed glucokinase activities

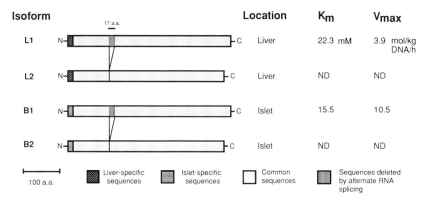

FIGURE 5.   Kinetics analysis of four different rat glucokinase isoforms produced by alternate RNA splicing. The coding nucleotide sequences for each of four different glucokinase isoforms were cloned into the eukaryotic expression vector pCMV4 and transfected into NIH-3T3 cells. The K_m and V_{max} for each isoform was determined measuring glucose phosphorylation at different glucose concentrations.[45] An endogenous low K_m hexokinase activity was present in the NIH-3T3 cells but did not affect the activity determinations for glucokinase. ND = not detected.

were then determined.[45] The results obtained (shown with the corresponding isoform in Figure 5) indicated that both the L1 and B1 isoforms were enzymatically active while the L2 and B2 isoforms were inactive. The K_ms of the enzymatically active L1 and B1 isoforms were comparable, but a more substantial difference was observed in the V_{max}s of the enzyme expressed in this manner. However, the differences in the V_{max} of the L1 and B1 isoforms could be due to the expression of different amounts of protein since the mRNAs encoding these isoforms have different sequences surrounding their translation initiation codons. The differences in the V_{max} and K_m of the L1 and B1 isoforms were small compared to the effect of the alternate splice acceptor site in the 4th exon of the gene. Selection of this splice site resulted in a glucokinase isoform without apparent catalytic activity. This was an important findng since the generation of aberrant glucokinase isoforms that were not functional could diminish glucose phosphorylation and glycolysis in the β cell and liver, thereby affecting glucose sensing and glucose usage, respectively.

B. ABERRANT GLUCOKINASE ISOFORMS IN LIVER AND ISLET

The identification of aberrant glucokinase isoforms due to the use of an alternate splice acceptor site in the 4th exon of the gene was only the first example of such events (see summary in Table 3). Subsequently, another aberrant glucokinase isoform, produced by splicing from an alternate donor site in the 2nd exon of the gene, has been identified. Hayzer and Iynedjian[59] cloned a glucokinase cDNA from a rat liver library that inserts sequence from

TABLE 3
Alternate RNA Splicing Events Involved in the Production
of Glucokinase mRNA that Have a Deleterious
Effect on Enzyme Activity

Location protein in gene (a.a.)	Tissues involved	No. nucleotides added or deleted	Frameshift	Size of product
Exon 2A	Liver[1]	+ 151	Yes	498
Exon 2	Pituitary[2]	− 25	Yes	68
Exon 2	Pituitary Liver[1]	− 52	Yes	58
Exon 4	β cells Liver Pituitary[2]	− 51	No	448

Note: The origin of each splicing variant is discussed in the text.

[1] or [2] indicates that these two alternative splice sites were utilized in a single mRNA.

From Magnuson, M. A., *J. Cell. Biochem.*, 48, 115, 1992. With permission.

a cassette exon (termed 2A) immediately before the sequence provided by exon 2. This exon is used in approximately 5% of the glucokinase mRNA present in liver. The use of the cassette exon, which adds 151 bases, occurred in conjunction with the use of an alternate splice donor site in exon 2 which removes 52 bases from the mRNA. Together, the insertion/deletion events add 87 new residues to the reading frame of the enzyme resulting in a new isoform of 498 amino acids. When this isoform was expressed in bacteria it did not phosphorylate glucose.[60]

C. MULTIPLE HUMAN LIVER GLUCOKINASE ISOFORMS

Recently, Tanizawa et al.[61] have cloned two human liver glucokinase cDNAs and found that a similar cassette exon is utilized to produce two different hepatic glucokinase isoforms, both of which are enzymatically active. The human liver glucokinase isoform designated hLGLK1[61] has 97% homology with rat liver glucokinase, although the 14 N terminal residues are entirely different due to the use of a different translation initiation codon in the cassette exon sequences. Although the insertion of a cassette exon at the junction of exons 1 and 2 is similar to the situation described in rat liver,[59] the size of the exon is different, 124 bp, and its sequence is not homologous to the rat exon 2A. The human liver isoform hLGLK2, which is produced from a mRNA lacking the cassette exon, is 98% homologous to rat liver glucokinase[61] and has sequences at the amino terminal end which are very similar to those in rat. The relative abundance of the two different human glucokinase isoforms has not been determined.

D. ABERRANT SPLICING OF GLUCOKINASE mRNA IN PITUITARY

The functional significance of aberrant glucokinase mRNAs in the islet and liver is unresolved. However, the finding of aberrant glucokinase mRNAs in pituitary may offer an explanation why such forms are found in the liver and islet. Historically, glucokinase activity has been identified only in the liver and in the pancreatic islet. Therefore, it was a surprise when Newgard and his colleagues found that AtT-20 cells, a mouse corticotroph-derived cell line, were a rich source of glucokinase mRNA.[42] This observation raised the issue of whether or not glucokinase mRNA and glucokinase activity were present in pituitary. Studies reported by two groups have addressed somewhat different aspects of this question.

Hughes et al.[42] analyzed anterior pituitary RNA for the presence of a glucokinase mRNA by Northern blot analysis and found a transcript that was of similar size to that in islet. They then amplified glucokinase cDNAs from AtT-20 cells, a corticotroph-derived cell line, and pituitary and determined the sequence of both. The glucokinase cDNAs obtained in their experiments showed that alternate RNA splicing of the glucokinase mRNA was occurring in the pituitary that utilized the alternate splice donor site in exon 2 which was described first by Hayzer and Iynedjian.[59] Of three AtT-20 cell PCR-derived clones studied, two utilized the alternate acceptor site while one did not. However, of three pituitary clones examined, all utilized the alternate acceptor site. The alternate splice donor site in exon 2, when used by itself, shifts the reading frame of the enzyme and predicts a peptide of only 58 amino acids, far too small to be catalytically active.

Liang et al.[45] addressed the question of glucokinase mRNA and activity in AtT-20 cells and pituitary both by measuring glucokinase activity and amplifying glucokinase mRNA from pituitary. In their studies no glucokinase activity could be detected in either AtT-20 cells or pituitary, even after partial purification by a chromatographic purification step. Amplification of a pituitary glucokinase cDNA and determination of its sequence showed a different pattern of alternate splice sites as shown in Figure 6. Of 3 PCR-derived cDNA clones that were sequenced, all showed the use of an alternate splice donor site in exon 2 that removes 25 bases and shifts the reading frame. In addition, the same clones all used the same alternate splice acceptor site in exon 4 that was identified previously as occurring in rat insulinoma tissue. The mRNA cognate to this cDNA clone, if translated, would produce only a 68 amino acid peptide. The absence of glucokinase in pituitary has also been addressed by immunohistochemical staining of pituitary for the protein using antibodies that detect the enzyme in liver and islet. In these studies, no immunoreactive glucokinase was detected.[62]

While the actual alternate splice sites identified by the different groups who have performed these studies differ, in no case has a glucokinase mRNA encoding a normal glucokinase isoform been identified from pituitary tissue.

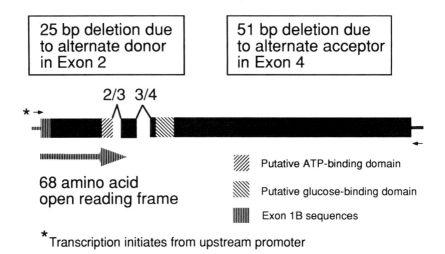

FIGURE 6. Pituitary glucokinase cDNA: alternate RNA splicing disrupts the reading frame for glucokinase. Glucokinase cDNAs from rat pituitary tissue were obtained by RNA-PCR as described previously.[45] Two different alternate splicing events were identified as illustrated. Either alternate splicing event, by itself, would be sufficient to prevent the production of an active glucokinase isoform.

However, the situation with the radiation-induced AtT-20 cells is more confusing but may simply reflect the phenotypic divergence of this cell line into different sublines.

VII. CONCLUSIONS

Biochemical studies over at least the last 25 years have established the importance of glucokinase in the metabolism of glucose by the liver and β cell. Studies within the last 5 years that have examined the molecular control points of this system have shown that glucokinase gene expression is under multiple levels of regulation, as is shown diagramatically in Figure 7. Most striking in this regard is the finding of alternate promoter regions in a single glucokinase gene that impart differential cell type-specific regulation to the enzyme. An upstream promoter region in the gene drives expression of glucokinase mRNA in both the pancreatic β cell and the pituitary while the downstream promoter region in the gene controls expression of glucokinase mRNA in the liver. This gene organization allows a different set of regulatory factors to be involved in both tissues, presumably allowing the tissue-specific control of glucokinase gene expression.

In addition to the use of alternate promoter regions in the glucokinase gene, alternate RNA processing contributes in an important way to the regulation of glucokinase activity. While glucokinase mRNAs are produced in

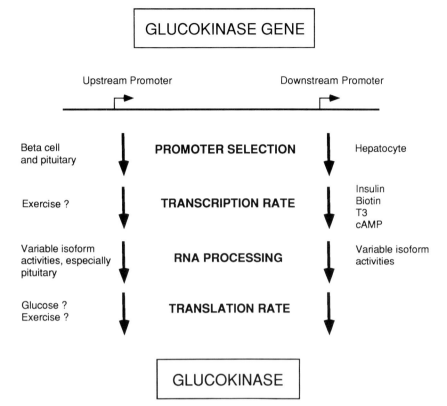

FIGURE 7. Multiple control points affecting glucokinase gene expression. A diagram showing the flow of information from the glucokinase gene to protein. The sites at which regulation of glucokinase gene expression has been identified are illustrated.

the liver, β cell, and pituitary, posttranscriptional regulation through the selection of alternate splice donor and acceptor sites, in addition to the use of a cassette exon, have the potential to alter the reading frame of the enzyme and lead either to early termination events or to the production of catalytically inactive isoforms. This effect is most pronounced in the pituitary where glucokinase protein and activity is not produced despite generation of a mRNA.

The mechanisms underlying the tissue-specific expression and regulation of the glucokinase gene expression are just beginning to be studied in depth. Undoubtedly studies of these mechanisms will provide a fertile area for future investigation. Since glucokinase is a key regulatory control point for the response of the β cell to glucose and for the utilization of glucose by the liver, an understanding of the mechanisms involved is important. As additional knowledge is obtained, it is likely that additional complexities will be discovered relating to the interplay of nutritional, neuroendocrine, and genetic factors that act to determine glucokinase gene expression.

ACKNOWLEDGMENTS

These studies were supported by grants from the National Institutes of Health and the Juvenile Diabetes Foundation. We thank our colleagues and collaborators for their contributions to the studies which are summarized here.

REFERENCES

1. **Granner, D. and Pilkis, S.,** The genes of hepatic glucose metabolism, *J. Biol. Chem.,* 265, 10173, 1990.
2. **Iynedjian, P. B. and Girard, J.,** Nomenclature for mammalian glucokinase, *Biochem. J.,* 275, 821, 1990.
3. **Ureta, T.,** The comparative isozymology of vertebrate hexokinases, *Comp. Bichem. Physiol.,* 71B, 549, 1982.
4. **Cornish-Bowden, A. and Cardenas, M. L.,** Hexokinase and glucokinase in liver metabolism, *Trends Biochem. Sci.,* 16, 281, 1991.
5. **Weinhouse, S.,** Regulation of glucokinase in liver, in *Current Topics in Cellular Regulation,* Vol. 11, Horecker, B. L. and Stadtman, E. R., Eds., Academic Press, New York, 1976, 1.
6. **Lawrence, G. M. and Trayer, I. P.,** Hexokinase isoenzymes: Antigenic cross-reactivities and amino acid copositional relatedness, *Comp. Biochem. Physiol.,* 79B, 233, 1984.
7. **Andreone, T. L., Printz, R. L., Pilkis, S. J., Magnuson, M. A., and Granner, D. K.,** The amino acid sequence of rat liver glucokinase deduced from cloned cDNA, *J. Biol. Chem.,* 264, 363, 1989.
8. **Kopetzki, E., Entian, K.-D., and Mecke, D.,** Complete nucleotide sequence of the hexokinase PI gene (HXK1) of *Saccharomyces cervisiae, Gene,* 39, 95, 1985.
9. **Stachelek, C., Stachelek, J., Swan, J., Botstein, D., and Konigsberg, W.,** Identification, cloning, and sequence determination of the genes specifying hexokinase A and B from yeast, *Nucleic Acids Res.,* 14, 1986.
10. **Nishi, S., Seino, S., and Bell, G. I.,** Human hexokinase: sequences of amino- and carboxyl-terminal halves are homologous, *Biochem. Biophys. Res. Commun.,* 157, 937, 1988.
11. **Schwab, D. A. and Wilson, J. E.,** Complete amino acid sequence of rat brain hexokinase, deduced from the cloned cDNA, and proposed structure of a mammalian hexokinase, *Proc. Natl. Acad. Sci. U.S.A.,* 86, 2563, 1989.
12. **Arora, K. K., Fanciulli, M., and Pedersen, P. L.,** Glucose phosphorylation in tumor cells, *J. Biol. Chem.,* 265, 6481, 1990.
13. **Bennett, W. S. and Steitz, T. A.,** Structure of a complex between yeast hexokinase A and glucose, *J. Mol. Biol.,* 140, 211, 1980.
14. **Lange, A. L., Xu, L. Z., Poelwijk, F. V., Lin, K., Granner, D. K., and Pilkis, S. J.,** Expression and site-directed mutagenesis of hepatic glucokinase, *Biochem. J.,* 277, 159, 1991.
15. **Iynedjian, P. B., Mobius, G., Seitz, H. J., Wollheim, C. B., and Renold, A. E.,** Tissue-specific expression of glucokinase: Identification of the gene product in liver and pancreatic islets, *Proc. Natl. Acad. Sci. U.S.A.,* 83, 1998, 1986.
16. **Thorens, B., Sarkar, H. K., Kaback, H. R., and Lodish, H. F.,** Cloning and functional expression in bacteria of a novel glucose transporter present in liver, intestine, kidney, and beta-pancreatic islet cells, *Cell,* 55, 281, 1988.

17. **Mueckler, M.,** Family of glucose-transporter genes. Implications for glucose homeostasis and diabetes, *Diabetes,* 39, 6, 1990.

18. **James, D. E., Brown, B., Navarro, J., and Pilch, P. F.,** Insulin-regulatable tissues express a unique insulin-sensitive glucose transport protein, *Nature (London),* 330, 183, 1988.

19. **Vilaro, S., Palacin, M., Pilch, P. F., Testar, X., and Zorzano, A.,** Expression of an insulin regulatable glucose carrier in muscle and fat endothelial cells, *Nature (London),* 342, 792, 1989.

20. **Matschinsky, F. M. and Ellerman, J. E.,** Metabolism of glucose in the islets of Langerhans, *J. Biol. Chem.,* 243, 2730, 1968.

21. **Malaisse, W. J., Sener, A., Herchuelz, A., and Hutton, J. C.,** Insulin release: the fuel hypothesis, *Metabolism,* 28, 373, 1979.

22. **Malaisse, W. J.,** Insulin release: the fuel concept, *Diabete Metab.,* 9, 313, 1983.

23. **Orci, L., Thorens, B., Ravazzola, M., and Lodish, H. F.,** Localization of the pancreatic beta cell glucose transporter to specific plasma membrane domains, *Science,* 245, 295, 1989.

24. **Chen, L., Alam, T., Johnson, J. H., Hughes, S., Newgard, C. B., and Unger, R. H.,** Regulation of β-cell glucose transporter gene expression, *Proc. Natl. Acad. Sci. U.S.A.,* 87, 4088, 1990.

25. **Giroix, M.-H., Sener, A., Pipeleers, D. G., and Malaisse, W. J.,** Hexose metabolism in pancreatic islets, *Biochem. J.,* 223, 447, 1984.

26. **Sener, A., Malaisse-Lagae, F., Giroix, M.-H., and Malaisse, W. J.,** Hexose metabolism in pancreatic islets: Compartmentation of hexokinase in islet cells, *Arch. Biochem. Biophys.,* 265, 61, 1986.

27. **Shimizu, T., Knowles, B. B., and Matschinsky, F. M.,** Control of glucose phosphorylation and glucose usage in clonal insulinoma cells, *Diabetes,* 37, 563, 1988.

28. **Shimizu, T., Parker, J. C., Najafi, H., and Matschinsky, F. M.,** Control of glucose metabolism in pancreatic β-cells by glucokinase, hexokinase, and phosphofructokinase, *Diabetes,* 37, 1524, 1988.

29. **Liang, Y., Najafi, H., and Matschinsky, F. M.,** Glucose regulates glucokinase activity in cultured islets from rat pancreas, *J. Biol. Chem.,* 265, 16863, 1990.

30. **Meglasson, M. D. and Matschinsky, F. M.,** New perspectives on pancreatic islet glucokinase, *Am. J. Physiol.,* 13, E1, 1984.

31. **Unger, R. H.,** Diabetic hyperglycemia: link to impaired glucose transport in pancreatic β cells, *Science,* 251, 1200, 1991.

32. **Ashcroft, F. M., Harrison, D. E., and Ashcroft, S. J. H.,** Glucose induces closure of single potassium channels in isolated rat pancreatic β-cells, *Nature (London),* 312, 446, 1984.

33. **Cook, D. L., Satin, L. S., Ashford, M. L. J., and Hales, C. N.,** ATP-sensitive K^+ channels in pancreatic β-cells, *Diabetes,* 37, 495, 1988.

34. **Boyd, A. E.,** Sulfonylurea receptors, ion channels, and fruit flies, *Diabetes,* 37, 847, 1988.

35. **Iynedjian, P. B., Pilot, P. R., Nouspikel, T., Milburn, J. L., Quaade, C., Hughes, S., Ucla, C., and Newgard, C. B.,** Differential expression and regulation of the glucokinase gene in liver and islets of Langerhans, *Proc. Natl. Acad. Sci. U.S.A.,* 86, 7838, 1989.

36. **Bedoya, F. J., Matschinsky, F. M., Shimizu, T., ONeil, J. J., and Appel, M. C.,** Differential regulation of glucokinase activity in pancreatic islets and liver of the rat, *J. Biol. Chem.,* 261, 10760, 1986.

37. **Magnuson, M. A.,** Glucokinase gene structure: functional implications of molecular genetic studies, *Diabetes,* 39, 523, 1990.

38. **Matschinsky, F. M.,** Glucokinase as glucose sensor and metabolic signal generator in pancreatic beta-cells and hepatocytes, *Diabetes,* 39, 647, 1990.

39. **Iynedjian, P. B., Ucla, C., and Mach, B.,** Molecular cloning of glucokinase cDNA, *J. Biol. Chem.,* 262, 6032, 1987.

40. **Chien, C.-T., Tauler, A., Lange, A. J., Chan, K., Printz, R. L., El-Maghrabi, M. R., Granner, D. K., and Pilkis, S. J.,** Expression of rat hepatic glucokinase in *Escherichia coli, Biochem. Biophys. Res. Commun.,* 165, 817, 1989.

41. **Magnuson, M. A. and Shelton, K. D.,** An alternate promoter in the glucokinase gene is active in the pancreatic beta cell, *J. Biol. Chem.,* 264, 15936, 1989.

42. **Hughes, S. D., Quaade, C., Milburn, J. L., Cassidy, L., and Newgard, C. B.,** Expression of normal and novel glucokinase mRNAs in anterior pituitary and islet cells, *J. Biol. Chem.,* 266, 4521, 1991.

43. **Magnuson, M. A., Andreone, T. L., Printz, R. L., Koch, S., and Granner, D. K.,** Rat glucokinase gene: Structure and regulation by insulin, *Proc. Natl. Acad. Sci. U.S.A.,* 86, 4838, 1989.

44. **Niswinder, K. and Magnuson, M. A.,** unpublished data.

45. **Liang, Y., Jetton, T. L., Zimmerman, E., Najafi, H., Matschinsky, F. M., and Magnuson, M. A.,** Effects of alternate RNA splicing on glucokinase isoform activities in the pancreatic islet, liver, and pituitary, *J. Biol. Chem.,* 266, 6999, 1991.

46. **Noguchi, T., Takenaka, M., Yamada, K., Matsuda, T., Hashimoto, M., and Tanaka, T.,** Characterization of the 5′ flanking region of rat glucokinase gene, *Biochem. Biophys. Res. Commun.,* 164, 1247, 1989.

47. **Iynedjian, P., Gjinovci, A., and Renold, A. E.,** Stimulation by insulin of glucokinase gene transcription in liver of diabetic rats, *J. Biol. Chem.,* 263, 740, 1988.

48. **Hoppner, W. and Seitz, H.-J.,** Effect of thyroid hormones on glucokinase gene transcription in rat liver, *J. Biol. Chem.,* 264, 20643, 1989.

49. **Narkewicz, M. R., Iynedjian, P. B., Ferre, P., and Girard, J.,** Insulin and triiodothyronine induce glucokinase mRNA in primary culture of neonatal rat hepatocytes, *Biochem. J.,* 271, 1990.

50. **Chauhan, J. and Dakshinamurti, K.,** Transcriptional regulation of the glucokinase gene by biotin in starved rats, *J. Biol. Chem.,* 266, 10035, 1991.

51. **Iynedjian, P. B., Jotterand, D., Nouspikel, T., Asfari, M., and Pilot, P.-R.,** Transcriptional induction of glucokinase gene by insulin in cultured liver cells and its repression by the glucagon-cAMP system, *J. Biol. Chem.,* 264, 21824, 1989.

52. **Shelton, K. D., Franklin, A. J., Khoor, A., Beechem, J., and Magnuson, M. A.,** Multiple elements in the upstream glukokinase promoter contribute to transcription in insulinoma cells, *Mol. Cell. Biol.,* in press.

53. **Jetton, T. L. and Magnuson, M. A.,** Heterogeneous expression of glucokinase among different pancreatic β-cells, *Proc. Natl. Acad. Sci. U.S.A.,* 89, 2619, 1992.

54. **Schuit, F. C., IntVeld, P. A., and Pipeleers, D. G.,** Glucose stimulates proinsulin biosynthesis by a dose-dependent recruitment of pancreatic beta cell, *Proc. Natl. Acad. Sci. U.S.A.,* 85, 3865, 1988.

55. **Solomon, D. and Meda, P.,** Heterogeneity and contact-dependent regulation of hormone secretion by individual beta cells, *Exp. Cell Res.,* 162, 507, 1986.

56. **Koranyi, L. I., Bourey, R. E., Slentz, C. A., Holloszy, J. O., and Permutt, M. A.,** Coordinate reduction of rat pancreatic islet glucokinase and proinsulin mRNA by exercise training, *Diabetes,* 40, 401, 1991.

57. **Newgard, C. B., Quaade, C., Hughes, S. D., and Milburn, J. L.,** Glucokinase and glucose transporter expression in liver and islets: implications for control of glucose homeostasis, *Biochem. Soc. Trans.,* 18, 851, 1990.

58. **Nakashima, R. A., Mangan, P. S., Columbini, M., and Pedersen, P. L.,** Hexokinase receptor complex in hepatoma mitochondria: Evidence from N,N′-dicyclohexylcarbodiimide-labeling studies for the involvement of the pore-forming protein VDAC, *Biochemistry,* 25, 1986.

59. **Hayzer, D. J. and Iynedjian, P. B.,** Alternative splicing of glucokinase mRNA in rat liver, *Biochem. J.,* 270, 261, 1990.

60. **Quaade, C., Hughes, S. D., Coats, W. S., Sestak, A. L., Iynedjian, P. B., and Newgard, C. B.,** Analysis of the protein products encoded by variant glucokinase transcripts via expression in bacteria, *FEBS Lett.,* 280, 47, 1991.
61. **Tanizawa, Y., Koranyi, L. I., Welling, C. M., and Permutt, M. A.,** Human liver glucokinase gene: cloning and sequence determination of two alternately spliced cDNAs, *Proc. Natl. Acad. Sci. U.S.A.,* 88, 7294, 1991.
62. **Jetton, T. L. and Magnuson, M. A.,** unpublished data.
63. **Magnuson, M. A.,** Tissue-specific regulation of glucokinase gene expression, *J. Cell. Biochem.,* 48, 115, 1992.

Chapter 7

DIETARY AND HORMONAL REGULATION OF L-TYPE PYRUVATE KINASE GENE EXPRESSION

Tamio Noguchi and Takehiko Tanaka

TABLE OF CONTENTS

0-8493-6961-4/93/$0.00 + $.50

169

I. OVERVIEW

Pyruvate kinase (PK), a key glycolytic enzyme, catalyzes formation of pyruvate and ATP from phosphoenol-pyruvate and ADP. Four isozymes of this enzyme have been identified in mammals.[1] They are named L-, R-, M_1-, and M_2-types, and differ in primary structure, kinetic properties, and tissue distribution.[1] However, the L- and R-types, and the M_1- and M_2-types are encoded by the same genes called the PKL[2] and PKM[3,4] gene, respectively. Both genes consist of 12 exons and 11 introns. The introns interrupt the protein coding sequences of the four isozymes at identical positions. The first (R) and second (L) exons of the PKL gene contain 5′ noncoding and amino terminal sequences specific to the R- and L-types, respectively, while the remaining downstream exons are common to the two isozymes. By use of two alternative promoters, which are located upstream of the type-specific exons, the R- and L-type isozymes are expressed in a tissue-specific manner. On the other hand, the 9th (M_1) and 10th (M_2) exons of the PKM gene encode sequences specific to the M_1- and M_2-types, respectively, while the remaining exons are common to the two isozymes. Thus, the M_1- and M_2-type isozymes are produced from the same transcript of the M gene by alternative RNA splicing, which occurs tissue specifically.

The L-type PK is largely expressed in the liver.[1] Gene expression of this enzyme is regulated by hormones and diet[1] like the expression of genes of other key enzymes involved in glycolysis and lipogenesis, including glucokinase (GK), glucose-6-phosphate dehydrogenase, malic enzyme, acetyl CoA carboxylase, and fatty acid synthase.[5] A high glucose diet or insulin induces expression of these enzymes, while starvation, diabetes, or glucagon suppresses their expression. Dietary fructose, which is known to stimulate lipogenesis more than dietary glucose,[6] also induces all these enzymes except glucokinase,[7-9] Its effect is observed even in diabetic liver in contrast to that of dietary glucose. Since fructokinase, which catalyzes the first step of fructose metabolism, is not regulated by hormones or diet,[10] the L-type PK is also a key enzyme of fructose metabolism as well as glucose metabolism. Thus, this enzyme is very important in not only glycolysis but also lipogenesis in the liver. In this chapter, we describe the molecular mechanism of induction of the L-type PK by insulin and dietary fructose.

II. REGULATION OF L-TYPE PK GENE EXPRESSION IN NORMAL AND DIABETIC RATS

The L-type PK mRNA level is markedly reduced in streptozotocin-induced diabetes in the liver of rats fed a high dextrose (glucose) diet.[11,12] Insulin administration to diabetic rats results in an increase in the level of the L-type PK mRNA in the liver.[11,12] As seen in Figure 1, this response shows

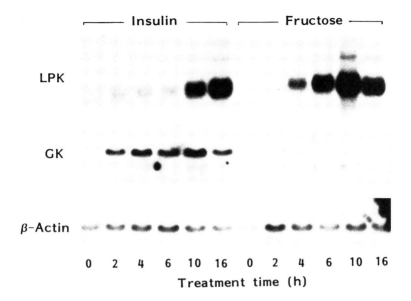

FIGURE 1. Time courses of induction by insulin or dietary fructose of L-type pyruvate kinase and glucokinase mRNAs in diabetic rat liver. LPK, L-type pyruvate kinase; GK, glucokinase.

a characteristic delay, the increase becoming detectable only after 6 h,[13] whereas induction of GK mRNA is nearly maximal 2 h after insulin injection.[14,18] These findings suggest that the mechanisms of induction of the two enzymes are different. Similar time courses of induction of the two mRNAs in normal rat liver have been observed after glucose feeding.[15,16] Dietary fructose also increases the level of the L-type PK mRNA in diabetic liver,[17] but not that of GK mRNA (Figure 1).[18] The induction of the L-type PK mRNA by fructose occurs faster than that by insulin. A similar response of the L-type PK mRNA to dietary fructose has been observed in normal rat liver.[16] The effect of fructose on the L-type PK seems to be directly on the liver because the possibility that hormones such as glucagon or adrenal or thyroid hormones mediate the fructose effect is excluded.[13] Dietary glycerol is also a potent inducer of the L-type PK in normal and diabetic rat liver (Table 1).[7,17,19] Thus, we consider that the effect of fructose is attributable to an intermediate common to the metabolism of both fructose and glycerol.

The L-type PK is expressed not only in the liver, but also in kidney and small intestine.[1] The major sites for the metabolism of fructose are known to be the liver, kidney, and small intestine,[6] whereas glycerol is metabolized appreciably only in the liver and kidney.[20] Therefore, it was of interest to examine the effects of various carbohydrates on the levels of the L-type PK mRNA in the kidney and small intestine of normal rats. As shown in Table 1, dietary fructose induced an increase in the L-type PK mRMA level in these

TABLE 1
Effects of Various Carbohydrates on the Level of L-Type Pyruvate Kinase mRNA in Liver, Kidney, and Small Intestine

Carbohydrate	Amount of mRNA (fold change)		
	Liver	Kidney	Small intestine
Control	1.0 ± 0.1	1.0 ± 0.6	1.0 ± 0.1
Fructose	5.1 ± 1.4	4.6 ± 1.9	3.0 ± 0.2
Glycerol	9.6 ± 6.0	3.2 ± 0.9	1.0 ± 0.2
Glucose	1.2 ± 0.2	1.0 ± 0.2	1.6 ± 0.1
Galactose	1.5 ± 0.5	0.7 ± 0.2	0.7 ± 0.1
Sorbitol	1.0 ± 0.5	0.6 ± 0.2	0.6 ± 0.2
Xylose	0.9 ± 0.2	0.5 ± 0.1	0.3 ± 0.1
Xylitol	0.9 ± 0.3	0.5 ± 0.3	0.5 ± 0.2

Note: Normal rats were used for determination of the mRNA in the kidney and small intestine, and diabetic rats for its determination in the liver. Rats were starved overnight and then given the indicated diets for 6 h. Control rats were killed at zero time. Data are normalized to the value for control rats.

tissues.[19] The time courses of these inductions were very similar to that observed in diabetic liver. Glycerol feeding also increased the mRNA level in the kidney after 6 h, but did not affect the mRNA level in the small intestine. Glucose induced a slight but significant (1.6-fold, $p < 0.001$) increase in the level of the L-type PK mRNA in the small intestine, but not in the kidney. No other carbohydrates, including galactose, sorbitol, xylose, and xylitol, were effective inducers of the L-type PK mRNA in the kidney or small intestine or in the liver of diabetic rats. Thus, the tissues in which fructose and glycerol induce the L-type PK mRNA correspond to the tissues in which these compounds are metabolized, supporting the metabolite hypothesis mentioned above. The candidate molecule may not accumulate significantly during metabolism of any other carbohydrate except glucose.

We examined the mechanism of induction of the L-type PK mRNA by insulin and dietary carbohydrates in diabetic rat liver using a nuclear transcription assay. No significant change in gene transcription was observed 3 h after insulin treatment.[13] As shown in Figure 2, however, insulin enhanced the rate of transcription of this gene about 28-fold after 12 h.[13,18] This stimulation was inhibited by cycloheximide, suggesting that insulin may stimulate transcription of the L-type PK gene by stimulating the synthesis of some unknown protein. On the other hand, only four- to sixfold stimulation of the gene transcription was observed 4 h after fructose or glycerol feeding.[18,19] On the contrary, Munnich et al. reported that dietary fructose stimulated transcription of the L-type PK gene in the liver of normal rats.[16] We tested the

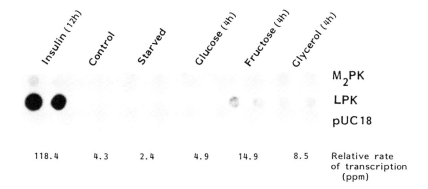

FIGURE 2. Effects of insulin, fructose, and glycerol on transcription of pyruvate kinase genes. LPK, L-type pyruvate kinase; M_2PK, M_2-type pyruvate kinase.

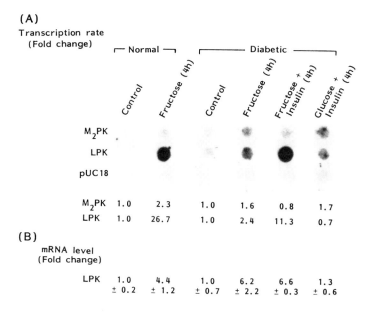

FIGURE 3. Effects of fructose on transcription of pyruvate kinase genes in normal and diabetic liver. Data are normalized to the value for control rats. LPK, L-type pyruvate kinase; M_2PK, M_2-type pyruvate kinase.

possibility that this difference was due to a difference in the condition of the animals used, that is, diabetic or normal.[19] Feeding fructose for 4 h resulted in only 2.4-fold stimulation of the transcription of the L-type PK gene in diabetic rats, but about 27-fold increase in its transcription in normal rats (Figure 3). In contrast, the extent of increase in the L-type PK mRNA levels was similar in these two conditions. To determine whether this difference in the responses of gene transcription to fructose was due to insulin, we injected insulin 30 min after the start of fructose or glucose feeding. Insulin treatment resulted in 4.7-fold further increase in transcription in fructose-fed diabetic rats after 4 h, but no change in the transcription in glucose-fed diabetic rats, consistent with results described above. On the contrary, insulin did not increase the mRNA level further in fructose-fed rats. These results indicate that the mechanism of induction of the L-type PK in the liver by fructose is dependent on the plasma insulin level. Dietary fructose acts mainly at the transcriptional level in normal liver, whereas it acts mainly at the posttranscriptional level in the liver in the insulin-deficient state. In the latter case, we suggest that fructose stabilizes nuclear RNA species of the L-type PK mRNA since increases in the levels of nuclear RNA precursors of this enzyme after fructose feeding precede changes in the levels of cytosolic L-type PK mRNA.[17]

Another group reported that glucocorticoids and thyroid hormones were involved in dietary induction of the L-type PK mRNA at a posttranscriptional level.[21] We did not, however, observe any significant role of these compounds in the induction of the mRNA by insulin and dietary fructose.[13] Glucagon or cyclic AMP inhibited transcriptional stimulation of the L-type PK gene by insulin or dietary fructose and also the posttranscriptional effect of fructose in diabetic liver.[13,21] These effects of glucagon may be related to inhibition of carbohydrate metabolism in the liver because its metabolism is involved in not only the effects of fructose but also the effect of insulin as discussed later.

Next, we examined the rates of transcription of the L-type PK gene in the kidneys of normal and diabetic rats fed a high fructose diet for 4 h (Figure 4).[19] Dietary fructose stimulated transcription only 1.8-fold and 2.2-fold in normal and diabetic kidney, respectively. However, the magnitude of these increases was much lower than that of the mRNA. Thus, in both normal and diabetic rats, dietary fructose acts mainly at the posttranscriptional level in the kidney, as in diabetic liver. These results are understandable since dietary glucose and insulin have no effect on the expression of the L-type PK gene in the kidney.[19]

III. REGULATION OF L-TYPE PK GENE EXPRESSION IN PRIMARY CULTURES OF RAT HEPATOCYTES

Based upon *in vivo* observations, we investigated the mechanisms of action of insulin and carbohydrates further using cultured hepatocytes.[22] Insulin caused time- and dose-dependent increases in the amount of the L-type PK mRNA in cultured hepatocytes maintained in a medium containing glucose (28 mM). The maximal response to 10 nM insulin was observed after 24 to 36 h (Figure 5). The effect of insulin was dependent on the nature and concentration of the carbohydrate used (Figure 6). The addition of glucose

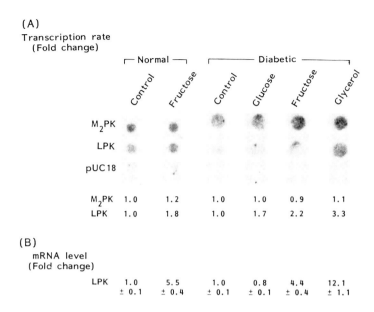

FIGURE 4. Effects of fructose and glycerol on transcription of pyruvate kinase genes in normal and diabetic kidney. Data are normalized to the control values. LPK, L-type pyruvate kinase; M_2PK, M_2-type pyruvate kinase.

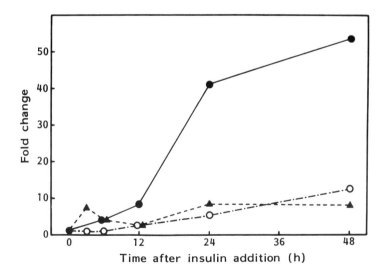

FIGURE 5. Time courses of induction by insulin of L-type pyruvate kinase mRNA in the presence of various carbohydrates. Hepatocytes were cultured with 0.1 μM insulin in the presence of 28 mM glucose (●), 5.6 mM fructose (○), or 5.4 mM glycerol (▲) for the indicated times. Data are expressed relative to the value at zero time.

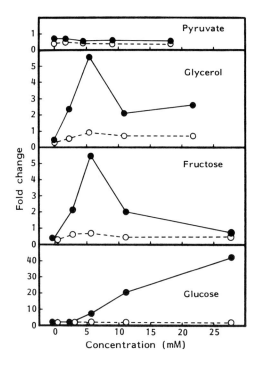

FIGURE 6.   Effects of carbohydrates at various concentrations on the level of L-type pyruvate kinase mRNA in cultured hepatocytes. Hepatocytes were incubated for 48 h with the indicated concentrations of carbohydrates in the presence (●) or absence (○) of 0.1 μ*M* insulin. Data are expressed relative to the value at zero time.

to the medium caused a marked, dose-dependent increase in the induction of the L-type PK mRNA by insulin. The additions of fructose and glycerol also enhanced the induction by insulin, but the maximal increases, observed at about 5.5 m*M* in both cases, were less than 20% of that with 28 m*M* glucose. No induction of the L-type PK mRNA by insulin was observed in hepatocytes cultured in the presence of pyruvate at any concentration examined. On the other hand, none of these carbohydrates induced the L-type PK mRNA in the absence of insulin. This was unexpected because fructose and glycerol are potent inducers of the mRNA even in diabetic liver.

The time course of induction of the L-type PK mRNA by insulin in the presence of fructose, but not of glycerol, was similar to that in the presence of glucose (Figure 5). In the presence of glycerol, the mRNA increased in a diphasic manner: the first increase reached a maximum after 3 h, whereas the second increase corresponded to the increase in the presence of glucose. The early effect of insulin/glycerol was inhibited by actinomycin D. Thus, this effect can be regarded as a partial reproduction of the fructose effect in normal liver, which occurred much faster than the insulin effect in diabetic

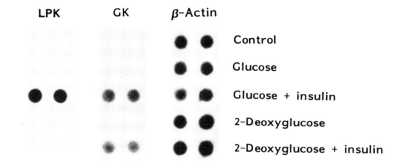

FIGURE 7.   Effects of 2-deoxyglucose on the induction by insulin of L-type pyruvate kinase and glucokinase mRNAs in cultured hepatocytes. Hepatocytes were incubated with glucose (28 m*M*) or 2-deoxyglucose (28 m*M*) in the presence or absence of insulin (0.1 μ*M*) for 36 h. Cells at zero time were used as controls. LPK, L-type pyruvate kinase; GK, glucokinase.

liver as described above. The reason why the effects of fructose in normal and diabetic liver were not reproduced in cultured hepatocytes is not known. Quantitative and/or qualitative reproduction of the *in vivo* metabolism of fructose and glycerol in a culture system may be difficult because the effect of fructose *in vivo* is attributable to a common metabolite of fructose and glycerol. In any case, these results suggest that the metabolism of glucose is required for the induction of the L-type PK mRNA by insulin. If so, the effect of insulin should not be observed in the presence of 2-deoxyglucose, a non-metabolizable glucose analog. As shown in Figure 7, this was, in fact, the case: the effect of insulin was not observed in hepatocytes cultured in a medium containing 2-deoxyglucose. This was not due to a nonspecific effect of 2-deoxyglucose because the induction of GK mRNA by insulin was not impaired by this glucose analog. In fact, the effect of insulin on GK mRNA did not require the presence of glucose itself. Thus, we suggest that some metabolite accumulating during the metabolism of glucose is necessary for the induction of the L-type PK mRNA by insulin and that this metabolite can be produced from fructose and glycerol to some extent, but not from pyruvate, in cultured hepatocytes. However, the role of insulin in the induction remains to be determined. Insulin may simply increase the concentration of this metabolite by stimulating glucose metabolism, or, in addition to this effect, another signal from insulin may also be necessary.

Consistent with *in vivo* studies, cycloheximide inhibited the induction of the L-type PK mRNA by insulin, suggesting that ongoing protein synthesis is required for the effect of insulin on the L-type PK. The addition of H-7, an inhibitor of protein kinase C, caused dose-dependent inhibition of the induction of L-type PK mRNA by insulin. However, phorbol 12-myristate 13-acetate did not increase the level of the mRNA. Thus, the effect of H-7 may not involve protein kinase C, although the possibility cannot be ruled

TABLE 2
Effects of Various Hormones on the Level
of L-Type Pyruvate Kinase mRNA in
Cultured Hepatocytes

Treatment	Amount of mRNA (fold change)	
	Pyruvate kinase	β-Actin
Control	1.0	1.0
Dexamethasone (1 μM)	0.1	1.3
Insulin (0.1 μM)	26.8	1.4
8-CPT-cAMP (0.1 mM)	0.8	2.8
T$_3$ (10 μM)	1.8	1.7
Dexamethasone + Insulin	76.9	1.5
Dexamethasone + Insulin + 8-CPT-cAMP	0.2	1.3
Dexamethasone + T$_3$	0.3	1.2
Dexamethasone + T$_3$ + Insulin	77.1	1.0

Note: Hepatocytes were incubated with various hormones at the concentrations indicated in parentheses for 36 h. The medium contained 28 mM glucose. Values are expressed relative to the control value at zero time.

out that kinase C may be involved in the action of insulin but not sufficiently so to induce the L-type PK mRNA by itself.

As mentioned above, another group reported that glucocorticoids and thyroid hormones are involved in dietary induction of the L-type PK mRNA by acting at a posttranscriptional level.[21] If this is the case in cultured hepatocytes, the additions of these hormones alone to cultured hepatocytes should cause some increase in the L-type PK mRNA level. As shown in Table 2, dexamethasone alone slightly decreased the mRNA level. This is consistent with the *in vivo* findings that adrenalectomy resulted in a slight elevation of the L-type PK mRNA level in the liver.[13] Interestingly, however, dexamethasone greatly enhanced the induction of the mRNA by insulin when added with insulin. As glucocorticoids are known to be necessary to maintain the differentiated function of hepatocytes *in vitro*,[23] this hormone may act to maintain the levels of hepatocyte-specific transcription factors *in vitro*. Consistent with *in vivo* findings, T$_3$ had no effect on the L-type PK mRNA level either alone or with dexamethasone or on the induction of the mRNA by insulin/dexamethasone. However, this hormone had a permissive effect on the induction of GK mRNA by insulin. 8-CPT-cAMP, a stable analog of cAMP, inhibited the induction of the mRNA by insulin, confirming the *in vivo* results.

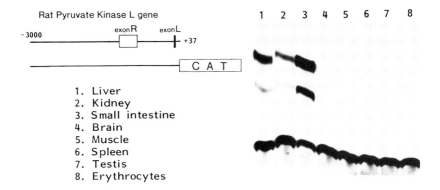

FIGURE 8. CAT activities in various tissues of transgenic mice.

IV. IDENTIFICATION OF *CIS*-ACTING REGULATORY REGIONS OF THE L-TYPE PK GENE

The *in vivo* and *in vitro* studies described above suggest that the mechanism of transcriptional stimulation of the L-type PK gene by insulin or dietary fructose is very complicated. In both cases, metabolism of carbohydrates is involved, and so a common metabolite could be responsible for the induction of the L-type PK. If this is the case, the same *cis*- and *trans*-acting factors are involved in both cases. To test this possibility, we tried to identify a *cis*-acting element responsible for transcriptional stimulation of the L-type PK gene by insulin or dietary fructose. First, we produced transgenic mice carrying the 5' flanking region of the L-type PK gene from about nucleotide -3000 to $+37$ linked to the chloramphenicol acetyltransferase (CAT) structural gene and determined the CAT activity in various tissues.[24] This activity was detected in the liver, kidney, and small intestine, but not in other tissues (Figure 8). Thus the tissues expressing CAT activity were the same as those expressing the endogenous L-type PK. Dietary glucose or insulin induced similar increases in the levels of CAT and L-type PK mRNAs in the liver of transgenic mice (Figure 9). Dietary fructose also caused a 15-fold increase in the level of CAT mRNA, but this increase extent was significantly less than that of the L-type PK mRNA (41-fold increase), suggesting that stabilization of the transcripts of the L-type PK gene plays a significant role in the induction by fructose even in normal liver. Thus, the sequence of about 3 kb upstream of the L-type PK gene contains all the *cis*-acting elements responsible for tissue-specific expression of this enzyme and its transcriptional stimulation by dietary carbohydrates and insulin. Dietary fructose, but not glucose, also induced a marked increase in the L-type PK mRNA level in the kidney, but induced only a marginal increase in the level of CAT mRNA. These results are consistent with our finding mentioned above that posttran-

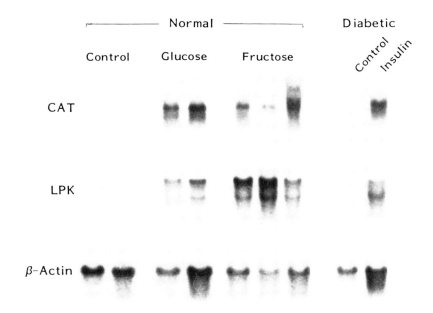

FIGURE 9.   Northern blot analysis of liver RNA of transgenic mice after administration of glucose or fructose diet or insulin for 16 h. LPK, L-type pyruvate kinase.

scriptional regulation is a major factor in the induction of the L-type PK in the kidney by fructose.

For further analysis of the *cis*-acting element, we introduced plasmid DNA containing 3.2 kb of the 5' flanking region of the L-type PK gene linked to the CAT gene into adult rat hepatocytes by electroporation and measured transient CAT expression.[25] Although hepatocytes transfected with this plasmid showed high CAT activity, the addition of insulin did not affect expression of the CAT gene.[26] This is in contrast with the results obtained in transgenic mice for some unknown reason. The results, however, indicated that strong promoter activity is present in the 3.2 kb upstream region. Therefore, we carried out functional analyses of a series of 5' and internal deletion constructs of the CAT fusion gene. In this way, we identified three positive regulatory regions required for expression of the L-type PK in hepatocytes. These regions, designated as PKL-I, PKL-II, and PKL-III, were located between nucleotides -76 and -94, -126 and -149, and -150 and -170, respectively (Figure 10). Then, we determined the individual activities of these elements. As shown in Figure 11, PKL-I showed enhancer-like activity alone, whereas PKL-II and PKL-III did not have any independent effects. Combinations of L-I + L-II and L-II + L-III, but not of L-I + L-III, showed synergistic enhancer activities when oriented in the same direction. The inclusion of all three elements oriented in the same direction had the maximum

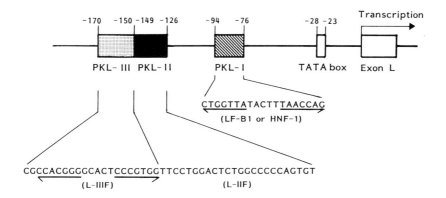

FIGURE 10. Schematic representation of the regulatory regions of the rat pyruvate kinase gene.

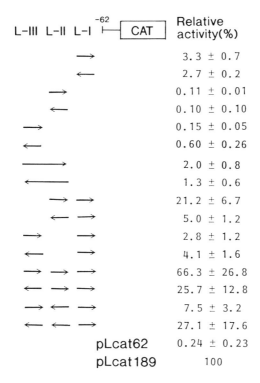

FIGURE 11. Interaction of three *cis*-acting elements of the L-type pyruvate kinase gene. A single copy of each double-stranded oligonucleotide was ligated to pLcat62 in the orientation shown by arrows on the left, which is the 5' to 3' direction. The relative activities of the constructs are expressed as percentages of that of pLcat189, which includes the region up to −189.

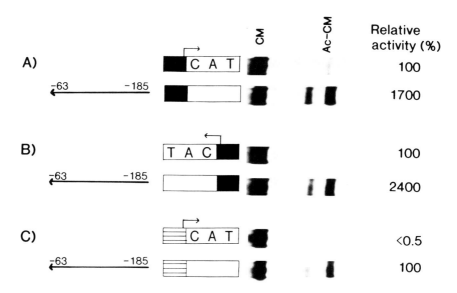

FIGURE 12. Interaction of the regulatory regions with heterologous promoter. A fragment containing the three elements was inserted upstream (A) or downstream (B) of the SV40 early promoter, or upstream of the pyruvate kinase M gene promoter (C) in the orientation shown by the arrows, which are the 5' to 3' direction.

synergistic effect, indicating that the three elements function as a unit. To examine whether this unit exerts its effect on heterologous promoters, a fragment containing all three elements was inserted upstream or downstream of the SV40 early promoter or the PKM gene promoter. In all cases, inclusion of the unit markedly stimulated expression of the CAT gene (Figure 12). Thus this unit enhanced expression from heterologous as well as homologous promoters in a manner that was independent of its orientation and position relative to the cap site. The activity of the unit was not detected in HeLa cells or K562 erythroleukemia cells, suggesting that this unit has cell-type specificity. Judging from this finding together with the results in transgenic mice, this unit seems very likely also to be responsible for expression of the L-type PK in the kidney and small intestine.

PKL-I contains a sequence homologous to the LF-B1[27] (also called HNF1[28] or HNF1α[29]) binding site. PKL-II and PKL-III include sequences homologous to the binding sites of LF-A1[30] and the adenovirus major late transcription factor (MLTF),[31] respectively. We carried out gel retardation assays to examine whether these nuclear proteins in fact bind to these three elements. The results indicated that the different *trans*-acting factors interacted with the three elements and that the *trans*-acting protein bound to PKL-I was in fact LF-B1. However, the *trans*-acting factors bound to PKL-II and PKL-III were different from LF-A1 and MLTF, respectively. We tentatively named these factors L-IIF and L-IIIF, respectively (Figure 10). Since the three elements had synergistic effects on the promoter activity, these proteins, bound to the elements, must interact with each other. Very recently, another *trans*-acting protein that binds to the LF-B1 binding site was identified and characterized. This protein called LF-B3,[32] vHNF1,[33] or HNF1α[29] is a transcriptional factor that binds to DNA as a dimer like LF-B1 and forms a heterodimer with LF-B1. The tissue distribution of LF-B3 resembles that of LF-B1 in adult rats: the two proteins are expressed in the liver, kidney, and small intestine, although their relative levels are different. Thus, LF-B3 also plays an important role in expression of the L-type PK gene in these tissues.

An important question is whether these *cis*- and *trans*-acting factors are involved in transcriptional stimulation of the L-type PK gene by insulin and dietary fructose. This problem could not be solved in our transfection system using cultured hepatocytes, as mentioned above. Therefore, we produced another type of transgenic mouse carrying a L-type PK/CAT fusion gene with a 5′ deletion up to nucleotide −610.[34] Expression of the CAT gene showed similar tissue-specificity to that of the endogenous L-type PK gene, strongly supporting the conclusion that the three *cis*-acting elements are important in tissue-specific expression of the L-type PK gene. However, preliminary experiments showed that dietary glucose and fructose caused only a slight change in the CAT mRNA level although they greatly increased the L-type PK mRNA level. Thus, the three elements may not alone be sufficient to confer insulin- and fructose-responsiveness on the CAT reporter gene. On the contrary, Thompson and Towle reported very recently that the 101-bp fragment from nucleotide −197 to −96 of the L-type PK gene conferred insulin/glucose-responsiveness on the CAT reporter gene with the thymidine kinase promoter when transfected into primary cultured hepatocytes using synthetic liposomes.[35] This region contained PKL-II and PKL-III. This is in contrast with our finding that insulin had no effect on expression of the CAT gene when the CAT fusion gene containing the upstream region of the L-type PK gene up to −3000 was introduced into hepatocytes by electroporation, as mentioned above. Another group also could not identify the regulatory region responsive to insulin by transfection of DNA into cultured hepatocytes by the calcium phosphate method.[36] These different results may be due to the different transfection procedures used. Further studies both *in vitro* and *in vivo* are necessary to clarify this discrepancy and elucidate the molecular mechanism underlying induction of the L-type PK by insulin and carbohydrates.

V. SUMMARY

Work to date in our laboratory as well as in others has demonstrated that the expression of genes coded for pyruvate kinase is influenced by hormonal and dietary factors. The four isozymes of this enzyme differ in their location in the body. In turn, the genes for these isozymes are different in their responsiveness to the kind of dietary carbohydrate and the hormones which influence its expression. The identification of the *cis*-acting regulatory region of the liver type pyruvate kinase gene has yet to be resolved.

REFERENCES

1. **Imamura, K., Noguchi, T., and Tanaka, T.,** Regulation of isozyme patterns of pyruvate kinase in normal and neoplastic tissues, in *Markers of Human Neuroectodermal Tumors,* Staal, G. E. J. and van Veelen, C. W. M., Eds., CRC Press, Boca Raton, FL, 1986, chap. 13.
2. **Noguchi, T., Yamada, K., Inoue, H., Matsuda, T., and Tanaka, T.,** The L-type and R-type isozymes of rat pyruvate kinase are produced from a single gene by use of different promoters, *J. Biol. Chem.,* 262, 14366, 1987.
3. **Noguchi, T., Inoue, H., and Tanaka, T.,** The M_1- and M_2-type isozymes of rat pyruvate kinase are produced from the same gene by alternative RNA splicing, *J. Biol. Chem.,* 261, 13807, 1986.
4. **Takenaka, M., Noguchi, T., Inoue, H., Yamada, K., Matsuda, T., and Tanaka, T.,** Rat pyruvate kinase M gene. Its complete structure and characterization of the 5'-flanking region, *J. Biol. Chem.,* 264, 2363, 1989.
5. **Goodridge, A. G.,** Dietary regulation of gene expression: enzymes involved in carbohydrate and lipid metabolism, *Annu. Rev. Nutr.,* 7, 157, 1987.
6. **Van Den Berghe, G.,** Metabolic effects of fructose in the liver, in *Current Topics in Cellular Regulation,* Vol. 13, Horecker, B. L. and Stadtman, E. R., Eds., Academic Press, New York, 1978, 97.
7. **Gunn, J. M. and Taylor, C. B.,** Relationships between concentration of hepatic intermediary metabolites and induction of the key glycolytic enzymes in vivo, *Biochem. J.,* 136, 455, 1973.
8. **Takeda, Y., Inoue, H., Honjo, K., Tanioka, H., and Daikuhara, Y.,** Dietary response of various key enzymes related to glucose metabolism in normal and diabetic rat liver, *Biochim. Biophys. Acta,* 136, 214, 1967.
9. **Fukuda, H., Iritani, N., and Tanaka, T.,** Effects of high-fructose diet on lipogenic enzymes and their substrate and effector levels in diabetic rats, *J. Nutr. Sci. Vitaminol.,* 29, 691, 1983.
10. **Adelman, R. C., Spolter, P. D., and Weinhouse, S.,** Dietary and hormonal regulation of enzymes of fructose metabolism in rat liver, *J. Biol. Chem.,* 241, 5467, 1966.
11. **Noguchi, T., Inoue, H., and Tanaka, T.,** Regulation of rat liver L-type pyruvate kinase mRNA by insulin and by fructose, *Eur. J. Biochem.,* 128, 583, 1982.
12. **Noguchi, T., Inoue, H., Chen, H., Matsubara, K., and Tanaka, T.,** Molecular cloning of DNA complementary to rat L-type pyruvate kinase mRNA. Nutritional and hormonal regulation of L-type pyruvate kinase mRNA concentration, *J. Biol. Chem.,* 258, 15220, 1983.
13. **Noguchi, T., Inoue, H., and Tanaka, T.,** Transcriptional and post-transcriptional regulation of L-type pyruvate kinase in diabetic rat liver by insulin and dietary fructose, *J. Biol. Chem.,* 260, 14393, 1985.
14. **Iynedjian, P. B., Gjinovci, A., and Renold, A. E.,** Stimulation by insulin of glucokinase gene transcription in liver of diabetic rats, *J. Biol. Chem.,* 263, 740, 1988.

15. **Iynedjian, P. B., Ucla, C., and Mach, B.,** Molecular cloning of glucokinase cDNA. Developmental and dietary regulation of glucokinase mRNA in rat liver, *J. Biol. Chem.,* 262, 6032, 1987.
16. **Munnich, A., Lyonnet, S., Chauvet, D., Van Schaftingen, E., and Kahn, A.,** Differential effects of glucose and fructose on liver L-type pyruvate kinase gene expression in vivo, *J. Biol. Chem.,* 262, 17065, 1987.
17. **Inoue, H., Noguchi, T., and Tanaka, T.,** Rapid regulation of L-type pyruvate kinase mRNA by fructose in diabetic rat liver, *J. Biochem. (Tokyo),* 96, 1457, 1984.
18. **Noguchi, T. and Tanaka, T.,** unpublished data, 1989.
19. **Matsuda, T., Noguchi, T., Takenaka, M., Yamada, K., and Tanaka, T.,** Regulation of L-type pyruvate kinase gene expression by dietary fructose in normal and diabetic rats, *J. Biochem. (Tokyo),* 197, 655, 1990.
20. **Vernon, R. G. and Walker, D. G.,** Glycerol metabolism in the neonatal rat, *Biochem. J.,* 118, 531, 1970.
21. **Vaulont, S., Munnich, A., Decaux, J.-F., and Kahan, A.,** Transcriptional and post-transcriptional regulation of L-type pyruvate kinase gene expression in rat liver, *J. Biol. Chem.,* 261, 7621, 1986.
22. **Matsuda, T., Noguchi, T., Yamada, K., Takenaka, M., and Tanaka, T.,** Regulation of the gene expression of glucokinase and L-type pyruvate kinase in primary cultures of rat hepatocytes by hormones and carbohydrates, *J. Biochem. (Tokyo),* 108, 778, 1990.
23. **Feliu, J. E., Coloma, J., Gomez-Lechon, M.-J., Garcia, M. D., and Baguena, J.,** Effect of dexamethasone on the isozyme pattern of adult rat liver parenchymal cells in primary cultures, *Mol. Cell. Biochem.,* 45, 73, 1982.
24. **Yamada, K., Noguchi, J., Miyazaki, J., Matsuda, T., Takenaka, M., Yamamura, K., and Tanaka, T.,** Tissue-specific expression of rat pyruvate kinase L/chloramphenicol acetyltransferase fusion gene in transgenic mice and its regulation by diet and insulin, *Biochem. Biophys. Res. Commun.,* 171, 243, 1990.
25. **Yamada, K., Noguchi, T., Matsuda, T., Takenaka, M., Monaci, P., Nicosia, A., and Tanaka, T.,** Identification and characterization of hepatocyte-specific regulatory regions of the rat pyruvate kinase L gene. The synergistic effects of multiple elements, *J. Biol. Chem.,* 265, 19885, 1990.
26. **Noguchi, T., Yamada, K., and Tanaka, T.,** unpublished data, 1989.
27. **Frain, M., Swart, G., Monaci, P., Nicosia, A., Stampfli, S., Frank, R., and Cortese, R.,** The liver-specific transcription factor LF-B1 contains a highly diverged homeobox DNA binding protein, *Cell,* 59, 145, 1989.
28. **Chouard, T., Blumenfeld, M., Bach, I., Vandekerhove, J., Cereghini, S., and Yaniv, M.,** A distal dimerization domain is essential for DNA binding by the atypical HNF1 homeodomain, *Nucleic Acids Res.,* 18, 5853, 1990.
29. **Mendel, D. B., Hansen, L. P., Graves, M. K., Conley, P. B., and Crabtree, G. R.,** HNF-1α and HNF1β (vHNF-1) share dimerization and homeodomains, but not activation domains, and form heterodimers in vitro, *Genes Dev.,* 5, 1042, 1991.
30. **Hardon, E. M., Frain, M., Paonessa, G., and Cortese, R.,** Two distinct factors interact with the promoter regions of several liver-specific genes, *EMBO J.,* 7, 1711, 1988.
31. **Chodosh, L. A., Carthew, R. W., Morgan, J. G., Crabtree, G. R., and Sharp, P. A.,** The adenovirus major late transcription factor activates the rat γ-fibrinogen promoter, *Science,* 238, 684, 1987.
32. **De Simone, V., De Magistris, L., Lazzaro, D., Gerstner, J., Monaci, P., Nicosia, A., and Cortes, R.,** LFB3, a heterodimer-forming homeoprotein of the LFB1 family, is expressed in specialized epithelia, *EMBO J.,* 10, 1435, 1991.
33. **Rey-Campos, J., Chouard, T., Yaniv, M., and Cereghini, S.,** vHNF1 is a homeoprotein that activates transcription and forms heterodimers with HNF1, *EMBO J.,* 10, 1445, 1991.

34. **Noguchi, T., Miyazaki, J., Yamamura, K., and Tanaka, T.,** unpublished data, 1991.

35. **Thomson, K. S. and Towle, H. C.,** Localization of the carbohydrate response element of the rat L-type pyruvate kinase gene, *J. Biol. Chem.,* 266, 8679, 1991.

36. **Cognet, M., Bergot, M.-O., and Kahn, A.,** cis-Acting elements regulating expression of the liver pyruvate kinase gene in hepatocytes and hepatoma cells. Evidence for tissue-specific activators and extinguisher, *J. Biol. Chem.,* 266, 7368, 1991.

Chapter 8

GLUCOSE-6-PHOSPHATE DEHYDROGENASE: DIET AND HORMONAL INFLUENCES ON *DE NOVO* ENZYME SYNTHESIS

Rolf F. Kletzien and Carolyn D. Berdanier

TABLE OF CONTENTS

0-8493-6961-4/93/$0.00 + $.50

I. INTRODUCTION

Among the many nutrient-gene interactions described in this volume, none have been described as often or as fully as that which occurs when animals are starved and realimented. Tepperman and Tepperman[1] described the response to intermittent starvation as adaptive hyperlipogenesis. Periodic famine even today afflicts human populations in third world countries and in countries facing catastrophic events such as war, earthquakes, crop failures, and economic failures. Those that can adapt and increase their metabolic efficiency survive; those that cannot die. Those that live pass on their genetic material to subsequent generations. Those that die do not. Truly, survival under adverse nutritional circumstances is the ultimate result of nutrient-gene interactions.

II. DIET EFFECTS ON G6PD ACTIVITY

Although survival and reproduction is the result of numerous nutrient-gene interactions, this chapter will discuss only one: the influence of nutritional state on the enzyme glucose-6-phosphate dehydrogenase G6PD (E.C. 1.1.1.49, D-glucose-6-phosphate: $NADP^+$ oxidoreductase). This enzyme catalyzes the first rate-limiting reaction in the pentose phosphate shunt. Together with the other pentose shunt dehydrogenase (6-phosphogluconate dehydrogenase E.C. 1.1.1.44), this reaction sequence provides approximately 50% of the reducing equivalents needed to support fatty acid synthesis. The remaining reducing equivalents are provided by the malic enzyme reaction ($\sim$40%) and the isocitrate dehydrogenase reaction ($\sim$10%). Glock and McLean[2] were among the first to report that diet restriction would result in a decrease in the activity of G6PD. This report was followed by others[3,4] that described the increase in enzyme activity in rats that were switched from a carbohydrate-poor diet to one rich in simple sugars. Even more enzyme activity was observed if the animals were first starved and then fed a sugar-rich diet.[1,5-15] There is a high correlation between the activity of G6PD and *de novo* fatty acid synthesis.[13] As shown in Table 1 the percent increment in enzyme activity depends on a variety of dietary and nondietary factors. Dietary factors include the type and amount of carbohydrate, the kind and amount of fat, and the level of protein. In Table 1 are presented some of the observations from a variety of studies using rats starved for 48 h and refed for 48 h. The literature is vast with respect to the responses of rats, chickens, cows, mice, pigs, fish, and humans. Only data from rats are presented in Table 1. Rats that are starved and refed diets rich in sugars such as sucrose generally have a higher G6PD response than rats refed an equivalent amount of starch or fructose.[13,14] Those refed fructose have a lower G6PD response than those refed glucose.[11] Those refed a glucose-menhaden oil diet have a lower response than those whose diet contained corn oil or coconut oil or beef tallow.[15,16] Note the rather

TABLE 1
Influence of Diet Composition on the Percent Increment in Hepatic G6PD Activity Due to Starvation-Refeeding

Diet	% Increment[a]	Reference
65% Glucose-5% corn oil	91–514	11, 13, 15, 16
65% Starch-5% corn oil	262–278	13, 14
65% Sucrose-5% corn oil	190–233	13–15
65% Fructose-5% corn oil	72	11
65% Glucose-5% menhaden oil		15, 16, 101–122
65% Glucose-5% beef tallow	270	16
65% Glucose-5% coconut oil	218	16
65% Starch-5% coconut oil	291	14
65% Sucrose-5% coconut oil	319	14
65% Protein-20% glucose-5% corn oil	154	13
40% Fat, 30% glucose, 20% protein	274	13

Note: SR, starved 48 h; refed 48 h.

[a] % increment = $\dfrac{\text{G6PD activity in S-R rats} - \text{G6PD activity in AL rats}}{\text{G6PD activity in AL rats}}$

broad range in values for some of the diets. In part, this is due to the variation within strains and in part to the variation in response between strains. Some rats, notably those of the Wistar strain, are relatively unresponsive to starvation-refeeding, whereas Sprague-Dawley rats are more responsive to this feeding paradigm. BHE rats are more responsive than Sprague-Dawley rats.[15] Not only are there differences due to the strain in enzyme activity, there are also differences attributable to the diet fed the rat prior to its use in a starvation-refeeding experiment, and, too, its sex and its age. If rats were fed a high fat diet prior to use, the increment in enzyme activity was twice as high as in rats fed a standard stock diet prior to use.[17] Middle-aged rats (~12 months of age) were similar to young rats[18,19] in their response to starvation-refeeding, while aged rats (22 months of age) were less responsive to this treatment than their younger (2 months old) cohorts.[20] Female rats are less responsive to starvation-refeeding than are male rats.[21] This gender difference in response is probably due to the fact that female rats have a higher G6PD activity in the basal *ad libitum* fed state. In part, this is due to the cyclical presence of estrogen which imposes a cyclical pattern of reductions and increases in food intake. Thus, postpubescent female rats experience a 5-day food intake decrease/increase cycle which, in effect, is a modified starve-refeed cycle. Prepubescent female rats do not evidence an estrus cycle and also do not have cyclical variations in food intake. Estrogen has also been shown to be essential to the induction of G6PD activity in the uterus.[22]

Variation in food supply, as well, can affect the G6PD response to starvation-refeeding. Rats that are deprived of food during shipping have a larger response to starvation-refeeding than rats obtained from a local source where the stress of shipping is minimized.[23,25] A larger response has also been observed in rats that have been starved and refed more than once.[10,25,26] The length of starvation also affects the response.[27] In rats starved for 3 to 6 days the enzyme response was greater than in rats starved for only 2 days prior to refeeding. All of these observations suggest that the increase in G6PD activity in animals refed following starvation is due to signals generated by the starvation or food restriction that stimulates G6PD.

III. SIGNALS THAT INCREASE G6PD ACTIVITY

At first, it was thought that the increase in G6PD activity in starved-refed rats was due to the fact that these rats overeat during the initial 2 days of refeeding.[28] However, Gimenez and Johnson showed that the G6PD enzyme response still occurred in rats not allowed to overeat.[29] A number of papers reported that the enzyme response could be suppressed if the antibiotic 8-azaguanine were administered during the refeeding period.[9,10,18,25-27] 8-Azaguanine is a compound that is incorporated into messenger RNA as it is being synthesized. However, because it is a substituted guanine, the messenger RNA cannot be translated. Rats that are starved and refed and given this drug will have enzyme activities similar to those found in *ad libitum* fed rats but will not evidence the large increase in activity typical of the response to starvation-refeeding. Thus, one can infer that the large increase in enyzme activity is due to *de novo* synthesis of mRNA coded for the G6PD enzyme.

Although the mechanism for the increase in enzyme protein synthesis was clarified by the use of 8-azaguanine, the signal or signals which stimulated the mRNA synthesis were unknown. Some suggested that insulin might serve as that signal since a high glucose diet was needed in order to demonstrate the enzyme response. Glucose is a potent insulin secretogogue and insulin does stimulate protein synthesis. However, a variety of experiments using intact animals failed to demonstrate unequivocally that insulin was the primary signal for the enzyme overshoot.[4,10,27,28,30,31]

The fact that starvation was an absolute requirement for the overshoot suggested that the stress response was involved in the generation of the enzyme overshoot. This was tested by removing the adrenals (ADX) from rats and then subjecting them to starvation-refeeding.[32] When refed, the ADX rats had enzyme activity levels similar to those of the *ad libitum* fed rats but no overshoot (Figure 1). When the glucocorticoids were replaced, the large increase in G6PD activity in the starved-refed rat was restored.[32-34] Using 8-azaguanine, it was shown that the glucocorticoid effect was that of stimulating G6PD mRNA synthesis.[31] This effect of glucocorticoid, although dependent to some extent on other hormones (insulin and thyroxine),[33,34] was indeed a direct effect of the hormone glucocorticoid on the synthesis of mRNA coding for G6PD.

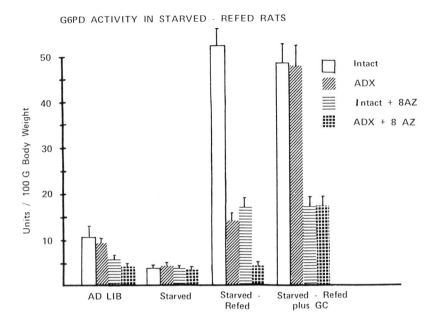

FIGURE 1. Hepatic G6PD activity in ad libitum fed and starved-refed Sprague-Dawley rats. Rats were fed or refed a 65% glucose diet. The starvation period and the refeeding period were 48 h each; 8 azaguanine was administered daily. Abbreviations: GC, glucocorticoid; 8 AZ, 8-azaguanine. Data are from References 33 and 34.

IV. USE OF CELLULAR MODELS AND MOLECULAR PROBES TO DEFINE THE SIGNALS RESPONSIBLE FOR REGULATION OF GLUCOSE-6-PHOSPHATE DEHYDROGENASE

The studies carried out with intact animals have demonstrated that a variety of nutrients and hormones are either directly or indirectly involved with the regulation of hepatic G6PD. However, it is often difficult to distinguish direct vs. indirect involvement of hormones or nutrients on gene expression in intact animals. For example, in attempting to determine whether insulin or glucose is a primary signal for induction of G6PD, experimentation in animals is compromised since blood glucose is maintained at 5 mM, and elevating it by feeding glucose causes insulin release. Eliminating insulin by making the animal diabetic results in high levels of glucagon which cause an elevation of hepatic cAMP and gluconeogenesis, conditions that might repress G6PD. The well-known induction of G6PD following injection of insulin into the animal could be the result of a positive influence of insulin on the G6PD gene or it could be result of derepression of the influence of glucagon. Thus, direct assignment of the responsibility for control of G6PD expression requires the use of less complex systems than the intact animal.

In order to sort out the various factors involved in the regulation of hepatic G6PD, we have employed primary cultures of liver parenchymal cells maintained in chemically defined, serum-free medium. In this system, cells retain the terminally differentiated phenotype for a discrete period of time (up to several weeks or months, depending on culture conditions),[37-39] in contrast to hepatoma cell lines which seldom display the full realm of differentiated functions expected of a hepatocyte. Thus, metabolic responses of the primary cultures to hormones or nutrients can be expected to mimic that observed *in vivo*. With this system it is then possible to add hormones or nutrients, individually or in combination, to define precisely the involvement of these factors in the control of G6PD expression. Several groups have employed the hepatocyte primary culture system in studies on G6PD regulation.[40-45] While culture conditions were not identical, some common themes have emerged and some areas of disagreement are apparent. This review will attempt to analyze and summarize results on nutrient and hormonal control of G6PD from the six laboratories[40-45] employing this culture system.

An additional development that has greatly aided analysis of G6PD expression is the availability of molecular probes for the mRNA encoding rat liver G6PD.[46] The complete sequence of the protein[47] and mRNA[48] has been established and genomic fragments have been isolated.[49] These sequences along with high titer antibodies against G6PD provide all of the necessary tools to analyze expression in either animals or in cells in culture. Thus, it is now possible to dissect in a molecular fashion the effects of the hormones or nutrients on the control of expression. This is particularly important in those situations where a hormone or nutrient might increase transcription of the G6PD gene, but a different hormone might be required for the optimum rate of translation of that mRNA species. If one simply measures enzyme activity or enzyme protein by immunochemical techniques, this sort of interaction would not be observed. Therefore, combining these molecular tools with the primary culture model should provide precise information on the control of G6PD expression.

V. HORMONAL REGULATION OF G6PD

Insulin has long been considered the primary signal for induction of G6PD in addition to the other lipogenic enzymes in animals. Kurts and Wells[45] and Nakamura et al.[41] were the first to demonstrate that insulin could induce G6PD in primary cultures of hepatocytes. Subsequently, others extended this observation to show that insulin increased the relative rate of G6PD synthesis[50] and mRNA encoding G6PD.[50,51] The ability of insulin to induce G6PD is not rapid in onset,[50] taking 36 to 48 h to reach its maximum effect. However, induction of G6PD *in vivo* following feeding a high carbohydrate diet is also a relatively slow process, taking 48 to 72 h to reach maximal levels of activity.[49,52] Thus, although the absolute change in enzyme activity observed in tissue culture following insulin treatment is not as great as that observed in the intact animal following carbohydrate refeeding, the time course is quite similar. This lends strength to the argument that insulin is the primary signal involved but suggests that other factors may be involved in amplification of the insulin response. The six laboratories[40-45] that have utilized primary cultures of hepatocytes are in agreement with the primary role of insulin in control of G6PD expression.

The precise level at which insulin regulates G6PD expression has not been established. The fact that insulin elevates G6PD mRNA in cultured hepatocytes suggests that transcriptional regulation of the G6PD gene may be a critical site of regulation. In addition, two studies employing transcription "run on" assays from rat liver nuclei isolated from fasted-refed animals have demonstrated that there is a transient activation of G6PD gene transcription.[49,53] However, it is also possible that insulin stabilizes the G6PD mRNA and/or causes a selective translation of it. Changes in G6PD mRNA stability following refeeding a high carbohydrate diet have been observed in rats.[49] The study of how insulin accomplishes regulation of G6PD and other insulin regulated genes remains as a major area of interest for molecular endocrinologists.

The glucocorticoids have a well established role in lipogenesis (for review see Reference 54), and, thus, the interaction of insulin and the glucocorticoids in regulation of G6PD has been explored in the hepatocyte primary culture model. The results from these studies have shown that some differences are apparent which are perhaps the result of culture conditions. Kurtz and Wells[45] first demonstrated that a synthetic glucocorticoid, dexamethasone (Dex), induced G6PD activity by itself and its effect was additive with insulin. In addition, the study[45] suggested that the glucocorticoid played a critical "permissive" role for insulin induction of G6PD since insulin failed to increase G6PD in hepatocytes from adrenalectomized rats unless Dex was added to the cultures. Kletzien and co-workers[50,51,55] concluded that while the glucocorticoids enhanced the insulin induction of G6PD activity, it exerted its main effect on mRNA levels encoding G6PD. Using molecular probes and a high titer, monospecific antibody against G6PD, they demonstrated that Dex appears to increase the mRNA encoding G6PD but translation of the active protein does not occur unless insulin is also present.[50,51] Holten and co-workers have also demonstrated that maximum induction of G6PD in the primary culture model requires the presence of a glucocorticoid.[42] Thus, it would seem that this class of hormones is involved in at least a permissive way, in regulating G6PD expression.

In contrast to the above results, two groups[41,43] have reported that the glucocorticoids are not involved in regulation of G6PD. However, Ichihara and colleagues[41] maintained their cultures in the continuous presence of glucocorticoids and serum. Thus, it is not surprising that addition of glucocorticoids failed to cause any further increase in enzyme activity.[41] It is more difficult to understand why Salati et al.[43] did not observe enhancement of insulin induction of G6PD activity, but it may be related to the use of serum in the tissue culture medium during the early phase of attachment of cells to the culture dish. Serum may contain sufficient glucocorticoids or other factors that exhibit long range effects on subsequent insulin stimulation of gene expression. Alternatively, many of these experiments[43] were conducted with vitamin E present in the culture medium which would be expected to decrease oxidant stress, a potential stimulant of G6PD. The glucocorticoids may exhibit different effects depending on the redox state of the cell.

Experiments conducted in animals some time ago have suggested that triiodothyronine (T3) may be involved in regulation of G6PD.[56] Little evidence from the primary culture model has been forthcoming to support a role for thyroxine. Fritz and Kletzien[57] demonstrated that T3 increased G6PD expression in the hypothyroid rat through a mechanism that did not involve an increase in G6PD mRNA. However, Kletzien and co-workers have not been able to demonstrate any direct effect of T3 on G6PD induction in the hepatocyte primary cultures, a result that is in concert with the other three groups that have directly examined this question.[42,43,45] Thus, it would appear that T3 has no immediate, direct role in regulation of hepatic G6PD expression.

Yoshimoto et al.[58] have shown that epidermal growth factor (EGF) is capable of inducing G6PD in primary cultures of hepatocytes and, in the presence of insulin, additivity in the response was noted. This effect appeared to be disconnected from lipogenesis since EGF repressed induction of malic enzyme and the insulin stimulated incorporation of acetate into triglycerides. EGF was shown to increase both the relative rate of G6PD synthesis and mRNA encoding the enzyme.[58] An EGF growth stimulus may be responsible for the increase in G6PD. Nevertheless, since it is sometimes recommended that EGF be included in the medium for hepatocyte primary cultures,[38] the influence of EGF or other growth or serum factors must be taken into account when attempting to ascertain direct vs. indirect effects of these agents on expression. Clearly, it would be ideal to always use a chemical defined, serum-free medium although optimum conditions may not yet be defined.

VI. NUTRIENT REGULATION OF G6PD

Investigation of the role of carbohydrate in the regulation of G6PD in rat hepatocytes in primary culture has led to the conclusion that carbohydrate does[40,43] or does not[42,45] influence induction of the enzyme by itself or in conjunction with hormones. In attempting to resolve this difference, it is clear that culture conditions can influence the response of cells to carbohydrate. Recent work from Holten's laboratory has shown that fructose, when present at 5 mM in the culture medium, increases expression in the presence of Dex and insulin, while at 10 m*M*, expression is decreased.[42] In the same study, these investigators found that the presence of D-glucose by itself or in combination with hormones did not influence expression of G6PD.[42] Kurtz and Wells[45] also showed that glucose did not enhance basal or hormone-induced expression of G6PD. In an equally clear fashion, two studies have demonstrated that glucose enhances basal and hormone-regulated expression[40,43] with the maximum observed effect of glucose being to increase basal expression fivefold.[43] In attempting to resolve this difference, one could consider the potential role of glucokinase expression. Both of the studies[40,43] that observed a glucose effect used a Waymouth medium while the other studies[42,45] did not. Since induction of glucokinase in primary cultures of hepatocytes is highly dependent on culture medium conditions,[59] it seems entirely possible that cells which lack glucokinase would not respond to glucose. The common observation that high density primary cultures of hepatocytes maintained in high glucose Waymouth medium produce excessive quantities of lactic acid clearly demonstrates the cells' ability to use glucose. Presumably, a metabolite of glucose is responsible for mediating the effect on gene expression. Thus, it is likely that both fructose and glucose are working through the same mechanism to produce the increased expression. No information is available as to the type of regulation that glucose or fructose are exerting.

Nutrient suppression of hepatic G6PD has also been reported. Results from studies in animals showed that a high level of fatty acid in the diet suppressed the basal level of G6PD activity and that which could be induced by a high carbohydrate diet.[52] Using primary cultures of hepatocytes, Salati et al.[43] have demonstrated that linoleate suppressed insulin induction of G6PD but not that elicited by glucose. Arachidonate was found to be a more effective inhibitor than linoleate, and both fatty acids caused a right shift in the insulin dose response curve. The insulin induction of two other lipogenic enzymes (acetyl CoA carboxylase and malic enzyme) was similarly affected,[43] suggesting that nutrient control of gene expression in this case may be linked to a mechanism involving desensitization of hepatocytes to the lipogenic action of insulin. Conversely, glucose enhancement of G6PD induction may represent sensitization of hepatocytes to insulin. Evidence has been presented to suggest that fatty acid suppression of G6PD occurs at the level of gene transcription.[53]

VII. OTHER FACTORS CAPABLE OF REGULATING G6PD

A variety of nonnutrient, nonhormonal agents are known to induce hepatic G6PD. The effect of ethanol, alone and in combination with hormones and glucose, has been extensively characterized.[40,55] Ethanol increases both the relative rate of G6PD synthesis and mRNA encoding the enzyme.[55] Particularly noteworthy is the ability of the drug to increase the glucocorticoid enhancement of G6PD induction. Recent work[60] in this area suggests that a metabolite of ethanol, acetaldehyde, may be the active species. Other agents, such as carbon tetrachloride and the drug Primaquine, can also induce the enzyme.[61] A common property of all these agents is that they are capable of producing oxidant stress in the hepatocyte. The participation of G6PD in controlling oxidant stress in a variety of tissues is well known.[61-63] The G6PD gene may be a sentinel for cellular oxidant stress and capable of rapidly responding to the need for cellular NADPH. If this notion is correct, it may also explain some of the differences observed in the primary culture model. For example, Manos et al.[42] reported that G6PD could be induced 12-fold in hepatocytes maintained in medium with only amino acids as an energy source. The lack of a readily metabolizable carbohydrate to run through the pentose phosphate pathway could result in considerable oxidant stress and induction of G6PD. Incubation of G6PD deficient erythrocytes in carbohydrate and pyruvate-free medium is known to accelerate lysis.[63]

VIII. RECAPITULATION OF WORK ON HEPATOCYTES

Use of primary cultures of hepatocytes to define precisely the hormonal and nutrient signals responsible for regulation of G6PD has demonstrated the central role that insulin occupies. Most investigators have also demonstrated glucocorticoid involvement, although this class of hormones seems to enhance insulin's effects rather than directly increasing expression. Participation of carbohydrate in the control of G6PD expression is more controversial with suggestions that glucose is involved,[40,43] is not involved,[42,45] or that fructose is involved.[42] Use of different culture conditions and culture media appear to be responsible for these discrepancies. These differences could also contribute to changes in the redox state of the cells which may influence G6PD activity. Fatty acid has been demonstrated to be a negative regulator of expression,[43,53] while hormones or conditions associated with fasting appear to not be involved in directly repressing G6PD expression.[41]

The level at which hormones and nutrients control G6PD expression has not yet been established. It is evident that mRNA encoding G6PD is elevated by insulin treatment[42,50,51] and the glucocorticoids enhance the level by an additional increment.[50,51] Thus, there is the suggestion that expression is controlled at the transcriptional level. This is supported by two recent studies[49,53] in which transcription "run off" assays were used to examine the effect of carbohydrate refeeding on hepatic G6PD in rats. Both studies demonstrated a transient increase in transcription within hours of feeding. However, one study also demonstrated that the G6PD mRNA half-life increased substantially (1.5 to 2 h for fasted animals vs. 9 to 11 h for refed) following refeeding of the high sucrose diet.[49] Therefore, there is the expectation that control of G6PD expression will be exerted at the transcriptional level, at the level of mRNA stability, and perhaps, as suggested earlier,[50,51] at the translational level. The availability of molecular probes and appropriate genomic fragments should permit conclusive experiments to be carried out in primary cultures of hepatocytes.

IX. RECENT INSIGHTS INTO G6PD GENE STRUCTURE: IMPLICATIONS FOR REGULATION OF EXPRESSION

G6PD is an X chromosome-linked gene in mammals, and the enzyme is subject to enormous genetic variability with over 300 variant forms described to date in man.[64,65] An unusual feature of the deficiency syndrome is that the loss of enzyme activity is manifest primarily in the red blood cell.[65] Since classical genetic linkage studies have shown that G6PD is encoded by a single allele, it is difficult to understand why only the red blood cell would exhibit the decreased enzyme activity.[65] However, recent studies on L- and R-type pyruvate kinase provide a possible framework for considering how this might occur.[66] The L- and R-type isozymes are encoded by the same gene but transcription is controlled by different promoters. Thus, the first exon in each mRNA is unique to the species resulting in amino terminal heterogeneity in the protein expressed in liver L-type vs. red blood cells (R-type).

Recently, partial G6PD protein sequence data for the rat liver protein has been published.[47] Comparison of this sequence with that of the human red blood cell protein[67] shows a very high degree of sequence homology from human and drosophilia sources and has shown a very high degree of conservation except at the 5' end of the cDNA.[68,69] Thus, it appears that the majority of the G6PD sequence has been conserved in evolution from drosophilia to man except the amino terminal sequence. Heterogeneity in the amino terminal sequence may reflect different tissue-specific requirements in the regulation of G6PD activity and/or control of gene expression as has been observed in the key glycolytic enzyme, pyruvate kinase.[66]

X. SEQUENCE AND STRUCTURE OF THE RAT G6PD GENE AND PSEUDOGENE

The consensus sequence for rat G6PD derived from cDNA and genomic clones is presented in Figure 2. We have previously shown that fragments of this sequence will hybrid select mRNA encoding G6PD[46]. Recently, partial protein sequence for rat liver G6PD has been published[47] allowing direct comparison with our nucleic acid sequence. For the 71 amino acid amino terminal sequence published by Jeffery et al.,[47] there exists 100% identity with our deduced amino acid sequence from the 5' coding region, unequivocally establishing this consensus sequence for rat G6PD. In addition, comparison of our consensus sequence with one of the published sequences for

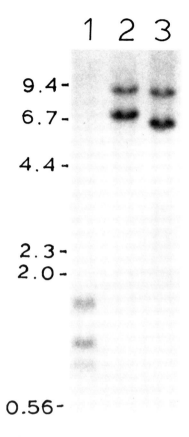

FIGURE 2. Southern blot of rat liver chromosomal DNA. Rat liver chromosomal DNA was digested with Pstl, Eco R1, and Bam H1 (Lanes 1, 2, and 3). The blot was probed with a nicknamed cDNA from pGDY-38[46] which is a 3′ untranslated region probe. The 9.4 and 9 kb (lanes 2 and 3) fragment is the functional gene and the 6.7 and 6.2 kb (lanes 2 and 3) fragment is the pseudogene.

human G6PD[70] revealed an overall sequence identity of 76.6%. If the individual regions of the two sequences are compared, there is 87.4% identity in the coding region while the 5′ and 3′ untranslated regions demonstrated a higher degree of divergence at 42.8% and 64.7% homology, respectively. The high degree of sequence conservation between the rat and human protein is demonstrated when the deduced protein sequences are compared (93.2% sequence identity), indicating that many of the nucleotide changes in the coding region are in the wobble position. While the 3′ untranslated flank was less homologous than the coding sequence, there were, nevertheless, areas within this region of near 100% homology. The sequences immediately following translation termination, in and around the unusual polyadenylation

TABLE 2
Comparison of Rat/Human Exon/Intron Boundaries

EXON	1	ACTAAATTCA CGA GCGCAG	GT	IVS	1	(>110 bp)	AG1	GAAAACATCA AC GCGTCAT
EXON	2	GGGTGCATCG	GT	IVS	2	(>500 bp)	AG	GGTGACCTGG
EXON	3	CTACCATCTG C	GT	IVS	3	(110 bp)	AG	GTGGCTGTTC
EXON	4	CTTCTTTAAA C G	GT	IVS	4	(>250 bp)	AG	GTCACTCCAG C C
EXON	5	TGAGTAGAC C T	GT	IVS	5	(>270 bp)	AG	AGGCTGGAAC
EXON	6	TGGTGCTGAG	GT	IVS	6	(228 bp)	AG	ATTTGCCAAC
EXON	7	GGATCATCAG C	GT	IVS	7	(306 bp)	AG	GGATGTCATG
EXON	8	TGATGAGAAG	GT	IVS	8	(128 bp)	AG	GTCAAAGTGT G
EXON	9	CGGTGGGATG A	GT	IVS	9	(270 bp)	AG	GAGTACCCTT G G
EXON	10	CAGATACAAG	GT	IVS	10	(119 bp)	AG	AATGTGAAGC C
EXON	11	TTGTCCGTAG C G C	GT	IVS	11	(96 bp)	AG	TGATGAACTC C C G
EXON	12	TCTATGGCAG T	GT	IVS	12	(101 bp)	AG	CCGAGGTCCC C
EXON	13	—						

Note: The rat sequence is shown on the top line, and in those positions where the human sequence is different, the base present in the human sequence is on the bottom line.

signal, and in two other areas within this region are highly conserved between rat and human.

While the human genome apparently does not contain a G6PD pseudo-gene,[71] a processed pseudogene for G6PD is present (Figure 2, 6.7 kb) in the rat genome and is more highly represented than the single copy of the functional gene (Figure 2, 9.4 kb). A genomic clone was isolated which contained a 6.7 Eco R1 fragment within which was a 3.2 kb sequence having the characteristics of a processed pseudogene for G6PD. Sequence analysis (Figure 3) revealed the absence of introns and the presence of a poly A tail, indicative of a pressed pseudogene. As in the case of other processed pseudogenes, a short direct repeat (GAAAAGACAA) which probably represents the site of insertion was located flanking the sequence. It occurs immediately 3' of the poly A tail and 65 base pairs 5' of the "translation start" site. Numerous base substitutions were found throughout the coding, 5' and 3' untranslated regions. The overall sequence identity between the pseudogene and consensus sequence is 95.8%.

```
                                                                        +1v
CONS                   GTGAACGTGTTTGGCAGCGGCAACTAAATTCAGAAAACATCATGGCAGAGCAGGTGGCTTTGAGCCGGACCCAGGTGTGTGGGATC 45
PG     GAAAAGACAAGAAACGGGCTATGCAATCCAGATCTGTGAACGTGTTTGGCAGCAGCAACTAAATTCAGAAAACATCATGGCAGAGCAGGTGGCTTTGAGCTGGACCCAGGTGTGTGGGATC

CONS   CTGAGGGAAGAGTTGTACCAGGGTGATGCCTTCCACCAAGCTGATACACACATATTTATCATCATGGGTGCATCGGGTGACCTGGCCAAGAAGAAGATTTATCCTACCATCTGGTGGCTG 165
PG     CTAAGGGAAGAGTTGTACCATGGTGATGCCTTCCACAAGGCTGATACACA--TATTTATCATCATGGGTGCATCGGGTGACCTGGCCAAGAAGAAGATTTATCCTACCATCTGGTGGCTG

CONS   TTCTGGGGATGGCCTTCTACCCGAAGACACCTTCATTGTAGGCTATGCCCGCTCACGACTCACAGTGGATGACATCCGCAAACAGAGTGAGCCCTTCTTTAAAGTCACTCCAGAAGAAAGA 285
PG     TTCCGGGGATGGCCTTTTACCCAAAGACACCTTCATTGTAGGCTATGCCTGCTCATGACTCACAGTGGATGACATCCGCAAACAGAGTGAGCCCTTCTTTAAAGTCACTCCAGAAGAAAGA

CONS   CCCAAGCTAGAGGAGTTCTTTGCCCGTAACTCCTATGTAGCTGGCCAGTATGATGATCCAGCCTCCTACAAGCACCTCAACAGCC-ACATGAATGCCCTGCACCAGGGAATGCAGGCCAA 405
PG     CCCAAGCTACAGGAGGAGTTCTTTGCCTGTAAACACCTATGTAGCTGGCCAGTATGATGATCCAGCCTTCTACAAGCACCTCAATAGCCTACATGAATGCCCTGCACCAGGGAATGCAGGCTAA

CONS   CCGTCTGTTCTACCTGGCCTTGCCCCCACTGTCTATGAAGCAGTCACCAAGAACCATTCAAGAGATCTGCATGAGTCAGACAGGCTGGAACCGCATCATAGTGGAGAAGCCCTTCGGGAG 525
PG     CCATCTATTCTATCTGCCCTTGCCCCCCACTGTCTATGAAGCAGTCACCAAGAACCATTCAAGAGATCTGTATGAGGTCAGACAGGCTGGAACCGCATCATAGTGGAGAAGCCCTTCGGGAG

CONS   AGACCTGCAGAGCTCCAATCAACTGTCGAACCACATCTCCTCTCTGTTTCGTGAGGACCAGATCTACCGCATTGACCACTACCTGGGCAAAGAGATGGTCCAGAACCTCATGGTGCTGAG 645
PG     AGACCTGCAGAGCTCCAATCAACTGTTGAACCACATCTCCTCTCTGTTTCGTGAGGACCAGATCTACCGCATTGACCACTACCTGG------------TCCAGAACCTCATGGTGCTGAG

CONS   ATTTGCCAACAGGATCTTTGGACCCATCTGGAATCGAGACAACATTCCTTGT--------GTGATCTTACATTTAAAGAGCCCTTTGGTACTGAGGGTCGTGGGGCTATTTTGATGAA 765
PG     ATTTGCCAACAGGACCTTTGGACCCATCTGGAATCGAGACAACATTCCTTGTGTCCATGCTGTGATCCTTACATTTAAAGAGCCCTTTGATACTGAGGGTTGTGGGGGCTATTTTGATGAA

CONS   TTTGGGATCATCAGGGGATGTCATGCAGAACCACCTCCTGCAGATGTTGTGTCTAGTGGGCCATGGAAAAGCCTGCCTCTACAGATTCAGATGATGTCCGTGATGAGAAGGTCAAAGTGTTA 885
PG     TTTAGGATCATAGGGGATGTTATACAGAACCACCTCCTGCAGATGTTGTGTCTAGTGGGCCATGGAAAAGCCTGCCTCTACAGATTCAAATGATGTCCGTGATGAGAAGGTCAAAGTGTTA

CONS   AAATGTATCTCAGAGTGGAAACTACAACGTGGTCCTTGGCCAGTATGTGGGGAACCCCAGTGGGAGAGGAGAGAGCTACCAATGGGTACTTAGATGACCCCACAGTACCCCATGGGTCT 1005
PG     AAATGTATCTCAGAGGTGGAAATTGGACAACGTGGTCCTTGGCCAGTATGTGGGGAACCCCAGTGGGAGAGGAGAGAGCTACCAATGGGTACTTAGATGACCCCACAGTACCCCATGGGTCCT

CONS   ACCACTGCTACCTTTGCAGCAGCTGTCCTCTATGTGGAGAATGAACGGTGGGATGGAGTACCCTTCATCCTGCGCTGTGGCAAAGCTCTGAATGAGCGCAAAGCTGAAGTGAGACTTCAG 1125
PG     ACCACTGCTACCTTAGCAGCAGCTGTCCTCTATGTGGGAGAATGAACAATGGGATGGAGTACCCTTCATCCTGCACTGTGGCAAAGCTCTGAATGAGTGCAAAGCTGAAGTGAAAACTTCAG

CONS   TTCCGCGATGTGGCAGGTGACATCTTCCACCAGCAGTGCAAGCGTAACGAGCTGGTCATCCGTGTGCAGCCCAATGAGGCGGTATACACCAAGATGATGACCAAGAAGCCTGGCATGTTC 1245
PG     TTCTGCGATGTGGCAGGTGACATCTTCCACCAGCAGTGCAAGTGTAACAAGCTGGTCATCTGTGTGTCA-----ATGAGGCGGTATACACCAAGATGATGACCAAGAACCCTGGCCTTGTTC

CONS   TTCAACCCTGAGGAGTCTGAGCTGGACCTAACCTATGGCAACAGATACAAGAATGTGAAGCTCCCTGATGCCTATGAACGCCTCATCCTGGATGTGTTCTGTGGGAGCCAAATGCACTTT 1365
PG     TTCAACCCTGAGGAGTCTGAGCTGGACCTAACCTATGGCAACAGA----------------------------------------------------------------------CTTT

CONS   GTCCGTAGTGATGAAACTCAGGGAAGCCTGGCGTATCTTCACACCATTGCTGCACAAGATTGATCGAGAGAGCCCCAGCCCATCCGTATGTCTATGGCAGCCGAGGTCCCACAGAGGCA 1485
PG     GTCCATAATGATGAAACTCGGCACACGCCTGGCATATCTTCACACATTGCTGCACAAGATTGGTGCAGAGAGAGCCCCAGCCCATTCCTTTATGTCTATGGCAGCCGAGGTCCCACAGAGGCA

CONS   GATGAGCTGATGAAGAAGAGTGGGCTTCCAGTATGAGGGTACCTACAAGTGGGTGAACCCTCACAAGCTCTGAGCCCTGGAAACTTACACCCATCTGCACTCTGCCTCTTCTGGCCACCCTT 1605
PG     GATGAGCTGATGAAGAAGAGTGGGCTTCCAGTATGAGGGT----------GGGTGAACCCTCACAAGCTCTGAGCCCTGGAAACTTACACCATCTGCATTCTGCCTCTTCTGGCCACCCTT

CONS   TCTGCATCTGCCCTTCTCACCATCTAACCCTCTATTAGGACTATTGACCTCATATTGGAAAGACTTTGGGACCCATAGGCCTTAGCTACACATTCTAGTCCTCTGGGCTTAGGCCACCATTC 1725
PG     TCTGCATCTGCCCTTCTCACCATCTAACCCTATATTAGGACTATTGACCTCATATTGGAAAGACTTTGGGACCCATAGGCCTTAGCTACACATTCTAGTCTCTGGGCTTAGGCCACCATTC

CONS   TGTCCTATGCTGCTGCCACTGCCACTACCACTAAACCCAGCTACATTCCTCAGATACCAGGCATTCAAAACGCATTGCAATGCTTTCAGGACCACCACTGTCCCTATCTGAGCCACCCAT 1845
PG     TGTCCTATGCTGCTGCCACTGCCACTACCACTAAACCCAGCTACATTCCTCAAATACCAGGCATTCAAAACACATTGCAATGCTTTCAGGACCACCACTGTCCCTATCTGAGCCACCCAT

CONS   CTTTCCACAAGACCTGAATCACCTCCTCCCCTCAATCCCCTGCAGAAAGAAGCGCTATCAGTCTGTCCCTGGACTCCTTAAGATAGGAGTTAGGAACAATTGGGAGGAGCCTTGGGCCTT 1965
PG     CTTTCCACAAGACCTGAATCACCTCCTCCCCTCCAACTCCTGCAGAATGAACGCCTATCAGTCTGTCCCTGGACTCCTTAAGATAGGAGTTAGGAACAATTGGGAGGAGCCTTGGGCCTT

CONS   GGAAGGGACAATGACCAAACCACACTTCCCTGAGACTGTGGGCAAGCTCCTCAAAACTTAAAGTGATCAAGGACACCCATCTGAGAGGGACCTGCCCATAGCCACACTAGCCTTAGTGCTAC 2085
PG     GGAGGGGACAATGACCAAACCACACTTCCCTGAGACTGTGGGCAAATTCCTCAAAACTTAAAGTGATGAAGGAGCACCCATTTGAGAGGGACCTGCCCATAGCCACACTAGCCTTAGTGCTAC

CONS   TTGACATTCCTCCTCACCAGTGGAAGAACTCTCATGCTGCCTAGCAATATTTTGGGGGCCATAGATATCTCCTAAACAATTCCATAGTCCATAGTCAGCCTCATCCAACCCATGGGCAG 2205
PG     TTGACATTCCTCCTCACCAGTGGAAGAAACCTCATGCTGCCTAGCGATATTTTGGGGGCCATAGATATCTCCTAAACAATTCCATAGTCCATAGTCAACCTCATCCAACCCATGGGCAG

CONS   CCTCCTTACCAAAGGAAGGTAAGAGCAGCAGCAGCTAGAATTTTCCTACCCCAACCCTGCCATTAAATCCTCAAAAAAAAAAAAAAAAAA 2292
PG     CCTCATTACCAAAGGAAGGTAAGAGCAGCAGCTAGAATTTTCCTACCCCAACCCTGCCATTAAATCATCAAAAAAGTTAAAAAAAAGAAAGAAAAAGACAA
```

The coding region and intron/exon boundaries were sequences from the genomic fragments containing the functional G6PD gene from rat liver (Table 2 and Figure 3). The structure of the rat liver gene was found to be remarkably similar to that of the human gene as reported by Martini et al.[71] The number of exons (13), the position of the introns in the coding sequence, and the size of the corresponding exons are identical in both genes. The intron boundaries (indicated by arrows in Figure 3) are highly conserved although intron size varies considerably (Table 2). IVS I does not intersect the coding region, and the exon/intron junction at this site in the 5′ untranslated region is not conserved. Exon 13 contains all of the 3′ untranslated region, the termination codon, and 88 base pairs of coding sequence.

Comparison of the rat consensus G6PD sequence with that of the human demonstrates a high degree of homology except at the sequences encoding the amino terminal end of the protein. However, two different sequences have been reported for the human amino terminal sequence of G6PD. One was determined by sequence analysis of the protein from red blood cells[67] and the other deduced from cDNAs produced from a variety of human cell types and transformed cell lines.[69,70] The sequence reported by Jeffery et al.[47] for the rat liver protein corresponds to that deduced from the human cDNAs. In order to determine the location of this difference in the sequence information, we compared cDNA and protein sequence data from rat and human with that of the G6PD gene. As shown in Figure 4, the point of divergence is localized to amino acid position #35 in the G6PD protein sequence. At the nucleic acid level, the site of sequence divergence is immediately 5′ of the IVS 2 region of the gene. IVS 2 is extremely large ($>$11 kb in the human gene) and is the first intron to intersect the coding sequence. In the case of the rat liver sequence, only 120 bases that encode protein are upstream of this intron. That this structural feature has been preserved in evolution is indicative of a possible important structural role in gene function. Either alternative splicing or use of different promoters as in the case of R-type vs. L-type pyruvate kinase could give rise to amino terminal heterogeneity for G6PD.

FIGURE 3. Comparison of the consensus coding sequence for rat G6PDH with a processed pseudogene. Nucleotide residues are numbered from the start of translation. The consensus sequence for G6PDH (upper, CONS) is from complete, overlapping cDNA clones from rat liver and brain libraries and from the coding region of the functional gene. Each clone was sequenced twice in both directions to verify the authenticity of the sequence. The pseudogene sequence (lower, PG) starts with the direct repeat (underlined) and ends with the same sequence immediately 3′ of the poly A tail. Bases that are different between the two sequences are indicated with a bar. The polyadenylation signal, the start, and termination of translation are denoted by the box. The positions of the 12 introns are denoted by arrows.

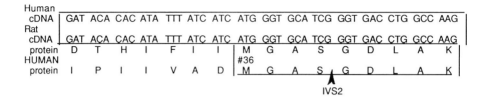

FIGURE 4. Amino terminal protein and 5′ coding sequence relationships between rat and human G6PDH. The rat protein sequence (1) and the human protein (19) and cDNA sequence (17) are from published reports. Identities in the sequences are posed. The position of the second intron of the G6PDH gene is noted by the arrow. The rat amino acid positions from 29 to 44 are shown and for the human, positions 46 to 61.

XI. CONCLUSIONS

In conclusion, genetic studies have established that G6PD is encoded by a single copy X-linked gene in mammals, and the expectation has been that the protein would be the same in all tissues. Comparison of the nucleic acid sequence encoding G6PD from drosophilia to man has revealed a high degree of sequence conservation except for that encoding the extreme amino terminal end of the protein. The conservation of the coding sequence highlights the importance of the protein and the reaction carried out by it for metabolism in complex biological systems. The amino terminal heterogeneity may be important in tissue-specific expression and control of G6PD activity. In this regard, G6PD occupies a delicate place at the interface of carbohydrate and lipid metabolism. While it is primarily a housekeeping enzyme in most tissues, expression is clearly regulatable in others. Liver is most notable since, as discussed above, a variety of hormones, nutrients, and other agents capable of producing oxidant stress regulate the activity of the enzyme. Thus, the cellular machinery responsible for controlling expression would be expected to be complex in order that regulation be accomplished in those tissues. Here it is inducible while maintaining a constitutive ''housekeeping'' level in other tissues. The current problems associated with isolating and characterizing the promoter for the G6PD gene may be a reflection of the complex nature of the putative control region.

REFERENCES

1. **Tepperman, H. M. and Tepperman, J.,** The hexose monophosphate shunt and adaptive hyperlipogenesis, *Diabetes,* 7, 478, 1958.
2. **Glock, G. E. and McLean, P.,** A preliminary investigation of the hormonal control of the hexose monophosphate oxidative pathway, *Biochem. J.,* 61, 390, 1955.
3. **Fitch, W. M. and Chaikoff, J. L.,** Extent and patterns of adaptation of enzyme activities in liver of normal rats fed diet high in glucose and fructose, *J. Biol. Chem.,* 235, 554, 1960.
4. **Rudack, D., Chisholm, E. M., and Holten, D.,** Rat liver glucose 6 phosphate dehydrogenase. Regulation by carbohydrate diet and insulin, *J. Biol. Chem.,* 246, 1249, 1971.
5. **Potter, V. R. and Ono, T.,** Enzyme patterns in rat liver and Morris hepatoma 5123 during metabolic transitions, *Cold Spring Harbor Symp. Quant. Biol.,* 26, 355, 1961.
6. **Tepperman, J. and Tepperman, H. M.,** Effects of antecedent food intake pattern on hepatic lipogenesis, *Am. J. Physiol.,* 193, 55, 1958.
7. **Tepperman, H. M. and Tepperman, J.,** On the response of hepatic glucose 6 phosphate dehydrogenase activity to changes in diet composition and food intake pattern, *Adv. Enzyme Reg.,* 1, 121, 1963.
8. **Sassoon, H. F., Dror, Y., Watson, J. J., and Johnson, B. C.,** Dietary regulation of liver glucose 6 phosphate dehydrogenase in the rat: starvation and dietary carbohydrate induction, *J. Nutr.,* 103, 321, 1973.
9. **Szepesi, B. and Freedland, R. A.,** Differential requirement for de novo RNA synthesis in the starved refed rat; inhibition of the overshoot by 8-azaguanine after refeeding, *J. Nutr.,* 99, 449, 1969.
10. **Szepesi, B. and Berdanier, C. D.,** Time course of the starve-refeed response in rats; the possible role of insulin, *J. Nutr.,* 101, 1563, 1971.
11. **Szepesi, B. and Michaelis, O. E.,** Comparison of the fructose effect and the starve-refeed response of rat liver enzymes, *Life Sci.,* 11, 113, 1972.
12. **Michaelis, O. E. and Szepesi, B.,** Effect of various sugars on hepatic glucose 6 phosphate dehydrogenase, malic enzyme and total lipid and the rat, *J. Nutr.,* 103, 697, 1973.
13. **Williams, B. H. and Berdanier, C. D.,** Effects of diet composition and adrenalectomy on the lipogenic responses of rats to starvation-refeeding, *J. Nutr.,* 112, 534, 1982.
14. **Baltzell, J. K. and Berdanier, C. D.,** Effect of the interaction of dietary carbohydrate and fat on the responses of rats to starvation-refeeding, *J. Nutr.,* 115, 104, 1985.
15. **Berdanier, C. D., Johnson, B., and Buchanan, M.,** Interacting effects of menhaden oil and sucrose on the lipogenic response to starvation-refeeding, *Nutr. Res.,* 9, 1167, 1989.
16. **Johnson, B. J. and Berdanier, C. D.,** Effect of menhaden oil on the responses of rats to starvation-refeeding, *Nutr. Rep. Int.,* 36, 809, 1987.
17. **Bue, J. and Berdanier, C. D.,** Effect of antecedent diet on the responses of BHE rats to starvation-refeeding, *Nutr. Rep. Int.,* 35, 791, 1987.
18. **Szepesi, B., Berdanier, C. D., and Michaelis, O. E.,** Effect of age on the responses to starvation-refeeding in the rat, *Proc. Soc. Exp. Biol. Med.,* 147, 195, 1974.
19. **Kim, M.-J., Pan, J.-S., and Berdanier, C. D.,** Age and thyroid hormone as factors in the responses of BHE rats to starvation-refeeding, *Proc. Soc. Exp. Biol. Med.,* 183, 273, 1986.
20. **Bois-Joyeux, B., Chanez, M., Aranda-Haro, F., and Peret, J.,** Age dependent hepatic lipogenic enzyme activities in starved-refed rats, *Diabetes Metab.,* 16, 290, 1990.
21. **Berdanier, C. D.,** Effects of estrogen on the responses of male and female rats to starvation refeeding, *J. Nutr.,* 111, 1425, 1981.
22. **Smith, E. R. and Barker, K. L.,** Effects of estradiol and NADP on the rate of synthesis of uterine glucose 6 phosphate dehydrogenase, *J. Biol. Chem.,* 249, 6541, 1974.

23. **Szepesi, B. and Michaelis, O. E.,** Problems in the demonstration of the disaccharide effect due to stress during shipment of rats, *Nutr. Rep. Int.,* 12, 299, 1975.

24. **Michaelis, O. E., Scholfield, D. J., Nace, C. S., and Reiser, S.,** Demonstration of the disaccharide effect in nutritionally stressed rats, *J. Nutr.,* 108, 919, 1978.

25. **Szepesi, B.,** Metabolic memory: effect of antecedent dietary manipulations on subsequent diet induced responses of rats. I. Effects on body weight, food intakes, glucose 6 phosphate dehydrogenase and malic enzyme, *Can. J. Biochem.,* 51, 1604, 1973.

26. **Szepesi, B.,** Effect of previous caloric restriction on two NADP-linked dehydrogenases of rat liver, *Nutr. Rep. Int.,* 10, 127, 1974.

27. **Szepesi, B., Berdanier, C. D., Diachenko, S. K., and Moser, P. B.,** Effect of length of starvation, refeeding and 8-azaquanine on serum insulin and NADP-linked dehydrogenases of rat liver, *J. Nutr.,* 101, 1147, 1971.

28. **Homan, H. D. and Berdanier, C. D.,** Effect of adrenalectomy on the enzyme responses of starved-refed rats, *Nutr. Rep. Int.,* 22, 213, 1980.

29. **Gimenez, R. S. and Johnson, B. C.,** Pair feeding in the dietary control of G6PD, *J. Nutr.,* 111, 260, 1981.

30. **Freedland, R. A., Cunliffe, T. L., and Zinkl, J. G.,** Effect of insulin on enzyme adaptations to diets and hormones, *J. Biol. Chem.,* 241, 5448, 1966.

31. **Szepesi, B., Vegors, R., and DeMosey, J. M.,** On the possible role of insulin in the starve-refeed response, *Nutr. Rep. Int.,* 5, 281, 1972.

32. **Berdanier, C. D., Wurdeman, R., and Tobin, R. B.,** Further studies on the role of the adrenal hormones in the responses of rats to meal feeding, *J. Nutr.,* 106, 1791, 1976.

33. **Wurdeman, R., Berdanier, C. D., and Tobin, R. B.,** Enzyme overshoot in starved-refed rats: role of adrenal glucocorticoid, *J. Nutr.,* 108, 1457, 1978.

34. **Bouillon, D. J. and Berdanier, C. D.,** Role of glucocorticoid in adaptive hyperlipogenesis in the rat, *J. Nutr.,* 110, 286, 1980.

35. **Berdanier, C. D. and Shubeck, D.,** Interaction of glucocorticoid and insulin in the responses of rats to starvation-refeeding, *J. Nutr.,* 109, 1766, 1979.

36. **Nessmith, S., Baltzell, J., and Berdanier, C. D.,** Interaction of glucocorticoid and thyroxine in the responses of rats to starvation-refeeding, *J. Nutr.,* 113, 2260, 1983.

37. **Kletzien, R. F., Pasiza, M. W., Becker, J. E., Potter, V. R., and Butcher, F. R.,** Induction of amino acid transport in primary cultures of adult rat liver parenchymal cells by insulin, *J. Biol. Chem.,* 251, 3014, 1976.

38. **Inoue, C., Yamamoto, H., Nakamura, T., Ichihara, A., and Okamoto, H.,** Nicotinamide prolongs survival of primary cultured hepatocytes without involving loss of hepatocyte-specific functions, *J. Biol. Chem.,* 264, 4747, 1989.

39. **Dunn, J. C. Y., Yarmush, M. L., Koebe, H. G., and Tompkins, R. G.,** Hepatocyte function and extracellar matrix geometry: long-term culture in a sandwich configuration, *Fed. Proc.,* 3, 174, 1989.

40. **Kelley, D. S. and Kletzien, R. F.,** Ethanol modulation of the hormonal and nutritional regulation of glucose-6-phosphate dehydrogenase activity in primary cultures of rat hepatocytes, *Biochem. J.,* 217, 543, 1984.

41. **Nakamura, T., Yoshimoto, K., Aoyama, K., and Ichihara, A.,** Hormonal regulations of glucose-6-phosphate dehydrogenase and lipogenesis in primary cultures of rat hepatocytes , *J. Biochem.,* 91, 681, 1982.

42. **Manos, P., Nakayama, R., and Holten, D.,** Regulation of glucose-6-phosphate dehydrogenase synthesis and mRNA abundance in cultured rat hepatocytes, *Biochem. J.,* 276, 245, 1991.

43. **Salati, L. M., Adkins-Finke, B., and Clarke, S. D.,** Free fatty acid inhibition of the insulin induction of glucose-6-phosphate dehydrogenase in rat hepatocyte monolayers, *Lipids,* 23, 36, 1988.

44. **Kelley, D. S., Nelson, G. J., and Hunt, J. E.,** Effect of prior nutritional status on the activity of lipogenic enzymes in primary monolayer cultures of rat hepatocytes, *Biochem. J.,* 235, 87, 1986.

45. **Kurtz, J. W. and Wells, W. W.,** Induction of glucose-6-phosphate dehydrogenase in primary cultures of adult rat hepatocytes: requirement for insulin and dexamethasone, *J. Biol. Chem.,* 256, 10870, 1981.

46. **Kletzien, R. F., Prostko, C. R., Stumpo, D. J., McClung, J. K., and Dreher, K. L.,** Molecular cloning of DNA sequences complementary to rat liver glucose-6-phosphate dehydrogenase mRNA: nutritional regulation of mRNA levels, *J. Biol. Chem.,* 260, 5621, 1985.

47. **Jeffery, J., Soderling-Barros, J., Murray, L. A., Hansen, R. J., Szepesi, B., and Jornvall, H.,** Molecular diversity of glucose-6-phosphate dehydrogenase: rat enzyme structure identifies NH_2-terminal segment, *Proc. Natl. Acad. Sci. U.S.A.,* 85, 7840, 1988.

48. **Ho, Y.-S., Howard, A. J., and Crapo, J. D.,** Cloning and sequence of a cDNA encoding rat glucose-6-phosphate dehydrogenase, *Nucleic Acids Res.,* 16, 7746, 1988.

49. **Prostko, C. R., Fritz, R. S., and Kletzien, R. F.,** Nutritional regulation of hepatic glucose-6-phosphate dehydrogenase: transient activation of transcription, *Biochem. J.,* 258, 295, 1989.

50. **Stumpo, D. J. and Kletzien, R. F.,** Regulation of glucose-6-phosphate dehydrogenase mRNA by insulin and the glucocorticoids in primary cultures of rat hepatocytes, *Eur. J. Biochem.,* 144, 497, 1984.

51. **Fritz, R. S., Stumpo, D. S., and Kletzien, R. F.,** Glucose-6-phosphate dehydrogenase mRNA sequence abundance in primary cultures of rat hepatocytes: effect of insulin and dexamethasone, *Biochem. J.,* 237, 637, 1986.

52. **Kelley, D. S., Nelson, G. J., Serrato, C. M., and Schmidt, P. C.,** Nutritional regulation of hepatic lipogenesis in the rat, *Nutr. Res.,* 7, 509, 1987.

53. **Katsurada, A., Iritani, N., Fukuda, H., Matsumura, Y., Noguchi, T., and Tanaka, T.,** Effects of nutrients and insulin on transcriptional and post-transcriptional regulation of glucose-6-phosphate dehydrogenase synthesis in rat liver, *Biochim. Biophys. Acta,* 1006, 104, 1989.

54. **Berdanier, C. D.,** Role of glucocorticoids in the regulation of lipogenesis, *Fed. Proc.,* 3, 2179, 1989.

55. **Stumpo, D. J. and Kletzien, R. F.,** The effect of ethanol, alone and in combination with the glucocorticoids and insulin, on glucose-6-phosphate dehydrogenase synthesis and mRNA in primary cultures of hepatocytes, *Biochem. J.,* 226, 123, 1985.

56. **Mariash, C. N., Kaiser, F. E., Schwartz, H. L., Towle, H. C., and Oppenheimer, J. H.,** Synergism of thyroid hormone and high carbohydrate diet in the induction of lipogenic enzymes in the rat, *J. Clin. Invest.,* 65, 1126, 1980.

57. **Fritz, R. S. and Kletzien, R. F.,** Regulation of glucose-6-phosphate dehydrogenase by diet and thyroid hormone, *Mol. Cell Endo.,* 51, 13, 1987.

58. **Yoshimoto, K., Nakamura, T., and Ichihara, A.,** Reciprocal effects of epidermal growth factor on key lipogenic enzymes in primary cultures of adult rat hepatocytes, *J. Biol. Chem.,* 258, 12355, 1983.

59. **Iynedjian, P. B., Jotterand, D., Nouspikel, T., Asfari, M., and Pilot, P.,** Transcriptional induction of glucokinase gene by insulin in cultured liver cells and its repression by the glucagon-cAMP system, *J. Biol. Chem.,* 264, 21824, 1989.

60. **Teel, J. F., Kletzien, R. F., Ginsberg, L. C., Stevens, G. J., and Stapleton, S. R.,** Acetaldehyde increases mRNA levels of glucose-6-phosphate dehydrogenase in rat hepatocytes in culture, *Fed. Proc.,* 5, 310 (Abstr.), 1991.

61. **Tarlou, A. R., Brewer, G. J., Carson, P. E., and Alving, A. S.,** Primaquine sensitivity, *Arch. Intern. Med.,* 109, 137, 1962.

62. **Shapiro, B. M.,** The control of oxidant stress at fertilization, *Science,* 252, 533, 1991.

63. **Beutler, E., Robson, M., and Buttenwies, E.,** The mechanism of glutathione destruction and protection in drug-sensitive and non-sensitive erythrocytes, *J. Clin. Invest.,* 36, 617, 1957.

64. **Kirkman, H. N. and Hendrickson, E. M.,** Sex-linked electrophoretic differences in glucose-6-phosphate dehydrogenase, *Am. J. Human Genet.,* 15, 241, 1963.

65. **Luzzato, L.,** Glucose-6-phosphate dehydrogenase: genetic and haematological aspects, *Cell. Biochem. Function,* 5, 101, 1987.

66. **Noguchi, T., Yamada, K., Inoue, H., Matsuda, T., and Tanaka, T.,** The L- and R-type isozymes of rat pyruvate kinase are produced from a single gene by use of different promoters, *J. Biol. Chem.,* 262, 14366, 1987.

67. **Jakizawa, T., Huang, T.-Y., Ikuta, T., and Yoshida, A.,** Human glucose-6-phosphate dehydrogenase: primary structure and cDNA cloning, *Proc. Natl. Acad. Sci. U.S.A.,* 83, 4157, 1986.

68. **Fouts, D., Ganguly, R., Gutierrez, A. G., Lucchesi, J. C., and Manning, J. E.,** Nucleotide sequence of the Drosophilia G6PD gene (ZW+) and comparison with the homologous human gene, *Gene,* 63, 261, 1988.

69. **Vulliamy, T. J., D-Urso, M., Battistuzzi, G., Estrada, M., Foulkes, N. S., Martini, G., Calabro, V., Poggi, V., Giordano, R., Town, M., Luzzatto, L., and Persico, M. G.,** Diverse point mutations in the human glucose-6-phosphate dehydrogenase gene cause enzyme deficiency and mild or serve hemolytic anemia, *Proc. Natl. Acad. Sci. U.S.A.,* 85, 5171, 1988.

70. **Persico, M. G., Vigletto, G., Martini, G., Toniolo, D., Paonessa, G., Moscatelli, C., Dono, R., Vulliamy, T., Luzzatto, L., and D-Urso, M.,** Isolation of human glucose-6-phosphate dehydrogenase cDNA clones: primary structure of the protein and unusual 5' non-coding region, *Nucleic Acids Res.,* 14, 2511, 1986.

71. **Martini, G., Toniolo, D., Vulliamy, T., Luzzatto, L., Dono, R., Viglietto, G., D'Urso, M., and Persico, M. G.,** Structural analysis of the X-linked gene encoding human glucose-6-phosphate dehydrogenase, *EMBO J.,* 5, 1849, 1986.

Chapter 9

NUTRITIONAL AND HORMONAL REGULATION OF GENES ENCODING ENZYMES INVOLVED IN FAT SYNTHESIS

Hei Sook Sul, Naima Moustaid, Kenji Sakamoto, Cynthia Smas, Nick Gekakis, and Ann Jerkins

TABLE OF CONTENTS

207

I. INTRODUCTION

Fatty acid and triacylglycerol synthesis for energy storage is known to be regulated in response to the nutritional and hormonal state of the animal. Subjecting rats to a few days of fasting results in the inhibition of lipogenesis; when fasted animals are subsequently fed a diet high in carbohydrate and low in fat, there is a prompt rise in the production of fatty acids and triacylglycerol to levels above those seen in normally fed rats.[1,2] Under these conditions, excess glucose in the cell is first converted to pyruvate via glycolysis. Pyruvate is converted to acetyl CoA which is used for the synthesis of long chain fatty acids, primarily palmitate. NADPH required for fatty acid synthesis is produced by the malic enzyme or via the pentose phosphate pathway. The fatty acids produced are then used for esterification onto glycerol-3-phosphate to generate triacylglycerol. Many key enzymes involved in these metabolic pathways are induced during the refeeding period.[1] The increase in the concentration of enzymes, such as of fatty acid synthase, has been shown to be due to the increased synthesis of enzyme protein.[3]

It may be that expression of genes encoding some of these inducible enzymes is regulated by nutritional and hormonal stimuli via common mechanisms. Changes in nutrient intake bring changes in circulating glucose and amino acid concentrations, which, in turn, signal secretion of hormones. It is generally accepted that insulin in the circulation is elevated during feeding and has positive effects, whereas the glucagon level is decreased and has negative effects on the expression of enzymes participating in fatty acid and triacylglycerol synthesis.[4] Circulating T3 is also known to decrease during fasting and increase during feeding. In addition, a high concentration of glucose is known to be required for the induction of these enzymes. The actual molecular mechanisms underlying this adaptive metabolic process, however, are, as yet, poorly understood. It is, in general, accepted that the thyroid hormone-receptor complex itself interacts with a thyroid hormone response element to activate gene transcription.[5] In cAMP mediated transcriptional activation, a nuclear protein is phosphorylated by the cAMP-dependent protein kinase and interacts with the cAMP response element. This protein is called the cAMP response element binding protein (CREB).[6] However, transcriptional regulation by the major anabolic hormone insulin is not well understood. Moreover, how a combination of hormones, which regulate specific genes coding for enzymes participating in fat synthesis, brings about an appropriate response to dietary stimuli is also not known.

This review will discuss the regulation of expression of genes for three enzymes participating in fatty acid and fat synthesis, liver phosphofructokinase (E.C. 2.7.1.11), fatty acid synthase (E.C. 2.3.1.85), and p90 (E.C. 2.3.1.15). These three proteins have been selected for investigation in our laboratory to compare their regulation at the gene level by nutritional and hormonal stimuli.

Each catalyzes an important step in different pathways through which metabolic flux increases during fat synthesis for energy storage. Phosphofructokinase is a key regulatory enzyme in glycolysis and exists in three isozymes specific to the tissues in which each are found: liver, muscle, and brain. It has been shown that the concentration of the major liver isozyme, a tetramer of the liver type subunit, was sensitive to nutritional and hormonal changes in that it was decreased by starvation or by insulin deprivation.[9,10] Fatty acid synthase, a dimer of 250 kDa subunit, carries out all the steps for synthesis of fatty acids from acetyl CoA and malonyl CoA. The fatty acid synthase activity is regulated by changes in enzyme concentration. Allosteric effectors or covalent modifications of this enzyme are not known.[2] p90 has been tentatively identified as glycerol-3-phosphate acyltransferase, an enzyme which catalyzes the esterification of glycerol-3-phosphate to 1-monoacylglycerol-3-phosphate (lysophosphatidate), the committed step in triacylglycerol biosynthesis. Glycerol-3-phosphate acyltransferase activity is found in both microsomal and mitochondrial membranes, and it is thought that the mitochondrial form in particular is decreased by starvation and increased by insulin.[7,8]

II. CLONING OF NUTRITIONALLY REGULATED cDNA SEQUENCES

To study the regulatory mechanisms of gene expression, we first isolated cDNA sequences coding for these enzymes using the following strategies. We isolated liver poly(A^+) RNA from mice under the condition of high fat synthesis by first fasting and then feeding a diet high in carbohydrate but devoid of fat. Since fatty acid synthase has a large subunit, we enriched poly(A^+) RNA for higher size messages by sucrose density gradient centrifugation before cDNA library construction and screening.[11] Replica filters of this library were hybridized to the total cDNA synthesized by reverse transcribing mRNAs prepared from fasted or refed mouse liver; clones that hybridized preferentially to the refed liver cDNA were picked and characterized. The cloned cDNA sequence for murine fatty acid synthase hybridized to a single 8.2 kb mRNA. This is in contrast to avian and rat tissues where two species of mRNA for fatty acid synthase are detected.[12,13] In addition to the fatty acid synthase cDNA clone, we have also isolated cDNA sequences to three other mRNAs of unknown identity. The open reading frame of one such mRNA with a size of 6.8 kb (p90 mRNA) showed sequence homology to *E. coli* glycerol-3-phosphate acyltransferase and was tentatively identified as the murine enzyme.[14] The observation that glycerol-3-phosphate acyltransferase activity in rat liver, especially in the mitochondrial fraction, was increased by feeding a high carbohydrate diet supports this tentative identification. For isolation of the cDNA sequence for liver phosphofructokinase, an enzyme known to be present in very low concentration, the polysome immunopre-

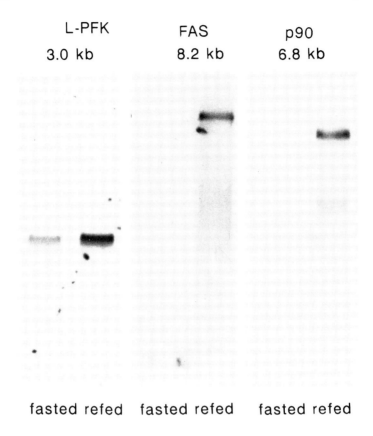

L-PFK   FAS   p90
3.0 kb   8.2 kb   6.8 kb

fasted refed   fasted refed   fasted refed

FIGURE 1. Relative levels of liver (B-type) phosphofructokinase, fatty acid synthase, and p90 mRNAs during fasting and refeeding of a high carbohydrate, fat-free diet. Poly (A +) RNA (4 μg per lane) prepared from livers of mice starved 48 h or previously starved mice refed for 16 h were subjected to 0.7% agarose gel electrophoresis, transferred to nitrocellulose, and probed with ³²P-labeled cDNA inserts.

cipitation method was employed for enrichment of its message before preparation of a cDNA library in λ gt11 and screening with liver phosphofructokinase isozyme-specific antibodies.[15]

Northern blot analysis using these cDNA sequences showed that the fatty acid synthase and p90 mRNAs were present in high concentration in adipose tissue and liver.[11,14] Liver phosphofructokinase mRNA, on the other hand was detected in all tissues examined although liver showed the highest concentration.[15] Fatty acid synthase and p90 mRNAs were not detectable in fasted mouse liver and increased by two orders of magnitude when mice were refed a high carbohydrate, fat-free diet.[11] In contrast, the liver phosphofructokinase mRNA was easily detectable in fasted mice liver and increased fivefold during refeeding (Figure 1). Dunaway and co-workers previously reported a decrease in liver phosphofructokinase concentration in starved or insulin-deprived animals that was primarily due to an enhanced rate of enzyme degradation.[10] Our results indicate that liver phosphofructokinase synthesis is regulated by changes in mRNA level. However, it is clear that the glycolytic enzyme liver phosphofructokinase, although regulated in the same fashion as the lipogenic enzymes,[4,11,16,17] is altered to a much smaller extent. It is worth noting that in the liver, glycolysis is expected to occur even during gluconeogenic periods. Moreover, since phosphofructokinase, unlike fatty acid synthase, is highly regulated by metabolites, which act as allosteric effectors, control at the gene level may not play a primary role in regulation of phosphofructokinase activity. This is in contrast to the regulation of a well-characterized glycolytic enzyme, pyruvate kinase; the mRNA level for pyruvate kinase seems to be altered during fasting/refeeding to a similar order of magnitude as those for fatty acid synthase and p90.[18]

III. *IN VIVO* NUTRITIONAL AND HORMONAL REGULATION OF GENE EXPRESSION

The roles for specific hormones in controlling expression of fatty acid synthase during fasting/refeeding have been suggested in various studies. Such nutritional manipulation causes an increase in circulating insulin and a decrease in glucagon levels. In addition, the increase in synthesis of fatty acid synthase in rat liver caused by refeeding is not observed in diabetic animals. Insulin specifically restores this increase in fatty acid synthase synthesis.[3] The lipolytic hormone glucagon decreases the concentrations of lipogenic enzymes and also suppresses glycolysis. Evidence for the role of glucagon was also demonstrated by an inhibition of the effects of refeeding on enzyme activities.[3]

With the cloned cDNA sequences, we were able to examine expression of these nutritionally regulated genes at the pretranslational level. In normal animals, there was a steady increase in both fatty acid synthase and p90 mRNAs for 16 h of refeeding after a lag period of a few hours. The lag was probably due to the digestion and absorption processes (Figure 2). However, in streptozotocin-diabetic animals, the mRNA levels approximated those of fasted animals when measured 8 h after refeeding. This indicated that insulin is required for the fasting/refeeding induction of the mRNAs for fatty acid synthase and p90. Dibutyryl cAMP, administered at the time of refeeding to normal fasted animals, inhibited the induction of the two mRNAs by 90% (Figure 2). Because glucagon stimulates production of cAMP in liver, these data suggest that the barely detectable levels of these mRNAs during fasting may be partly due to the increased plasma glucagon level.

The effect of insulin was further examined by measuring steady-state mRNA levels after insulin administration to streptozotocin-diabetic animals fed a high carbohydrate, fat free diet (Figure 3). The induction of fatty acid synthase and p90 mRNAs by insulin is rapid and marked. The mRNA levels

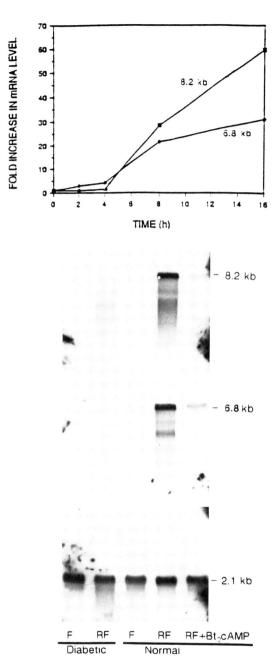

FIGURE 2. Effects of streptozotocin-induced diabetes and cAMP on fatty acid synthase and p90 mRNA induction by fasting/refeeding. Poly (A +) RNA was prepared from livers of diabetic and normal mice fasted (F) or refed (RF) for 8 h. Dibutyryl cAMP and theophylline were injected at the time of refeeding in normal, fasted mice. Northern analyses were carried out as in Figure 1. 2.1 kb mRNA represents α-actin.

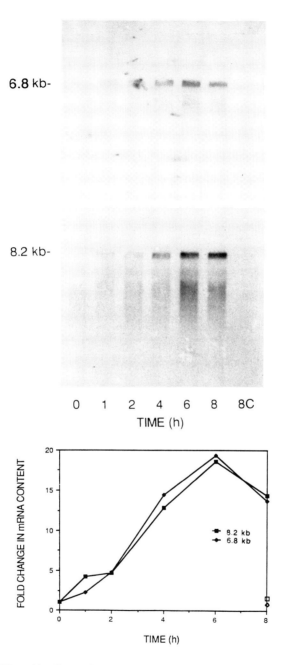

FIGURE 3. Effect of insulin on mRNA levels for p90 and fatty acid synthase in streptozotocin-diabetic mice. Poly (A +) RNA was prepared from livers of streptozotocin-diabetic mice killed at the indicated times after administration of combined dose of regular insulin (3 U/100 g) intraperitoneally and Lente Insulin (30 U/100 g) subcutaneously. Saline-injected mice were also used at 8 h as a control (8C). Northern analyses were done as in Figures 1 and 2.

for fatty acid synthase and p90 increased four- and twofold 1 h after intra-peritoneal insulin injection, respectively, and achieved maximal increases of about 20-fold after 6 h. However, it is known that glucagon levels are decreased when insulin is administered to diabetic animals and this may be a factor. In cultured avian embryo hepatocytes, Goodridge and co-workers reported that insulin only amplified the effect of T3 on fatty acid synthase gene expression.[19,20] It would be surprising if there is a difference in regulation of avian and murine fatty acid synthase by insulin.

IV. REGULATION OF TRANSCRIPTION

mRNA concentrations can be, in theory, controlled by the rate of specific transcription, processing, or turnover. The first approach would be to ask if the altered transcription can quantitatively account for the observed effects of fasting/refeeding, insulin, and cAMP on mRNA levels for fatty acid synthase and p90. Therefore, transcription run-on analyses were carried out in isolated liver nuclei from normal and diabetic animals treated as described above.[14,20] The transcription rates of the fatty acid synthase and p90 genes were increased when previously fasted mice were refed a high carbohydrate diet (Figure 4). The maximum increase in transcription rate of 39-fold for fatty acid synthase was attained at 6 h after refeeding and was maintained up to 16 h, whereas the transcriptional increase for the p90 gene was substantially slower in that it was 2.5-fold at 6 h, 7-fold at 9 h, and reached 22-fold after 16 h of refeeding. However, there was no detectable transcription of fatty acid synthase and p90 genes in fasted or fasted-refed streptozotocin-diabetic animal liver, indicating a requirement of insulin for transcriptional induction by fasting/refeeding. Furthermore, there was a rapid and marked increase in transcription rates of fatty acid synthase and p90 genes when insulin was given to diabetic mice; transcription rates of both of these genes increased approximately fourfold after 30 min and attained a maximum increase of seven- to eightfold at 2 h (Figure 5). The results demonstrate that these genes are highly regulated at the transcriptional level by nutritional and hormonal stimuli. We then asked whether the insulin effect on specific transcription requires ongoing protein synthesis and/or a regulatory protein with a short half-life. In the presence of cycloheximide, the increases in mRNA levels and transcriptional rates for fatty acid synthase and p90 caused by insulin in diabetic mice were not observed. Thus, it appears insulin is required for ongoing protein synthesis and that a similar mechanism may be involved in the insulin regulation of these two genes.

The effect of cAMP on fatty acid synthase and p90 gene transcription was also examined. When dibutyryl cAMP was given at the start of refeeding of normal animals, there was no increase in transcription of either fatty acid synthase or p90 genes by fasting/refeeding, indicating a negative transcrip-

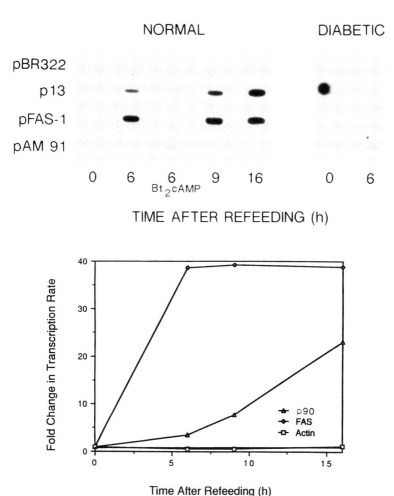

FIGURE 4. Effects of fasting/refeeding, cAMP, and streptozotocin-diabetes on fatty acid synthase and p90 gene transcription. Normal or streptozotocin-diabetic mice were treated as described in Figure 1. Nuclei were isolated from livers and run-on transcription analyses were carried out. Plasmids containing cDNA inserts coding for fatty acid synthase (pFAS-1), p90 (p13), and actin (pAM91) were used. pBR322 is a negative control.

tional control of these genes by cAMP (Figure 4). Similar effects of cAMP were observed in liver phosphofructokinase gene expression. The increase in mRNA level for liver phosphofructokinase in refed mice was reduced by 50% by cAMP administration. Likewise, the four- to fivefold increase in transcription rate of the phosphofructokinase gene observed during refeeding was totally abolished by cAMP administration.

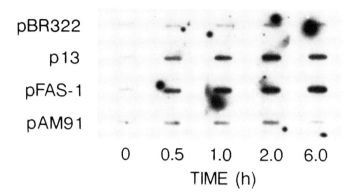

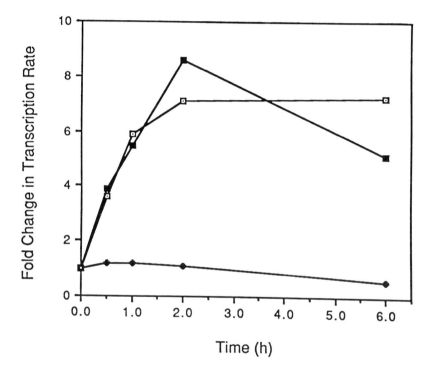

FIGURE 5.   Effect of insulin on transcriptional rates of p90 and fatty acid synthase genes in streptozotocin-diabetic mice liver. Nuclei were isolated from diabetic mice treated with insulin as in Figure 3 and run-on assays were carried out. (■) p13; (□) pFAS-1; (♦) pAM91.

V. DIFFERENTIATION-DEPENDENT EXPRESSION IN 3T3-L1 CELLS

The murine 3T3-L1 cells, upon differentiation, acquire adipocyte morphology, accumulate large amounts of triacylglycerol, and become responsive

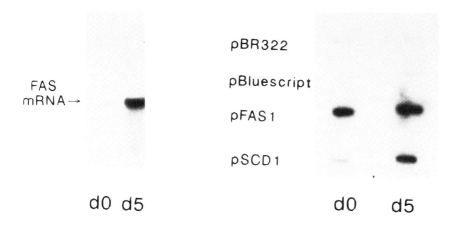

FIGURE 6. Comparisons of mRNA levels and transcription rates for fatty acid synthase gene in 3T3-L1 preadipocytes and adipocytes. RNA (5 μg per lane) prepared from confluent 3T3-L1 cells (d0) and from cells differentiated by treatment of methylisobutylxanthine and dexamethasone (d5) were used for Northern blot. Nuclei isolated from preadipocytes and adipocytes were used for run-on transcription and hybridized with 10 μg each of fatty acid synthase (pFAS-1), stearoyl CoA desaturase (pSCD1), and control plasmids fixed on nitrocellulose.

to lipolytic and lipogenic hormones. The concentrations of several lipogenic enzymes, along with other differentiation-dependent proteins, increase concomitantly with differentiation. These cells have been used as a model system for studying the biochemical and molecular mechanisms underlying hormonal regulation of lipogenic enzyme genes and adipose tissue development.[22-24]

As with other lipogenic enzymes, fatty acid synthase activity, mass, and rate of synthesis increase by an order of magnitude during conversion of 3T3-L1 preadipocytes to adipocytes.[25-27] We recently showed that mRNA levels for both fatty acid synthase and p90 increased about 20-fold during differentiation. The increase in fatty acid synthase mRNA level can account for the increase in fatty acid synthase synthesis observed, demonstrating regulation at a pretranslational level. Interestingly, however, run-on assays showed that fatty acid synthase gene transcription was significant in preadipocytes and the transcription increased only 1.5-fold during adipocyte differentiation (Figure 6). Increase in the transcription rate for the p90 gene was three- to fourfold. In contrast, transcription of the stearoyl-CoA desaturase gene increased markedly during adipose conversion as reported by others.[28,29] The discrepancy between the induced level of fatty acid synthase and p90 mRNAs and transcription rates strongly suggests that not only an increase in transcription but altered RNA stability and/or processing may contribute to the increase in mRNA levels for fatty acid synthase and p90 during adipocyte conversion. To test whether changes in fatty acid synthase mRNA turnover rates during differentiation could account for the induction, fatty acid synthase

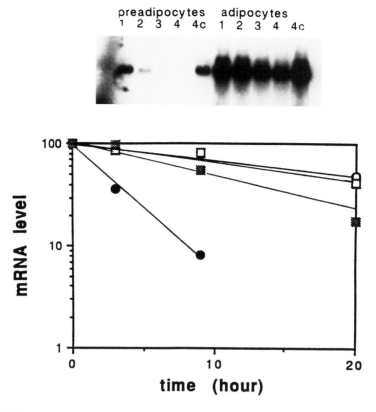

FIGURE 7. Comparison of mRNA stability in 3T3-L1 preadipocytes and adipocytes. Preadipocytes (day 0) and adipocytes (day 5) were treated with actinomycin D (10 μg/ml). RNA were then prepared from control cells (lane 1) or from cells 3 h (lane 2), 9 h (lane 3), or 20 h (lane 4) after actinomycin treatment. Lane 4c represents control cells maintained for 20 h with no actinomycin D treatment. Northern blot shown is for the fatty acid synthase mRNA. Quantification by densitometric scanning for relative decrease in fatty acid synthase and actin mRNAs in preadipocytes (○, □) or adipocytes (●, ■) were shown.

mRNA abundance was measured as a function of time after the addition of a transcription inhibitor, actinomycin, in preadipocytes and adipocytes (Figure 7). The decrease in fatty acid synthase mRNA occurred more rapidly in preadipocytes compared to adipocytes; after 3 h of actinomycin treatment, fatty acid synthase mRNA levels decreased by 64% in preadipocytes while there was no change in adipocytes. Apparent half-lives of fatty acid synthase mRNAs were estimated to be 2.5 h in preadipocytes vs. 20 h in adipocytes. The eightfold increase in fatty acid synthase mRNA half-life in adipocytes, along with the 1.5-fold increase in the transcription rate of the gene, accounts for the 10- to 15-fold increase in fatty acid synthase mRNA observed during differentiation. Therefore, the increase in fatty acid synthase mRNA during adipose conversion is due to both transcriptional activation of the gene and posttranscriptional stabilization of the mRNA. The increase observed in most of the developmentally regulated mRNAs during adipocyte differentiation, such as glycerol-3-phosphate dehydrogenase, has been attributed to the transcriptional activation of their gene.[30,31] In the case of adipsin, mRNA stability was also suggested to contribute to the increase in RNA abundance during the differentiation of 3T3-F442A adipocytes. However, the transcription rates or the mRNA stability for adipsin mRNA could not be compared due to its low detectability.[30]

It has been previously reported that, in addition to the activities of lipogenic enzymes, the activities of several glycolytic enzymes are increased during adipocyte differentiation.[32,33] Therefore, we have examined the rate of synthesis of liver phosphofructokinase during 3T3-L1 differentiation. We observed that the rate of synthesis of liver phosphofructokinase increased fivefold during adipose conversion as measured by pulse-chase experiments. In addition, the mRNA level for this enzyme was also increased an order of magnitude during differentiation, indicating pretranslational control. We have also found that 3T3-L1 cells expressed high level of muscle phosphofructokinase mRNA during the fibroblastic stage but that during adipocyte differentiation, the muscle isoform is downregulated as the liver phosphofructokinase mRNA level increases (unpublished results). It is clear that isozyme switching from muscle to liver phosphofructokinase occurs during adipocyte differentiation. It may be that all three isoforms are expressed in embryonic tissue and, as cells differentiate to become specialized, phosphofructokinase isozyme(s) with regulatory properties suited for the cell function is expressed. We also have previously shown that myoblasts express all three isozymes but that during differentiation to myotubes, expression of the muscle isozyme predominates.[34]

VI. HORMONAL REGULATION OF GENE EXPRESSION IN 3T3-L1 ADIPOCYTES

Mature 3T3-L1 adipocytes, which acquire responsiveness to lipolytic and lipogenic hormones during the differentiation process, were utilized to determine independent hormone effects on gene expression. When mature 3T3-L1 adipocytes were treated with insulin for 16 h in a serum-free medium, both fatty acid synthase and p90 mRNA levels increased approximately threefold (Figure 8). However, the increase in expression of fatty acid synthase and p90 by insulin in streptozotocin-diabetic mice was much higher than the increase observed in cultured adipocytes. Treatment of 3T3 adipocytes with dibutyryl cAMP caused a 60 and 80% decrease in fatty acid synthase and p90 mRNA levels, respectively. Overall, the effects of insulin and cAMP on these two mRNAs in cultured adipocytes are in agreement with those obtained

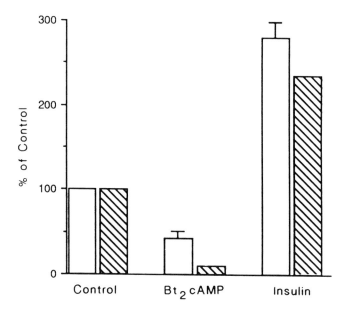

FIGURE 8.   Effect of insulin and cAMP on mRNA levels for fatty acid synthase p90 in 3T3-L1 adipocytes. 3T3-L1 cells were differentiated, maintained in serum-free medium for 24 h, and then treated either with 1 μg/ml of insulin or 1 mM dibutyryl cAMP plus 1 mM theophylline for 16 h. Poly (A+) RNA prepared were used in Northern blot analyses for fatty acid synthase (□) and p90 (▨).

in mouse liver described in the previous section, indicating independent effects of these agents.

The role of thyroid hormone in regulation of fatty acid synthase activity is well documented in studies which utilized animals of different thyroid status. Liver fatty acid synthase activity was reduced in hypothyroidism,[35] and thyroid hormone injection for 7 days increased fatty acid synthase activity two- to threefold in liver.[36,37] Possible synergistic effects of thyroid hormones and a high carbohydrate diet have also been suggested in the regulation of fatty acid synthase activity.[38] In cultured primary rat hepatocytes, treatment with T3 for 24 h has been shown to produce a two- to threefold increase in lipogenic enzyme activities, including fatty acid synthase.[39] Therefore, we examined the effect of T3 on fatty acid synthase in mature 3T3-L1 adipocytes. The fatty acid synthase mRNA level increased 2.4-fold at 12 h of 10 nM T3 treatment and further increased to 4.5-fold at 24 h. A similar increase in p90 mRNA level was observed after T3 treatment. The increase in fatty acid synthase mRNA level was correlated with the increase in synthesis of fatty acid synthase. It was next examined if the increase in fatty acid synthase mRNA was due to an increase in the transcription rate of this gene. Treatment with 10 nM T3 for 6 h increased fatty acid synthase gene transcription to a

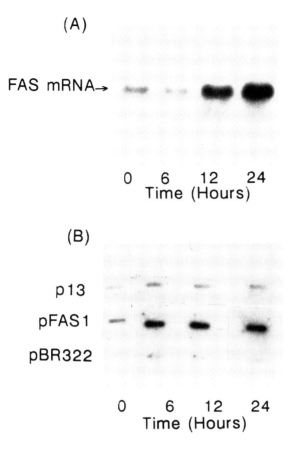

(A)

FAS mRNA→

0   6   12   24
Time (Hours)

(B)

p13

pFAS1

pBR322

0   6   12   24
Time (Hours)

FIGURE 9. Effect of T3 on mRNA level and transcription rate for fatty acid synthase. Differentiated 3T3-L1 adipocytes were treated with T3 for indicated times. RNA and nuclei were prepared and used for Northern blot (A) and transcription run-on assays (B), respectively.

maximum of fourfold (Figure 9). The increase in transcription preceded the increase in mRNA levels. In contrast to the fatty acid synthase gene, transcription of the p90 gene was not appreciably affected by T3, indicating possible regulation at the posttranscriptional level. These two genes seem to be regulated by thyroid hormone by different mechanisms.

VII. CONCLUSIONS

The *in vivo* and *in vitro* experiments described above led us to conclude that mammalian fatty acid synthase and p90 genes are under transcriptional control by insulin and cAMP, and that insulin upregulates and cAMP downregulates these genes. Liver phosphofructokinase is regulated in a similar

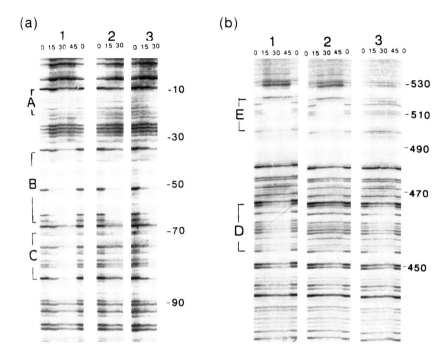

FIGURE 10. Binding of nuclear proteins to the mouse liver phosphofructokinase promoter region. DNA fragments containing the mouse liver phosphofructokinase promoter regions. $+87$ to -169 (a) and -351 to -699 (b), were labeled, incubated with indicated amounts (μg) of liver nuclear extracts, and digested with DNase 1. The regions of protection are demarcated: (a) 1, no competing oligonucleotides; 2, 60-fold excess of oligonucleotides (-65 to -97); 3, 60-fold excess of AP-1 consensus sequence and (b) 1, no competing oligonucleotides; 2, excess of oligonucleotides (-445 to -475); 3, excess of oligonucleotides (-485 to -522).

fashion. Nevertheless, these genes may be regulated independently and different mechanisms may be involved.

At present, in order to identify hormone response elements and nuclear factors which interact with these regulatory sequences, we have isolated genomic clones coding for these enzymes and defined transcription units and 5' and 3' flanking regions. We have also examined by DNase 1 footprinting assays sequences at 5' flanking regions which may be potentially important for transcriptional regulation. As shown in Figure 10, liver nuclear proteins were found to interact with several regions of the liver phosphofructokinase gene promoter *in vitro*.[40] Interestingly, multiple GC boxes which are usually present in "housekeeping" genes and are probably Sp1 binding sites, showed protection by liver nuclear proteins. Also protected was an AP-1-like sequence. Competition experiments showed that a distinct protein was involved in binding. In addition, there were two unique sequences at the distal region of the liver phosphofructokinase gene which showed protection from DNase digestion. The fatty acid synthase gene promoter contains multiple GC boxes and several unique regions including possible hormone response elements which show interaction with liver nuclear proteins (unpublished results). The conserved 8-bp motif (CRE) present in a number of genes upregulated by cAMP is not present in either liver phosphofructokinase or fatty acid synthase gene promoter regions.

To determine the functional significance of these sequences in the hormonal regulation of these genes, we have constructed fusion genes linking promoters to the chloramphenicol acetyltransferase reporter gene and transiently transfected 3T3-L1 adipocytes. To date, we found that 2.1 kb and 0.7 kb of the immediate 5′ flanking sequences of the fatty acid synthase and liver phosphofructokinase genes, respectively, contain *cis*-acting sequences for the upregulation by insulin. The studies of these genes will enable us to dissect nutritional and hormonal regulation of gene expression during energy storage. Moreover, these studies will allow us to define *cis*-acting elements and *trans*-acting factors which interact with these sequences for the transcriptional activation by insulin. Unlike the transcriptional regulation by T3 and cAMP, the mechanism of insulin regulation of gene transcription is not well understood, although some progress has been made in characterizing sequences which may be important for insulin regulation in other genes, such as phosphoenolpyruvate carboxykinase and glyceraldehyde-3-phosphate dehydrogenase.[41,42] Identification and studies of *cis-trans* components for insulin regulation may also lead to understanding a signal transduction pathway for insulin. In addition, by characterizing *cis-trans* components for various hormone responsiveness, it will be possible to examine potential interactions of *trans*-acting factors in achieving multihormonal regulation of these genes during changes in the nutritional status of the animal.

ACKNOWLEDGMENTS

We thank former members of this laboratory, J. D. Paulauskis, S. C. Gehnrich, and P. Rongnoparut whose efforts contributed to the studies reviewed here. The research was supported by National Institutes of Health Grants DK-36264 and DK-40518 and the Juvenile Diabetes Foundation (H. S. S.). N. M. is the recipient of Juvenile Diabetes Foundation Postdoctoral Research Fellowship.

REFERENCES

1. **Gibson, D. M., Lyons, R. T., Scott, D. F., and Muto, Y.,** Synthesis and degradation of the lipogenic enzymes of rat liver, *Adv. Enzyme Regul.,* 10, 187, 1972.
2. **Volpe, J. J. and Vagelos, P. R.,** Mechanisms and regulation of biosynthesis of saturated fatty acids, *Physiol. Rev.,* 56, 339, 1976.
3. **Lakshmanan, M. R., Nepokroeff, C. M., and Porter, J. W.,** Control of the synthesis of fatty acid synthetase in rat liver by insulin, glucagon, and adenosine 3':5'cyclic monophosphate, *Proc. Natl. Acad. Sci. U.S.A.,* 69, 3516, 1972.
4. **Goodridge, A. G.,** Dietary regulation of gene expression: Enzymes involved in carbohydrate and lipid metabolism, *Annu. Rev. Nutr.,* 7, 157, 1987.
5. **Brent, G. A., Moore, D. D., and Larsen, P. R.,** Thyroid hormone regulation of gene expression, *Annu. Rev. Physiol.,* 53, 17, 1991.
6. **Gonzalez, G. A., Yamamoto, K. K., Fischer, W. H., Karr, D., Menzel, P., Biggs, W., Vale, W. W., and Montiminy, M. R.,** A cluster of phosphorylation sites on the cAMP-regulated nuclear factor CREB predicted by its sequence, *Nature (London),* 337, 749, 1989.
7. **Bates, E. J. and Saggerson, E. D.,** A study of the glycerol phosphate acyltransferase and dihydroxyacetone phosphate acyltransferase activities in rat liver mitochondrial and microsomal fractions, *Biochem. J.,* 182, 751, 1979.
8. **Saggerson, E. D. and Carpenter, C. A.,** Effects of streptozotocin-diabetes and insulin administration in vivo or in vitro on the activities of five enzymes in the adipose-tissue triacylglycerol-synthesis pathway, *Biochem. J.,* 243, 289, 1987.
9. **Dunaway, G. A. and Weber, G.,** Effects of hormonal and nutritional changes on rates of synthesis and degradation of hepatic phosphofructokinase isozymes, *Arch. Biochem. Biophys.,* 162, 629, 1974.
10. **Dunaway, G. A., Leung, G. L.-Y., Thrasher, J. R., and Cooper, M. D.,** Turnover of hepatic phosphofructokinase in normal and diabetic rats. Role of insulin and peptide stabilizing factor, *J. Biol. Chem.,* 253, 7460, 1978.
11. **Paulauskis, J. D. and Sul, H. S.,** Cloning and expression of mouse fatty acid synthase and other specific mRNAs. Development and hormonal regulation in 3T3-L1 cells, *J. Biol. Chem.,* 263, 7049, 1988.
12. **Yan, C., Wood, E. A., and Porter, J. W.,** Characterization of fatty acid synthetase cDNA clone and its mRNA, *Biochem. Biophys. Res. Commun.,* 126, 1235, 1985.
13. **Back, D. W., Goldman, M. J., Fisch, J. E., Ochs, R. S., and Goodridge, A. G.,** The fatty acid synthase gene in avian liver. Two mRNAs are expressed and regulated in parallel by feeding, primarily at the level of transcription, *J. Biol. Chem.,* 261, 15179, 1986.
14. **Shin, D. H., Paulauskis, J. D., Moustaid, N., and Sul, H. S.,** Transcriptional regulation of p90 with sequence homology to E. coli glycerol-3-phosphate acyltransferase, *J. Biol. Chem.,* 266, 23 834, 1991.
15. **Gehnrich, S. C., Gekakis, N., and Sul, H. S.,** Liver (B-type) phosphofructokinase mRNA. Cloning, structure, and expression, *J. Biol. Chem.,* 263, 11755, 1988.
16. **Raymondjean, M., Vaulont, S., Cognet, M., Decaux, J. F., Puzenat, N., Bergot, M. O., and Kahn, A.,** Positive and negative regulation of gene expression by insulin and glucagon: the model of L-type pyruvate kinase gene, *Biochimie,* 73, 41, 1991.
17. **Sul, H. S., Wise, L. S., Brown, M. L., and Rubin, C. S.,** Cloning of cDNA sequences for murine ATP-citrate lyase. Construction of recombinant plasmids using an immuno-purified mRNA template and evidence for the nutritional regulation of ATP-citrate lyase mRNA content in mouse liver, *J. Biol. Chem.,* 259, 1201, 1984.
18. **Sul, H. S., Wise, L. S., Brown, M. L., and Rubin, C. S.,** Cloning of cDNA sequences for murine malic enzyme and the identification of aberrantly large malic enzyme mRNA in MOD-1 null mice, *J. Biol. Chem.,* 259, 555, 1984.

19. **Wilson, S. B., Back, D. W., Morris, S. M., Jr., Swierczynski, J., and Goodridge, A. G.,** Hormonal regulation of lipogenic enzymes in chick embryo hepatocytes in culture. Expression of the fatty acid synthase gene is regulated at both translational and pretranslational steps, *J. Biol. Chem.,* 261, 15179, 1986.

20. **Stapleton, S. R., Mitchell, D. A., Salati, L. M., and Goodridge, A. G.,** Triiodothyronine stimulates transcription of the fatty acid synthase gene in chick embryo hepatocytes in culture. Insulin and insulin-like growth factor amplify that effect, *J. Biol. Chem.,* 265, 18442, 1990.

21. **Paulauskis, J. D. and Sul, H. S.,** Hormonal regulation of mouse fatty acid synthase gene transcription in liver, *J. Biol. Chem.,* 264, 574, 1989.

22. **Green, H. and Kehinde, O.,** Sublines of mouse 3T3 cells that accumulate lipid, *Cell,* 1, 113, 1974.

23. **Rubin, C. S., Hirsch, A., Fung, C., and Rosen, O. M.,** Development of hormone receptors and hormonal responsiveness in vitro, *J. Biol. Chem.,* 253, 7570, 1978.

24. **Sul, H. S.,** Adipocyte differentiation and gene expression, *Curr. Opinion Cell Biol.,* 1, 1116, 1989.

25. **Weiss, G. H., Rosen, O. M., and Rubin, C. S.,** Regulation of fatty acid synthetase concentration and activity during adipocyte differentiation. Studies on 3T3-L1 cells, *J. Biol. Chem.,* 255, 4751, 1980.

26. **Student, A. K., Hsu, R. Y., and Lane, M. D.,** Induction of fatty acid synthetase synthesis in differentiating 3T3-L1 preadipocytes, *J. Biol. Chem.,* 255, 4745, 1980.

27. **Wise, L. S., Sul, H. S., and Rubin, C. S.,** Coordinate regulation of the biosynthesis of ATP-citrate lyase and malic enzyme during adipocyte differentiation. Studies on 3T3-L1 cells, *J. Biol. Chem.,* 259, 4827, 1984.

28. **Moustaid, N. and Sul, H. S.,** Regulation of expression of the fatty acid synthase gene in 3T3-L1 cells by differentiation and triiodothyronine, *J. Biol. Chem.,* 266, 18550, 1991.

29. **Bernlohr, D. A., Bolanowski, M. A., Kelly, T. J., and Lane, M. D.,** Evidence for an increase in transcription of specific mRNAs during differentiation of 3T3-L1 preadipocytes, *J. Biol. Chem.,* 260, 5563, 1985.

30. **Cook, D. S., Hunt, C. R., and Spiegelman, B. M.,** Developmentally regulated mRNAs in 3T3-adipocytes: analysis of transcriptional control, *J. Cell. Biol.,* 100, 514, 1985.

31. **Djian, P., Phillips, M., and Green, H.,** The activation of specific gene transcription in the adipose conversion of 3T3 cells, *J. Cell. Physiol.,* 124, 554, 1985.

32. **Mackall, J. C. and Lane, M. D.,** Role of pyruvate carboxylase in fatty acid synthesis: alterations during preadipocyte differentiation, *Biochem. Biophys. Res. Commun.,* 79, 720, 1977.

33. **Spiegelman, B. M. and Green, H.,** Control of specific protein biosynthesis during the adipose conversion of 3T3 cells, *J. Biol. Chem.,* 255, 8811, 1980.

34. **Gekakis, N., Gehnrich, S. C., and Sul, H. S.,** Phosphofructokinase isozyme expression during myoblast differentiation, *J. Biol. Chem.,* 264, 3658, 1989.

35. **Volpe, J. J. and Kishimoto, Y.,** Fatty acid synthetase of brain: development, influence of nutritional and hormonal factors and comparison with liver enzyme, *J. Neurochem.,* 19, 737, 1972.

36. **Mariash, C. N., Kaiser, F. E., and Oppenheimer, J. H.,** Comparison of the response characteristics of four lipogenic enzymes to 3,5,3'-triiodothyronine administration: evidence for variable degrees of amplification of the nuclear 3,5,3'-triiodothyronine signal, *Endocrinology,* 106, 22, 1980.

37. **Diamant, S., Gorin, E., and Shafrir, E.,** Enzyme activities related to fatty acid synthesis in liver and adipose tissue of rats treated with triiodothyronine, *Eur. J. Biochem.,* 26, 553, 1972.

38. **Mariash, C. N., Kaiser, F. E., Schwartz, H. L., and Towle, H. C.,** Synergism of thyroid hormone and high carbohydrate diet in the induction of lipogenic enzymes in the rat. Mechanisms and implications, *J. Clin. Invest.,* 65, 1126, 1980.

39. **Spence, J. T. and Pitot, H. C.,** Induction of lipogenic enzymes in primary cultures of rat hepatocytes. Relationship between lipogenesis and carbohydrate metabolism, *Eur. J. Biochem.,* 36, 15, 1982.

40. **Rongnoparut, P., Verdon, C. P., Gehnrich, S. C., and Sul, H. S.,** Isolation and characterization of the transcriptionally regulated mouse liver (B-type) phosphofructo-kinase gene and its promoter, *J. Biol. Chem.,* 266, 8086, 1991.

41. **O'Brien, R. M., Lucas, P. C., Forest, C. D., Magnuson, M. A., and Granner, D. K.,** Identification of a sequence in the PEPCK gene that mediates a negative effect of insulin on transcription, *Science,* 249, 533, 1990.

42. **Nasrin, N., Ercolani, L., Denaro, M., Kong, X. F., Kang, I., and Alexander, M.,** An insulin response element in the glyceraldehyde-3-phosphate dehydrogenase gene binds a nuclear protein induced by insulin in cultured cells and by nutritional manipulations in vivo, *Proc. Natl. Acad. Sci. U.S.A.,* 87, 5273, 1990.

Chapter 10

REGULATION OF HEPATIC GENE EXPRESSION BY DIETARY FATS: A UNIQUE ROLE FOR POLYUNSATURATED FATTY ACIDS

Steven D. Clarke and Donald B. Jump

TABLE OF CONTENTS

This chapter is dedicated to Dr. Sandy Abraham whose tireless and frontiering efforts in the area of polyunsaturated fatty acid metabolism and the control of gene expression served as the foundation for the work presented here.

I. INTRODUCTION

Numerous proteins involved in carbohydrate and lipid metabolism (Table 1) are regulated by dietary fat, but it has become increasingly clear over the last decade that regulation of the expression of these proteins is differentially affected by saturated and polyunsaturated fats.[1-6] Using the fatty acid biosynthetic pathway as a model, Gibson and his colleagues discovered that only polyunsaturated fatty acids inhibited hepatic fatty acid biosynthesis.[1] Dietary saturated fats such as tallow were found to have little or no ability to suppress hepatic lipogenesis.[1-5] This unique characteristic of dietary polyunsaturated fats has led us to propose that at least part of the hypotriglyceridemic action of dietary polyunsaturated fats is due to the inhibition of *de novo* fatty acid synthesis which in turn reduces substrate availability for triglyceride synthesis, and hence decreases the hepatic output of triglycerides. Consequently, the amount of circulating plasma triglyceride is decreased.

The suppression of hepatic lipogenesis by dietary polyunsaturated fatty acids appears to require that the fatty acid contain a minimum of 18-carbons and two double bonds in the 9 and 12 positions, and at least one of these bonds must be of the cis configuration (Figure 1). In addition, the inhibition of fatty acid biosynthesis by polyunsaturated fats is: (1) specific for hepatic lipogenesis,[4-7] (2) mediated by the fatty acid component of the dietary lipid,[4] (3) dependent upon the desaturation of the 18-carbon fatty acid by the delta-6 desaturase,[8,9] (4) not mediated by a prostaglandin or thromboxane metabolite,[6,9] and (5) related to the suppression of genes coding for proteins directly involved fatty acid biosynthesis.[10-12] The objective of this review is to trace the evolution of our understanding of the mechanism by which polyunsaturated fatty acids uniquely regulate the expression of lipogenic proteins, notably fatty acid synthase, and to speculate as to the potential molecular mechanisms by which polyunsaturated fatty acids regulate nuclear transcriptional events. It should be noted that the data to be discussed are derived primarily from studies with rodents, and that the unique effects of polyunsaturated fatty acids may not include all species, e.g., polyunsaturated fats do not selectively inhibit lipogenic enzymes in birds.[13]

II. POST-TRANSLATIONAL REGULATION OF LIPOGENIC ENZYMES

The classic studies of Chaikoff and his colleagues[14] had indicated that fat ingestion would suppress acetate incorporation into hepatic lipids. It was

TABLE 1
Dietary Fatty Acid Effects on Rat Liver Proteins

Enzyme	Fat	Amount	mRNA	Transcription	Ref.
Glucokinase	Saf oil	−30%			5, 47, 49
Phosphofructokinase	18:2	nc			5
Pyruvate kinase	18:2	nc			5
Glucose-6-phosphate dehydrogenase	16:0/18:1	nc			4, 48
	18:2/18:3	−60%			
	Saf oil	−42%	−52%		22
Malic enzyme	16:0/18:1	nc			4, 30
	18:2/corn	−45%	nc		
Citrate lyase	18:2	−55%			47
Acetyl-CoA carboxylase	16:0/Tallow	nc	5		
	18:2/Saf	−60%			5
Fatty acid synthase	16:0	nc	nc	nc	11, 12
	18:1	nc	+30%		11
	18:2	60%	−70%		6, 11, 20
	Fish oil	−80%	−75%	−94%	11, 12
Delta-9 desaturase	18:2/18:3	−50%			49
	Fish oil	−85%			
Delta-6 desaturase	18:2/18:3	+150%			49
	Fish oil	−20%			
Delta-5 desaturase	18:2/18:3	+350%			49
lipase	Fish oil	nc			
	Lard	nc			50
	Corn oil	+155%			
	Fish oil	+105%			
apo A I	High P/S	+22–50%		51	
S14	16:0		nc	nc	11, 12
	18:1		−17%	nc	
	18:2		−80%		
	Fish oil		−84%	−80%	
β-Actin	18:1		nc	nc	12
PEPCK	Fish oil	nc			

proposed that fatty acids or their CoA esters functioned as negative allosteric modifiers of key glycolytic and lipogenic enzymes, e.g., phosphofructokinase and acetyl-CoA carboxylase.[14,15] In our early studies, the putative fatty acid allosteric binding sites of hepatic phosphofructokinase and/or acetyl-CoA carboxylase were hypothesized to possess a higher affinity for polyunsaturated fatty acyl-CoAs than for the saturated fatty acid CoA esters. Thus, polyunsaturated fatty acids could selectively inhibit glucose flux to fatty acids by selectively binding to regulatory enzymes and inhibiting their catalytic effi-

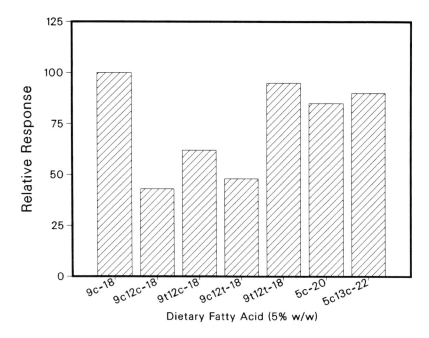

FIGURE 1. Structural requirements for a fatty acid to suppress hepatic lipogenesis. Mice (n = 4) were fed for 5 d a high glucose diet containing 5% (w/w) of the respective fatty acids cited in the figure. *In vivo* fatty acid synthesis was determined by tritiated water incorporation into fatty acids. Relative response is defined relative to mice fed a high glucose fat-free diet (SEM < 15%).

ciency.[16,17] Recently, we reported that polyunsaturated fats were in fact better acute inhibitors of hepatic fatty acid synthesis and malonyl-CoA accumulation than were saturated fats.[16] However, the acute suppression of fatty acid synthesis by the ingestion of a single meal containing polyunsaturated fat was modest when compared to the 60 to 90% reduction in fatty acid synthesis and lipogenic enzymatic activities brought on by the prolonged (3 to 4 d) ingestion of polyunsaturated fat (Figure 4).[1-12] The acute suppression of lipogenesis by safflower oil was not the consequence of reduced catalytic efficiency or capacity of pyruvate kinase or phosphofructokinase, i.e. glycolytic activity was unaffected.[5] In addition, the quantity of catalytically functional acetyl-CoA carboxylase (as determined by increased resistance to avidin activation)[18,19] was not reduced by the ingestion of saturated or polyunsaturated fats (Figure 5). Subsequent analysis of the time course of inhibition of fatty acid synthesis vs. acetyl-CoA carboxylase and fatty acid synthase activities revealed that the polyunsaturated fatty acid suppression of hepatic fatty acid biosynthesis paralleled the decline in the amount of acetyl-CoA carboxylase and fatty acid synthase.[5] Thus, polyunsaturated fatty acids appeared to exert their greatest impact on the fatty acid biosynthetic capacity via a coordinate suppression of the amount of lipogenic enzymes.[5,20-22] In light of this conclusion, our attention focused on the events which regulate enzyme synthesis and gene expression.

III. POLYUNSATURATED FATTY ACID SUPPRESSION OF FATTY ACID SYNTHASE AND S14 mRNA ABUNDANCE

Inhibition of enzyme protein synthesis is the major reason for the reduction in hepatic activity of lipogenic enzymes (e.g., fatty acid synthase) caused by the ingestion of polyunsaturated fatty acids.[6,20] Two mechanisms may explain this inhibition of enzyme synthesis: (1) interference with translation rate of the mRNA and/or (2) reduction in the amount of mRNA. To address this issue we have focused our efforts on two genes: fatty acid synthase and the putative lipogenic protein, S14.[10-12]

When young growing rats were fed for 6 d a diet containing 10% safflower oil, the abundance of fatty acid synthase and S14 mRNA was decreased to levels that were 30 and 50%, respectively, of the values observed with a fat-free high glucose diet.[11] Comparable amounts of dietary monounsaturated or saturated fat (e.g., triolein) had no suppressive effect.[11] Polyunsaturated fatty acids not only impaired fatty acid synthase and S14 expression in the growing and adult rodents, but also blocked the 20- to 30-fold rise in mRNA levels associated with weaning rats onto a high carbohydrate diet.[11] The weanling animal appeared to be more sensitive to the polyunsaturated fat inhibition than was the growing or adult rat, e.g., 3% 18:2 (n-6) suppressed fatty acid synthase expression 80 to 90% in the 30-d-old rat, whereas 10% 18:2 (n-6) suppressed fatty acids synthase mRNA only 65% in the 3-month-old rat.[11] The regulation of gene expression by dietary polyunsaturated fats is not unique to fatty acid synthase and S14, but also extends to the suppression of acetyl CoA carboxylase[21] and glucose-6-phosphate dehydrogenase.[22] Thus, dietary polyunsaturated fats appear to coordinately suppress the level of transcripts coding for lipogenic enzymes, and in this way decrease the synthesis of these enzymes which in turn results in a reduced lipogenic capacity.[4,5,11,12] However, each gene possesses its own regulatory sequences, and expression of each gene coding for a protein in the fatty acid biosynthetic pathway is a summation of integrated hormonal and nutrient signals. Consequently, individual proteins may vary in the degree of response to dietary polyunsaturated fatty acids, e.g., S14 mRNA is less dramatically reduced by polyunsaturated fats than is fatty acid synthase mRNA.[11] Nevertheless, polyunsaturated fatty acids clearly possess a selective ability to reduce the amount of protein and mRNA for several enzymes of fatty acid biosynthesis.

While dietary fat rich in 18:2 (n-6) suppressed the expression of enzymes involved in fatty acid biosynthesis, the level of mRNA and protein for the liver fatty acid binding protein (l-FABP) is induced by dietary fats.[23] A

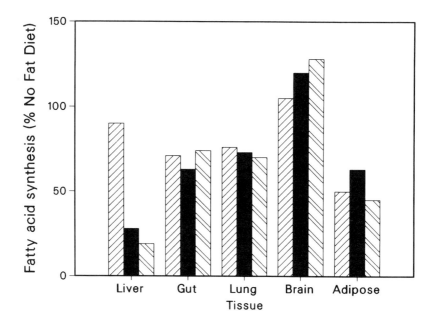

FIGURE 2. Tissue-specific inhibition of lipogenesis by polyenoic fatty acids. Mice (n = 4) were fed for 5 d a high glucose diet containing 10% hydrogenated cottonseed oil or triolein (single slash bars), 10% corn oil (solid bars), or 10% menhaden fish oil (double slash bars). *In vivo* fatty acid synthesis was measured by tritiated water incorporation (SEM < 15%).

selective induction of 1-FABP due to polyunsaturated fats has not been firmly established, but studies with rat liver cells in primary culture indicate 18:2 (n-6) directly enhanced 1-FABP synthesis, and resulted in higher levels of 1-FABP protein than did 16:0.[24] Clearly, polyunsaturated fatty acids uniquely regulate hepatic gene expression, and this effect may extend to genes coding for proteins other than enzymes in fatty acid biosynthesis.

The polyunsaturated fatty acid control of lipogenic proteins is predominately a liver-specific phenomenon (Figures 2 and 3).[25] Fatty acid biosynthesis and the activities of lipogenic enzymes within rodent lung, brain, and small intestine were unaffected by a diet that contained 20 to 30% kcal as hydrogenated cottonseed oil, corn oil, and menhaden oil (Figure 2).[9,25] Comparable levels of dietary fat reduced *de novo* fatty acid biosynthesis in mammary and epididymal fat pads, but saturated and polyunsaturated fats were equipotent (Figure 2). Interestingly, fatty acid synthase mRNA levels were unaffected by either dietary saturated or polyunsaturated fat (Figure 3),[9] which is in sharp contrast to the hepatic responses to these dietary lipids.[11,23] Since dietary fats suppress adipose fatty acid synthase activity and rates of fatty acid biosynthesis,[15] the data suggest fatty acid synthase synthesis in adipose tissue may be regulated at the level of translation as well as at the point of transcription.

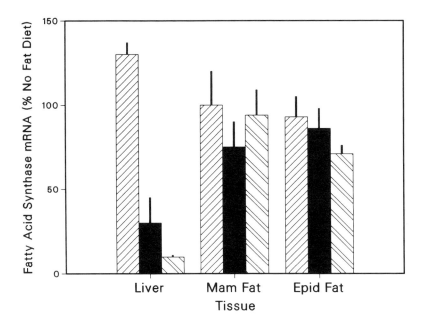

FIGURE 3. Tissue-specific suppression of fatty acid synthase by dietary polyunsaturated fats. Mice (n = 4) were fed for 5 d a high glucose diet containing 10% hydrogenated cottonseed oil (single slash bars), 10% corn oil (solid bars), or 10% menhaden fish oil (double slashed bars). Fatty acid synthase mRNA abundance was determined by DNA/RNA hybridization.[11]

Regardless of the mechanism of regulation, clearly the unique control of fatty acid synthase expression by polyunsaturated fats noted for hepatocytes does not extend to the adipocyte.

IV. POTENCY OF FATTY ACID INHIBITORS

As might be expected the inhibitory potency of dietary polyunsaturated fatty acids varies depending upon the degree of unsaturation and chain length (Figures 1 through 4).[2-5,11] In general, 18:3 (n-3) is more potent than 18:2 (n-6), but, most importantly, products of the delta-6 desaturase are 2 to 4 times more potent suppressors than are the respective fatty acid precursors (Figure 6).[8,9,22] Obviously, since dietary oils vary in their polyunsaturated fatty acid content, they also differ in their capacity to inhibit lipogenesis and the expression of lipogenic enzymes. Thus, olive oil (rich in 18:1) has no suppressive effect, safflower oil (75% 18:2 [n-6]) is more potent than corn oil (50% 18:2 [n-6]), and menhaden fish oil is two- to threefold more potent than safflower oil. The greater potency of fish oil is observed as a twofold greater maximal suppression of fatty acid synthase expression, and as a threefold lower re-

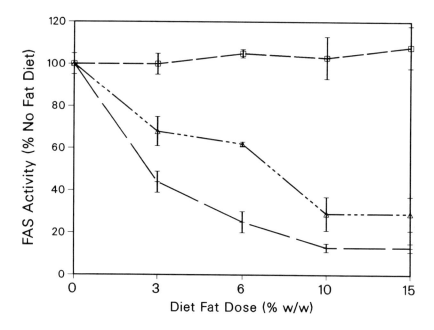

FIGURE 4. Response of hepatic fatty acid synthase activity to varying doses and types of dietary fat. Mice (n = 4) were fed a high glucose diet for 5 d containing the level of fat (w/w) cited in the figure. Hydrogenated cottonseed oil, corn oil, and menhaden fish oil are represented by the open squares, open triangles, and crosses, respectively. Data are expressed as means ± SEM for n = 4 mice per point.

quirement for fish oil to achieve 50% of maximum inhibition (Figure 4).[9] A comparison of 18:3 (n-6) and 20:4 (n-6) with fish oil has shown that the key difference in potency between fish oil and vegetable oil lies in the conversion rate of 18:2 (n-6) to 18:3 (n-6), i.e., 20:4 (n-6) is equipotent to fish oil (Figure 6). Thus, the greater potency of fish oil as an inhibitor of fatty acid synthase and S14 gene expression is due primarily to its high proportion (>30%) of 20- and 22-carbon polyenoic fatty acids.

V. HEPATIC-SPECIFIC INHIBITION

Dietary studies cannot determine if fatty acids directly inhibit hepatocyte gene expression or indirectly function via peripheral release of a fatty acid metabolite or hormonal factor. Results with rat hepatocytes in primary culture indicate that polyunsaturated fatty acids do not require peripheral tissue metabolism nor cause the release of a peripheral tissue factor or hormone.[26,27] The hormonal induction of acetyl CoA carboxylase, malic enzyme, and glucose-6-phosphate dehydrogenase activities as well as the induction of fatty acid synthase mRNA is suppressed by the addition of albumin-bound poly-

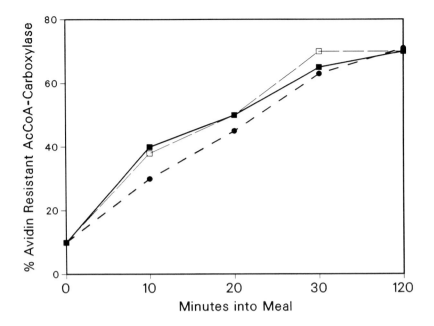

FIGURE 5. The effect of dietary saturated and polyunsaturated fats on the activation of acetyl CoA carboxylase during meal ingestion. Rats were trained to eat their entire daily food allotment during a 3-h meal. The fraction of catalytically active acetyl CoA carboxylase present in hepatic tissue during the ingestion of a single meal of 10% tripalmitin (solid circles), 10% safflower oil (open squares), or no fat (solid squares) was determined using the avidin inactivation procedure.[18,19]

enoic (n-6) fatty acids to the media (Figure 7).[26,27] As occurs *in vivo,* 20:4 (n-6) was more potent than 18:2 (n-6). Approximately 100 μM 20:4 (n-6) was required to suppress fatty acid synthase mRNA induction 50%.[26] Admittedly, circulating levels of 20:4 (n-6) above 50 μM are unlikely. However, hepatocytes in culture rapidly utilize 20:4 (n-6) for triglyceride and phospholipid synthesis.[28] Thus, the media concentration of 20:4 (n-6) must be maintained at a high level in order to provide an effective concentration sufficient to exert an inhibitory action. As would be expected from the *in vivo* responses, as much as 500 μM 20:1 (n-9) did not inhibit fatty acid synthase expression, and in fact, increased synthase mRNA levels twofold (Figure 7).[26] One possible explanation for the enhancement of fatty acid synthase gene expression by 20:1 (n-9) may be that 20:1 (n-9) displaced the inhibitor 20:4 (n-6).

Interestingly, as the media content of 20:4 (n-6) was increased from 50 to 500 μM, the level of β-actin mRNA rose 70 to 275%, but 20:1 (n-9) had no effect on β-actin expression (Figure 7).[26] β-Actin expression is used as an indicator of hepatocyte differentiation status, i.e., high levels of β-actin mRNA may be indicative of a loss of adult differentiation traits.[29] It is curious

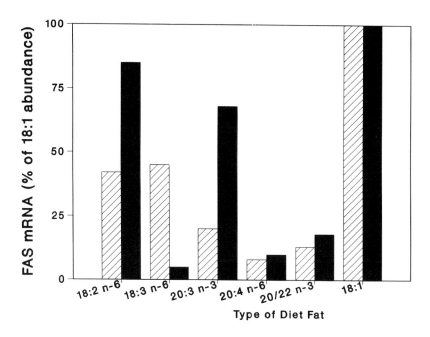

FIGURE 6.   Delta-6 desaturation of linoleate and α-linolenate is required to suppress fatty acid synthase gene expression. Mice were fed a high glucose diet for 5 days containing various fatty acids with (solid bars) and without (slashed bars) the delta-6 desaturase inhibitor eicosatetraynoic acid (ETYA). The level of fat supplementation to the diet was 5% for 18:2, 18:3, and 18:1, and 3% for 20:3 and 20:4. The 20:5/22:6 column represents 10% dietary menhaden fish oil which provides 3% 20:5/22:6 (n-3).

as to what effects polyunsaturated fatty acids may exert on adult hepatic function and differentiation characteristic. In any case, it appears that polyunsaturated fatty acids specifically regulate gene expression events of the hepatocyte and that these events may extend to hepatic genes other than those coding for lipogenic proteins.

VI. STRUCTURAL REQUIREMENTS AND FATTY ACID POTENCY

The basic requirement for a dietary fatty acid to inhibit expression of lipogenic proteins is that it contain 18 carbons and possess at least 2 conjugated double bonds in the 9,12-positions.[7-9] Thus, polyenoic fatty acids of the (n-9) family as well as unusual fatty acids such as 5c,13c-22:2 do not suppress fatty acid biosynthesis (Figure 1). Similarly, loss of one of the double bonds by hydroxylation renders the fatty acid inactive.[30] However, a fatty acid may contain additional double bonds and still retain inhibitory potency. For example, dietary columbinic acid (5t,9c,12c-18:3) and alpha-linolenic acid

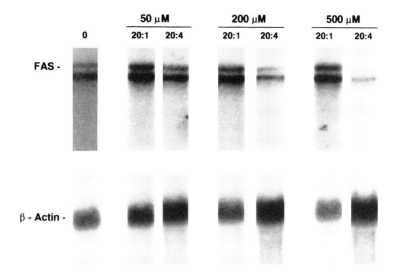

FIGURE 7.   Northern analysis demonstrating suppression of fatty acid synthase gene expression in primary cultures of rat liver cells by albumin-bound arachidonate. Rat liver cells were maintained in primary culture[38] and incubated for 36 h with the levels of albumin-bound 20:4 (n-6) or 20:1 (n-9) cited in the figure.

(9c,12c,15c-18:3) both reduce the abundance of fatty acid synthase mRNA and the rate of lipogenesis.[11,20] With respect to double bond conformation, one of the two double bonds may be in the trans configuration, but 9t,12t-18:2 does not lower the level of hepatic fatty acid synthase mRNA (Figure 7).[7]

While dietary 18:2 (n-6) and 18:3 (n-3) markedly reduce the abundance of hepatic fatty acid synthase and S14 mRNAs,[11] the actual intracellular modulator of gene expression is not 18:2 (n-6) or 18:3 (n-3) per se.[9] In order for these fatty acids to suppress the expression of genes coding for lipogenic enzymes, they must first undergo desaturation by the delta-6 desaturase. When delta-6 desaturation was blocked by the fatty acid analog, eicosatetraynoic acid (ETYA), the suppression of fatty acid synthase mRNA level by dietary safflower oil, corn oil, or 18:2 (n-6) was completely prevented (Figure 6), but the hepatic concentration of 18:2 (n-6) dramatically increased.[8,9] Adding the product of the delta-6 desaturase (e.g., 6c,9c,12c-18.3) completely reinstated the inhibition of gene expression (Figure 6). Under these conditions, 18:3 (n-6) and 20:3 (n-6) content of hepatic lipids increased, but conversion of 20:3 (n-6) to 20:4 (n-6) continued to be impaired because ETYA also inhibited the delta-5 desaturase. Thus, while 20:4 (n-6) and 20:5 (n-3) are potent suppressors of fatty acid synthase gene expression, the formation of these fatty acids is not an *a priori* requirement for the polyenoic fatty acid regulation of gene expression to occur.

The potent suppression of fatty acid synthase and S14 mRNA levels by fish oil rich in (n-3) fatty acids combined with the observation that formation of 20:4 (n-6) was not required to observe the regulation of gene expression. This suggests that prostanoids may not be involved in the polyunsaturated fatty acid mechanism. Consistent with this conclusion is the failure of several prostanoid inhibitors to prevent the suppression of fatty acid synthase gene expression.[6,8,31] For example, the prostaglandin inhibitors indomethacin and ibuprofin[6,31] and the leukotriene inhibitors nordihydroquaiaretic acid, quercitin, and phenidone did not prevent dietary corn oil or safflower oil from suppressing hepatic fatty acid synthase and glucose-6-phosphate dehydrogenase activities.[9] Similarly, columbinic acid, which is not a substrate for prostanoid synthesis, continued to suppress fatty acid synthase enzyme activity.[20] However, the inability to detect an effect does not mean an effect does not exist. Selective regulation of gene transcription by a fatty acid may require high affinity binding of a polyunsaturated fatty acid mediator to a nuclear *trans*-acting protein. Such selectivity will likely require modification of the fatty acid by mechanisms such as hydroxylation or oxidation.[32] Since numerous possibilities exist for this hypothetical fatty acid metabolite, elucidation of the nuclear proteins which regulate fatty acid synthase and S14 gene expression will greatly aid in identifying the fatty acid metabolite ligand responsible for controlling gene expression.

VII. TRANSCRIPTIONAL REGULATION BY POLYUNSATURATED FATS

The suppression of fatty acid synthase and S14 mRNA abundance in hepatic tissue caused by the ingestion of polyunsaturated fatty acids could result from: (1) an inhibition of gene transcription, (2) an interference with the maturation of the nascent transcript, and/or (3) an acceleration of the rate of mRNA degradation. Nuclear run-on assays with nuclei prepared from livers of rats fed a diet containing 10% menhaden fish oil vs. tripalmitin or triolein demonstrated that polyunsaturated fats determine the level of fatty acid synthase and S14 mRNAs by controlling the rate of gene transcription.[12] For example, menhaden oil suppressed fatty acid synthase gene transcription 94% and reduced the level of its mRNA 80%.[12] Similarly, hepatic S14 gene transcription and mRNA levels were both suppressed 60 to 70% by dietary fish oil.[12] Since the transcription rates for phosphoenolpyruvate carboxykinase and β-actin were unchanged, the effects of fish oil appear to be specific for genes coding for lipogenic proteins. These results do not completely eliminate the possibility that polyunsaturated fatty acids could subtly influence transcript maturation processes and/or mRNA degradation rates, but certainly the predominate control mechanism by which polyunsaturated fatty acids regulate the hepatic expression of lipogenic proteins appears to be via inhibition of gene transcription. Malic enzyme may be an exception to this conclusion,[30,33,34] but this protein does not play a pivotal role as a determinant of flux in the fatty acid biosynthetic pathway.

In light of the potent suppression of fatty acid synthase transcription by polyunsaturated fatty acids, we have speculated that polyunsaturated fatty acids or a metabolite regulate the activity of a specific transacting factor which acts on a specific *cis*-acting element associated with the fatty acid synthase and S14 genes. In order to address this question we have cloned from a rat genomic DNA cosmid library a 39-kb insert that contains the entire fatty acid synthase gene. Based upon a recently published sequence for the putative start site of fatty acid synthase transcription,[35] our clone extends 2.5 kb upstream from the start site of transcription. The region of −2495 to +1500 of the fatty acid synthase gene has been fully sequenced and contains several nucleotide motifs which are characteristic of a promoter region. First, TATA and inverted CAAT boxes were located within the first 100-bp 5′ to the putative transcription start site. The TATA and CAAT boxes were separated by a 60-bp sequence that was comprised of more than 75% G-C nucleotides. In addition, two G-C boxes with consensus sequences characteristic of a binding site for transcription factor Sp1 (i.e., 5′-GGGCGG-3′) were located within 150 bp of the CAAT box. Also contained within the +1 to −2452 region were 2 sets of glucocorticoid-like response elements (GREs) which were located at −1534 and −2000, as well as a prospective thyroid response element (TRE) located at −650.

Using the *in vitro* transcription initiation assay and transient transfection of liver cells, we have found the region −2195 to +65 of the fatty acid synthase gene does in fact contain a transcriptionally active promoter and certain hormonal regulatory elements characteristic of the endogenous gene. Nuclear protein extracts from hepatic tissue, but not spleen or kidney, have been found to drive the *in vitro* transcription of an artificial G-free gene.[36] These responses required the 5′ to 3′ orientation. When the −2195 to +65 region of the fatty acid synthase gene was linked to the chloramphenicol acetyltransferase reporter and rat liver cells in primary culture were transiently transfected, CAT activity was expressed. Moreover, the CAT expression was stimulated by glucocorticoids (Figure 8). The glucocorticoid-dependent CAT expression is consistent with the glucocorticoid-dependent induction of the endogenous fatty acid synthase gene (Figure 9) and consistent with the presence of the two GREs within this region. Neither T_3 nor insulin enhanced CAT expression (Figure 8). Since expression of fatty acid synthase is an insulin-dependent process (Figure 9),[37] it would appear that the −2195 region does not contain all regulatory nucleotide sequences characteristic of the fatty acid synthase gene. Thus, while the −2195 to +65 region contains an active promoter for fatty acid synthase which is capable of initating transcription in a tissue-specific manner, it appears that additional sequences are essential to understanding *in vivo* expression in this gene. The successful cloning of the

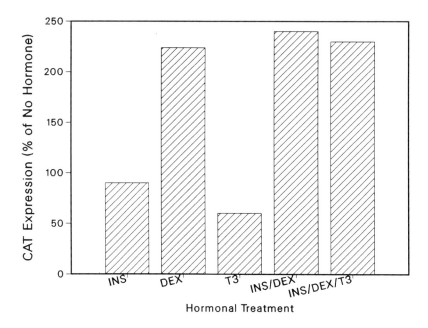

FIGURE 8. Stimulation of fatty acid synthase promoter activity by glucocorticoids. Rat liver cells in primary culture[38] were transiently transfected by lipofectin with a pSV2 vector containing −2190 to +65 of the fatty acid synthase gene, the chloramphenicol acetyltransferase reporter, and the SV40 enhancer 5′ to the promoter. The hormone concentrations were 0.1 μM and SEM < 15%.

fatty acid synthase gene and its promoter, coupled with the extensive characterization of the S14 gene, will permit us to identify potential *cis*-elements and *trans*-acting nuclear proteins which may be under polyunsaturated fatty acid control.

VIII. MECHANISMS OF TRANSCRIPTIONAL CONTROL

Our early understanding of the mechanisms by which polyunsaturated fats suppressed lipogenesis and of the activities of lipogenic enzymes suggested that the process was a long-term adaptation that required peripheral metabolism of the ingested polyunsaturated fatty acids to an active derivative or required the release of a hormone which suppressed hepatic lipogenesis. However, these early kinetic studies relied heavily upon changes in the activities of lipogenic enzymes.[1-5] Since the average half-life of a lipogenic enzyme is 36 to 48 h, we could not evaluate the early changes which might occur in gene transcription or mRNA abundance as a result of polyunsaturated fat ingestion. Thus, the inhibitory effects of polyunsaturated fats would (by

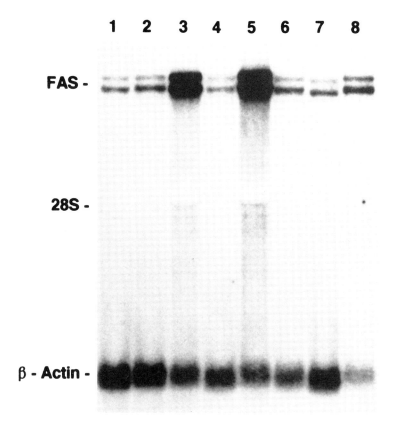

FIGURE 9. Hormonal induction of fatty acid synthase gene expression in primary cultures of rat liver cells. Rat liver cells were maintained in primary culture[38] and treated with the hormones cited in the figure. RNA was extracted and subjected to Northern analyses. The hormonal concentrations were 0.1 μM. Lane 1, no hormone; lane 2, insulin; lane 3, insulin/dexamethasone; lane 4, dexamethasone; lane 5, insulin/dexamethasone/T_3; lane 6, T_3; lane 7, T_3/insulin; and lane 8, T_3/dexamethasone.

definition) appear to involve long-term events. Through the use of primary rat hepatocytes in monolayer culture,[26,27,38] we have demonstrated that polyunsaturated fatty acids directly regulate hepatocyte gene expression and do not require their peripheral metabolism to an active metabolite (Figure 7).

However, one obvious and frequently argued adaptation is that ingestion of oils rich in polyenoic fatty acids results in the enrichment of hepatic membrane phospholipids with polyenoic fatty acids, and that the accompanying rise in membrane fluidity causes an increase hepatocyte responsivity to β-agonists (e.g., glucagon) and/or decreased responsivity to lipogenic agonists (e.g., insulin). Such a shift in hormonal balance could lead to increased cAMP-dependent protein kinase activity and subsequently suppression of fatty acid synthase and S14 transcription. There is little question that dietary polyunsaturated fats alter hormone binding and membrane signaling mechanisms,[39-42] but the hormonal responsivity changes induced by polyunsaturated fatty acids do not present a pattern consistent with suppression of fatty acid synthase and S14 gene expression. For example, although dietary polyunsaturated fats increase glucagon-stimulated adenylate cylase activity twofold,[39] they also significantly increase tissue sensitivity to insulin.[40,42] Moreover, the dose of corn oil required to elicit maximal increases in glucagon sensitivity was only 2% of the diet,[39] while inhibition of fatty acid synthase gene expression by corn or safflower oil continues to be expressed at doses up to 10% of the diet (Figure 4).[11,37] The strongest argument against the idea that fatty acid membrane compositional changes are responsible for the control of gene expression lies in the kinetics of transcriptional inhibition vs. membrane fatty acid changes. Removing fish oil from the diet for simply one 3-h meal completely reverses the near 90% suppression of fatty acid synthase gene expression.[11] Similarly, transcription of acetyl CoA carboxylase was suppressed 50% within 6 h of adding corn oil to the diet.[21] In our opinion, 3 to 6 h is insufficient time for a dietary lipid to achieve significant changes in membrane phospholipid fatty acid composition. Rather, the kinetics of transcriptional regulation are more consistent with direct control of gene transcription by polyunsaturated fatty acids or their metabolites.

The final piece of evidence against a change in cAMP balance caused by dietary polyunsaturated fats is derived from the effects of these fatty acids on gene transcription.[12] If the polyunsaturated fatty acid suppression of fatty acid synthase and S14 gene transcription were to involve a cAMP-dependent process, then one would expect that the transcription rate of phosphoenolpyruvate carboxykinase (PEPCK) would be increased by dietary polyunsaturated fats. However, PEPCK transcription, which is exquisitely sensitive to stimulation by cAMP and inhibition by insulin,[43] was not affected by dietary polyunsaturated fat.[12] Thus, collectively the available data suggest that the regulation of gene expression by polyunsaturated fats does not involve shifts in the balance of hormones which modulate the cAMP second message system.

While enrichment of membrane lipids with polyunsaturated fats appears not to be the mechanism by which polyunsaturated fatty acids regulate fatty acid synthase and S14 gene expression, it is feasible that long chain polyenoic fatty acids interfere with nuclear events regulated by lipogenic hormones. For example, fatty acids or nuclear fatty acid binding proteins might potentially interact with T_3 or glucocorticoid receptors in such a way as to interfere with receptor/DNA interactions and thereby alter transcriptional events. While such an idea is purely speculation, dietary polyunsaturated fats were found to interfere with T_3 induction of hepatic fatty acid synthase, malic enzyme, and glucose-6-phosphate dehydrogenase.[44] The effect of a diet containing 10% safflower oil was to shift the dose-response curve for T_3 to the right,[44] but not to reduce the maximum activity induced by high doses of T_3. Such a shift in T_3 responsivity suggests dietary polyunsaturated fatty acids may function as competitive inhibitors of the T_3 induction of lipogenic enzymes. The mechanism of such a shift is unclear, but interference with T_3 receptor binding and/or receptor/DNA interactions may be potential sites of polyenoic fatty acid action.

IX. SUMMARY AND SPECULATION

The tissue-specific response of the fatty acid synthase gene to polyunsaturated fat inhibition, and the rapid transcriptional suppression of genes coding for lipogenic enzymes, has led us to speculate that the nucleus of hepatocytes contains a nuclear protein (e.g., nuclear fatty acid binding protein) which selectively binds 18:3 (n-6) and 18:4 (n-4) polyenoic fatty acids and their metabolites. Upon binding of the polyenoic fatty acid ligand, the putative *trans*-acting protein would bind to a *cis*-acting element linked to the fatty acid synthase (or S14) gene which would result in suppression of gene transcription. Such a *cis*-element could be specific for the proposed nuclear fatty acid binding protein or could involve interaction with a defined hormonal response element, e.g., thyroid response element. Alternatively, transcriptional regulation could also be regulated via fatty acid ligand binding to nuclear fatty acid binding protein which would allow the protein to interact with other nuclear hormonal receptors, such as the glucocorticoid or thyroid receptors, and thereby alter the hormonal action of these signals. A single nuclear fatty acid binding protein which bound to a specific nucleotide response sequence that was common to all genes coding for lipogenic enzymes would explain the apparent coordinate suppression of lipogenic enzymes by dietary polyunsaturated fats. Obviously at this time our hypothesis is simply speculation, but precedent for the concept is derived from the mechanism by which cholesterol regulates HMG-CoA reductase and the LDL receptor,[45] as well as from the recent demonstration of a nuclear binding protein for clofibrate and its involvement in the induction of peroxisomes.[46]

ACKNOWLEDGMENTS

The work presented within this review was supported in part by the Upjohn Company, Kalamazoo, MI and the National Institutes of Health—DK 39302 (SDC) and GM 36851 (DBJ). The authors wish to thank Ormond MacDougal, William Blake, Meera Kamdar, and Ana Mildner for the technical assistance and thoughtful discussions. Dr. Clarke is the recipient of the Lillian Fountain Smith Endowed Professorship of Human Nutrition at Colorado State University.

REFERENCES

1. **Allmann, D. W. and Gibson, D. W.,** Fatty acid synthesis during early linoleic acid deficiency in the mouse, *J. Lipid Res.,* 6, 51, 1969.
2. **Sabine, J. R., McGrath, H., and Abraham, S.,** Dietary fat and the inhibition of hepatic lipogenesis in the mouse, *J. Nutr.,* 98, 312, 1969.
3. **Musch, K., Ojakian, M. S., and Williams, M. A.,** Comparison of α-linolenate and oleate in lowering activity of lipogenic enzymes in rat liver: evidence for a greater effect of dietary linolenate independent of food and carbohydrate intake, *Biochim. Biophys. Acta,* 337, 343, 1974.
4. **Clarke, S. D., Romsos, D., and Leveille, G. A.,** Differential effects of dietary methyl esters of long-chain saturated and polyunsaturated fatty acids on rat liver and adipose tissue lipogenesis, *J. Nutr.,* 107, 1170, 1977.
5. **Toussant, M. J., Wilson, M. D., and Clarke, S. D.,** Coordinate suppression of liver acetyl-CoA carboxylase and fatty acid synthase by polyunsaturated fat, *J. Nutr.,* 111, 146, 1981.
6. **Flick, P. K., Chen, J., and Vagelos, P. R.,** Effect of dietary linoleate on synthesis and degradation of fatty acid synthetase from rat liver, *J. Biol. Chem.,* 252, 4242, 1977.
7. **Emken, E. A., Abraham, S., and Lin, C. Y.,** Metabolism of cis-12-octadecenoic acid and trans-9, trans-12 octadecadienoic acid and their influence on lipogenic enzyme activities in mouse liver, *Biochim. Biophys. Acta,* 919, 111, 1987.
8. **Clarke, B. A. and Clarke, S. D.,** Suppression of rat liver fatty acid synthesis by eicosa-5,8,11,14-tetraynoic acid without a reduction in lipogenic enzymes, *J. Nutr.,* 112, 1212, 1982.
9. **Clarke, S. D. and Abraham, S.,** Inhibition of fatty acid synthase gene expression by polyunsaturated fat: a requirement for delta-6 desaturation, submitted.
10. **Clarke, S. D., Armstrong, M. K., and Jump, D. B.,** Nutritional control of rat liver fatty acid synthase and S14 mRNA abundance, *J. Nutr.,* 120, 218, 1990.
11. **Clarke, S. D., Armstrong, M. K., and Jump, D. B.,** Dietary polyunsaturated fats uniquely suppress rat liver fatty acid synthase and S14 mRNA content, *J. Nutr.,* 120, 225, 1990.
12. **Blake, W. L. and Clarke, S. D.,** Suppression of hepatic fatty acid synthase and S14 gene transcription by dietary polyunsaturated fat, *J. Nutr.,* 120, 1727, 1990.
13. **Hillard, B. L., Lundin, P., and Clarke, S. D.,** Essentiality of dietary carbohydrate for maintenance of liver lipogenesis in the chick, *J. Nutr.,* 110, 1533, 1980.
14. **Bortz, W. M., Abraham, S., and Chaikoff, I. L.,** Localization of the block in lipogenesis resulting from feeding fat, *J. Biol. Chem.,* 238, 1266, 1963.
15. **Romsos, D. R. and Leveille, G. A.,** Effects of diet on activity of enzymes involved in fatty acid and cholesterol synthesis, *Adv. Lipid Res.,* 12, 97, 1974.
16. **Wilson, M. D., Blake, W. L., Salati, L. M., and Clarke, S. D.,** Potency of polyunsaturated and saturated fats as short term inhibitors of hepatic lipogenesis in rats, *J. Nutr.,* 120, 544, 1990.
17. **Clarke, S. D. and Hillard, B. L.,** Suppression of hepatocyte fatty acid synthesis by albumin-bound linoleate involves depolymerization of acetyl-CoA carboxylase filaments, *Lipids,* 15, 207, 1981.
18. **Ashcraft, B. A., Fillers, W. S., Augustine, S. L., and Clarke, S. D.,** Polymer-protomer transition of acetyl-CoA carboxylase occurs *in vivo* and varies with nutritional conditions, *J. Biol. Chem.,* 255, 10033, 1980.
19. **Clarke, B. A. and Clarke, S. D.,** Polymer-protomer transition of acetyl-CoA carboxylase as a regulator of lipogenesis in rat liver, *Arch. Biochem. Biophys.,* 218, 92, 1982.
20. **Schwartz, R. S. and Abraham, S.,** Effect of dietary polyunsaturated fatty acids on the activity and content of fatty acid synthetase in mouse liver, *Biochim. Biophys. Acta,* 711, 316, 1982.

21. **Katsurada, A., Iritani, N., Fukuda, H., Matsumura, Y., Nishimoto, N., Noguchi, T., and Tanaka, T.,** Effects of nutrients and hormones on transcriptional and post-transcriptional regulation of acetyl-CoA carboxylase in rat liver, *Eur. J. Biochem.,* 190, 435, 1990.

22. **Tomlinson, J. E., Nakayama, R., and Holten, D.,** Repression of pentose phosphate pathway dehydrogenase synthesis and mRNA by dietary fat in rats, *J. Nutr.,* 118, 408, 1988.

23. **Dempsey, M. E., Hargis, P. S., McGuire, D. M., McMahon, A., Olson, C. D., Salati, L. M., Clarke, S. D., and Towle, H. C.,** Role of sterol carrier protein in cholesterol metabolism, *Chem. Phys. Lipids,* 38, 223, 1985.

24. **Salati, L. M., Hargis, P. S., Olson, C. D., Clarke, S. D., and Dempsey, M. E.,** Rapid regulation of sterol carrier protein synthesis, secretion, and turnover in rat hepatocytes, *Fed. Proc.,* 45 (Abstr.), 1581, 1986.

25. **Clarke, S. D., Wilson, M. D., and Ibnoughazala, T.,** Resistance of lung fatty acid synthesis to inhibition by dietary fat in the meal-fed rat, *J. Nutr.,* 114, 598, 1984.

26. **Armstrong, M. K., Blake, W. L., and Clarke, S. D.,** Arachidonic acid suppression of fatty acid synthase gene expression in cultured rat hepatocytes, *Biochem. Biophys. Res. Commun.,* 177, 1056, 1991.

27. **Salati, L. M., Adkins-Finke, B. A., and Clarke, S. D.,** Fatty acid regulation of glucose-6-phosphate dehydrogenase in rat liver cell monolayers, *Lipids,* 23, 36, 1988.

28. **Odin, R. S., Adkins-Finke, B. A., Blake, W. L., Phinney, S. D., and Clarke, S. D.,** Membrane phospholipid enrichment of cultured rat hepatocytes with omega-3 and omega-6 fatty acids is associated with decreased triglyceride production and secretion, *Biochim. Biophys. Acta,* 921, 378, 1987.

29. **Reid, L. M., Narita, M., Fujita, M., and Rosenberg, L.,** *Isolated and Cultured Hepatocytes,* Guillouzo, A. and Gugen-Guillouzo, C., Eds., Hohn Libbey, London, 1986, 225.

30. **Schwartz, R. S. and Abraham, S.,** Effect of dietary fat on the activity, content, rates of synthesis, and degradation and translation of messenger RNA coding for malic enzyme, *Arch. Biochem. Biophys.,* 221, 206, 1983.

31. **Szepesi, B., Kamara, A. K., and Clarke, S. D.,** Lack of specificity of polyunsaturated fats in the inhibition of rat liver glucose-6-phosphate dehydrogenase, *J. Nutr.,* 119, 161, 1989.

32. **Nebert, D. W.,** Proposed role of drug-metabolizing enzymes: regulation of steady state levels of the ligands that effect growth, homeostasis, differentiation, and neuroendocrine functions, *Mol. Endocrinol.,* 5, 1203, 1991.

33. **Dozin, B., Rall, J., and Nikodem, V. M.,** Tissue-specific control of rat malic enzyme activity and messenger RNA levels by a high carbohydrate diet, *Proc. Natl. Acad. Sci. U.S.A.,* 83, 4705, 1986.

34. **Katsurada, A., Iritani, N., Fukuda, H., Noguchi, T., and Tanaka, T.,** Influence of diet on the transcriptional and post-transcriptional regulation of malic enzyme induction in the rat liver, *Eur. J. Biochem.,* 168, 487, 1987.

35. **Amy, C. M., Williams-Ahlf, B., Naggert, J., and Smith, S.,** Molecular cloning of the mammalian fatty acid synthase gene and identification of the promoter region, *Biochem. J.,* 271, 675, 1990.

36. **MacDougal, O., Clarke, S. D., and Jump, D. B.,** Unpublished observations.

37. **Jump, D. B., Bell, A., Lepar, G., and Hu, D.,** Insulin rapidly induces rat liver S14 gene transcription, *Mol. Endocrinol.,* 4, 1655, 1990.

38. **Salati, L. M. and Clarke, S. D.,** Fatty acid inhibition of the hormonal induction of acetyl-CoA carboxylase in hepatocyte monolayers, *Arch. Biochem. Biophys.,* 246, 82, 1986.

39. **Dax, E. M., Partilla, J. S., Pineyro, M. A., and Gregerman, R. I.,** Altered glucagon-and catecholamine hormone-sensitive adenylyl cyclase responsiveness in rat liver membranes induced by manipulation of dietary fatty acid intake, *Endocrinology,* 127, 2236, 1990.

40. **Field, C. J., Ryan, E. A., Thomson, A. B. R., and Clandinin, M. T.,** Diet fat composition alters membrane phospholipid composition insulin binding, and glucose metabolism in adipocytes from control and diabetic animals, *J. Biol. Chem.,* 265, 11143, 1990.

41. **Blake, W. L. and Clarke, S. D.,** Type of dietary fat alters rat liver cell responsiveness to vasopressin, *J. Nutr. Biochem.,* 2, 87, 1991.

42. **Clandinin, M. T., Cheema, S., Field, C. J., Garg, M. L., Vendatraman, J., and Clandinin, T. R.,** Dietary fat: exogenous determination of membrane structure and cell function, *FASEB J.,* 5, 2761, 1991.

43. **O'Brien, R. M., Lucas, P. C., Forest, C. D., Magnuson, M. A., and Granner, D. K.,** Identification of a sequence in the PEPCK gene that mediates a negative effect of insulin on transcription, *Science,* 249, 533, 1990.

44. **Clarke, S. D. and Hembree, J.,** Dietary fat suppresses the triiodothyronine induction of rat liver lipogenic enzymes, *J. Nutr.,* 120, 625, 1990.

45. **Dawson, P. A., Hofmann, S. L., Van der Westhuyzen, D. R., Sudhof, T. C., Brown, M. S., and Goldstein, J. L.,** Sterol-dependent repression of low density lipoprotein receptor promoter mediated by 16-base pair sequence adjacent to binding site for transcription factor Sp1, *J. Biol. Chem.,* 263, 3372, 1988.

46. **Issemann, I. and Green, S.,** Activation of a member of the esteroid hormone receptor superfamily by peroxisome proliferators, *Nature (London),* 347, 645, 1990.

47. **Clarke, S. D., Romsos, D. R., and Leveille, G. A.,** Influence of dietary fatty acids on liver and adipose tissue lipogenesis and on liver metabolites in meal-fed rats, *J. Nutr.,* 107, 1277, 1977.

48. **Herzberg, G. R.,** The influence of dietary fatty acid composition on lipogenesis, *Adv. Nutr. Res.,* 5, 221, 1983.

49. **Christiansen, E. N., Lund, J. S., Rortveit, T., and Rustan, A. C.,** Effect of dietary n-3 and n-6 fatty acids on fatty acid desaturation, *Biochim. Biophys. Acta,* 1082, 57, 1991.

50. **Baltzell, J. K., Wooten, J. T., and Otto, D. A.,** Lipoprotein lipase in rats fed fish oil: apparent relationship to plasma insulin levels, *Lipids,* 26, 289.

51. **Miller, J. C. E., Barth, R. K., Shaw, P. H., Elliot, R. W., and Hastie, N. D.,** Identification of a cDNA clone for mouse apoprotein A-1 (apo A 1) and its use in characterization of apo A-1 mRNA expression in liver and small intestine, *Proc. Natl. Acad. Sci. U.S.A.,* 80, 1511, 1983.

Chapter 11

DIETARY FAT, GENE EXPRESSION, AND CARCINOGENESIS

Howard P. Glauert

TABLE OF CONTENTS

0-8493-6961-4/93/$0.00 + $.50
© 1993 by CRC Press, Inc.

I. INTRODUCTION

The amount and type of fat consumed in the diet may be important in the development of human cancer. Several organizations have advocated decreasing the fat content of the diet as a means of preventing the development of cancer. For example, the American Cancer Society[1] advises people to "cut down on total fat intake", and the American Institute for Cancer Research[2] advises people to "reduce the intake of total dietary fat from the current average of approximately 37% to a level of no more than 30% of total calories and, in particular, reduce the intake of saturated fat."

In this chapter, the role of dietary fat on the development of human and experimental cancer will be discussed. The role of gene expression as a mechanism by which dietary fat may alter the development of cancer will be examined. Cancer clearly is a disease of alterations both in genetic structure and in genetic expression, both of which can be affected by dietary fat.

II. MECHANISMS OF CARCINOGENESIS

A. CHEMICAL CARCINOGENESIS

At present, epidemiologists believe that 60 to 90% of human cancers are caused by environmental agents.[3] Worldwide, the incidence of cancer in different tissues and organs varies widely from country to country.[4] When people migrate from one country to another, their descendants tend to develop the same types of cancers as the native population.

The main environmental agents which are thought to cause cancer are viruses, radiation, and chemicals.[5] The induction of cancer by chemicals was first documented in 1775 by Percival Pott, who found that men who worked as chimney sweeps had a high incidence of scrotal cancer.[6] Many other human carcinogens have since been identified, including azo dyes, benzene, vinyl chloride, and cigarette smoke.[3] In animals, chemical carcinogenesis was first observed in the early 1900s, when the application of coal tar to skin was found to induce tumors.[7,8] Subsequently, chemicals isolated from coal tar, the polycyclic aromatic hydrocarbons (PAH), were found to induce skin tumors in mice.[9,10] Several other classes of chemicals have since been found to induce tumors, including aromatic amines, dialkylnitrosamines, and alkylating agents.[3]

In order to induce cancer, most chemical carcinogens have to be metabolized to some other form. The procarcinogens are generally metabolized by mixed-function oxidases in the smooth endoplasmic reticulum to electrophilic forms (the ultimate carcinogens).[3] These electrophiles then react with nucleophilic sites in the cell, such as the nitrogen and oxygen atoms in DNA. The binding of an ultimate carcinogen to DNA can lead to mispairing during DNA replication and thus to the formation of mutations in DNA. Chemical carcinogens or their reactive metabolites have been shown to be mutagenic in bacterial and mammalian mutagenesis systems.[11,12]

B. MULTISTAGE CARCINOGENESIS

After the administration of a chemical carcinogen, there is a latency period before the actual appearance of a neoplasm. The latency period may vary depending on the carcinogen used, its dosage, and the tissue or organ being investigated, but it may be considered to be a general feature of the pathogenesis of neoplasia.[13] Concurrently with early investigations into the latency phenomenon, Rous and Kidd[14] provided experimental evidence suggesting a two-stage mechanism for carcinogenesis in skin. They found that coal tar application could "initiate" tumor formation while wounding could "promote" this process. Subsequently, Berenblum and Shubik[15] clearly demonstrated that skin carcinogenesis could be divided into an initiation stage followed by a longer promotion stage. These authors and many others have demonstrated that administering a low dose of an initiating agent (which may be a chemical, UV light, or ionizing radiation) to mouse skin normally results in few or no tumors being formed during the animal's lifetime. Subsequent repeated application of a promoting agent (such as croton oil or its most potent component, 12-O-tetradecanoylphorbol-13-acetate [TPA]) over several weeks or months will, however, induce the formation of skin papillomas and carcinomas.

Subsequent studies have demonstrated that carcinogenesis in other tissues—such as the liver, mammary gland, bladder, and pancreas—can also be divided into the two stages of initiation and promotion.[5] Initiation and promotion, regardless of the tissue in question, have certain common properties. In two-stage carcinogenesis, the administration of a single subcarcinogenic dose of a carcinogen (i.e., initiation) is followed by the long-term administration of a promoter.[16] Initiation is not reversible, whereas promotion is. Neither the administration of an initiator alone, repetitive doses of the promoter alone, nor the administration of the promoter before the initiator will lead to tumor production: the application of the initiator must be followed by repeated doses of the promoter.[16] Mechanistically, an initiator has been defined as an agent which irreversibly alters the native molecular structure of DNA (i.e., produces a mutation), whereas a promoter has been defined as an agent that alters the expression of genetic information in the cell.[13]

An additional model for multistage carcinogenesis has been proposed by Moolgavkar and colleagues.[17] In this model, two mutations must occur in order for a cell to become neoplastic. The rate at which a tumor appears can be influenced by two factors: an increase in the mutation rate, which could be brought about by a genotoxic carcinogen, or an increase in the rate of cell proliferation, which would expand the population of normal and intermediate cells (i.e., cells with one mutation), and increase the likelihood of a mutation occurring. A tumor-promoting agent could bring about this second mechanism.

C. GENES IMPORTANT IN CARCINOGENESIS

The identity of the genes which are mutated by chemical carcinogens or radiation is of great importance in understanding the carcinogenic process. Dietary fat likely influences carcinogenesis by bringing about an alteration in the expression of one or more of these genes. Cellular oncogenes or tumor suppressor genes are the most probable targets whose mutation would lead to the development of tumors. Cellular oncogenes, or proto-oncogenes, are the cellular analogs of viral oncogenes, which have the ability to transform cells and produce tumors. Over 50 viral oncogenes have been discovered. When mutations are produced in specific locations in cellular oncogenes, they gain the ability to transform cells. Cellular oncogenes containing mutations in specific codons have been demonstrated in several model systems.[18] In mouse skin, papillomas and carcinomas induced by initiation/promotion protocols contain activated *ras* oncogenes. Activated *ras* oncogenes are also seen in mouse liver tumors induced by several chemicals.

Another probable target is one or more of the tumor suppressor genes. These genes are normally expressed in nontransformed cells and appear to play a role in chromosome stability or in the regulation of cellular differentiation or cellular proliferation.[19] Certain types of genetic human tumors, such as retinoblastoma, involve the loss of specific genes. The loss of tumor suppressor genes may involve point mutation in the specific gene or deletion of part of the chromosome containing the specific gene. Mutations and allelic losses of the p53 tumor suppressor gene have been observed in several human tumors.[20-22]

III. EPIDEMIOLOGICAL STUDIES

Epidemiological studies do not provide definitive conclusions as to the effect of dietary fat on human cancer. Prospective studies are few in number and are not in agreement. Several epidemiological studies have examined the relationship between the intake of dietary fat or meat and the development of colon cancer.[23,24] Several correlational and case-control epidemiological studies have found a positive association between dietary fat and colon cancer, but others have not. Prospective epidemiological studies, however, have found no effect in three studies, a positive association in one study, and an actual protective effect of high fat intakes in two studies.[22,24] The interpretation of these studies may be complicated by the total energy intake;[23,25,26] energy intake has been correlated to colon cancer risk in several correlational and case-control studies.[23,26]

Numerous epidemiological studies have attempted to identify factors which influence breast cancer risk in humans. Established breast cancer risk factors include age of first pregnancy, body build, age at menarche or menopause, and the amount of radiation received by the chest.[27] The effect of dietary fat has been studied in correlational, case-control, and prospective epidemiological studies. Studies examining international correlations between dietary fat intake and breast cancer risk and migrant studies have reported a positive association between dietary fat intake and breast cancer risk.[27] Case-control studies generally have also supported a connection between total fat intake and breast cancer risk.[28] Recently, a combined analysis of 12 case-control studies found a highly significant positive association between breast cancer risk and saturated fat intake.[29] Two prospective studies, however, did not find any link between dietary fat intake and the development of breast cancer,[30,31] whereas another found a positive correlation between fat intake and the development of breast cancer.[32]

In the pancreas, dietary fat consumption has been found to be associated with an increased human cancer risk in the few correlational and case-control epidemiological studies that have been published.[33-35] In correlational epidemiological studies, high dietary fat intakes are also associated with a higher incidence of ovarian and prostatic cancer.[36]

IV. EXPERIMENTAL CARCINOGENESIS STUDIES

A. NUTRITIONAL CONSIDERATIONS

In experiments examining the effect of dietary fat, fat has to be substituted for carbohydrate isocalorically in order to ensure that fat is the only experimental variable. In some studies, fat was added to a nutritionally complete diet, such as a ground unrefined diet. Since animals consume approximately the same number of calories, animals on such a high fat diet would consume less of all other nutrients in the diet. Therefore, any changes seen in the high fat group could not strictly be attributed to a higher consumption of fat. In other studies, fat was substituted for a carbohydrate (such as corn starch or dextrose) on a gram-for-gram basis. Since fat contains 9 kcal per gram and carbohydrate contains only 4 kcal per gram,[37] such a protocol would also result in lower consumption of other nutrients in the diet. In order to substitute fat for carbohydrate on a metabolizable energy basis, 4 g of fat have to be substituted for 9 g of carbohydrate.

Another issue in studies examining dietary fat and carcinogenesis is whether the enhancing effect of dietary fat, when it is seen, is due to dietary fat per se (for example, by an effect on bile acid metabolism in colon carcinogenesis) or rather to a more efficient utilization of calories. Even when the ratio of calories to nutrients remains the same when the dietary fat content of the diet is raised, animals frequently gain more weight than animals on a low fat diet. When fat is substituted for carbohydrate isocalorically, the substitution is done on the basis of metabolizable energy rather than net energy. Donato and Hegsted,[38] however, have demonstrated that fat is used more efficiently than carbohydrate, so that equating their metabolizable energy values may be inaccurate.

B. SKIN CARCINOGENESIS

Mouse skin is one of the oldest and most widely used systems for studying chemical carcinogenesis, including multistage carcinogenesis. As discussed earlier, two-stage carcinogenesis (initiation-promotion) was first observed in mouse skin and involves initiation by a subcarcinogenic dose of radiation or of a chemical, such as PAH, followed by the long-term administration of croton oil or its active ingredient TPA.

Most studies examining dietary fat have studied complete carcinogenesis by PAH or ultraviolet light. Early studies demonstrated that high fat diets enhanced skin carcinogenesis induced by tar[39] or PAH.[40-46] In studies where skin tumors were induced by ultraviolet light, Mathews-Roth and Krinsky[47] found that high fat diets increased skin carcinogenesis, whereas Black et al.[48] found that high fat diets did not increase skin carcinogenesis, but that feeding a saturated fat inhibited tumorigenesis.

The effect of fatty acids on the initiation and promotion of skin carcinogenesis has also been studied. Certain fatty acids—oleic acid and lauric acid—have promoting activity when applied daily to mouse skin after a single application of 7,12-dimethylbenz(a)anthracene (DMBA); stearic acid and palmitic acid do not have any effect.[49] When diets varying in their fat content are fed during the initiation or promotion stages of DMBA-initiated, TPA-promoted skin carcinogenesis, high fat diets enhance the promotion of skin carcinogenesis and slightly inhibit initiation.[50,51]

C. HEPATOCARCINOGENESIS

Many early studies of dietary fat and cancer used the liver as the target organ. In these studies, aromatic amines and azo dyes were frequently used to induce hepatocellular carcinomas. In later studies, effects of dietary fat on initiation and promotion in the liver were examined. In initiation-promotion protocols, the administration of a single subcarcinogenic dose of a carcinogen (such as diethylnitrosamine [DEN] or DMBA) along with a proliferative stimulus (such as partial hepatectomy) followed by the long-term feeding of chemicals such as phenobarbital, 2,3,7,8-tetrachlorodibenzo-*p*-dioxin, or polyhalogenated biphenyls, leads to a high incidence of hepatocellular adenomas and carcinomas.[13,52] In addition, foci of putative preneoplastic hepatocytes appear before the development of gross tumors. These foci, known as altered hepatic foci or enzyme-altered foci, contain cells which exhibit qualitatively altered enzyme activities or alterations in one or more cell functions.[53] The enzymes most frequently studied include γ-glutamyltranspeptidase (GGT) and placental glutathione-*S*-transferase, which are normally not present in adult liver, but which are often present in foci, and ATPase and glucose-6-phosphatase, which are normally present but which are frequently missing from foci.[54,55] Altered hepatic foci can also be identified on hematoxylin and eosin-stained tissue.[56,57] The appearance of foci has been correlated with the later development of malignant neoplasms.[58,59]

The first studies examined the effect of dietary fat on the induction of hepatocellular carcinomas by complete hepatocarcinogens. In the liver, increasing the fat content of the diet enhances the development of 2-acetylaminofluorene (AAF)-, *p*-dimethylaminoazobenzene (DAB)-, and aflatoxin B_1 (AFB)-induced tumors and GGT-positive foci in rats.[60-63] Furthermore, hepatocarcinogenesis by DAB is enhanced by feeding a diet which contains a greater proportion of polyunsaturated fatty acids.[64,65] In these studies, however, the diets were administered at the same time as the carcinogen injections, so that the stage of carcinogenesis which was affected could not be determined.

More recent studies have examined whether this enhancement of hepatocarcinogenesis is caused by an effect on the initiation of carcinogenesis, the promotion of carcinogenesis, or both. Misslbeck et al.[66] found that increasing the corn oil content of the diet after the administration of 10 doses of aflatoxin increased the number and size of GGT-positive foci, but Baldwin and Parker,[67] using a similar protocol, found no effect of dietary corn oil. Glauert and Pitot[68] similarly found that increasing the safflower oil or palm oil content of the diet did not promote DEN-induced GGT-positive foci or greatly affect phenobarbital promotion of GGT-positive foci. The promotion of GGT-positive foci by dietary tryptophan also is not affected by dietary fat.[69] Newberne et al.[70] found that increasing dietary corn oil (but not beef fat) during and after the administration of AFB increased the incidence of hepatic tumors, but not when the diets were fed only after AFB administration. Baldwin and Parker[67] also found that increasing the corn oil content of the diet before and during AFB administration increased the number and volume of GGT-positive foci. Recently, it has been found that when rats are fed diets high in polyunsaturated fatty acids (but not in saturated fatty acids) before receiving the hepatocarinogen DEN, they develop more GGT-positive and ATPase-negative foci than rats fed low fat diets.[71] The results of these studies suggest that the enhancement of hepatocarcinogenesis by dietary fat is primarily due to an effect on initiation, and that polyunsaturated fats have a greater effect than do saturated fats.

D. COLON CARCINOGENESIS

Studies in experimental animals have produced differing results. Because no animal models develop colon tumors spontaneously, a variety of chemicals have been used to induce colon tumors, usually in rats or mice. These include 1,2-dimethylhydrazine (DMH) and its metabolites azoxymethane (AOM) and methylazoxymethanol (MAM); 3,2'-dimethyl-4-aminobiphenyl (DMAB); methylnitrosourea (MNU); and *N*-methyl-*N'*-nitro-*N*-nitrosoguanidine (MNNG).[72-75] DMH and AOM have been used most frequently to study nutritional effects. Both can induce colon tumors by single[76-80] or multiple[81-85] injections.

Animal studies examining the effect of dietary fat have used a variety of protocols, and the results obtained often have been dependent on the investigator's protocol. In several studies, fat was either added to a chow diet or was substituted for carbohydrate on a weight basis, so that the ratio of calories to essential nutrients was altered.[83,84,86-91] These studies all found that increasing dietary fat enhanced colon carcinogenesis, but the effect could have been due to a lower consumption of essential nutrients rather than to an effect of fat. In other studies, rats or mice were subjected to multiple doses of a colon carcinogen, with the dietary fat content being varied (isocalorically) during, and frequently before or after, the carcinogen injections. Some of these studies found an enhancement when the dietary fat content of the diet was increased, but most saw no effect and one saw an inhibition of tumor development.[73,75,81,82,92-95] Feeding unsaturated rather than saturated fat in the diet also was found to increase colon carcinogenesis induced by multiple injections of AOM.[96] More recent studies have examined the effect of the amount and type of dietary fat on the initiation and promotion of colon carcinogenesis induced by one, two, or three injections of AOM. Reddy and colleagues found that increasing the unsaturated fat, but not the saturated fat, content of the diet during the promotional phase enhanced colon carcinogenesis; the initiation of colon carcinogenesis by AOM, however, was not affected by the level of unsaturated fat, but was slightly enhanced by increased levels of saturated fat.[97-99] Other investigators, however, did not see an enhancement of the promotional phase by dietary polyunsaturated fat.[100,101]

E. PANCREATIC CARCINOGENESIS

Dietary fat has been studied extensively in animal models, primarily rats and hamsters. In rats, azaserine is the chemical carcinogen usually used.[102] Single or multiple doses of this chemical induce carcinomas derived from acinar cells. Putative preneoplastic lesions—basophilic and acidophilic foci—have also been described in the rat pancreas.[102] The carcinogen *N*-nitrosobis(2-oxopropyl)amine (BOP) has been used primarily to induce pancreatic tumors in hamsters, although it also has been shown to be effective in rats.[102] In hamsters, tumors are derived from ductal cells, as are the majority of human pancreatic tumors. The hamster model, therefore, may be more relevant to human pancreatic cancer.

Dietary fat has been found to influence tumorigenesis in both animal models. In rats, feeding high-fat diets after, or during and after, the injection of azaserine enhances the development of pancreatic tumors and putative preneoplastic lesions.[103-111] Pancreatic carcinogenesis induced by *N*-nitroso(2-hydroxypropyl) (2-oxopropyl)amine in rats is also enhanced by feeding high fat diets.[112] In several of these studies, the effect cannot be attributed unequivocally to dietary fat because fat was substituted for carbohydrate on a weight basis. In hamsters, BOP-induced pancreatic carcinogenesis is also increased by feeding high-fat diets.[108-110,113-116] Roebuck and colleagues[104,105,111] found that polyunsaturated fat, but not saturated fat, enhanced pancreatic carcinogenesis, and that a certain level of essential fatty acids are required for the enhancement of pancreatic carcinogenesis. Birt et al.,[116] however, found that feeding a saturated fat (beef tallow) enhanced pancreatic carcinogenesis in hamsters greater than a polyunsaturated fat (corn oil).

Finally, it has been observed in 2-year carcinogenesis studies in which corn oil gavage has been used as the vehicle for the carcinogen that a higher incidence of pancreatic acinar cell adenomas is present in corn oil gavage-treated male Fischer-344 control rats than in untreated controls.[117,118] This association was not observed in female rats or in male or female B6C3F$_1$ mice. The mechanism of this effect is not known.

F. MAMMARY CARCINOGENESIS

The effect of dietary fat on mammary carcinogenesis in experimental animals has been examined extensively: over 100 experiments have been conducted.[119] Two animal models are primarily used: a rat model (usually the Sprague-Dawley strain) in which mammary tumors are induced by DMBA or MNU and a mouse model, in which tumors develop spontaneously. The use of these models is advantageous because tumor latency, tumor size, and tumor progression can easily be quantified by palpation of mammary tumors as they appear. In both models, increasing the fat content of the diet clearly enhances the development of mammary tumors.[119] In the rat model, a high fat diet increases tumorigenesis both when it is fed during and after carcinogen administration, and when it is fed only after carcinogen injection.

G. OTHER SITES

Dietary fat has also been studied for its effect on experimental carcinogenesis in other organs. In the lung, dietary fat enhanced BP- or BOP-induced carcinogenesis in hamsters,[114,120] whereas in mice, a high fat diet did not affect spontaneous carcinogenesis in one study, but the feeding of egg extracts enhanced it in another.[44,121] Spontaneous brain tumors in mice and prostate tumors in rats were increased by the feeding of high fat diets, as was the induction of BOP-induced renal tumors, but DMH- or MAM-induced renal or ear duct tumors were not affected by dietary fat.[87,114,122,123]

H. EFFECT OF n–3 FATTY ACIDS

Feeding diets high in n–3 fatty acids generally inhibits chemical carcinogenesis; in most of these studies, oils high in n–3 fatty acids are substituted for corn oil.[124] In the pancreas, feeding diets high in menhaden oil decreases the development of azaserine-induced preneoplastic lesions.[125,126] In the colon, feeding menhaden oil in place of corn oil, or eicosapentaenoic acid in place of linoleic acid, decreases the development of DMH- or AOM-induced colon tumors, but adding menhaden oil to a low-fat diet does not affect colon carcinogenesis.[99,127-129] In the mammary gland, feeding a diet high in menhaden oil instead of a high corn oil diet decreases the incidence of DMBA-induced tumors in rats.[130,131] Locniskar et al.,[132] however, found that substituting menhaden oil for corn oil or coconut oil did not affect skin tumor promotion by TPA.

V. MECHANISMS BY WHICH DIETARY FAT MAY INFLUENCE CARCINOGENESIS

A. MEMBRANE FLUIDITY

An important function of dietary fatty acids is their presence in membrane lipids. Altering the fatty acid content of the diet alters the composition of membrane lipids, particularly in certain tissues; feeding diets high in n–6 or n–3 fatty acids increases the concentrations of these fatty acids in membrane lipids.[133] The activities of membrane-bound enzymes are increased in membranes that are more fluid, i.e., that have a higher content of polyunsaturated fatty acids.[133] Higher amounts of unsaturated fatty acids in cell membranes are also associated with a higher rate of cell proliferation.[134] The alteration by dietary fatty acids of the catalytic abilities of membrane-bound enzymes, such as cytochrome P-450, may play an important role in carcinogenesis.

B. TOXICITY

One possible mechanism by which dietary fat may enhance carcinogenesis is by the toxicity of fatty acids or of metabolites that increase after the feeding of high fat diets. Such toxicity would bring about a proliferative response in the tissue to replace lost cells. Cellular genes involved in cell proliferation, including cellular oncogenes, would likely be increased.

In the colon, toxicity may play a role in the enhancement of carcinogenesis by dietary fat. One hypothesis for the effect of dietary fat is that dietary fat increases the concentration of metabolites with carcinogenic or promoting activity in the fecal stream. Bile acids, particularly secondary bile acids, have promoting activity in the colon;[135-137] their concentration in the feces has been found to be increased by dietary fat in some but not all studies.[83,138-141] Bile acids function as detergents; therefore, high concentrations may be toxic to epithelial cells in the colon. Increasing the concentration of bile acids as well as fatty acids in the colon has been demonstrated to increase colon epithelial cell proliferation.[142-146] It has been further demonstrated that the increased expression of protein kinase C may play a role in bile acid-induced cell proliferation.[147]

C. EICOSANOID METABOLISM

Another mechanism by which dietary fat may influence carcinogenesis is by altering the synthesis of eicosanoids. Fatty acids that are consumed in the diet can be metabolized to a variety of other compounds, including longer and more unsaturated fatty acids, prostaglandins, leukotrienes, thromboxanes, hydroperoxyeicosatetraenoic acids, and hydroxyeicosatetraenoic acids.[148] Altering the type of fatty acid in the diet changes the composition of the eicosanoids that are produced by the body.[149] For example, increasing the n−3 content of the diet will alter the type of eicosanoids that are produced as well as their function.[150] Specific eicosanoids bind to receptors and cause specific alterations in gene expression and cellular function,[151] some of which may be related to carcinogenesis. It has been found that inhibition of eicosanoid synthesis inhibits DNA synthesis and tumor promotion in both the skin and liver.[152-154]

D. CALORIC EFFECTS

The issue of whether the enhancing effect of fat in carcinogenesis is due to higher consumption of calories or more efficient utilization of energy has been examined in several tissues. The earliest study was conducted by Boutwell et al.[46] using the mouse skin carcinogenesis system; they attributed most of the enhancing effect of dietary fat to an increased consumption of calories. Birt et al.,[51] however, found that the promotion of skin carcinogenesis was enhanced even though the high-fat diets were pair-fed. The greater caloric density of fat has also been proposed to play a role in colon tumorigenesis. Caloric restriction inhibits chemically induced colon carcinogenesis, even if the percentage of dietary fat in the diet is greatly increased.[155,156] In the pancreas, the enhancement by dietary fat appears to be an effect of dietary fat rather than of an increased consumption of calories, as pair-feeding does not inhibit the enhancing effect of dietary fat in hamsters.[115] Several studies have suggested that the enhancement of mammary carcinogenesis by dietary fat may be caused, at least in part, by an alteration in the efficiency of energy utilization.[157-159] Recently, Freedman et al.,[119] using a combined statistical analysis of over 100 animal experiments, concluded that the enhancing effect of high-fat diets was caused in part by a higher caloric intake and in part by some other effect of dietary fat. Finally, caloric restriction has been found to inhibit tumorigenesis in many tissues in experimental animals.[160]

Several mechanisms have been proposed to explain how caloric consumption affects carcinogenesis. These include effects on growth factors, immunity, neuroendocrine function, and metabolic regulation.[160,161] Any of these effects could be brought about by alterations in gene expression.

E. EFFECT ON INITIATION

Dietary fat may also affect the initiation stage of carcinogenesis. Since initiation involves the mutation of DNA, its alteration (by dietary fat or other agents) would mainly affect the structure of genes rather than their expression. In the liver, dietary fat appears to enhance carcinogenesis primarily by an effect on initiation. In other tissues, many protocols have varied the levels of dietary fat during the time of carcinogen injections; therefore dietary fat may be affecting some aspect of initiation in these studies. Higher levels of dietary fat may enhance initiation of carcinogenesis by several mechanisms, including alterations in absorption of the carcinogen from the gut, transport to the target organ, uptake by the target organ, metabolism by cytochrome P-450 or other drug-metabolizing enzymes to a form which can react with DNA, and DNA repair. Any of these processes could be altered by changes in the expression of the appropriate gene. But dietary fat could affect these processes by mechanisms independent of gene expression. Several of these processes occur in membranes, whose lipid composition can be altered by changing the amount or type of dietary fat.[162,163] Increasing the fat content of the diet increases cytochrome P-450 and related activities.[164-167] Western blot analysis has indicated that higher amounts of enzyme protein are present after feeding diets high in polyunsaturated fat.[168] Therefore dietary fat may be affecting both gene expression and the surrounding matrix necessary for optimum enzyme activity. The metabolism of several chemicals, including hexobarbital, aniline, ethylmorphine, benzo[a]pyrene (BP), and dimethylnitrosamine, is also enhanced by feeding diets high in polyunsaturated fatty acids.[164,165,169-171]

F. INDUCTION OF SPECIFIC GENES

Altering the level of dietary fat changes the expression of many genes. Most of the genes studied, however, are related to carbohydrate or lipid metabolism and are not likely to play a role in carcinogenesis.[172-175] As mentioned above, the expression of cytochrome P-450 genes is influenced by the level of dietary fat.[164-168]

Very little work has been published on the effect of dietary fat on the expression of genes which are likely to be important in carcinogenesis: oncogenes and tumor suppressor genes. In the colon, dietary fat did not affect the expression of either c-*myc* or c-H-*ras* oncogenes in normal or tumor tissue.[101] In the mammary gland, substituting 1–3 fatty acid-rich oils for corn oil decreased the expression of the H-*ras* oncogene in one study but had no effect in the other.[176,177]

Another approach has been to examine for genes whose expression is altered greatly when the level of dietary fat is changed. Using subtractive hybridization, Elliott et al.[178] isolated a clone that is expressed fivefold higher in the mammary gland of mice fed low fat diets. This gene was found to have over 80% homology to a gene induced by UV irradiation (LFM 1).[179] The relationship of this gene to tumorigenesis is currently unknown.

VI. SUMMARY AND CONCLUSIONS

Clearly there is much variability in studies of dietary fat and cancer, both in epidemiological and experimental studies. In epidemiological studies, a relationship between dietary fat and colon, pancreatic, and mammary cancer has been found in correlational and case-control studies. Few prospective studies, however, have found an enhancement of these cancers by dietary fat, and, for colon cancer, two studies have found that a high-fat diet is protective.

In experimental studies, dietary fat generally enhances chemically induced skin, liver, pancreatic, and mammary carcinogenesis, whereas conflicting results have been seen in colon carcinogenesis. Dietary fat appears to act primarily during the promotional stage of carcinogenesis in all of these models except the liver, where the effect of dietary fat is primarily on initiation. Because of the variability seen in studies of dietary fat and cancer (particularly prospective epidemiological studies), it cannot be stated unequivocally at this time that human cancer can be prevented by decreasing the fat content of the diet.

The mechanisms by which high-fat diets enhance experimental carcinogenesis are unclear, but probably involve both a higher intake of net energy and other effects, some of which are organ-specific. Nearly all of the mechanisms by which dietary fat may influence carcinogenesis involve alterations in genetic expression. Because of the variability that is seen in studies of dietary fat and cancer, it is critical that basic scientific studies be conducted to elucidate the mechanisms by which dietary fat influences carcinogenesis. At present, few studies have been conducted which have examined the effect of dietary fat on genes which are important in the carcinogenic process. The determination of genes which are turned on or off by dietary fat will likely provide answers as to the role of dietary fat in carcinogenesis.

REFERENCES

1. American Cancer Society, *Cancer Facts and Figures 1991,* American Cancer Society, Atlanta, 1991.
2. American Institute for Cancer Research, *Dietary Guidelines to Lower Cancer Risk,* American Institute for Cancer Research, Washington, D.C., 1990.
3. **Miller, E. C.,** Some current perspectives on chemical carcinogenesis in humans and experimental animals, *Cancer Res.,* 38, 1479, 1978.
4. **Armstrong, B. and Doll, R.,** Environmental factors and cancer incidence and mortality in different countries with special references to dietary practices, *Br. J. Cancer,* 15, 617, 1975.
5. **Pitot, H. C.,** *Fundamentals of Oncology,* 3rd ed., Marcel Dekker, New York, 1986.
6. **Pott, P.,** *Chirurgical observations relative to cancer of the scrotum,* Reprinted in *Natl. Cancer Inst. Monogr.,* 10, 7, 1963.
7. **Yamagiwa, K. and Ichikawa, K.,** Experimental study of the pathogenesis of carcinoma, *J. Cancer Res.,* 3, 1, 1918.
8. **Tsutsui, H.,** Über das kunstlich erzeugte Cancroid bei der Maus, *Gann,* 12, 17, 1918.
9. **Cook, J. W., Hewett, C. L., and Heiger, I.,** The isolation of cancer-producing hydrocarbon from coal tar, *J. Chem. Soc.,* 395.
10. **Kennaway, E. L. and Hieger, I.,** Carcinogenic substances and their fluorescence spectra, *Br. Med. J.,* ii, 1044, 1930.
11. **McCann, J., Choi, E., Yamasaki, E., and Ames, B. N.,** The detection of carcinogens as mutagens in the Salmonella/microsome test: assay of 300 chemicals, *Proc. Natl. Acad. Sci. U.S.A.,* 72, 5135, 1975.

12. **DeMarini, D. M., Brockman, H. E., de Serres, F. J., Evans, H. H., Stankowski, C. F., and Hsie, A. W.,** Specific-locus mutations induced in eukaryotes (especially mammalian cells) by radiation and chemicals: a perspective, *Mutat. Res.,* 220, 11, 1989.

13. **Pitot, H. C. and Sirica, A. E.,** The stages of initiation and promotion in hepatocarcinogenesis, *Biochem. Biophys. Acta,* 605, 191, 1980.

14. **Rous, P. and Kidd, J. G.,** Conditional neoplasms and subthreshold neoplastic states, *J. Exp. Med.,* 73, 365, 1941.

15. **Berenblum, I. and Shubik, P.,** The role of croton oil applications, associated with a single painting of a carcinogen, in tumour induction of the mouse's skin, *Br. J. Cancer,* 1, 379, 1947.

16. **Diamond, L., O'Brien, T. G., and Baird, W. M.,** Tumor promoters and the mechanism of tumor promotion, *Adv. Cancer Res.,* 32, 1, 1980.

17. **Moolgavkar, S. H. and Knudson, A. G.,** Mutation and cancer: a model for human carcinogenesis, *J. Natl. Cancer Inst.,* 66, 1037, 1981.

18. **Stowers, S. J., Maronpot, R. R., Reynolds, S. H., and Anderson, M. W.,** The role of oncogenes in chemical carcinogenesis, *Environ. Health Perspect.,* 75, 81, 1987.

19. **Sager, R.,** Tumor suppressor genes: the puzzle and the promise, *Science,* 246, 1406, 1989.

20. **Smith, M. L., Yeleswarapu, L., Locker, J., and Lombardi, B.,** p53 mutation(s) in diethylnitrosamine-induced foci of putative preneoplastic hepatocytes in male Fischer 344 rats, *Proc. Am. Assoc. Cancer Res.,* 32, 136, 1991.

21. **Hsu, I. C., Metcalf, R. A., Sun, T., Welsh, J. A., Wang, N. J., and Harris, C. C.,** Mutational hotspot in the p53 gene in human hepatocellular carcinomas, *Nature (London),* 350, 427, 1991.

22. **Okamoto, A., Sameshima, Y., Yokoyama, S., Terashima, Y., Sugimura, T., Terada, M., and Yokota, J.,** Frequent allelic losses and mutations of the p53 gene in human ovarian cancer, *Cancer Res.,* 51, 5171, 1991.

23. **Kolonel, L. N.,** Fat and colon cancer: how firm is the epidemiologic evidence?, *Am. J. Clin. Nutr.,* 45, 336, 1987.

24. **Willett, W. C., Stampfer, M. J., Colditz, G. A., Rosner, B. A., and Speizer, F. E.,** Relation of meat, fat, and fiber intake to the risk of colon cancer in a prospective study among women, *New Engl. J. Med.,* 323, 1664, 1990.

25. **Willett, W. C.,** Implications of total energy intake for epidemiologic studies of breast and large-bowel cancer, *Am. J. Clin. Nutr.,* 45, 354, 1987.

26. **Lyon, J. L., Mahoney, A. W., West, D. W., Gardner, J. W., Smith, K. R., Sorenson, A. W., and Stanish, W.,** Energy intake: its relationship to colon cancer risk, *J. Natl. Cancer Inst.,* 78, 853, 1987.

27. **Kelsey, J. L. and Berkowitz, G. S.,** Breast cancer epidemiology, *Cancer Res.,* 48, 5615, 1988.

28. **Prentice, R. L., Pepe, M., and Self, S. G.,** Dietary fat and breast cancer: a quantitative assessment of the epidemiological literature and a discussion of methodological issues, *Cancer Res.,* 49, 3147, 1989.

29. **Howe, G. R., Hirohata, T., Hislop, T. G., Iscovich, J. M., Yuan, J. M., Katsouyanni, K., Lubin, F., Marubini, E., Modan, B., Rohan, T., Toniolo, P., and Shunzhang, Y.,** Dietary factors and risk of breast cancer: combined analysis of 12 case-control studies, *J. Natl. Cancer Inst.,* 82, 561, 1990.

30. **Willett, W. C., Stampfer, M. J., Colditz, G. A., Rosner, B. A., Hennekens, C. H., and Speizer, F. E.,** Dietary fat and the risk of breast cancer, *New Engl. J. Med.,* 316, 22, 1987.

31. **Jones, D. Y., Schatzkin, A., Green, S. B., Block, G., Brinton, L. A., Ziegler, R. G., Hoover, R., and Taylor, P. R.,** Dietary fat and breast cancer in the National Health and Nutrition Examination Survey I. Epidemiological followup study, *J. Natl. Cancer Inst.,* 79, 465, 1987.

32. **Howe, G. R., Friedenreich, C. M., Jain, M., and Miller, A. B.,** A cohort study of fat intake and risk of breast cancer, *J. Natl. Cancer Inst.,* 83, 336, 1991.

33. **Wynder, E. L.,** An epidemiological evaluation of the causes of cancer of the pancreas, *Cancer Res.,* 35, 2228, 1975.

34. **Durbec, J. P., Chevilotte, G., Bidart, J. M., Berthezene, P., and Sarles, H.,** Diet, alcohol, tobacco and risk of cancer of the pancreas: a case-control study, *Br. J. Cancer,* 47, 463, 1983.

35. **Norell, S. E., Ahlbom, A., Erwald, R., Jacobson, G., Lindberg-Navier, I., Okin, R., Törnberg, B., and Wiechel, K. L.,** Diet and pancreatic cancer: a case-control study, *Am. J. Epidemiol.,* 124, 894, 1986.

36. **Carroll, K. K.,** Lipids and carcinogenesis, *J. Environ. Pathol. Toxicol.,* 3, 253, 1980.

37. **National Research Council,** *Recommended Dietary Allowances, 10th edition,* National Academy Press, Washington, D.C., 1989.

38. **Donato, K. and Hegsted, D. M.,** Efficiency of utilization of various sources of energy for growth, *Proc. Natl. Acad. Sci. U.S.A.,* 82, 4866, 1985.

39. **Watson, A. F. and Mellanby, E.,** Tar cancer in mice. II. The condition of the skin when modified by external treatment or diet, as a factor in influencing the cancerous reaction, *Br. J. Exp. Pathol.,* 11, 311, 1930.

40. **Baumann, C. A., Jacobi, H. P., and Rusch, H. P.,** The effect of diet on experimental tumor production, *Am. J. Hygiene,* 30A, 1, 1939.

41. **Jacobi, H. P. and Baumann, C. A.,** The effect of fat on tumor formation, *Am. J. Cancer,* 39, 338, 1940.

42. **Lavik, P. S. and Baumann, C. A.,** Dietary fat and tumor formation, *Cancer Res.,* 1, 181, 1941.

43. **Lavik, P. S. and Baumann, C. A.,** Further studies on the tumor-promoting action of fat, *Cancer Res.,* 3, 749, 1943.

44. **Tannenbaum, A.,** The genesis and growth of tumors. III. Effects of a high fat diet, *Cancer Res.,* 2, 468, 1942.

45. **Tannenbaum, A.,** The dependence of the genesis of induced skin tumors on the fat content of the diet during different stages of carcinogenesis, *Cancer Res.,* 4, 683, 1944.

46. **Boutwell, R. K., Brush, M. K., and Rusch, H. P.,** The stimulating effect of dietary fat on carcinogenesis, *Cancer Res.,* 9, 741, 1949.

47. **Mathews-Roth, M. M. and Krinsky, N. I.,** Effect of dietary fat level on UV-B induced skin tumors, and anti-tumor action of beta-carotene, *Photochem. Photobiol.,* 40, 671, 1984.

48. **Black, H. S., Lenger, W., Phelps, A. W., and Thornby, J. I.,** Influence of dietary lipid upon ultraviolet-light carcinogenesis, *Nutr. Cancer,* 5, 59, 1983.

49. **Holsti, P.,** Tumor promoting effects of some long chain fatty acids in experimental skin carcinogenesis in the mouse, *Acta Pathol. Microbiol. Scand.,* 46, 51, 1959.

50. **Birt, D. F., Pelling, J. C., Tibbels, M. G., and Schweickert, L.,** Acceleration of papilloma growth in mice fed high-fat diets during promotion of two-stage skin carcinogenesis, *Nutr. Cancer,* 12, 161, 1989.

51. **Birt, D. F., White, L. T., Choi, B., and Pelling, J. C.,** Dietary fat effects on the initiation and promotion of two-stage skin tumorigenesis in the SENCAR mouse, *Cancer Res.,* 49, 4170, 1989.

52. **Silberhorn, E. M., Glauert, H. P., and Robertson, L. W.,** Carcinogenicity of poly-halogenated biphenyls: PCBs and PBBs, *Crit. Rev. Toxicol.,* 20, 439, 1990.

53. **Beer, D. G. and Pitot, H. C.,** Biological markers characterizing the development of preneoplastic and neoplastic lesions in rodent liver, *Arch. Toxicol.,* 10 (Suppl.), 68, 1987.

54. **Pitot, H. C., Glauert, H. P., and Hanigan, M.,** The significance of biochemical markers in the characterization of putative initiated cell populations in rodent liver, *Cancer Lett.,* 29, 1, 1985.

55. **Hendrich, S., Campbell, H. A., and Pitot, H. C.,** Quantitative stereological evaluation of four histochemical markers of altered foci in multistage hepatocarcinogenesis in the rat, *Carcinogenesis,* 8, 1245, 1987.
56. **Harada, T., Maronpot, R. R., Morris, R. W., Stitzel, K. A., and Boorman, G. A.,** Morphological and stereological characterization of hepatic foci of cellular alteration in control Fischer 344 rats, *Toxicol. Pathol.,* 17, 579, 1989.
57. **Bannasch, P., Enzmann, H., Klimek, F., Weber, E., and Zerban, H.,** Significance of sequential cellular changes inside and outside foci of altered hepatocytes during hepatocarcinogenesis, *Toxicol. Pathol.,* 17, 617, 1989.
58. **Kunz, H. W., Tennekes, H. A., Port, R. E., Schwarz, M., Lorke, D., and Schaude, G.,** Quantitative aspects of chemical carcinogenesis and tumor promotion in liver, *Environ. Health Perspect.,* 50, 113, 1983.
59. **Emmelot, P. and Scherer, E.,** The first relevant cell stage in rat liver carcinogenesis. A quantitative approach, *Biochim. Biophys. Acta,* 605, 247, 1980.
60. **Kline, B. E., Miller, J. A., Rusch, H. P., and Baumann, C. A.,** Certain effects of dietary fats on the production of liver tumors in rats fed p-dimethylaminoazobenzene, *Cancer Res.,* 6, 5, 1946.
61. **Sugai, M., Witting, L. A., Tsuchiyama, H., and Kummerow, F. A.,** The effect of heated fat on the carcinogenic activity of 2-acetylaminofluorene, *Cancer Res.,* 22, 510, 1962.
62. **McCay, P. B., King, M., Rikans, L., and Pitha, J. V.,** Interactions between dietary fats and antioxidants on DMBA-induced hyperplastic nodules and hepatomas, *J. Environ. Pathol. Toxicol.,* 3, 451, 1980.
63. **Baldwin, S. and Parker, R. S.,** The effect of dietary fat and selenium on the development of preneoplastic lesions in rat liver, *Nutr. Cancer,* 8, 273, 1986.
64. **Miller, J. A., Kline, B. E., Rusch, H. P., and Baumann, C. A.,** The carcinogenicity of p-dimethylaminoazobenzene in diets containing hydrogenated coconut oil, *Cancer Res.,* 4, 153, 1944.
65. **Miller, J. A., Kline, B. E., Rusch, H. P., and Baumann, C. A.,** The effect of certain lipids on the carcinogenicity of p-dimethylaminoazobenzene, *Cancer Res.,* 4, 756, 1944.
66. **Misslbeck, N. G., Campbell, T. C., and Roe, D. A.,** Effect of ethanol consumed in combination with high or low fat diets on the post initiation phase of hepatocarcinogenesis, *J. Nutr.,* 114, 2311, 1984.
67. **Baldwin, S. and Parker, R. S.,** Influence of dietary fat and selenium in initiation and promotion of aflatoxin B_1-induced preneoplastic foci in rat liver, *Carcinogenesis,* 8, 101, 1987.
68. **Glauert, H. P. and Pitot, H. C.,** Effect of dietary fat on the promotion of diethylnitrosamine-induced hepatocarcinogenesis in female rats, *Proc. Soc. Exp. Biol. Med.,* 181, 498, 1986.
69. **Sidransky, H., Verney, E., and Wang, D.,** Effects of varying fat content of a high tryptophan diet on induction of gamma-glutamyltranspeptidase positive foci in the livers of rats treated with hepatocarcinogen, *Cancer Lett.,* 31, 235, 1986.
70. **Newberne, P. M., Weigert, J., and Kula, N.,** Effects of dietary fat on hepatic mixed-functions oxidases and hepatocellular carcinoma induced by aflatoxin B1 in rats, *Cancer Res.,* 39, 3986, 1979.
71. **Glauert, H. P., Lay, L. T., Kennan, W. S., and Pitot, H. C.,** Effect of dietary fat on the initiation of hepatocarcinogenesis by diethylnitrosamine or 2-acetylaminofluorene in rats, *Carcinogenesis,* 12, 991, 1991.
72. **Druckrey, H.,** Production of colonic carcinomas by 1,2-dialkyhydrazines and azoalkanes, in *Carcinoma of the Colon and Antecedent Epithelium,* Burdette, W. J., Ed., Springfield, IL, Thomas, 1970, 267.
73. **Reddy, B. S. and Ohmori, T.,** Effect of intestinal microflora and dietary fat on 3,2′-dimethyl-4-aminobiphenyl-induced colon carcinogenesis in F344 rats, *Cancer Res.,* 41, 1363, 1981.

74. **Rogers, A. E. and Nauss, K. M.,** Rodent models for carcinoma of the colon, *Dig. Dis. Sci.,* 30, 87S, 1985.

75. **Nauss, K. M., Locniskar, M., Sondergaard, D., and Newberne, P. M.,** Lack of effect of dietary fat on N-nitrosomethylurea (NMU)-induced colon tumorigenesis in rats, *Carcinogenesis,* 5, 255, 1984.

76. **Glauert, H. P. and Weeks, J. A.,** Dose- and time-response of colon carcinogenesis in Fischer-344 rats after a single dose of 1,2-dimethylhydrazine, *Toxicol. Letters,* 48, 283, 1989.

77. **Karkare, M. R., Clark, T. D., and Glauert, H. P.,** Effect of dietary calcium on colon carcinogenesis induced by a single injection of 1,2-dimethylhydrazine in rats, *J. Nutr.,* 121, 568, 1991.

78. **Schiller, C. M., Curley, W. H., and McConnell, E. E.,** Induction of colon tumors by a single oral dose of 1,2-dimethylhydrazine, *Cancer Lett.,* 11, 75, 1980.

79. **Decaens, C., Gautier, R., Bara, J., and Burtin, P.,** Induction of rat intestinal carcinogenesis with single doses, low and high repeated doses of 1,2-dimethylhydrazine, *Carcinogenesis,* 10, 69, 1989.

80. **Ward, J. M.,** Dose response to a single injection of azoxymethane in rats, *Vet. Pathol.,* 12, 165, 1975.

81. **Glauert, H. P., Bennink, M. R., and Sander, C. H.,** Enhancement of 1,2-dimethylhydrazine-induced colon carcinogenesis in mice by dietary agar, *Food Cosmetics Toxicol.,* 19, 281, 1981.

82. **Nauss, K. M., Locniskar, M., and Newberne, P. M.,** Effect of alterations in the quality and quantity of dietary fat and 1,2-dimethylhydrazine-induced colon carcinogenesis in rats, *Cancer Res.,* 43, 4083, 1983.

83. **Reddy, B. S., Weisburger, J. H., and Wynder, E. L.,** Effects of dietary fat level and dimethylhydrazine on fecal bile acid and neutral sterol excretion and colon carcinogenesis in rats, *J. Natl. Cancer Inst.,* 52, 507, 1974.

84. **Bull, A. W., Soullier, B. K., Wilson, P. S., Hayden, M. T., and Nigro, N. D.,** Promotion of azoxymethane-induced intestinal cancer by high fat diet in rats, *Cancer Res.,* 39, 4956, 1979.

85. **Sakaguchi, M., Minuora, T., Hiramatsu, T., et al.,** Effects of dietary saturated and unsaturated fatty acids on bile acids and colon carcinogenesis induced by azoxymethane in rats, *Cancer Res.,* 46, 61, 1986.

86. **Nigro, N. D., Singh, D. V., Campbell, R. L., and Pak, M. S.,** Effect of dietary fat on intestinal tumor formation by azoxymethane in rats, *J. Natl. Cancer Inst.,* 54, 439, 1975.

87. **Reddy, B. S., Watanabe, K., and Weisburger, J. H.,** Effect of high-fat diet on colon carcinogenesis in F344 rats treated with 1,2-dimethylhydrazine, methylazoxymethanol acetate, or methylnitrosourea, *Cancer Res.,* 37, 4156, 1977.

88. **Bull, A. W., Bronstein, J. C., and Nigro, N. D.,** The essential fatty acid requirement for azoxymethane-induced intestinal carcinogenesis in rats, *Lipids,* 24, 340, 1989.

89. **Reddy, B. S., Narisawa, T., Vukusich, D., Weisburger, J. H., and Wynder, E. L.,** Effect of quality and quantity of dietary fat and dimethylhydrazine on colon carcinogenesis in rats, *Proc. Soc. Exp. Biol. Med.,* 151, 237, 1976.

90. **Bansal, B. R., Rhoads, J. E. J., and Bansal, S. C.,** Effects of diet on colon carcinogenesis and the immune system in rats treated with 1,2-dimethylhydrazine, *Cancer Res.,* 38, 3293, 1978.

91. **Schmaehl, D., Habs, M., and Habs, H.,** Influence of a non-synthetic diet with a high fat content on the local occurrence of colonic carcinomas induced by N-nitroso-acetoxymethylmethylamine (AMMN) in Sprague-Dawley rats, *Hepato-gastroenterol.,* 30, 30, 1983.

92. **Locniskar, M., Nauss, K. M., Kaufmann, P., and Newberne, P. M.,** Interaction of dietary fat and route of carcinogen administration on 1,2-dimethylhydrazine-induced colon tumorigenesis in rats, *Carcinogenesis,* 6, 349, 1985.

93. **Nauss, K. M., Bueche, D., and Newberne, P. M.,** Effect of beef fat on DMH-induced colon tumorigenesis: influence of rat strain and nutrient composition, *J. Nutr.,* 117, 739, 1987.

94. **Pence, B. C. and Buddingh, F.,** Inhibition of dietary fat-promoted colon carcinogenesis in rats by supplement calcium or vitamin D_3, *Carcinogenesis,* 9, 187, 1988.

95. **Hardman, W. E., Heitman, D. W., and Cameron, I. L.,** Suppression of the progression of 1,2 dimethylhydrazine (DMH) induced colon carcinogenesis by 20% dietary corn oil in rats supplemented with dietary pectin, *Proc. Am. Assoc. Cancer Res.,* 32, 131, 1991.

96. **Sakaguchi, M., Hiramatsu, Y., Takada, H., Yamamura, M., Hioki, K., Saito, K., and Yamamoto, M.,** Effect of dietary unsaturated and saturated fats on azoxymethane-induced colon carcinogenesis in rats, *Cancer Res.,* 44, 1472, 1984.

97. **Reddy, B. S. and Maeura, Y.,** Tumor promotion by dietary fat in azoxymethane-induced colon carcinogenesis in female F344 rats: influences of amount and source of dietary fat, *J. Natl. Cancer Inst.,* 72, 745, 1984.

98. **Reddy, B. S. and Maruyama, H.,** Effect of different levels of dietary corn oil and lard during the initiation phase of colon carcinogenesis in F344 rats, *J. Natl. Cancer Inst.,* 77, 815, 1986.

99. **Reddy, B. S. and Sugie, S.,** Effect of different levels of omega-3 and omega-6 fatty acids on azoxymethane-induced colon carcinogenesis in F344 rats, *Cancer Res.,* 48, 6642, 1988.

100. **Wargovich, M. J., Allnutt, D., Palmer, C., Anaya, P., and Stephens, L. C.,** Inhibition of the promotional phase of azoxymethane-induced colon carcinogenesis in the F344 rat by calcium lactate: effect of simulating two human nutrient density levels, *Cancer Lett.,* 53, 17, 1990.

101. **Guillem, J. G., Hsieh, L. L., O'Toole, K. M., Forde, K. A., LoGerfo, P., and Weinstein, I. B.,** Changes in expression of oncogenes and endogenous retroviral-like sequences during colon carcinogenesis, *Cancer Res.,* 48, 3964, 1988.

102. **Longnecker, D.,** Experimental pancreatic cancer: role of species, sex and diet, *Bull. Cancer,* 77, 27, 1990.

103. **Roebuck, B. D., Yager, J. D., Longnecker, D. S., and Wilpone, S. A.,** Promotion by unsaturated fat of azaserine-induced pancreatic carcinogenesis in the rat, *Cancer Res.,* 41, 3961, 1981.

104. **Roebuck, B. D., Longnecker, D. S., Baumgartner, K. J., and Thron, C. D.,** Carcinogen-induced lesions in the rat pancreas: effects of varying levels of essential fatty acid, *Cancer Res.,* 45, 5252, 1985.

105. **Roebuck, B. D.,** Effects of high levels of dietary fats on the growth of azaserine-induced foci in the rat pancreas, *Lipids,* 21, 281, 1986.

106. **Roebuck, B. D., Kaplita, P. V., Edwards, B. R., and Praissman, M.,** Effects of dietary fats and soybean protein on azaserine-induced pancreatic carcinogenesis and plasma cholecystokinin in the rat, *Cancer Res.,* 47, 1333, 1987.

107. **O'Connor, T. P., Roebuck, B. D., and Campbell, T. C.,** Dietary intervention during the postdosing phase of 1-azaserine-induced preneoplastic lesions, *J. Natl. Cancer Inst.,* 75, 955, 1985.

108. **Woutersen, R. A. and van Garderen-Hoetmer, A.,** Inhibition of dietary fat promoted development of (pre)neoplastic lesions in exocrine pancreas of rats and hamsters by supplemental vitamins A, C and E, *Cancer Lett.,* 41, 179, 1988.

109. **Woutersen, R. A., van Garderen-Hoetmer, A., Bax, J., and Scherer, E.,** Modulation of dietary fat-promoted pancreatic carcinogenesis in rats and hamsters by chronic ethanol ingestion, *Carcinogenesis,* 10, 453, 1989.

110. **Woutersen, R. A., van Garderen-Hoetmer, A., Bax, J., and Scherer, E.,** Modulation of dietary fat-promoted pancreatic carcinogenesis in rats and hamsters by chronic coffee ingestion, *Carcinogenesis,* 10, 311, 1989.

111. **Roebuck, B. D., Yager, J. D. J., and Longnecker, D. S.,** Dietary modulation of azaserine-induced pancreatic carcinogenesis in the rat, *Cancer Res.,* 41, 888, 1981.

112. **Longnecker, D. S., Roebuck, B. D., and Kuhlmann, E. T.,** Enhancement of pancreatic carcinogenesis by a dietary unsaturated fat in rats treated with saline or N-nitroso(2-hydroxypropyl) (2-oxopropyl)amine, *J. Natl. Cancer Inst.,* 74, 219, 1985.

113. **Birt, D. F., Salmasi, S., and Pour, P. M.,** Enhancement of experimental pancreatic cancer in Syrian golden hamsters by dietary fat, *J. Natl. Cancer Inst.,* 67, 1327, 1981.

114. **Birt, D. F. and Pour, P. M.,** Increased tumorigenesis induced by N-nitrosobis(2-oxopropyl)amine in Syrian golden hamsters fed high-fat diets, *J. Natl. Cancer Inst.,* 70, 1135, 1983.

115. **Birt, D. F., Julius, A. D., White, L. T., and Pour, P. M.,** Enhancement of pancreatic carcinogenesis in hamsters fed a high-fat diet ad libitum and at a controlled calorie intake, *Cancer Res.,* 49, 5848, 1989.

116. **Birt, D. F., Julius, A. D., Dwork, E., Hanna, T., White, L. T., and Pour, P. M.,** Comparison of the effects of dietary beef tallow and corn oil on pancreatic carcinogenesis in the hamster model, *Carcinogenesis,* 11, 745, 1990.

117. **Eustis, S. L. and Boorman, G. A.,** Proliferative lesions of the exocrine pancreas: relationship to corn oil gavage in the National Toxicology Program, *J. Natl. Cancer Inst.,* 75, 1067, 1985.

118. **Haseman, J. K., Huff, J. E., Rao, G. N., Arnold, J. E., Boorman, G. A., and McConnell, E. E.,** Neoplasms observed in untreated and corn oil gavage control groups of F344/N rats and (C57BL/6N X C3H/HeN)F$_1$ (B6C3F$_1$) mice, *J. Natl. Cancer Inst.,* 75, 975, 1985.

119. **Freedman, L. S., Clifford, C., and Messina, M.,** Analysis of dietary fat, calories, body weight, and the development of mammary tumors in rats and mice: a review, *Cancer Res.,* 50, 5710, 1990.

120. **Beems, R. B. and van Beek, L.,** Modifying effect of dietary fat on benzo(a)pyrene-induced respiratory tract tumors in hamsters, *Carcinogenesis,* 5, 413, 1984.

121. **Szepsenwol, J.,** Carcinogenic effect of ether extract of whole egg, alcohol extract of egg yolk, and powdered egg free of the ether extractable part in mice, *Proc. Soc. Exp. Biol. Med.,* 116, 1136, 1964.

122. **Szepsenwol, J.,** Brain nerve cell tumors in mice on diets supplemented with various lipids, *Pathol. Microbiol.,* 34, 1, 1969.

123. **Pollard, M. and Luckert, P. H.,** Promotional effects of testosterone and dietary fat on prostate carcinogenesis in genetically susceptible rats, *Prostate,* 6, 1, 1985.

124. **Cave, W. T.,** Dietary n-3 polyunsaturated fatty acid effects on animal tumorigenesis, *FASEB J.,* 5, 2160, 1991.

125. **O'Connor, T. P., Roebuck, B. D., Peterson, F., and Campbell, T. C.,** Effect of dietary intake of fish oil and fish protein on the development of L-azaserine-induced preneoplastic lesions in the rat pancreas, *J. Natl. Cancer Inst.,* 75, 959, 1985.

126. **O'Connor, T. P., Roebuck, B. D., Peterson, F. J., Lokesh, B., Kinsella, J. E., and Campbell, T. C.,** Effect of dietary omega-3 and omega-6 fatty acids on development of azaserine-induced preneoplastic lesions in rat pancreas, *J. Natl. Cancer Inst.,* 81, 858, 1989.

127. **Nelson, R. L., Tanure, J. C., Andrianopoulos, G., Souza, G., and Lands, W. E. M.,** A comparison of dietary fish oil and corn oil in experimental colorectal carcinogenesis, *Nutr. Cancer,* 11, 215, 1988.

128. **Minoura, T., Takata, T., Sakaguchi, M., Takada, H., Yamamura, M., Hioki, K., and Yamamoto, M.,** Effect of dietary eicosapentaenoic acid on azoxymethane-induced colon carcinogenesis in rats, *Cancer Res.,* 48, 4790, 1988.

129. **Reddy, B. S. and Maruyama, H.,** Effect of fish oil on azoxymethane-induced colon carcinogenesis in male F344 rats, *Cancer Res.,* 46, 3367, 1986.

130. **Abou-El-Ela, S. H., Prasse, K. W., Carroll, R., Wade, A. E., Dharwadkar, S., and Bunce, O. R.,** Eicosanoid synthesis in 7,12-dimethylbenz(a)anthracene-induced mammary carcinomas in Sprague-Dawley rats fed primrose, menhaden or corn oil diets, *Lipids,* 23, 948, 1988.

131. **Abou-El-Ela, S. H., Prasse, K. W., Farrell, R. L., Carroll, R. W., Wade, A. E., and Bunce, O. R.,** Effects of D,L-2-difluoromethylornithine and indomethacin on mammary tumor promotion in rats fed high n-3 and/or n-6 fat diets, *Cancer Res.,* 49, 1434, 1989.

132. **Locniskar, M., Belury, M. A., Cumberland, A. G., Patrick, K. E., and Fischer, S. M.,** Lack of a protective effect of menhaden oil on skin tumor promotion by 12-O-tetradecanoylphorbol-13-acetate, *Carcinogenesis,* 11, 1641, 1990.

133. **Murphy, M. G.,** Dietary fatty acids and membrane protein function, *J. Nutr. Biochem.,* 1, 68, 1990.

134. **Welsch, C. W.,** Enhancement of mammary tumorigenesis by dietary fat: review of potential mechanisms, *Am. J. Clin. Nutr.,* 45, 192, 1987.

135. **Narisawa, T., Magadia, N. E., Weisburger, J. H., and Wynder, E. L.,** Promoting effect of bile acids on colon carcinogenesis after intrarectal instillation of N-methyl-N′-nitro-N-nitrosoguanidine in rats, *J. Natl. Cancer Inst.,* 53, 1093, 1974.

136. **Reddy, B. S., Narisawa, T., Weisburger, J. H., and Wynder, E. L.,** Promoting effect of sodium deoxycholate on colon adenocarcinomas in germ-free rats, *J. Natl. Cancer Inst.,* 56, 441, 1976.

137. **Reddy, B. S., Watanabe, K., Weisburger, J. H., and Wynder, E. L.,** Promoting effect of bile acids in colon carcinogenesis in germ-free and conventional F344 rats, *Cancer Res.,* 37, 3238, 1977.

138. **Reddy, B. S., Mangat, S., Sheinfil, A., Weisburger, J. H., and Wynder, E. L.,** Colon carcinogenesis. Effect of type and amount of dietary fat and 1,2-dimethylhydrazine on biliary bile acids and neutral sterols in rats, *Cancer Res,* 37, 2132, 1977.

139. **Glauert, H. P. and Bennink, M. R.,** Influence of diet or intrarectal bile acid injections on colon epithelial cell proliferation in rats previously injected with 1,2-dimethylhydrazine, *J. Nutr.,* 113, 475, 1983.

140. **Reddy, B. S., Hanson, D., Mangat, S., Mathews, L., Sbaschnig, M., Sharma, C., and Simi, B.,** Effect of high-fat, high-beef diet and of mode of cooking of beef in the diet on fecal bacterial enzymes and fecal bile acids and neutral sterols, *J. Nutr.,* 110, 1880, 1980.

141. **Gallaher, D. D. and Franz, P. M.,** Effects of corn oil and wheat brans on bile acid metabolism in rats, *J. Nutr.,* 120, 1320, 1990.

142. **Deschner, E. E., Cohen, B. I., and Raicht, R. F.,** Acute and chronic effect of dietary cholic acid on colonic epithelial cell proliferation, *Digestion,* 21, 290, 1981.

143. **Cohen, B. I., Raicht, R. F., Deschner, E. E., Takahashi, M., Sarwal, A. N., and Fazzini, E.,** Effect of cholic acid feeding on N-methyl-N-nitrosourea-induced colon tumors and cell kinetics in rats, *J. Natl. Cancer Inst.,* 64, 573, 1980.

144. **Skraastad, O. and Reichelt, K. L.,** An endogenous colon mitosis inhibitor and dietary calcium inhibit increased colonic cell proliferation induced by cholic acid, *Scand. J. Gastroenterol.,* 23, 801, 1988.

145. **Wargovich, M. J., Eng, V. W. S., Newmark, H. L., and Bruce, W. R.,** Calcium ameliorates the toxic effect of deoxycholic acid on colonic epithelium, *Carcinogenesis,* 4, 1205, 1983.

146. **Wargovich, M. J, Eng, V. W. S., and Newmark, H. L.,** Calcium inhibits the damaging and compensatory proliferative effects of fatty acids on mouse colon epithelium, *Cancer Lett.,* 23, 253, 1984.

147. **Craven, P. A., Pfanstiel, J., and DeRubertis, F. R.,** Role of activation of protein kinase C in the stimulation of colonic epithelial proliferation and reactive oxygen formation by bile acids, *J. Clin. Invest.,* 79, 532, 1987.

148. **Rosenthal, M. D.,** Fatty acid metabolism of isolated mammalian cells, *Prog. Lipid Res.,* 26, 87, 1987.

149. **Lands, W. E. M.,** Biosynthesis of prostaglandins, *Annu. Rev. Nutr.,* 11, 41, 1991.

150. **Dyerberg, J.,** Linolenate-derived polyunsaturated fatty acids and prevention of athero-sclerosis, *Nutr. Rev.,* 44, 125, 1986.

151. **Nicosia, S. and Patrono, C.,** Eicosanoid biosynthesis and action: novel opportunities for pharmacological intervention, *FASEB J.,* 3, 1941, 1989.

152. **Fischer, S. M., Baldwin, J. K., Jasheway, D. W., Patrick, K. E., and Cameron, G. S.,** Phorbol ester induction of 8-lipoxygenase in inbred SENCAR (SSIN) but not C57BL/6J mice correlated with hyperplasia, edema, and oxidant generation but not or-nithine decarboxylase induction, *Cancer Res.,* 48, 658, 1988.

153. **Fischer, S. M., Fürstenberger, G., Marks, F., and Slaga, T. J.,** Events associated with mouse skin tumor promotion with respect to arachidonic acid metabolism: a com-parison between SENCAR and NMRI mice, *Cancer Res.,* 47, 3174, 1987.

154. **Denda, A., Ura, H., Tsujiuchi, T., Tsutsumi, M., Eimoto, H., Takashima, Y., Kitazawa, S., Kinugasa, T., and Konishi, Y.,** Possible involvement of arachidonic acid metabolism in phenobarbital promotion of hepatocarcinogenesis, *Carcinogenesis,* 10, 1929, 1989.

155. **Reddy, B. S., Wang, C. X., and Maruyama, H.,** Effect of restricted caloric intake on azoxymethane-induced colon tumor incidence in male F344 rats, *Cancer Res.,* 47, 1226, 1987.

156. **Klurfeld, D. M., Weber, M. M., and Kritchevsky, D.,** Inhibition of chemically induced mammary and colon tumor promotion by caloric restriction in rats fed increased dietary fat, *Cancer Res.,* 47, 2759, 1987.

157. **Boissonneault, G. A., Elson, C. E., and Pariza, M. W.,** Net energy effects of dietary fat on chemically induced mammary carcinogenesis in F344 rats, *J. Natl. Cancer Inst.,* 76, 335, 1986.

158. **Kritchevsky, D., Weber, M. M., and Klurfeld, D. M.,** Dietary fat versus caloric content in initiation and promotion of 7,12-dimethylbenz(a)anthracene-induced mammary tumorigenesis in rats, *Cancer Res.,* 44, 3174, 1984.

159. **Welsch, C. W., House, J. L., Herr, B. L., Eliasberg, S. J., and Welsch, M. A.,** Enhancement of mammary carcinogenesis by high levels of dietary fat: a phenomenon dependent on ad libitum feeding, *J. Natl. Cancer Inst.,* 82, 1615, 1990.

160. **Boissonneault, G. A.,** Calories and carcinogenesis: modulation by growth factors, in *Nutrition, Toxicity, and Cancer,* Rowland, I. R., Ed., CRC Press, Boca Raton, FL, 1991, 413.

161. **Snyder, D. L.,** *Dietary Restriction and Aging,* Alan R. Liss, New York, 1989.

162. **Baldwin, S. and Parker, R. S.,** Effects of dietary fat level and aflatoxin B_1 treatment on rat hepatic lipid composition, *Food Chem. Toxicol.,* 23, 1049, 1985.

163. **Neelands, P. J. and Clandinin, M. T.,** Diet fat influences liver plasma-membrane lipid composition and glucagon-stimulated adenylate cyclase activity, *Biochem. J.,* 212, 573, 1983.

164. **Wade, A. E., Norred, W. P., and Evans, J. S.,** Lipids in drug detoxification, in *Nutrition and Drug Interrelations,* Hathcock, J. N. and Coon, J., Eds., Academic Press, New York, 1978, 475.

165. **Hammer, C. T. and Wills, E. D.,** Dependence of the rate of metabolism of benzo(a)pyrene on the fatty acid composition of the liver endoplasmic reticulum and on dietary lipids, *Nutr. Cancer,* 2, 113, 1980.

166. **Rutten, A. A. J. J. L. and Flake, H. E.,** Influence of high dietary levels of fat on rat hepatic phase I and II biotransformation enzyme activities, *Nutr. Rep. Int.,* 36, 109, 1987.

167. **Cassanol, P., Bonnamour, D., Grolier, P., Pelissier, M. A., Amelizad, Z., Albrecht, R., and Narbonne, J. F.,** The effect of dietary imbalances on the activation of benzo[a]pyrene by the metabolizing enzymes from rat liver, *Mutat. Res.,* 191, 67, 1987.

168. **Kim, H. J., Choi, E. S., and Wade, A. E.,** Effect of dietary fat on the induction of hepatic microsomal cytochrome P450 isozymes by phenobarbital, *Biochem. Pharmacol.,* 39, 1423, 1990.

169. **Lam, T. C. L. and Wade, A. E.,** Influence of dietary lipid on the metabolism of hexobarbital by the isolated, perfused rat liver, *Pharmacology,* 21, 64, 1980.

170. **Lam, T. C. L. and Wade, A. E.,** Effect of dietary lipid on benzo(a)pyrene metabolism by perfused rat liver, *Drug Nutr. Interact.,* 1, 31, 1981.

171. **Wade, A. E., Harley, W., and Bunce, O. R.,** The effects of dietary corn oil on the metabolism and mutagenic activation of N-nitrosodimethylamine (DMN) by hepatic microsomes from male and female rats, *Mutat. Res.,* 102, 113, 1982.

172. **Clarke, S. D., Armstrong, M. K., and Jump, D. B.,** Dietary polyunsaturated fats uniquely suppress rat liver fatty acid synthase and S14 mRNA content, *J. Nutr.,* 120, 225, 1990.

173. **Brannon, P. M.,** Adaptation of the exocrine pancreas to diet, *Annu. Rev. Nutr.,* 10, 85, 1990.

174. **Ribeiro, A., Mangeney, M., Cardot, P., Loriette, C., Chambaz, J., Rayssiguier, Y., and Bereziat, G.,** Nutritional regulation of apolipoprotein genes: effect of dietary carbohydrates and fatty acids, *Adv. Exp. Med. Biol.,* 285, 407, 1991.

175. **Goodridge, A. G.,** Dietary regulation of gene expression: enzymes involved in carbohydrate and lipid metabolism, *Annu. Rev. Nutr.,* 7, 157, 1987.

176. **Karmali, R. A., Chao, C. C., Basu, A., and Modak, M., II,** Effect of n-3 and n-6 fatty acids on mammary H-ras expression and PGE-2 levels in DMBA-treated rats, *Anticancer Res.,* 9, 1169, 1989.

177. **Ronai, Z., Lau, Y. Y., and Cohen, L. A.,** Dietary n-3 fatty acids do not affect induction of Ha-ras mutations in mammary glands of NMU-treated rats, *Mol. Carcinog.,* 4, 120, 1991.

178. **Elliott, T. S., Swartz, D. A., Visek, W. J., and Kaput, J.,** Molecular analyses of differences in gene expression due to changes in the level of dietary fat, *FASEB J.,* 4, A916, 1990.

179. **Swartz, D. A., Visek, W. J., and Kaput, J.,** Genomic organization of a fat regulated gene, *FASEB J.,* 5, A1649, 1991.

Chapter 12

REGULATION OF ACETYL CoA CARBOXYLASE AND GENE EXPRESSION

John B. Allred and Diana F. Bowers

TABLE OF CONTENTS

0-8493-6961-4/93/$0.00 + $.50
© 1993 by CRC Press, Inc.

269

I. OVERVIEW

Acetyl CoA carboxylase is most often considered in its role in *de novo* fatty acid biosynthesis. In that context, acetyl CoA carboxylase is usually cited as the key enzyme which regulates fatty acid biosynthesis,[1] but it is only one of several enzymes required for the process *in vivo*.[2] As shown in Figure 1, the "lipogenic" enzymes, in addition to acetyl CoA carboxylase, include ATP-citrate lyase, fatty acid synthase, glucose-6-phosphate dehydrogenase, and malic enzyme, all of which are active in cytosol where fatty acid biosynthesis occurs.[1] When experimental animals are subjected to enhanced lipogenic conditions (e.g., refeeding a high carbohydrate diet to previously fasted rats), there is an increase in activity of acetyl CoA carboxylase in liver with a concomitant, coordinated rise in the activities of the other lipogenic enzymes.[2] At the same time the product of acetyl CoA carboxylase, malonyl CoA, can inhibit carnitine palmitoyltransferase, a key enzyme necessary for the transport of long-chain fatty acyl CoA esters into mitochondria. This inhibition is important in the regulation of β-oxidation of long-chain fatty acids, as discussed in Chapter 14.[3] In the simplest interpretation, this inhibition by the product of acetyl CoA carboxylase may be a mechanism to minimize the possibility that synthesis of long-chain fatty acids and their degradation do not occur at the same time.

In addition to its role in *de novo* fatty acid biosynthesis, acetyl CoA carboxylase is also required for elongation of long-chain fatty acids. While the initial product of the fatty acid synthesis system is palmitate, part of it is elongated to form stearate and oleate before exiting the liver as a triglyceride in the form of VLDL.[4] The conversion of linoleic acid to arachidonic acid by the microsomal elongation system requires malonyl CoA produced by acetyl CoA carboxylase.[5] Such a conversion is the reason that linoleic, but not arachidonic acid, is a dietary essential. Arachidonic acid is an integral constituent of membranes and is necessary for the synthesis of an important family of prostaglandins.[6]

The fact that acetyl CoA carboxylase has a dual role in metabolism is important in terms of potential dietary/hormonal regulation of the enzyme. Early work showed that fatty acid biosynthesis in animal models increased in response to a diet high in carbohydrate.[2] The two hormones most closely associated with changes in dietary carbohydrate level, insulin and glucagon, have opposing effects on lipogenesis and the activity of acetyl CoA carboxylase[7] in isolated rat hepatocytes. On the other hand, even when carbohydrate consumption is limited, the elongation system and consequently a basal level of the activity of acetyl CoA carboxylase is still necessary.

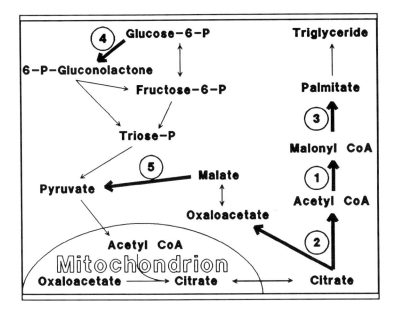

FIGURE 1. Fatty acid biosynthesis from glucose.[2] Acetyl CoA carboxylase catalyzes the formation of malonyl CoA (Reaction 1) from acetyl CoA produced in cytosol by the action of ATP-citrate lyase (Reaction 2). Conversion of malonyl CoA to palmitate is catalyzed by fatty acid synthase (Reaction 3), using NADPH produced by glucose-6-phosphate dehydrogenase (Reaction 4) and malic enzyme (Reaction 5).

II. IS CARBOHYDRATE CONVERTED TO FATTY ACIDS IN HUMANS?

In the context of human nutrition, the question of the importance of *de novo* fatty acid biosynthesis must be considered. There is no doubt that humans, in common with other higher animals, have acetyl CoA carboxylase as well as the other necessary enzymes for the elongation of linoleic to arachidonic acid. In fact, the gene for acetyl CoA carboxylase has been shown to be present in human genetic material, reportedly located on chromosome 17.[8] Further, an inborn error of metabolism in which acetyl CoA carboxylase is missing is known to exist in humans.[9]

While acetyl CoA carboxylase is clearly functional as a part of the long-chain fatty acid elongation system in human tissue, how important is it and the other lipogenic enzymes in *de novo* fatty acid synthesis from carbohydrate? It is generally accepted that when *de novo* fatty acid synthesis happens in humans, the process occurs in liver and not adipose tissue.[10] Still, there have been a number of reports over the years which have questioned the importance of the conversion of carbohydrate to fat in humans. Early work[11] measured the respiratory quotients of subjects who were given an oral carbohydrate load and found RQ values less than 1, whereas the synthesis of fatty acids from carbohydrate should give values greater than 1. A subsequent study[12] confirmed that after a single, high-carbohydrate meal equivalent to 479 g of starch, nonprotein-RQ values did not exceed 1 over the next 8 h. The authors concluded that the ingested carbohydrate was not converted to fat under these conditions.

The question was recently addressed more directly by administering acetate labeled with the nonradioactive, heavy isotope of carbon, ^{13}C, to human volunteers who were given either an infused or oral glucose load.[4] While it was demonstrated that the administered [^{13}C]acetate did label the liver cytosolic acetyl CoA pool, surprisingly little isotope was found in fatty acids of triglycerides of blood VLDL. It was calculated that less than 2% of the VLDL palmitate and less than 1% of the VLDL stearate came from *de novo* fatty acid biosynthesis. Thus, the authors concluded that the glucose load failed to stimulate *de novo* fatty acid biosynthesis.

If this conclusion is extrapolated to the general case, it raises a number of questions. Not the least of these is, if excessive carbohydrate consumption does not increase *de novo* fatty acid synthesis and deposition, what happens to the calories? Further, since feeding diets high in carbohydrate for extended periods is known to stimulate fatty acid synthesis in the tissue of a number of animal models, including rats,[13,14] mice,[15] pigs,[16] and even ruminants,[17] the lack of an effect of dietary carbohydrate would imply that humans have a markedly different lipogenic regulatory system than that found in other higher animals. It seems much more likely that the apparent lack of a carbohydrate-stimulated increase in lipogenesis in humans in these three reports[4,11,12] was due to the lack of subject adaptation to an increased carbohydrate intake. That is, in each of these human studies, subjects were allowed to consume their ''normal'' diet (presumably containing substantially more fat and less carbohydrate than the test diet) prior to the experiment and measurements were made in the few hours following a single carbohydrate load. Numerous studies have made it clear that while some of the regulatory mechanisms that control lipogenesis in animals occur very rapidly, others require substantially more time. One such time-dependent, adaptive mechanism is the regulation of the quantity of the active cytosolic form of acetyl CoA carboxylase. Much of the remainder of this chapter is devoted to a discussion of the time-dependent adaptive mechanisms which regulate acetyl CoA carboxylase activity and thus lipogenesis.

III. THE GENE FOR ACETYL CoA CARBOXYLASE

Insight into the molecular biology of acetyl CoA carboxylase has a relatively short history. Kim's laboratory was the first to purify a functional acetyl CoA carboxylase mRNA from mammalian tissue (rat mammary gland)[18] and to obtain acetyl CoA carboxylase c-DNA[19] which could be used as a probe to study the molecular biology of the rat enzyme. Using these tools in subsequent experiments, overlapping c-DNA clones were found which covered the entire coding sequence of rat acetyl CoA carboxylase mRNA.[20] The sequence of the coding section, consisting of 7035 bases, was used to predict the primary structure of rat acetyl CoA carboxylase. By this analysis, rat acetyl CoA carboxylase contained 2345 amino acids and had a molecular weight of 265,220. Although the entire predicted primary structure of rat acetyl CoA carboxylase has been published,[20] Figure 2 shows only selected segments of the enzyme to illustrate some features which may be important in its synthesis and regulation.

A c-DNA probe for chicken acetyl CoA carboxylase has also been prepared[24] and used[25] to determine the sequence of bases in the coding section of DNA for acetyl CoA carboxylase mRNA, from which the primary amino acid sequence of the chicken enzyme was deduced. Chicken acetyl CoA carboxylase was predicted to contain 21 fewer amino acid residues than the rat enzyme and have a molecular weight of 262,706. Not surprisingly, acetyl CoA carboxylase from the two sources have very similar primary structures. It is somewhat more surprising that both the rat[20] and chicken[25] liver enzyme exhibit a high degree of homology with carbamoyl phosphate synthetase at the amino terminal end of the protein.

IV. RAT ACETYL CoA CARBOXYLASE mRNA

The use of a c-DNA probe to study rat acetyl CoA carboxylase mRNA has provided some very interesting observations and, at the same time, showed that the transcriptional process and its regulation are far from simple. The basic technique usually used to study mRNA is called Northern blotting. It involves the electrophoretic separation by size of RNA extracted from a given tissue and then transferring the RNA from the electrophoresis gel to a solid support, usually a nylon membrane. The membrane is subsequently probed with a c-DNA which has been made radioactive by the incorporation of [32]P (nick-translated). Any mRNA on the membrane whose base sequence is complementary to that of the c-DNA will bind the radioactive c-DNA probe, in proportion to the quantity of mRNA. When the membrane is placed in contact with X-ray film, the film is exposed wherever radioactivity occurs. From the developed film, the approximate size(s) of any mRNA which bound the c-DNA probe can be estimated. The relative quantity of each mRNA species, over a limited concentration range, can be estimated by scanning the film with a densitometer.

Northern blot analysis of mRNA from rat liver[19] showed that the acetyl CoA carboxylase c-DNA probe hybridized to a mRNA of varying sizes, but two major bands were observed. The larger of the two contained approximately 10 kilobases (kb) while the smaller, more diffuse band was estimated to contain about 3 kb. Since acetyl CoA carboxylase is such a large protein

```
  1   Met-Asp-Glu-Pro-Ser-Pro-Leu-Ala-Lys-Thr-Leu-Glu-Leu-Asp-Gln-His-Ser-Arg-Phe-Ile   20
                                 ↓       ↓
 21   Ile-Gly-Ser-Val-Ser-Glu-Asp-Asn-Ser-Glu-Asp-Glu-Ile-Ser-Asn-Leu-Val-Lys-Leu-Asp   40


 61   Leu-Gly-Ile-Ser-Ala-Leu-Gln-Asp-Gly-Leu-Ala-Phe-His-Met-Arg-Ser-Ser-Met-Ser-Gly   80
                                                             ↓
 81   Leu-His-Leu-Val-Lys-Gln-Gly-Arg-Asp-Arg-Lys-Ile-Asp-Ser-Gln-Arg-Asp-Phe-Thr      100


                            Biotin
781   Ile-Glu-Val-Met-Lys-Met-Val-Met-Thr-Leu-Thr-Ala-Val-Glu-Ser-Gly-Cys-Ile-His-Tyr  800


                                        |--------Deletion--------|
1181  Phe-Met-Leu-Pro-Thr-Ser-His-Pro-Asn-Arg-Gly-Asn-Ile-Pro-Thr-Leu-Asn-Arg-Met-Ser  1200
                                                                          ↓
1201  Phe-Ala-Ser-Asn-Leu-Asn-His-Tyr-Gly-Met-Thr-His-Val-Ala-Ser-Val-Ser-Asp-Val-Leu  1220


2321  Met-Thr-Gln-His-Ile-Ser-Pro-Thr-Gln-Arg-Ala-Glu-Val-Val-Arg-Ile-Leu-Leu-Ser-Thr-Met  2340
2341  Asp-Ser-Pro-Ser-Thr.                                                              2345
```

FIGURE 2. Selected segments of the primary structure of acetyl CoA carboxylase, based on the amino acid sequence predicted from analysis of c-DNA clones of the enzyme.[20] The arrows (↓) show the eight serine residues which can be phosphorylated.[21] The lysine residue in the indicated (Biotin) tetrapeptide is biotinylated in the holoenzyme[20,22] and the indicated octapeptide (----Deletion----) is deleted in a putative isozymic form of the enzyme.[23]

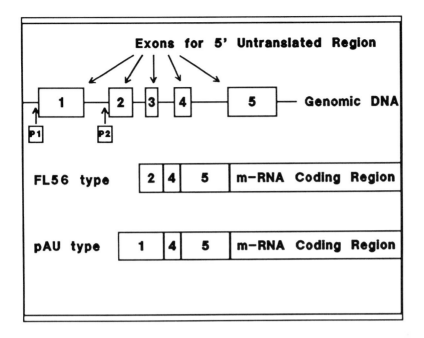

FIGURE 3. Reported[27] exon-intron structure of genomic DNA coding for the 5' untranslated region of acetyl CoA carboxylase, showing differential splicing of exons to produce heterogeneous acetyl CoA carboxylase mRNA. The protein synthesis initiation codon (AUG) is located in exon 5. P1 and P2 show the approximate locations of two putative promoters of transcription.

that mRNA coding for it must contain at least 7 kb,[20] it has been concluded that the 10-kb band represented translatable acetyl CoA carboxylase mRNA while the 3-kb band represented degradative products.[18]

Although electrophoresis can separate species of mRNA which have substantial differences in molecular weight, detection of more subtle structural differences require more sophisticated techniques. Using c-DNA probes complementary to specific base sequences in acetyl CoA carboxylase mRNA for Northern blot analysis and primer extension analysis, acetyl CoA carboxylase mRNA has been examined in mammary gland from lactating rats[26,27] and in rat liver.[28] These analyses established that there was substantial heterogeneity of the mRNA within the 10-kb band in both tissues. Five different species of mRNA were found in rat mammary gland all of which had the same base sequence in the coding region but differed in length and/or base sequence of the 5' untranslated region. The coding region of genomic DNA for the 5' untranslated portion of mRNA was found in five exons containing a total of 645 nucleotides, scattered over a 50-kb region. The authors concluded that these heterogeneous species of mRNA occurred because of differential splicing of the transcripts of these exons (Figure 3). For example, it was proposed that the 5' untranslated region of the most abundant specie of mRNA in rat mammary gland (FL56) is formed by splicing mRNA transcripts from exons 2, 4, and 5 without incorporation of the transcripts of exons 1 and 3. The 5' untranslated region of the second most abundant form of mRNA in rat mammary gland (FL63) is formed by splicing the transcripts of exons 2, 3, 4, and 5 but not that of exon 1.

Acetyl CoA carboxylase mRNA in rat liver was found to be even more heterogeneous than mammary gland mRNA, again attributable to differential splicing of the transcripts of the five exons coding for the 5' untranslated region of mRNA.[28] The most abundant mRNA species in the liver of rats fed laboratory chow was the same as the most abundant type (FL56) found in mammary gland in lactating rats. However, when rats were fasted and then refed a diet high in carbohydrate to stimulate hepatic lipogenesis, a type of mRNA (pAU) not found in the mammary gland (regardless of diet) became the most abundant species in liver although the FL56 type was still present. As with various mRNA species in mammary gland, the pAU type of mRNA did not differ in the coding region from other forms, but it did differ from other species in the length and, in part, base sequence of the 5' untranslated region. Analysis indicated that in the pAU type of mRNA, the 5' untranslated region was made by splicing the transcripts of exons 1, 4, and 5 (Figure 3). That is, the transcript of exon 2 found in the FL56 type was replaced in the pAU type by the transcript of exon 1.

These interesting observations led to the hypothesis that transcription of exons for the 5' untranslated region of acetyl CoA carboxylase mRNA is regulated by two different gene promoters.[27,28] It was proposed that Promoter I is located upstream from exon 1 while Promoter II is after exon 1 but ahead of exon 2 (Figure 3). The presence of two promoters, differentially activated depending upon physiological conditions, could account for the formation of dissimilar mRNA species. For example, the 5' untranslated region of the major species of mRNA (FL56 = ACC[2:4:5] mRNA) in liver of rats fed laboratory chow and in mammary gland of lactating rats contained the transcript of exon 2, indicating that Promoter II (but not I) was active. In contrast, when rats were fasted and then refed a diet high in carbohydrate, the 5' untranslated region of the most abundant species of mRNA (pAU = ACC[1:4:5] mRNA) contained the transcript of exon 1, indicating that Promoter I was activated under these conditions. In adipose tissue, pAU is by far the major species of acetyl CoA carboxylase mRNA, regardless of whether rats are fed a chow diet or refed a diet high in carbohydrate.

A more detailed analysis of the physiological regulation of these two promoters, using primer extension analysis, has appeared.[29] While this method provides only an indirect measure of promoter activity, the results indicated that when rats were fed laboratory chow *ad libitum*, only basal levels of Promoter I in adipose tissue and Promoter II in liver were found. After rats were fasted and then refed a diet high in carbohydrate, both promoters were activated in liver but only Promoter I was activated in adipose tissue. The activities of promoters in both liver (Promoter I and II) and adipose tissue (Promoter I) were depressed in streptozotocin-diabetic rats. Administration of insulin rapidly increased Promoter I activity in adipose tissue but, surprisingly, had no effect on the activity of either promoter in liver of the same rats even after a period of 24 h where insulin was administered every 6 h.

It has been proposed[29] that acetyl CoA carboxylase mRNA species can be classified as either class 1 (represented by the pAU type of mRNA in Figure 3) or class 2 (represented by the FL56 type of mRNA in Figure 3), depending upon whether they were produced in response to activation of Promoter I or II, respectively. Activation of Promoter I in liver occurs only under increased lipogenic conditions while Promoter II is active in liver of chow fed rats as well as in other tissue, presumably to maintain a basal level of acetyl CoA carboxylase. In fact, it has been suggested that Promoter II is a "house-keeping" promoter.[30] This scenario implies that the presence of two different promoters may be related to the dual function of acetyl CoA carboxylase as discussed in previous sections. That is, the class 1 mRNA transcript produced following the activation of Promoter I may be involved in increased *de novo* synthesis of fatty acids destined for stored triglyceride while the type of mRNA (class 2) produced from activation of Promoter II may be involved in long chain fatty acyl CoA elongation and/or synthesis of fatty acids for membrane lipids.

While this scenario represents an interesting hypothesis, it may be an oversimplification because the reported[29] promoter activity of mammary gland and adipose tissue does not precisely fit. That is, since initiation of lactation is accompanied by a substantial increase in *de novo* fatty acid biosynthesis in the mammary gland, it might be expected on the basis of the proposed model that Promoter I would become activated during lactation, but Promoter I was not found to be active in mammary gland under any condition. Instead, Promoter II activity increased during lactation. In adipose tissue, only Promoter I (and not Promoter II) activity has been found, even under nonlipogenic conditions. Further, the expression of Promoter II has been found to require enhancer elements.[30] Enhancer sequences, which activate transcription, are not usually found in "house-keeping" genes.[30] Finally, additional information on Promoter II regulation has been obtained by the study of insulin-dependent differentiation of 30A5 preadipocytes into adipocytes.[31] Differentiation of these cells, which contain only Class 2 acetyl CoA carboxylase mRNA species whose production is under the control of Promoter II, is accompanied by an increased activity/quantity of acetyl CoA carboxylase and the concentration of acetyl CoA carboxylase mRNA.[32] Insulin apparently activates Promoter II, but such acute hormonal regulation of "house-keeping genes" would not be expected.

Surprisingly, insulin induction of acetyl CoA carboxylase and cell differentiation required prior exposure of the 30A5 preadipocytes to cyclic-AMP.[31] Specific regions of genomic DNA were shown to be required for this cyclic-AMP effect. It is not clear that these results are relevant to the regulation of acetyl CoA carboxylase mRNA production in cells of growing and adult rats because hormonal effects during differentiation may be quite different than their effects on differentiated cells. Semenkovich et al.[33] have shown that the primary effect of insulin on lipoprotein lipase during the differentiation of 3T3 preadipocytes into adipocytes was at the transcriptional level, but in differentiated adipocytes, insulin had no effect on transcription. Rather, in differentiated cells, exposure to insulin affected lipoprotein lipase production only at the translational and posttranslational levels. Still, Swierczynski et al.[34] showed that triiodothyroxine-stimulated accumulation of acetyl CoA carboxylase mRNA in cultured chicken embryo hepatocytes, as well as increased concentrations of mRNA of other lipogenic enzymes, was blocked by protein kinase inhibitors. The authors concluded that ongoing protein phosphorylation was specifically required for triiodothyroxine to stimulate transcription of genes of lipogenic enzymes.

It is clear from these reports that regulation of acetyl CoA carboxylase mRNA synthesis is not simple. Understanding of this complex process has been made even more difficult by evidence that there are isozymic forms of acetyl CoA carboxylase.

V. EVIDENCE OF ISOZYMIC FORMS OF ACETYL CoA CARBOXYLASE

While the heterogeneous forms of acetyl CoA carboxylase mRNA found in rat tissue discussed above differ in the noncoding region, Kong et al.[23] found rat tissue acetyl CoA carboxylase mRNA which does differ in base composition in the coding region. Specifically, c-DNA clones were isolated with and without 24 bases which code for 8 amino acids corresponding to residues 1189 through 1196 (Figure 2). Further, it was shown that the quantity of mRNA coding for the ''long'' (2345 amino acid residues) and ''short'' (2337 amino acids) forms of acetyl CoA carboxylase varied among tissues and physiological conditions. The ''short'' form of mRNA accounted for about 10% of the total acetyl CoA carboxylase mRNA in liver whether rats were fed laboratory chow or fasted rats were refed a diet high in carbohydrate. In adipose tissue from rats fed a chow diet, the ''short'' form was 20% of the total acetyl CoA carboxylase mRNA but increased to about 36% of the total adipocyte acetyl CoA carboxylase mRNA in response to refeeding the diet high in carbohydrate. In the mammary gland of lactating rats, the ''short'' form of mRNA accounted for about 70% of total acetyl CoA carboxylase mRNA. Assuming that the two types of acetyl CoA carboxylase mRNA are translated with somewhat equal efficiency *in vivo*, these results indicate that rat tissues contain isozymic forms of acetyl CoA carboxylase.

Work in our laboratory has shown that cytosolic rat liver acetyl CoA carboxylase in purified[35,36] as well as crude[36] preparations occurs as a double band after SDS-PAGE, indicating the presence of two forms of the enzyme which differ in subunit molecular weight. That observation has been confirmed by Thampy[37] who concluded that the two forms were isozymes on the basis of antibody reactivity and tissue distribution. More recently, Bianchi et al.[38] also concluded that these two subunit molecular weight forms of acetyl CoA carboxylase represent isozymes of the enzyme, based on differences in binding by two different monoclonal antibody preparations. It was reported that while liver contained both subunit forms, heart and skeletal muscle contained only the larger subunit form and adipose tissue contained only the smaller form.

While the reports by Thampy,[37] Kong et al.,[23] and Bianchi et al.[38] all concluded that there are isozymic forms of acetyl CoA carboxylase, a more detailed analysis of the observations shows very little agreement on how the putative isozymes differ from each other. First, our observations[36] indicated that the two high molecular weight forms of acetyl CoA carboxylase in liver cytosol differed in mass by about 13,000 daltons (13 kDa) while Bianchi et al.[38] and Thampy[37] estimated a 15 kDa difference. In contrast, the mass of isozymes with and without the 8 amino acids as shown in Figure 2 would differ by less than 1 kDa, an amount too small to be detected by SDS-PAGE in such a large protein. Second, tissue distribution of the isozymic forms were markedly different. For example, while Bianchi et al.[38] found only one isozymic form in rat adipose tissue, Kong et al.[23] reported that adipocytes contained substantial amounts of mRNA for both isozymes. Finally, the monoclonal antibodies used by Bianchi et al.[38] to differentiate the two isozymic forms of the enzyme were raised against synthetic peptides containing only the 15 amino acids in the N terminal or C terminal fragments of acetyl CoA carboxylase where the amino acid composition of the two isozymes reported by Kong et al.[23] do not differ (Figure 2). In spite of the lack of agreement on specifics, the available evidence does suggest that rat tissues contain isozymic forms of acetyl CoA carboxylase which may have different tissue distribution and regulatory properties.

VI. POSSIBLE PHYSIOLOGICAL RELEVANCE OF MULTIPLE FORMS OF mRNA CODING FOR ACETYL CoA CARBOXYLASE

Although it is clear that there are multiple forms of acetyl CoA carboxylase mRNA in tissue, much work remains to be done before the physiological relevance of them can be understood. In the case of the species of acetyl CoA carboxylase mRNA which differ only in the 5′ untranslated region, translation of each of them would produce enzymes with identical primary structure. Since translation of mRNA occurs from the 5′ to 3′ direction, these 5′ regions are upstream from the coding section. In fact, the protein synthesis initiation codon, AUG, is located within exon 5 (Figure 3). The function of the 5′

untranslated region of mRNA is not known but it is possible that the length and base sequence of this section, while not affecting amino acid sequence of the protein, may affect interaction with ribosomes. In the case of acetyl CoA carboxylase mRNA species, differences in the 5′ untranslated region were found to affect translation in an *in vitro* rabbit reticulocyte protein synthesis assay.[29] With that system, translation was about 10 times better with the pAU type of mRNA than with the FL56 type.

If this difference in translation efficiency occurs *in vivo*, the cell could effectively control the rate of translation simply by regulating relative amounts of different types of mRNA produced. Further, from an investigative viewpoint, this would mean that experiments to determine the relative importance of transcriptional vs. posttranscriptional processes in the control of enzyme synthesis are difficult to interpret unless relative amounts of acetyl CoA carboxylase mRNA differing in the 5′ untranslated region are determined. That is, a multifold increase in the quantity of class 2 type of acetyl CoA carboxylase mRNA might produce no more enzyme protein per unit of time than would be obtained from a much smaller increase in the quantity of class 1 type of acetyl CoA carboxylase mRNA. Unfortunately, simple Northern blot analysis which is most often used for determining dietary/hormonal effects on acetyl CoA carboxylase mRNA levels does not differentiate between class 1 and class 2 types of mRNA.

Similarly, simple Northern blot analysis does not differentiate between the two acetyl CoA carboxylase mRNAs which code for isozymic forms of the enzyme. Yet, the relative amounts of the two forms differ among tissues and their relative concentrations vary under different physiological conditions.[23] The ratio of the two isozymic forms of acetyl CoA carboxylase may be very important because their regulatory properties differ. It has been shown that deletion of the eight amino acids from the larger form of acetyl CoA carboxylase allowed cyclic-AMP-dependent protein kinase to catalyze the phosphorylation of the serine residue located at position 1200 which makes the enzyme more dependent upon citrate for activation.[23]

VII. REGULATION OF ACETYL CoA CARBOXYLASE: RELATIVE IMPORTANCE OF TRANSCRIPTION, TRANSLATION, AND POSTTRANSLATIONAL PROCESSING

There is an abundance of evidence indicating that the quantity of the active forms of acetyl CoA carboxylase increases in liver cytosol under increased lipogenic conditions, e.g., when fasted rats are refed a diet high in carbohydrate. Indeed, over two decades ago, Nakanishi and Numa[14] showed that fasting of rats decreased acetyl CoA carboxylase quantity and its rate of synthesis, as determined by the incorporation of injected [^{3}H]-leucine into immunologically precipitated cytosolic acetyl CoA carboxylase, relative to that of fed control rats. Refeeding a diet high in carbohydrate markedly increased both the quantity and the apparent rate of synthesis of the cytosolic enzyme.

Dietary/hormonal regulation of the synthesis of active acetyl CoA carboxylase could occur at any of several points in the multistep process. Regulation could occur either at the level of acetyl CoA carboxylase mRNA synthesis (transcription), synthesis of the nascent protein (translation), or during subsequent steps which converts the nascent protein to the active holoenzyme. The simplest hypothesis to account for the increased quantity of enzyme in liver of fasted/refed rats is that dietary/hormonal conditions which increase lipogenesis result in a higher rate of transcription, leading to a higher steady state concentration of acetyl CoA carboxylase mRNA. This hypothesis assumes that the production of acetyl CoA carboxylase mRNA is rate limiting and that translation and posttranslational processing are relatively fast.

Data from reports where the activity/quantity of cytosolic acetyl CoA carboxylase and mRNA concentration are measured at a point in time after refeedings are consistent with the hypothesis that regulation of acetyl CoA carboxylase can occur, at least in part, at the transcriptional level. Pape et al.[39] confirmed previous reports that fasting (48 h) decreased and refeeding a fat-free diet (72 h) markedly increased acetyl CoA carboxylase activity in liver cytosol relative to that found in liver of fed controls. Northern blot analysis of RNA from liver of the same rats, using a nick-translated c-DNA probe for acetyl CoA carboxylase mRNA, showed very little of acetyl CoA carboxylase mRNA in liver of chow-fed rats, no detectable acetyl CoA carboxylase mRNA in liver of fasted rats, and relatively large amounts in liver of rats refed the fat-free diet. While the paper did not report quantitative amounts of acetyl CoA carboxylase mRNA under any dietary conditions (e.g., from densitometer scanning of X-ray film from Northern blots), the authors concluded that while regulation of acetyl CoA carboxylase synthesis may occur in part by transcriptional control, posttranscriptional processing may also be involved because acetyl CoA carboxylase activity was only three times greater in liver of refed than fed rats while the increase in acetyl CoA carboxylase mRNA concentration was much larger.

An almost identical experiment was conducted by Batenburg and Whitsett[40] using an acetyl CoA carboxylase c-DNA probe provided by Kim (i.e., the same probe as used by Pape et al.[39]), except that the relative concentration of acetyl CoA carboxylase mRNA was determined by densitometer scanning of the X-ray film from Northern blots. In this study, acetyl CoA carboxylase mRNA was found to increase by only 2.5-fold in liver of rats refed a fat-free diet (72 h) relative to that in liver of fed control rats, whereas Pape et al.[39] concluded that there was a much larger increase in acetyl CoA carboxylase mRNA in response to refeeding a similar diet for the same time period. Further, in contrast to the results reported by Pape et al. who found that acetyl

CoA carboxylase mRNA was not detectable in liver of fasted (48 h) rats,[39] Batenburg and Whitsett[40] reported that fasting (48 h) resulted in only a small, statistically insignificant decrease in rat liver acetyl CoA carboxylase mRNA relative to that in liver from fed control animals. Unfortunately, Batenburg and Whitsett[40] did not report changes in acetyl CoA carboxylase quantity or activity in response to fasting, but Pape et al.[39] reported that fasting rats for 48 h resulted in a 96% decrease in both acetyl CoA carboxylase quantity and activity in liver relative to that in liver of fed control rats. Thus, it is highly unlikely that the level of acetyl CoA carboxylase mRNA regulates enzyme synthesis during fasting, but the data in these reports are compatible with the hypothesis that refeeding increases transcription of the gene for acetyl CoA carboxylase, although neither report indicated which of the multiple forms of acetyl CoA carboxylase mRNA were affected.

Data in these reports[39,40] showed that both acetyl CoA carboxylase mRNA and enzyme activity/quantity were increased after rats were refed a diet high in carbohydrate. In each case, however, acetyl CoA carboxylase mRNA was measured only at a single point in time (72 h) after refeeding. Such an experiment does not indicate whether the rise in acetyl CoA carboxylase activity/quantity was due to increased acetyl CoA carboxylase mRNA. The relative importance of each potential regulatory mechanism can better be assessed by determining the time-dependent relationships of transcription rate, acetyl CoA carboxylase mRNA levels, and quantity of active enzyme. When Katsurada et al.[41] conducted such an experiment, the results (Figure 4) were not consistent with the hypothesis of simple transcriptional control. They found that the rate of acetyl CoA carboxylase mRNA synthesis increased significantly ($p < 0.05$) within an hour after refeeding a fat-free, high carbohydrate diet and remained at a similarly high rate until 8 h after refeeding. Meanwhile, acetyl CoA carboxylase mRNA concentration steadily increased over the first 8 h and then remained relatively constant through 24 h of refeeding. Over the next 24 and 48 h, acetyl CoA carboxylase mRNA concentration steadily dropped to reach a level at 72 h after refeeding that was significantly ($p < 0.05$) below the 8- to 24-h plateau. In contrast, cytosolic acetyl CoA carboxylase activity (and quantity as judged by immunoreactive protein) increased at a much slower rate than the rate of accumulation of mRNA and did not reach a plateau until 48 h after refeeding. It should be noted that at a point in time (72 h) after refeeding, both acetyl CoA carboxylase mRNA and acetyl CoA carboxylase activity/quantity were elevated relative to that in liver of fasted rats, in agreement with the results of Pape et al.[39] and Batenburg and Whitsett,[40] but measurement of changes in these parameters over time leads to a different conclusion. That is, since there was a time lag of several hours between measurable increased acetyl CoA carboxylase mRNA and the increased enzyme activity/quantity, the data are not consistent with the hypothesis that formation of cytosolic enzyme is limited simply by acetyl CoA carboxylase mRNA template availability. Rather, posttranscriptional

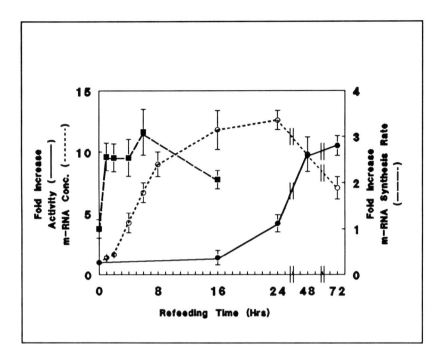

FIGURE 4.   Effect of refeeding a high carbohydrate diet on acetyl CoA carboxylase activity, mRNA synthesis rate, and mRNA levels in rat liver, adapted from Katsurada et al.[41] Enzyme activity was shown to be directly proportional to quantity of immunoprecipitated enzyme.

processes appeared to be of primary importance in determining the quantity of cytosolic acetyl CoA carboxylase in liver of fasted-refed rats.

Determination of the temporal relationships of acetyl CoA carboxylase mRNA synthesis rates, acetyl CoA carboxylase mRNA concentration, and acetyl CoA carboxylase activity/quantity in liver of insulin-treated strepto-zotocin-diabetic rats led to essentially the same result as refeeding a high-carbohydrate diet.[41] That is, acetyl CoA carboxylase mRNA synthesis rate rapidly increased after insulin injection which led to a steady increase in acetyl CoA carboxylase mRNA concentration. Acetyl CoA carboxylase mRNA concentration reached a plateau in 8 to 16 h after insulin injection and remained at this elevated concentration through the 72-h experiment. In contrast, acetyl CoA carboxylase activity/quantity did not change for the first 24 h postinjection, after which time it rose linearly for the next 48 h. Again the time lag between the rise in acetyl CoA carboxylase mRNA concentration and the increase in enzyme quantity indicated that insulin-dependent formation of cytosolic acetyl CoA carboxylase is not regulated simply by the availability of the mRNA template but rather by posttranscriptional processes.

While the studies described in the preceding paragraphs dealt with intact rats, a number of investigations on hormonal regulation of acetyl CoA carboxylase have been conducted with primary cultures of hepatocytes. Use of such cell cultures has an advantage over studies with intact animals in that the substrate and hormonal milieu of the cells can be controlled, but have the disadvantage that it is not certain whether results from the use of these systems accurately reflect *in vivo* regulatory mechanisms. With that caveat, it appears that insulin stimulates the synthesis of acetyl CoA carboxylase in cultured rat liver hepatocytes. Giffhorn and Katz[42] found that the activity of acetyl CoA carboxylase per mg DNA was increased 24 h after insulin and glucose were added to cultured hepatocytes. Acetyl CoA carboxylase synthesis, as measured by incorporation of [^{35}S]-methionine into immunoprecipitated enzyme, was also stimulated by a combination of insulin (.01 μM) and glucose (20 mM).

Spence et al.[43] conducted a more in-depth study of regulation of acetyl CoA carboxylase in cultured rat hepatocytes in which they determined the time-dependent changes in acetyl CoA carboxylase activity and synthesis rates. Enzyme activity was measured[44] by a [^{14}C]-bicarbonate fixation assay in cytosol after preincubation with 20 mM citrate for 30 min at 37°. Protein synthesis rates were determined at each time point by measuring the amount of [^{3}H]-leucine in 20 min into protein precipitated by polyclonal antiacetyl CoA carboxylase antibody. Enzyme activity began to increase immediately after insulin was added and reached a plateau after 6 h, but the rate of protein synthesis increased much more slowly (Figure 5). Although the quantity of acetyl CoA carboxylase was not measured in this study, preincubation of the enzyme preparation with citrate has been found by others[39] to negate the effects of any short-term regulatory mechanisms, in which case activity is directly proportional to quantity. Therefore, it is likely that the increase in acetyl CoA carboxylase activity well before synthesis of the enzyme was increased indicates that the initial effects of insulin were on posttranslational processes rather than transcription or translation.

Instead of measuring acetyl CoA carboxylase mRNA by Northern blot, Spence et al.[43] determined time-dependent changes in acetyl CoA carboxylase mRNA template activity. Template activity ("translatable acetyl CoA carboxylase mRNA") was measured at each time point by determining the relative synthesis rates of immunoprecipitated acetyl CoA carboxylase vs. synthesis rates of total soluble protein in a rabbit reticulocyte-lysate assay, using total RNA extracted from hepatocytes as template. Following insulin exposure, the acetyl CoA carboxylase mRNA activity increased at a substantially slower rate than enzyme activity but at a considerably faster rate than enzyme synthesis (Figure 5). The temporal relationship between enzyme synthesis rates and acetyl CoA carboxylase mRNA template activity was much closer in cultured hepatocytes (Figure 5) than that found between enzyme quantity and acetyl CoA carboxylase mRNA determined by Northern blot analysis in liver of intact rats (Figure 4). Although this may be due, at least

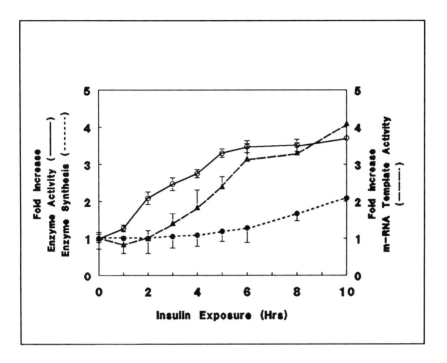

FIGURE 5.   Activity, synthesis, and mRNA template activity of acetyl CoA carboxylase over time after exposure of cultured rat hepatocytes to insulin, adapted from Spence et al.[43] Activity is very likely directly proportional to enzyme quantity since it was determined by the bicarbonate fixation assay after preincubation with 20 mM citrate. Protein synthesis rates were determined at each time point by measuring the amount of [³H]-leucine in 20 min into protein precipitated by polyclonal antiacetyl CoA carboxylase antibody. Template activity of mRNA was measured at each time point by determining the relative synthesis rates of immunoprecipitated acetyl CoA carboxylase vs. synthesis rates of total soluble protein in a rabbit reticulocyte-lysate assay, using total RNA extracted from hepatocytes as template.

in part, to the fact that these are two different systems, it could also be argued that measurement of acetyl CoA carboxylase mRNA template activity gives a more accurate assessment of the relative importance of transcription and translation in regulating the synthesis of acetyl CoA carboxylase. Unlike results from standard Northern blot analysis, measurement of acetyl CoA carboxylase mRNA template activity would reflect both the quantity of mRNA and variations in translation efficiency of the heterogeneous forms of acetyl CoA carboxylase mRNA. Thus, the data in Figure 5 could be interpreted to mean that acetyl CoA carboxylase synthesis over several hours may be regulated by an insulin effect, at least in part, at the transcriptional level, but again there was a substantial time lag between the rise in acetyl CoA carboxylase mRNA and increased acetyl CoA carboxylase synthesis.

Using similar primary cultures of rat hepatocytes, Spence et al.[43] also made the interesting observation that incubation of cultured cells with 25 mM

fructose (but not glucose), in the absence of insulin, had an effect on enzyme activity, rates of enzyme synthesis, and acetyl CoA carboxylase mRNA template activity almost identical to the insulin response, except for an initial 2-h time lag following fructose addition before any of the three parameters changed. Combination experiments showed that insulin and fructose effects on all three parameters were additive. The results of this study indicate that effects of fructose, like insulin, can be observed in primary cultures of rat hepatocytes on both transcription and posttranscriptional processes. Studies with intact rats have similarly indicated that feeding diets high in fructose (but not glucose) to streptozotocin-diabetic rats had almost the same effect on acetyl CoA carboxylase activity/quantity,[41,43] acetyl CoA carboxylase mRNA template activity,[43] acetyl CoA carboxylase mRNA concentration,[41] and acetyl CoA carboxylase synthesis rates[41] in liver as insulin injection. The mechanism by which dietary fructose affects these parameters has not been established.

In addition to fructose, the effects of other dietary components on regulation of acetyl CoA carboxylase have been studied. Katsurada et al.[45] compared the effects in liver on activity of lipogenic enzymes, including acetyl CoA carboxylase, after fasted rats were refed a diet containing 85% sucrose without protein vs. rats refed a diet composed of 67% sucrose/18% casein. In each case, the remainder of the diet was composed of cellulose (9.9%), vitamins, and minerals. Acetyl CoA carboxylase activity increased at the same rate for the first 2 days of refeeding and remained at this high level (about sixfold greater than that of fasted rats) until day 4 regardless of protein content of the diet. In contrast, maximal increased activity of fatty acid synthase, glucose-6-phosphate dehydrogenase and malic enzyme was observed only when the diet contained protein. At the end of the 4 days, [^{3}H]-leucine was injected 2 h before rats were killed, and radioactivity in immunoprecipitated cytosolic enzymes was determined as a measure of enzyme synthesis rates. While the presence or absence of dietary protein had no effect on acetyl CoA carboxylase synthesis rates, the rates of synthesis of both glucose-6-phosphate dehydrogenase and malic enzyme were significantly lower ($p < 0.01$) in liver of rats refed the protein-free diet. The levels of mRNA template activity for glucose-6-phosphate dehydrogenase and malic enzyme[45] were unaffected by dietary protein, indicating that dietary protein affected the synthesis of these two enzymes at posttranscriptional level. Subsequent experiments,[41] in which fasted rats were refed these identical diets, confirmed that the presence or absence of dietary protein had no effect on either the increase in acetyl CoA carboxylase activity/quantity, acetyl CoA carboxylase mRNA synthesis rates, or acetyl CoA carboxylase mRNA concentration.

The effects of dietary triglycerides on acetyl CoA carboxylase depend upon the chain length and degree of unsaturation of the fatty acids present. For example, after rats were fed a diet in which 32% of total calories were derived from medium chain triglycerides (predominant fatty acids = $C_{8:0}$ and $C_{10:0}$) for 21 d, acetyl CoA carboxylase activity in liver was not different than that in liver of rats fed a high carbohydrate (corn starch), low fat (1% corn oil) diet.[46] In contrast, when an equivalent amount of energy was derived from corn oil, liver acetyl CoA carboxylase activity was reduced by a factor of 2.

Other studies have shown that dietary polyunsaturated fatty acids are much more effective than saturated or monounsaturated fatty acids in depressing lipogenesis and acetyl CoA carboxylase activity. For example, dietary safflower oil had a greater inhibitory effect on rat liver acetyl CoA carboxylase activity than tallow[47] or palmitate.[47,48] Emken et al.[49] compared the effect of feeding various long chain fatty acids to mice on the activity in liver of acetyl CoA carboxylase and other lipogenic enzymes. They found that linoleate dramatically decreased the activity of all of the lipogenic enzymes, including acetyl CoA carboxylase, but neither palmitate nor stearate had any effect on the activity of acetyl CoA carboxylase or any of the other lipogenic enzymes.

The mechanism by which polyunsaturated fatty acids decrease acetyl CoA carboxylase activity remains unknown although Katsurada et al.[41] reported that when rats were fed a diet containing 57% sucrose, 18% casein, and 10% corn oil, acetyl CoA carboxylase activity, acetyl CoA carboxylase mRNA synthesis, and acetyl CoA carboxylase mRNA concentration in liver were all significantly lower ($p < 0.001$) than when rats were fed a diet containing 67% sucrose and 18% casein but no fat. While these data might suggest that dietary corn oil affected transcription, time-dependent changes in these parameters were not reported. The authors suggested that dietary polyunsaturated fatty acids might influence insulin secretion or insulin binding to cellular receptors.

Salati et al.,[50] who studied the effects of long chain fatty acids on induction of acetyl CoA carboxylase and other lipogenic enzymes in rat hepatocyte cultures, similarly concluded that fatty acids might affect insulin interaction with the hepatocytes. It should be noted, however, that the effect of feeding long chain fatty acids to rats is different than the effects observed when primary cultures of hepatocytes are exposed to them. When Salati and Clarke[51] determined the effects of adding either 0.2 mM palmitate, oleate, linoleate, or arachidonate to the media of primary cultures of rat hepatocytes, they found that each of them caused a small (30 to 35%) albeit statistically significant reduction in acetyl CoA carboxylase activity. But, in contrast to results from studies with intact rats, the effect did not vary with either chain length or degree of unsaturation.

VIII. SUBCELLULAR DISTRIBUTION OF ACETYL CoA CARBOXYLASE AS A POTENTIAL POSTTRANSLATIONAL REGULATORY MECHANISM

Witters et al.[52] published the first report in recent years which suggested that there may be particulate forms of acetyl CoA carboxylase. They found substantial acetyl CoA carboxylase activity was precipitated when rat liver homogenates were subjected to high speed centrifugation and concluded that acetyl CoA carboxylase was associated with microsomes, the extent of which depended upon dietary conditions. Imesch et al.[53] drew a similar conclusion from the study of rat adipose tissue. Subsequently, Evans and Witters[54] separated proteins in rat liver preparations by SDS-PAGE and transferred them to nitrocellulose (Western blot) which was then probed with polyclonal and monoclonal antiacetyl CoA carboxylase antibody. They found that only 57% of the enzyme was soluble. Nine percent of the total acetyl CoA carboxylase was found in low-speed (20,000 $\times$ g) and 34% was found in high-speed (105,000 $\times$ g) precipitates. It was concluded that much of the enzyme was microsomal although no marker enzymes were measured to confirm that conclusion. More recently, a report from the same laboratory[55] concluded that there were no particulate forms of acetyl CoA carboxylase in rat liver, based upon the observation that neither mitochondrial nor microsomal forms of the enzyme were detectable on Western blots of rat liver preparations using monoclonal antiacetyl CoA carboxylase antibodies raised against fragments of acetyl CoA carboxylase.[38]

In the meantime, it was discovered that radioactive avidin would bind to biotinylated proteins and remain bound through SDS-PAGE.[36] That observation has been confirmed by others.[38] Radioactive bands could be found by exposing the dried gel to X-ray film, and the quantity of each biotinyl enzyme applied to the gel could be determined by measuring the amount of radioactivity in each band.[56] Using this method, it was concluded[57] that as much as three fourths of the total acetyl CoA carboxylase found in liver of fed rats was particulate and that it was associated with the outer mitochondrial membrane. Only a trace amount of enzyme was found in the microsome fraction. There were two mitochondrial forms of acetyl CoA carboxylase, both of which were precipitated by polyclonal antiacetyl CoA carboxylase antibody. The larger of the two forms had a molecular mass similar to that of cytosolic acetyl CoA carboxylase while the smaller was about 30 kDa less. Both of these mitochondrial biotinyl proteins were larger by at least 100 kDa than the three known mitochondrial biotinyl enzymes, propionyl CoA carboxylase, pyruvate carboxylase, and 3-methyl crotonyl CoA carboxylase.

The discovery that the amount of the active cytosolic forms of acetyl CoA carboxylase vs. the amount of the relatively inactive mitochondrial forms varied as a function of diet led to the hypothesis that the mitochondrial forms represented a storage reservoir of acetyl CoA carboxylase which could be mobilized/activated under increased lipogenic conditions.[57] When rats were fasted (48 h), the quantity of active cytosolic acetyl CoA carboxylase decreased while the quantity of mitochondrial forms increased, with no change in the total quantity of acetyl CoA carboxylase. When fasted rats were refed a high-carbohydrate diet, the quantity of the active cytosolic forms increased, beginning about 8 h after refeeding[58] while mitochondrial forms decreased, again with no change in the total quantity of acetyl CoA carboxylase. The study of acute (3 d) alloxan-diabetic rats[59] indicated that mobilization/activation of mitochondrial forms of acetyl CoA carboxylase was insulin dependent. Examination of the subcellular distribution of acetyl CoA carboxylase in the liver of Zucker rats[60] indicated that obese rats had no more enzyme than that of lean litter mates, but rather more of the enzyme was in the active cytosolic form.

Evidence of particulate forms of two other lipogenic enzymes which are normally active in cytosol has been found. A form of ATP-citrate lyase has been shown to be associated with mitochondria.[61] The distribution of this enzyme between cytosol and mitochondria is dietary dependent.[62] More recently, it was concluded[63] that epidermal growth factor increased the activity of glucose-6-phosphate dehydrogenase of rat renal cortical cells by causing its release from a particulate form within the cell into the cytosol.

Although the mechanism by which acetyl CoA carboxylase might be bound to the outer mitochondrial membrane has not been established, several enzymes have now been shown to be attached to the cellular plasma membrane via a glycosylated phosphatidyl inositol linkage.[64] The discovery that acetyl CoA carboxylase is a glycoprotein[65,66] raises the possibility that it may be attached to the mitochondrial membrane by a similar mechanism. Analysis of the carbohydrates present in purified acetyl CoA carboxylase by GC-MS confirmed a previous report[67] that the enzyme contained inositol and established that it also contained mannose and galactose, as well as trace amounts of *N*-acetyl glucosamine, fucose, and 6-deoxy mannose. It was calculated that the combined mass of the carbohydrates amounted to about 6 kDa. While glycosylation is most often associated with exported proteins,[64] it has been established that another enzyme intimately involved in lipid metabolism, HMG CoA reductase, is a glycoprotein.[68] HMG CoA reductase remains within the cell in which it is produced and, unlike exported glycoproteins, is not processed in the Golgi apparatus.[69]

If rat liver contains mitochondrial storage forms of acetyl CoA carboxylase which can be mobilized under lipogenic conditions, it could provide a mechanism by which cells could more rapidly respond to dietary/hormonal changes than *de novo* enzyme synthesis might allow. Such a mechanism could account for the rapid rise in enzyme quantity implied by the data in Figure 5 well before acetyl CoA carboxylase synthesis is increased after cultured rat hepatocytes are exposed to insulin. In this scenario, insulin stimulation of acetyl CoA carboxylase transcription and translation would still be necessary in order to replace mobilized mitochondrial enzyme. While the existence of storage forms of acetyl CoA carboxylase represents a plausible hypothesis, definitive proof of the physiological significance of particulate forms of the enzyme has not been obtained.

IX. BIOTINYLATION

Acetyl CoA carboxylase is synthesized as an apoenzyme which must be biotinylated to form the holoenzyme. It is generally assumed that the biotinylation occurs rapidly except in biotin-deficient animals and therefore does not play a regulatory role.[70] It is known that in the active form of acetyl CoA carboxylase, biotin is covalently bonded to the epsilon amino group of lysine in the tetrapeptide [val-met-lys-met] located toward the amino terminal end of the protein at position 783 to 786 (Figure 2).[21] The biotinylation site of chicken liver acetyl CoA carboxylase contains the same tetrapeptide.[24] In contrast, in all other biotinyl enzymes present in tissues of higher animals (propionyl CoA carboxylase, pyruvate carboxylase, and 3-methyl crotonyl CoA carboxylase), the tetrapeptide sequence contains alanine instead of valine[71] and is located very near the carboxyl end of the protein.[72] While it has generally been assumed that all four biotin enzymes are biotinylated by the same enzyme,[70] studies involving structural alterations of acetyl CoA carboxylase suggest that this may not be the case. When site-directed mutagenesis was used to change the coding sequence of c-DNA so that alanine was substituted for the valine located at position 783, biotinylation of the protein expressed in a bacterial system was abolished.[22] When the amino acids surrounding the [ala-met-lys-met] site of acetyl CoA carboxylase were further altered to make a protein more closely resembling the structure of the other biotinyl enzymes, biotinylation of the lysine did occur. This suggests that acetyl CoA carboxylase and the other biotinyl enzymes may be biotinylated by distinct enzymes, differing in specificity.

X. SHORT-TERM REGULATORY MECHANISMS

A number of short-term mechansism which may regulate the activity of acetyl CoA carboxylase have been found, including covalent modification, citrate activation, and enzyme polymeric state. Most of the information about these mechanisms lie outside the scope of this review with the exception that there is evidence of interaction of these regulatory mechanisms with gene expression.

Shortly after the initial report that acetyl CoA carboxylase activity could be affected by its phosphorylation state,[73] it was discovered that derivatives of cyclic-AMP rapidly inhibited lipogenesis in rat[74] and chicken[75] liver slices and that the addition of cyclic-AMP decreased the activity of acetyl CoA carboxylase in rat liver homogenates.[76] These observations have been confirmed in numerous subsequent reports. It has now been shown that acetyl CoA carboxylase contains at least eight serine residues (Figure 2) which can be phosphorylated by various protein kinases.[21] Phosphorylation of the serine residue at position 79 is catalyzed by AMP-dependent (but not cyclic-AMP-dependent) protein kinase which inactivates the enzyme,[77] while phosphorylation of the serine at position 1200, catalyzed by cyclic-AMP-dependent protein kinase, does not inactivate acetyl CoA carboxylase directly but rather affects its sensitivity to citrate activation.[23] Cyclic-AMP-dependent protein kinase was found to catalyze the phosphorylation of serine-1200 only when the smaller (2337 amino acids) isozymic form of acetyl CoA carboxylase, in which 8 amino acids are deleted (Figure 2), served as the source of enzyme.[23] Thus, cyclic-AMP and citrate effects depend upon which isozyme is being studied. In turn, the polymeric state of acetyl CoA carboxylase depends upon phosphorylation state and citrate sensitivity.[78]

XI. CONCLUSION

It is clear that dietary/hormonal regulation of acetyl CoA carboxylase ultimately invovles gene expression. It is equally apparent, however, that regulation of the enzyme is remarkably complex. The available data are not compatible with the hypothesis that the quantity of active forms of the cytosolic enzyme is solely dependent upon the rate of gene transcription. Rather, dietary/hormonal regulation of acetyl CoA carboxylase can occur at the transcriptional, translational, and posttranslational levels.

REFERENCES

1. **Volpe, J. J. and Vagelos, P. R.,** Mechanisms and regulation of biosynthesis of saturated fatty acids, *Physiol. Rev.,* 56, 339, 1976.
2. **Gibson, D. M., Lyons, R. T., Scott, D. F., and Muto, Y.,** Synthesis and degradation of the lipogenic enzymes of rat liver, *Adv. Enzyme Regul.,* 10, 187, 1970.
3. **Brady, L.,** Regulation of hepatic carnitine palmitoyltransferase, *Nutrition and Gene Expression,* CRC Press, Boca Raton, FL, 1992.
4. **Hellerstein, M. K., Christiansen, M., Kaempfer, S., Kletke, C., Wu, K., Reid, J. S., Mulligan, K., Hellerstein, N. S., and Shackleton, C. H. L.,** Measurement of de novo hepatic lipogenesis in humans using stable isotopes, *J. Clin. Invest.,* 87, 1841, 1991.
5. **Mead, J. F., Alfin-Slater, R. B., Howton, D. R., and Popjak, G.,** *Lipids: Chemistry, Biochemistry and Nutrition,* Plenum Press, New York, 1986, chaps. 8 and 9.
6. **Needleman, P., Turk, J., Jakschik, B. A., Morrison, A. R., and Lefkowith, J. B.,** Arachidonic acid metabolism, *Annu. Rev. Biochem.,* 55, 69, 1986.
7. **Bijleveld, C., Vaartjes, W. J., and Geelen, M. J.,** Time course of hormonal effects on acetyl-CoA carboxylase as measured in digitonin-permeabilized rat hepatocytes, *Horm. Metab. Res.,* 21, 602, 1989.
8. **Milatovich, A., Plattner, R., Heerema, N. A., Palmer, C. G., Lopez-Casillas, F., and Kim, K.-H.,** Localization of the gene for acetyl-CoA carboxylase to human chromosome 17, *Cytogenet. Cell. Genet.,* 48, 190, 1988.
9. **Blom, W., DeMuinck-Keizer, S. M. P. F., and Scholte, H. R.,** Acetyl CoA carboxylase deficiency: an inborn error of *de novo* fatty acid synthesis, *New Engl. J. Med.,* 305, 465, 1981.

10. **Shrago, E., Spennetta, T., and Gordon, E.,** Fatty acid synthesis in human adipose tissue, *J. Biol. Chem.,* 244, 2761, 1969.

11. **Passmore, R. and Swindells, Y. E.,** Observations on the respiratory quotients and weight gain of man after eating large quantities of carbohydrate, *Brit. J. Nutr.,* 17, 331, 1963.

12. **Acheson, K. J., Flatt, J. P., and Jequier, E.,** Glycogen synthesis versus lipogenesis after a 500 gram carbohydrate meal in man, *Metabolism,* 31, 1234, 1982.

13. **Tepperman, H. M. and Tepperman, J.,** Adaptive hyperlipogenesis, *Fed. Proc.,* 23, 73, 1964.

14. **Nakanishi, S. and Numa, S.,** Purification of rat liver acetyl CoA carboxylase and immunochemical studies on its synthesis and degradation, *Eur. J. Biochem.,* 16, 161, 1970.

15. **Buckley, M. G. and Rath, E. A.,** Regulation of fatty acid synthesis and malonyl-CoA content in mouse brown adipose tissue in response to cold-exposure, starvation or refeeding, *Biochem. J.,* 243, 437, 1987.

16. **Mersmann, H. J. and Koong, L.-J.,** Effect of plane of nutrition on adipose tissue metabolism in genetically obese and lean pigs, *J. Nutr.,* 114, 862, 1986.

17. **Smith, S. B., Prior, R. L., Ferrell, C. L., and Mersmann, H. J.,** Interrelationships among diet, age, fat deposition and lipid metabolism in growing steers, *J. Nutr.,* 114, 153, 1984.

18. **Lopez-Casillas, F., Pape, M. E., Bai, D. H., Kuhn, D. N., Dixon, J. E., and Kim, K. H.,** Preparation of functional acetyl-CoA carboxylase mRNA from rat mammary gland, *Arch. Biochem. Biophys.,* 257, 63, 1987.

19. **Bai, D. H., Pape, M. E., Lopez-Casillas, F., Luo, X. C., Dixon, J. E., and Kim, K. H.,** Molecular cloning of cDNA for acetyl-coenzyme A carboxylase, *J. Biol. Chem.,* 261, 12395, 1986.

20. **Lopez-Casillas, F., Bai, D. H., Luo, X. C., Kong, I. S., Hermodson, M. A., and Kim, K. H.,** Structure of the coding sequence and primary amino acid sequence of acetyl-coenzyme A carboxylase, *Proc. Natl. Acad. Sci. U.S.A.,* 85, 5784, 1988.

21. **Davies, S. P., Sim, A. T., and Hardie, D. G.,** Location and function of three sites phosphorylated on rat acetyl-CoA carboxylase by the AMP-activated protein kinase, *Eur. J. Biochem.,* 187, 183, 1990.

22. **Bai, D. H., Moon, T. W., Lopez-Casillas, F., Andrews, P. C., and Kim, K. H.,** Analysis of the biotin-binding site on acetyl-CoA carboxylase from rat, *Eur. J. Biochem.,* 182, 239, 1989.

23. **Kong, I. S., Lopez-Casillas, F., and Kim, K. H.,** Acetyl-CoA carboxylase mRNA species with or without inhibitory coding sequence for Ser-1200 phosphorylation, *J. Biol. Chem.,* 265, 13695, 1990.

24. **Takai, T., Wada, K., and Tanabe, T.,** Primary structure of the biotin-binding site of chicken liver acetyl-CoA carboxylase, *FEBS. Lett.,* 212, 98, 1987.

25. **Takai, T., Yokoyama, C., Wada, K., and Tanabe, T.,** Primary structure of chicken liver acetyl-CoA carboxylase deduced from cDNA sequence, *J. Biol. Chem.,* 263, 2651, 1988.

26. **Lopez-Casillas, F., Luo, X. C., Kong, I. S., and Kim, K. H.,** Characterization of different forms of rat mammary gland acetyl-coenzyme A carboxylase mRNA: analysis of heterogeneity in the 5' end, *Gene,* 83, 311, 1989.

27. **Luo, X. C., Park, K., Lopez-Casillas, F., and Kim, K. H.,** Structural features of the acetyl-CoA carboxylase gene: mechanisms for the generation of mRNAs with 5' end heterogeneity, *Proc. Natl. Acad. Sci. U.S.A.,* 86, 4042, 1989.

28. **Lopez-Casillas, F. and Kim, K. H.,** Heterogeneity at the 5' end of rat acetyl-coenzyme A carboxylase mRNA. Lipogenic conditions enhance synthesis of a unique mRNA in liver, *J. Biol. Chem.,* 264, 7176, 1989.

29. **Lopez-Casillas, F., Ponce-Castaneda, M. V., and Kim, K. H.,** In vivo regulation of the activity of the two promoters of the rat acetyl coenzyme-A carboxylase gene, *Endocrinology,* 129, 1049, 1991.

30. **Luo, X. C. and Kim, K. H.,** An enhancer element in the house-keeping promoter for acetyl-CoA carboxylase gene, *Nucleic Acids Res.,* 18, 3249, 1990.

31. **Park, K. and Kim, K. H.,** Regulation of acetyl-CoA carboxylase gene expression: Insulin induction of acetyl-CoA carboxylase and differentiation of 30A5 preadipocytes require prior cAMP action on the gene, *J. Biol. Chem.,* 266, 12249, 1991.

32. **Pape, M. E. and Kim, K. H.,** Transcriptional regulation of acetyl coenzyme A carboxylase gene expression by tumor necrosis factor in 30A-5 preadipocytes, *Mol. Cell. Biol.,* 9, 974, 1989.

33. **Semenkovich, C. F., Wims, M., Noe, L., Etienne, J., and Chan, L.,** Insulin regulation of lipoprotein lipase activity in 3T3-L1 adipocytes is mediated at the posttranscriptional and posttranslational levels, *J. Biol. Chem.,* 264, 9030, 1989.

34. **Swierczynski, J., Mitchell, D. A., Reinhold, D. S., Salati, L. M., Stapleton, S. R., Klautky, S. A., Struve, A. E., and Goodridge, A. G.,** Triiodothyronine-induced accumulations of malic enzyme, fatty acid synthase, acetyl-coenzyme A carboxylase, and their mRNAs are blocked by protein kinase inhibitors. Transcription is the affected step, *J. Biol. Chem.,* 266, 17459, 1991.

35. **Allred, J. B., Harris, G. J., and Goodson, J.** Regulation of purified rat liver acetyl CoA carboxylase by phosphorylation, *J. Lipid Res.,* 24, 449, 1983.

36. **Goodson, J., Pope, T. S., and Allred, J. B.,** Molecular weights of subunits of acetyl CoA carboxylase in rat liver cytoplasm, *Biochem. Biophys. Res. Commun.,* 122, 694, 1984.

37. **Thampy, K. G.,** Formation of malonyl coenzyme A in rat heart. Identification and purification of an isozyme of acetyl CoA carboxylase from rat heart, *J. Biol. Chem.,* 264, 17631, 1989.

38. **Bianchi, A., Evans, J. L., Iverson, A. J., Nordlund, A. C., Watts, T. D., and Witters, L. A.,** Identification of an isozymic form of acetyl-CoA carboxylase, *J. Biol. Chem.,* 265, 1502, 1990.

39. **Pape, M. E., Lopez-Casillas, F., and Kim, K. H.,** Physiological regulation of acetyl-CoA carboxylase gene expression: effects of diet, diabetes, and lactation on acetyl-CoA carboxylase mRNA, *Arch. Biochem. Biophys.,* 267, 104, 1988.

40. **Batenburg, J. J. and Whitsett, J. A.,** Levels of mRNAs coding for lipogenic enzymes in rat lung upon fasting and refeeding and during perinatal development, *Biochim. Biophys. Acta.,* 1006, 329, 1989.

41. **Katsurada, A., Iritani, N., Fukuda, H., Matsumura, Y., Nishimoto, N., Noguchi, T., and Tanaka, T.,** Effects of nutrients and hormones on transcriptional and post-transcriptional regulation of acetyl-CoA carboxylase in rat liver, *Eur. J. Biochem.,* 190, 435, 1990.

42. **Giffhorn, S. and Katz, N. R.,** Glucose-dependent induction of acetyl-CoA carboxylase in rat hepatocyte cultures, *Biochem. J.,* 221, 343, 1984.

43. **Spence, J. T., Koudelka, A. P., and Tseng-Crank, J. C.,** Role of protein synthesis in the carbohydrate-induced changes in the activities of acetyl CoA carboxylase and hydroxymethylglutaryl CoA reductase in cultured rat hepatocytes, *Biochem. J.,* 227, 939, 1985.

44. **Craig, M. C., Dugn, R. E., Muesing, R. A., Slakey, L. L., and Porter, J. W.,** Comparative effects of dietary regimen on the levels of enzymes regulating the synthesis of fatty acids and cholesterol in rat liver, *Arch. Biochem. Biophys.,* 151, 128, 1972.

45. **Katsurada, A., Iritani, N., Fukuda, H., Noguchi, T., and Tanaka, T.,** Effects of dietary nutrients on lipogenic enzyme and mRNA activities in rat liver during induction, *Biochim. Biophys. Acta.,* 877, 350, 1986.

46. **Chanez, M., Bois-Joyeux, B., Arnaud, M. J., and Peret, J.,** Metabolic effects in rats of a diet with a moderate level of medium-chain triglycerides, *J. Nutr.,* 121, 585, 1991.

47. **Wilson, M. D., Hays, R. D., and Clarke, S. D.,** Inhibition of liver lipogenesis by dietary polyunsaturated fat in severely diabetic rats, *J. Nutr.,* 116, 1511, 1986.

48. **Clarke, S. D., Wilson, M. D., and Ibnoughazala, T.,** Resistance of lung fatty acid synthesis to inhibition by dietary fat in the meal-fed rat, *J. Nutr.,* 114, 598, 1984.
49. **Emken, E. A., Abraham, S., and Lin, C. Y.,** Metabolism of cis-12-octadecenoic acid and trans-9,trans-12-octadecadienoic acid and their influence on lipogenic enzyme activities in mouse liver, *Biochim. Biophys. Acta.,* 919, 111, 1987.
50. **Salati, L. M., Adkins-Finke, B., and Clarke, S. D.,** Free fatty acid inhibition of the insulin induction of glucose-6-phosphate dehydrogenase in rat hepatocyte monolayers, *Lipids,* 23, 36, 1988.
51. **Salati, L. M. and Clarke, S. D.,** Fatty acid inhibition of hormonal induction of acetyl-coenzyme A carboxylase in hepatocyte monolayers, *Arch. Biochem. Biophys.,* 246, 82, 1986.
52. **Witters, L. A., Friedman, S. A., and Bacon, G. W.,** Microsomal acetyl CoA carboxylase: evidence for association of enzyme polymer with liver microsomes, *Proc. Natl. Acad. Sci. U.S.A.,* 78, 3639, 1981.
53. **Imesch, E., Wolczunowicz, M., and Rous, S.,** Enzymatic activities of cytoplasmic and of microsomal acetyl CoA carboxylase of rat epididymal adipose tissue; different regulatory effects of a short-term fast and palmitoyl CoA on these two enzymes, *Int. J. Biochem.,* 15, 977, 1983.
54. **Evans, J. L. and Witters, L. A.,** Quantitation by immunoblotting of the in vivo induction and subcellular distribution of hepatic acetyl-CoA carboxylase, *Arch. Biochem. Biophys.,* 264, 103, 1988.
55. **Iverson, A. J., Bianchi, A., Nordlund, A. C., and Witters, L. A.,** Immunological analysis of acetyl-CoA carboxylase mass, tissue distribution and subunit composition, *Biochem. J.,* 269, 365, 1990.
56. **Roman-Lopez, C. R., Goodson, J., and Allred, J. B.,** Determination of the quantity of acetyl CoA carboxylase by [14C]methyl avidin binding, *J. Lipid Res.,* 28, 599, 1987.
57. **Allred, J. B. and Roman-Lopez, C. R.,** Enzymatically inactive forms of acetyl-CoA carboxylase in rat liver mitochondria, *Biochem. J.,* 251, 881, 1988.
58. **Roman-Lopez, C. R., Shriver, B. J., Joseph, C. R., and Allred, J. B.,** Mitochondrial acetyl-CoA carboxylase: time course of mobilization/activation in liver of refed rats, *Biochem. J.,* 260, 927, 1989.
59. **Roman-Lopez, C. R. and Allred, J. B.,** Acute alloxan diabetes alters the activity but not the total quantity of acetyl CoA carboxylase in rat liver, *J. Nutr.,* 117, 1976, 1987.
60. **Allred, J. B., Roman-Lopcz, C. R., Jurin, R. R., and McCune, S. A.,** Mitochondrial storage forms of acetyl CoA carboxylase: mobilization/activation accounts for increased activity of the enzyme in liver of genetically obese Zucker rats, *J. Nutr.,* 119, 478, 1989.
61. **Janski, A. M. and Cornell, N. W.,** Association of ATP citrate lysase with mitochondria, *Biochem. Biophys. Res. Commun.,* 92, 305, 1980.
62. **Cornell, N. W., Janski, A. M., and Rendon, A.,** Compartmentation of enzymes: ATP citrate lysase in hepatocytes from fed and fasted rats, *Fed. Proc.,* 44, 2448, 1985.
63. **Stanton, R. C., Seifters, J. L., Boxer, D. C., Zimmerman, E., and Cantley, L. C.,** Rapid release of bound glucose-6-phosphate dehydrogenase by growth factors. Correlation with increased enzymatic activity, *J. Biol. Chem.,* 266, 12442, 1991.
64. **Low, M. G.,** Biochemistry of the glycosyl-phosphatidylinositol membrane protein anchors, *Biochem. J.,* 244, 1, 1987.
65. **Bowers, D. F. and Allred, J. B.,** Acetyl CoA carboxylase is a glycoprotein, *FASEB J.,* 5, A462, 1991.
66. **Bowers, D. F.,** Identification and Characterization of Acetyl CoA Carboxylase as a Glycoprotein, Ph.D. dissertation, The Ohio State University, Columbus, 1991.
67. **Heger, H. W. and Peter, H. W.,** Phosphatidylinositol as an essential constituent of the acetyl CoA carboxylase from rat liver, *Int. J. Biochem.,* 8, 841, 1977.

68. **Liscum, L., Cummings, R. D., Anderson, R. G. W., DeMartino, G. N., Goldstein, J. L., and Brown, M. S.,** 3-Hydroxy-3-methylglutaryl CoA reductase: a transmembrane glycoprotein of the endoplasmic reticulum with N-linked "high-mannose" oligosaccharide, *Proc. Natl. Acad. Sci. U.S.A.,* 80, 7165, 1983.

69. **Haro, D., Marrero, P. F., Ayte, J., and Hegardt, F. G.,** Identification of a cholesterol-regulated 180-kDa microsomal protein in rat hepatocytes, *Eur. J. Biochem.,* 188, 123, 1990.

70. **Moss, J. and Lane, M. D.,** The biotin-dependent enzymes, *Adv. Enzymol.,* 35, 321, 1971.

71. **Shenoy, B. C., Paranjape, S., Murtif, V. L., Kumar, G. K., Samols, D., and Wood, H. G.,** Effect of mutations at met-88 and met-90 on the biotinylation of lys-89 of the apo 1.3S subunit of transcarboxylase, *FASEB J.,* 2, 2505, 1988.

72. **Lamhonwah, A.-M., Quan, F., and Gravel, R. A.,** Sequence homology around the biotin-binding site of human propionyl CoA carboxylase and pyruvate carboxylase, *Arch. Biochem. Biophys.,* 254, 631, 1987.

73. **Carlson, C. A. and Kim, K.-H.,** Regulation of hepatic acetyl CoA carboxylase by phosphorylation and dephosphorylation, *J. Biol. Chem.,* 248, 378, 1973.

74. **Allred, J. B. and Roehrig, K. L.,** Inhibition of hepatic lipogenesis by cyclic-3', 5'-nucleotide monophosphate, *Biochem. Biophys. Res. Commun.,* 46, 1135, 1972.

75. **Allred, J. B. and Roehrig, K. L.,** Metabolic oscillations and food intake, *Fed. Proc.,* 32, 1727, 1973.

76. **Allred, J. B. and Roehrig, K. L.,** Inhibition of rat liver acetyl CoA carboxylase in N^6, O^2-dibutyryl cyclic-3', 5'-adenosine monophosphate *in vitro, J. Biol. Chem.,* 248, 4131, 1973.

77. **Haystead, T. A., Moore, F., Cohen, P., and Hardie, D. G.,** Roles of the AMP-activated and cyclic-AMP-dependent protein kinases in the adrenaline-induced inactivation of acetyl-CoA carboxylase in rat adipocytes, *Eur. J. Biochem.,* 187, 199, 1990.

78. **Thampy, K. G. and Wakil, S. J.,** Regulation of acetyl-coenzyme A carboxylase. II. Effect of fasting and refeeding on the activity, phosphate content, and aggregation state of the enzyme, *J. Biol. Chem.,* 263, 6454, 1988.

Chapter 13

APOLIPOPROTEIN B mRNA EDITING

Nicholas O. Davidson

TABLE OF CONTENTS

0-8493-6961-4/93/$0.00 + $.50
© 1993 by CRC Press, Inc.

I. OVERVIEW AND DEFINITION

Apolipoprotein B (apo B) is a large hydrophobic protein synthesized in the liver and small intestine of mammals. Apo B serves an essential although incompletely understood role in the assembly and secretion of triglyceride-rich lipoproteins (chylomicrons and very low density lipoproteins) and also functions in the catabolic clearance of low density lipoprotein (LDL), the major transport vehicle of plasma cholesterol in humans. Mammalian apo B is the product of a single gene which maps to the p23-p24 region of chromosome $2^{1,2}$ (for detailed reviews see References 3 and 4). Apo B mRNA is expressed and processed in a tissue-specific fashion, the molecular and cellular basis for which will be the focus of this chapter. As currently viewed, one form of apo B is synthesized in the human liver as a protein of 4536 amino acids with a relative molecular mass of 512,000. This form of the protein is referred to on a centile scale as apo B100.[5-9] By contrast, the intestinal form of the protein contains 2152 amino acids, is colinear with the amino terminal half of apo B100, and is referred to as apo B48.[10,11] Several important structural domains have been identified for apo B based upon cDNA sequence and monoclonal epitope mapping. The most important of these are the LDL receptor binding domain and the attachment site of apolipoprotein (a) which both reside in the carboxyl terminal half of the protein.[12-14] The biological relevance of apo B48 may thus relate to the absence of the above domains which are present only in the full length protein. Apo B48 is found in the systemic circulation in association with intestinally derived lipoprotein particles, namely chylomicrons and chylomicron remnants. These particles are cleared from the plasma compartment via a (principally hepatic) receptor which has been partially characterized and is referred to as the LDL receptor-related protein or LRP.[15,16] This receptor recognizes apolipoprotein E as its major ligand.[17] Thus, lipoprotein particles containing apo B48 have a distinct catabolic fate from those with apo B100 which are cleared principally via the LDL receptor. The implications for this observation in terms of atherosclerosis susceptibility remain to be tested experimentally, but it suggests several models whereby such characteristics of an intestinal particle may have evolved to facilitate their function as an efficient delivery system for dietary triglyceride. By contrast, the finer regulation of plasma cholesterol homeostasis may be achieved through hepatic very low density lipoprotein (VLDL) secretion and ultimately LDL uptake.

As alluded to above, this chapter will focus on the posttranscriptional modification of apo B mRNA referred to as apo B mRNA editing. This process is distinct from other co- or posttranscriptional processing events such as capping, polyadenylation, and splicing and, until recently, was unique to apo B as an example of posttranscriptional modification of the coding information in a mammalian gene. Apo B mRNA editing was the first example in vertebrates of this departure from one of the central tenets of molecular biology that DNA encodes an RNA template which is identical and which subsequently specifies a predictable protein. Over the last few years, several examples of RNA editing have been described in lower eukaryotes, and these have been classified according to the underlying mechanism (for detailed reviews, see References 18 and 19). These mechanisms include the posttranscriptional insertion or deletion of uridine residues in trypanosome mitochondrial genes which results in the production of a translationally competent open reading frame. This form of RNA editing is of interest also because it is mediated by small guide RNAs.[20] Other forms of RNA editing include the posttranscriptional insertion of guanosine residues in paramyxovirus and cytidine residues in physarum polycephalum. These examples are discussed in detail in the reviews cited.[18,19]

Apo B mRNA editing was described in 1987 when several groups simultaneously reported the site-specific modification of apo B mRNA as the underlying mechansim for the production of distinct isoforms of the protein from the human liver and small intestine.[21-23] Powell et al.[21] prepared intestinal cDNAs and demonstrated that nucleotide 6666 in human and rabbit intestinal apo B cDNA was changed from the genomically templated cytidine to a uridine residue. This change modified codon 2153 from a CAA which encodes glutamine to UAA which specifies an in-frame stop codon. Chen et al.[22] reported the identical conclusion but used a strategy consisting of tryptic fragment analysis and direct RNA sequencing. These findings indicated that intestinal apo B was the product of a single apo B gene in which codon 2153 is altered to produce a translational stop codon and thereby specify a truncated apo B (apo B48) as the primary translation product. Recently, a second example of mammalian RNA editing was described in which certain subunits of the brain glutamate receptor transcripts undergo a single base change in which a glutamine codon (CAG) is changed to an arginine codon (CGG).[24] The significance of the observation that both forms of RNA editing involve a glutamine codon may be fortuitous since the base involved is different (cytidine vs. adenosine) although the mechanism may involve a deamination in both instances. The consequences following editing of the brain glutamate receptor transcripts are profound since the arginine residue regulates the functional characteristics of these gated receptor channels. Thus, in both situations currently known of mammalian RNA editing, the process has enormous functional consequences for gene expression. It is likely that an understanding of RNA editing mechanisms and their evolution will provide important insight into a novel area of regulation of gene expression and will reveal new functions for this process.

II. TISSUE-SPECIFIC DISTRIBUTION AND DEVELOPMENTAL REGULATION

A. MAMMALIAN APO B mRNA EDITING

Several studies have addressed the tissue-specific distribution of apo B mRNA editing. The original descriptions[21-23] were of intestinal apo B mRNA editing since this was central to an understanding of the production of apo B48 by the small intestinal enterocyte. Following this description, Davidson et al.[25] cloned and sequenced the region of rat apo B cDNA corresponding to the 3' terminus of apo B48 and demonstrated that apo B mRNA editing occurs in both the rat liver and small intestine. This finding was confirmed by others[26] and raises an important and presently unresolved question—namely, the evolutionary and biological advantage of editing both the intestinal and hepatic apo B transcripts. The major functional consequence of this observation is that the rat and mouse liver synthesize and secrete both apo B100 and B48. As far as currently known, mammalian hepatic apo B synthesis and secretion in all other species, including humans, is exclusively apo B100. A single report[27] of apo B mRNA editing in the human liver has not been confirmed by others and the validity of this observation is in doubt since subsequent, independent investigation of this issue has determined that human hepatic apo B mRNA is essentiallly all unedited.[28] Apo B mRNA editing in the mammalian enterocyte occurs as a developmentally regulated event in human,[28] rat,[29] and pig small intestine.[30] Teng et al.[28] demonstrated that human intestinal apo B was more than 90% unedited in fetal small intestinal RNA from late first trimester samples. This finding confirmed earlier preliminary work suggesting that human fetal intestine synthesized both forms of apo B.[31] Teng et al.[28] also demonstrated that the early gestation fetal small intestine synthesizes and secretes both apo B100 and apo B48, indicating that the unedited form of the apo B transcript is translationally competent in the small intestine and, furthermore, leads to apo B100 secretion. As the small intestine undergoes maturation during the late first and early second trimester, the proportion of edited apo B mRNA increases such that at 19 to 20 weeks gestation, small intestinal apo B mRNA is approximately 80 to 90% edited. Adult small intestine was found to contain a variable quantity of unedited apo B mRNA, varying from 3 to 19% in one series. Other studies have independently established that the human small intestine indeed synthesizes apo B100.[32,33] Of interest in this regard is the observation that although adult small intestine synthesizes and secretes apo B100, the proportions of newly synthesized apo B100 to apo B48 in the tissue or the culture media do not approach the proportions of unedited to edited apo B mRNA—there being far less than 3% apo B100 detectable as a percentage of total immunoreactive apo B species. The implications of this are that the unedited transcript is translationally active but that the determinants of apo B100 synthesis and secretion from the small intestine involve other, co- , or posttranslational events which may influence the stability of the nascent apo B peptide. Such events have been postulated to play a major role in the regulation of apo B secretion from Hep G2 cells particularly in relation to altered lipid flux.[34]

Other studies have established that rat apo B mRNA editing is developmentally regulated in both the small intestine and liver. Work by Wu et al.[29] demonstrated that the temporal sequence of the developmental changes in apo B mRNA editing is distinct for the liver and small intestine with a striking increase in intestinal editing prenatally while the hepatic transcript is largely unedited until postnatal day 20. Thus, the emergence of edited apo B mRNA in both the rat liver and intestine appears to coincide with developmental changes in triglyceride metabolism. Parallel conclusions cannot be proposed for either human or porcine intestine since fetal hepatic and intestinal triglyceride metabolism have yet to be characterized in these species. In regard to the observation that apo B mRNA editing is developmentally induced in the fetal small intestine, studies conducted in our laboratory have addressed the question of whether apo B mRNA editing changes in the adult as enterocytes migrate up the villus during their 3- to 5-day life span. Using adult rats, sequential enterocyte isolation from villus tip to crypt was undertaken by citrate-EDTA chelation and RNA extracted from isolated cells whose position was inferred from sucrase activity enrichment (Figure 1). Apo B mRNA editing was determined by reverse-transcription and coupled polymerase chain reaction amplification of apo B cDNA followed by single nucleotide discrimination using differential oligonucleotide hybridization. As can be seen in Figure 1, the small intestinal enterocyte edits apo B mRNA as soon as it is transcribed in cells emerging from the crypt.

As a model of the developing fetal small intestine, workers have used a human colon cancer-derived cell line (Caco-2) which, in culture, undergoes a form of "spontaneous differentiation" and displays certain phenotypic characteristics of developing enterocytes. Teng et al.[28] used this cell line to study apo B mRNA editing and found that during the course of differentiation from pre- to late postconfluency, apo B mRNA abundance increased 20-fold but the proportions of edited to unedited transcript remained unaltered at less than 5% at all times studied. Thus, in this cell line apo B mRNA abundance appears to be regulated by mechanisms distinct from those which influence apo B mRNA editing. Other investigators[36] using this cell line have found that apo B mRNA editing increases when the cells are grown on semipermeable filters rather than plastic. The specific cues which mediate this response are presently unknown but may provide insight into the mechanisms underlying developmental changes in human small intestinal apo B mRNA editing.

B. APO B mRNA EDITING IN NON-APO B-PRODUCING CELLS AND TISSUES

Mammalian apo B is expressed in a tissue-specific manner with gene transcription and protein synthesis predominantly confined to the adult liver and small intestine. In the fetus, however, apo B mRNA and protein biosynthesis have been found to occur in a number of extraintestinal, extrahepatic sites. In the rat[29,37] these include the placental and fetal membranes, while in humans, we and others have found that numerous tissues, including lung, kidney, stomach, colon, adrenal, and fetal membranes were all competent to synthesize and secrete apo B.[28,38] These latter studies demonstrated the pres-

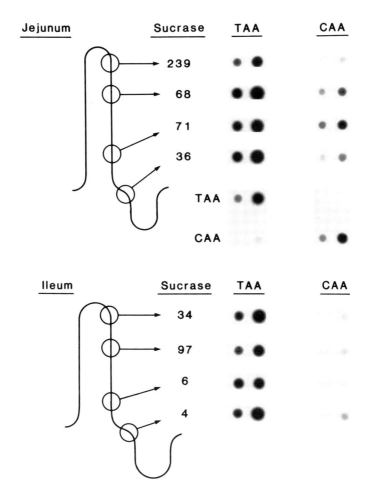

FIGURE 1. Gradients of apo B mRNA editing in rat small intestine. Isolated rat enterocytes were prepared from both jejunum and ileum by citrate-EDTA chelation with estimates of the relative position of each fraction on the villus-crypt axis determined by sucrase-specific activity. RNA was prepared from each fraction and subjected to reverse-transcription and polymerase chain reaction amplification of apo B cDNA followed by differential oligonucleotide hybridization to detect the single nucleotide substitution. It is evident that intestinal apo B mRNA is extensively edited in all regions of the small intestine with no horizontal or vertical gradient detectable.

ence of apo B mRNA using reverse transcription and coupled polymerase chain reaction amplification (RT-PCR) of low levels of endogenous apo B mRNA. Of equal significance was the observation that a number of simultaneously analyzed tissue RNA samples, such as placenta, brain, and heart, were negative. Analysis of these PCR amplified cDNA samples indicated that apo B was edited to a varying extent (10 to 50%) in all the fetal tissues examined with the notable exception of the liver, and, in all these sites, apo

B cDNA was edited to a progressively greater extent during development. Of importance was the observation that despite the presence of edited transcript in these extraintestinal sites, the form of apo B synthesized and secreted was exclusively apo B100. Thus, the regulation of apo B48 synthesis involves other possibly co- or posttranslational events.

A different approach to this question was used by Bostrom et al.[39,40] who utilized a chimeric apo B expression construct to determine the presence of apo B mRNA editing in a variety of cell lines. The advantage of this approach is the ability to detect apo B processing events in the transfected construct distinct from the endogenous transcript. These workers used either a 63-, 186-, or a 354-base pair (bp) region of human apo B cDNA flanking the edited base (6507 to 6860) annealed in midreading frame to an apo E cDNA under the transcriptional control of the Moloney murine leukemia virus long terminal repeat. Previous studies[39] by these workers demonstrated that the 354- but not the 63-bp construct was competent to undergo editing in Caco-2 cells, suggesting that the requisite length of apo B mRNA sequence to direct editing of codon 2153 was over 63 bps. These workers transfected 18 different cell lines with the 354-bp construct, using cells that contained endogenous apo B mRNA, such as Caco-2 and various rat and mouse liver cell lines, in which the transcript is edited albeit to varying degrees, and HepG2 cells where apo B mRNA is expressed but not edited. In addition, these workers transfected cells from a variety of embryonal origins where no endogenous apo B mRNA is expressed. The findings of this study indicate that the 354-bp construct was edited in cells that edit endogenous apo B mRNA, although details on the relative efficiency of exogenous vs. endogenous apo B mRNA editing and corresponding synthesis rates for apo B100 vs. apo B48 were not provided. Of greater interest was the observation that a number of cells of different embryonic origin, none of which express endogenous apo B, were competent in editing the 354-bp construct and in some instances, to secrete a truncated apo B protein. The implication of this finding is that numerous cells possess the editing machinery, and that apo B mRNA editing may represent one example of a process which is more widespread than previously thought.

C. TOPOLOGY OF APO B mRNA EDITING

The mechanism involved in apo B mRNA editing is unknown but several possibilities exist. The ready explanation for this single nucleotide change is that it involves a site-specific cytosine deamination to a uracil. Other possibilities include some form of base exchange reaction or modification such as occurs in certain tRNAs or the posttranscriptional insertion and deletion of residues which occurs in trypanosomes. Bostrom et al.[40] demonstrated that a synthetic radiolabeled apo B cRNA could be edited *in vitro* and the resulting RNA digested to its 5′-monophosphates and resolved by two-dimensional thin layer chromatography. The finding that the edited base comigrated with authentic uridine 5'-monophosphate indicates strongly that the process involves a cytosine deaminase. The question of where in the cell this modification occurs was addressed in an imaginative series of experiments from Lau et al.[41] These workers addressed this issue using the rat liver which is known to contain both edited and unedited apo B mRNA. The authors first prepared nuclear and cytoplasmic RNA and demonstrated that nuclear RNA contained less than 20% edited apo B mRNA while cytoplasmic apo B mRNA—either total or polysomal—contained approximately 60% edited species. Lau et al.[41] then went on to fractionate nuclear RNA into polyA + and polyA − RNA in order to ascertain the relationship between polyadenylation and editing of apo B mRNA. They found that polyA + RNA contained approximately 50% edited apo B mRNA while polyA − RNA contained only 10% edited species. These workers then identified an intron in the rat apo B genomic DNA sequence 5% to the edited base in a similar position to that reported for the human apo B gene. Using a primer downstream of the edited base and another primer at the 3' terminus of the upstream exon 25, Lau et al.[41] were able to use the polymerase chain reaction to amplify approximately 3 kb of apo B flanking the intron-exon junction in both genomic DNA and unspliced RNA. Splicing was shown to remove the 500-bp intron, and thus the amplification of a 2.5-kb fragment could be distinguished from the 3-kb fragment alluded to above. These workers demonstrated that unspliced nuclear pre-RNA contained less than 10% edited apo B mRNA and that unspliced polyA − RNA contained essentially no edited apo B mRNA. Thus, the conclusion from these experiments is that apo B mRNA editing is not a cotranscriptional process, but occurs predominantly in the nucleus and is completed by the time the RNA is spliced and polyadenylated. These findings have important implications for the identity and characteristics of the postulated editing machinery which should presumably be nuclear, although it is well recognized that passage of proteins and other macromolecules occurs through nuclear pores, which thus does not preclude the possibility of the editing machinery originating in the cytoplasmic compartment. The topology of apo B mRNA editing in other cell types, particularly the small intestine, has not been evaluated to date, at least in part because of difficulties associated with the preparation of intestinal nuclei and intact nuclear RNA.

III. MODULATION OF APO B mRNA EDITING BY HORMONAL AND NUTRITIONAL FACTORS

A. MODULATION BY THYROID HORMONE

As detailed above, studies by many workers demonstrated that the rat liver synthesizes and secretes both forms of apo B, but Davidson et al.[42] were the first to demonstrate that the proportions of these isoforms was subject to modulation following alterations in thyroid hormone status. Using rats made hypothyroid following dietary exposure to propylthiouracil, Davidson et al.[42]

treated groups of these animals with increasing doses of T3 and examined apolipoprotein gene expression in the liver and small intestine. These studies revealed a tissue-specific effect of thyroid hormone with regard to the expression of apolipoprotein A-I and A-IV genes which were transcriptionally upregulated in the liver but not the small intestine. Of interest was the effect of T3 administration on apo B synthesis in the liver. In contrast to the euthyroid and hypothyroid animals, animals treated with T3 had no detectable apo B100 synthesis following immunoprecipitation of radiolabeled hepatic cytosolic ($225,000 \times g$) supernatants. Following up on these observations, Davidson et al.[25] used a reverse-transcription/polymerase-chain reaction based strategy to amplify the region corresponding to the 3' terminus of rat apo B48 and demonstrated, by differential oligonucleotide hybridization, that the mechanism underlying the absence of hepatic apo B100 synthesis in the T3 treated animals was a switch from approximately 60/40 edited to unedited apo B mRNA to greater than 90% edited apo B mRNA. Thus apo B mRNA editing was demonstrated to be both a developmentally regulated event and to be modulated in the adult animal following hormonal manipulation. It bears emphasis that changes in rat hepatic apo B mRNA editing occur with only minor (less than twofold) changes in apo B mRNA abundance in all the settings that we have encountered. Subsequent studies[43] were undertaken to establish the functional consequences for hepatic triglyceride-rich lipoprotein assembly of the deletion of the carboxyl terminal half of apo B. Animals were prepared as before and groups of euthyroid, hypothyroid, and hyperthyroid rats used to prepare hepatic Golgi for analysis of the content lipoproteins. Using nascent intracellular Golgi VLDLs prepared from control, euthyroid animals demonstrated a polydisperse morphology with sizes averaging 419 Å while the analogous particles isolated from hypothyroid animals averaged only 160 Å. Previous work[42] suggested that the hypothyroid rat has a relative block in the secretion of hepatic triglyceride, and the findings of this study extend these general observations by demonstrating that lipoproteins isolated from a presecretory intracellular organelle are smaller and presumably underlipidated. By contrast, Golgi lipoproteins from hyperthyroid animals where apo B mRNA editing was greater than 90% demonstrated a normal morphology and size distribution with an average diameter of 328 Å. Analysis[43] of the protein components of these Golgi VLDLs by Western blotting revealed the presence of essentially only apo B48 in the hyperthyroid animals while both hypothyroid and control euthyroid animals demonstrated a large preponderance of apo B100. Thus, as would be anticipated from the results of studies of intestinal lipoprotein synthesis and secretion (where apo B48 is synthesized almost exclusively), apo B48 is competent to direct the synthesis and secretion of hepatic triglyceride-rich lipoproteins.

In regard to the mechanism of the T3-dependent modulation of hepatic apo B mRNA editing in the rat, other studies have attempted to address the role of the nuclear thyroid hormone receptor and potential intermediates of

T3 action, such as growth hormone, as the proximate mediators of the alterations of apo B mRNA editing. These were important considerations in view of the observation that, following T3 administration to hypothyroid rats, there was a lag period of 12 to 18 h before any changes were detectable in either hepatic apo B mRNA editing or apo B100 synthesis. Studies were conducted in hypothyroid rats treated with one of two synthetic analogs of T3.[44] These analogs were developed as hypocholesterolemic agents since it appeared that they were able to lower serum cholesterol levels in rats without precipitating the cardiac side effects common with authentic T3.[45] Both analogs chosen were able to bind to the hepatic (but not cardiac) nuclear thyroid hormone receptor with reduced but comparable affinity to native T3. However, one analog demonstrated approximately 20 times the bioactivity of the other. Following administration of these agents to hypothyroid rats, hepatic apo B mRNA editing was examined and found to be increased with both the "active" and "inactive" analog. Thus, binding to the nuclear thyroid hormone receptor was considered an important component of the induction of apo B mRNA editing following thyroid hormone administration.[44] Induction of apo B mRNA editing following T3 administration was accompanied by coordinate pretranslational increases in hepatic apo A-I and A-IV gene expression. Administration of T3 analogs produced a less marked increase in hepatic mRNA abundance for these genes (3- to 5-fold compared to 6- to 14-fold for native T3) and a variable increase in hepatic mRNA abundance for several prototypic T3-responsive genes, such as malic enzyme and spot 11. Thus the induction of hepatic apo B mRNA editing occurs somewhat independently from the hormonal induction of other classes of T3 responsive genes.[44] When hypothyroid animals were treated with pharmacologic doses of growth hormone there was no change in the proportions of edited and unedited apo B mRNA.[44] This was despite effective reversal of the growth arrest which accompanies hypothyroidism in the rat. Thus growth hormone was considered, by itself, insufficient to mediate the hormonal modulation of apo B mRNA editing. Recent studies[46] using hypophysectomized rats treated with various combinations of growth hormone and T4 have suggested a facilitative role for growth hormone in mediating apo B mRNA editing both in the whole animal and in isolated hepatocytes, but clearly more work is needed in this area. A possible clue in this regard may come from preliminary work[47] where the nuclear thyroid hormone receptor was transfected into Caco-2 cells and apo B mRNA editing found to be modulated. However, this work requires additional confirmation. Finally, the role of thyroid hormone in the developmental regulation of both intestinal and hepatic apo B mRNA editing needs clarification. In the rat, hepatic apo B mRNA editing is predominantly a postnatal event,[29] and it is tempting to speculate that thyroid hormone may undergo developmental increases in parallel with the induction of apo B mRNA editing. This remains to be demonstrated, however. It is unlikely that thyroid hormone plays a role in the development or maintenance of intestinal apo B mRNA editing since the changes in apo B mRNA editing in both the rat and human small intestine precede any developmental increase in fetal thyroid hormone activity and maternal T3 does not cross the placenta. Additionally in the adult rat small intestine, there was no change in apo B mRNA editing following changes in thyroid hormone status.[25]

B. MODULATION OF APO B mRNA EDITING FOLLOWING CHANGES IN HEPATIC LIPOGENESIS

Thyroid hormone modulation of hepatic apo B mRNA editing was accompanied by striking alterations in hepatic lipid mobilization.[42] Since there are well characterized examples of genes which display response to both T3 and high carbohydrate intake, such as malic enzyme and spot 11 and spot 14 (reviewed in Reference 44), it was of interest to examine the possibility that the gene(s) responsible for mediating apo B mRNA editing in the rat liver may represent an example of this general class of response. Against this idea is the previous observation that changes in apo B mRNA editing show a longer lag period following T3 administration (12 to 18 h) than is generally encountered with this class of responsive genes (which often show responses within minutes to a few hours), and, secondly, that apo B mRNA editing has not been reproduced *in vitro* following T3 administration. Baum et al.[48] studied groups of rats fasted for 24 or 48 h and other groups fasted for 48 h and subsequently fed a high carbohydrate, fat-free diet for 24 or 48 h. This maneuver produced a 30-fold range of hepatic triglyceride concentration from a nadir at 48 h fasting to a peak at 48 h refeeding a high carbohydrate diet. In association with these changes, apo B synthesis rates were determined following intraportal vein administration of tritiated leucine and quantitative immunoprecipitation of apo B. In animals fasted for 48 h there was a decrease in the ratio of apo B48 to apo B100 synthesis and a corresponding decrease in hepatic apo B mRNA editing. In animals fasted for 48 h and then refed a high carbohydrate diet for either 24 or 48 h, there was no apo B100 synthesis detectable and hepatic apo B mRNA was greater than 90% edited. In association with these findings, serum apo B isomorphs as demonstrated on Western blots, were found to be altered in parallel such that control animals and animals fasted for 24 or 48 h demonstrated mostly apo B100 in their serum while animals fasted and refed a high carbohydrate diet demonstrated essentially only apo B48. This latter finding[48] clearly reflects a number of events including synthetic and catabolic processes, but as a maneuver to alter circulating levels of apo B100, these observations suggest an important avenue of investigation. When all animals were included in an analysis of the impact of apo B mRNA editing on hepatic apo B100 synthesis rates, there was a striking correlation found, suggesting that hepatic apo B mRNA editing is an extremely important determinant of this process *in vivo*. Studies[48] were extended to an evaluation of the regulation of other hepatic apolipoprotein genes in the setting of the changes in hepatic triglyceride metabolism alluded to above. It was found that apo A-I and A-IV mRNA abundance and protein biosynthesis were increased approximately two- to fourfold in the animals refed a high carbohydrate diet for 48 h. These changes are thus of a lesser magnitude but in the same direction as encountered[42] with T3 treatment. The time course of these changes has not been fully characterized since the earliest time point examined was 24 h after exposure to a high-carbohydrate diet and the experiment only continued up to 48 h of refeeding. Thus, important questions still remain, such as the earliest point that a response to carbohydrate is detectable and whether equivalent responses can be obtained with different carbohydrate sources such as fructose and glucose. Additionally, it will be of interest to examine longer periods of exposure to a high-carbohydrate diet to determine whether the liver is capable of effectively unloading the accumulated lipid after sustained intake of this modified isocaloric diet. Finally it will be important to evaluate the synthesis of nascent intrahepatic lipoproteins by isolating Golgi VLDLs and examining their morphology, lipid content, and apo B isomorph distribution. These studies are currently underway in the laboratory.

C. MODULATION OF APO B mRNA EDITING BY OTHER MANIPULATIONS

Since the first report that hepatic apo B mRNA editing was modulated by thyroid hormone, several investigators have examined other potential routes of manipulating this activity. Two recent reports[49,50] are of interest in this regard. Work from Glickman's laboratory[49] has examined the regulation of both hepatic and intestinal apolipoprotein gene expression following the administration of ethinyl estradiol to rats. This agent produces a profound hypocholesterolemia in rats which is thought to be secondary to overexpression of hepatic LDL receptors. These workers found that administration of this agent to rats for 5 d produced a decrease in hepatic but not small intestinal apo B mRNA editing and a corresponding increase in the relative synthesis rates of apo B100. In keeping with the findings of Baum et al.,[48] alluded to above, these workers found a strong correlation between hepatic apo B mRNA editing and apo B100 synthesis. The possible contribution of prolonged fasting to the decrease observed in hepatic apo B mRNA editing in this model was explored and found not to be a factor in this response although the precise mechanism remains obscure. A second report[50] was presented in abstract form recently and describes the changes in hepatic apo B mRNA editing in noninsulin-dependent diabetes mellitus. This report describes a decrease in hepatic apo B mRNA editing in uncontrolled noninsulin-dependent diabetes in rats. The report raises the larger and more interesting possibility that insulin may play a role in the regulation of hepatic apo B mRNA editing distinct from its well characterized effects on the posttranslational processing of nascent apo B100 and B48.

IV. *IN VITRO* EDITING OF APO B mRNA

A. METHODOLOGICAL CONSIDERATIONS

The major question for investigators examining the biology of apo B mRNA editing is the underlying mechanism of this interesting process. As indicated above the most likely explanation is a site-specific deamination of the cytosine at nucleotide 6666, but the basis for this conclusion was attained experimentally as a result of a technical advance pioneered by Driscoll et al.,[51] which described an *in vitro* system to examine apo B mRNA editing. This sytem used cytoplasmic S100 supernatants prepared from McA7777 cells, a rat hepatoma line which synthesizes and secretes both apo B100 and apo B48 and which therefore provides a ready source of "editing activity" to examine possible *in vitro* RNA modification. Driscoll et al.[51] incubated McA7777 extracts with various lengths of synthetic apo B RNA prepared from the region flanking the edited base. In order to detect the single nucleotide change, these workers devised an ingenious scheme whereby an antisense apo B oligonucleotide (35-mer) is annealed downstream from the edited base at position 6666 and extended with reverse transcriptase in the presence of dATP, dCTP, dTTP, and dideoxy-GTP as a chain terminator. There being no other "C" residues between the 5' end of the antisense oligonucleotide and the edited base, the primer is extended until it reaches the first upstream "C". If nucleotide 6666 is unedited the primer undergoes chain termination and produces an extension product of 42 bp. If nucleotide 6666 is edited to a "U" then the primer extends to the next upstream "C", at 6655 in human apo B cDNA and 6661 in rat and mouse apo B cDNA. The extended and terminated primers are resolved effectively by urea-acrylamide electrophoresis. In the original description by Driscoll et al.,[51] *in vitro* editing was demonstrated with S100 extracts from McA7777 cells, but the efficiency of editing was less than 5% of the input RNA. The reasons for the low activity are of interest since endogenous apo B mRNA from these cells is edited approximately 20%. One possibility is that subtle sequence differences between human and rat apo B cDNA may be of importance (see below and Figure 2). There may be additional factors, however, since an intriguing observation is that McA7777 cells transfected with a human full-length apo B cDNA edit a substantial proportion of the human apo B mRNA (greater than 50%) while the endogenous transcript is edited less than half of this value.[52,53] Thus issues relating to differential accessibility and transcript stability require consideration. The relatively low level of *in vitro* apo B mRNA editing observed with McA7777 cells has been observed by others in nuclear extracts prepared from rat liver.[41] The question arises as to whether this particular source (i.e., rat hepatic or hepatoma-derived nuclear or S100 extract) contains relatively low levels of editing activity or whether the purification scheme used is inappropriate. Investigators using rat liver nuclear extracts to examine *in vitro* editing have had to optimize their methods for

A. Aligned nucleotide sequences

```
        6482                                        6522
Human   CCCACAGCAAGCTAATGATTATCTGAATTCATTCAATTGGGAGAGACAAGTTTCACATGC
Pig     -----------T---------------G-A------G------A----------TGAG---
Rabbit  -----------T---------------G-A---------------------------CAG---
Rat     T--T-----GAT-C--------------G---CTG-C-------------AG-TGG---
Mouse   T--T-----GAT-C--C-----------G---CTG-C-------------AG-TGG---

        6542                                        6582
Human   CAAGGAGAAACTGACTGCTCTCACAAAAAAGTATAGAATTACAGAAAATGATATACAAAT
Pig     ---AA------AT-G--A-T---TGG--G-T---------------------G---G---
Rabbit  -----------A---A--T----------T----A---------G-------------C
Rat     ------A---T-A---T--T---TGG----C---------------T------G---T---
Mouse   ------A---A-A---T--T---TGG----T---------------T------G---T---

        6602                                        6642
Human   TGCATTAGATGATGCCAAAATCAACTTTAATGAAAAACTATCTCAACTGCAGACATATAT
Pig     ---------CA--------------C--------------A----G--A--A------G-
Rabbit  ------G---A-----------------A-----------G----------T---------G-
Rat     ---C-----AG--------------G----------C--------TG-------CGC
Mouse   ---CA-----AG--------------C----------C--------TG-------CGC

        6662                                        6702
Human   GATACAATTTGATCAGTATATTAAAGATAGTTATGATTTACATGATTTGAAAATAGCTAT
Pig     -----------------------------A-----------------T--G-C------
Rabbit  ---------------------------A--T----C----------T----------
Rat     -----------------------G-----A--------GC---G--C--A----G-A----
Mouse   ---------------------------A-------CC------C--A----G-A----

        6722                                        6762
Human   TGCTAATATTATTGATGAAATCATTGAAAAATTAAAAAGTCTTG
Pig     -----GG-------------------C--C---G----T-----
Rabbit  A----G------A---C-------G-------------T-----
Rat     ----C-G---------AG------------GC------TG----
Mouse   ----G-G---------CG-----------G-------TG----
```

B. Aligned amino acid sequences

```
        2092                                        2132
Human   PQQANDYLNSFNWERQVSHAKEKLTALTKKYRITENDIQIALDDAKINFNEKLSQLQTYM
Pig     ---V----ST-S-----LS--K-HSDFMED-------VR----N----L----T----V
Rabbit  ---V----ST--------S------TF--N-K-------T---N----L----------V
Rat     ---IH----ASD-----AG------SFMEN----D--VL----S----L-------E--A
Mouse   ---IHH---ASD-----AG----I-SFMEN----D--VL -I-S-----------E--A

        2152                                        2182
Human   IQFDQYIKDSYDLHDLKIAIANIIDEIIEKLKSL
Pig     ---------N-----F-T---R------AT--I-
Rabbit  ---------NF----F-----S---Q-M----I-
Rat     -------R-N--AQ---RT--Q---R------M-
Mouse   ---------N--P----RT--E---R------M-
```

FIGURE 2. (A) Aligned nucleotide sequence of apo B cDNA flanking the edited base. Four mammalian apo B cDNAs are contrasted to the human sequence with identity represented by a hyphen.[30] (B) Aligned amino acid sequence deduced from four mammalian apo B cDNAs in comparison to human apo B flanking the edited base. The single letter code is employed for amino acid identity.[30]

reliable detection of low levels of apo B mRNA editing which in some cases are less than 1 to 2% of input RNA.[54] Studies from a number of labs indicate that the yield and functional activity of small intestinal S100 extracts is considerably higher than that reported for the rat liver. Using isolated rabbit,[40] rat,[55] or baboon[56] enterocyte S100 extracts, workers have reported *in vitro* conversion rates of ~10 to 40% of input RNA. Whether this editing activity is developmentally induced in the small intestine awaits further study. There appear to be some inconsistencies between different reports concerning the optimal conditions for *in vitro* editing with the original report describing a cryolabile activity which required 50 mM EDTA.[51] Later reports have found a number of differences from the earlier study, and most recently Greeve et al.[55] have detailed extensive findings in regard to the characterization of rat enterocyte S100 editing activity. The major conclusions emerging from this and other reports is that the editing activity is most likely a protein with no cofactor or additional RNA requirements. Attempts to purify this editing activity from crude enterocyte S100 extracts have resulted in severalfold enrichment, but no consensus has emerged as to its apparent molecular mass. Driscoll has suggested that baboon enterocyte S100 extracts contain an editing activity of approximately 125 kDa.[56] Another controversial aspect of the *in vitro* editing methodology is whether apo B mRNA assembles into a higher order molecular complex following incubation with a source of editing activity. Following an initial report of this phenomenon,[57] others[55] have failed to confirm the existence of a so-called "editosome". Distinct from the editing activity, there appears to be an apo B mRNA binding activity in rat liver nuclear extracts.[58] This binding activity has an apparent molecular weight of 40 kDa following UV cross-linking and specifically binds a 26-nucleotide apo B RNA fragment spanning the edited base. It will be of interest to investigate the relationship of this apparent RNA binding protein to the editing machinery by way of mutagenesis studies in which editing is abolished. Species considerations may be relevant in this regard also, since studies[59] from our laboratory have indicated that chick small intestinal enterocytes contain an editing enhancing factor(s), and studies are currently underway to purify and compare these factor(s) to the binding activity found in rat tissue.

B. SEQUENCE REQUIREMENTS FOR APO B mRNA EDITING

Several studies have addressed this issue following the technical advances alluded to above. A general point of interest is that the nucleotide sequence spanning the edited base at position 6666 is highly conserved in mammalian apo B cDNAs (Figure 2). In regard to the minimal length of sequence information, there is some confusion in the literature most likely reflecting experimental differences. Davies et al.[60] determined that a 26-nucleotide region of human apo B RNA flanking the edited base was as efficiently edited as the endogenous (rat) transcript when transfected into McA7777 cells. This result is at variance with data suggesting that Caco-2 cells edit a 354- and 186- but not a 63 bp chimeric construct following stable transfection.[39,40] The reasons for this discrepancy are unclear. Recent work from Scott's laboratory[61] has demonstrated that a human apo B RNA sequence of 26 nucleotides can be edited *in vitro* using S100 extracts prepared from rat enterocytes. This is of interest since earlier studies from this group[51] indicated that a 55- but not a 26-nucleotide fragment could be edited *in vitro* using S100 extracts prepared from McA7777 cells. Thus it is likely that there are true biological differences in editing activity within different cell types which may, additionally, be more or less preserved during the preparation of S100 extracts. In regard to the nucleotide specificity for apo B mRNA editing, recent studies have provided some important advances. Chen et al.[54] used a coupled transcription editing reaction to demonstrate that in a series of mutations in a 9-nucleotide sequence flanking the edited base, in only 2 of 22 such mutants (a transversion mutant changing the wild type ATACAATTT to ATTCTATTT and a deletion mutant ATAC-TTTT) was editing abolished. The authors contend that this indicates a relatively lax sequence specificity for apo B mRNA editing. An important limitation of this study was the low level of editing demonstrated by rat liver nuclear extracts which was less than 2% of the input apo B RNA. In order to investigate whether more distant sequence information conferred important specificity Shah et al.[61] systematically mutagenized a 55-bp region of human apo B cDNA and demonstrated that the minimal sequence information required for *in vitro* editing by rat enterocyte S100 extracts is contained with a 26-nucleotide stretch flanking the edited base. However, these workers found an 11-nucleotide region downstream of position 6666 in which all but one point mutations abolished editing. This result is of interest since recent work from our laboratory[59] has demonstrated that chick apo B mRNA is not edited *in vivo* by small intestinal enterocytes, and that the chick apo B cDNA sequence is divergent from mammalian apo B cDNAs with 3 changes from the conserved mammalian sequence noted in the 11-nucleotide cassette described by Shah et al.[61] These changes would be predicted to abolish *in vitro* apo B RNA editing according to the data of Shah et al., and, indeed, no editing could be detected when chick apo B RNA was incubated *in vitro* with S100 extracts prepared from rat enterocytes while human, rat, and porcine apo B RNAs were all efficiently edited. A further point of interest in regard to the sequence specificity of apo B RNA editing concerns the demonstration that a second edited base is detectable in human apo B RNA at position 6802 which changes threonine (ACA) to isoleucine (AUA) at codon 2198.[62] This is of no functional consequence since the transcript is terminated upstream of this nucleotide in all instances examined. The finding is of interest in view of the sequence homology flanking both edited sites, and it reinforces the hypothesis that the requisite information for editing is contained within the 26 nucleotides flanking position 6666.

V. SUMMARY AND CONCLUSIONS

Apo B mRNA editing is the first example of a process which expands the repertoire of vertebrate gene regulation. It is likely that the basis for this interesting process will be of importance in other contexts particularly in view of the demonstration that apo B mRNA editing is detectable in cells which contain only trivial amounts of apo B mRNA. Areas that will provide interesting information in regard to the biology of apo B mRNA editing will be to determine the tissue- and cell-specific cues which regulate this process in response to development and to hormonal and nutritional modulation. These issues are currently under investigation and will form the focus of future reports.

ACKNOWLEDGMENTS

Work cited from the author's laboratory was supported by Grants HL-38180, HL KO-4 02166 from the NHLBI and DK-42086 from NIDDK. I would like to recognize the contributions of BaBie Teng, Charles Baum, Ruth Carlos, and Annalise Hausman to the work presented in this review. This chapter is dedicated to my father, Jack Davidson, M.B. Ch.B., on the occasion of his 75th birthday.

REFERENCES

1. **Chan, L., VanTuinen, P., Ledbetter, D. H., Daiger, S. P., Gotto, A. M., and Chen, S. H.,** The human apolipoprotein B-100 gene: a highly polymorphic gene that maps to the short arm of chromosone 2, *Biochem. Biophys. Res. Comm.,* 133, 248, 1985.
2. **Mehrabian, M., Sparkes, R. S., Mohandas, T., Klisak, I. J., Schumaker, V. N., Heinzmann, C., Zollman, S., Ma, Y., and Lusis, A. J.,** Human apolipoprotein B: chromosmal mapping and DNA polymorphisms of hepatic and intestinal species, *Somatic Cell Mol. Genet.,* 12, 245, 1986.
3. **Luo, C. C., Li, W. H., Moore, M. N., and Chan, L.,** Structure and evolution of the apolipoprotein multigene family, *J. Mol. Biol.,* 187, 325, 1986.
4. **Li, W. H., Tanimura, M., Luo, C. C., Datta, S., and Chan, L.,** The apolipoprotein multigene family: biosynthesis, structure, structure-function relationships, and evolution, *J. Lipid Res.,* 29, 245, 1988.
5. **Knott, T. J., Pease, R. J., Powell, L. M., Wallis, S. C., Rall, S. C., Innerarity, T. L., Blackhart, D., Taylor, W. H., Marcel, Y., Milne, R., Johnson, D., Fuller, M., Lusis, A. J., McCarthy, B. J., Mahley, R. W., Levy-Wilson, B., and Scott, J.,** Complete protein sequence and identification of structural domains of human apolipoprotein B, *Nature (London),* 323, 734, 1986.
6. **Yang, C. Y., Chen, S. H., Gianturco, S. H., Bradley, W. A., Sparrow, J. T., Tanimura, M., Li, W. H., Sparrow, D. A., DeLoof, H., Roseneu, M., Lee, F. S., Gu, Z. W., Gotto, A. M., and Chan, L.,** Sequence, structure, receptor-binding domains and internal repeats of human apolipoprotein B-100, *Nature (London),* 323, 738, 1986.
7. **Chen, S. H., Yang, C. Y., Chen, P. F., Setzer, D., Tanimura, M., Li, W. H., Gotto, A. M., and Chan, L.,** The complete cDNA and amino acid sequence of human apolipoprotein B-100, *J. Biol. Chem.,* 261, 12918, 1986.
8. **Cladaras, C., Hadzopoulou-Cladaras, M., Nolte, R. T., Atkinson, D., and Zannis, V. I.,** The complete sequence and structural analysis of human apolipoprotein B-100: relationship between apoB-100 and apoB-48 forms, *EMBO J.,* 5, 3495, 1986.
9. **Ludwig, E. H., Blackhart, B. D., Pierotti, V. R., Caiati, L., Fortier, C., Knott, T., Scott, J., Mahley, R. W., Levy-Wilson, B., and McCarthy, B. J.,** DNA sequencing of the human apolipoprotein B gene, *DNA,* 6, 363, 1987.

10. **Innerarity, T. L., Young, S. G., Poksay, K. S., Mahley, R. W., Smith, R. S., Milne, R. W., Marcel, Y. L., and Weisgraber, K. H.,** Structural relationship of human apolipoprotein B48 to apolipoprotein B100, *J. Clin. Invest.,* 80, 1794, 1987.

11. **Hardman, D. A., Protter, A. A., Chen, G. C., Schilling, J. W., Sato, K. Y., Lau, K., Yamanaka, M., Mikita, T., Miller, J., Crisp, T., McEnroe, G., Scarborough, R. M., and Kane, J. P.,** Structural comparison of human apolipoproteins B-48 and B-100, *Biochemistry,* 26, 5478, 1987.

12. **Yang, C. Y., Kim, T. W., Weng, S. A., Lee, B., Yang, M., and Gotto, A. M., Jr.,** Isolation and characterization of sulfhydryl and disulfide peptides of human apolipoprotein B-100, *Proc. Natl. Acad. Sci. U.S.A.,* 87, 5523, 1990.

13. **Law, A. and Scott, J.,** A cross-species comparison of the apolipoprotein B domain that binds to the LDL receptor, *J. Lipid Res.,* 31, 1109, 1990.

14. **Pease, R. J., Milne, R. W., Jessup, W. K., Law, A., Provost, P., Fruchart, J. C., Dean, R. T., Marcel, Y. L., and Scott, J.,** Use of bacterial expression cloning to localize the epitopes for a series of monoclonal antibodies against apolipoprotein B-100, *J. Biol. Chem.,* 265, 553, 1990.

15. **Herz, J., Hamann, U., Rogne, S., Myklebost, O., Gausepohl, H., and Stanley, K. K.,** Surface location and high affinity for calcium of a 500-kd liver membrane protein closely related to the LDL-receptor suggest a physiological role as lipoprotein receptor, *EMBO J.,* 7, 4119, 1988.

16. **Kowal, R. C., Herz, J., Goldstein, J. L., Esser, V., and Brown, M. S.,** Low density lipoprotein receptor-related protein mediates uptake of cholesterol esters derived from apoprotein E-enriched lipoproteins, *Proc. Natl. Acad. Sci. U.S.A.,* 86, 5810, 1989.

17. **Beisegel, U., Weber, W., Ihrke, G., Herz, J., and Stanley, K. K.,** The LDL-receptor-related protein, LRP, is an apolipoprotein E-binding protein, *Nature (London),* 341, 162, 1989.

18. **Cattaneo, R.,** Different types of messenger RNA editing, *Annu. Rev. Genet.,* 25, 71, 1991.

19. **Sollner-Webb, B.,** RNA editing, *Curr. Op. Cell Biol.,* 3, 1056, 1991.

20. **Blum, B., Bakalara, N., and Simpson, L.,** A model for RNA editing in kinetoplastid mitochondria: "Guide" RNA molecules transcribed from maxicircle DNA provide the edited information, *Cell,* 60, 189, 1990.

21. **Powell, L. M., Wallis, S. C., Pease, R. J., Edwards, Y. H., Knott, T. J., and Scott, J.,** A novel form of tissue-specific RNA processing produces apolipoprotein B48 in intestine, *Cell,* 50, 831, 1987.

22. **Chen, S. H., Habib, G., Yang, C. Y., Gu, Z. W., Lee, B. R., Weng, S., Silberman, S. R., Cai, S. J., Deslypere, J. P., Rosseneu, M., Gotto, A. M., Li, W. H., and Chan, L.,** Apolipoprotein B-48 is the product of a messenger RNA with an organ-specific in-frame stop codon, *Science,* 238, 363, 1987.

23. **Hospattanker, A. V., Higuchi, K., Law, S. W., Meglin, N., and Brewer, H. B.,** Identification of a novel in-frame translational stop codon in human intestine ApoB mRNA, *Biochem. Biophys. Res. Commun.,* 148, 279, 1987.

24. **Sommer, B., Kohler, M., Sprengel, R., and Seeburg, P. H.,** RNA editing in brain controls a determinant of ion flow in glutamate-gated channels, *Cell,* 67, 11, 1991.

25. **Davidson, N. O., Powell, L. M., Wallis, S. C., and Scott, J.,** Thyroid hormone modulates the introduction of a stop codon in rat liver apolipoprotein B messenger RNA, *J. Biol. Chem.,* 263, 13482, 1988.

26. **Tennyson, G. E., Sabatos, C. A., Higuchi, K., Meglin, N., Brewer, H. B.,** Expression of apolipoprotein B mRNAs encoding higher- and lower-molecular weight isoproteins in rat liver and intestine, *Proc. Natl. Acad. Sci. U.S.A.,* 86, 500, 1989.

27. **Higuchi, K., Hospattankar, A. V., Law, S. W., Meglin, N., Cortright, J., and Brewer, H. B.,** Human apolipoprotein B (apo B) mRNA: identification of two distinct apo B mRNAs, an mRNA with the apo B-100 sequence and an apoB mRNA containing a premature in-frame translational stop codon, in both liver and intestine, *Proc. Natl. Acad. Sci. U.S.A.,* 85, 1772, 1988.

28. **Teng, B., Verp, M., Salomon, J., and Davidson, N. O.,** Apolipoprotein B messenger RNA editing is developmentally regulated and widely expressed in human tissues, *J. Biol. Chem.,* 265, 20616, 1990.

29. **Wu, J. H., Semenkovich, C. F., Chen, S. H., Li, W. H., and Chan, L.,** Apolipoprotein B mRNA editing: validation of a sensitive assay and developmental biology of RNA editing in the rat, *J. Biol. Chem.,* 265, 12312, 1990.

30. **Teng, B., Black, D. D., and Davidson, N. O.,** Apolipoprotein B messenger RNA editing is developmentally regulated in pig small intestine: nucleotide comparison of apolipoprotein B editing regions in five species, *Biochem. Biophys. Res. Comm.,* 173, 74, 1990.

31. **Glickman, R. M., Rogers, M., and Glickman, J. N.,** Apolipoprotein B synthesis by human liver and intestine in vitro, *Proc. Natl. Acad. Sci. U.S.A.,* 83, 5296, 1986.

32. **Levy, E., Rochette, C., Londono, I., Roy, C. C., Milne, R. W., Marcel, Y. L., and Bendayan, M.,** Apolipoprotein B-100: immunolocalization and synthesis in human intestinal mucosa, *J. Lipid Res.,* 31, 1937, 1990.

33. **Hoeg, J. M., Sviridov, D. D., Tennyson, G. E., Demosky, S. J., Meng, M. S., Bojanovski, D., Safonova, I. G., Repin, V. S., Kuberger, M. B., Smirnov, V. N., Higuchi, K., Gregg, R. E., and Brewer, H. B.,** Both apolipoproteins B-48 and B-100 are synthesized and secreted by the human intestine, *J. Lipid Res.,* 31, 1761, 1990.

34. **Dixon, J. L., Furukawa, S., and Ginsberg, H. N.,** Oleate stimulates secretion of apolipoprotein B-containing lipoproteins from Hep G2 cells by inhibiting early intracellular degradation of apolipoprotein B, *J. Biol. Chem.,* 266, 5080, 1991.

35. **Coleman, R. A., Haynes, E. B., Sand, T. M., and Davis, R. A.,** Developmental coordinate expression of triacylglycerol and small molecular weight apoB synthesis and secretion by rat hepatocytes, *J. Lipid Res.,* 29, 33, 1988.

36. **Jiao, S., Moberly, J. B., and Schonfeld, G.,** Editing of apolipoprotein B messenger RNA in differentiated Caco-2 cells, *J. Lipid Res.,* 31, 695, 1990.

37. **Demmer, L. A., Levin, M. S., Elovson, J., Reuben, M. A., Lusis, A. J., and Gordon, J. I.,** Tissue-specific expression and developmental regulation of the rat apolipoprotein B gene, *Proc. Natl. Acad. Sci. U.S.A.,* 83, 8102, 1986.

38. **Hopkins, B., Brice, A. L., Schofield, P. N., Baralle, F. E., and Graham, C. F.,** Identity of cells containing apolipoprotein B messenger RNA, in 6- to 12-week postfertilization human embryos, *Development,* 100, 83, 1987.

39. **Bostrom, K., Lauer, S. J., Poksay, K. S., Garcia, Z., Taylor, J. M., and Innerarity, T. L.,** Apolipoprotein B48 RNA editing in chimeric apolipoprotein EB mRNA, *J. Biol. Chem.,* 264, 15701, 1989.

40. **Bostrom, K., Garcia, Z., Poksay, K. S., Johnson, D. F., Lusis, A. J., and Innerarity, T. L.,** Apolipoprotein B mRNA editing: direct determination of the edited base and occurrence in non-apolipoprotein B producing cell lines, *J. Biol. Chem.,* 265, 22446, 1990.

41. **Lau, P. P., Xiong, W., Zhu, H. J., Chen, S. H., and Chan, L.,** Apolipoprotein B mRNA editing is an intracellular event that occurs posttranscriptionally coincident with splicing and polyadenylation, *J. Biol. Chem.,* 266, 20550, 1991.

42. **Davidson, N. O., Carlos, R. C., Drewek, M. J., and Parmer, T. G.,** Apolipoprotein gene expression in the rat is regulated in a tissue-specific manner by thyroid hormone, *J. Lipid Res.,* 29, 1511, 1988.

43. **Davidson, N. O., Carlos, R. C., Sherman, H. L., and Hay, R. V.,** Modulation of apolipoprotein B-100 mRNA editing: effects on hepatic very low density lipoprotein assembly and intracellular apoB distribution in the rat, *J. Lipid Res.,* 31, 899, 1990.

44. **Davidson, N. O., Carlos, R. C., and Lukaszewicz, A. M.,** Apolipoprotein B mRNA editing modulated by thyroid hormone analogs but not growth hormone administration in the rat, *Mol. Endocrinol.,* 4, 779, 1990.

45. **Underwood, A. H., Emmett, J. C., Ellis, D., Flynn, S. B., Leeson, P. D., Benson, G. M., Novelli, R., Pearce, N. J., and Shah, V. P.,** A thyromimetic that decreases plasma cholesterol levels without increasing cardiac activity, *Nature (London),* 324, 425, 1986.

46. **Sjoberg, A., Oscarsson, J., Bostrom, K., Innerarity, T. L., Eden, S., and Olofsson, S. O.,** Growth hormone participates in the regulation of the editing of apolipoprotein B mRNA in the rat liver, *Arterioscler. Thromb.,* 11, 1402a, 1991.

47. **Patterson, A. P., Eggerman, T. L., Demosky, S. J., and Brewer, H. B.,** T_3 Nuclear receptor and growth hormone modulate apolipoprotein B editing, *Arteriosclerosis and Thrombosis,* 11, 1403a, 1991.

48. **Baum, C. L., Teng, B., and Davidson, N. O.,** Apolipoprotein B messenger RNA editing in the rat liver: modulation by fasting and refeeding a high carbohydrate diet, *J. Biol. Chem.,* 265, 19263, 1990.

49. **Seishima, M., Bisgaier, C. L., Davies, S. L., and Glickman, R. M.,** Regulation of hepatic apolipoprotein synthesis in the 17 alpha-ethinyl estradiol-treated rat, *J. Lipid Res.,* 32, 941, 1991.

50. **Jiao, S., Yanagi, K., Shimomura, I., Kobatake, T., Keno, Y., Kubo, M., Tokunaga, K., Matsuzawa, Y., and Tarui, S.,** Enhancement of apoB mRNA editing in the liver of non-insulin-dependent diabetic rats, *Arteriosclerosis and Thrombosis,* 11, 1424a, 1991.

51. **Driscoll, D. M., Wynne, J. K., Wallis, S. C., and Scott, J.,** An *in vitro* system for the editing of apolipoprotein B mRNA, *Cell,* 58, 519, 1989.

52. **Blackhart, B. D., Yao, Z., and McCarthy, B. J.,** An expression system for human apolipoprotein B100 in a rat hepatoma cell line, *J. Biol. Chem.,* 265, 8358, 1990.

53. **Yao, Z., Blackhart, B. D., Linton, M. F., Taylor, S. M., Young, S. G., and McCarthy, B. J.,** Expression of carboxyl-terminally truncated forms of human apolipoprotein B in rat hepatoma cells: Evidence that the length of apolipoprotein B has a major effect on the buoyant density of the secreted lipoproteins, *J. Biol. Chem.,* 266, 3300, 1991.

54. **Chen, S. H., Li, X., Liao, W. S., Wu, J. H., and Chan, L.,** RNA editing of apolipoprotein B mRNA: sequence specificity determined by in vitro coupled transcription editing, *J. Biol. Chem.,* 265, 6811, 1991.

55. **Greeve, J., Navaratnam, N., and Scott, J.,** Characterization of the apolipoprotein B mRNA editing enzyme: no similarity to the proposed mechanism of RNA editing in kinetoplasid protozoa, *Nucleic Acids Res.,* 19, 3569, 1991.

56. **Driscoll, D. M. and Casanova, E.,** Characterization of the apolipoprotein B mRNA editing activity in enterocyte extracts, *J. Biol. Chem.,* 265, 21401, 1990.

57. **Smith, H. C., Kuo, S. R., Backus, J. W., Harris, S. G., Sparks, C. E., and Sparks, J. D.,** In vitro apolipoprotein B mRNA editing: Identification of a 27S editing complex, *Proc. Natl. Acad. Sci. U.S.A.,* 88, 1489, 1991.

58. **Lau, P. P., Chen, S. H., Wang, J. C., and Chan, L.,** A 40 kilodalton rat liver nuclear protein binds specifically to apolipoprotein B mRNA around the RNA editing site, *Nucleic Acids Res.,* 18, 5817, 1991.

59. **Teng, B. B. and Davidson,** Evolution of intestinal apolipoprotein B mRNA editing: chicken apolipoprotein B mRNA is not edited but chicken enterocytes contain *in-vitro* editing enhancement factor(s), *J. Biol. Chem.,* in press, 1992.

60. **Davies, M. S., Wallis, S. C., Driscoll, D. M., Wynne, J. K., Williams, G. W., Powell, L. M., and Scott, J.,** Sequence requirements for apolipoprotein B RNA editing in transfected rat hepatoma cells, *J. Biol. Chem.,* 264, 13395, 1989.

61. **Shah, R. R., Knott, T. J., Legros, J. E., Navaratnam, N., Greeve, J. C., and Scott, J.,** Sequence requirements for the editing of apolipoprotein B mRNA, *J. Biol. Chem.,* 266, 16301, 1991.

62. **Navaratnam, N., Patel, D., Shah, R. R., Greeve, J. C., Powell, L. M., Knott, T. J., and Scott, J.,** An additional editing site present in apolipoprotein B mRNA, *Nucleic Acids Res.,* 19, 1741, 1991.

Chapter 14

REGULATION OF THE 68-kDa HEPATIC CARNITINE PALMITOYLTRANSFERASE

Paul S. Brady, Roderick A. Barke, and Linda J. Brady

TABLE OF CONTENTS

0-8493-6961-4/93/$0.00 + $.50
© 1993 by CRC Press, Inc.

I. INTRODUCTION

A. GENERAL ASPECTS OF FATTY ACID OXIDATION IN LIVER

When a long-chain fatty acid enters the liver, a number of metabolic fates are available to it (Figure 1). The major fate of a long-chain fatty acid, such as palmitic acid, is activation by esterification with coenzyme A. This is catalyzed by one member of the family of acyl CoA synthetases. A minor fate of long chain fatty acids, which apparently does not require this initial CoA activation, is microsomal ω-oxidation. The main pathways of acyl CoA metabolism are β-oxidation in the mitochondria or esterification to either triglyceride or phospholipid in the cytosol.

The mitochondrial β-oxidative pathway accounts for 90% of total cellular β-oxidation of long chain fatty acids. Under most conditions, peroxisomal chain shortening of long chain fatty acids represents a small fraction of long chain fatty acid β-oxidation. However, under certain conditions, such as high dietary fat intake or administration of peroxisome proliferating drugs, the peroxisomal contribution to long chain fatty acid β-oxidation may increase to 20 to 25% or more of total β-oxidation.[1]

The carnitine acyltransferases are a family of enzymes involved in fatty acid transport within the cell. The nomenclature of the carnitine acyltransferases relies on the purported chain length specificity of the various enzymes. Generally, short (carnitine acetyltransferases), medium (carnitine octanoyltransferases), and long chain (carnitine palmitoyltransferases) acyltransferases are recognized. However, the chain length specificity of the medium and long chain enzymes is quite broad, and the specificity overlaps to such an extent that chain length specificity per se cannot be used to distinguish these enzymes.[2,3]

A long chain carnitine acyltransferase, *carnitine palmitoyltransferase* (CPT), is located in the mitochondrion only. CPT regulates the flux of long chain fatty acids into the mitochondria. This enzyme is the focus of the present review. However, long to medium chain acyltransferases, carnitine *octanoyltransferases*, are located in peroxisomes and endoplasmic reticulum. Short chain acyltransferases, *carnitine acetyltransferases*, are located in mitochondria and peroxisomes. Bieber has reviewed the functions and regulation of these latter two acyltransferases.[2]

B. GENERAL ASPECTS OF CPT

Carnitine palmitoyltransferase (CPT; E.C. 2.3.1.23) catalyzes the reversible reaction:

long chain acyl-CoA
$$+ \text{l-carnitine} \leftrightarrow \text{long chain acyl-l-carnitine} + \text{CoASH} \quad (1)$$

The inner mitochondrial membrane is impermeable to long chain acyl CoAs, while long chain acylcarnitines can readily traverse this barrier. Thus, CPT

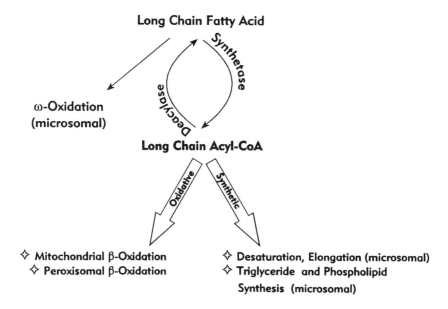

FIGURE 1. Pathways of hepatic fatty acid utilization. When a long chain fatty acid (here represented as palmitic acid) enters the liver, two major metabolic fates are available, oxidation via the mitochondria or peroxisomes and reesterification to triglyceride or phospholipid.

modulates the flux of long chain acyl CoA into the mitochondria. Figure 2 presents a *classical* view of CPT function within the mitochondria. It is clear that CPT *activity* is found on the inner surface of the mitochondrial inner membrane. This activity has been referred to as CPT-B, CPT-II, CPT_2, or CPT_i (for CPT inner). It is also clear that CPT activity is found on the cytosolic side of the mitochondrion. The outer activity has been referred to as CPT-A, CPT-I, CPT_1, or CPT_o (for CPT outer). Further, the outer CPT is inhibited by malonyl CoA, a substrate in the fatty acid synthetic pathway. Unfortunately, while localization of *activities* is possible, it remains very difficult to determine how many CPT *proteins* are involved in producing these CPT *activities*. The problem of definition of the number of distinct CPT proteins, as well as the problems mentioned in the following section, has confounded efforts to study the regulation of CPT at the gene level. Only recently has a 68-kDa CPT been purified, cloned, and its regulation studied.[4-6]

C. PROBLEMS IN THE STUDY OF CPT

Mitochondrial carnitine palmitoyltransferase activities can be demonstrated both on the cytosolic and the matrix faces of the mitochondrion. Recently, it has been demonstrated that the outer CPT activity may be associated with the inner face of the outer mitochondrial membrane.[7] However, it has been impossible to determine whether these topologically distinct ac-

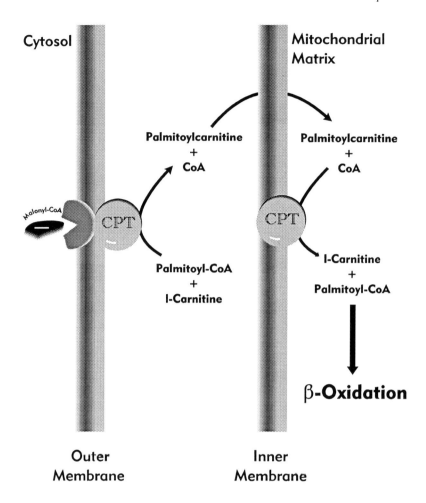

Cytosol

Mitochondrial
Matrix

Palmitoylcarnitine
+
CoA

Palmitoylcarnitine
+
CoA

Malonyl-CoA

CPT

CPT

Palmitoyl-CoA
+
l-Carnitine

l-Carnitine
+
Palmitoyl-CoA

β-Oxidation

Outer
Membrane

Inner
Membrane

FIGURE 2. Mitochondrial carnitine palmitoyltransferase. Mitochondrial carnitine palmitoyl-transferase activities are found on the outer and inner surfaces of the mitochondrion. The outer activity, *in situ*, exhibits sensitivity to malonyl CoA inhibition.

tivities represent discrete proteins or are the product of a single protein in different membrane environments. Studies with intact mitochondria have shown that the apparent K_ms and V_{max}s are quite similar, given the constraint that both long chain acyl CoAs and long chain acylcarnitines are detergents and, therefore, cannot be used at concentrations high enough to calculate accurate kinetic constants.[2] The outer CPT *activity* is inhibited by malonyl CoA, while the inner *activity* is not.

The first major difficulty encountered in defining the number of CPTs is that there is no intrinsic difference in the reaction catalyzed. The reaction (1) is freely reversible, and the kinetic constants observed on either face of the mitochondria are not sufficiently different to distinguish two enzymes. Further, the mitochondrial inner membrane has an asymmetric lipid distribution.[8] The specific lipid composition of the membrane face contributes significantly to the measured characteristics of the enzyme activity.[9-12] Historically, a number of investigators have attempted to attribute the CPT activities to separate proteins based on differential sensitivity to a variety of inhibitors or differential solubilization by detergents.[13,14] However, such approaches generally compare intact mitochondria to mitochondria treated with detergent or other mechanism to expose the inner CPT. Invariably, this leads to some form of disruption of the mitochondrial membrane, which, itself, alters CPT K_ms, V_{max}, and sensitivity to malonyl CoA.

Some years ago we employed the approach of inverting mitochondrial vesicles. This approach avoids the more obvious artifacts. In this type of study, the inner surface of the mitochondria exhibits sensitivity to bromopalmitoyl CoA and 2-tetradecylglycidyl CoA, inhibitors commonly proposed to distinguish the two CPT proteins. In fact, the purified 68-kDa CPT protein, which is clearly at least the inner CPT (whether it is the one and *only* CPT is subject to debate), can be induced to recover at least partial 2-tetradecylglycidyl CoA sensitivity by the addition of dimethyl sulfoxide.[12] Still, under no experimental conditions has any investigator reported malonyl CoA sensitivity of the purified 68-kDa CPT alone.

The second key problem is that removal of the protein from the membrane alters the protein's characteristics. To date, only one CPT protein has been isolated from mitochondria.[2,15,16] The protein has a M_r of 68 to 69 kDa and lacks malonyl CoA sensitivity. Antibody generated to this protein reacts with a 68 to 69-kDa band on Western blotting and precipitates only a single mRNA translation product. All of the CPT activity solubilized from hepatic mitochondria by either Tween 20 or by sonication is precipitable by antibody generated to this protein.[17,18] Unfortunately (for the sake of simplicity), this does not preclude the possibility that another CPT protein, the outer, malonyl CoA-sensitive CPT, is lost due to the various solubilization schemes. Indeed, this possibility has been proposed.

D. WHICH CPT ARE WE DEALING WITH?

To date a single mitochondrial CPT has been cloned, the precursor to the 68-kDa CPT. This protein appears to correspond to the inner CPT. A second protein, which binds malonyl CoA or 2-tetradecylglycidyl CoA, has been isolated from mitochondrial outer membrane.[19-21] This 85- to 95-kDa protein has not been demonstrated to have catalytic activity. In studies described below, combining this protein with the 68-kDa CPT results in malonyl CoA-sensitive CPT activity.[22,23]

Ghadiminejad and Saggerson[24] have shown *cross-reactivity* of the 68-kDa CPT and a larger molecular weight protein associated with the mitochondrial inner membrane on Western blots. Our own blots of total mitochondria often show two bands, particularly where CPT activity is induced. Calculation of M_r led us to conclude that the less abundant, upper band represented the 68-kDa precursor. In support of this, mRNA yields a single immunoprecipitable band with M_r corresponding to the higher weight band observed on Western blots. Hoppel recently presented the possibility that outer membrane malonyl CoA binding protein and inner membrane 68-kDa CPT are associated at mitochondrial membrane pores or junctions.[25] Such junctions are known to serve as sites for protein import. His suggestion may, indeed, explain the enrichment of the outer mitochondrial membrane with a 68-kDa CPT precursor. Can we, then, be sure that there is a single mitochondrial CPT protein associated with a malonyl CoA-binding regulatory protein? The answer remains an unfortunate no. The good fortune in this is that the activity of the 68-kDa CPT is regulated. If this is the only mitochondrial CPT, we are well on our way to understanding its regulation. If there is a second mitochondrial CPT, the data available from *in situ* studies indicate that it too is regulated; that is, outer CPT activity changes. It would, then, be important to determine how this change in activity is mediated. Such data are currently available for the 68-kDa CPT.

Regulation of the activity of any enzyme can be divided into a number of events (Figure 3). In the most general sense, these events can be subdivided into (1) short-term events which alter catalytic efficiency but do not alter the quantity of protein, and (2) longer term events which specifically alter protein quantity. If protein quantity is regulated, this regulation can be accomplished at a number of levels. As with many regulatory enzymes, CPT exhibits both forms of regulation. The specific difficulty which arises with CPT is that the majority of *in situ* studies examine the outer CPT activity, while all studies of purified CPT involve the 68-kDa protein.

II. REGULATION OF CATALYTIC EFFICIENCY OF CPT

Several forms of regulation of catalytic activity of carnitine palmitoyltransferase have been reported. Competitive inhibition (with palmitoyl CoA) by malonyl CoA of carnitine palmitoyltransferase was the first reported mechanism of control of this enzyme.[26] More recently, other events, such as phosphorylation and membrane fluidity, have been suggested to exert some level of control. Unfortunately, the impact of each event on the overall flux through CPT is undefined.

A. REGULATION BY MALONYL CoA

Control of carnitine palmitoyltransferase by malonyl CoA inhibition is the oldest reported mechanism of control. However, in the 14 years or so since the initial report, the mechanism remains controversial.

Malonyl CoA sensitivity has been presumed to be a characteristic of the outer CPT activity. However, all studies employed some means of disrupting

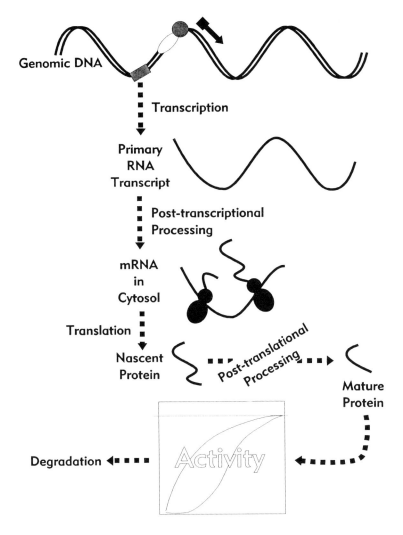

FIGURE 3. Control of protein activity. Control of carnitine palmitoyltransferase activity (or the activity of any enzyme) can be controlled at a variety of levels, from the rate of transcription of RNA to interaction of the protein with specific inhibitors.

the mitochondrion, be it detergent lysis or sonication, in demonstrating that the inner CPT activity is not inhibited by malonyl CoA. Our own studies with inverted mitochondrial vesicles, which should represent the least disrupted case, also failed to show malonyl CoA sensitivity of the inner CPT activity.[9] It may be that the outer and inner surfaces *talk* to each other through some physical connection. In unpublished studies, exposure of intact mitochondria to malonyl CoA, followed by inversion and washing, resulted in some (25 to 50%) decline in the inner CPT activity. We had previously suggested such a connection in studies with 2-tetradecylglycidyl CoA (TDGA-CoA).[12]

2-Tetradecylglycidyl CoA is presumed to inhibit only the outer CPT activity.[27] However, if one exposes the intact mitochondria to TGDA-CoA, washes them, and then inverts them, the inner CPT activity is inhibited.[12] Still, we also reported that the inner CPT activity is, of itself, sensitive to this inhibitor. Further, the purified 68-kDa CPT protein can be rendered TGDA-CoA-sensitive under proper conditions. Thus, there was always the possibility that TDGA-CoA leaked from the outer to the inner surface.

Current and ongoing work from two groups[22,23] demonstrates that the combining of 68-kDa CPT and outer mitochondrial membrane or malonyl CoA-binding protein, under defined conditions, results in a malonyl CoA-sensitive CPT activity. These same conditions neither result in CPT activity of the outer membrane preparation in the absence of the 68-kDa CPT nor malonyl CoA sensitivity of the 68-kDa CPT in the absence of the outer membrane preparation. Whether they are observing regained CPT activity of the outer membrane malonyl CoA-binding protein or recovered malonyl CoA sensitivity of the 68-kDa CPT is unclear. Putting aside the question of one vs. two CPT proteins, these data would serve to confirm the observation that mitochondrial outside and inside could communicate relative to a malonyl CoA inhibition of CPT activities on both faces.

While it remains uncertain how malonyl CoA sensitivity is conferred, a recurring theme in CPT regulation has been that when CPT activity increases, malonyl CoA sensitivity declines. This effect is likely quite important in governing total flux through CPT.

B. REGULATION BY PHOSPHORYLATION

Harano and co-workers have reported that CPT is phosphorylated in response to cAMP in cultured cells.[28] To our knowledge, this initial observation has not been pursued. It remains unclear if the phosphorylation has physiologic significance, but the possibility of this form of regulation will need to be further investigated.

C. REGULATION BY ALTERATION OF LIPID DOMAIN

As mentioned above, the mitochondrial membranes represent a highly complex, highly structured, and incredibly asymmetric domain. That CPT activity is regulated is unquestioned. However, we tend to neglect the membrane association of the activities. Or, if we do consider membrane association, it is as an inconvenience in the purification of the protein.

The membrane characteristics do appear to alter CPT activities on both inner and outer mitochondrial surfaces. This can be demonstrated by two means. First, altering membrane fluidity, with agents such as ethanol or benzoyl alcohol, results in increased CPT activity.[9] As mentioned above, the mitochondrial inner membrane has a very asymmetrical phospholipid distribution, with a very high cardiolipin content on the inner face. Exposure of the outer face to cardiolipin also results in increased CPT activity.[11] Second, a number of physiologic states where CPT activity is elevated, such as starvation or severe streptozotocin-diabetes in the rat, also produce increased mitochondrial fluidity.[12] Thus, membrane fluidity can and does have an effect on CPT activity without an effect on malonyl CoA sensitivity. The extent to which membrane fluidity does contribute to overall CPT regulation remains uncertain. More recent observations on the regulation of the 68-kDa CPT regulation show that a primary mechanism of control is via the control of protein synthesis. At the same time, the largest observed increase in 68-kDa CPT protein is fivefold with a two- to threefold increase a more typical observation. In assessing the overall control of flux through CPT, the combination of short-term and long-term effects must be considered.

III. REGULATION OF MASS OF 68-kDa CPT

A. PURIFICATION AND ANTIBODY GENERATION

To date, only a single 68-kDa mitochondrial carnitine palmitoyltransferase has been isolated. This protein is generated from a larger precursor, as is typical of proteins inserted into the mitochondrial membrane.[29,30] The purification of the protein has allowed the generation of antibody, which, in turn, has allowed quantification of CPT protein. Antibody to the 68-kDa protein has been generated in a number of laboratories, and the exchange of antibody has introduced some uniformity into the study of CPT.[15-17] Antibody to the 68-kDa CPT is effective in precipitating the protein for turnover and translation studies, and in quantifying the protein by Western blot.[17,29] The polyclonal antibody, regardless of source, is not inhibitory to catalytic activity of the CPT. This indicates that the catalytic site is not available, probably because it represents a hydrophobic pocket.

Increases in CPT activity have been shown to closely follow CPT quantity. Consequently, we believe that, although other factors such as phosphorylation or changes in the lipid environment have a limited effect, the *major* mechanism of control of the 68-kDa CPT must be protein synthesis. The generation of an effective antibody has allowed us to directly measure protein turnover, as well as protein quantity. In a number of studies, we have now shown that the synthesis of the 68-kDa carnitine palmitoyltransferase is rapidly increased by a variety of stimuli, including diabetes *in vivo*[31] and cAMP administration in cell systems.[32] Alterations in degradation are not observed. This increase in synthesis can be linked directly to increases in translation rate.

If communication of malonyl CoA sensitivity to the inner enzyme is proven possible, or if the 68-kDa CPT proves to be the only mitochondrial CPT protein, then malonyl CoA may share an equal role in the overall regulation of this enzyme.

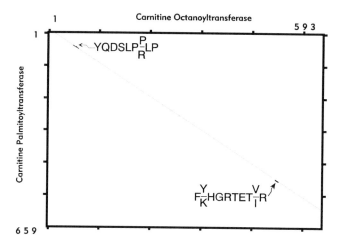

FIGURE 4. Relationship of carnitine palmitoyltransferase and carnitine octanoyltransferase amino acid sequences. The amino acid sequences of CPT and COT were compared without allowing conservative amino acid substitution and looking for 8 amino acid segments showing at least 75% conservation. The specific amino acid sequences are shown.

B. CLONING AND SEQUENCING

The rat liver 68-kDa carnitine palmitoyltransferase has now been cloned by two groups[4,5] and the sequence of the corresponding human enzyme has also been reported by a third.[6] All three clones are in generally good agreement, although we observe a somewhat longer rat CPT sequence than do Woeltje et al. The rat and human enzymes show a high degree of homology.[6] The peroxisomal carnitine octanoyltransferase has also been cloned and sequenced.[33] The peroxisomal enzyme is not membrane associated. The comparison of the amino acid homology and hydrophobicity for these two amino acid sequences are presented (Figure 4). The sequences are not strikingly similar. There are three small regions with strong homology: an early region of 26 amino acids, a small central region of 6 amino acids, and a region toward the carboxy terminal of 9 amino acids. This is not too surprising, since antibodies to these proteins do not cross-react. The hydrophobicity plots (Kyte-Doolittle) are quite similar, reflecting the observation that the 68-kDa carnitine palmitoyltransferase membrane protein is not strikingly hydrophobic. The value of a cDNA is twofold. First, as seen above, we can predict an amino acid sequence, make structural predictions, and begin to evaluate the protein configuration. The value of this is that we can assess the specific critical catalytic components. Second, we can employ the cDNA as a tool to examine changes in mRNA levels and transcriptional activity in various states. As we have seen, CPT activity, CPT protein level, and CPT translation rate are increased in response to starvation,[4,33] diabetes,[31] and peroxisome proliferator administration,[17,30] and decreased in response to sepsis.[34] We can proceed to determine if these changes are the product of alterations in the quantity of mRNA and, ultimately, the product of alterations in transcription rates.

C. REGULATION OF CPT GENE EXPRESSION

It has been possible to determine CPT activity *in situ* for approximately 20 years. It is only recently that we have been able to determine the specific effects of various perturbations on the inner CPT protein through the use of antibody. The inner CPT level responds to a wide variety of agents. Agents which increase inner CPT protein include starvation and diabetes, mediated through glucagon/cAMP and insulin. However, in addition to this relatively well-studied control axis, inner CPT protein is also increased in response to a variety of agents which proliferate peroxisomes or which interfere with normal fatty acid oxidation (e.g., clofibrate treatment,[30] riboflavin deficiency).[4] CPT activity and protein levels are decreased in sepsis.[34] The septic state represents a complex set of hormone-mediated responses, not only to glucagon/insulin, but potentially to cytokines, prostaglandins, and other metabolic mediators.

The determination of the cDNA sequence for the 68-kDa CPT allows direct quantification of the mRNA levels for this protein (by dot or Northern blot), as well as the transcription rate (by nuclear run-on assay). A selection of available data is presented in Table 1. In all cases studied to date, changes in CPT activity are directly associated with changes in the quantity of protein. The quantity of protein is a product of the level of mRNA, which, in turn, is a product of the nuclear transcription rate. Simply stated, the regulation of 68-kDa protein levels must be controlled at the transcriptional level. Subsequent analysis is aimed at determining how regulation is accomplished.

D. PROMOTER ANALYSIS

The regulation of the 68-kDa CPT protein must, therefore, involve interaction of specific nuclear regulatory proteins with the promoter region for the CPT gene. Such genomic regulatory elements are generally found *upstream* or 5′ to the structural elements. Therefore, we have undertaken the isolation of promoter regions for both the rat and human 68-kDa CPTs.[35] Given the high degree of homology of the cDNA for these two proteins, the probability of successful simultaneous isolation seemed high. Using polymerase chain reaction approaches to the isolation of these promoters, we were able to initially isolate regions of approximately 500 bp upstream from the mRNA coding region. The 68-kDa CPT, and CPT activity, in general, are quite interesting because not only do they respond to expected stimuli, such as cAMP-mediated induction, but, also, they respond to apparently unrelated stimuli, such as the peroxisomal proliferating agents. We have taken three simultaneous approaches to the study of the promoter. First, we have examined the promoter region directly using sequence comparison to known consensus nuclear protein binding elements, footprinting with purified nuclear binding

Nutrition and Gene Expression

TABLE 1
Regulation of the Hepatic Mitochondrial 68-kDa Carnitine Palmitoyltransferase

State	Model	68-kDa CPT Protein (immunoreactive)	68-kDa CPT mRNA	68-kDa CPT Transcription	Ref.
Starvation	Rat	↑	↑	↑	17
Diabetes	Rat	↑↑	↑↑	↑↑	17,31
Riboflavin deficiency	Rat	↑↑	↑↑	ND	4
cAMP	Rat	↑↑	↑↑	↑↑	36
	H4IIE Cell			ND	32
Sepsis	Rat	↓	↓	ND	34
Interleukin-6	H4IIE Cell	↑↑↑	↑↑↑	ND	37
DEHP	Rat	↑↑↑	↑↑↑	↑↑↑	5,15,17
Clofibrate	Rat	↑↑	↑↑	↑	38

Note: The quantity of 68 kDa carnitine palmitoyltransferase protein is regulated primarily at the transcriptional level. Increases in the protein, determined by densitometric scanning of Western blots, increases in mRNA, determined by blot or translational activity, and transcription rates correlate quite well. The table illustrates the approximate degree of response (↑ = 1.5- to 2-fold increase; ↑↑ = 2- to 3-fold increase; ↑↑↑ ≥ 4-fold; ↓ = 50% decrease; ND = not determined). The table also indicates something of the variety of stimuli to which the 68-kDa CPT responds.

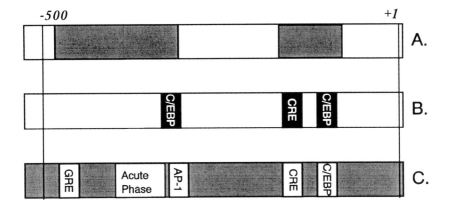

FIGURE 5. Mapping the 68-kDa carnitine palmitoyltransferase promoter. We have isolated an approximately 600-bp region of the CPT promoter: (A) shows the areas of protection by rat liver nuclear extract in footprint analysis; (B) represents regions of protection by purified nuclear binding proteins CREB, C/EBPα, and C/EBPβ; (C) shows the predicted binding regions for a variety of binding proteins based on consensus nucleotide sequences in the promoter.

proteins where available, and gel mobility shift assays. These studies have indicated that the 68-kDa CPT promoter contains a single cAMP response element (CRE), binding sites for C/EBPα (CCAAT/enhancer binding protein), and for C/EBPβ (the DBP-IL6 dependent binding protein). Studies with rat liver nuclear extracts also define a large region of binding which incudes a C/EBPα and a C/EBP binding region, but one which is much broader and more complex, containing consensus binding elements for other proteins, which we have not directly tested. A map of the CPT promoter is presented (Figure 5).

The second major approach to promoter analysis is functional. The CPT promoter has been introduced into a plasmid containing a reporter gene, chloramphenicol acetyltransferase. The CPT promoter is then introduced into HepG2 cells and its response to various stimuli examined. These can include hormonal factors as cAMP or thyroid hormone, peroxisomal proliferators, such as clofibrate, or expression vectors for nuclear regulatory proteins, such as c-*fos*, c-*jun*, or C/EBPα.[35]

The final approach represents a descriptive analysis of changes in the level of nuclear binding proteins. Rat liver or cell nuclear extracts are prepared following stimulus with cAMP or peroxisomal proliferators. The changes in specific nuclear proteins can be assayed by Western blotting. Preliminary data indicate that c-*fos*, c-*jun*, and C/EBPβ increase in sepsis.

The difficulty with the analysis of any promoter is that what *may* bind can be easily defined. What *does* bind is more difficult to determine. Current and ongoing work in this area is aimed at answering this question.

IV. FUTURE DIRECTIONS TO DEFINE THE REGULATION OF FATTY ACID OXIDATION

In our initial enthusiasm, many of us had assumed that the isolation of the cDNA for the 68-kDa CPT would allow us to resolve the question of how many mitochondrial CPTs truly exist. So far, this has not been the case. What we have resolved is that changes in 68-kDa CPT activity can be directly attributed to changes in transcription of 68-kDa CPT mRNA. However, if a second mitochondrial CPT exists, it may be sufficiently different from the 68-kDa CPT that the antibody to one does not cross-react with the other. This degree of difference exists between the 68-kDa mitochondrial CPT and the peroxisomal COT, and molecular probes for one protein do not detect the other protein. Have the tools of molecular biology let us down in our desire to distinguish the events which produce outer and inner CPT activity? We would hold that this is not so. We have not yet asked the correct question for the available tools; we need to begin to apply the tools that we have. A number of groups remain in the search for the still elusive outer CPT protein. However, the immediately approachable question is whether the 68-kDa CPT represents the only mitochondrial CPT protein. Since we cannot yet find the other CPT protein, perhaps the approach should be to remove the protein that we do recognize and to see what activity remains. Methods are available for this, using antisense probes to the 68-kDa CPT. While we do not yet have an answer, we are currently using this approach to contribute to our understanding.

In a much broader sense, the mitochondrial β-oxidative pathway must interact with peroxisomal and microsomal fatty acid oxidative events. CPT activities are part of a family of enzymes including several other long chain and short chain carnitine acyltransferase activities located in these other subcellular compartments. The long chain carnitine acyltransferases are all malonyl CoA-sensitive and are all increased by starvation and diabetes. However, the metabolic reasons for, and the consequences of, this subcellular compartmentation are not known. The proteins have not all been successfully isolated. Nonetheless, isolation and subsequent cloning appear imminent. With the development of molecular tools for this entire cast of carnitine acyltransferases, we may be able to examine their interaction by selective overexpression or suppression. That is, by placing the coding region for the specific protein under the control of a strong transfectable promoter, we can generate high levels of protein in cell systems. Alternatively, use of antisense probes may allow us to eliminate the activity of a specific protein in cell systems. Such approaches will allow us to examine both the response of the other related enzymes to this perturbation and the response of the overall metabolic processes of fatty acid oxidation and esterification. Through such approaches we can begin to study the role of each enzyme in the whole organism.

REFERENCES

1. **Kondrup, J. and Lazarow, P. B.,** Peroxisomal β-oxidation in intact rat hepatocytes: quantitation of its flux, *Ann. N.Y. Acad. Sci.,* 386, 404, 1982.
2. **Bieber, L. L.,** Carnitine, *Annu. Rev. Biochem.,* 57, 261, 1988.
3. **Ramsay, R. R.,** Microsomal, mitochondrial and peroxisomal carnitine palmitoyltransferase activities change differently in starvation and diabetes, *Biochim. Biophys. Acta,* (submitted.)
4. **Brady, P. S., Feng, Y.-X., and Brady, L. J.,** Transcriptional regulation of carnitine palmitoyltransferase synthesis in riboflavin deficiency in rats, *J. Nutr.,* 118, 1128, 1988.
5. **Woeltje, K. F., Esser, V., Weis, B. C., Sen, A., Cox, W. F., McPhaul, M. J., Slaughter, C. A., Foster, D. W., and McGarry, J.D.,** Cloning, sequencing, and expression of a cDNA encoding rat liver mitochondrial carnitine palmitoyltransferase II, *J. Biol. Chem.,* 265, 10720, 1990.
6. **Finocchiaro, G., Taroni, F., Rocchi, M., Martin, A. L., Colombo, I., Tarelli, G. T., and DiDonato, S.,** cDNA cloning, sequence analysis, and chromosomal localization of the gene for human carnitine palmitoyltransferase, *Proc. Natl. Acad. Sci. U.S.A.,* 88, 661, 1991.
7. **Murthy, M. S. R. and Pande, S. V.,** Malonyl-CoA binding site and the overt carnitine palmitoyltransferase activity reside on the opposite sides of the outer mitochondrial membrane, *Proc. Natl. Acad. Sci. U.S.A.,* 84, 378, 1987.
8. **Nicolay, K., Timmers, R., Spoelstra, E., VanderNeut, R., Fok, J., Huigen, Y., Verkleij, A., and deKruiff, B.,** *Biochim. Biophys. Acta,* 778, 359, 1984.
9. **Brady, L. J., Silverstein, L. J., Hoppel, C. L., and Brady, P. S.,** Hepatic mitochondrial inner membrane properties and carnitine palmitoyltransferase A and B: effect of diabetes and starvation, *Biochem. J.,* 232, 445, 1985.
10. **Brady, L. J., Hoppel, C. L., and Brady, P. S.,** Hepatic mitochondrial inner membrane properties, β-oxidation and carnitine palmitoyltransferases A and B, *Biochem. J.,* 233, 427, 1986.
11. **Brady, L. J. and Brady, P. S.,** Hepatic and cardiac carnitine palmitoyltransferase activity: effects of adriamycin and galactosamine, *Biochem. Pharmacol.,* 36, 3419, 1987.
12. **Brady, P. S. and Brady, L. J.,** Action in vivo and in vitro of 2-tetradecylglycidic acid, 2-tetradecylglycidyl-CoA and 2-tetradecylglycidylcarnitine on hepatic carnitine palmitoyltransferase, *Biochem J.,* 238, 801-809, 1986.
13. **Yates, D. W. and Garland, P. B.,** Carnitine palmitoyltransferase activities (EC 2.3.1.-) of rat liver mitochondria, *Biochem. J.,* 119, 547, 1970.
14. **Hoppel, C. L. and Brady, L.,** Carnitine palmitoyltransferase and transport of fatty acids, in *The Enzymes of Biological Membranes,* Martonosi, A., Ed., Vol. 2, Plenum Press, NY, 1985, 139.
15. **Miyazawa, S., Ozasa, H., Osumi, T., and Hashimoto, T.,** Purification and properties of carnitine octanoyltransferase and carnitine palmitoyltransferase from rat liver, *J. Biochem. (Tokyo),* 94, 529, 1983.
16. **Clarke, P. R. H. and Bieber, L. L.,** Isolation and purification of mitochondrial carnitine octanoyltransferase from beef heart, *J. Biol. Chem.,* 256, 9861, 1981.
17. **Brady, P. S. and Brady, L. J.,** Hepatic carnitine palmitoyltransferase turnover and translation rates in fed, starved, streptozotocin-diabetic and diethylhexylphthalate-treated rats, *Biochem. J.,* 246, 641, 1987.
18. **Kerner, J. and Bieber, L. L.,** Studies on the association of b-oxidation enzymes with CPTi and CPTo of heart mitochondria, *FASEB J.,* 4 (Abstr. 3104), A802, 1990.
19. **Declerq, P. E., Falck, J. R., Kuwajima, M., Tyminski, H., Foster, D. W., and McGarry, J. D.,** Characterization of the mitochondrial carnitine palmitoyltransferase enzyme system: use of inhibitors, *J. Biol. Chem.,* 262, 9812, 1987.

20. **Woeltje, K. F., Kuwajima, M., Foster, D. W., and McGarry, J. D.,** Characterization of the mitochondrial carnitine palmitoyltransferase enzyme system: use of detergents and antibodies, *J. Biol. Chem.,* 262, 9822, 1987.

21. **Zammit, V. A., Corstorphine, C. G., and Kolodziej, M. P.,** Evidence for distinct functional molecular sizes of carnitine palmitoyltransferases I and II in rat liver mitochondria, *Biochem. J.,* 263, 89, 1989.

22. **Ghagiminejad, I. and Saggerson, E. D.,** Carnitine palmitoyltransferase (CPT2) from mitochondrial inner membrane becomes inhibitable by malonyl-CoA if reconstituted with outer membrane malonyl-CoA binding protein, *FEBS,* 269, 406, 1990.

23. **Chung, C., Woldegiorgis, G., and Bieber, L. L.,** Restoration of malonyl-CoA sensitivity to purified rat heart mitochondrial carnitine palmitoyltransferase by addition of protein fraction(s) from an 86 kD malonyl-CoA binding affinity column, *FASEB J.,* 5 (Abstr. 1290), A592, 1990.

24. **Ghagiminejad, I. and Saggerson, E. D.,** The relationship of rat liver overt carnitine palmitoyltransferase to the mitochondrial malonyl-CoA binding entity and to the latent palmitoyltransferase, *Biochem. J.,* 270, 787, 1990.

25. **Hoppel, C. L.,** Carnitine palmitoyltransferases, in *Current Concepts in Carnitine Research,* Carter, A., Ed., CRC Press, Boca Raton, FL, 1992, 153.

26. **McGarry, J. D., Leatherman, G. F., and Foster, D. W.,** Carnitine palmitoyltransferase. I, *J. Biol. Chem.,* 253, 4128, 1978.

27. **Kiorpes, T. C., Hoerr, D., Ho, W., Weaner, L. E., Inman, M. G., and Tutwiler, G. F.,** Identification of 2-tetradecylglycidyl coenzyme A as the active form of methyl 2-tetradecylglycidate (methyl palmoxirate) and its characterization as an irreversible, active site-directed inhibitor of carnitine palmitoyltransferase A in isolated rat liver mitochondria, *J. Biol. Chem.,* 259, 9750, 1984.

28. **Harano, Y., Kashiwago, H., Kojima, H., Suzuki, M., Hashimoto, T., and Shigeta, Y.,** Phosphorylation of carnitine palmitoyltransferase and activation by glucagon in isolated rat hepatocytes, *FEBS Lett.,* 188, 267-272, 1985.

29. **Ozasa, H., Miyazawa, S., and Osumi, T.,** Biosynthesis of carnitine octanoyltransferase and carnitine palmitoyltransferase, *J. Biochem. (Tokyo),* 94, 543, 1983.

30. **Brady, P. S., Marine, K. A., Brady, L. J., and Ramsay, R. R.,** Co-ordinate induction of hepatic mitochondrial and peroxisomal carnitine acyltransferase synthesis by diet and drugs, *Biochem. J.,* 260, 93, 1989.

31. **Brady, L. J. and Brady, P. S.,** Regulation of carnitine palmitoyltransferase synthesis in spontaneously diabetic BB Wistar rats, *Diabetes,* 38, 65,

32. **Wang, L., Brady, P. S., and Brady, L. J.,** Turnover of carnitine palmitoyltransferase mRNA and protein in H4IIE cells: effect of cyclic AMP and insulin, *Biochem. J.,* 263, 703, 1989.

33. **Chatterjee, B., Song, C. S., Kim, J. M., and Roy, A. K.,** Cloning, sequencing and regulation of rat liver carnitine octanoyltransferase: transcriptional stimulation of the enzyme during peroxisome proliferation, *Biochemistry,* 27, 9000, 1988.

34. **Birklid, S. D., Brady, P. S., Brady, L. J., and Barke, R. A.,** Hepatic gene expression in a rat intra-abdominal sepsis model: fatty acid oxidation, *FASEB J.,* 4 (Abstr. 2250), A654, 1990.

35. **Brady, P. S., Liu, J.-S., Park, E. A., Hanson, R. W., and Brady, L. J.,** Isolation and characterization of the promoter for the gene coding for carnitine palmitoyltransferase, *Biochem. J.,* 286, 779, 1992.

36. **Brady, L. J. and Brady, P. S.,** Regulation of carnitine palmitoyltransferase by glucagon and insulin, *Biochem. J.,* 258, 677, 1989.

37. **Barke, R. A., Brady, L. J., and Brady, P. S.,** The Ca2+ second messenger system and IL-1-alpha modulation of hepatic gene transcription and mitochondrial fatty acid oxidation, *Surgery,* 110, 285, 1991.

38. **Brady, L. J., Ramsay, R. R., and Brady, P. S.,** Regulation of carnitine acyltransferase synthesis in lean and obese Zucker rats by DHEA and clofibrate, *J. Nutr.,* 121, 525, 1991.

Chapter 15

LONG-TERM REGULATION OF HEPATIC GLUTAMINASE AND THE UREA CYCLE ENZYMES

Malcolm Watford

TABLE OF CONTENTS

I. INTRODUCTION

Glutamine is not a dietary essential amino acid, but it does play an important role within the body, and recent developments in the treatment of several pathological states have resulted in the provision of large amounts of dietary glutamine, either enterally or parenterally.[1-3] A major fate of glutamine within the body is catabolism via the hepatic pathways of gluconeogenesis and urea synthesis. This review focuses on the long-term regulation of glutamine catabolism within the liver through changes in the activities of phosphate-activated glutaminase and the urea cycle enzymes.

II. SOURCES AND ROLES OF GLUTAMINE

It is widely believed that ingested glutamine, at normal dietary intake levels ($\sim$4 g per day), is metabolized by the small intestine and therefore does not enter the bloodstream intact.[4,5] Feeding of large (up to 40 g per day) quantities of glutamine in pathological and experimental situations possibly overloads the capacity of the intestine and may result in dietary glutamine reaching the portal blood. Little is known regarding the fate of such excess glutamine, but it is likely that most of it is taken up by the liver and catabolized to urea and glucose.

Glutamine is the most abundant amino acid in the body, being present at >20 mM in human skeletal muscle[6] and 0.5 to 0.8 mM in plasma;[6-8] furthermore, this glutamine pool is turning over rapidly, at a rate of 130 μmol/min per 100-g body weight, in the adult rat.[9] This implies a continuous and high rate of glutamine production within the body. The major sites of glutamine synthesis are skeletal muscle,[10] lungs,[11,12] and adipose tissue.[13] The liver can also be a site of net glutamine production,[8,14,15] and in the rat, and some herbivorous species, the kidneys may exhibit net glutamine synthesis.[16]

For this chapter the utilization of glutamine will be reviewed without regard to the site of production. Possibly the major, quantitative site of glutamine utilization is the small intestine where glutamine is an important respiratory fuel for enterocytes, the end products being ammonia, citrulline, alanine, proline, lactate, and carbon dioxide.[4,17] Another major site of glutamine utilization is a group of "active" cells, such as lymphocytes, thymocytes, macrophages, reticulocytes, and tumor cells.[5,18,19] Again, the role of glutamine in such cells is as a respiratory substrate and these cells may represent the largest net "organ" of glutamine catabolism within the body. Other sites of net glutamine utilization are the kidneys in metabolic acidosis,[16] the mammary gland during lactation,[20,21] and under some conditions the liver.[8,10,15,22,23]

The liver possesses both the potential for glutamine utilization and glutamine syntheses. However, net exchange of glutamine across the liver is not always observed. There is extensive evidence that both synthesis and hydrolysis of glutamine occur in the liver concomitantly and the net flux is, therefore, the difference between these two pathways.[24] Although species differences exist, it appears that the liver is a site of glutamine synthesis in the normal physiological conditions[14] but changes to a site of net utilization in diabetes,[25,26] high protein feeding,[15,22] exercise,[23] and during some stages of feeding and starvation.[27] This is in keeping with its role as a direct substrate for gluconeogenesis and urea synthesis. Glutamine also serves as an important precursor for liver urea synthesis by the indirect pathway of glutamine catabolism, in extrahepatic cells, such as enterocytes, to alanine, ammonia, and other amino acids which are then taken up by the liver.[8]

III. PATHWAYS OF GLUTAMINE METABOLISM

A. GLUTAMINE SYNTHETASE
Glutamine synthesis is catalyzed by glutamine synthetase (EC 6.3.1.2, L-glutamate:ammonia ligase [ATP]) which utilizes ATP to add ammonia to glutamate (Equation 1). The activity was first described by Krebs[28] in 1935 and has recently been reviewed.[29] It will not be discussed further in this chapter.

$$\text{Glutamate} + NH_3 + ATP \rightarrow \text{Glutamine} + ADP + Pi \qquad (1)$$

B. GLUTAMINASE
The degradation of glutamine can occur by many reactions, including many amidotransferases. However, there is only one true glutaminase, giving rise to stoichiometric amounts of glutamate and ammonia.[30] Unfortunately, many enzymes will produce ammonia from glutamine and therefore reliable methods for the assay of glutaminase activity in nonpurified preparations must rely on the measurement of glutamate production. Phosphate activated glutaminase, sometimes known as phosphate-dependent glutaminase or glutaminase 1 (EC 3.5.1.2, L-glutamine amidohydrolase), is the only true glutaminase found in mammalian tissues (Equation 2). It is the major enzyme responsible for glutamine catabolism in the tissues mentioned in this review, with the possible exception of lactating mammary gland.[21,31] It is hereafter referred to simply as glutaminase. However, there are two isozymes of glutaminase, described here as liver-type and kidney-type; the latter is sometimes referred to as brain-type in the literature.

$$\text{Glutamine} + H_2O \rightarrow \text{Glutamate} + NH_4^+ \qquad (2)$$

Krebs[28] was the first to describe the glutamine-hydrolyzing activity of mammalian tissues. He also noted the existence of the two isozymes differing in pH optima and sensitivity to glutamate inhibition. Kidney-type glutaminase has a relatively low K_m for glutamine, a low affinity for phosphate, and is subject to inhibition by low concentrations of glutamate.[21,32,33] Liver-type glutaminase exhibits a higher K_m for glutamine, higher affinity for phosphate, and is not affected by low concentrations of glutamate but has an absolute requirement for activation by ammonia (half maximal at 0.25 to 0.5 mM).[32-35] Both enzymes are located in the mitochondrial matrix in loose association with the inner mitochondrial membrane.[35,36] There is evidence that the kinetic properties of the enzymes are considerably altered by association with the inner mitochondrial membrane and therefore the properties listed cannot be assumed to be maintained *in vivo*.[37]

Kidney-type glutaminase has been purified from kidney,[38,39] brain,[40] and Ehrlich ascites tumor cells.[41] Antibodies raised against the purified kidney enzyme cross-react with the enzyme in kidney, brain, and small intestine but not with that of adult rat liver.[42] Purification procedures developed for the kidney enzyme were based on the reversible polymerization of the protein in the presence of phosphate or borate. Since the liver-type enzyme does not undergo such polymerization, alternative purification methods were developed.[43-45] The liver enzyme appears to possess only one type of polypeptide, 58 kDa in size, but the size of the active enzyme is not known.[45] The kidney enzyme is active as a heterotetramer comprised of three 65-kDa subunits and one 68-kDa subunit.[46] In addition, antibodies raised to liver-type glutaminase show very low reactivity towards the enzyme of brain, kidney, and small intestine.[45] The recent cloning of cDNAs to both kidney- and liver-type glutaminases[46-48] has revealed other differences (see Section V).

The two isozymes are found in different tissues.[32] The liver type is only seen in liver parenchymal cells after birth and is not found in any other tissue.[45,48] Kidney-type glutaminase is present in all other glutamine-utilizing tissues with the possible exception of mammary gland.[21,31] By characterization of different kinetic properties, Linder-Horowitz[49] examined the appearance of the two types of glutaminase activity in liver during development. Fetal liver contains only kidney-type activity until about 2 days before birth at which time this activity decreases sharply and is not detectable in postnatal liver. Conversely liver-type activity first appears around birth and rapidly (within 2 to 4 d) reaches adult levels. The recent availability of antibody[45] and cDNA[48] probes to the glutaminases has extended these results to show that liver-type glutaminase protein and mRNA first appear about 1 day before birth and reach adult levels within 2 days after birth.

C. HEPATIC GLUTAMINE METABOLISM AND THE UREA CYCLE

Glutamine is catabolized by the liver by the pathway outlined in Figure 1. Glutamine is taken up by parenchymal cells by transport system N.[50] There is also evidence of a facilitative system,[15,51] perhaps system L,[52] which transports glutamine out of the cell. Glutamine then enters the mitochondria to be

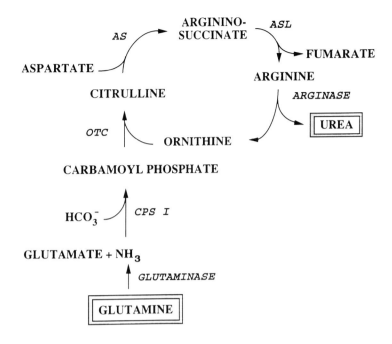

FIGURE 1. Hepatic glutaminase and the urea cycle.

irreversibly hydrolyzed by phosphate-activated glutaminase. The ammonia and glutamate formed can then be used for urea synthesis and gluconeogenesis. Little, or no, glutamine is completely oxidized[53,54] and evidence from a careful analysis of dietary protein suggests that this is true for most amino acids.[54] This review takes the opinion that glutamine catabolism and urea synthesis within the liver are part of the same pathway and focuses on the coordinate regulation of hepatic glutaminase with the urea cycle enzymes. The urea cycle is therefore described in some detail.

The urea cycle is usually described as beginning within the mitochondrial matrix where NH_3 and bicarbonate condense to form carbamoyl phosphate via the action of carbamoyl phosphate synthetase 1 (CPS 1). Ornithine trans-carbamoylase (OTC) then catalyzes the combination of carbamoyl phosphate with ornithine to form citrulline. Citrulline exits the mitochondria to the cytosol where argininosuccinate synthetase (AS) adds an aspartate moiety to form argininosuccinate. This is then hydrolyzed by argininosuccinate lyase (ASL) to give arginine and fumarate. The arginine is hydrolyzed by arginase to urea and at the same time regenerates ornithine.[55] There is considerable evidence that the cycle efficiently channels intermediates, and this possibly extends to the channeling of substrates into the cycle itself.[55-59] For example Meijer[60] has shown that ammonia generated by the action of glutaminase is preferentially used for the carbamoyl phosphate synthetase reaction in liver mitochondria.

The function of the urea cycle is ammonia detoxification but since it also removes bicarbonate it has been proposed to play a significant role in the regulation of systemic pH. The complete cycle is found only in the liver although partial reactions are found in the kidney and small intestine constituting tissue-compartmented system for arginine synthesis.[61] These aspects have been the subject(s) of many recent reviews.[55,62-65]

IV. REGULATION OF GLUTAMINASE ACTIVITY

A. SHORT-TERM REGULATION

Both hepatic glutamine metabolism and the urea cycle are subject to regulation by both short-term and long-term mechanisms. Short-term regulation can be inferred from the observation that the concentration of glutamine within the liver is held constant at about 5 μmoles/g wet weight. The concentrations of urea cycle substrates, ammonia, and bicarbonate are also not reduced to zero and do not vary widely. A number of key regulatory points have been identified in the pathway of hepatic glutamine metabolism, transport into the cell, glutaminase and α-ketoglutarate dehydrogenase activities, and entry into the urea cycle.[34,35,52] A variety of evidence has been cited to demonstrate a regulatory role for hepatic glutaminase activity. Recently Pogson et al.[52] have applied the calculations of quantitative regulatory theory to this problem in isolated rat hepatocytes. They report that glutaminase exhibits a control strength of close to 1.0 (indicative of a major regulatory role), while that of transport system N is 0.31 and system L is -0.4 (indicating a negative effect). Despite problems with the application of this theory to enzymes within the mitochondrial matrix the results are further indication that glutaminase activity is of regulatory significance.

Although glutamine at high concentrations (>2 mM) is readily utilized by experimental systems *in vitro*, such as the perfused liver and isolated hepatocytes, no net glutamine utilization is observed at physiological glutamine concentrations (<1 mM).[8] A number of substances will stimulate glutamine utilization in such systems, for example, glucagon, ammonia, leucine, cysteine, bicarbonate, low pH, cAMP, catecholamines, vasopressin, angiotensin II, and thyroxine.[34,66-69] The mechanisms by which these agents exert their effects is not well understood and their relevance *in vivo* is not clear. They may act by changing the interaction of glutaminase with the inner-mitochondrial membrane.[69] Recent work has suggested that hormonal regulation of cell (and mitochondrial) volume may play a role.[70,71] Ochs[72] has proposed that glucagon and cAMP act at the level of glutaminase activity but that hormones acting via changes in calcium (α-adrenergic, vasopressin, and angiotensin II) act at the level of α-ketoglutarate dehydrogenase. As mentioned in Section III, hepatic glutaminase activity is activated by one of its products, ammonia. This was first described by Charles[73] and subsequently shown to be an absolute requirement for the enzyme.[34,74] Welbourne[75] has proposed an intricate regulatory system involving the feed-forward activation of hepatic glutaminase by ammonia produced from intestinal glutamine catabolism.

Kaiser et al.[76] have found flux through liver glutaminase to be increased sixfold in human cirrhotic liver, and in rats bearing ascites tumor cells, Quesada et al.[77] report a dramatic decrease in hepatic glutaminase activity within 24 h of injection of the cells into the peritoneal cavity. It is not clear what mechanism is responsible for these effects, but they probably reflect acute regulation rather than long-term changes in the amount of enzyme protein.

Glutamine is also reported to affect a number of hepatic pathways, i.e., glycogen synthesis, proteolysis, and glycolysis. Many, if not all, of these effects are due to cell volume (swelling) changes resulting from the cotransport of sodium with glutamine.[70,71,78]

Kidney-type glutaminase has also been proposed to be of regulatory importance in the short-term regulation of renal ammoniagenesis. This has been extensively reviewed[79,80] recently and will not be considered further here. Short-term regulation of kidney-type glutaminase in other tissues has not been extensively characterized.

The short-term regulation of the urea cycle is proposed to be exerted primarily at the level of carbamoyl phosphate synthetase I which is subject to allosteric activation by *N*-acetyl glutamate. The hydrolysis of glutamine to glutamate, which can be used for *N*-acetyl glutamate synthesis, is in accord with the major fate of hepatic glutamine catabolism being urea synthesis. Meijer et al.[55] have recently reviewed the role of carbamoyl phosphate synthetase I, and other enzymes, in the acute regulation of the cycle.

B. LONG-TERM REGULATION

Since the pathways of glutamine catabolism and urea synthesis within the liver are tightly regulated in the short-term, the capacity of these pathways is, by definition, in excess under normal circumstances. But there is also extensive evidence of long-term regulation, although for hepatic glutaminase this has only recently been investigated in depth.

As pointed out by Folin[81] in 1905, urinary urea represents the most variable nitrogenous excretion product. In what must now be considered a potentially dangerous series of experiments, Folin showed that urea production was variable and regulated by diet. He fed himself, and others, a very low protein diet of arrowroot starch and cream for 14 d and found that urea excretion fell to low levels, but it rose again within 24 h of resuming his normal diet. Although based on anecdotal evidence, this downregulation of urea production has been reported to be of pathological importance in the refeeding of high protein diets to severely malnourished prisoners at the end of World War II.[82]

The long-term regulation of hepatic glutaminase has only been recently recognized. Some earlier reports indicated that the activity was increased after feeding high protein diets[83] and decreased on feeding low protein diets.[84]

TABLE 1
Long-Term Regulation of Glutaminase in Rat Liver

	Activity[86]	mRNA[48]	Transcription rate[88]
Control	100 ± 10	100 ± 28	100 ± 29
Diabetes (6 days)	402 ± 12	476 ± 36	448 ± 160

Note: All values reported as % of control values and are means ± SEM
for 4 to 6 animals in each study. Control activity = 4.1 μmol/
min/g wet wt. Other values are based on arbitrary densitometry
units.

These results were based on questionable assay methods but did reflect changes in activity although the absolute values were incorrect. Examination of the regulation of hepatic glutaminase required the establishment of an accurate and reliable assay procedure based on that of Curthoys and Lowry.[85] The original method was modified to accommodate the kinetic properties of the hepatic enzyme. This meant the inclusion of high substrate concentration (100 mM glutamine), limited phosphate (50 mM), and the inclusion of ammonium chloride (2 mM) to ensure complete activation of the enzyme.[86] Using this method, it has been shown that hepatic glutaminase activity is increased during diabetes,[86] high protein feeding,[87] and starvation,[87] and decreased during low protein feeding[88] (Table 1). These changes are unrelated to changes in acid-base status, the major determinant of renal glutaminase activity. Using a protein-free diet, McGivan et al.[89] found reduced activity of hepatic glutaminase. Similar results are seen with a 5% protein diet[88] which gives a more consistent response since rats consuming a protein-free diet lose weight and are obviously in a catabolic state. The observed pattern of regulation of hepatic glutaminase activity is similar to the pattern observed for net hepatic glutamine catabolism *in vivo*.

Long-term changes in hepatic glutaminase activity parallel those observed in the activity of all five urea cycle enzymes and in the activity of a key regulatory enzyme of gluconeogenesis, phosphoenolpyruvate carboxykinase.[65] Thus changes in the maximal capacity of hepatic glutaminase are in keeping with the role of glutamine as a substrate for urea synthesis and gluconeogenesis. It is well established that hormones, such as glucagon, insulin, and glucocorticoids, are the agents acting at the level of the liver cell to regulate expression of the gluconeogenic and urea cycle enzymes.[65,90] To date the question of hormonal regulation of liver glutaminase has received little attention. In rat hepatocytes in primary culture, the activity is reported to increase in response to glucagon and to exogenous ammonia.[89] However, the activity decreases in cells in primary culture in the absence of hormones, and even with hormones the activity does not rise above that seen in freshly isolated hepatocytes. It is therefore not clear what mechanisms are involved in this response or if it has any bearing on the situation *in vivo*.

Kidney-type glutaminase is subject to long-term regulation, but this has only been extensively characterized in kidney, with regard to metabolic acidosis. The activity is increased three- to fourfold in the kidney of acidotic rats,[85,91] but changes in the glutaminase activity in other tissues, lung, skeletal muscle, adipose tissue, and intestine are not seen.[91] There is a large increase observed in glutaminase activity of the small intestine during streptozotocin diabetes in the rat,[86,92,93] but this is the result of an increased number of enterocytes, not to changes in glutaminase activity per enterocyte.[87]

V. REGULATION OF EXPRESSION OF GLUTAMINASE GENES

A. GLUTAMINASE cDNAs

Liver glutaminase is encoded by a mRNA of approximately 2.8 kb in length which is expressed only in postnatal liver.[48] It is absent until a few hours prior to birth but rapidly reaches adult levels within 2 days of birth.[48] As mentioned elsewhere, kidney-type glutaminase is comprised of two different sized subunits. Curthoys[46] has recently shown that both subunits are the product of the same gene, located on chromosome 9 in the rat.[94] Transcription of the gene gives rise to two species of mRNA, 6.0 and 3.3 kb in length. Both encode a 72-kDa precursor polypeptide which is then differentially processed to give the 65- and 68-kDa subunits.[46]

Comparison of the known cDNA sequences of liver- and kidney-type glutaminases shows little similarity at the nucleotide level. When the predicted amino acid sequences are compared a region of ~150 residues with high identity (>85%) has been found.[46,48] The complete sequence of liver-type glutaminase has not yet been determined. The results to date indicate that the two isozymes are the products of different but evolutionary related genes. Interestingly, the family of CPS enzymes has been shown to have evolved from the fusion of a glutaminase with a synthetase gene,[95] and many of the amidotransferases exhibit a conserved active site triad of cysteine/histidine/aspartate.[96] The presence of such a site has not been detected in the mammalian glutaminase sequences determined to date[46,48] and may therefore indicate a separate line of evolution.

B. HETEROGENEITY OF EXPRESSION IN THE LIVER

Liver possesses both the capacity for glutamine utilization (glutaminase) and glutamine synthesis (glutamine synthetase). It has been shown that these enzymes are simultaneously active and that the net utilization or production of glutamine by the liver is the difference in activity.[24,64] The elegant immunochemical studies of Gebhardt and Mecke[97] showed that glutamine synthetase protein was located in a restricted population of cells perhaps only 1 to 3 cells in thickness surrounding the venous exit. This has been confirmed in numerous publications and is now the standard index for identity of perivenous cells.[55,98,99] Haussinger and co-workers,[64] using the isolated perfused rat liver, proposed an intercellular glutamine cycle, with glutamine being degraded in the periportal cells and synthesized in perivenous cells. The model required that glutaminase be located predominantly in periportal cells. Studies[100] of enzyme activity and mRNA abundance in isolated periportal and perivenous cells show glutaminase to be mainly located in the periportal cells along with other enzymes of glutamine catabolism, phosphoenolpyruvate carboxykinase, and the urea cycle enzymes. The pattern of expression of the carbamoyl synthetase I changes during different physiological conditions,[101] and the availability of cDNA probes will allow experiments to examine if this is also true for liver-type glutaminase.

C. REGULATION OF EXPRESSION

Long-term increases in the activity of enzymes may be the result of the modification of existing protein or due to changes in the amount of the protein. In the rat the activity of hepatic glutaminase is increased fourfold in streptozotocin diabetes.[86] Using equivalence point titration and Western blotting, the increase in glutaminase activity during diabetes was shown to be completely accounted for by an increase in the amount of glutaminase protein with no change in the specific activity of the protein.[45] The availability of a cDNA has allowed determination of the relative abundance of glutaminase mRNA and this too increases fourfold.[48] Recent work has established that this is due to a fourfold increase in the rate of transcription of the hepatic glutaminase gene (Table 1).[88] The abundance of the mRNA is also increased in response to injection of large doses of dibutyryl cAMP.[101] After high (60%) protein feeding or starvation, there is an increase in liver glutaminase activity, mRNA abundance, and rate of transcription.[88] Conversely, low (0 or 5%) protein feeding results in decreases in glutaminase activity, mRNA abundance, and transcription rate.[88] These results establish that liver glutaminase is regulated in the long-term primarily by changes in the rate of transcription of the gene, and no changes in protein or mRNA turnover need be postulated.

This is in striking contrast to the expression of kidney-type glutaminase in rat kidney. Expression is regulated by changes in the amount of enzyme and rate of enzyme synthesis and this correlates with changes in the relative abundance of the mRNA.[103] However, it appears that mRNA turnover, rather than changes in transcription rate, is the predominant mechanism of regulation.[103,104] In metabolic acidosis the abundance of the message increases five-to sevenfold and appears to be very stable. On recovery from acidosis, the abundance of the message falls rapidly and no changes in the rate of transcription are seen. In LLPC-PK$_1$-F$^+$ cells in culture the abundance of the mRNA is increased >20-fold at pH 6.9 compared to pH 7.4, and again no changes in the rate of transcription of the gene are seen. The apparent half-life of the mRNA changes from 2.5 h at pH 7.4 to 15 h at pH 6.9 and thus accounts for the rise in abundance.[104] The long (6 kb) kidney-type glutaminase mRNA contains many sequences in the noncoding region which could be involved in mRNA stabilization.

D. COORDINATE EXPRESSION WITH THE UREA CYCLE ENZYME GENES

The enzymes of the urea cycle are considered as classics for the study of enzyme protein turnover and adaptation. In the 1930s it was demonstrated that arginase activity changed with dietary protein intake, and in the 1960s Schimke[105] demonstrated coordinate changes in the activity of all five enzymes. He extended the field by using the response to develop quantitative aspects of enzyme protein turnover and adaptation which became the basis for all subsequent studies.

The five urea cycle enzymes are subject to coordinate regulation, as is the activity of *N*-acetyl glutamate synthetase.[55,65] The enzymes are all located primarily in the periportal region of the liver acinus.[55] Activities are highest under conditions of high protein feeding or starvation and lowest in low protein feeding. Changes also occur in uncontrolled diabetes and during hypercatabolic states such as sepsis and trauma.[65] The apparent half-lives of the proteins are long (3 to 6 d), and therefore changes in the amount of enzyme activity are not manifested rapidly and several days may be required to reach new steady states. Since all of the enzymes respond to hormones, such as glucagon, glucocorticoids, thyroxine and insulin, it is likely that the changes observed with diet are mediated by hormonal signals.[65,90,105] Changes in activity appear to be due to increased amounts of enzyme protein resulting from increased abundance of the specific mRNAs and increased rates of transcription of the genes. Exceptions are an increase in arginase activity in diabetes without a change in the amount of enzyme,[107] changes in the rate of arginase enzyme protein degradation during starvation,[108] and evidence of translational regulation of carbamoyl phosphate synthetase I.[109] Furthermore, in contrast to the other enzymes, ornithine transcarbamoylase does not show dramatic responses to glucagon, cAMP, or glucocorticoids in tissue culture and short-term experiments *in vivo*.[110]

Morris,[65] and others,[111,112] have reviewed the molecular biology of the long-term regulation of the urea cycle enzymes, and it will not be covered in great detail here. The evidence is not complete. Much of the work is from computer analysis of DNA sequence or is inferred from the pattern of responses seen *in vivo* and *in vitro*. Very little work has been done with reporter genes fused to urea cycle enzyme gene promoter sequences. The pattern that is emerging is that the urea cycle enzyme genes possess 5′ regulatory sequences which allow regulation of expression by cAMP and glucocorticoids. In addition, argininosuccinate synthase is subject to regulation by arginine which acts to inhibit transcription through a region 5′ to the gene.[113,114] There is also evidence of the presence of tissue-specific elements.[65] It is likely, given the coordinate regulation of expression, that the genes possess similar regulatory elements. For CPS I and OTC and arginase, the presence of elements recognized by C/EBP or related factors has been demonstrated, and it is likely that CRE and GRE will be present.[65] A unique sequence has been found, "UCE" (urea cycle element),[65,115] in the regions 5' to at least 4 of the genes (OTC, AS, ASL, and arginase) and this may play a role in the coordinate regulation. It has been hypothesized that this element may be involved in the coordinate expression of other genes involved in nitrogen metabolism.

The common expression of glutaminase with urea cycle enzymes (and gluconeogenic enzymes) in periportal cells, and the similar ontogenic development pattern, appearing around birth in the liver, is in keeping with the role of hepatic glutamine metabolism. However, differences do exist. The urea cycle enzymes are expressed in other tissues, CPS I and OTC in the intestine, and ASS and ASL in kidney, while hepatic glutaminase is only expressed at measurable levels in postnatal liver. Evidence from experiments *in vivo* and *in vitro* indicate that glutaminase will have similar elements for the tissue-specific and hormonal regulation of expression. The presence or absence of these, and the UCE, await the availability of genomic clones and sequence analysis.

VI. SUMMARY

This chapter has centered on the long-term regulation of expression of hepatic glutaminase gene and how this relates to expression of the urea cycle. These enzymes are proteins with relatively long (days) half-lives, and changes in activity take some time to be manifested. This could mean that the changes are simply adaptive in nature and do not themselves play any role in the regulation of metabolism. This may be true in subjects eating regular diets. However, the decrease in activity observed in times of low rates of hepatic amino acid catabolism certainly limits the capacity of the body to deal with a sudden nitrogen load, while the increased capacity observed in hypercatabolic states allows the body to deal with the high rates of nitrogen disposal.

There is considerable evidence that the urea cycle acts as a metabolon; its substrates and intermediates are efficiently channeled along the pathway. This concept may even extend to the enzymes involved in the provision of substrates, such as glutaminase. Many enzymes involved in the degradation of excess amino acids are also subject to coordinate long-term regulation with those of the urea cycle. Perhaps this reflects the integrated nature of the system. Application of molecular biological techniques to the comparison of expression of these various hepatic genes is leading to an understanding of how hepatic amino acid metabolism is regulated.

ACKNOWLEDGMENTS

Work done in the author's laboratory was supported by Grant DK-37301 from the National Institutes of Health. The author is grateful to Drs. Brosnan, Curthoys, and Morris for providing manuscripts in press.

REFERENCES

1. **Lacey, J. M. and Wilmore, D. W.**, Is glutamine a conditionally essential amino acid?, *Nutr. Rev.*, 48, 297, 1990.
2. **Lowe, D. K., Benfell, K., Smith, R. J., Jacobs, D. O., Murawski, B., Ziegler, T. R., and Wilmore, D. W.**, Safety of glutamine-enriched parenteral nutrient solutions in humans, *Am. J. Clin. Nutr.*, 52, 1101, 1990.
3. **Furst, P., Alpers, S., and Stehle, P.**, Dipeptides in clinical nutrition, *Proc. Nutr. Soc.*, 49, 343, 1990.
4. **Windmueller, H. G. and Spaeth, A. E.**, Uptake and metabolism of plasma glutamine by the small intestine, *J. Biol. Chem.*, 249, 5070, 1974.
5. **Newsholme, E. A. and Parry-Billings, M.**, Properties of glutamine release from skeletal muscle and its importance to the immune system, *J. Parent. Ent. Nutr.*, 14, 63S, 1990.
6. **Souba, W. W.**, Glutamine. A key substrate for the splanchnic bed, *Annu. Rev. Nutr.*, 11, 285, 1991.
7. **Souba, W. W., Smith, R. J., and Wilmore, D. W.**, Glutamine metabolism by the intestinal tract, *J. Parent. Ent. Nutr.*, 9, 608, 617, 1985.
8. **Lund, P. and Watford, M.**, Glutamine as a precursor of urea, in *The Urea Cycle*, Grisolia, S., Baguena, R., and Mayor, F., Eds., John Wiley & Sons, New York, 1976, 479.
9. **Squires, E. J. and Brosnan, J. T.**, Measurement of the turnover rate of glutamine in normal and acidotic rats, *Biochem. J.*, 210, 277, 1983.
10. **Lund, P. and Williamson, D. H.**, Inter-tissue nitrogen fluxes, *Brit. Med. Bull.*, 41, 251, 1985.
11. **Welbourne, T. C.**, Role of the lung in glutamine homeostatis, *Contr. Nephrol.*, 63, 178, 1988.
12. **Souba, W. W., Herkowitz, K., and Plumley, D. A.**, Lung glutamine metabolism, *J. Parent. Ent. Nutr.*, 14, 68S, 1990.
13. **Frayn, K. N., Khan, K., Coppack, S. W., and Elia, M.**, Amino acid metabolism in human subcutaneous adipose tissue, *Clin. Sci.*, 80, 471, 1991.
14. **Aikawa, T., Matsuka, H., Yamamoto, H., Okuda, T., Ishikawa, E., Kawano, T., and Matsumura, E.**, Gluconeogenesis and amino acid metabolism. Interorganal relations and roles of glutamine and alanine in the amino acid metabolism of fasted rats, *J. Biochem.*, 74, 1003, 1974.
15. **Fafournoux, P., Remesy, C., and Demigne, C.**, Fluxes and membrane transport of amino acids in rat liver under different protein diets, *Am. J. Physiol.*, 259, E614, 1990.
16. **Tannen, R. L.**, Ammonia metabolism, *Am. J. Physiol.*, 235, F265, 1978.
17. **Watford, M., Lund, P., and Krebs, H. A.**, Isolation and metabolic characteristics of rat and chicken enterocytes, *Biochem. J.*, 178, 589, 1979.
18. **Ardawi, M. S. M. and Newsholme, E. A.**, Metabolism in lymphocytes and its importance in the immune response, *Essays Biochem.*, 21, 1, 1985.

19. **Newsholme, E. A., Crabtree, B., and Ardawi, M. S. M.,** The role of high rates of glycolysis and glutamine metabolism in rapidly dividing cells, *Biosci. Rep.,* 5, 393, 1985.

20. **Watford, M., Erbelding, E. J., and Smith, E. M.,** Glutamine metabolism in rat small intestine: response to lactation, *Biochem. Soc. Trans.,* 14, 1058, 1986.

21. **Horowitz, M. L. and Knox, W. E.,** A phosphate activated glutaminase in rat liver different from that in kidney and other tissues, *Enzymol. Biol. Clin.,* 9, 214, 1968.

22. **Yamamoto, H., Aikawa, T., Matsumura, H., Okuda, T., and Ishikawa, E.,** Interorganal relationships of amino acid metabolism in fed rats, *Am. J. Physiol.,* 244, 1428, 1974.

23. **Wasserman, D. H., Greer, R. J., Williams, P. E., Becker, T., Lacy, D. B., and Abumrad, N. N.,** Interaction of gut and nitrogen metabolism during exercise, *Metabolism,* 40, 307, 1991.

24. **Vincent, N., Martin, G., and Baverel, G.,** Simultaneous synthesis and degradation of glutamine in isolated rat liver cells. Effect of vasopressin, *Biochim. Biophys. Acta,* 1014, 184, 1989.

25. **Schrock, H. and Goldstein, L.,** Interorgan relationships for glutamine metabolism in normal and acidotic rats, *Am. J. Physiol.,* 240, E519, 1981.

26. **Brosnan, J. T., Man, K.-C., Hall, D. E., Colbourne, S. A., and Brosnan, M. E.,** Interorgan metabolism of amino acids in streptozotocin-diabetic ketoacidotic rat, *Am. J. Physiol.,* 244, E151, 1983.

27. **Yamamoto, H., Aikawa, T., Matsutaka, H., Okuda, T., and Ishikawa, E.,** Interorganal relationships of amino acid metabolism in fed rats, *Am. J. Physiol.,* 226, 1428, 1974.

28. **Krebs, H. A.,** Metabolism of amino acids. The synthesis of glutamine from glutamic acid and ammonia, and the enzymic hydrolysis of glutamine in animal tissues, *Biochem. J.,* 29, 1951, 1935.

29. **Cooper, A. J. L.,** Glutamine synthetase, in *Glutamine and Glutamate in Mammals,* Kvamme, E., Ed., CRC Press, Boca Raton, FL, 1988, 7.

30. **Meister, A.,** Function of glutathione in kidney via the gamma-glutamyl cycle, *Med. Clin. N. Am.,* 59, 649, 1975.

31. **Erbelding, E. J.,** Glutamine Metabolism in the Small Intestine, MS thesis, Cornell University, Ithaca, NY, 1985.

32. **Huang, Y.-Z. and Knox, W. E.,** A comparative study of glutaminase isozymes in rat tissues, *Enzyme,* 21, 408, 1976.

33. **Curthoys, N. P.,** Cellular distribution and induction of the enzymes of renal ammoniagenesis and gluconeogenesis, in *pH Homeostasis,* Haussinger, D., Ed., Academic Press, New York, 323, 1988.

34. **McGivan, J. D.,** Metabolism of glutamine and glutamate in liver—regulation and physiological significance, in *Glutamine and Glutamate in Mammals,* Kvamme, E., Ed., CRC Press, Boca Raton, FL, 1988, 183.

35. **Kovacevic, Z. and McGivan, J. D.,** Mitochondrial metabolism of glutamine and glutamate and its physiological significance, *Physiol. Rev.,* 63, 547, 1983.

36. **Kalra, J. and Brosnan, J. T.,** Localization of glutaminase in rat liver, *FEBS Lett.,* 32, 325, 1973.

37. **McGivan, J. D., Lacey, J. H., and Joseph, S. K.,** Localization and some properties of phosphate-dependent glutaminase in disrupted liver mitochondria, *Biochem. J.,* 192, 537, 1980.

38. **Kvamme, E., Tveit, B., and Svenneby, G.,** Glutaminase from pig renal cortex. I. Purification and general properties, *J. Biol. Chem.,* 245, 1871, 1970.

39. **Curthoys, N. P., Kuhlenschmidt, T., and Godfrey, S. S.,** Regulation of renal ammoniagenesis. Purification and characterization of phosphate-dependent glutaminase from rat kidney, *Arch. Biochem. Biophys.,* 174, 82, 1976.

40. **Svenneby, G.,** Pig brain glutaminase: purification and identification of different enzyme forms, *J. Neurochem.,* 17, 1591, 1970.

41. **Quesada, A. R., Sanchez-Jimenez, F., Perez-Rodriguez, J., Marquez, J., Medina, M. A., and Nunez de Castro, I.,** Purification of phosphate-dependent glutaminase from isolated mitochondria of Ehrlich ascites-tumour cells, *Biochem. J.,* 255, 1031, 1988.

42. **Curthoys, N. P., Kuhlenschmidt, K., Godfrey, S. S., and Weiss, R. F.,** Phosphate-dependent glutaminase from rat kidney. Cause of increased activity in response to acidosis and identity with glutaminase from other tissues, *Arch. Biochem. Biophys.,* 172, 162, 1976.

43. **Patel, M. and McGivan, J. D.,** Partial purification and properties of rat liver glutaminase, *Biochem. J.,* 220, 583, 1984.

44. **Heini, H. G., Gebhardt, R., and Mecke, D.,** Purification and characterization of rat liver glutaminase, *Eur. J. Biochem.,* 162, 541, 1987.

45. **Smith, E. M. and Watford, M.,** Rat hepatic glutaminase: purification and immunochemical characterization, *Arch. Biochem. Biophys.,* 260, 740, 1988.

46. **Shapiro, R. A., Farrell, L., Srinivasan, M., and Curthoys, N. P.,** Isolation, characterization, and in vitro expression of a cDNA that encodes the kidney isoenzyme of the mitochondrial glutaminase, *J. Biol. Chem.,* 266, 18792, 1991.

47. **Banner, C., Hwang, J.-J., Shapiro, R. A., Wenthold, R. J., Nakatani, Y., Lampel, K. A., Thomas, J. W., Huie, D., and Curthoys, N. P.,** Isolation of a cDNA for rat brain glutaminase, *Mol. Brain Res.,* 3, 247, 1988.

48. **Smith, E. M. and Watford, M.,** Molecular cloning of a cDNA for rat hepatic glutaminase. Sequence similarity to kidney-type glutaminase, *J. Biol. Chem.,* 265, 10631, 1990.

49. **Linder-Horowitz, M.,** Changes in glutaminase activities of rat liver and kidney during pre- and post-natal development, *Biochem. J.,* 114, 65, 1969.

50. **Kilberg, M. S., Handlogten, M. E., and Christensen, H. N.,** Characteristics of an amino acid transport system in rat liver for glutamine, asparagine, histidine and closely related analogs, *J. Biol. Chem.,* 255, 4011, 1980.

51. **Farfournoux, P., Demigne, C., Remesy, C., and Le Cam, A.,** Bidirectional transport of glutamine across the cell membrane in rat liver, *Biochem. J.,* 216, 401, 1983.

52. **Pogson, C. I., Low, S. Y., Knowles, R. N., Salter, M., and Rennie, M. J.,** Application of metabolic control theory to amino acid metabolism in liver, *Alfred Benson Symp.,* 30, 262, 1991.

53. **Blackburn, E. H. and Hird, F. J. R.,** Metabolism of glutamine and glutamate by rat liver mitochondria, *Arch. Biochem. Biophys.,* 152, 258, 1972.

54. **Jungas, R. L., Halperin, M. L., and Brosnan, J. T.,** A quantitative analysis of amino acid oxidation and related gluconeogenesis in man, *Physiol. Rev.,* in press.

55. **Meijer, A. J., Lamers, W. H., and Chamuleau, R. A. F. M.,** Nitrogen metabolism and ornithine cycle function, *Physiol. Rev.,* 70, 701, 1990.

56. **Cohen, N. S., Cheung, C.-W., and Raijman, L.,** Channeling of extramitochondrial ornithine to matrix ornithine transcarbamoylase, *J. Biol. Chem.,* 262, 203, 1987.

57. **Cheung, C.-W., Cohen, N. S., and Raijman, L.,** Channeling of urea cycle intermediates in situ in permeabilized hepatocytes, *J. Biol. Chem.,* 264, 4038, 1989.

58. **Watford, M.,** Channeling in the urea cycle: a metabolon spanning two compartments, *Trends Biochem. Sci.,* 14, 313, 1989.

59. **Watford, M.,** The urea cycle: a two-compartment system, *Essays Biochem.,* 26, 49, 1990.

60. **Meijer, A. J.,** Channeling of ammonia from glutaminase to carbamoyl-phosphate synthetase in liver mitochondria, *FEBS Lett.,* 191, 249, 1985.

61. **Dhanakoti, S. N., Brosnan, J. T., Herzberg, G. R., and Brosnan, M. E.,** Renal arginine synthesis: studies in vitro and in vivo, *Am. J. Physiol.,* 259, E437, 1990.

62. **Atkinson, D. E. and Bourke, E.,** The role of ureagenesis in pH homeostasis, *Trends Biochem. Sci.,* 9, 291, 1984.

63. **Brosnan, J. T., Lowry, M., Vinay, P., Gougoux, A., and Halperin, M. L.,** Renal ammonium production—une vue canadienne, *Can. J. Physiol. Pharamacol.,* 65, 489, 1987.

64. **Haussinger, D.**, Nitrogen metabolism in liver: structural and functional organization and physiological relevance, *Biochem. J.*, 267, 281, 1990.

65. **Morris, S. M.**, Regulation of enzymes of urea and arginine synthesis, *Annu. Rev. Nutr.*, 12, in press.

66. **Corvera, S. and Garcia-Sainz, J. A.**, Hormonal stimulation of mitochondrial glutaminase. Effects of vasopressin, angiotensin II, adrenaline and glucagon, *Biochem. J.*, 210, 957, 1983.

67. **Kashiwagura, T., Erecinska, M., and Wilson, D. F.**, pH Dependence of hormonal regulation of gluconeogenesis and urea synthesis from glutamine in suspensions of hepatocytes, *J. Biol. Chem.*, 260, 407, 1985.

68. **Verhoeven, A. J., Estrela, J. M., and Meijer, A. J.**, Adrenergic stimulation of glutamine metabolism in isolated rat hepatocytes, *Biochem. J.*, 230, 457, 1985.

69. **McGivan, J. D., Vadher, M., Lacey, J., and Bradford, N.**, Rat liver glutaminase. Regulation by reversible interaction with the mitochondrial membrane, *Eur. J. Biochem.*, 148, 323, 1985.

70. **Haussinger, D. and Lang, F.**, The mutual interaction between cell volume and cell function: a new principle of metabolic regulation, *Biochem. Cell. Biol.*, 69, 1, 1991.

71. **vonDahl, S., Hallbrucker, C., Lang, F., and Haussinger, D.**, Regulation of liver cell volume and proteolysis by glucagon and insulin, *Biochem. J.*, 278, 771, 1991.

72. **Ochs, R. S.**, Glutamine metabolism in isolated hepatocytes. Evidence for catecholamine stimulation of α-ketoglutarate dehydrogenase, *J. Biol. Chem.*, 259, 13004, 1984.

73. **Charles, R.**, Mitochondriale Citrulline Synthese: een Ammonik Fixerend en ATP Verbruikend Proces, Ph.D. thesis, University of Amsterdam, Rototype, Amsterdam, 1968.

74. **McGivan, J. D. and Bradford, N. M.** Characteristics of the activation of glutaminase by ammonia in sonicated rat liver mitochondria, *Biochim. Biophys. Acta*, 759, 241, 1983.

75. **Welbourne, T. C.**, Interorgan glutamine flow in metabolic acidosis, *Am. J. Physiol.*, 253, F1069, 1987.

76. **Kaiser, S., Gerok, W., and Haussinger, D.**, Ammonia and glutamine metabolism in human liver slices. New aspects of the pathogenesis of hyperammonemia in chronic liver disease, *Eur. J. Clin. Invest.*, 18, 535, 1988.

77. **Quesada, A. R., Medina, M. A., Marquez, J., Sanchez-Jimenez, F. M., and Nunez de Castro, I.**, Contribution by host tissue to circulating glutamine in mice inoculated with Ehrlich ascites tumor cells, *Can. Res.*, 48, 1551, 1988.

78. **Watford, M.**, A swell way to regulate metabolism, *Trends Biochem. Sci.*, 15, 329, 1990.

79. **Tannen, R. L. and Sastrasinh, S.**, Response of ammonia metabolism to acute acidosis, *Kidney Int.*, 25, 1, 1984.

80. **Schoolwerth, A. C. and LaNoue, K.**, Transport of metabolic substrates into renal mitochondria, *Annu. Rev. Physiol.*, 47, 143, 1985.

81. **Folin, O.**, Laws governing the chemical composition of urine, *Am. J. Physiol.*, 13, 66, 1905.

82. **Freedland, R. A. F. and Briggs, S.**, *A Biochemical Approach to Nutrition*, Chapman and Hall, London, 1977, 46.

83. **Freedland, R. A. F. and Taylor, A. R.**, Studies on glucose-6-phosphatase and glutaminase in rat liver and kidney, *Biochim. Biophys. Acta*, 92, 567, 1964.

84. **Simell, O.**, Effect of low protein diet and hyperammonemia on liver glutaminase activity in the rat, *Experientia*, 30, 324, 1974.

85. **Curthoys, N. P. and Lowry, O. H.**, The distribution of glutaminase isoenzymes in the various structures of the nephron in normal, acidotic, and alkalotic rat kidney, *J. Biol. Chem.*, 248, 162, 1973.

86. **Watford, M., Smith, E. M., and Erbelding, E. J.**, Regulation of phosphate-activated glutaminase activity and glutamine metabolism in the streptozotocin-diabetic rat, *Biochem. J.*, 224, 207, 1984.

87. **Watford, M., Erbelding, E. J., Shapiro, A. C., Zakow, A. C., and Smith, E. M.,** The adaptive response of phosphate-activated glutaminase in the rat, *Contr. Neph.*, 47, 140, 1985.

88. **Watford, M., Zhan, Z., and Vincent, N. C.,** unpublished.

89. **McGivan, J. D., Boon, K., and Doyle, F. A.,** Glucagon and ammonia influence the long-term regulation of phosphate-dependent glutaminase activity in primary cultures of rat hepatocytes, *Biochem. J.,* 274, 103, 1991.

90. **Snodgrass, P. J., Lin, R. C., Muller, W. A., and Aoki, T. T.,** Induction of urea cycle enzymes of rat liver by glucagon, *J. Biol. Chem.,* 253, 2748, 1978.

91. **Watford, M.,** Regulation of expression of the genes for glutaminase and glutamine synthetase in the acidotic rat, *Contr. Neph.,* 92, 211, 1991.

92. **Watford, M., Erbelding, E. J., and Smith, E. M.,** The regulation of glutamine and ketone body metabolism in the small intestine of the long-term (40-day) streptozotocin-diabetic rat, *Biochem. J.,* 242, 61, 1987.

93. **Nagy, L. E. and Kretchmer, N.,** Effect of diabetic ketosis on jejunal glutaminase, *Arch. Biochem. Biophys.,* 248, 80, 1986.

94. **Mock, B., Kozak, C., Seldin, M. F., Ruff, N., D'Hoostelaere, L., Szpirer, C., Seuanez, H., O'Brien, S., and Banner, C.,** A glutaminase (Gls) gene maps to mouse chromosome 1, rat chromosome 9, and human chromosome 2, *Genomics,* 5, 291, 1989.

95. **Nyunoya, H., Broglie, K. E., and Lusty, C. J.,** The gene coding for carbamoyl-phosphate synthetase I was formed by fusion of an ancestral glutaminase gene and a synthetase gene, *Proc. Natl. Acad. Sci. U.S.A.,* 82, 2244, 1985.

96. **Zalkin H., Argos, P., Narayana, S. V. L., Tiedeman, A. A., and Smith, J. M.,** Identification of a trpG-related glutamine amide transfer domain in *Escherichia coli* GMP synthetase, *J. Biol. Chem.,* 260, 3350, 1985.

97. **Gebhardt, R. and Mecke, D.,** Heterogeneous distribution of glutamine synthetase among rat liver parenchymal cells in situ and in primary culture, *EMBO J.,* 2, 567, 1983.

98. **Jungermann, K. and Katz, N.,** Functional specialization of different hepatocyte populations, *Physiol. Rev.,* 69, 708, 1989.

99. **Gaasbeek-Janzen, J. W., Lamers, W. H., Moorman, A. F. M., deGraaf, A., Los, J. A., and Charles, R.,** Immunohistochemical localization of carbamoylphosphate synthetase (ammonia) in adult rat liver. Evidence for a heterogeneous distribution, *J. Histochem. Cytochem.,* 32, 557, 1984.

100. **Watford, M. and Smith, E. M.,** Distribution of hepatic glutaminase activity and mRNA in perivenous and periportal hepatocytes, *Biochem. J.,* 267, 265, 1990.

101. **Moorman, A. F. M., deBoer, P. A. J., Charles, R., and Lamers, W. H.,** Diet and hormone induced reversal of the carbamoylphosphate synthetase mRNA gradient in the rat liver lobulus, *FEBS Lett.,* 276, 9, 1990.

102. **Watford, M. and Smith, E. M.,** Regulation of hepatic glutaminase mRNA levels in the rat, *Biochem. Soc. Trans.,* 17, 175, 1988.

103. **Tong, J., Shapiro, R. A., and Curthoys, N. P.,** Changes in the levels of translatable glutaminase mRNA during onset and recovery from metabolic acidosis, *Biochemistry,* 26, 2773, 1987.

104. **Kaiser, S. and Curthoys, N. P.,** Effect of pH and bicarbonate on phosphoenolpyruvate carboxykinase and glutaminase mRNA levels in cultured renal epithelial cells, *J. Biol. Chem.,* 266, 9397, 1991.

105. **Schimke, R. and Doyle, D.,** Control of enzyme levels in animal tissues, *Annu. Rev. Biochem.,* 39, 929, 1970.

106. **Morris, S. M., Moncman, C. L., Rand, K. D., Dizikes, G. J., Cederbaum, S. D., and O'Brien, W. E.,** Regulation of mRNA levels for five urea cycle enzymes in rat liver by diet, cyclic AMP and glucocorticoids, *Arch. Biochem. Biophys.,* 256, 343, 1987.

107. **Bond, J. S., Failla, M. L., and Unger, D. F.,** Elevated manganese concentration and arginase activity in livers of streptozotocin-induced diabetic rats, *J. Biol. Chem.,* 258, 8004, 1983.

108. **Schimke, R. T.,** The importance of both synthesis and degradation in the control of arginase levels in rat liver, *J. Biol. Chem.,* 239, 3808, 1964.
109. **vanRoon, M. A., Zonneveld, D., Charles, R., and Lamers, W. H.,** Accumulation of carbamoylphosphate synthetase and phosphoenolpyruvate carboxykinase mRNA in embryonic rat hepatocytes. Evidence for translational control during the initial phases of hepatocyte specific gene expression *in vitro, Eur. J. Biochem.,* 178, 191, 1988.
110. **Nebes, V. L. and Morris, S. M.,** Regulation of messenger ribonucleic acid levels for five urea cycle enzymes in cultured rat hepatocytes. Requirements for cyclic adenosine monophosphate, glucocorticoids and ongoing protein synthesis, *Mol. Endocrinol.,* 2, 444, 1988.
111. **Jackson, M. J., Beaudet, A. L., and O'Brien, W. J.,** Mammalian urea cycle enzymes, *Annu. Rev. Genet.,* 20, 431, 1986.
112. **Takiguchi, M., Matsubasa, T., Amaya, Y., and Mori, M.,** Evolutionary aspects of urea cycle genes, *BioEssays,* 10, 163, 1989.
113. **Boyce, F. M., Anderson, G. M., Rusk, C. D., and Freytag, S. O.,** Human argininosuccinate synthetase minigenes are subject to arginine mediated repression but not to trans induction, *Mol. Cell. Biol.,* 6, 1244, 1986.
114. **Jackson, M. J., O'Brien, W. E., and Beaudet, A. L.,** Arginine-mediated regulation of an argininosuccinate synthetase minigene in normal and canavanine resistant human cells, *Mol. Cell. Biol.,* 6, 2257, 1986.
115. **Engelhardt, J. F., Steel, G., and Valle, D.,** Transcriptional analysis of the ornithine aminotransferase promoter, *J. Biol. Chem.,* 266, 752, 1990.

Chapter 16

EFFECTS OF DIETARY PROTEIN ON GENE EXPRESSION

David K.-C. Chan and James L. Hargrove

TABLE OF CONTENTS

0-8493-6961-4/93/$0.00 + $.50

353

I. OPPOSING EFFECTS OF DIETARY PROTEIN AND CARBOHYDRATE

When meals that contain protein are ingested, the liver is exposed to elevated concentrations of amino acids and hormones in the blood of the hepatic portal vein. This produces two important effects that alter the regulation of specific genes. The first of these effects is an increase in the size of the liver and kidneys. Hepatic and renal cells proliferate, and this involves an increase in DNA synthesis, RNA synthesis, and the activation of enzymes needed for synthesis of cellular constituents.[1,2] In addition to cell division, cell growth occurs with accumulation of protein. The second effect is an increase in the concentration of enzymes needed to catabolize the excess amino acids, dispose of the unneeded amino group, and utilize the resulting carbon chain.[3-6] Due to this adaptation, the concentration of amino acids in the general circulation does not increase to the degree noted in the portal circulation.

Although amino acids can be converted to glucose through the processes of *trans-* or deamination and gluconeogenesis, fatty acids do not fuel gluconeogenesis efficiently in liver. This causes dietary protein and carbohydrate to have a special, opposing relationship on the synthesis of enzymes involved in amino acid catabolism. Fasting or consuming diets that contain relatively large amounts of protein causes adaptive increases in the concentrations of enzymes in liver and kidney that convert amino acids to precursors for synthesis of glucose and fatty acids;[4-6] conversely, dietary carbohydrate decreases the activity of enzymes that participate in gluconeogenesis and amino acid catabolism.[7,8] The ability of carbohydrates to decrease the synthesis of enzymes that participate in catabolism of amino acids is superficially similar to catabolite repression in bacteria and has been referred to as glucose repression. These responses have been characterized best in animals fed experimental diets that provide extreme levels of protein or carbohydrate and produce large changes in enzyme levels. Although these diets are extreme in comparison to the mixed diets typical of North Americans, adaptive changes in gene expression probably occur in response to diets of any composition. In the words of Helen Teppermann et al.,[9] "the business of differentiation in the liver is eternally unfinished", and this comment no doubt applies to cells of the endocrine system and central nervous system that contribute to nutrient-initiated signaling.

The means by which the diet affects genetic activity probably differs for individual organs, and depends on the duration of any prior fast and the composition of the ensuing meals. Levels of nutrients in the blood plasma are significant even during starvation, and their concentrations increase only moderately during the absorptive phase. For example, it is unusual for the concentration of glucose or amino acids in the general circulation to increase by more than 50% after a meal in nondiabetic subjects.[10-13] The concentration in the splanchnic circulation may increase two- or threefold. Similarly, eating a single, high-fat meal increases the level of fatty acids in the blood of human subjects by 20 to 40% compared to fasting levels.[14] The intestinal secretion of chylomicrons causes turbidity of the blood plasma to increase noticeably, even though the degree of lipemia is not extreme. Therefore, changes in gene expression in response to dietary modification take place during periods when levels of nutrients differ from values found in fasted subjects by relatively small amounts, but the expression of particular genes may increase or decrease by factors of tenfold or more as a consequence of signal amplification within responsive cells. This chapter recounts how dietary protein and carbohydrate affect hormonal secretion and genes that encode several inducible enzymes.

II. DIETARY PROTEIN AS A STIMULUS TO ORGAN GROWTH

The liver and kidneys both play important roles in regulating levels of amino acids in the blood plasma and in synthesizing glucose from excess amino acids. Perhaps for this reason, dietary protein promotes growth of these organs more strongly than it stimulates growth of the rest of the body. This topic has received considerable attention due to the concern that excessive amino acids may cause kidney function to deteriorate;[15] in addition, excess amino acids stimulate kidney growth, and this could be beneficial in some cases.[2] Depending on the level of protein, the weight of these organs in rats may increase from 10 to 50% over a period of several days, and the rate of DNA replication increases, as judged by incorporation of radiolabeled deoxythymidine.[16] The amount of protein relative to DNA also increases, suggesting that protein or amino acids promote both hypertrophy and hyperplasia; hyperplasia predominates in young rats and hypertrophy predominates in older ones.[17] In addition to their effects on organ growth, amino acids rapidly increase the rate of renal blood flow (hyperemia) and glomerular filtration rate (GFR).[15,18] When the subjects are refed diets containing lower levels of protein, the blood flow and organ weights return to normal values.

Amino acids are thought to affect the function of kidney and liver by direct and indirect mechanisms. Mixtures of amino acids that cause renal hyperemia also stimulate secretion of insulin, glucagon, and growth hormone.[18] When infused alone, glucagon increases GFR. However, a mixture of branched-chain amino acids increased secretion of insulin and growth hormone without affecting GFR.[18] This evidence suggests that glucagon participates in the hyperemic response to dietary protein. In addition to the hormonal responses, amino acids may generate other mediators that affect organ function. Arginine not only stimulates glucagon secretion, but is converted to nitric oxide by the vascular endothelium. Nitric oxide increases blood flow and alters levels of cyclic guanosine $3',5'$-monophosphate in glomeruli.[19] It counteracts the action of vasoconstrictors, such as the local vascular hormone, endothelin. Other mediators may also be involved, for example, the prostaglandins. Inhibitors of cyclooxygenase prevent the synthesis of prostaglandins, which can dilate renal blood vessels. In the relative absence of eicosanoids, amino acids do not induce hyperemia. Moreover, dietary protein activates the renin-angiotensin system by increasing plasma renin activity; the quantity of mRNA encoding renin increases in kidney but the message for angiotensinogen is unchanged.[20]

What promotes the growth response of liver and kidney to amino acids? The answer is probably complex, because both organs contain receptors for a variety of steroidal and peptide hormones, as well as for growth factors and cytokines. Stimuli that are associated with cell growth almost invariably increase the activity of ornithine decarboxylase (ODC), and dietary protein potently stimulates activity of this enzyme in both organs. Farwell et al. observed that diets that contained 26% casein caused a 20-fold increase in ODC activity in liver and a 10-fold increase in kidneys in comparison to responses of rats fed a protein-free diet.[16] Peak values occurred during the nocturnal feeding period and declined with continued feeding of the protein-containing diet. The protein-containing diet stimulated hepatic DNA synthesis, but this repsonse was prevented by propane-1,3-diamine, an inhibitor of ODC activity. Kaysen et al.[21] observed that the increase in glomerular filtration rate occurred before the ODC response, and that ODC activity returned to low values after three days of feeding a high protein diet. The changes in GFR and ODC activity preceded changes in kidney weight and were implicated in the cell proliferation which was needed for kidney growth. The effect of amino acids on ODC activity could be reproduced by injecting casein hydrolyzate or mixtures of amino acids, and a mixture of glutamate, aspartate, and alanine was the smallest group that gave the full response.[22] All effective mixtures increased the content of polyamines in kidney, and it was postulated that effective groups also stimulated ureagenesis. Although other genes in the kidney also respond to dietary protein, the response is much less than the magnitude noted for ODC activity. For instance, the mRNAs for arginosuccinate synthetase and lyase increase only about threefold in animals fed a 60% protein diet.[23] Ornithine decarboxylase activity is regulated at multiple levels, and the points affected by dietary protein have not yet been explored in the kidney. It will be interesting to learn more about mediators of the growth-promoting effects of dietary protein on these two organs.

III. HORMONAL SECRETION DURING FASTING AND REFEEDING

Many nutritional and biochemical studies concerning expression of genes for enzymes have utilized an experimental design in which animals that have been starved for 2 or 3 d are then refed a diet of specified composition. This starvation-refeeding paradigm produces reproducible changes in hormone se-

TABLE 1
Effects of Energy-Yielding Nutrients on Hormone Secretion

Organ/Hormone	Glucose	Amino acids	Fatty acids	Ref.
Gut				
Neurotensin	0[a]	0	+ +	25
Peptide YY	+	+	+ +	26
Gastric inhibitory peptide	+ +	0/ +	+ +	27
Cholecystokinin	0/ +	+ +	+ +	28
Pancreas				
Insulin	+ +	+	+ /0	29–32
Glucagon	− [b]	+ +	−	24, 32–34
Pancreatic polypeptide	+ / − [b]	+ +	−	26
Anterior pituitary				
Growth hormone	− [b]	+ +	−	23, 34–37
Adrenocorticotropin	0/ − [b]	+ +	0	38–42
Thyroid stimulating hormone		−	−	43
Adrenal				
Cortisol	+	+ +	0	32, 39–41
Thyroid gland				
T_3	+	−	0	32, 40, 44
T_4	−	+	0	44
T_3/T_4	+	−	0	

[a] Definition of symbols: 0, no effect; +, stimulatory effect; −, inhibitory effect; + /0 or + / − , variable.

[b] Hypoglycemia is a strong stimulus for secretion of these hormones.

cretion during the initial phase, but changes that occur after refeeding depend on composition of the diet, and the hormones participate in activating specific genes. Table 1 summarizes effects of glucose, amino acids, and fatty acids on secretion of a variety of hormones from the gut, pancreas, and anterior pituitary gland. It is evident that each class of energy-yielding macronutrient produces a distinct pattern of hormone secretion from the intestinal mucosa, pancreas, and pituitary. Peptide hormones are gene products of the endocrine system, and nutrients frequently affect gene expression in hormone-secreting cells; glucose regulation of transcription from the insulin gene is an important example.[24]

A. PEPTIDES FROM THE GUT AND PANCREAS

Peptide hormones from the enteroendocrine system, such as cholecystokinin, gastric inhibitory peptide, and pancreatic polypeptide, are secreted into the hepatic portal circulation in response to nutrients present in the lumen of the gut (Table 1),[25-44] and starvation causes their output to decline. Glucose stimulates secretion of gastic inhibitory peptide and insulin with weaker effects on other hormones. Amino acids stimulate secretion of pancreatic polypeptide from the intestine and pancreas, and augment secretion of glucagon and insulin; dietary protein or amino acids also increase secretion of growth hormone, adrenocorticotropin (ACTH), and cortisol. Fatty acids have minor effects on insulin and glucagon but strongly stimulate secretion of neurotensin, peptide YY, gastic inhibitory peptide, and cholecystokinin.

Tissues that contain receptors for these peptide hormones may respond to diet-induced secretion of the gut peptides. For example, the endocrine and exocrine pancreas contains receptors for many peptides.[45] Insulin output is increased by gastric inhibitory peptide, pancreatic polypeptide, neurotensin, and bombesin, and may be decreased by galanin. The liver contains receptors for secretin, pancreatic polypeptide, and vasoactive intestinal peptide,[46] and starvation would be expected to decrease the activity of signaling pathways linked to these receptors. At present, it is not clear to what extent peptides from the intestinal mucosa affect gene expression in the liver. However, vasoactive intestinal peptide activates adenylate cyclase,[47] and cyclic adenosine-3'5'-monophosphate (cAMP) regulates numerous hepatic genes. During the initial phase of starvation in humans and rats, secretion of insulin from the pancreas declines as levels of glucose and other secretagogues decrease in the blood plasma.[24,48-51] The output of pancreatic glucagon changes less than insulin, but fasting causes an initial increase in glucagon secretion in the human.[50] In both species, the ratio of insulin to glucagon in the blood plasma declines. Fasting increases the activity of the sympathetic nervous system and leads to increased secretion of epinephrine from the adrenal medulla.[52] Rising levels of catecholamines and ACTH stimulate output of glucocorticoids from the adrenal cortex and antagonize the action of insulin while enhancing the action of glucagon on expression of genes that participate in gluconeogenesis.

Studies done with experimental animals suggest that the relative levels of carbohydrate and protein in the diet modify the ratio of insulin to glucagon secreted into the plasma.[53] Glucose is the primary stimulus to insulin secretion and is necessary for secretion in response to most amino acids; in contrast, glucose inhibits glucagon secretion, whereas amino acids stimulate this process.[33] In portal blood, the ratio of glucagon to insulin was correlated with the concentrations of total amino acids and branched-chain plus aromatic amino acids.[53] Diets containing high carbohydrates produce different metabolic effects on liver than diets with high protein content. With mixed or high carbohydrate diets, the level of glycogen in liver increases during the feeding period, and cyclic AMP levels remain low. However, diets containing elevated protein stimulate production of cAMP, and little glycogen is deposited.[54] This is confirmed by measurements of enzyme activities that are regulated in opposing fashion by insulin compared to glucagon or cAMP. Diets containing high protein caused phosphoenolpyruvate carboxykinase (PEPCK) levels to increase, whereas pyruvate kinase declined, and carbohydrates promoted the opposite effect.[55]

The ability of different amino acids to promote hormone secretion differs among hormones and across species. In general, amino acids are weaker secretagogues for insulin than is glucose, but leucine, glycine, serine, and alanine all enhanced insulin secretion after intravenous administration to sheep.[56] While sheep are ruminants, monogastric animals have similar responses with respect to insulin release. In many species, arginine stimulates insulin secretion more potently than other amino acids. Secretion of glucagon in sheep was elevated most by alanine, glycine, serine, and arginine, whereas aspartic acid, phenylalanine, and glutamate were the most effective releasers of growth hormone (GH or somatotropin).[56] The hormonal response caused by refeeding diets that contain elevated protein may depend on the relative level of different amino acids in the protein source.

B. PITUITARY HORMONES, THYROXINE, AND GLUCOCORTICOIDS

Nutrient status strongly modifies hormonal secretion from the pituitary gland, and this effect is thought to be mediated by altered synthesis and secretion of neurotransmitters and releasing factors from specific areas in the hypothalamus. Fasting in humans and experimental animals causes secretion of most pituitary hormones to decline, and brings about a reversible condition with symptoms similar to effects of hypophysectomy.[48-51] Growth is inhibited, reproductive cycles cease, and the metabolic rate declines because the secretion of thyroid-stimulating hormone, prolactin, luteinizing hormone, and follicle-stimulating hormone is depressed. Decreased secretion of gonadotropic hormones represses the preovulatory surge of LH secretion in the female and causes a decline in output of gonadal steroids in both sexes. Sex steroids are anabolic agents, and their decreased secretion reduces the capacity to maintain synthesis of proteins that are essential to normal functions of muscle and bone. In the rat, secretion of GH declines, whereas in humans output of GH increases during fasting or hypoglycemia.

The response of the thyroid gland to stimulation by thyrotropin diminishes with food deprivation in man and experimental animals.[44,57] Although plasma concentrations of thyroid-stimulating hormone (TSH) may not differ greatly, levels of $3,5,3'$-triiodothyronine (T_3) and T_4 decline during long-term food deprivation, and TSH levels do not rise to compensate. Pituitary function is affected, as indicated by a lower level of mRNA for thyroid-stimulating hormone in the anterior pituitary gland of rats. The mRNA for thyrotropin-releasing hormone in the hypothalamus declines in rats.[58]

Acute hypoglycemia, which can be caused by injecting insulin, increases the secretion of counter-regulatory hormones from the pituitary gland, including ACTH, GH, and prolactin. This effect has been postulated to involve glucoreceptors in specific areas of the brain, and is thought to reflect the lack of glucose and not the presence of insulin. For instance, administering sufficient glucose to prevent hypoglycemia inhibited the release of ACTH and

GH caused by insulin.[38] During starvation, elevated output of ACTH causes the weight of the adrenal glands to increase, and increases plasma levels of corticosterone three- to fivefold in rats.[39-41]

Effects of GH on metabolism partly result from increased synthesis and secretion of the somatomedin, insulin-like growth factor I (IGF-I) from liver and other tissues, which is initiated through binding of GH to its receptor. In rats, the level of IGF-I declines during fasting or restriction of dietary protein, and this is paralleled by decreased levels of IGF-I mRNA in liver.[59] Fasting also causes resistance to the action of GH, and this may be a consequence of decreased levels of the GH receptor in liver.[60] Fasting causes the concentration of this receptor and its mRNA to decline in liver.

The secretion of GH is controlled by GH-releasing factor (GHRF), somatostatin, and neurotransmitters such as dopamine,[61] and the cause for the changes due to fasting and refeeding have not been fully explored. However, Bruno et al.[62,63] have found that food deprivation in rats decreased the concentration of the mRNA for prepro-GHRF in the hypothalamus without affecting the mRNA for somatostatin. Upon refeeding, low levels of protein (4 to 12% of the diet by weight) were found to increase the concentration of prepro-GHRF mRNA. Feeding meals containing protein increases the concentration of amino acids in the brain,[64] and GH secretion is stimulated by an intravenous infusion of amino acids in man and other animals.[36,65] In sheep, the response of GH to arginine and leucine was smaller than responses to the acidic amino acids. Several amino acids caused a mild hyperglycemia, with glycine, asparagine, serine, alanine, threonine, glutamine, arginine, methionine, proline, and phenylalanine being most effective. Leucine caused plasma glucose levels to decline. In humans, the amino acids that best stimulated secretion of growth hormone were arginine, histidine, lysine, leucine, and phenylalanine.[65]

During refeeding, metabolic effects produced by the diet are most extreme when the majority of calories are derived from one nutrient. Matzen et al.[36] compared effects in human subjects who ingested 500 kcal of carbohydrate (dextrin-maltose) or an isocaloric diet with 80% of calories from protein. Plasma glucose and insulin did not rise in subjects receiving the protein diet, but levels of GH and glucagon were both elevated, and TSH gradually declined. During this period, levels of T_3, reverse T_3, and T_4 did not change significantly. In evaluating studies that make use of the fasting-refeeding paradigm, it is important to be mindful of these complex effects of nutrients on hormonal secretion.

IV. PHYSIOLOGICAL EFFECTS OF AMINO ACIDS

Although amino acids serve a common purpose during protein synthesis, each amino acid has unique properties and functions in interorgan metabolism, some of which may influence gene expression. For example, alanine and glutamine are released from extrahepatic tissues (mainly skeletal muscle) under gluconeogenic conditions.[5,6] Alanine is an important glucose precursor through the glucose-alanine cycle, whereas glutamine serves as an important energy source for the intestine and kidney.[66] Besides serving as a glucose precursor, alanine facilitates gluconeogenesis by inhibiting pyruvate kinase,[67,68] thus preventing a futile cycle from developing between glycolysis and gluconeogenesis. In addition, alanine inhibits proteolysis and stimulates protein synthesis in liver,[69,70] which may help prevent excessive loss of protein during starvation. Finally, alanine serves as a potent stimulus to release of glucagon,[71] which is elevated relative to insulin by fasting, diabetes, or ingestion of diets containing protein. Leucine, or a product of leucine metabolism, inhibits protein degradation in muscle and liver,[72] and is a potent secretagogue for glucagon and insulin.[73] Aspartic acid and glutamate are excitatory neurotransmitters,[74] and both stimulate secretion of GH from the anterior pituitary after intravenous infusion. It is feasible that some central effects of acidic amino acids could be mediated through receptors in circumventricular organs. In addition to serving as a precursor for neurotransmitter synthesis, tryptophan exerts numerous metabolic effects including stabilization of several hepatic enzymes that do not participate in its metabolism, including tyrosine aminotransferase and phosphoenolpyruvate carboxykinase.[75] Tryptophan or a related metabolite activates PEP carboxykinase by a mechanism that may require glutathione peroxidase.[76,77] Pharmacological doses of tryptophan stimulate protein synthesis, and evidence suggests that it binds to a receptor on the nuclear envelope that may facilitate translocation of mRNA into the cytoplasm.[78] Arginine stimulates secretion of insulin, glucagon, and somatotropin, and is the substrate used for synthesis of nitric oxide, a relaxant of smooth muscle in the vascular endothelium.[79]

After protein-containing meals have been ingested, the level of amino acids increases in the portal and systemic blood. The partial list of functions described above suggests that hyperaminoacidemia may have numerous metabolic consequences that are unrelated to protein synthesis, but that may be mediated by altered hormone secretion, activation of the autonomic nervous system, formation of ammonia, or by several other processes. Therefore, it is essential that the level of catabolic enzymes for amino acids adapt to the dietary intake and prevent large changes in their concentration.

V. REGULATION OF ENZYME SYNTHESIS BY DIETARY PROTEIN

A. EFFECTS OF DIETARY COMPOSITION ON ENZYME RESPONSES

The metabolic response of the liver to refeeding diets with different compositions has been studied extensively, and it is clear that nutrient levels as well as hormonal effects are important in the induction or repression of

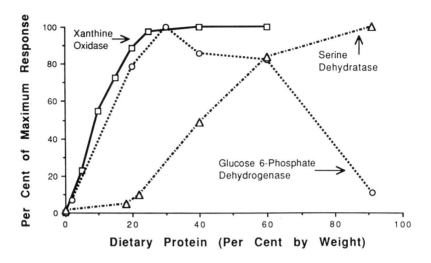

FIGURE 1. Patterns of induction by dietary protein differ for various enzymes and peptides. The threshold of induction is quite low for xanthine oxidase (squares) and glucose-6-phosphate dehydrogenase (circles), but is substantially higher for serine dehydratase (triangles) and most transaminases (not shown). Maximal expression of xanthine oxidase occurs at a dietary casein level of about 20%, whereas serine dehydratase continues to increase in proportion to protein consumed, and glucose-6-phosphate declines at high levels of protein. Data were replotted from articles by Harper,[1] Potter and Ono,[81] and Ogawa et al.[82]

various enzymes. The synthesis of lipogenic enzymes is increased by diets containing high carbohydrate content,[80] and these conditions suppress synthesis of several transaminases, dehydratases, and enzymes of the urea cycle that are sensitive to dietary amino acids. As a group, the concentrations of these enzymes increase with dietary protein and decrease with dietary carbohydrate.[4-8] However, the level of protein required to induce a response differs for various enzymes. As shown in Figure 1, refeeding diets that contain as little as 5% protein by weight causes induction of xanthine oxidase and glucose-6-phosphate dehydrogenase,[81] whereas a threshold of about 18% protein is required to induce serine dehydratase and its mRNA.[82] Glucose-6-phosphate dehydrogenase is interesting in the requirement for both amino acids and carbohydrate to achieve full induction, and dietary fat inhibits induction of the enzyme.[83] Whereas xanthine oxidase reaches a plateau beyond which it is not further induced at about 20% protein, and very high levels of protein cause glucose-6-phosphate dehydrogenase to decline,[81] serine dehydratase continues to increase in proportion to the amount of amino acids that is ingested.

It is interesting to compare the response of gluconeogenic enzymes during fasting and refeeding to that of glucokinase. During starvation, the liver must maintain euglycemia, and glucokinase hinders this by trapping glucose in the liver as glucose-6-phosphate. Levels of glucokinase decline during starvation,

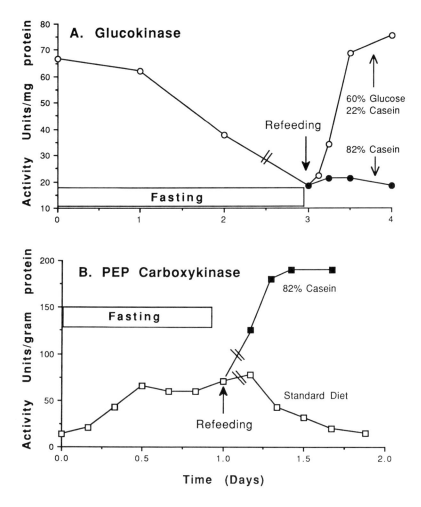

FIGURE 2. Enzyme induction in response to refeeding depends on composition of the diet. Glucokinase concentration in liver declines during fasting and is induced to high levels in response to mixed diets that contain a large proportion of carbohydrate, but not by diets that contain extreme levels of protein (Panel A). PEP carboxykinase levels increase during fasting in support of gluconeogenesis, and diets containing high levels of carbohydrate rapidly inhibit mRNA transcription and cause the enzyme concentration to decline (Panel B, open squares). However, refeeding diets that contain high levels of protein increase synthesis of PEP carboxykinase, and the enzyme concentration increases. The figure was prepared from data reported by Blumenthal et al.,[84] Hopgood et al.,[85] and Seitz et al.[86]

whereas levels of PEP carboxykinase increase (Figure 2).[84,85] After feeding, the level of glucokinase increases and PEP carboxykinase decreases in proportion to dietary carbohydrate. In contrast, dietary protein fails to produce an increase in glucokinase in fasted rats, and increases the activity of PEP carboxykinase (Figure 2). The decline in PEP carboxykinase activity results from the ability of insulin and carbohydrate to inhibit mRNA production.[87] Recent studies in transgenic mice have shown that the stimulatory effect of protein on PEPCK expression in liver and kidney is also due to increased mRNA accumulation.[88]

When high-protein diets are first introduced, the level of amino acids in the bloodstream increases greatly after feeding. However, continual feeding of diets with elevated protein leads to adaptive increases in the capacity of the liver to metabolize amino acids. As a result, the concentration of amino acids that are degraded in the liver actually decreases in well-adapted animals. No comparable increase occurs for the branched-chain amino acid aminotransferase, which is located in the mitochondria of skeletal muscle and liver, but the keto acid dehydrogenase activity increases in proportion to dietary protein.[89] Therefore, the level of leucine, valine, and isoleucine increases linearly as dietary protein content increases. This may serve several useful purposes, such as permitting the adaptation to high levels of protein to continue despite the increased capacity for degradation of amino acids. When high protein diets are fed continually, the activity of enzymes involved in lipogenesis and glycolysis decreases and remains at a low level.

B. DIETARY REGULATION OF TYROSINE AMINOTRANSFERASE

Tyrosine aminotransferase (TyrAT; E.C. 2.6.1.5) provides a useful example of a mammalian gene that is regulated by dietary protein and carbohydrate. This enzyme is found only in liver, and the gene is inactive during most of fetal life. Transcription does not begin until the last third of gestation in the rat (Figure 3A). Just prior to birth, the promoter becomes active in hepatocytes but not elsewhere, and the level of TyrAT reaches a value that is less than one tenth of adult levels.[90,91] In the earliest period, the rate of transcription appears to be fixed at a low level, and no response is obtained for hormones that augment transcription. However, just prior to birth this changes, as the gene acquires the ability to respond weakly to glucagon or cyclic AMP, and marginally to hydrocortisone.

Just after birth, the response of TyrAT changes impressively. At this time, animals that fail to replenish their stores of blood glucose must initiate gluconeogenesis or risk death, and synthesis of a set of enzymes that supports this process begins at a high rate. The enzyme levels increase manyfold within a few hours of birth, surge past the adult levels, and then fall back to typical values that are roughly twentyfold higher than were found in the fetal liver.[90,91] Not only does transcription from the promoter increase to a higher "basal" rate,[92] but the capacity to respond to glucocorticoid hormones becomes prominent (Figure 3A). Whereas the fetal hepatocyte responds with at best a twofold increase in enzyme after steroid treatment, in the adult, responses of tenfold are usual at optimal doses of hormone.

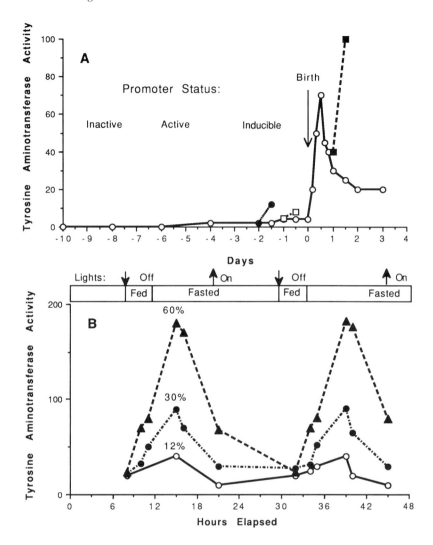

FIGURE 3. Regulation of the gene encoding the enzyme, tyrosine aminotransferase, is typical of mammalian genes that are induced by dietary protein. Panel A: Acquisition of the capacity to synthesize this enzyme and its mRNA occurs in liver during the latter third of gestation. Enzyme activity is indicated by the open circles. At the time of birth (arrow), synthesis begins at a high rate in support of gluconeogenesis, and the capacity to respond to diet and hormones begins. Symbols indicate the following treatments: injection of fetuses with cyclic AMP (closed circles) of hydrocortisone (open squares); injection of neonates with hydrocortisone (closed squares). Panel B: The quantity of enzyme in adult liver undergoes a diurnal cycle because mRNA synthesis responds to feeding and the level of protein in the diet. Tissue from animals that are fed a diet containing 12% protein contains very little TyrAT, and the diurnal variation is small (open circles). The variation becomes much larger when dietary protein makes up 30% (closed circles) or 60% (triangles) of the diet by weight. Diurnal rhythms of this magnitude require that the enzyme and its mRNA are both unstable, as predicted by the kinetic theory discussed in Chapter 1. The figure is a composite drawn from data reported in References.[90-93]

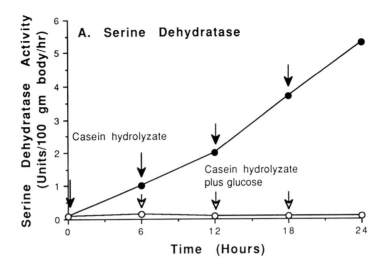

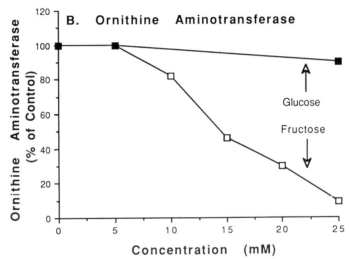

FIGURE 4.   Glucose interferes with the ability of protein to increase the synthesis of several enzymes of amino acid catabolism in liver. (A) Serine dehydratase is induced by dietary protein or by hydrolyzed casein, administered by gastric incubation to intact rats at the times indicated by arrows (closed circles). If 1 g of glucose is included along with the casein, the induction is prevented (open circles). (B) Carbohydrate repression of ornithine aminotransferase can be observed in isolated cells. Hepatocytes were prepared from rats fed a low protein diet and cultured in the presence of hexoses with insulin added to the culture medium. After 28 h, fresh medium containing the hexoses and 0.1 mM dibutyryl cyclic AMP was added, and samples were taken for assay at 52 h. (Data replotted from Peraino et al.[96] and Merrill and Pitot[101].)

A second prominent feature of the gene for TyrAT in adult liver is that it generates enzyme in a diurnal rhythm that is driven by the feeding cycle. Shown in Figure 4B are results of allowing rats access to diets of different protein content (12, 30, or 60% by weight) under an inverted light/dark cycle.[93] The enzyme level begins to rise at the onset of darkness after feeding is initiated, but begins to decline soon after food is removed. This type of circadian rhythm is a prominent feature of short-lived gene products that respond to nutrients. In the case of TyrAT, the immediate stimulus that causes the increased rate of synthesis is not totally clear; it is likely to involve changes in levels of insulin, glucagon, and adrenal steroids in the blood plasma in addition to elevated amino acid content. Recent evidence suggests that the nutrient-related induction of TyrAT that occurs during feeding is a result of altered production of mRNA, as judged by quantitative Northern analysis.[94] The low amplitude of the rhythm observed when animals are fed low protein diets may indicate that carbohydrates decrease the rate at which this enzyme is synthesized; this effect is noted for a variety of enzymes that initiate degradation of amino acids.[7]

C. REGULATION OF SERINE-THREONINE DEHYDRATASE BY PROTEIN AND CARBOHYDRATE

Serine Dehydratase (SDH; E.C. 4.2.1.13) is a pyridoxal phosphate-dependent enzyme that catalyzes deamination of serine and threonine. SDH expression is limited to the liver and kidney cortex and is developmentally controlled. Levels of enzyme begin to increase in liver shortly after birth, but adult levels are not attained until the end of the first month in the rat.[95] Dietary protein and carbohydrate regulate its synthesis in opposing fashion, with protein causing a large increase in the rate of enzyme synthesis, whereas carbohydrate inhibits enzyme synthesis.[96-98] The concentration of SDH begins to increase after feeding of high protein diets is initiated, and increases up to 100-fold or more, with the final concentration almost directly proportional to the proportion of protein in the diet. By comparison, most transaminases increase no more than tenfold under this regimen. Synthesis of SDH in isolated hepatocytes is under the control of glucagon and glucocorticoids, which must be present simultaneously for optimal induction to occur.[99] Insulin strongly inhibits synthesis of the enzyme in cultured cells. Hormones do not affect the stability of the mRNA or the enzyme, which have half-lives of about 5 to 10 h and 13 h, respectively.

The opposing effects of dietary protein and carbohydrate on the synthesis of SDH and ornithine aminotransferase can be observed after intubation of hydrolyzed casein into the stomach at 6 h intervals (Figure 4A).[96] This causes a large increase in enzyme activity as a result of increased synthesis of protein. However, the induction is prevented if glucose or fructose is administered with the casein hydrolyzate. Insulin secretion does not fully account for the ability of glucose to inhibit induction by dietary protein because glucose represses the synthesis of SDH in livers of streptozotocin-diabetic rats whereas insulin had relatively little effect.[7] Glucose repression of SDH is due to inhibition of transcription, for SDH mRNA virtually disappears when rats are fed a protein free diet.[82] Although the intracellular mediators of this effect are unknown, Sudilovsky et al. showed that intubation of glucose inhibited induction of SDH and TyrAT by glucagon without preventing the increase in cAMP caused by this hormone.[100]

Repression of enzyme synthesis by carbohydrates can be demonstrated in cultured cells. Merrill and Pitot[101] isolated hepatocytes from livers of rats that had previously been fed a low-protein diet. The dibutyryl analog of cAMP increased the synthesis and concentration of ornithine aminotransferase sev-eralfold in the presence of culture medium containing insulin and glucose. However, addition of fructose (Figure 4B), sorbose, sorbitol, or glycerol to the culture medium prevented the induction. The inhibitory effect of these carbohydrates required their metabolism and was associated with a decrease in mRNA concentration. This evidence suggested that part of the inhibitory effect of carbohydrates on enzyme synthesis was due to a direct action of the sugar or a metabolite of the sugar on production of specific mRNAs. Direct effects of glucose in stimulating mRNA production in hepatocytes has been observed by Towle et al. for pyruvate kinase and a protein called S14.[102] In these cases, glucose causes a concentration-dependent increase in mRNA levels and transcription rate when insulin is present at a fixed concentration.

Ogawa et al.[82] showed that the concentration of SDH mRNA achieved after feeding levels of dietary casein ranging from 0 to 91% of total calories was proportional to the amount of enzyme activity, and presumably protein concentration. When the protein-fed animals were switched back to a protein-free diet, enzyme activity declined with a half-life of about 18 h, and mRNA declined with a half-life of about 3 h. It was not clear whether half-life of the mRNA was affected by diet, but earlier evidence suggested that glucose might destabilize the mRNA.

Recent studies have demonstrated that when the gene for SDH is activated by protein, an upstream region becomes sensitive to digestion *in vitro* with the nuclease, DNAse I. Four DNAse I-hypersensitive sites (I to IV in Figure 5) were identified in the region located within 6 kb upstream from the structural gene. Sites II and III were especially sensitive in samples from rats fed the high-protein diet, indicating that diet altered structure of the chromatin in these areas. This change in chromatin structure occurred only in liver, although a second hypersensitive area near the start site of transcription was observed in kidney (Figure 5). Another interesting feature observed in this study was that diet contributed to a pronounced, diurnal cycle in mRNA production that was not as evident for the enzyme level, possibly because of the mRNA had a much shorter half-life. Region II was most affected by the diurnal cycle, and sensitivity in region I decreased. Ogawa et al.[82] concluded that nutritional status could activate or inactivate transcription of the SDH gene by controlling regulatory factors that act in these regions of the SDH gene.

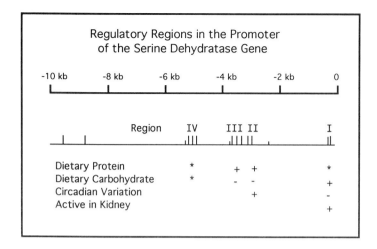

FIGURE 5.   DNA elements that respond to dietary protein and that promote circadian variation in mRNA synthesis have been identified in the promoter for the gene encoding serine dehydratase by testing for their sensitivity to digestion by DNAse I. DNA that was prepared from rats fed diets containing elevated casein was readily digested by the nuclease at sites that mapped to about -3200 and -3800 nucleotides 5′ to the start site of transcription (regions II and III). These sites were not sensitive to nuclease digestion in DNA isolated from animals fed a protein-free diet, but a site located at positions -100 to -200 was more sensitive to digestion (region I). Sensitivity of DNA in regions I and II varied in circadian fashion, with region II being more sensitive in the nocturnal feeding period. The figure is based on data reported by Ogawa et al.[82]

VI. CONCLUSIONS

Even though fasting causes reproducible changes in hormone secretion and gene expression, the consequences of refeeding depend on the composition of the diet. During fasting, the rate of gluconeogenesis is elevated, and this becomes inappropriate if an animal is refed a diet that contains abundant carbohydrates. However, if an animal is refed a diet containing high protein or lipid but little carbohydrate, that animal must continue to produce glucose. Therefore, it is not surprising that hormonal responses differ according to dietary content of the energy-yielding nutrients.

Studies of changes in gene expression after fasting and refeeding have already yielded several striking results. First, fasting and refeeding alter the structure of chromatin in regions that flank structural genes involved in metabolic regulation, and the alterations depend on the amount of carbohydrate, protein, and fat in the diet.[82,103] This confirms the idea that second messenger systems are activated in specific cells in response to dietary constituents, either directly or indirectly as a result of diet-related hormone secretion. It should prove very interesting to determine the components of these transducing systems and compare them with systems used in microorganisms. Second, some effects of nutrients take place in the absence of hormones, whereas others require hormones or are secondary to hormone secretion. In most cases, nutrients and hormones must both be present to obtain an optimum response. This makes sense because the point of modifying gene transcription in such cases is to permit adaptation to nutrient supply. Third, it is likely that the genetic apparatus responds more to nutritional supply than is apparent from measures based on enzyme activities. Regulated mRNAs generally have shorter half-lives and much lower concentrations than proteins, and are much more sensitive indices of responses to diet. It will be important to survey changes in mRNA levels in various organs and determine the extent to which nutrients reprogram the genetic apparatus not just in the liver and pancreas, but also in the brain, kidney, skeletal muscles, and other organs.

REFERENCES

1. **Harper, A. E.,** Effect of variations in protein intake on enzymes of amino acid metabolism, *Can. J. Biochem. Physiol.,* 43, 1589, 1965.
2. **Fine, L. G.,** The role of nutrition in hypertrophy of renal tissue, *Kidney Int.,* 32 (Suppl. 22), S2, 1987.
3. **Szepesi, B. and Freedland, R. A.,** Dietary effects on rat liver enzymes in meal-fed rats, *J. Nutr.,* 96, 382, 1968.
4. **Peters, J. C. and Harper, A. E.,** Adaptation of rats to diets containing different levels of protein: effects on food intake, plasma and brain amino acid concentrations and brain neurotransmitter metabolism, *J. Nutr.,* 115, 382, 1985.
5. **Munro, H. N. and Crim, M. C.,** The proteins and amino acids, in *Modern Nutrition in Health and Disease,* Shils, M. E. and Young, V. R., Eds., 7th ed., Lea and Febiger, Philadelphia, 1988, 1.
6. **Young, V. R. and Marchini, J. S.,** Mechanisms and nutritional significance of metabolic responses to altered intakes of protein and amino acids, with reference to nutritional adaptation in humans, *Am. J. Clin. Nutr.,* 51, 270, 1990.
7. **Soling, H., Kaplan, J., Erbstoeszer, M., and Pitot, H. C.,** The role of hormones in glucose repression in rat liver, *Adv. Enzyme Regul.,* 7, 171, 1969.
8. **Goldberg, M. L.,** The glucose effect: carbohydrate repression of enzyme induction, RNA synthesis, and glucocorticoid activity—a role for cyclic AMP and cyclic GMP, *Life Sci.,* 17, 1747, 1975.
9. **Teppermann, H. M., Teppermann, J., Pownall, J. D., and Branch, A.,** On the response of hepatic glucose-6-phosphate dehydrogenase activity to changes in diet composition and food intake pattern, *Adv. Enzyme Regul.,* 1, 121, 1963.
10. **Harper, A. E.,** Diet and plasma amino acids, *Am. J. Clin. Nutr.,* 21, 358, 1968.
11. **Ashley, D. V., Barclay, D. V., Chauffard, F. A., Moennoz, D., and Leathwood, P. D.,** Plasma amino acid responses in humans to evening meals of differing nutritional composition, *Am. J. Clin. Nutr.,* 36, 143, 1982.
12. **Eisenstein, A. B., Srack, I., Gallo-Torres, H., Georgiadis, and Miller, O. N.,** Increased glucagon secretion in protein-fed rats: lack of a relationship to plasma amino acids, *Am. J. Physiol.,* 236, E20, 1979.
13. **Felig, P., Wahren, J., and Hendler, R.,** Influence of oral glucose ingestion on splanchnic glucose and gluconeogenic substrate metabolism in man, *Diabetes,* 24, 468, 1975.

14. **Dole, V. P., James, A. T., Webb, P. W., Rizack, M. A., and Sturman, M. F.,** The fatty acid patterns of plasma lipids during alimentary lipemia, *J. Clin. Invest.,* 38, 1544, 1959.

15. **Klahr, S.,** Effects of protein intake on the progression of renal disease, *Annu. Rev. Nutr.,* 9, 87, 1989.

16. **Farwell, D. C., Miguez, J. B., and Herbst, E. J.,** Ornithine decarboxylase and po-lyamines in liver and kidneys of rats on cyclical regimens of protein-free and protein-containing diets, *Biochem. J.,* 168, 49, 1977.

17. **Jakobsson, B., Celsi, G., Lindblad, B., and Aperia, A.,** Influence of different protein intake on renal growth in young rats, *Acta Paediatr. Scand.,* 76, 293, 1987.

18. **Wada, L., Don, B. R., and Schambelan, M.,** Hormonal mediators of amino acid-induced glomerular hyperfiltration in humans, *Am. J. Physiol.,* 260, F787, 1991.

19. **King, A., Troy, J., Anderson, S., Neuringer, J., Gunning, M., and Brenner, B. M.,** Nitric oxide: a potential mediator of renal hyperemia and hyperfiltration, *J. Am. Soc. Nephrol.,* 1, 1271, 1991.

20. **Rosenberg, M. E., Chmielewski, D., and Hostetter, T. H.,** Effect of dietary protein on rat renin and angiotensinogen gene expression, *J. Clin. Invest.,* 85, 1144, 1990.

21. **Kaysen, G. A., Rosenthal, C., and Hutchinson, F. N.,** GFR increases before renal mass or ODC activity increase in rats fed high protein diets, *Kidney Int.,* 36, 441, 1989.

22. **Sens, D. A., Levine, J. H., and Buse, M. G.,** Stimulation of hepatic and renal ornithine decarboxylase activity by selected amino acids, *Metabolism,* 32, 787, 1983.

23. **Morris, S. M., Jr., Moncman, C. L., Holub, J. S., and Hod, Y.,** Nutritional and hormonal regulation of mRNA abundance for arginine biosynthetic enzymes in kidney, *Arch. Biochem. Biophys.,* 273, 230, 1989.

24. **Efrat, S., Surana, M., and Fleischer, N.,** Glucose induces insulin gene transcription in a murine pancreatic β-cell line, *J. Biol. Chem.,* 266, 11141, 1991.

25. **Ferris, C. F.,** Neurotensin, in *Handbook of Physiology,* Section 6, The Gastrointestinal System, Vol. II, Neural and Endocrine Biology, Schultz, S. G., Makhlouf, G. M., and Rauner, B. B., Eds., Waverly Press, Inc., Baltimore, MD, 1989, 559.

26. **Taylor, I. L.,** Pancreatic polypeptide family: pancreatic polypeptide, neuropeptide Y, and peptide YY, *Op. cit.,* 1989, 475.

27. **Brown, J. C., Buchan, A. M. J., Mcintosh, C. H. S., and Pederson, R. A.,** Gastric inhibitory polypeptide, *Op. cit.,* 1989, 403.

28. **Rehfeld, J. F.,** Cholecystokinin, *Op. cit.,* 1989, 337.

29. **Mayhew, D. A., Wright, P. H., and Ashmore, J.,** Regulation of insulin secretion, *Pharmacol. Rev.,* 21, 183, 1969.

30. **Martin, C. R.,** *Endocrine Physiology,* Oxford University Press, Oxford, 1985, 159.

31. **Unger, R. H., Dobbs, R. E., and Orci, L.,** Insulin, glucagon, and somatostatin secretion in the regulation of metabolism, *Annu. Rev. Physiol.,* 40, 307, 1978.

32. **Edozien, J. C., Niehaus, N., Mar, M.-H., Makoui, T., and Switzer, B. S.,** Diet-hormone interrelationships in the rat, *J. Nutr.,* 108, 1767, 1978.

33. **Rorsman, P., Ashcroft, F. M., and Berggen, P.-O.,** Regulation of glucagon release from pancreatic A-cells, *Biochem. Pharmacol.,* 41, 1783, 1991.

34. **Castellino, P., Giordano, C., Perina, A., and DeFronzo, R. A.,** Effects of plasma amino acid and hormone levels on renal hemodynamics in humans, *Am. J. Physiol.,* 255, F444, 1988.

35. **Pecile, A. and Muller, E. E.,** Control of growth hormone secretion, in *Neuroendocri-nology,* Vol. 1, Martini, L. and Ganong, W., Eds., Academic Press, New York, 1966, 537.

36. **Matzen, L. E., Andersen, B. B., Jensen, B. G., Gjessing, H. J., Sindrup, S. H., and Kvetny, J.,** Different short-term effect of protein and carbohydrate intake on TSH, growth hormone, (GH), insulin, C-peptide, and glucagon in humans, *Scand. J. Clin. Lab. Invest.,* 50, 801, 1990.

37. **Alvarez, C. V., Mallo, F., Burguera, B., Cacicedo, L., Dieguez, C., and Casanueva, F. F.,** Evidence for a direct pituitary inhibition by free fatty acids of in vivo growth hormone responses to growth hormone-releasing hormone in the rat, *Neuroendocrinology,* 53, 185, 1991.

38. **Vigas, M., Tatar, P., Jurcovicova, J., and Jezlova, D.,** Glucoreceptors located in different areas mediate the hypoglycemia-induced release of growth hormone, prolactin, and adrenocorticotropin in man, *Neuroendocrinology,* 51, 365, 1990.

39. **Munro, H. N., Steele, M. H., and Hutchison, W. C.,** Action of dietary proteins and amino acids on the rat adrenal gland, *Br. J. Nutr.,* 19, 137, 1965.

40. **Rabolli, D. and Martin, R. J.,** Effects of diet composition on serum levels of insulin, thyroxine, triiodothyronine, growth hormone, and corticosterone in rats, *J. Nutr.,* 107, 1068, 1977.

41. **Slag, M. F., Ahmed, M., Gannon, M. C., and Nuttall, F. Q.,** Meal stimulation of cortisol secretion: a protein induced effect, *Metabolism,* 30, 1104, 1981.

42. **Aizawa, T., Yasuda, N., and Greer, M. A.,** Hypoglycemia stimulates ACTH secretion through a direct effect on the basal hypothalamus, *Metabolism,* 30, 996, 1981.

43. **Jepson, M. M., Bates, P. C., and Millward, D. J.,** The role of insulin and thyroid hormones in the regulation of muscle growth and protein turnover in response to dietary protein in the rat, *Br. J. Nutr.,* 59, 397, 1988.

44. **Danforth, E., Jr. and Burger, A. G.,** The impact of nutrition on thyroid hormone physiology and action, *Annu. Rev. Nutr.,* 9, 201, 1989.

45. **Gardner, J. D. and Jensen, R. T.,** Receptors for gut peptides and other secretagogues on pancreatic acinar cells, in *The Gastrointestinal System, Handbook of Physiology,* Section 6, Vol. 2, Schultz, S. G., Makhlouf, G. M., and Rauner, B. M., Eds., American Physiological Society, 1989, 171.

46. **Rosselin, G.,** Liver receptors for regulatory peptides, *Op. cit.,* 245.

47. **Sanchez, V., Governa, R., and Calvo, J. K.,** Glycogenolytic effect of vasoactive intestinal peptide in the rat in vivo, *Experientia,* 47, 625, 1991.

48. **Becker, D. J.,** The endocrine responses to protein calorie malnutrition, *Annu. Rev. Nutr.,* 3, 187, 1983.

49. **Goodman, M. N., Larsen, P. R., Kaplan, M. M., Aoki, T. T., Young, V. R., and Ruderman, N. B.,** Starvation in the rat. II. Effect of age and obesity on protein sparing and fuel metabolism in the rat, *Am. J. Physiol.,* 239, E277, 1980.

50. **Xie, Q.-W.,** Experimental studies on changes of neuroendocrine functions during starvation and refeeding, *Neuroendocrinology,* 53 (Suppl. 1), 52, 1991.

51. **Thomas, G. B., Mercer, J. E., Karalis, T., Rao, A., Cummins, J. T., and Clarke, I. J.,** Effect of restricted feeding on the concentrations of growth hormone, gonadotropins, and prolactin in plasma, and on the amounts of messenger ribonucleic acid for GH, gonadotropin subunits, and PRL in the pituitary glands of adult ovariectomized ewes, *Endocrinology,* 126, 1361, 1990.

52. **Sakaguchi, T., Takahashi, M., and Bray, G. A.,** Diurnal changes in sympathetic activity, *J. Clin. Invest.,* 82, 282, 1988.

53. **Jarrousse, C., Lardeux, B., Bourdel, G., Girard-Globa, A., and Rosselin, G.,** Portal insulin and glucagon in rats fed proteins as a meal: immediate variations and circadian modulations, *J. Nutr.,* 110, 1764, 1980.

54. **Tiedgen, M. and Seitz, H. J.,** Dietary control of circadian variations in serum insulin, glucagon, and hepatic cyclic AMP, *J. Nutr.,* 110, 876, 1980.

55. **Peret, J., Foustock, S., Chanez, M., Bois-Joyeux, B., and Assani, R.,** Plasma glucagon and insulin concentrations and hepatic phosphoenolpyruvate carboxykinase and pyruvate kinase activities during adaptation of rats to a high protein diet, *J. Nutr.,* 111, 1173, 1981.

56. **Kuhara, T., Ikeda, S., Ohneda, A., and Sasaki, Y.,** Effects of intravenous infusion of 17 amino acids on the secretion of GH, glucagon, and insulin in sheep, *Am. J. Physiol.,* 260, E21, 1991.

57. **Hugues, J.-N., Enjalbert, A., Burger, A., Voirol, M.-J., Sebaoun, J., and Epelbaum, J.,** Sensitivity of thyrotropin (TSH) secretion to 3,5,3'-triiodothyronine and TSH-releasing hormone in rat during starvation, *Endocrinology,* 119, 253, 1986.

58. **Blake, N. G., Eckland, D. J. A., Foster, O. J. F., and Lightman, S. L.,** Inhibition of hypothalamic thryrotropin-releasing hormone messenger ribonucleic acid during food deprivation, *Endocrinology,* 129, 2714, 1991.

59. **Straus, D. S. and Takemoto, C. D.,** Effect of fasting on insulin-like growth factor-I (IGF-1) and growth hormone receptor mRNA levels and IGF-1 gene transcription in rat liver, *Mol. Endocrinol.,* 4, 91, 1990.

60. **Thissen, J. P., Triest, S., Underwood, L. E., Maes, M., and Ketelslegers, J. M.,** Divergent responses of serum insulin-like growth factor-I and liver growth hormone receptors to exogenous GH in protein-restricted rats, *Endocrinology,* 126, 908, 1990.

61. **Plotsky, P. M. and Vale, W. W.,** Patterns of growth hormone-releasing factor and somatostatin secretion into the hypophyseal-portal circulation of the rat, *Science,* 230, 461, 1986.

62. **Bruno, J. F., Olchovsky, D., White, J. D., Leidy, J. W., Song, J., and Berelowicz, M.,** Influence of food deprivation in the rat on hypothalamic expression of growth hormone-releasing factor and somatostatin, *Endocrinology,* 127, 2111, 1990.

63. **Bruno, J. F., Song, J., and Berelowitz, M.,** Regulation of rat hypothalamic prepro-growth hormone-releasing factor messenger ribonucleic acid by dietary protein, *Endocrinology,* 129, 1226, 1991.

64. **Glaeser, B. S., Maher, T. J., and Wurtman, R. J.,** Changes in brain levels of acidic, basic, and neutral amino acids after consumption of single meals containing various proportions of protein, *J. Neurochem.,* 41, 1016, 1983.

65. **Knopf, R. F., Conn, J. W., Fajans, S. S., Floyd, J. C., Guntsche, E. M., and Rull, J. A.,** Plasma growth hormone response to intravenous amino acids, *Am. J. Endocrinol.,* 26, 1140, 1965.

66. **Felig, P.,** Progress in endocrinology and metabolism. The glucose-alanine cycle, *Metabolism,* 22, 179, 1973.

67. **Kayne, F. J. and Price, N. C.,** Amino acid effector binding to rabbit muscle pyruvate kinase, *Arch. Biochem. Biophys.,* 159, 292, 1973.

68. **Dong, F. M. and Freedland, R. A.,** Effects of alanine on gluconeogenesis in isolated rat hepatocytes, *J. Nutr.,* 110, 2341, 1980.

69. **Perez-Sala, D., Parrila, R., and Ayuso, M. S.,** Key role of L-alanine in the control of hepatic protein synthesis, *Biochem. J.,* 241, 491, 1987.

70. **Poso, A. R. and Mortimore, G. E.,** Requirement for alanine in the amino acid control of hepatic protein degradation, *Proc. Natl. Acad. Sci. U.S.A.,* 81, 2039, 1984.

71. **Muller, W. A., Faloona, G. R., and Unger, R. H.,** The effect of alanine on glucagon secretion, *J. Clin. Invest.,* 50, 2215, 1971.

72. **Walser, M.,** Role of branched-chain ketoacids in protein metabolism, *Kidney Int.,* 38, 595, 1990.

73. **MacDonald, M. J., McKenzie, D. I., Kaysen, J. H., Walker, T. M., Moran, S. M., Fahien, L. A., and Towle, H. C.,** Glucose regulates leucine-induced insulin release and the expression of the branched chain ketoacid dehydrogenase E1α subunit gene in pancreatic islets, *J. Biol. Chem.,* 266, 1335, 1991.

74. **Garthwaite, J. and Garthwaite, G.,** Mechanisms of excitatory amino acid neurotoxicity in rat brain slices, *Adv. Exp. Biol. Med.,* 268, 505, 1990.

75. **Pestana, A.,** Dietary and hormonal control of enzymes of amino acid catabolism in liver, *Eur. J. Biochem.,* 11, 400, 1969.

76. **Lardy, H. A. and Merryfield, M. L.,** Ferroactivator and the regulation of gluconeogenesis, *Curr. Top. Cell. Reg.,* 18, 243, 1981.

77. **Punekar, N. S. and Lardy, H. A.,** Phosphoenolpyruvate carboxykinase ferroactivator 1. Mechanism of action and identity with glutathione peroxidase, *J. Biol. Chem.,* 262, 6714.

78. **Kurl, R. N., Verney, E., and Sidransky, H.,** Identification and immunohistochemical localization of tryptophan binding protein in nuclear envelopes of rat liver, *Arch. Biochem. Biophys.,* 265, 286, 1988.

79. **Palmer, R. M. J., Ashton, D. S., and Moncada, S.,** Vascular endothelial cells synthesize nitric oxide from L-arginine, *Nature (London),* 333, 664, 1989.

80. **Goodridge, A. G.,** Dietary regulation of gene expression: enzymes involved in carbohydrate and lipid metabolism, *Annu. Rev. Nutr.,* 7, 157, 1987.

81. **Potter, V. R. and Ono, T.,** Enzyme patterns during metabolic transitions in rat liver and Morris hepatoma 5123, *Cold Spring Harbor Symp. Quant. Biol.,* 26, 355, 1961.

82. **Ogawa, H., Fujioka, M., Su, Y., Kanamoto, R., and Pitot, H. C.,** Nutritional regulation and tissue-specific expression of the serine dehydratase gene in rats, *J. Biol. Chem.,* 266, 20412, 1991.

83. **Perez, N., Clark-Turri, L., Rabajille, E., and Niemeyer, H.,** Regulation of rat liver enzymes by natural components of the diet, *J. Biol. Chem.,* 239, 2420, 1964.

84. **Blumenthal, M. D., Abraham, S., and Chaikoff, I. L.,** Dietary control of liver glucokinase activity, *Arch. Biochem. Biophys.,* 104, 215, 1964.

85. **Hopgood, M. F., Ballard, F. J., Reshef, L., and Hanson, R. W.,** Synthesis and degradation of phosphoenolpyruvate carboxylase in rat liver and adipose tissue. Changes during a starvation-re-feeding cycle, *Biochem. J.,* 134, 445, 1975.

86. **Seitz, H. J., Tiedgen, M., and Tarnowski, W.,** Regulation of hepatic phosphoenolpyruvate carboxykinase (GTP). Role of dietary proteins and amino acids in vivo and in the isolated perfused rat liver, *Biochim. Biophys. Acta,* 632, 473, 1980.

87. **Beale, E. G., Hartley, J. L., and Granner, D. K.,** $N^6,O^{2'}$-dibutyryl cyclic AMP and glucose regulate the amount of messenger RNA coding for hepatic phosphoenolpyruvate carboxykinase (GTP), *J. Biol. Chem.,* 257, 2022, 1982.

88. **Short, M. K., Clouthier, D. E., Schaefer, I. M., Hammer, R. E., Magnuson, M. A., and Beale, E. G.,** Tissue-specific, developmental, hormonal, and dietary regulation of rat phosphoenolpyruvate-human growth hormone fusion genes in transgenic mice, *Mol. Cell. Biol.,* 12, 1007, 1992.

89. **Block, K. P., Aftring, R. P., and Buse, M. G.,** Regulation of rat liver branched-chain α-keto acid dehydrogenase activity by meal frequency and dietary protein, *J. Nutr.,* 120, 793, 1990.

90. **Sereni, F., Kenney, F. T., and Kretschmer, N.,** Factors influencing the development of tyrosine α-ketoglutarate transaminase activity in rat liver, *J. Biol. Chem.,* 234, 609, 1959.

91. **Greengard, O.,** Enzymic differentiation in mammalian liver, *Science,* 163, 891, 1969.

92. **Johnson, A. C., Lee, K.-L., Isham, K. R., and Kenney, F. T.,** Gene-specific acquisition of hormonal responsiveness in rat liver during development, *J. Cell. Biochem.,* 37, 243, 1988.

93. **Watanabe, M., Potter, V. R., and Pitot, H. C.,** Systematic oscillations in tyrosine transaminase and other metabolic functions in liver of normal and adrenalectomized rats on controlled feeding schedules, *J. Nutr.,* 95, 207, 1968.

94. **Bartels, H., Herbort, H., and Jungermann, K.,** Predominant periportal expression of the phosphoenolpyruvate carboxykinase and tyrosine aminotransferase genes in rat liver. Dynamics during the daily feeding rhythm and starvation-refeeding cycle demonstrated by in situ hybridization, *Histochemistry,* 95, 637, 1990.

95. **Snell, K.,** Enzymes of serine metabolism in normal, developing, and neoplastic rat tissues, *Adv. Enz. Regul.,* 22, 325, 1984.

96. **Peraino, C., Lamar, C., Jr., and Pitot, H. C.,** Studies on the induction and repression of enzymes in rat liver. IV. Effects of cortisone and phenobarbital, *J. Biol. Chem.,* 241, 2944, 1966.

97. **Anderson, H. L., Benevenga, N. J., and Harper, A. E.,** Associations among food and protein intake, serine dehydratase, and plasma amino acids, *Am. J. Physiol.,* 214, 1008, 1968.

98. **Jost, J. P., Khairallah, E. A., and Pitot, H. C.,** Studies on the induction and repression of enzymes in rat liver. Regulation of the rate of synthesis and degradation of serine dehydratase by dietary amino acids and glucose, *J. Biol. Chem.,* 243, 3057, 1968.

99. **Noda, C., Yakiyama, M., Nakamura, T., and Ichihara, A.,** Requirements of both glucocorticoids and glucagon as co-inducers for activation of transcription of the serine dehydratase gene in cultured rat hepatocytes, *J. Biol. Chem.,* 263, 14764, 1988.

100. **Sudilovsky, O., Pestana, A., Hinderaker, P. H., and Pitot, H. C.,** Cyclic adenosine 3',5'-monophosphate during glucose repression in rat liver, *Science,* 174, 142, 1971.

101. **Merrill, M. J. and Pitot, H. C.,** Inhibition of cyclic AMP-dependent induction of ornithine aminotransferase by simple carbohydrates in cultured hepatocytes, *Arch. Biochem. Biophys.,* 259, 250, 1987.

102. **Towle, H. C., Thompson, K. S., Liu, J., and Shih, H.-M.,** Carbohydrate metabolism regulates gene transcription in liver: identification of response elements and nuclear factors, *FASEB J.,* 6, A402, 1992.

103. **Castro, E. C.,** Nutrient effects on DNA and chromatin structure, *Annu. Rev. Nutr.,* 7, 407, 1987.

Chapter 17

VITAMIN D-DEPENDENT CALCIUM BINDING PROTEIN, CALBINDIN-D: REGULATION OF GENE EXPRESSION

Rajbir K. Gill and Sylvia Christakos

TABLE OF CONTENTS

0-8493-6961-4/93/$0.00 + $.50

I. INTRODUCTION

The vitamin D endocrine system is a major regulator of calcium home-ostasis. The primary function of vitamin D in maintaining calcium homeostasis is through direct stimulation of intestinal transport, effects on calcium reabsorption in the kidney, and mobilization of mineral from bone.[1,2] The biological responses of the hormonally active form of vitamin D, 1,25-dihydroxyvitamin D_3 (1,25$(OH)_2D_3$), are initiated at its target tissues by the interaction of the hormone with its specific high affinity intracellular receptor and the association of the steroid receptor complex with specific sequences of DNA nucleotides resulting in the activation and expression of genes.[1,2] The protein in intestine and kidney whose synthesis is induced by 1,25$(OH)_2D_3$ is the calcium-binding protein, calbindin.[3] It has been suggested that the renal and intestinal calbindin are involved in vitamin D-dependent intracellular translocation of calcium ions.[3-5] In addition, calbindin has been reported in many other tissues including bone (chrondrocytes of rat growth plate cartilage and ameloblasts of rodent teeth)[6-10] and tissues that are not regulators of serum calcium, such as pancreas,[11-15] uterus, placenta,[19,20] and brain,[21-25] and in a variety of species,[3,26,27] thus suggesting multiple effects of the vitamin D endocrine system. Biochemically the calbindins are divided into two classes of proteins which do not share amino acid homology; the 28,000 M_r protein (calbindin-D_{28k}, which has four high affinity calcium binding sites, which is conserved during evolution, and is present in avian intestine and in avian and mammalian kidney, pancreas, bone, and brain) and the 9000 M_r protein (calbindin-D_{9k} which has two calcium binding domains, is less conserved evolutionarily, and is present in mammalian intestine, placenta, uterus, bone and in mouse and neonatal rat kidney).[3] Although the exact role of the calbindins in many of these tissues remains to be determined, the importance of the discovery of the calbindins is that key advances in our understanding of the diversity of the vitamin D endocrine system have been made through the study of the tissue distribution of the calbindins and their colocalization with the 1,25$(OH)_2D_3$ receptor.[28] In addition the biosynthesis of calbindin has provided an important approach for studying the molecular mechanism of 1,25$(OH)_2D_3$ action in major target tissues such as intestine and kidney. This chapter will focus on how an understanding of the regulation of calbindin (regulation by 1,25$(OH)_2D_3$, tissue-specific regulation and regulation by factors other than 1,25$(OH)_2D_3$) has resulted in a better understanding of the possible functional significance of calbindin and has enabled us to ask appropriate questions so as to define specific genetic regulatory elements in the calbindin promoter.

II. REGULATION OF INTESTINAL AND RENAL CALBINDIN GENE EXPRESSION

In the intestine, calbindin (calbindin-D_{28k} in avian and calbindin-D_{9k} in mammals) is localized in columnar epithelial cells[29,30] and has been postulated to facilitate intracellular calcium translocation.[4] Avian and mammalian renal calbindin-D_{28k} is present in the principal cells of the distal tubule in which selective reabsorption of calcium is known to occur.[31-34] In order to obtain a better understanding of the molecular mechanisms of $1,25(OH)_2D_3$ action in these tissues, DNAs complementary to both rat and chick intestinal calbindins[35-37] and to mammalian calbindin-D_{28k}[38,39] have been cloned, the deduced amino acid sequences reported, and the probes have been used to examine the regulation of calbindin gene expression.[40-52] Administration of $1,25(OH)_2D_3$ to vitamin D-deficient animals has been shown to induce rapid transcription of the calbindin gene in chick[42] and rat intestine[43] and in rat kidney.[46] The peak of calbindin gene transcription occurs at 1 to 3 h which precedes the peak of accumulation of calbindin mRNA (12 h) and calbindin protein (48 h) (Figure 1). These results indicate that the early increases in calbindin mRNA and calbindin protein may be due to the stimulation of transcription. However, the long lag time between transcription and the peak of mRNA and protein accumulation reflects the involvement of posttranscriptional mechanisms. The accumulation of calbindin-D mRNA long after $1,25(OH)_2D_3$ injection suggests an effect on stabilization of RNA by hormone. Recent *in vitro* studies using actinomycin D and cycloheximide have also suggested that posttranscriptional mechanisms are involved in the regulation of the renal and intestinal calbindins by $1,25(OH)_2D_3$.[53,54] Thus, calbindin mRNA is regulated by the D hormone at the transcriptional and posttranscriptional level and stabilization of transcribed RNA may be an important mechanism in the regulation of calbindin gene expression by $1,25(OH)_2D_3$. Developmental studies have indicated that renal calbindin-D_{28k} mRNA increases sharply between birth and 1 week of age, the period of rapid nephron differentiation, suggesting a possible role for renal calbindin in the differentiation process.[47] In rat intestine, calbindin-D_{9k} mRNA is induced in the third postnatal week, coinciding with weaning and the period of increased duodenal active transport of calcium (Figure 2).[47] Thus, calbindin may be an essential factor involved in the development of intestinal active calcium transport. It is of interest that the developmental induction of $1,25(OH)_2D_3$ receptor (VDR) mRNA in intestine (Figure 2) and kidney coincides with the induction of calbindin mRNA in these tissues, indicating that the induction of VDR mRNA has a role in regulating the developmental expression of the calbindin gene.[47]

Further studies in intestine and kidney have provided evidence that the calbindin gene is not always regulated by $1,25(OH)_2D_3$ and that factors other than $1,25(OH)_2D_3$ can modulate calbindin gene expression. Glucocorticoid

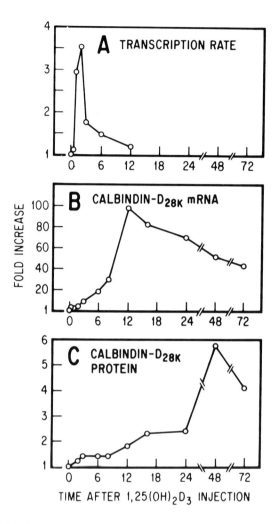

FIGURE 1. Analysis of rat renal calbindin-D_{28k} transcription, mRNA, and protein levels in kidney after $1,25(OH)_2D_3$ treatment. (A) Calbindin-D gene transcription at various times after $1,25(OH)_2D_3$ administration (200 ng/100 g body weight to vitamin D-deficient rats). The transcription rate was measured using the nuclear transcription assay[46] and expressed as fold increase compared to control. The results shown are the mean of four different experiments; (B) The time course of increase in calbindin-D_{28k} mRNA levels following $1,25(OH)_2D_3$ administration. Northern blots of poly(A^+)RNA isolated from kidneys of vitamin D-deficient rats at various times after single injection of $1,25(OH)_2D_3$ were quantitated by densitometric analysis.[44] The signal intensities were adjusted for variations in β-actin densities and the fold increase in mRNA levels compared to control was calculated. The values shown are the mean of three separate Northern hybridization experiments: (C) The time course of calbindin-D_{28k} accumulation after $1,25(OH)_2D_3$ treatment. The levels of calbindin-D at various times were measured by immunobinding assay[44] and expressed as fold increase compared to control. Each value represents the mean of three separate groups of rats.

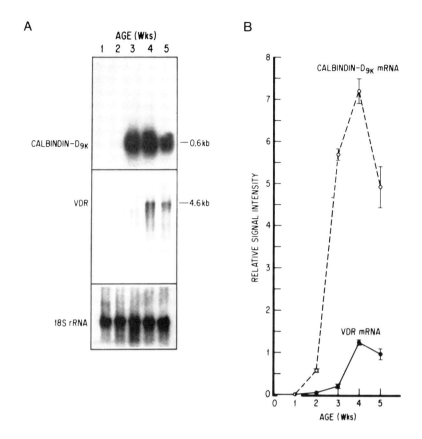

FIGURE 2. Developmental changes in intestinal calbindin-D_{9k} mRNA and VDR mRNA assessed by Northern blot hybridization. (A) Rat intestinal poly(A^+)RNA (5 μg per lane) was fractionated through a formaldehyde-agarose gel, transferred onto a nylon membrane, and hybridized with calbindin-D_{9k}, VDR, and 18S rRNA probes; (B) densitometric quantitation of Northern analysis. The results represent the mean ± SE of three separate Northern hybridization analyses from different groups of rats. The data were normalized based on results obtained upon rehybridization with 18S rRNA cDNA. (From Huang, et al., *J. Biol. Chem.*, 264, 17454, 1989. With permission.)

administration has been shown to result in an inhibition of rat and mouse intestinal calbindin-D_{9k} gene expression[47,52] which may be related to the previously reported glucocorticoid-mediated inhibition of intestinal calcium absorption.[55] Hall et al.[56] and Corradino et al.[57] have noted stimulatory as well as inhibitory effects of dexamethasone on chick intestinal calbindin-D_{28k}. Renal calbindin-D_{28k} is unaffected by glucocorticoids. Other tissues are, however, suggesting that tissue-specific regulation by glucocorticoids of calbindin gene expression exists.[47,52] Studies concerning the regulation of renal calbindin have suggested modulation by calcium of calbindin gene expression.[45,53] In addition, regulation of calbindin-D_{28k} by estrogen (in mouse and chick uterus) and regulation of neuronal calbindin-D_{28k} by retinoic acid have also been observed.[58-60] These findings suggest multiple, tissue-specific interactions in the regulation of calbindin gene expression. Similar to calbindin, a number of other steroid responsive genes have also been reported to be under the control of more than one steroid. 1,25-Dihydroxyvitamin D_3, dexamethasone, and retinoic acid have been reported to decrease procollagen gene expression, and receptors for both vitamins A and D recognize a common 11 nucleotide response element in the human osteocalcin gene.[61,62] The DNA binding domain of the steroid receptor mediates the specific recognition of the hormone response elements (HREs) and includes a cysteine-rich region that comprises two zinc finger-like structures. The first three amino acids in the stem of the first zinc finger (P box) specify the half-site sequence.[63] Based on the P box sequence, receptors are classified as members of either the glucocorticoid receptor (including glucocorticoid, mineralocorticoid, androgen, and progesterone receptors) or the estrogen/thyroid hormone receptor subfamily (including estrogen, thyroid hormone, retinoic acid, and vitamin D receptors).[63-65] On the basis of its distinct P box, the members of each class display cross-recognition of HRE. Umesono et al. recently reported that the retinoic acid receptor, the thyroid hormone receptor, and the vitamin D receptor recognize a common consensus half-site sequence and that this site is also the binding site for the estrogen receptor (but in the form of an inverted repeat).[66] It is of interest that 1,25(OH)$_2$D$_3$, estrogen, and retinoic acid (members of the estrogen receptor/thyroid receptor subfamily) have all been reported to regulate calbindin-D_{28k}. Thus, it is possible that receptors for these ligands may recognize a common sequence in the calbindin-D_{28k} promoter. It is possible that the glucocorticoid regulation of calbindin involves the recognition of a different sequence. Understanding the multiple steroid interactions in the regulation of calbindin gene expression should help clarify the mechanism of regulation of calbindin function and should aid in our understanding of the molecular interactions of the 1,25(OH)$_2$D$_3$ receptor with other steroid hormone receptors in the control of gene expression.

III. REGULATION OF CALBINDIN GENE EXPRESSION IN BONE

The calbindins have been localized immunocytochemically in the chondrocytes of the growth plate cartilage of the rat and chick and in the ameloblasts of rodent teeth.[7-10] The presence of the calbindins in chondrocytes suggests that these proteins may be involved in the transport of calcium toward extracellular sites of calcification in the growth plate. The presence of the calbindins in ameloblasts suggests a role for these proteins in mineralization. From the immunocytochemical studies it had not been clear whether the calbindins in chondrocytes and ameloblasts were vitamin D dependent. Recent studies by Berdal et al., using Northern analysis of total RNA from microdissected epithelial mesenchymal tooth germ of erupting rat incisor, indicated the presence of both calbindin-D_{9k} and calbindin-D_{28k} mRNA in this tissue.[10] The mRNA for both calbindins was nearly undectectable in vitamin D deficient rats but was upregulated by $1,25(OH)_2D_3$ injection. These findings, although suggestive of a direct role of $1,25(OH)_2D_3$ regulated calbindin gene expression in the mineralization process, are still preliminary. Further studies are needed to verify the vitamin D dependence of calbindin in chondrocytes and ameloblasts.

IV. REGULATION OF CALBINDIN GENE EXPRESSION IN THE NERVOUS SYSTEM

Calbindin-D_{28k} is one of the major calcium binding proteins in brain. It is present in some cells but not others throughout the brain and constitutes between 1 and 2% of the total soluble protein in certain brain areas.[12,13,21-25] Studies by Stumpf et al. demonstrated specific nuclear retention of [^{3}H]$1,25(OH)_2D_3$ in many of the same areas of rat brain in which calbindin was localized immunocytochemically, suggesting central modulation by $1,25(OH)_2D_3$.[67,68] It should be noted however that there is only a partial correlation between calbindin-D_{28k} and $1,25(OH)_2D_3$ receptors in the brain and that $1,25(OH)_2D_3$ administration has not been found to regulate calbindin or calbindin gene expression in this tissue. It is possible that $1,25(OH)_2D_3$ has a role in the regulation of other neuronal-related processes such as biorhythms, behavior, stress, and reproduction.[68] Although neuronal calbindin is not altered by $1,25(OH)_2D_3$ administration, specific decreases in calbindin gene expression have been observed in brain areas particularly affected in aging and in neurodegenerative disease processes (substantia nigra, Parkinson's disease; nucleus basalis, Alzheimers disease, and corpus striatum, Huntington's disease).[49,51] In the aging cerebellum, there is a 73% reduction in calbindin mRNA and only a 15 to 20% loss in calbindin containing Purkinje cells, suggesting that the reduction in calbindin mRNA is primarily due to a decrease of calbindin transcription with age.[51] These findings suggest that a decrease in calbindin may lead to a failure of calcium buffering or intraneuronal calcium homeostasis which may contribute to calcium-mediated irreversible cytotoxic events during aging and the pathological process. This hypothesis is supported by our recent findings using cell cultures of embryonic rat hippocampus.[69] A direct relationship between calbindin immunoreactivity and resistance to neurotoxicity induced by glutamate or calcium ionophore was documented. In addition calbindin positive neurons were better able to reduce free intracellullar calcium levels than calbindin negative neurons. These studies indicate excitoprotective and calcium-reducing roles for calbindin and are one of the first studies providing evidence for the functional significance of neuronal calbindin. Due to the significance of calbindin in the nervous system, it was important to determine the factors which could modulate the expression of this protein in brain. Recently, we reported an 80 to 85%

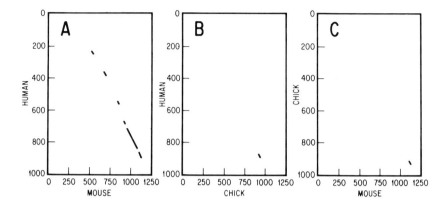

FIGURE 3. Dot matrix analysis showing the homologies between human $(-840/+154)$, mouse $(-1075/+154)$, and chick $(-850/+154)$ calbindin-D_{28k} gene promoter regions. Each dot indicates a minimum of 17 base pair matches out of 20. The coding regions are in the lower right corner.

decrease in the levels of calbindin and calbindin mRNA in the hippocampus of adrenalectomized rats.[50] Administration of corticosterone to normal and adrenalectomized rats resulted in the upregulation of calbindin-D_{28k} protein and mRNA in hippocampus. These studies present the first evidence of a regulator of calbindin gene expression in brain. Additionally, we recently found that calbindin-D_{28k} can be induced by retinoic acid in neuroblastoma and medulloblastoma cells.[60] Thus, we are only now beginning to understand the regulation of neuronal calbindin. If, as our results indicate, brain calbindin plays a critical role in neuronal calcium homeostasis, then future studies using the chromosomal gene for calbindin will be important for understanding the basic mechanisms involved in regulating neuronal calcium homeostasis.

V. ANALYSIS OF CALBINDIN PROMOTERS

To elucidate the molecular mechanism of calbindin gene regulation and to identify *cis*-acting elements in the promoter region, the chromosomal genes for human,[73] mouse,[74] and chick calbindin-D_{28k}[75-77] and rat calbindin-D_{9k}[78] have been isolated. The chicken calbindin-D_{28k} gene (18.5 kb) is split into 11 coding exons by 10 intervening sequences.[76] The calbindin-D_{9k} gene (2.5 kb) contains three exons interrupted by two introns. In the promoter region, a TATA box sequence is found 30 bp upstream from the Cap site of the rat calbindin-D_{9k} and the chick, mouse, and human calbindin-D_{28k} genes.[73-78] The consensus for the CAT box is present upstream from the TATA box at various lengths. The distal promoter (surrounding the TATA box) for each of these genes is highly GC rich.[73-78] Determination of the sequence homology among human, mouse, and chick calbindin-D_{28k} promoters was done by dot matrix analysis (Figure 3). When considering comparisons between chick-human and chick-mouse promoters there was no strongly conserved sequence (although avian and mammalian calbindin-D_{28k} share 79% amino acid identity). However, comparison between human-mouse (which share 98% amino acid sequence identity) indicated that the distal promoter was strongly conserved while the proximal part of the promoter diverged in the two species as shown in Figure 3. The lack of conserved sequences between the avian and mammalian promoters may be due to the evolutionary difference between the species as well as to the difference in tissue distribution (calbindin-D_{28k} is present in avian intestine but not in mammalian intestine).

Functional studies have been done by linking the mouse calbindin-D_{28k} promoter to a reporter gene coding for bacterial chloramphenicol acetyltransferase (CAT).[79] A fragment of the mouse calbindin-D_{28k} promoter with coordinates $-1075/+34$ was capable of promoting the expression of CAT in rat osteosarcoma cells (Ros 17/2.8), insulinoma cells (β cell line RIN_R; clone 1046-38, in which calbindin is present endogenously), and in LLC-PK_1 cells (pig kidney cells). The highest levels of expression were in the Ros 17/2.8 cells which contained the highest concentration of $1,25(OH)_2D_3$ receptors. These cells were used for all further studies. Deletion mutant analysis of the promoter indicated that the level of basal CAT expression increased fivefold with deletion of 587 bp of upstream sequences. This suggests the presence of an upstream repressor sequence. Upon treatment with sodium butyrate (2 mM) there occurred an eightfold increase in basal CAT activity. One segment located between -702 and -488 was found to confer a $1,25(OH)_2D_3$ dependent increase in CAT activity in the presence of butyrate (incubation of cells in the presence of $1,25(OH)_2D_3$ and butyrate [0.4 and 0.8 mM] resulted in a three- to fourfold stimulation in CAT activity above the stimulation observed with butyrate alone). These results suggest the presence of a regulatory repressor which needs to be released before activation by $1,25(OH)_2D_3$ can be observed or that butyrate is activating some other factor which is involved in the observed vitamin D mediated response. In addition, another segment located between -390 and -175 consistently contributed a significant twofold induction of CAT activity in the presence of $1,25(OH)_2D_3$ alone, without butyrate. Sequence $-198/-169$ has homology to the rat osteocalcin vitamin D response element (VDRE) (Figure 4).[80-81] To further characterize this downstream region gel-retardation assays were done using a ^{32}P labeled probe ($-248/-78$) and nuclear extract from $1,25(OH)_2D_3$-treated Ros 17/2.8 cells. Incubation of the probe with the nuclear extract resulted in a specific DNA-protein interaction as indicated by a band of retarded mobility. Addition of cold oligonucleotide representing the rat osteocalcin VDRE specifically competed with the probe, and this competition was demonstrated by abolishing the band of retarded mobility. These initial findings represent the first analysis of the mouse calbindin-D_{28k} promoter. It should be noted that, at this time, little is understood about sequences important in the regulation of the calbindin gene. With the availability of the calbindin gene we can now initiate studies

RAT OSTEOCALCIN   −458 GGTGA ATGA GGACA TTAC TGACC
MOUSE CALBINDIN−D$_{28k}$ −198 GGGGA TGTGA GGAGA 8bp TGAGC

FIGURE 4. Sequence homology between the VDRE of the rat osteocalcin gene and sequence −198/169 of the mouse calbindin-D$_{28k}$ promoter.

concerning the basic mechanisms, whereby calbindin is trancribed in only specific cells and the multiple interactions involved in the regulation of the calbindin gene. In addition, in order to obtain an understanding of the multiple mechanisms of 1,25(OH)$_2$D$_3$ action, the study not only of transcriptional but also of posttranscriptional mechanisms is needed. These studies should provide new insight concerning 1,25(OH)$_2$D$_3$-regulated gene expression and should help clarify the molecular interaction of the 1,25(OH)$_2$D$_3$ receptor with other factors in the control of gene expression.

REFERENCES

1. **DeLuca, H. F.,** The vitamin D story: a collaborative effort of basic science and clinical medicine, *FASEB J.,* 2, 224, 1988.
2. **Henry, H. L. and Norman, A. W.,** Vitamin D: metabolism and biological action, *Annu. Rev. Nutr.,* 4, 493, 1984.
3. **Christakos, S., Gabrielides, C., and Rhoten, W. B.,** Vitamin D-dependent calcium binding proteins: chemistry, distribution, functional considerations and molecular biology, *Endocrine Rev.,* 10, 3, 1989.
4. **Feher, J. J.,** Facilitated calcium diffusion by intestinal calcium binding protein, *Am. J. Physiol.,* 244, C303, 1983.
5. **Bronner, F., Pansu, D., and Stein, W. D.,** An analysis of intestinal calcium-transport across the rat intestine, *Am. J. Physiol.,* 250, G561, 1986.
6. **Christakos, S. and Norman, A. W.,** A vitamin D-dependent calcium binding protein in bone, *Science,* 202, 70, 1978.
7. **Balmain, N., Brehier, A., Cuisinier-Gleizes, P., and Mathieu, H.,** Evidence for the presence of calbindin-D$_{28k}$ (CaBP$_{28k}$) in the tibial growth cartilage of rats, *Cell Tissue Res.,* 245, 331, 1986.
8. **Zhou, X. Y., Dempster, D. W., Marion, S. L., Pike, J. W., Haussler, M. R., and Clemens, T. L.,** Bone vitamin D-dependent calcium binding protein is localized in chondrocytes of growth plate cartilage, *Calcif. Tissue Int.,* 38, 244, 1986.
9. **Elms, T. N. and Taylor, A. N.,** Calbindin-D$_{28k}$ localization in rat molars during odentogenesis, *J. Dent. Res.,* 66, 1431, 1987.
10. **Berdal, A., Dupret, J. M., Pike, J. W., and Mathieu, H.,** Vitamin D controlled expression of calbindin genes in teeth, *Abstracts, 8th Workship on Vitamin D,* 1991, 206.
11. **Morrisey, R., Bucci, T., Empson, R., and Lufkin, E.,** Calcium binding protein: its cellular localization in jejunum, kidney and pancreas, *Proc. Soc. Exp. Biol. Med.,* 149, 56, 1975.

12. **Christakos, S., Friedlander, E. J., Frandsen, B. R., and Norman, A. W.,** Development and application of a radioimmunoassay for chick intestinal calcium binding protein and tissue distribution, *Endocrinology,* 104, 1495, 1979.

13. **Sonnenberg, J., Pansini, A. R., and Christakos, S.,** Vitamin D-dependent rat renal calcium binding protein: development of a radioimmunoassay tissue distribution and immunological identification, *Endocrinology,* 115, 640, 1984.

14. **Roth, J., Bonner-Weir, S., Norman, A. W., and Orci, L.,** Immunocytochemistry of vitamin D-dependent calcium binding protein in chick pancreas: exclusive localization in β cells, *Endocrinology,* 111, 2216, 1982.

15. **Rhoten, W. B. and Christakos, S.,** Vitamin D dependent calcium binding protein is present in mammalian β cells, *Diabetes,* 32, 130A, 1983.

16. **Corradino, R. A., Wasserman, R. H., Pubolos, M. H., and Chang, S. I.,** Vitamin D_3 induction of a calcium binding protein in the uterus of the laying hen, *Arch. Biochem. Biophys.,* 125, 378, 1968.

17. **Delorme, A. C., Dana, J. L., Acker, M. G., Ripoche, M. A., and Mathieu, H.,** In rat uterus, 17β estradiol stimulates a calcium binding protein similar to duodenal vitamin D dependent calcium binding protein, *Endocrinology,* 113, 1340, 1983.

18. **Bruns, M. E., Overpeck, J. G., Smith, G. C., Hirsch, G. N., Mills, S. E., and Bruns, D. E.,** Vitamin D-dependent calcium binding protein in rat uterus: differential effects of estrogen, tamoxifen, progesterone and pregnancy on accumulation and cellular localization, *Endocrinology,* 122, 2371, 1988.

19. **Delorme, A. C., Cassier, P., Geny, B., and Mathieu, H.,** Immunocytochemical localization of vitamin D-dependent calcium binding protein in the yolk sac of the rat, *Placenta,* 4, 263, 1983.

20. **Bruns, M. E., Kleeman, E., and Bruns, D. E.,** Vitamin D-dependent calcium binding protein in mouse yolk sac: biochemical and immunochemical properties and response to 1,25-dihydroxycholecalciferol, *J. Biol. Chem.,* 261, 7485, 1986.

21. **Taylor, A. N.,** Chick brain calcium binding protein: response to cholecalciferol and some developmental aspects, *J. Nutr.,* 107, 480, 1977.

22. **Jande, S. S., Maler, L., and Lawson, D. E. M.,** Immunohistochemical mapping of vitamin D dependent calcium binding protein in brain, *Nature,* 294, 765, 1981.

23. **Roth, J., Baeten, D., Norman, A. W., and Garcia-Segura, L. M.,** Specific neurons in chick central nervous system stain with an antibody against intestinal vitamin D-dependent calcium binding protein, *Brain Res.,* 222, 452, 1981.

24. **Feldman, S. C. and Christakos, S.,** Vitamin D-dependent calcium binding protein in rat brain: biochemical and immunocytochemical characterization, *Endocrinology,* 112, 290, 1983.

25. **Celio, M.,** Calbindin-D_{28k} and parvalbumin in the rat nervous system, *J. Neuroscience,* 35, 375, 1990.

26. **Parmentier, M., Ghysens, M., Rysen, F., Lawson, D. E. M., Pasteels, J. L., and Pochet, R.,** Calbindin in vertebrate classes: immunohistochemical localization and Western blot analysis, *Gen. Comp. Endocrinol.,* 65, 399, 1987.

27. **Christakos, S., Malkowitz, L., Sori, A., Sperduto, A., and Feldman, S. C.,** Calcium binding protein in squid brain: biochemical similarity to the 28,000 M_r vitamin D dependent calcium binding protein, *J. Neurochem.,* 49, 1427, 1987.

28. **Christakos, S. and Norman, A. W.,** Vitamin D dependent calcium binding protein and its relation to 1,25-dihydroxyvitamin D receptor localization and concentration, in *Calcium Binding Proteins and Calcium Function,* Siegel, F. et al., Eds., Elsevier, New York, 1980, 371.

29. **Thorens, B., Roth, J., Norman, A. W., Perrelet, A., and Orci, L.,** Immunocytochemical localization of vitamin D-dependent calcium binding protein in chick duodenum, *J. Cell Biol.,* 94, 201, 1982.

30. **Taylor, A. N., Gleason, W. A., and Lankford, G. L.,** Vitamin D dependent calcium binding. Immunocytochemical localization of the rat intestinal protein, *J. Histochem. Cytochem.*, 32, 153, 1984.

31. **Christakos, S., Brunette, M. G., and Norman, A. W.,** Localization of immunoreactive vitamin D-dependent calcium binding protein in chick nephron, *Endocrinology,* 109, 322, 1981.

32. **Roth, J., Thorens, B., Hunziker, W., Norman, A. W., and Orci, L.,** Vitamin D-dependent calcium binding protein: immunocytochemical localization in chick kidney, *Science,* 214, 197, 1981.

33. **Rhoten, W. B. and Christakos, S.,** Immunocytochemical localization of vitamin D-dependent calcium binding protein in mammalian nephron, *Endocrinology,* 109, 981, 1981.

34. **Roth, J., Brown, D., Norman, A. W., and Orci, L.,** Localization of the vitamin D-dependent calcium binding protein in mammalian kidney, *Am. J. Physiol.,* 243, F243, 1982.

35. **Desplan, C., Thomasset, M., and Moukhatar, M. S.,** Synthesis, molecular cloning and restriction analysis of DNA complementary to vitamin D dependent calcium binding protein mRNA from rat intestine, *J. Biol. Chem.,* 258, 2762, 1983.

36. **Hunziker, W., Siebert, P., King, M., Strucki, P., Dugaiczk, A., and Norman, A. W.,** Molecular cloning of a vitamin D-dependent calcium binding protein mRNA sequence from chick intestine, *Proc. Natl. Acad. Sci. U.S.A.,* 80, 4228, 1983.

37. **Hunziker, W.,** The 28kDa vitamin D dependent calcium binding protein is a six domain structure, *Proc. Natl. Acad. Sci. U.S.A.,* 83, 7578, 1986.

38. **Wood, T. L., Kobayashi, Y., Frantz, G., Varghese, S., Christakos, S., and Tobin, A. J.,** Calcium binding protein (calbindin-D_{28k}): expression of calbindin-D_{28k} RNAs in rodent kidney and brain, *DNA,* 7, 585, 1988.

39. **Hunziker, W. and Schrickel, S.,** Rat brain calbindin-D_{28k}: six domain structure and extensive amino acid homology with chicken calbindin-D_{28k}, *Mol. Endocrinol.,* 2, 465, 1988.

40. **Perret, C., Desplan, C., Brehier, A., and Thomasset, M.,** Characterization of rat 9-kDa cholecalcin (CaBP) messenger RNA using a complementary DNA, *Eur. J. Biochem.,* 148, 61, 1983.

41. **Theofan, G. and Norman, A.,** Effects of α amanitin and cycloheximide on 1,25-dihydroxyvitamin D_3 dependent calbindin-D_{28k} and its mRNA in vitamin D_3 replete chick intestine, *J. Biol. Chem.,* 261, 7311, 1986.

42. **Theofan, G., Nguyen, A. P., and Norman, A. W.,** Regulation of calbindin-D_{28k} gene expression by 1,25-dihydroxyvitamin D_3 is correlated to receptor occupancy, *J. Biol. Chem.,* 261, 16943, 1986.

43. **Dupret, J., Brun, P., Perret, C., Lomri, N., Thomasset, M., and Cuisinier-Gleizes, P.,** Transcriptional and posttranscriptional regulation of vitamin D-dependent calcium binding protein in rat duodenum by 1,25-dihydroxycholecalciferol, *J. Biol. Chem.,* 262, 16553, 1987.

44. **Varghese, S., Lee, S., Huang, Y.-C., and Christakos, S.,** Analysis of rat vitamin D-dependent calbindin-D_{28k} gene expression, *J. Biol. Chem.,* 263, 9776, 1988.

45. **Huang, Y.-C. and Christakos, S.,** Modulation of rat calbindin-D_{28k} gene expression by 1,25-dihydroxyvitamin D_3 and dietary alteration, *Mol. Endocrinol.,* 2, 928, 1988.

46. **Varghese, S., Deaven, L. L., Huang, Y. C., Gill, R., Iacopino, A. M., and Christakos, S.,** Transcriptional regulation and chromosomal assignment of the mammalian calbindin-D_{28k} gene, *Mol. Endocrinol.,* 3, 495, 1989.

47. **Huang, Y.-C., Lee, S., Stolz, R., Gabrielides, C., Pansini-Porta, A., Bruns, M. E., Bruns, D., Mifflin, T., Pike, J. W., and Christakos, S.,** Effect of hormones and development on the expression on the rat 1,25-dihydroxyvitamin D_3 receptor gene and comparison to calbindin gene expression, *J. Biol. Chem.,* 264, 17454, 1989.

48. **Rhoten, W. B. and Christakos, S.,** Cellular gene expression for calbindin-D$_{28k}$ in mouse kidney, *Anat. Rec.,* 227, 145, 1990.

49. **Iacopino, A. M. and Christakos, S.,** Reduction of neuronal calcium binding protein (calbindin-D$_{28k}$) gene expression in aging and neurodegenerative diseases, *Proc. Natl. Acad Sci. U.S.A.,* 87, 4078, 1990.

50. **Iacopino, A. M. and Christakos, S.,** Corticosterone regulates calbindin-D$_{28k}$ mRNA and protein levels in rat hippocampus, *J. Biol. Chem.,* 265, 10177, 1990.

51. **Iacopino, A. M. and Christakos, S.,** Calcium binding protein (calbindin-D$_{28k}$) gene expression in the developing and aging mouse cerebellum, *Mol. Brain Res.,* 8, 283, 1990.

52. **Li, H. and Christakos, S.,** Differential regulation by 1,25 dihydroxyvitamin D$_3$ of calbindin-D$_{9k}$ and calbindin-D$_{28k}$ gene expression in mouse kidney, *Endocrinology,* 128, 2844, 1991.

53. **Enomoto, H., Hendy, G. N., and Clemens, T. L.,** Regulation of calbindin-D$_{28k}$ (CaBP) mRNA in chick kidney cells: critical importance of posttranscriptional mechanisms and calcium ion concentration, *J. Bone Min. Res.,* 6 (Suppl. 1), abstract 390, S183, 1991.

54. **Meyer, Galligan, M., Jones, G., Komm, B. S., and Haussler, M. R.,** 1,25(OH)$_2$D$_3$ dependent regulation of calbindin-D$_{28k}$ mRNA apparently requires the synthesis of an uncharacterized protein in chick duodenal organ culture, *J Bone Min. Res.,* 6 (Suppl. 1), abstract 406, S185, 1991.

55. **Kimberg, D. V., Baerg, R. D., Gershone, E., and Graudusius, R. T.,** Effect of cortisone treatment on the active transport of calcium by the small intestine, *J. Clin. Invest.,* 50, 1309, 1971.

56. **Hall, A. K., Bishop, J. E., and Norman, A. W.,** Inhibitory and stimulatory effects of dexamethasone and 1,25dihydroxyvitamin D$_3$ on chick intestinal calbindin-D$_{28k}$ and its mRNA, *Mol. Cell. Endocrinol.,* 51, 25, 1987.

57. **Corradino, R. A. and Fullmer, C. S.,** Positive cotranscriptional regulation of intestinal calbindin-D$_{28k}$ gene expression by 1,25dihydroxyvitamin D$_3$ and glucocorticoids, *Endocrinology,* 128, 944, 1991.

58. **Navikis, R. J., Katzenellenbogen, B. S., and Nalbandov, A. V.,** Effects of the sex steroids and vitamin D$_3$ on calcium binding protein in the chick shell gland, *Biol. Reprod.,* 21, 1153, 1979.

59. **Opperman, L. A., Sandes, T. J., Mills, S. E., Christakos, S., Bruns, D. E., and Bruns, M. E.,** The control of calbindin-D$_{9k}$ and calbindin-D$_{28k}$ in mouse uterus by 1,25(OH)$_2$D$_3$ and estrogen, *J. Bone Min. Res.,* 5 (Suppl. 2), S264, 1990.

60. **Hall, A. K., Wang, Y.-Z., Gill, R. K., and Christakos, S.,** Modulation of calbindin-D$_{28k}$-like gene expression by retinoic acid in neuronal cells overexpressing a retinoic acid receptor, *J. Bone Min. Res.,* 6 (Suppl. 1), S184, 1991.

61. **Kim, H. T. and Chen, T. L.,** 1,25Dihydroxyvitamin D$_3$ interaction with dexamethasone and retinoic acid: effects on procollagen messenger ribonucleic acid levels in rat osteoblast like cells, *Mol. Endocrinol.,* 3, 97, 1989.

62. **Schule, R., Umesono, K., Mangelsdorf, D. J., Bolado, J., Pike, J. W., and Evans, R. M.,** Jun-fos and receptors for vitamin A and D recognize a common response element in the human osteocalcin gene, *Cell,* 61, 497, 1990.

63. **Umesono, K. and Evans, R. M.,** Determinants of target gene specificity for steroid/ thyroid hormone receptor, *Cell,* 57, 1139, 1990.

64. **Mader, S., Kuman, U., de Verneuil, H., and Chambon, P.,** Three amino acids of the esterogen receptor are essential to its ability to distinguish an estrogen from glucocorticoid response element, *Nature,* 338, 271, 1989.

65. **Danielsen, M., Hinck, L., and Ringold, A. M.,** Two amino acids within the knuckle of the first zinc finger specify DNA response element activation by glucocorticoid receptor, *Cell,* 57, 1131, 1989.

66. **Umesono, K., Murakami, K. K., Thompson, C., and Evans, R. M.,** Direct repeats as selective response elements for the thyroid hormone, retinoic acid and vitamin D_3 receptors, *Cell,* 65, 1255, 1991.

67. **Stumpf, W. E. and O'Brien, L. P.,** 1,25(OH)$_2$ Vitamin D_3 sites of action in the brain, *Histochemistry,* 87, 393, 1987.

68. **Stumpf, W. E. and Bidman, H.-J.,** Sites and mechanisms of action of vitamin D in the nervous system, *Abstracts, Eighth Workshop on Vitamin D,* 23, 1991, 23.

69. **Mattson, M. P., Rychlik, B., Chu, C., and Christakos, S.,** Evidence for calcium reducing and excitoprotective roles for calcium binding protein, calbindin-D_{28k} in cultured hippocampal neurons, *Neuron,* 6, 41, 1991.

70. **Christakos, S. and Norman, A. W.,** Evidence for a specific high affinity binding protein for 1,25 dihydroxyvitamin D_3 in chick kidney and pancreas, *Biochem. Biophys. Res. Commun.,* 89, 56, 1979.

71. **Christakos, S. and Norman, A. W.,** Biochemical characterization of 1,25dihydroxyvitamin D_3 receptors in chick pancreas and kidney cytosol, *Endocrinology,* 108, 140, 1981.

72. **Clark, S. A., Stumpf, W. E., Sar, M., DeLuca, H. F., and Tanaka, Y.,** Target cells for 1,25dihydroxyvitamin D_3 in rat pancreas, *Cell Tissue Res.,* 109, 515, 1980.

73. **Parmentier, M., DeViglder, J. J. M., Muri, E., Szisirer, C., Islam, M. Q., Guerts Van Kessel, A., Lawson, D. E. M., and Vassart, G.,** The human calbindin 27 kDa gene: structural organization of the 5′ and 3′ region, chromosomal assignment and restriction fragment length polymorphism, *Genomics,* 4, 309, 1989.

74. **Gill, R. K. and Christakos, S.,** Isolation of genomic clones harboring sequences homologous to mammalian calbindin-D_{28k} and their characterization, *J. Bone Mineral Res.,* 4, 1989, S291.

75. **Minghetti, P. P., Cancela, L., Fujisawa, Y., Theofan, G., and Norman, A. W.,** Molecular structure of the chicken vitamin-D induced calbindin-D_{28k} gene reveals eleven exons, six Ca^{2+} binding domains and numerous promoter regulatory elements, *Mol. Endocrinol.,* 2, 355, 1988.

76. **Wilson, P. W., Rogers, J., Harding, M., Pohl, V., Pattyn, G., and Lawson, D. E. M.,** Structure of chick chromosomal gene for calbindin and calretinin, *J. Mol. Biol.,* 200, 615, 1988.

77. **Ferrari, S., Drusian, E., Baltini, R., and Fregni, M.,** Nucleotide sequence of the promoter region of the gene encoding chicken calbindin-D_{28k}, *Nucleic Acid Res.,* 16, 353, 1988.

78. **Perret, C., Lomri, N., Gouhier, N., Auffrey, C., and Thomasset, M.,** The rat vitamin-D dependent calcium binding protein (9-KDa CaBP) gene: complete nucleotide sequence and structural organization, *Eur. J. Biochem.,* 172, 43, 1988.

79. **Gill, R. and Christakos, S.,** Identification of sequence elements in the mouse calbindin-D_{28k} gene which confer basal activation and hormonal inducible response, *Abstracts, Eighth Workshop on Vitamin D,* 1991, 81.

80. **Markose, E. R., Stein, J. L., Stein, G. S., and Lian, J. B.,** Vitamin D-mediated modifications in protein-DNA interaction at two promoter elements of the osteocalcin gene, *Proc. Natl. Acad. Sci. U.S.A.,* 87, 1701, 1990.

81. **Demay, M. B., Gerardi, J. M., DeLuca, H. F., and Kronenberg, H. M.,** DNA sequence in the rat osteocalcin gene that bind the 1,25 dihydroxyvitamin D_3 receptor and confer responsiveness to 1,25-dihydroxyvitamin D_3, *Proc. Natl. Acad. Sci. U.S.A.,* 87, 369, 1990.

Chapter 18

VITAMIN D REGULATION OF OSTEOBLAST GROWTH AND DIFFERENTIATION

Jane B. Lian and Gary S. Stein

TABLE OF CONTENTS

0-8493-6961-4/93/$0.00 + $.50
© 1993 by CRC Press, Inc.

I. INTRODUCTION

Vitamin D is a steroid (pro)-hormone that undergoes several transformations to the active hormone 1,25-dihydroxyvitamin D_3 ($1,25(OH)_2D_3$). The active 1,25-dihydroxy D_3 is produced in the kidney and has three major target tissues (intestine, kidney, and bone) that contribute to the regulation of calcium homeostasis. Figure 1 schematically illustrates the vitamin D endocrine system. At the physiologic level, in response to a fall in circulating calcium levels, parathyroid hormone induces synthesis of the renal hydroxylase enzyme (25-hydroxycholecalciferol 1-hydroxylase). Then $1,25(OH)_2D_3$ promotes the absorption of calcium and phosphorus across the intestinal epithelial cells and mobilizes mineral from bone. Normal serum calcium levels are maintained, since high serum calcium indirectly and $1,25(OH)_2D_3$ directly suppresses parathyroid hormone secretion and function.[1] Calcitonin, also secreted in response to hypercalcemia, inhibits osteoclastic activity. At the cellular level, the interaction of $1,25(OH)_2D_3$ and its receptor is analogous to the mode of action of other steroid hormones and involves changes in gene expression in the target cells;[2] however, rapid ''non-genomic'' actions of vitamin D, possibly potentiating the regulation of calcium metabolism, are also being observed.[3-5]

For several decades, vitamin D has been recognized as functionally important to regulatory mechanisms operative in skeletal development (cartilage and bone formation) and the lifetime remodeling of bone involving resorption and bone formation. The active metabolite $1,25(OH)_2D_3$ mediates growth and differentiation of chondrocytes,[6-8] osteoblasts,[9,10] and progenitor cells that develop the osteoclast phenotype, both directly and/or as a consequence of the effects of the hormone on a broad spectrum of integrative cellular signaling mechanisms (reviewed in References 11, 12). Not to be dismissed are the well documented physiological influences of vitamin D on calcium metabolism that result in modifications of the development and activities of skeletal cells (reviewed in References 10, 13). In addition to these central roles of vitamin D in bone cell development and maintenance of functional properties, it is well established that vitamin D-related aberrations in metabolism of the hormone and/or receptor binding and the association of hormone-receptor complexes with vitamin D-responsive elements in gene regulatory sequences are the principal components of skeletal disorders.[14] Since vitamin D levels are a determinant of the physiological regulation and pathological status of bone, the nutritional contribution to skeletal development and maintenance must be recognized. Here, the dietary levels of vitamin D and factors that affect the metabolism or utilization of the hormone are key considerations.

This chapter focuses on the combined application of molecular, biochemical, and morphologic approaches that have contributed to defining the multiple effects of vitamin D on bone cell growth and differentiation. The osteocalcin gene is representative of a series of genes that are vitamin D-

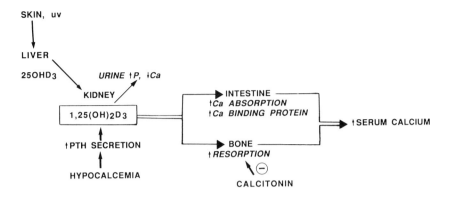

FIGURE 1. Interrelationships of the vitamin D endocrine system in response to low serum calcium. The final step in synthesis of the active hormone metabolite, $1,25(OH)_2D_3$, occurs in the kidney and can be stimulated by parathyroid hormone (PTH). Organs responsive to $1,25(OH)_2D_3$ to increase serum calcium are kidney, intestine, and bone. Calcitonin will inhibit excessive osteoclastic resorption in bone.

responsive, and stringently regulated during progressive development of the osteoblast phenotype. We will, therefore, address the influences of the hormone at the transcriptional and at several posttranscriptional levels. At each of these levels, osteocalcin expression can be modulated. Osteocalcin expression depends on the developmental stage of the osteoblast, both independently and synergistically or antagonistically, with other physiological effectors of this expression. It is also important to complement assessments of transcriptional activity or mRNA levels with *in situ* hybridization and immunohistochemistry where gene expression can be established at the single cell level. We will illustrate how such experimental approaches have been effectively integrated to address the structural and functional properties of vitamin D regulation of osteocalcin in bone.

II. VITAMIN D AND BONE PHYSIOLOGY

Many advances in our understanding of vitamin D regulation of skeletal tissues are attributable to availability of *in vivo* and *in vitro* model systems that support bone growth and differentiation. The foundation for our present knowledge of bone cell growth and differentiation under conditions where cell and tissue relationships are maintained, was provided by *in vivo* studies of growth and formation of bone tissue principally utilizing histology and histochemistry to examine skeletal development. Similarly, the effects of vitamin D depletion and repletion on bone growth were studied *in vivo*, where changes in serum parameters were correlated with perturbations in bone morphology, histochemistry and composition during avian and rodent skeletal development.[15-20]

Both inadequate or excessive vitamin D intakes have dramatic effects on bone morphology and metabolism. Vitamin D deficiency increases the amount of nonmineralized bone matrix (osteoid tissue), resulting in rickets in children and osteomalacia in adults. Administration of $1,25(OH)_2D_3$ to vitamin D-deficient animals stimulates osteoclast bone resorption, then cures rickets and osteomalacia.[21] Experimental studies in rodents have demonstrated that bone mineralization is not dependent upon $1,25(OH)_2D_3$ and can be maintained if an adequate supplementation of calcium and phosphate in the diet is used.[22-25] However, other studies have demonstrated that vitamin D can stimulate bone mineralization independently of changes in serum mineral concentrations.[21] Yet, pharmacologic doses of $1,25(OH)_2D_3$ administered *in vitro* in primary osteoblast cultures[9] and *in vivo* can inhibit mineralization of the newly formed bone matrix.[19,26] *In vivo* morphometric studies have demonstrated that high doses of $1,25(OH)_2D_3$ decrease the amount and rate of formation of new bone matrix in mice.[17] Thus, the effects of this metabolite on skeletal tissue are indeed extremely complex, encompassing multiple parameters related to several osteoblast gene products that are involved in tissue development and mineralization.[9,13]

Vitamin D's primary action in bone is to stimulate bone resorption as a calcitropic regulating hormone, and this has been well documented *in vitro*[27] and *in vivo*.[28] *In vivo* repletion of hormone to vitamin D-deficient rats results in an initial osteoclastic bone resorption. The target cell for this primary function of the hormone is the osteoblast, which contains receptors for dihydroxyvitamin D_3.[29,30] It has been proposed[31] that the osteoblast secretes factors that stimulate osteoclastic activity in response to hormone.[32] Vitamin D also contributes to bone resorption by promoting differentiation of the mononuclear progenitor cell to form the multinucleated osteoclast.[33,34] Thus, the hormone appears to have both anabolic effects and antianabolic effects on osteoblasts to promote bone resorption. Indeed, this dual function likely relates to the balance necessary to maintain the coupling of bone resorption and bone formation. Only recently have the responses of isolated osteoblasts and cocultures of these cells with osteoclasts to vitamin D been characterized; these studies show that the multiple actions of this hormone are highly complex.

III. VITAMIN D MODULATION OF OSTEOBLAST CELL GROWTH AND DIFFERENTIATION

From the earliest *in vivo* studies, it appeared that vitamin D receptors were confined to the osteoprogenitor cells that differentiate to osteoblasts and chondrocytes.[35] Several *in vitro* cellular systems are currently being utilized to examine the cascade of events in gene expression that supports the interdependent sequence of biochemical and structural aspects of growth and differentiation that are mediated by $1,25(OH)_2D_3$ during the progressive development of the osteoblast.

A. ROLE OF 1,25(OH)₂D₃ IN COMMITMENT TO CELL DIFFERENTIATION

Bone marrow cultures offer the possibility of examining events associated with stem cell differentiation to osteoblasts.[36,37] Clonal lines from single cells derived, for example, from fetal rat calvaria also provide an opportunity for examining selective expression of genes during commitment to the bone cell lineage.[38,39] These are promising systems, offering the potential to study growth and differentiation of osteoprogenitor cells. To date, our knowledge of this aspect of osteoblast differentiation is incomplete, largely due to the limited characterization of such model systems. The marrow tissue comprises the stroma which can be differentiated to osteoblast-like cells largely under the influence of glucocorticoids (e.g., dexamethasone). Glucocorticoids can regulate and increase the availability of vitamin D receptors in some osteoblast cell lines.[40] Marrow also contains cells of the hematopoietic lineage which differentiate into osteoclasts, and this occurs largely under the influence of 1,25(OH)₂D₃. In early studies of marrow cell cultures treated with vitamin D, both increases in multinucleated cells (reflecting osteoclasts) and the presence of alkaline phosphatase positive cells (reflecting osteoblasts) were observed.[10]

B. ROLE OF 1,25(OH)₂D₃ ON OSTEOBLAST ACTIVITY AND BONE MATRIX FORMATION

1. The Normal Sequelae of Osteoblast Differentiation

Osteoblastic cells released from bone as outgrowth cultures or by a sequential series of combined trypsin and collagenase digestions[41,42] are a viable cellular model for studying osteoblast growth and differentiation. The bone-derived cells are enriched in cells committed to the osteoblast lineage, exhibiting properties which include PTH and vitamin D responsiveness and synthesis of bone-related proteins. Although this precludes investigation of the process by which commitment to the osteoblast phenotype is regulated, the primary diploid osteoblast cultures offer an effective system to examine osteoblastic properties from a proliferating cell to a cell which becomes enveloped in a mineralized extracellular bone-like matrix. Normal diploid osteoblasts undergo an ordered development over several weeks in culture, whether isolated from human bone[43,44] or fetal calvaria from 21-d rat bone,[45,46] mouse,[47] 16-d embryonic chick[48] or bovine bone-derived cells.[49] Proliferating cells differentiate to osteoblasts forming nodules of multilayered cells with a mineralized, type I collagen extracellular matrix (Figure 2).[46,50,51] A bone-tissue-like organization of these nodules formed in culture is supported by a comparison of the ultrastructure of the mineralized regions of the culture to sections through an intact 21-d fetal rat calvarium; both exhibit ordered deposition of crystals within and between the orthogonally organized bundles of collagen fibrils (Figure 2). No evidence for cell necrosis or intracellular calcification is indicated in the cultures, particularly where osteoblasts are embedded in mineralized matrix.[46,51-53]

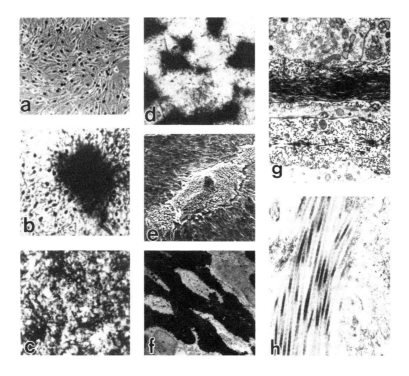

FIGURE 2. Morphology of normal rat diploid osteoblast cultures. (A through E) Light microscopy of cultured osteoblasts isolated from 21-d fetal calvaria by sequential collagenase digestion.[50] (A and B) Cells labeled with ^{3}H thymidine show greater than 95% of the cells are proliferating after plating (A, day 5). The cells multilayer, forming nodules (B, day 12), and the onset of expression of the osteoblast phenotype is indicated by alkaline phosphatase staining of cells where proliferation has ceased in the center of the nodule (C, day 18 alkaline phosphatase histochemistry). Every cell has ceased to proliferate and greater than 90% of the cells are alkaline phosphatase positive (D, day 28). It is within the multilayered nodule that mineralization is initiated; von Kossa silver staining shows mineralized nodules throughout the dish. A through E, photographed at 100×. (F) Scanning electron micrograph of a nodule on day 28 at 20×, with mineralized matrix enveloping osteoblasts. (G and H): Transmission electron microscopy (bars are 5 and 2.5 microns). (G) shows orthogonal organization of the collagen matrix in thick layers between the osteoblasts *in vitro*. (H) Higher magnification of the collagen bundles shows mineral deposition within the collagen fibrils and absence of intracellular calcification.

In these cultures of mammalian diploid osteoblasts, cell growth and tissue-specific gene expression have been mapped during the progressive development of the bone cell phenotype from a proliferating cell to a mature osteocytic-like cell within the context of a bone tissue-like organization (Figure 3).[50-53] Temporally, genes encoding osteoblast phenotype markers are expressed during three distinct periods, including a growth period (proliferation), a period of matrix development, and a mineralization period. During the first 10 to 12 d, the period of active proliferation is reflected by mitotic activity

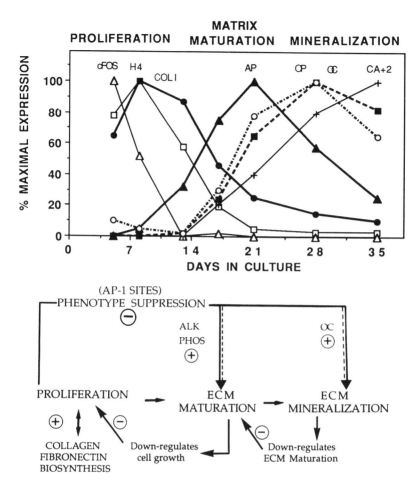

FIGURE 3. Temporal expression of cell growth and osteoblast phenotype related genes during the development of *in vitro* formed bone-like tissue by normal diploid rat osteoblasts. Top panel: isolated primary cells were initially cultured in MEM with 10% FCS, then after confluence, in BGJb medium supplemented with 10% FCS, 50 μg/ml ascorbic acid and 10 mM β-glycerol phosphate. Cellular RNA was isolated at the times indicated (3, 5, 7, 10, 12, 16, 21, 28, and 35 d) during the differentiation time course and assayed for the steady-state levels of various transcripts by Northern blot analysis. The resulting blots were quantitated by scanning densitometry and the results plotted relative to the maximal expression of each transcript. Three periods of gene expression are represented: (A) Genes characterizing the proliferation period include H4 histone (which reflects DNA synthesis), c-*myc* and c-*fos*, plus genes expressed for formation of the type I collagen, fibronectin (FN), and transforming growth factor-β (TGFβ); (B) alkaline phosphatase (AP) is expressed postproliferatively and associated with extracellular matrix maturation; (C) genes induced to high levels with onset of extracellular matrix mineralization; represented are osteopontin (OP) and osteocalcin (OC). Calcium (Ca^{2+}) accumulation is indicated. This temporal pattern defines three periods, proliferation, matrix maturation, and mineralization reflecting a developmental sequence of osteoblast maturation. Lower panel: the interrelationship and feedback signals between the periods are indicated. AP-1 activity in proliferating osteoblast, that is, fos-jun protein binding to AP-1 sites, suppresses activity of postproliferative expressed genes.

with expression of cell cycle (e.g., histone) and cell growth (e.g., c-*myc*, c-*fos* and AP-1 activity) regulated genes. It is fundamental to development of the bone cell phenotype that several genes associated with formation of the extracellular matrix (type I collagen, fibronectin, and TGFβ) are actively expressed in this proliferation period. Immediately following the downregulation of proliferation, as reflected by the decline in DNA synthesis (^{3}H-thymidine incorporation) and histone gene expression, the expression of alkaline phosphatase (enzyme activity and mRNA) increases greater than tenfold. At this time, a differentiation-specific histone gene is expressed.[54,55] During this period (from 12 to 18 d) the extracellular matrix undergoes a series of modifications in composition and organization that renders it competent for mineralization. During this matrix maturation phase, every cell has become alkaline phosphatase positive as the cultures progress into the mineralization stage. In heavily mineralized cultures, cellular levels of alkaline phosphatase mRNA decline.

With the onset of extracellular matrix mineralization, other bone-specific genes are induced. For example, osteopontin or SPPI, a 60-kDa calcium binding protein enriched in *O*-phosphoserine,[56] and osteocalcin, a γ-carboxyglutamate-containing protein,[57,58] are upregulated in concert with the accumulation of mineral. Notably, osteopontin is expressed during the period of active proliferation (at 25% of maximal levels), consistent with its high level of expression following oncogene transformation or phorbal ester treatments of cells.[59] Osteopontin mRNA is decreased postproliferatively and then exhibits an induction at the onset of mineralization achieving peak levels of expression during the mineralization period (days 16 to 20). In contrast to osteopontin, the vitamin K-dependent protein,[60] osteocalcin, is expressed only postproliferatively with the onset of nodule formation.[51] This 5.7-kDa calcium binding protein is characterized by 3 γ-carboxyglutamic acid residues (Gla), binds tightly to hydroxyapatite, and is maximally expressed during active mineralization of the extracellular matrix (ECM) *in vivo*[57] and *in vitro*.[51] A related vitamin K-dependent protein in bone, matrix Gla protein (MGP),[61] is characterized by the presence of 5 Gla residues in a 10-kDa molecule. MGP is present in proliferating cells and is increased in relation to accumulation of collagen, but levels of expression are not correlated with the mineralization phase.[62] Several manipulations of the culture system[51] demonstrate that induction of high levels of osteocalcin and osteopontin mRNA are dependent upon formation of a mineralized extracellular matrix whereas MGP levels are sustained independent of mineralization.[62] Both osteocalcin and osteopontin are expressed in mature osteoblasts within the nodules[63] and the proteins are localized to the mineralized bone extracellular matrix.[64] Osteocalcin has been shown to contribute to regulation of the mineral phase in bone both *in vitro* as a potential inhibitor of mineral nucleation[65] and *in vivo* as a bone matrix signal that promotes osteoclast differentiation and activation.[16,66,67] Osteopontin has been localized to the region of osteoclast contact.[68] More recently, others have used synthetic peptides containing osteopontin's Arg-Gly-Asp sequence region, to show stimulation of osteoclast activity.[69] While the induction of osteopontin and osteocalcin during the period of active mineral accumulation may support the onset and progression of extracellular matrix mineralization, these proteins may also function in the regulation of turnover of the mineral phase of bone *in vivo*. This potential function would be consistent with vitamin D regulation of these proteins during bone resorption.

The observed temporal expression of osteoblast phenotype markers during the developmental sequence is a reflection of functional activities necessary for the progressive formation of bone tissue. The expression of alkaline phosphatase mRNA and enzyme activity prior to the initiation of osteoblast mineralization suggests that alkaline phosphatase may be involved in the preparation of the extracellular matrix for the ordered deposition of mineral. In the normal temporal expression of osteoblast genes, alkaline phosphatase peak levels precede maximal deposition of mineral and in heavily mineralized cultures, alkaline phosphatase mRNA levels and enzyme activity are downregulated. This is consistent with *in vivo* observations of vitamin D deficiency where serum and tissue alkaline phosphatase levels are high in the absence of mineralized matrix in rachitic bone. In this situation, with the absence of matrix mineralization, alkaline phosphatase is apparently not downregulated. The coinduction of genes, such as osteocalcin and osteopontin with active accumulation of mineral, may support the onset and progression of extracellular matrix mineralization. Alternatively, the induction of these mineralization-associated genes may reflect an acquisition of osteoblast properties necessary for signaling bone turnover *in vivo* as discussed above. The sequence of expression of genes that encode osteoblast phenotype markers in culture follows a pattern of gene expression and tissue distribution that occurs in neonatal bones determined by *in situ* hybridization[70,71] and during fetal calvarial development *in vivo*,[72] supporting the biological relevance of the osteoblast culture system as a model for intramembranous bone differentiation. Taken together, the patterns of expression of these genes and the synthesis of the encoded proteins demonstrate that a temporal sequence of gene expression during the culture period is associated with development of the extracellular matrix and reflects maturation of the osteoblast phenotype *in vitro*.

2. Vitamin D Modification of Osteoblast Growth and Differentiation

Vitamin D affects numerous osteoblast products,[11] including cell growth related genes,[73-75] nuclear and cytoskeletal proteins,[76,77] structural proteins involved in bone matrix formation, and mineralization,[9] enzymes,[78] and plasma membrane proteins[79] such as alkaline phosphatase.[80,81] Many responsive proteins have not yet been characterized.[82] Vitamin D gene regulation has been studied for only a few proteins, but these represent some of the major constituents of the bone matrix. These include, for example, collagen,[83] osteopontin,[84] osteocalcin,[85-87] and matrix Gla protein.[61,62] The development of

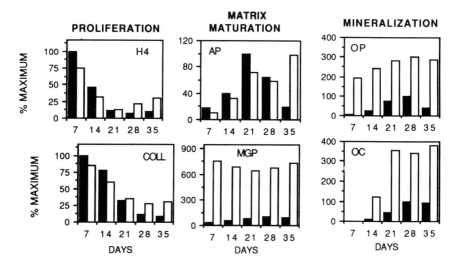

FIGURE 4. Examples of vitamin D-regulated genes during the three periods of osteoblast phenotype development. Solid bars are controls and open bars are cells treated 48 h prior to harvest with (10^{-8} *M*) 1,25(OH)$_2$D$_3$. Samples during the time course were hybridized together for quantitation of the control and hormone-treated samples as percent maximal expression for each gene. Note the pleiotropic effects on proliferation related genes. Histone (H4) and type I collagen (COLL) are inhibited early (day 8 through 16), then stimulated in mature osteoblasts in a mineralized matrix (day 28 through 35); alkaline phosphatase (AP) is also stimulated.

normal diploid osteoblast cultures that produce a bone-like mineralized extracellular matrix has allowed for studies that can evaluate the effects of the hormone at several stages of osteoblast differentiation from the proliferating cell to the mature osteocyte-like cell enveloped in a mineralized matrix. Vitamin D influences the extent to which osteoblast genes are expressed, in most cases modulating the extent of expression, rather than inducing or suppressing expression (Figure 4). For example, osteocalcin is not expressed during proliferation and is not induced by vitamin D. Other genes that are expressed in proliferating cells, such as OP and MGP, are vitamin D responsive, indicating that the vitamin D receptor is available, along with a subset of accessory factors that is required for hormone binding and activity at vitamin D responsive promoter elements.

Biphasic effects of vitamin D on gene expression are observed (Figure 4) that reflect differences in regulation dependent upon the stage of osteoblast maturation.[9,88,89] For example, in normal diploid osteoblast cultures, vitamin D inhibits proliferation when the cells are actively dividing, but stimulates proliferation in quiescent osteocyte-like cells in a mineralized matrix.[9] The antiproliferative or proliferative effects may be a consequence of vitamin D regulation of growth factors, such as skeletal growth factor,[73,74] epidermal growth factor, and TGFβ.[75] Genes, such as TGFβ and collagen, are expressed at high levels during proliferation and are downregulated by vitamin D in proliferating rat osteoblasts,[83,90] but are upregulated in mature osteoblasts.[9,43] TGFβ levels are also stimulated when expressed at low levels and inhibited when expressed at high levels (unpublished observations, this laboratory). Alkaline phosphatase also exhibits biphasic responsiveness to vitamin D consistent with both positive or negative regulation observed in other cell culture systems.[43,80,91-93] Such biphasic responsiveness can be a consequence of the maturation state and matrix environment of the osteoblast. The inhibition of collagen production and alkaline phosphatase enzyme activity by vitamin D *in vitro* may relate to the pharmacologic effect in animals where an inhibition of mineralization and a decrease in bone formation are observed.[19-21,26] Vitamin D both positively and negatively regulates expression of genes encoding osteoblast phenotype markers as a function of the differentiated stage of the cell and basal levels of expression. The hormone can thus serve as a physiological mediator of bone formation and remodeling, which are functionally related to bone matrix competency and the differentiated state of the osteoblasts.

A reciprocal and functionally coupled relationship between the expression of growth-related and differentiation-related genes exists, as schematically illustrated in Figure 5 for development of the osteoblast phenotype. Several lines of evidence support this growth-differentiation relationship; e.g., growth supports collagen biosynthesis, collagen matrix formation contributes to cessation of proliferation, and this is required for increased expression of alkaline phosphatase and osteocalcin.[51,53] The acute effect of $1,25(OH)_2D_3$, if added to the culture for periods up to 48 h, is to shift the profile of expressed genes to one that represents a more differentiated cell, while maintaining the growth-differentiation relationship. This may be a consequence of its antiproliferative effect since the downregulation of proliferation promotes expression of phenotypic genes.[53] However, if 10^{-8} M $1,25(OH)_2D_3$ is present throughout the culture period (continuously added from day 5, proliferating period), the resulting inhibition of collagen and alkaline phosphatase gene expression prevents formation of a mineralized extracellular matrix.[9] In conditions where mineralization is prevented, there is no induction of osteopontin and osteocalcin by $1,25(OH)_2D_3$. This defines the second transition point and illustrates the role of the mineralized extracellular matrix in promoting osteoblast differentiation. Thus, the program of bone cell maturation is blocked because of the inhibition of proliferation and collagen synthesis. Although vitamin D is well documented to increase osteocalcin transcription and synthesis,[86,94] osteocalcin cannot be expressed nor modulated by vitamin D if osteoblast differentiation is blocked by high concentrations and long exposure of proliferating cells to vitamin D in this system. This concept is further supported by experiments in which the cultures were treated with $1,25(OH)_2D_3$ continuously at various times after formation of the mineralized matrix.[9] Since the OC gene is expressed under these conditions, stimulation by hormone is

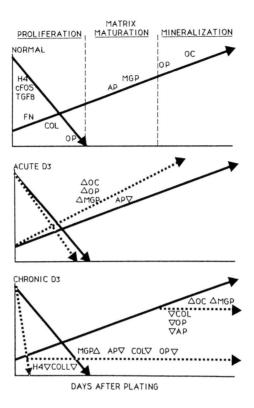

FIGURE 5. Model of the reciprocal relationship between proliferation and differentiation in normal diploid cells during the rat osteoblast developmental sequence (top panel) modifications in vitamin D treated cells with acute (middle) and chronic (lower) treatment. The proliferation-differentiation relationships are schematically illustrated as arrows representing changes in expression of cell cycle and cell growth regulated genes (proliferation arrow) and genes associated with the maturation (differentiation arrow) of the osteoblast phenotype as the extracellular matrix develops and mineralizes in normal diploid cell cultures (top panel). Here, the three principal periods of the osteoblast developmental sequence are designated within broken vertical lines (proliferation, matrix development and maturation, and mineralization). These broken lines indicate the two experimentally established principal transition points in the developmental sequence exhibited by normal diploid osteoblasts during the progressive acquisition of the bone cell phenotype: the first at the completion of proliferation when genes associated with matrix development and maturation are upregulated, and the second at the onset of extracellular matrix mineralization. The middle panel schematically illustrates the acceleration of the relationship between growth and differentiation by vitamin D in short-term treatment to 48 h. Hormone accelerates proliferation and differentiation (broken arrows) by changing levels of gene expression in one period relative to that found in the following period. In the lower panel the inhibitory effects of vitamin D on differentiation of osteoblasts *in vitro* (broken arrows) are illustrated for continuous hormone treatment initiated on day 5 or day 20 reflected by inhibition (day 5) or cessation (from day 20) of the formation and growth of mineralized nodules. H4, histone; COL-I, type αI collagen; FN, fibronectin; ALK PHOS, alkaline phosphatase; MGP, matrix Gla protein; OP, osteopontin; OC, osteocalcin.

observed related to extent of mineralized matrix that was formed at the time of vitamin D exposure (Figure 5, lower panel). Continuous presence of hormone blocked further maturation of the cultures (e.g., growth and mineralization of nodules were inhibited). It has been observed that chronic exposure of rat osteosarcoma cells (Ros 17/2.8) to 1,25(OH)$_2$D$_3$ can result in altered phenotypic expression. After 5 d, osteocalcin mRNA was decreased and MGP expression, which is not normally detected in this Ros line, was induced.[61]

One factor contributing to the variability of response of the osteoblast to 1,25(OH)$_2$D$_3$ is potentially related to changes in the receptor levels of vitamin D. Receptor level is regulated by proliferative status of the cell, presence of glucocorticoids, and homologous upregulation of its receptor, as shown in the Ros 17/2.8 rat osteosarcoma osteoblast-like cell line.[29,95] However, it can be seen in Figure 4 that genes expressed both during proliferation and differentiation (e.g., OP and MGP) are regulated by the hormone throughout the developmental sequences. Thus, receptor availability is not rate limiting. Calcium is yet another contributing factor to the regulation of the vitamin D receptor in osteoblast-like cells. This was suggested by the report[96] that the upregulation of the vitamin D receptors in UMR 106 cells is inhibited by two different calcium channel blockers, nitrendipine and verapamil. Such observations raise the possibility that osteoblasts within a matrix that is undergoing extracellular mineralization, may show variable levels of vitamin D responsiveness if the receptor level is closely regulated by the extracellular calcium concentration.

In addition to the direct modulation of osteoblast gene activity that results from interaction of the vitamin D receptor complex with specific gene sequences, new evidence has accumulated to indicate several nongenomic effects. Changes in cellular calcium occur in response to 1,25(OH)$_2$D$_3$ in a matter of seconds and these have been classified as "rapid" or nongenomic effects. 1-α-25(OH)$_2$D$_3$ rapidly increases cytosolic calcium and alters membrane phospholipid metabolism in hepatocytes,[97] cultured osteoblasts,[3,4,98] chondrocytes[7] and isolated enterocytes.[4] The rapid transmembrane influx of extracellular calcium after stimulation by 1,25(OH)$_2$D$_3$ is mediated largely by the opening of voltage-gated calcium channels.[3,99] The rapid, "nongenomic" effects of vitamin D include prolonged opening of the voltage-sensitive calcium channels present in the plasma membrane,[99] phosphoinositide breakdown,[4] altered enzymatic activity in matrix vesicles,[7] and calcium transport in chick intestine.[100] A fundamental question raised by these observations is whether the responses are mediated by a different type of receptor for 1,25(OH)$_2$D$_3$[4,101] and whether there are interactions with the steroid-receptor signaling pathway.

Thus, many aspects of osteoblast metbolism are regulated by vitamin D. Several factors, however, cloud an understanding of the precise effects of vitamin D on the osteoblast in relation to bone formation and resorption. First, other than an understanding of collagen as a bone matrix structural

TABLE 1
Sequence Comparison of Known Vitamin D-Responsive Elements in the Promoters of Vitamin D-Regulated Genes

	AP1				
	−512				−483
Human osteocalcin	5′ GGTGACTCA	CC	GGGTGA	ACG	GGGGCA TT 3′

	AP1				
	−470				−443
Rat osteocalcin	5′ TGCCCTGCA	CT	GGGTGA	ATG	AGGACA TT 3′

	−761			−741
Mouse osteopontin	5′ AC	AA GGTTCA	CGA	GGTTCA CG 3′

	−198		−180
Mouse calbindin-D_{28k}	5′ GGGGGA	TGTG	AGGAGA 3′

	−489		−475
Rat calbindin-D_{9k}	5′ GGGTCT	CGG	AAGCCC 3′

Note: Homologous half steroid elements are underlined.

protein, the functions of other products that are regulated by vitamin D in osteoblasts are not clearly defined.[65] Second, the transcriptional effects of the hormone on osteoblast expressed genes are not straightforward. As indicated above, studies have reported either stimulation or inhibition of gene expression and protein synthesis dependent upon the level of vitamin D *in vivo* or concentration of hormone *in vitro*. Effects vary with the cell culture system being used, the density of the cells, and the duration of hormone treatment. From our most recent studies in normal osteoblasts[9] and numerous reports in transformed cells, one explanation for the apparent discrepancies is that the ability of $1,25(OH)_2D_3$ to modulate gene transcription depends upon the basal levels of expression of the gene, which in turn relates to the maturational state of the cell being studied. The question arises as to whether changes in these osteoblast parameters are primary or secondary to the hormone's effect on changes in the differentiated state of the cell or the proliferative capacity of the cell, since a change in proliferation alters the levels of nuclear proto-oncogenes and other factors that regulate gene transcription. An analysis of the vitamin D responsive element (VDRE) in the genes regulated by vitamin D should lead to clarification of these complexities. To date, VDREs have been identified in only five genes (Table 1): mouse osteopontin,[84] calbindin-D_{28k},[102] calbindin-D_{9k},[103] and human and rat osteocalcins.[85,87,94,104]

C. INFLUENCE OF $1,25(OH)_2D_3$ ON OSTEOCLAST DIFFERENTIATION AND BONE RESORPTION

When vitamin D-deficient animals are treated with $1\text{-}\alpha\text{-}25(OH)_2D_3$, osteoclastic bone resorption is enhanced and the number of osteoclasts increases.[21] Osteoclastic bone resorption is induced by two different mechanisms: first the recruitment of osteoclast mononuclear progenitors and their transition to multinucleated osteoclasts and second, the activation of quiescent osteoclasts.[10] Numerous studies have established that osteoclasts are formed by the differentiation and fusion of progenitors derived from hematopoietic stem cells.[105,106] Equally important, many studies have demonstrated that $1,25(OH)_2D_3$ is a potent mediator of differentiation of such progenitors; for example, the human leukemia HL60 cell line can be induced to differentiate to adherent macrophages by vitamin D.[107] As mentioned above, marrow mononuclear cells can be promoted to differentiate into cells having osteoclast characteristics such as ruffled borders and clear zones, tartrate resistant acid phosphatase activity (TRAP), calcitonin receptors, and the ability to resorb bone matrix.[34] Thus, while the fully differentiated multinucleated osteoclast itself does not have receptors for $1,25(OH)_2D_3$, the hormone can promote formation of this bone resorbing cell.

Vitamin D also contributes to activation of the mature osteoclast via its effects on osteoblasts. Indeed morphologic changes in osteoclasts are evident in response to $1,25(OH)_2D_3$ in cultured long bones.[108] Osteoclast-like multinucleated cells can be formed in cocultures of spleen cells together with osteoblastic cells derived from the calvaria in the presence of $1,25(OH)_2D_3$.[109] TRAP is a marker of active osteoclasts. In mouse marrow cultures, TRAP positive multinucleated cells were formed only near clusters of alkaline phosphatase-positive stromal cells, indicative of an osteoblast-like cell.[110] Thus, osteoblastic cells appear to be involved in osteoclast formation induced by $1,25(OH)_2D_3$. These and several other studies have provided evidence consistent with the model of Rodan and Martin[31] which proposed that the effects of $1,25(OH)_2D_3$ on bone resorption occurs by a direct effect on bone that could be elicited by an initial effect on the osteoblast to produce a local humoral mediator which would then activate the osteoclast. While many osteoblast products, such as osteopontin and osteocalcin, are secreted into the extracellular matrix in response to vitamin D, the specific protein or factor that interacts with the osteoclast to promote activation to a resorbing cell is not yet defined. Thus, vitamin D induces bone resorption through differentiation of osteoclast progenitors in a manner that involves marrow-derived cells and by activation of existing osteoclasts mediated by its effects on the osteoblast.

The effect of vitamin D on osteoblast characteristics is also consistent with the hormone's ability to stimulate bone resorption. Collagen synthesis in mature osteoblasts is inhibited by $1,25(OH)_2D_3$ *in vitro* and parallels the effects for bone resorption. This downregulation of the major structural component of bone tissue, but the upregulation of noncollagenous proteins, osteocalcin, and osteopontin, may relate to resorptive activities in bone. Yet anabolic effects of vitamin D on collagen synthesis are also known. These appear to be limited to the periosteal cells or cloned osteosarcoma lines (e.g.,

MG63 cells) that exhibit phenotypic properties of immature osteoblasts indicated by significant levels of type III collagen synthesis.[111] Here, the effect of hormone to stimulate proliferation and collagen production in osteoblast progenitor-like cells may serve as a coupling phenomenon to balance bone resorption. Thus, it is possible for $1,25(OH)_2D_3$ to meet the demands required to maintain serum calcium by replacing the resorbed bone with newly differentiated osteoblasts that have been stimulated to produce matrix.

IV. VITAMIN D REGULATION OF THE OSTEOCALCIN GENE

It is well known that vitamin D anabolically and catabolically modulates bone cell metabolic activities. More recently, it has become apparent that this occurs through selective expression of a series of vitamin D responsive genes. Therefore, several fundamental questions involve: (1) the identification and characterization of the regulatory elements in skeletal-associated gene promoters that respond to the hormone; (2) the mechanisms by which vitamin D-responsive sequences of specific skeletal genes are selectively rendered competent to bind the vitamin D receptor complex in a cell and tissue specific manner and/or at particular stages of differentiation; and (3) the combined influence of multiple physiological mediators on vitamin D responsiveness at the level of transcriptional control. The VDRE of the osteocalcin gene was the first vitamin D receptor binding sequence to be identified; and, as we gain further insight into the structural and functional properties of the osteocalcin gene VDRE, the complexity of regulatory events at this transcription control element is becoming increasingly apparent. Perhaps this is an indication of the basis for involvement of vitamin D as a mediator of osteocalcin gene expression within the context of a broad spectrum of physiological responses of the osteoblast.

Osteocalcin is regarded as a bone-specific protein. The 5.7-kDa polypeptide (46 to 50 amino acids depending on the species) is synthesized by osteoblasts as a 10,000-kDa precursor that undergoes posttranscriptional conversion of glutamic acids 17, 21, and 24 to the calcium binding γ-carboxyglutamic acid that is catalyzed by a vitamin K-dependent carboxylase. While the functions of osteocalcin remain to be definitively established, and the protein may indeed have multiple biological activities, several lines of evidence convincingly support involvement of osteocalcin both in skeletal development and in bone resorption. Here, it should be noted that while osteocalcin is primarily associated with the bone extracellular matrix, serum levels of osteocalcin provide an extremely important clinical indication of osteoblast activity resulting from bone turnover (resorption). This is reflected by changes in serum osteocalcin levels in several skeletal disorders.[57] But perhaps more importantly, serum osteocalcin levels are modified in response to and/or causally related to a broad spectrum of physiologic circumstances where homeostatic mechanisms controlling calcium metabolism are operative. As a consequence, the calcitropic hormones (vitamin D, parathyroid hormone) participate in regulation of cellular osteocalcin levels and the extent to which osteocalcin is secreted.[112] Osteocalcin synthesis and mRNA levels are altered in response to vitamin D,[85-87,94,104] estrogen,[113,114] glucocorticoids,[104,115] retinoic acid,[116,117] TGF β,[118] cyclic nucleotides,[119,120] and γ-interferon.[121] The picture that is emerging is one of a protein whose regulation must be mediated by multiple factors and signaling mechanisms to accommodate a diverse series of physiological requirements.

A. SEQUENCE ORGANIZATION OF THE OSTEOCALCIN GENE REFLECTS VITAMIN D-MEDIATED CONTROL

Understanding the mechanisms controlling vitamin D responsiveness of the osteocalcin gene necessitates examining hormone-mediated regulation during the progressive development of the osteoblast phenotype. In rat osteoblast cultures where the influence of vitamin D on gene expression has been systematically studied (Figure 4), vitamin D upregulates osteocalcin gene transcription only by enhancing basal levels of expression that are initiated in mature osteoblasts during extracellular matrix mineralization. Vitamin D stimulation of osteocalcin was not observed in proliferating osteoblasts. Thus, vitamin D does not function as an inducer of osteocalcin gene transcription, but as an enhancer.

Sequence analysis of the osteocalcin gene provides an indication of the complexity of transcriptional and posttranscriptional regulation of the gene (Figure 6). The sequence organization of the mRNA coding region of the gene reflects the biochemical events associated with processing of the initially synthesized 10,000-kDa precursor. The complexity of vitamin D-mediated control of osteocalcin transcription is illustrated by direct interactions of the vitamin D receptor with the VDRE in the osteocalcin gene promoter, which is accompanied by and functionally related to modifications in protein/DNA interactions at other regulatory sequences of the osteocalcin gene. Identification and characterization of proteins that interact in a sequence-specific manner with their cognate regulatory elements in the osteocalcin gene promoter further establishes the physiological mediators of osteocalcin gene transcription and the circumstances under which they participate in regulation. Additionally, the cloned gene permits direct quantitation of the rate of osteocalcin gene transcription and cellular levels of osteocalcin mRNA, permitting a direct assessment of the involvement of transcriptional and posttranscriptional control during development, tissue maintenance, and in skeletal disorders.

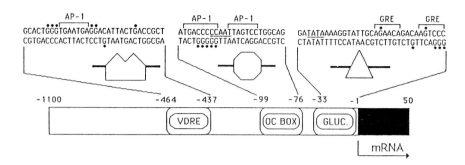

FIGURE 6.   Rat osteocalcin gene promoter indicating sequences for physiological regulatory elements. Structural organization of the rat osteocalcin gene. (A) Sequences of those regulatory elements in the proximal promoter that have been defined and partially characterized. These include the vitamin D responsive element (VDRE); the osteocalcin box (OC box), which is a primary proximal transcription regulatory element containing the CCAAT motif as a central core; and the glucocorticoid response elements (GLUC) in the rat gene[173] or overlapping the TATA motif as defined in the human gene.[121] Within the OC box and VDRE elements are also found active AP-1 sites which bind the oncogene-encoded Fos/Jun protein complex. The solid circles above or below G residues indicate vitamin D receptor protein-DNA interactions defined at single nucleotide resolution within the vitamin D responsive element.[94] (B) With reference to the OC box and TATA promoter elements, intron and exon organization and location of the pre-, pro-peptide, and mature osteocalcin forms are illustrated.

In contrast to other bone-related genes (e.g., type I collagen and alkaline phosphatase), only a single mRNA transcript has been observed from the osteocalcin gene. It appears that 1 transcription initiation site is utilized, and the 4 splicing events that fuse exons 1 through 4 into an mRNA encoding the osteocalcin prepropeptide are not known to vary. The presence of only a single osteocalcin mRNA additionally supports one functional poly A site and a fixed length to the 3' poly A sequence. Although regulation of osteocalcin expression does not appear to be modulated by changes in the organization of the mRNA transcripts, this does not preclude the presence of sequences in the transcribed region of the osteocalcin gene which contribute to the regulation of transcription.

Consensus sequences for responsiveness to several physiological media-tors of osteocalcin expression reside in the osteocalcin gene promoter. In addition to a VDRE, sequences bearing homology to other steroid receptor binding complexes, for example, estrogen, glucocorticoids, retinoic acid,[86,116,121] and regulatory factors such as γ-interferon[122] and cyclic AMP, are found in the 5' flanking sequences of the osteocalcin gene. As with many genes transcribed by RNA polymerase II, a TATA motif is located at -42 to -39, a CCAAT element at -92 to -88, and AP-1 and AP-2 sites are present at several sites in the osteocalcin gene promoter. However, while such consensus sequences provide an indication of potential regulatory mechanisms that may be operative, several lines of evidence are necessary to establish function, among which are (1) demonstration of an influence on transcriptional activity by deletion, substitution, or site-specific mutagenesis; (2) identification and characterization of sequence-specific regulatory element occupancy by cognate transcription factors; and (3) modifications in protein/ DNA interactions as a function of biological activity.

In the case of the osteocalcin gene promoter, to date only four regulatory sequences, which are schematically illustrated in Figure 6, have been established by the above criteria. Interestingly, they all contribute to vitamin D-mediated transcriptional control of osteocalcin gene expression. The "osteocalcin box", a 24-nucleotide element with a CCAAT motif as the central core, is a highly conserved regulatory sequence that is required for basal expression. The TATA sequence region is also required for rendering the osteocalcin gene transcribable. For both of these elements, a direct involvement in transcriptional control has been provided by experimental approaches *in vivo*, and sequence-specific protein/DNA interactions have been demonstrated which exhibit variations with modifications in biological activity, notably during development of the osteoblast phenotype and/or following vitamin D treatment. The VDRE (-466 to -437) and multiple glucocorticoid regulatory elements (GRE), one associated with the TATA domain, modulate steroid hormone effects on osteocalcin gene transcriptional activity; both elements exhibit protein/DNA interactions that influence vitamin D-mediated control of transcription from the osteocalcin gene. Not to be overlooked is evidence for potential interaction of these regulatory elements. Dexamethasone can, for example, inhibit or enhance the vitamin D stimulation of OC gene transcription.[115] This may occur through protein-protein interactions of transcription factors at each element or indirectly by either hormone inducing synthesis of a transcription factor that resides at the DNA binding domain of the other hormone. Preliminary evidence for both mechanisms is indicated. The first can occur if the three-dimensional structure of the promoter changes, bringing different regulatory elements into close proximity for protein-protein interactions. Recent studies[123] have shown the presence of nucleosomes between the VDRE and proximal promoter elements (TATA region and OC Box). Evidence for the second mechanism is indicated by the enhancement of protein-DNA interactions in the TATA region when nuclear extracts from vitamin D treated bone cells are examined.[124]

B. REGULATION OF VITAMIN D RESPONSIVE ELEMENT (VDRE) ACTIVITY

The VDRE of the rat[87,94,125] and human[85,104] osteocalcin genes have been identified and characterized by two approaches: (1) the determination of the ability of systematically introduced nucleotide deletions or substitutions to influence vitamin D upregulation of osteocalcin gene transcription[85,87] and (2) the definition at single nucleotide resolution of the binding of vitamin D receptor complexes.[94] As shown in Table 1, the osteocalcin gene VDRE of several mammals contains steroid half-element motifs with a 3-nucleotide spacer. Of interest, the osteopontin VDRE contains a minimal 9-bp motif, AGGTTCACG, that can confer vitamin D responsiveness[84] compared to the 24-bp osteocalcin VDRE within the osteocalcin gene promoter. Additionally, consensus sequences for another vitamin-derived steroid hormone, retinoic acid,[126] and for the nuclear proto-oncogene-encoded Fos and Jun proteins, have been identified.[82] Results suggest that retinoic acid[116,117] and the Fos-Jun complex[127] modulate osteocalcin gene transcription, thus providing a functional basis for synergistic and/or antagonistic control of multiple regulatory activities within the VDRE. Such activities within the VDRE reflect the contribution of multiple promoter binding factors competing for or complexing with and thereby promoting and/or inhibiting interactions at a single regulatory element. These relationships may provide an explanation for positive or negative activity of a single regulatory sequence under different biological conditions.

Competency for occupancy of the VDRE by the vitamin D - vitamin D receptor complex is requisite for the enhancement of osteocalcin gene transcription by the hormone. Equally important is the mechanism by which the VDRE is rendered refractory to binding of the regulatory complex in immature osteoblasts not expressing osteocalcin and in nonosteoblastic cells. The answers to these long-standing biological questions may be provided by the complement of proteins that complex with the VDRE. Protein-DNA interactions and protein-protein interactions must both be considered. Markose et al.[94] mapped the protein DNA interactions at the VDRE at specific guanine residues using nuclear extracts from osteosarcoma cells that express osteocalcin under the influence of vitamin D. These results provided the first indication of the complexity of transcription factor binding at the VDRE. Subsequent studies[128,129] demonstrated the requirement of an accessory factor for binding of the vitamin D receptor complex to the osteocalcin gene VDRE. Additionally, the finding that vitamin D receptor phosphorylation is essential for hormone-receptor complex formation[130,131] and upregulation of transcription[103] reflects the complexity of events associated with activity of the VDRE. Modifications that mediate or reflect activity at the osteocalcin gene VDRE under various biological circumstances are suggested by striking differences in VDRE protein-DNA interactions in normal diploid osteoblasts compared with osteosarcoma cells and during progression of osteoblast differentiation (Figure 7). Such differences in the VDR complex may relate to mechanisms by which osteocalcin synthesis and regulation by vitamin D can occur in proliferating transformed cells but is prevented in proliferating normal diploid osteoblasts.

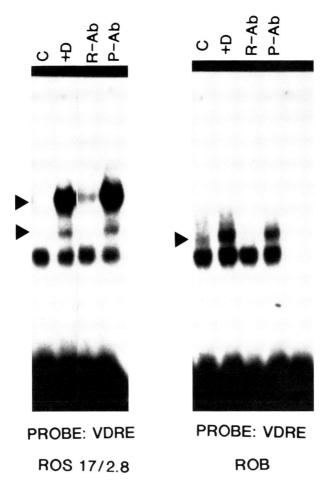

PROBE: VDRE

ROS 17/2.8

PROBE: VDRE

ROB

FIGURE 7. Characterization of vitamin D-mediated enhancement of protein-DNA interactions at the rat OC gene VDRE in normal osteoblasts and Ros 17/2.8 cells. Gel mobility shift assays using nuclear proteins from untreated (C) or 10^{-8} M vitamin D-treated (D) confluent (day 8) Ros 17/2.8 cells or mature (day 32) normal diploid osteoblasts show enhancement of protein-DNA interactions with a ^{32}P-labeled rat osteocalcin VDRE oligonucleotide DNA probe following treatment of the cells with vitamin D receptor (VDR) binding is indicated by arrows. Monoclonal antibodies against the porcine VDR which detect only the porcine VDR (P-Ab) or which cross-react with the rat VDR (R-Ab) were preincubated with the nuclear proteins for 20 min prior to DNA probe addition.[124] These antibodies demonstrate that the vitamin D enhanced protein-DNA interactions observed in both cell types are dependent on the binding of the vitamin D receptor. Note that two VDR-DNA complexes are formed in the osteosarcoma cells and only the fast-mobility complex is detectable in differentiated ROB cells expressing osteocalcin.

C. PHENOTYPE SUPPRESSION: A POSTULATED MECHANISM FOR THE COORDINATE REGULATION OF VITAMIN D-MEDIATED ENHANCEMENT AND BASAL TRANSCRIPTION OF THE OSTEOCALCIN GENE BY FOS-JUN INTERACTIONS AT AP-1 SITES WITHIN THE VDRE AND OTHER PROMOTER ELEMENTS

A mechanism by which vitamin D modulates the extent to which the osteocalcin is transcribed postproliferatively in osteoblasts may involve regulation of Fos-Jun binding at the VDRE and at the OC box. We identified a series of consensus sequences for transcription factor AP-1[86,94,127] and demonstrated sequence-specific interactions of these promoter binding sequences with a stable heterodimeric Fos-Jun complex.[127] The possibility therefore arises that Fos and Jun proteins expressed in proliferating osteoblasts could suppress osteocalcin gene transcription until late in the development of the bone cell phenotype.

Several experimental results support the concept of coordinate occupancy of the Fos-Jun protein complex at the VDRE and OC Box regulatory elements, providing a potential molecular mechanism to account for the absence of osteocalcin gene expression in proliferating osteoblasts. Expression of c-fos and c-jun occurs primarily during the proliferative period of the osteoblast developmental sequence.[51,54] Second, AP-1 binding activity is observed primarily in proliferating osteoblasts and dramatically decreases after the downregulation of proliferation and the initiation of extracellular matrix maturation and mineralization at which time osteocalcin gene transcription is initiated.[127] Clearly, distinct patterns of protein-DNA interactions occur at the VDRE as well as the OC Box, with nuclear protein extracts from proliferating (day 5) osteoblasts which do not express osteocalcin compared to differentiated cells where AP-1 activity has significantly declined. In tumor cells (Ros 17/2.8), osteocalcin expression is deregulated, in that transcription continues in proliferating cells. Even so, a reciprocal relationship between AP-1 activity and OC gene expression is maintained. In confluent osteoblast cultures AP-1 activity is markedly diminished and osteocalcin mRNA and transcription increase tenfold. The loss of AP-1 activity during the developmental sequence of osteoblast maturation coincides with marked changes in protein-DNA interactions in osteocalcin gene promoter elements having AP-1 sites. For example, we observed that protein-DNA contacts do not occur at G residues within the AP-1 consensus sequences of the VDRE and OC Box when the gene is actively transcribed.[94] Additionally, transfection of c-fos and c-jun into cells expressing osteocalcin resulted in the downregulation of osteocalcin gene transcription and further supports a Fos-Jun mediated suppression of the osteocalcin gene.[116] These results are consistent with a model in which coordinate occupancy of the AP-1 sites in the VDRE and osteocalcin box in proliferating osteoblasts may suppress both basal and vitamin D-enhanced osteocalcin gene transcription. This phenomenon, in which osteocalcin gene expression is prevented during proliferation, is called *phenotype suppression*.[127,132]

AP-1 SITES WITHIN THE HUMAN VDRE

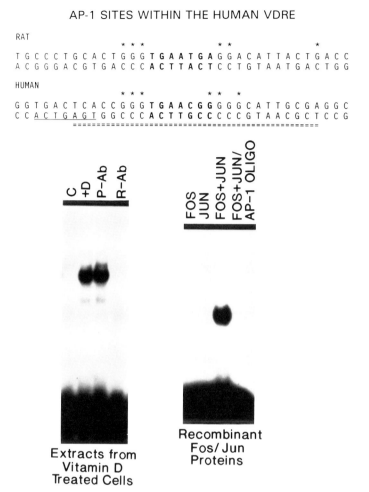

FIGURE 8. Vitamin D receptor and Fos-Jun protein complex binding to the human VDRE sequences. Nuclear extracts from 48-h vitamin D-treated Ros cells were used in gel mobility shift assays (left panel) with an oligonucleotide probe to the human VDRE as indicated by the dotted underline in the top panel sequence. The vitamin D receptor complex is blocked by preincubation with antibody recognizing the rat receptor (R-AB) but not the pig-specific antibody (9P-AB), provided by H. DeLuca, Madison, WI. The probe used for these protein-DNA interactions has the classic AP-1 consensus deleted (solid underline) but retains the AP-1 site homologous to the rat VDRE (bold letters) that resides within the core of this regulatory element. The right panel demonstrates that this is an active AP-1 site by binding of only the Fos-Jun protein complex which can be competed by the AP-1 oligonucleotide.

Mutual exclusion of occupancy by fos-jun and element-specific factors at the VDRE and OC Box is further supported by interactions of fos-jun binding using nuclear extracts from osteoblasts expressing the OC gene only when mutations are introduced that preclude VDR and CAAT motif factor interactions.[174] Using both recombinant fos and jun proteins,[127] and nuclear extracts, we have demonstrated fos-jun complex binding (Figure 8) to the internal AP-1 site (bold letters) in the human osteocalcin gene VDRE.[132] The human VDRE, in contrast to the rat VDRE, has an additional upstream AP-1 site (solid underline in Figure 8). While the internal AP-1 site may function in suppression of transcription, the upstream AP-1 site in the human VDRE has been shown to enhance vitamin D-stimulated gene transcription, but is not necessary for vitamin D regulation.[128] This suggests that responsiveness of the human and rat osteocalcin genes to factors that influence AP-1 activity may differ. Yet, the similar organization of the internal AP-1 sites of the VDRE and the identical AP-1 organization in the osteocalcin box for the human and rat osteocalcin genes suggests that there are some similar functional properties of the elements, perhaps in regulating expression in relation to the proliferative state of the cells.

A remaining question is the mechanism by which the osteocalcin gene is rendered transcribable and vitamin D responsive following the downregulation of proliferation. Here, the possibilities include:

1. Release of the Fos-Jun complex from the AP-1 sites to permit sequences to be available for occupancy by the vitamin D receptor complex and/ or by tissue-specific osteocalcin box transcription factors;
2. Modifications of the Fos-Jun complex that facilitate binding of activation related factors;
3. Association with other DNA binding proteins or protein-protein interactions.

With respect to the latter possibility, binding of the Fos-Jun complex may play a dual positive and negative role in the regulation of transcription. The Fos-Jun complex may suppress osteocalcin gene transcription when proliferation is ongoing by directly or indirectly modulating sequence-specific interactions at the vitamin D receptor binding domain. Then postproliferatively, the Fos-Jun complex may facilitate vitamin D receptor binding to support the sequential upregulation of osteocalcin and other vitamin D responsive genes. Such regulation could be accomplished by different fos-jun protein complexes. Both fos and jun are members of a family of related proteins, and a recent study indictes differential stimulation of c-jun and jun-B by $1,25(OH)_2D_3$ in MC3T3-E1 osteoblastic cells.[133]

Further experimental results are necessary to determine whether the organization of steroid receptor binding domains and AP-1 sites within promoters of genes that are hormone responsive can provide a general mechanism for the phenotype suppression and/or activation in different periods of cell and tissue differentiation. However, support for such a model is provided by the association of AP-1 sites within consensus sequences for other steroid responsive elements, for example, glucocorticoids.[116,127,134] Undoubtedly the phenotype suppression model represents a simplification of an extremely complex series of protein-DNA interactions whereby multiple physiological signals are transduced to the nucleus resulting in modifications in transcription that progressively alter phenotypic properties of cells leading to structural and functional events associated with differentiation. However, the strength of such a model is that it provides a basis for experimentally addressing the functional significance of organization of regulatory elements for phenoptypic marker genes of cell differentiation within the context of gene expression associated with proliferation.

It is necessary to account not only for precise control of basal transcription from the osteocalcin gene, but also to explain the sensitivity of vitamin D modulation to basal levels of expression. In focusing on this component of transcriptional regulation, we obtained results consistent with the possibility that the TATA box sequences (the primary transcriptional regulatory element) and the VDRE may interact to control the level of expression, possibly by cooperative interactions. We have shown sequence-specific, vitamin D-dependent, protein-DNA interactions at both the VDRE and TATA box using nuclear extracts from both Ros 17/2.8 cells[94] and normal diploid osteoblast cultures.[124,135] Thus, vitamin D-mediated regulation at the VDRE is integrated with activities at other basal regulatory elements via vitamin D-dependent protein-DNA or protein-protein complexes.

V. POTENTIAL THERAPEUTIC APPLICATIONS OF VITAMIN D

An increased understanding of molecular mechanisms by which vitamin D-mediated genes are responsive to the hormone can provide the basis for a broad spectrum of therapeutic uses. Here, knowledge of how VDREs of specific genes and tissues are selectively rendered operative will allow for development of treatment regimens that incorporate targeted responses. Specifically, we should take advantage of the complexity of regulatory events that include a series of protein-DNA interactions at the VDRE which render the hormone responsive element competent for binding the vitamin D receptor and, equally important, the observed and potential variations in responsiveness elicted by metabolites and analogs of the hormone.

Clinical disorders related to vitamin D can arise from multiple factors, either (1) altered availability of the parent molecule vitamin D (e.g., by dietary deficiency); (2) altered conversion of vitamin D to its principal active metabolite $1,25(OH)_2D_3$, produced in the kidney, (3) conditions that may be due to variations in organ responsiveness to the active metabolites; or (4) pertur-

TABLE 2
Examples of Human Diseases Related to
Disturbances in the Vitamin D Endocrine System

Anticonvulsant therapy
Cirrhosis
Diabetes
Drug-induced metabolism
Glucocorticoid antagonism
Hyperparathyroidism
Hypoparathyroidism
Leukemia
Malabsorption
Osteomalacia
Osteoporosis
Psoriasis
Renal disease
Sarcoidosis
Steatorrhea

bations in the integrated actions of these metabolites on selected responsive genes. Table 2 indicates some of the major disorders associated with vitamin D abnormalities or disturbances, including those secondary to alterations of PTH and/or calcitonin as components of the vitamin D endocrine system, or secondary to drug induced inactivation of vitamin D. Even today some groups are at risk for vitamin D deficiency, including breast-fed infants who do not receive vitamin D supplementation, and people in an environment with reduced exposure to sun and inadequate intake of fortified foods, e.g., Asian women (particularly primigravidas) who wear veils and who may also consume nonsupplemented foods. Disorders of the intestine, such as malabsorption syndrome, steatorrhea or tropical sprue, or liver disorders (e.g., cirrhosis or obstructive jaundice), lead to vitamin D deficiency and thereby altered bone formation.

Several hereditary disorders of the vitamin D endocrine system are known.[136] One group, vitamin D-dependent Rickets Type II, involves mutations that cause the vitamin D receptor gene to code for an abnormal receptor and lead to defects that disturb all tissues responsive to vitamin D. In these patients there is a marked heterogeneity of clinical manifestations.[137] This is likely a consequence of at least five phenotypically different defects which are related to variations in hormone-receptor binding capacity, affinity or nuclear localization and defects in binding of the hormone-receptor complex to DNA. In another inborn error of vitamin D metabolism, Vitamin D-dependent Rickets Type I, the autosomal recessive defect is presumed to involve the renal $25(OH)D_1$ hydroxylase enzyme. These patients can be treated with the active metabolite $1,25(OH)_2D_3$ for normalization of disturbed parameters of the vitamin D endocrine system.[138] In an X-linked dominant disorder, Hypophosphatemic Vitamin D Resistant Rickets of unknown etiology, high doses of phosphate and $1,25(OH)_2D_3$ have allowed for fairly normal growth of these children.[139] In addition to these inborn errors of vitamin D metabolism are those disorders involving a tissue-selective resistance to $1,25(OH)_2D_3$— but here there is, as yet, no understanding of the cellular basis which relates to a direct effect of vitamin D actions either at the genomic or nongenomic levels. For example, elevated $1,25(OH)_2D_3$ levels occur in human and rat osteopetrosis, a hereditary disorder having a spectrum of osteoclast abnormalities.[140] The hormone level may be elevated to maintain normal serum calcium in the absence of active bone resorption. Alterations in some vitamin D-regulated proteins (osteocalcin, osteopontin, collagen, alkaline phosphatase) indicate perturbations in bone responsiveness.[141]

Vitamin D $(1,25(OH)_2D_3)$ is being considered for the treatment of osteoporosis,[142] a growing problem.[143] Type I postmenopausal osteoporosis occurs predominantly in women within 20 years after the menopause and is related to estrogen deficiency. It is characterized by accelerated bone loss which leads to suppressed PTH secretion and a trend toward low serum $1,25(OH)_2D$ concentrations and inefficient calcium absorption.[144-146] In clinical trials in which calcium intake was limited and dosages of $1,25(OH)_2D_3$ of 0.5 μg/d were employed, hypercalcemia and hypercalcuria occurred uncommonly.[147] Short-term rather than continuous treatment preferentially stimulates osteoblast function but not osteoclast activity.[148] Type II is the age related osteoporosis resulting in defects in the vitamin D endocrine system that includes primary failure by the kidney and defective $25(OH)D1\alpha$-hydroxylase activity,[149] a decrease in vitamin D receptor concentration in intestinal mucosa and possibly in bone cells, and, in some elderly patients, vitamin D deficiency. With age there is decreased intestinal absorption of calcium, and this is a major determinant in the etiology of the disease.[144]

Although some investigators have documented reduced levels of $1,25(OH)_2D_3$ in osteoporotic subjects when compared to age matched controls, several studies document vitamin D levels in the normal range.[150] Thus, the value of calcitriol treatment for established postmenopausal osteoporosis remains controversial. Findings of several major clinical trials that have used calcitriol in the treatment of osteoporosis are conflicting. There are studies, e.g., Aloia et al.[151] and Caniggia et al.,[152] that indicate calcitriol treatment reduces bone loss in women with postmenopausal osteoporosis by increasing calcium absorption and reducing bone resorption. Gallagher et al.[153] has recently demonstrated that the treatment of postmenopausal osteoporotic women with calcitriol for 2 years was associated with an increase in spine density and total body calcium when compared to placebo. In the recent Tilyard study[176] (n = 600), a 50% reduction of spinal deformation compared to calcium-treated controls was observed. However, the Falch et al.[154] and the Ott and Chestnut[155] studies indicate that calcitriol is not effective in the treatment of established postmenopausal osteoporosis. The different observations may relate to the patient population and/or perhaps the stage of the disease.

More promising, however, are applications of the antiproliferative actions of $1,25(OH)_2D_3$, particularly effective in cells of the immune system[156] and keratinocytes, for the treatment of autoimmune disorders,[157] psoriasis,[158-160] cancer, and graft vs. host reactions.[161] However, the calcemic activity of the hormone, that is, its effects on bone, kidney, and intestine in elevating serum calcium, can be a serious problem. It is for this reason that analogs of $1,25(OH)_2D_3$ are being developed that have specific activities in particular target tissues and cells, e.g., leukemic cells, with little effect on bone calcium mobilization.[162,163] One compound, calcipotriol (MC903, Leo Pharmaceutical Products Company, DK-2750, Ballerup, Denmark) has low calcemic activity[164] and trial studies have demonstrated that it may be useful for treating psoriasis.[158] The reason for its low calcium mobilizing effect *in vivo* is not clear. Studies indicate that MC903 has good binding activity to the vitamin D receptor but low affinity for the vitamin D binding protein in the serum.[165] When tested in human osteoblast culture systems, one study found that calcipotriol and $1,25(OH)_2D_3$ resulted in similar stimulation of alkaline phosphatase and osteocalcin.[166] However, studies of Marie et al.[167] indicate that MC903 was less effective than $1,25(OH)_2D_3$ dependent upon the basal level of expression of osteocalcin. Another analog, 22-oxa-calcitriol (OCT), also promotes differentiation without the calcemia. However, the anticalcemic effect is not related to its inability to stimulate osteoblast activity. In two studies, in human MG63 cells[168] and in rat osteosarcoma Ros 17/2.8 cells,[169] the response of hormone and analog to osteocalcin, alkaline phosphatase, and induction of vitamin D receptors was similar. However, OCT was reported *not* to mediate the rapid nongenomic effects such as increasing cytosolic calcium.[3]

These observations lead one to several speculations regarding the mechanism of the differential actions of the analogs on specific gene products or cellular responses. Table 3 illustrates the variable degrees of mRNA stimulation of two vitamin D-regulated genes by several $1,25(OH)_2D_3$ analogs. The properties of the hormone-receptor complex's interaction with the vitamin D responsive gene segment may be altered. This hypothesis is supported by observations of the fluorinated analogs that are more potent than $1,25(OH)_2D_3$ *in vivo*[170] and *in vitro* in osteoblasts, with respect to alkaline phosphatase and osteocalcin stimulation and inhibition of DNA synthesis.[166] Studies show, however, that the increased potency is not due to receptor binding, as the fluorinated derivatives have less or similar affinity constants as $1,25(OH)_2D_3$.[171,172] Thus, it may be the specific interaction of the hormone-receptor complex which promotes interactions with other transactivating factors and the DNA sequences of the genes that accounts for gene activation or responsiveness. For example, PTH is downregulated by vitamin D, and the conformation of the receptor-hormone analog complex may promote an interaction with a particular transactivating factor that is specific for one gene and not another. Based upon the knowledge that the $1,25(OH)_2D_3$ receptor

TABLE 3
Stimulation of Osteocalcin (OC) and
Matrix Gla Protein (MGP) mRNA by
Vitamin D Analogs in Osteoblasts

		Ratio of Vitamin D/control[a]	
		OC	MGP
RO-5535	$1,25(OH)_2D_3$	5.3	5.2
RO-6218	$1,25(OH)_2D_2$	4.6	2.4
RO-6709	$1,25(OH)_2$ 24-epi$_\Delta$-$D_{22}3$	7	5
RO-8525	$1,25R,26(OH)_3D_3$	6	1.5
RO-6890	$1,25(OH)_2$-$26F_3D_3$	3.4	9.8

[a] Rat-derived osteoblasts were cultured 16 d for the development of mineralized bone nodules. Cells were harvested 24 h after exposure to 10^{-8} M metabolite for preparation of total cellular RNA and quantitation of mRNA. Note differences in the extent of stimulation among metabolities and between the two genes.[175]

complex can either up- or downregulate different genes in association with other gene regulatory elements and/or transcription factors, a similar mechanism may explain differential effects of the analogs.

VI. SUMMARY

The nutritional aspects of vitamin D on bone formation and turnover are appreciated within the context of the active metabolite of vitamin D, $1,25(OH)_2D_3$, that functions as a steroid hormone in modulating cellular differentiation and the production of specific gene products of bone cells. The effects of vitamin D at the genomic level and nongenomic levels are discussed in relation to the central role of vitamin D in maintaining normal bone metabolism. The concept of vitamin D as a physiologic factor in both promoting bone resorption and bone formation is presented in relation to positive and negative regulatory effects on the transcription of osteoblastic genes. Using the bone-specific protein, osteocalcin, as a model of a gene regulated by the hormone, the vitamin D responsive element is analyzed. Several concepts of transcriptional regulation are presented to account for (1) vitamin D modulating gene expression as an enhancer of basal, transcriptional activity, and (2) the synergistic and/or antagonistic interactions of vitamin D with other regulatory factors. In light of the very recent discoveries on gene regulation by vitamin D and consequent effects on bone formation and resorption, we have addressed current understanding of abnormalities of the vitamin D endocrine system and therapeutic applications of synthetic analogs of $1,25(OH)_2D_3$ that produce highly selective effects.

REFERENCES

1. **Naveh-Many, T., Marx, R., Keshet, E., Pike, J. W., and Silver, J.,** Regulation of 1,25-dihydroxyvitamin D3 receptor gene expression by 1,25-dihydroxyvitamin D3 in the parathyroid in vivo, *J. Clin. Invest.,* 86, 1968, 1990.
2. **Ozono, K., Sone, T., and Pike, J. W.,** The genomic mechanism of action of 1,25-dihydroxyvitamin D$_3$, *J. Bone Min. Res.,* 6, 1021, 1991.
3. **Civitelli, R., Kim, Y. S., Gunsten, S. L., Fujimori, A., Huskey, M., Avioli, L. V., and Hruska, K. A.,** Nongenomic activation of the calcium message system by vitamin D metabolites in osteoblast-like cells, *Endocrinology,* 127, 2253, 1990.
4. **Lieberherr, M., Grosse, B., Duchambon, P., and Drueke, T.,** A functional cell surface type receptor is required for the early action of 1,25-dihydroxyvitamin D$_3$ on the phosphoinositide metabolism in rat enterocytes, *J. Biol. Chem.,* 264, 20403, 1989.
5. **Baran, D. T., Sorensen, A. M., Owen, T., Shalhoub, V., Stein, G., and Lian, J.,** 1α,25-Dihydroxyvitamin D$_3$ rapidly increases cytosolic calcium in clonal rat osteosarcoma cells lacking the vitamin D receptor, *J. Bone Min. Res.,* 6, 1269, 1991.
6. **Harmond, M. F., Thomasset, M., Rouais, F., and Ducassou, D.,** In vitro stimulation of articular chondrocyte differentiated function by 1,25-dihydroxycholecalciferal or 24,R,25-dihydroxycholecalciferol, *J. Cell. Physiol.,* 119, 359, 1984.
7. **Boyan, B. D., Schwartz, I., Carnes, D. L., and Ramirez, V.,** The effects of vitamin D metabolites on the plasma and matrix vesicle membranes of growth and resting cartilage cells *in vitro, Endocrinology,* 122, 2851, 1988.
8. **Gerstenfeld, L. C., Kelly, C. M., von Deck, M., and Lian, J. B.,** Effect of 1,25-dihydroxyvitamin D$_3$ on induction of chondrocyte maturation in culture: extracellular matrix gene expression and morphology, *Endocrinology,* 126, 1599, 1990.
9. **Owen, T. A., Aronow, M. A., Barone, L. M., Bettencourt, B., Stein, G., and Lian, J. B.,** Pleiotropic effects of vitamin D on osteoblast gene expression are related to the proliferative and differentiated state of the bone cell phenotype: dependency upon basal levels of gene expression, duration of exposure and bone matrix competency in normal rat osteoblast cultures, *Endocrinology,* 128, 1496, 1991.
10. **Suda, T., Shinki, T., and Takahashi, N.,** The role of vitamin D in bone and intestinal cell differentiation, *Annu. Rev. Nutr.,* 10, 195, 1990.
11. **DeLuca, H. F.,** The vitamin D story: a collaborative effort of basic science and clinical medicine, *FASEB J.,* 2, 224, 1988.
12. **Minghetti, P. P. and Norman, A. W.,** 1,25(OH)$_2$-vitamin D$_3$ receptors: gene regulation and genetic circuitry, *FASEB J.,* 2, 3043, 1988.
13. **Stern, P. H.,** Vitamin D and bone, *Kidney Int.,* 29, S17, 1990.
14. **Marx, S. J. and Barsony, J.,** Tissue-selective 1,25-dihydroxyvitamin D$_3$ resistance: novel applications of calciferols, *J. Bone Min. Res.,* 3, 481, 1988.
15. **Toverud, S. U., Hammarstrom, L. E., and Kristoffersen, U. M.,** Quantitative studies on acid phosphatase in developing rat bones and teeth during hypervitaminosis D, *Arch. Oral Biol.,* 20, 175, 1975.
16. **Lian, J. B., Carnes, D. L., and Glimcher, M. J.,** Bone and serum concentrations of osteocalcin as a function of 1,25 (OH)$_2$D$_3$ circulating levels in rats with bone disorders, *Endocrinology,* 120, 2123, 1987.
17. **Marie, P. J., Hott, M., and Garba, M.-T.,** Contrasting effects of 1,25-dihydroxyvitamin D$_3$ on bone matrix and mineral appositional rates in the mouse, *Metabolism,* 34, 777, 1985.
18. **Weisbrode, S. E., Capen, C. C., and Norman, A. W.,** Ultrastructural evaluation of the effect of 1,25-dihydroxyvitamin D$_3$ on bone of thyroparathyroidectomized rats fed a low-calcium diet, *Am. J. Pathol.,* 92, 459, 1978.
19. **Hock, J. M., Gunnes-Hey, M., Poser, J., Olson, H., Bell, N. H., and Raisz, L. G.,** Stimulation of undermineralized matrix formation by 1,25 dihydroxyvitamin D$_3$ in long bones of rats, *Calcif. Tissue Int.,* 28, 79, 1986.

20. **Gallagher, J. A., Beneton, M., Harvey, L., and Lawson, D. E. M.,** Response of rachitic rat bones to 1,25-dihydroxyvitamin D_3: biphasic effects on mineralization and lack of effect on bone resorption, *Endocrinology,* 119, 1603, 1986.

21. **Marie, P. J. and Travers, R.,** Continuous infusion of 1,25-dihydroxyvitamin D_3 stimulates bone turnover in the normal young mouse, *Calcif. Tissue Int.,* 35, 418, 1983.

22. **Underwood, J. L. and DeLuca, H. F.,** Vitamin D is not directly necessary for bone growth and mineralization, *Am. J. Physiol.,* 246, E493, 1984.

23. **Holtrop, M. E., Cox, K. A., Carnes, D. L., and Holick, M. F.,** Effects of serum calcium and phosphorus on skeletal mineralization in vitamin D-deficient rats, *Am. J. Physiol.,* 251, E234, 1986.

24. **Clark, S. A., Boass, A., and Toverud, S. U.,** Effects of high dietary contents of calcium and phosphorus on mineral metabolism and growth of vitamin D-deficient suckling and weaned rats, *Bone Min.,* 2, 257, 1987.

25. **Kollenkirchen, U., Walters, M. R., and Fox, J.,** Plasma Ca influences vitamin D metabolite levels as rats develop vitamin D deficiency, *Am. J. Physiol.,* 260, E447, 1991.

26. **Wronski, T. J., Halloran, B. P., Bikle, D. D., Globus, R. K., and Morey-Holton, E. R.,** Chronic administration of 1,25-dihydroxyvitamin D_3: increased bone but imparied mineralization, *Endocrinology,* 119, 2580, 1986.

27. **Raisz, L. G., Trummel, C. L., Holick, M. F., and DeLuca, H. F.,** 1,25-dihydroxycholecalciferol, a potent stimulator of bone resorption in tissue culture, *Science,* 175, 768, 1972.

28. **Tanaka, Y. and DeLuca, H. F.,** Bone mineral mobilization activity of 1,25-dihydroxycholecalciferol, a metabolite of vitamin D., *Arch. Biochem. Biophys.,* 146, 574, 1971.

29. **Chen, T. L., Cohn, C. M., Morey-Holton, E., and Feldman, D.,** $1\alpha,25$-Dihydroxyvitamin D_3 receptors in cultured rat osteoblast-like cells, *J. Biol. Chem.,* 258, 4350, 1983.

30. **Chen, T. L., Li, J. M., van Ye, T., Cone, C. M., and Feldman, D.,** Hormonal responses to 1,25-dihydroxyvitamin D_3 in cultured mouse osteoblast-like cells-modulation by changes in receptor level, *J. Cell. Physiol.,* 126, 21, 1986.

31. **Rodan, G. A. and Martin, T. J.,** Role of osteoblasts in hormonal control of bone resorption—a hypothesis, *Calcif. Tissue Int.,* 33, 349, 1981.

32. **McSheehy, P. M. J. and Chambers, T. J.,** 1,25-Dihydroxyvitamin D_3 stimulates rat osteoblastic cells to release a soluble factor that increases osteoclastic bone resorption, *J. Clin. Invest.,* 80, 425, 1987.

33. **Abe, E., Miyaura, C., Tanaka, H., Shina, Y., Kuribayashi, T., Suda, S., Nishii, Y., DeLuca, H. F., and Suda, T.,** $1\alpha,25$-Dihydroxyvitamin D_3 promotes fusion of mouse alveolar macrophages both by a direct mechanism and by a spleen cell-mediated indirect mechanism, *Proc. Natl. Acad. Sci. U.S.A.,* 80, 5583, 1983.

34. **Roodman, G. D., Ibbotson, K. J., MacDonald, B. R., Kuehl, T. J., and Mundy, G. R.,** 1,25-Dihydroxyvitamin D_3 causes formation of multinucleated cells with several osteoclast characteristics in cultures of primate marrow, *Proc. Natl. Acad. Sci. U.S.A.,* 82, 8213, 1985.

35. **Narbaitz, R., Stumpf, W. E., Sar, M., Huang, S., and DeLuca, H. F.,** Autoradiographic localization of target cells for $1\alpha,25$-dihydroxyvitamin D_3 in bones from fetal rats, *Calcif. Tissue Int.,* 35, 177, 1983.

36. **Owen, M. and Friedenstein, A. J.,** Stromal stem cells: marrow-derived osteogenic precursors, in *Cell and Molecular Biology of Vertebrate Hard Tissues,* Evered, D. and Harnett, S., Eds., John Wiley & Sons, Chichester, 1988, 42.

37. **Benayhahu, D., Kletter, Y., Zipori, D., and Weintraub, S.,** Bone marrow-derived stromal cell line expressing osteoblastic phenotype in vitro and osteogenic capacity in vivo, *J. Cell. Physiol.,* 140, 1, 1989.

38. **Bernier, S. M., Desjardina, J., Sullivan, A. K., and Goltzman, D.,** Establishment of an osseous cell line from fetal rat calvaria using an immunocytolytic method of cell selection: characterization of the cell line and of derived clones, *J. Cell. Physiol.,* 145, 274, 1990.

39. **Grigoriadis, A. E., Heersche, J. N. M., and Aubin, J. E.,** Differentiation of muscle, fat, cartilage, and bone from progenitor cells present in a bone-derived clonal cell population: effect of dexamethasone, *J. Cell Biol.,* 106, 2139, 1988.

40. **Chen, T. L., Cone, C. M., Morey-Holton, E., and Feldman, D.,** Glucocorticoid regulation of $1,25(OH)_2D_3$-vitamin D_3 receptors in cultured mouse bone cells, *J. Biol. Chem.,* 257, 13564, 1982.

41. **Peck, W. A., Birge, S. J., and Fedak, S. A.,** Bone cells: biochemical and biological studies after enzymatic isolation, *Science,* 146, 1476, 1964.

42. **Wong, G. L., Luben, R. A., and Cohn, D. V.,** 1,25-Dihydroxycholecalciferol and parathormone: effects on osteoclast-like and osteoblast-like cells, *Science,* 197, 663, 1977.

43. **Beresford, J. N., Gallagher, J. A., and Russell, R. G. G.,** 1,25-Dihydroxyvitamin D_3 and human bone-derived cells *in vitro*: effects on alkaline phosphatase, Type I collagen and proliferation, *Endocrinology,* 119, 1176, 1986.

44. **Robey, P. G. and Termine, L. J.,** Human bone cells *in vitro, Calc. Tissue Int.,* 37, 453, 1985.

45. **Bellows, C. G., Aubin, J. E., Heersche, H. N. M., and Antosz, M. E.,** Mineralized bone nodules formed *in vitro* from enzymatically released rat calvaria cell populations, *Calcif. Tissue Int.,* 38, 143, 1986.

46. **Bhargava, U., Bar-Lev, M., Bellows, C. G., and Aubin, J. E.,** Ultrastructural analysis of bone nodules formed in vitro by isolated fetal rat calvaria cells, *Bone,* 9, 155, 1988.

47. **Ecarot-Charrier, B., Glorieux, F. H., van der Rest, M., and Pereira, G.,** Osteoblasts isolated from mouse calvaria initiate matrix mineralization in culture, *J. Cell Biol.,* 96, 639, 1983.

48. **Gerstenfeld, L. C., Chipman, S. D., Glowacki, J., and Lian, J. B.,** Expression of differentiated function by mineralizing cultures of chicken osteoblasts, *Dev. Biol.,* 122, 49, 1987.

49. **Ibaraki, K., Termine, J. D., Whitson, S. W., and Young, M. F.,** Bone matrix mRNA expression in differentiating fetal bovine osteoblasts, *J. Bone Miner. Res.,* 7, 743, 1992.

50. **Aronow, M. A., Gerstenfeld, L. C., Owen, T. A., Tassinari, M. S., Stein, G. S., and Lian, J. B.,** Factors that promote progressive development of the osteoblast phenotype in cultured fetal rat calvaria cells, *J. Cell Physiol.,* 143, 213, 1990.

51. **Owen, T. A., Aronow, M., Shalhoub, V., Barone, L. M., Wilming, L., Tassinari, M. S., Kennedy, M. B., Pockwinse, S., Lian, J. B., and Stein, G. S.,** Progressive development of the rat osteoblast phenotype *in vitro*: reciprocal relationships in expression of genes associated with osteoblast proliferation and differentiation during formation of the bone extracellular matrix, *J. Cell. Physiol.,* 143, 420, 1990.

52. **Gerstenfeld, L. C., Chipman, S. D., Kelly, C. M., Hodgens, K. J., Lee, D. D., and Landis, W. J.,** Collagen expression, ultrastructural assembly, and mineralization in cultures of chicken embryo osteoblasts, *J. Cell Biol.,* 106, 979, 1988.

53. **Stein, G. S., Lian, J. B., and Owen, T. A.,** Relationship of cell growth to the regulation of tissue-specific gene expression during osteoblast differentiation, *FASEB J.,* 4, 3111, 1990.

54. **Collart, D., Ramsey-Ewing, A., Bortell, R., Lian, J., Stein, J., and Stein, G.,** Isolation and characterization of a cDNA from a human histone H2B gene which is reciprocally expressed in relation to replication-dependent H2B histone genes during HL60 cell differentiation, *Biochemistry,* 30, 1610, 1991.

55. **Shalhoub, V., Gerstenfeld, L. C., Lian, J. B., Stein, J. L., and Stein, G. S.,** Down-regulation of cell growth and cell cycle regulated genes during chick osteoblast differentiation with the reciprocal expression of histone gene variants, *Biochemistry,* 28, 5318, 1989.

56. **Oldberg, A., Franzén, A., and Heinegård, D.,** Cloning and sequence analysis of rat bone sialoprotein (osteopontin) cDNA reveals an Arg-Gly-Asp cell-binding sequence, *Proc. Natl. Acad. Sci. U.S.A.,* 83, 8819, 1986.

57. **Hauschka, P. V., Lian, J. B., Cole, D. E. C., and Gundberg, C. M.,** Osteocalcin and matrix Gla protein: vitamin K-dependent proteins in bone, *Physiol. Rev.,* 69, 990, 1989.

58. **Price, P. A.,** Role of vitamin K-dependent proteins in bone metabolism, *Annu. Rev. Nutr.,* 8, 565, 1988.

59. **Craig, A. M., Smith, J. H., and Denhardt, D. T.,** Osteopontin, a transformation-associated cell adhesion phosphoprotein, is induced by 12-0-tetradecanoylphorbol 13-acetate in mouse epidermis, *J. Biol. Chem.,* 264, 9682, 1989.

60. **Lian, J. B. and Friedman, P. A.,** The vitamin K-dependent synthesis of gamma-carboxyglutamic acid by bone microsomes, *J. Biol. Chem.,* 253, 6623, 1978.

61. **Fraser, J. D. and Price, P. A.,** Induction of matrix Gla protein synthesis during prolonged 1,25-dihydroxyvitamin D_3 treatment of osteosarcoma cells, *Calcif. Tissue Int.,* 46, 270, 1990.

62. **Barone, L. M., Owen, T. A., Tassinari, M. S., Bortell, R., Stein, G. S., and Lian, J. B.,** Developmental expression and hormonal regulation of the rat matrix Gla protein (MGP) gene in chondrogenesis and osteogenesis, *J. Cell. Biochem.,* 46, 000-000, 1991.

63. **Pockwinse, S., Wilming, L., Conlon, D., Stein, G. S., and Lian, J. B.,** Expression of cell growth and bone specific genes at single cell resolution during development of bone tissue-like organization in primary osteoblast cultures, submitted.

64. **Nagata, T., Bellows, C. G., Kasugai, S., Butler, W. T., and Sodek, J.,** Biosynthesis of bone proteins [SPP-1 (secreted phosphoprotein-1, osteopontin), BSP (bone sialoprotein) and SPARC (osteonectin)] in association with mineralized-tissue formation by fetal rat calvarial cells in culture, *Biochem. J.,* 274, 513, 1991.

65. **Boskey, A. L., Wians, F. H., Jr., and Hauschka, P. V.,** The effect of osteocalcin on *in vitro* lipid-induced hydroxyapatite formation and seeded hydroxyapatite growth, *Calc. Tissue Int.,* 37, 57, 1985.

66. **Lian, J. B., Tassinari, M., and Glowacki, J.,** Resorption of implanted bone prepared from normal and warfarin-treated rats, *J. Clin. Invest.,* 73, 1223, 1984.

67. **Glowacki, J. and Lian, J.,** Impaired recruitment and differentiation of osteoclast progenitors by osteocalcin-depleted bone implants, *Cell Differ.,* 21, 247, 1987.

68. **Reinholt, F. P., Hultenby, K., Oldberg, Ă., and Heinegård, D.,** Osteopontin—a possible anchor of osteoclasts to bone, *Proc. Natl. Acad. Sci. U.S.A.,* 87, 4473, 1990.

69. **Miyauchi, A., Alvarez, J., Greenfield, E. M., Teti, A., Grano, M., Colucci, S., Zambonin-Zallone, A.., Ross, F. P., Teitelbaum, S. L., Cheresh, D., and Hruska, K. A.,** Recognition of osteopontin and related peptides by an $\alpha_v\beta_3$ integrin stimulates immediate cell signals in osteoclasts, *J. Biol. Chem.,* 266, 20369, 1991.

70. **Weinreb, M., Shinar, D., and Rodan, G. A.,** Different pattern of alkaline phosphatase, osteopontin, and osteocalcin expression in developing rat bone visualized by in situ hybridization, *J. Bone Min. Res.,* 5, 831, 1990.

71. **Sandberg, M., Autio-Harmainen, H., and Vuorio, E.,** Localization of the expression of types I, III, and IV collagen, TGF-β1 and c-fos genes in developing human calvarial bones, *Dev. Biol.,* 130, 324, 1988.

72. **Yoon, K., Buenaga, R., and Rodan, G. A.,** Tissue specificity and developmental expression of osteopontin, *Biochem. Biophys. Res. Comm.,* 148, 1129, 1987.

73. **Scharla, S. H., Strong, D. D., Mohan, S., Baylink, D. J., and Linkhart, T. A.,** 1,25-Dihydroxyvitamin D_3 differentially regulates the production of insulin-like growth factor I (IGF-I) and IGF-binding protein-4 in mouse osteoblasts, *Endocrinology,* 129, 3139, 1991.

74. **Kurose, H., Yamaoka, K., Okada, S., Nakajima, S., and Seino, Y.,** 1,25-Dihydroxyvitamin D_3 [1,25-$(OH)_2D_3$] increases insulin-like growth factor I (IGF-I) receptors in clonal osteoblastic cells. Study on interaction of IGF-I and 1,25-$(OH)_2D_3$, *Endocrinology,* 126, 2088, 1990.

75. **Petkovich, P. M., Wrana, J. L., Grigoriadis, A. E., Heersche, J. N. M., and Sodek, J.,** 1,25-Dihydroxyvitamin D_3 increases epidermal growth factor receptors and transforming growth factor β-like activity in a bone-derived cell line, *J. Biol. Chem.,* 262, 13424, 1987.

76. **Gronowicz, G., Egan, J. J., and Rodan, G. A.,** The effect of 1,25-dihydroxyvitamin D_3 on the cytoskeleton of rat calvaria and rat osteosarcoma (ROS 17/2.8) osteoblastic cells, *J. Bone Min. Res.,* 1, 441, 1986.

77. **Lomri, A. and Marie, P. J.,** Changes in cytoskeletal proteins in response to parathyroid hormone and 1,25-dihydroxyvitamin D in human osteoblastic cells, *Bone Min.,* 10, 1, 1990.

78. **Hong, M. H., Jin, C. H., Toshiyuki, S., Ishimi, Y., Abe, E., and Suda, T.,** Transcriptional regulation of the production of the third component of complement (C3) by 1α,25-dihydroxyvitamin D_3 in mouse marrow-derived stromal cells (ST2) and primary osteoblastic cells, *Endocrinology,* 129, 2774, 1991.

79. **Shull, S., Tracy, R. P., and Mann, K. G.,** Identification of a vitamin D-responsive protein on the surface of human osteosarcoma cells, *Proc. Natl. Acad. Sci. U.S.A.,* 86, 5405, 1989.

80. **Majeska, R. J. and Rodan, G. A.,** The effect of 1,25$(OH)_2D_3$ on alkaline phosphatase in osteoblastic osteosarcoma cells, *J. Biol. Chem.,* 257, 3362, 1982.

81. **Mulkins, M. A., Manolagas, S. C., Deftos, L. H., and Sussman, H. H.,** Dihydroxyvitamin D_3 increases bone alkaline phosphatase enzyme levels in human osteogenic sarcoma cells, *J. Biol. Chem.,* 258, 6219, 1983.

82. **Murray, E. J., Murray, S. S., and Manolagas, S. C.,** Two-dimensional gel autoradiographic analyses of the effects of 1,25-dihydroxycholecalciferol on protein synthesis in clonal rat osteosarcoma cells, *Endocrinology,* 126, 2679, 1990.

83. **Harrison, J. R., Petersen, D. N., Lichtler, A. C., Mador, A. T., Rowe, D. W., and Kream, B. E.,** 1,25-Dihydroxyvitamin D_3 inhibits transcription of type I collagen genes in the rat osteosarcoma cell line ROS 17/2.8, *Endocrinology,* 125, 327, 1989.

84. **Noda, M., Vogel, R. L., Craig, A. M., Prahl, J., DeLuca, H. F., and Denhardt, D. T.,** Identification of a DNA sequence responsible for binding of the 1,25-dihydroxyvitamin D_3 receptor and 1,25-dihydroxyvitamin D_3 enhancement of mouse secreted phosphoprotein 1 (Spp-1 or osteopontin) gene expression, *Proc. Natl. Acad. Sci. U.S.A.,* 87, 9995, 1990.

85. **Kerner, S. A., Scott, R. A., and Pike, J. W.,** Sequence elements in the human osteocalcin gene confer basal activation and inducible response to hormonal vitamin D_3, *Proc. Natl. Acad. Sci. U.S.A.,* 86, 4455, 1989.

86. **Lian, J. B., Stewart, C., Puchacz, E., Mackowiak, S., Shalhoub, V., Collart, D., Zambetti, G., and Stein, G.,** Structure of the rat osteocalcin gene and regulation of vitamin D-dependent expression, *Proc. Natl. Acad. Sci. U.S.A.,* 86, 1143, 1989.

87. **Demay, M. B., Gerardi, J. M., DeLuca, H. F., and Kronenberg, H. M.,** DNA sequences in the rat osteocalcin gene that bind the 1,25-dihydroxyvitamin D_3 receptor and confer responsive to 1,25-dihydroxyvitamin D_3, *Proc. Natl. Acad. Sci. U.S.A.,* 87, 369, 1990.

88. **Spiess, Y. H., Price, P. A., Deftos, J. L., and Manolagas, S. C.,** Phenotype-associated changes in the effects of 1,25-dihydroxyvitamin D_3 on alkaline phosphatase and bone GLA-protein of rat osteoblastic cells, *Endocrinology,* 118, 1340, 1986.

89. **Kim, Y. S., Birge, S. J., Avioli, L. V., and Miller, R.,** Cell density-dependent vitamin D effects on calcium accumulation in rat osteogenic sarcoma cells (ROS 17/2), *Calcif. Tissue Int.,* 41, 218, 1987.

90. **Canalis, E. and Lian, J. B.,** 1,25-dihydroxyvitamin D_3 effects on collagen and DNA synthesis in periosteum and periosteum-free calvaria, *Bone,* 6, 457, 1985.

91. **Franceschi, R. T. and Young, J.,** Regulation of alkaline phosphatase by 1,25-dihydroxyvitamin D_3 and ascorbic acid in bone-derived cells, *J. Bone Min. Res.,* 5, 1157, 1990.

92. **Kyeyune-Nyombi, E., Lau, W. K.-H., Baylink, D. J., and Strong, D. D.,** Stimulation of cellular alkaline phosphatase activity and its messenger RNA level in a human osteosarcoma cell line by 1,25-dihydroxyvitamin D_3, *Arch. Biochem. Biophys.,* 275, 363, 1989.

93. **Canalis, E.,** Effect of hormones and growth factors on alkaline phosphatase activity and collagen synthesis in cultured rat calvariae, *Metabolism,* 32, 14, 1983.

94. **Markose, E. R., Stein, J. L., Stein, G. S., and Lian, J. B.,** Vitamin D-mediated modifications in protein-DNA interactions at two promoter elements of the osteocalcin gene, *Proc. Natl. Acad. Sci. U.S.A.,* 87, 1701, 1990.

95. **Pan, L. and Price, P. A.,** Ligand-dependent regulation of the 1,25-dihydroxyvitamin D_3 receptor in rat osteosarcoma cells, *J. Biol. Chem.,* 262, 4670, 1987.

96. **van Leeuwen, J. P. T. M., Birkenhäger, J. C., Buurman, C. J., Schilte, J. P., and Pols, H. A. P.,** Functional involvement of calcium in the homologous up-regulation of the 1,25-dihydroxyvitamin D_3 receptor in osteoblast-like cells, *FEBS,* 270, 165, 1990.

97. **Baran, D. T., Sorensen, A. M., Honeyman, T. W., Ray, R., and Holick, M. F.,** $1\alpha,25$-dihydroxyvitamin D_3-induced increments in hepatocyte cytosolic calcium and lysophosphatidylinositol: inhibition by pertussis toxin and $1\alpha,25$-dihydroxyvitamin D_3, *J. Bone Min. Res.,* 5, 517, 1990.

98. **Leiberherr, M.,** Effects of vitamin D_3 metabolites on cytosolic free calcium in confluent mouse osteoblasts, *J. Biol. Chem.,* 262, 13168, 1987.

99. **Caffrey, J. M. and Farach-Carson, M. C.,** Vitamin D_3 metabolites modulate dihydropyridine-sensitive calcium currents in clonal rat osteosarcoma cells, *J. Biol. Chem.,* 264, 20265, 1989.

100. **Nemere, I. and Norman, A. W.,** Rapid action of 1,25-dihydroxyvitamin D_3 on calcium transport in perfused chick duodenum: effect of inhibitors, *J. Bone Min. Res.,* 2, 99, 1987.

101. **Farach-Carson, M. C., Sergeev, I., and Norman, A. W.,** Nongenomic actions of 1,25-dihydroxyvitamin D_3 in rat osteosarcoma cells: structure-function studies using ligand analogs, *Endocrinology,* 129, 1876, 1991.

102. **Gill, R. K. and Christakos, S.,** Identification of sequence elements in the mouse calbindin-D_{28k} gene which confer basal activation and hormone inducible response, in *Vitamin D Gene Regulation Structure-Function Analysis and Clinical Application,* Norman, A. W., Bouillon, R., and Thomasset, M., Eds., Walter de Gruyter, Berlin, 1991, 36.

103. **Darwish, H. M. and DeLuca, H. F.,** Identification of a 1,25-dihydroxyvitamin D_3-response element in the 5'-flanking region of the rat calbindin D-9k gene, *Proc. Natl. Acad. Sci. U.S.A.,* in press.

104. **Morrison, N. A., Shine, J., Fragonas, J.-C., Verkest, V., McMenemy, L., and Eisman, J. A.,** 1,25-Dihydroxyvitamin D-responsive element and glucocorticoid repression in the osteocalcin gene, *Science,* 246, 1158, 1989.

105. **Burger, E. H., Van Der Meer, J. W. M., Van De Gevel, J. S., Gribnau, J. C. Thesingh, C. W., and Van Furth, R.,** In vitro formation of osteoclasts from long-term culture of bone marrow mononuclear phagocytes, *J. Exp. Med.,* 156, 1604, 1982.

106. **Walker, D. G.,** Control of bone resorption by haematopoietic tissue. The induction and reversal of congenital osteopetrosis in mice through use of bone marrow and splenic transplants, *J. Exp. Med.,* 142, 651, 1975.

107. **Bar-Shavit, Z., Teitelbaum, S. L., Reitsma, P., Hall, A., Pegg, L. E., Trial, J., and Kahn, A.,** Induction of monocytic differentiation and bone resorption by 1,25-dihydroxyvitamin D$_3$, *Proc. Natl. Acad. Sci. U.S.A.,* 80, 5907, 1983.

108. **Holtrop, M. E. and Raisz, L. G.,** Comparison of the effects of 1,25-dihydroxycholecalciferol, prostaglandin E$_2$ and osteoclast-activating factor with parathyroid hormone on the ultrastructure of osteoclasts in cultured long bones of fetal rats, *Calc. Tissue Int.,* 29, 201, 1979.

109. **Takahashi, N., Akatsu, T., Sasaki, T., Yamaguchi, A., Kodama, H., Matin, T. J., and Suda, T.,** Osteoblastic cells are involved in osteoclast formation, *Endocrinology,* 123, 2600, 1988.

110. **Udagawa, N., Takahashi, N., Akatsu, T., Sasaki, T., Yamaguchi, A., Kodama, H., Matin, T. J., and Suda, T.,** The bone marrow-derived stromal cell lines MC3T3-G2/PA6 and ST2 support osteoclast-like cell differentiation in co-culture with mouse spleen cells, *Endocrinology,* 125, 1805, 1989.

111. **Franceschi, R. T., Romano, P. R., and Park, K.-Y.,** Regulation of type I collagen synthesis by 1,25-dihydroxyvitamin D$_3$ in human osteosarcoma cells, *J. Biol. Chem.,* 263, 18938, 1988.

112. **Lian, J. B., Couttes, M. C., and Canalis, E.,** Studies of hormonal regulation of osteocalcin synthesis in cultured fetal rat calvariae, *J. Biol. Chem.,* 60, 8706, 1985.

113. **Aronow, M. A., Owen, T. A., Stein, G. S., and Lian, J. B.,** Estrogen inhibition of osteocalcin mRNA expression occurs in cultured rat osteoblasts only after formation of a mineralized extracellular matrix, *J. Bone Min. Res.,* 5, S273, 1990.

114. **Turner, R. T., Colvard, D. S., and Spelsberg, T. C.,** Estrogen inhibition of periosteal bone formation in rat long bones: down-regulation of gene expression for bone matrix proteins, *Endocrinology,* 127, 1346, 1990.

115. **Schepmoes, G., Breen, E., Owen, T. A., Aronow, M. A., Stein, G. S., and Lian, J. B.,** Influence of dexamethasone on the vitamin D-mediated regulation of osteocalcin gene expression, *J. Cell. Biochem.,* 47, 184, 1991.

116. **Schüle, R., Umesono, K., Mangelsdorf, D. J., Bolado, J., Pike, J. W., and Evans, R. M.,** Jun-Fos and receptors for vitamins A and D recognize a common response element in the human osteocalcin gene, *Cell,* 61, 497, 1990.

117. **Nishimoto, S. K., Salka, C., and Nimni, M. E.,** Retinoic acid and glucocorticoids enhance the effect of 1,25-dihydroxyvitamin D$_3$ on bone γ-carboxyglutamic acid protein synthesis by rat osteosarcoma cells, *J. Bone Min. Res.,* 2, 571, 1987.

118. **Noda, M.,** Transcriptional regulation of osteocalcin production by transforming growth factor-β in rat osteoblast-like cells, *Endocrinology,* 124, 612, 1989.

119. **Noda, M., Yoon, K., and Rodan, G. A.,** Cyclic AMP-mediated stabilization of osteocalcin mRNA in rat osteoblast-like cells treated with parathyroid hormone, *J. Biol. Chem.,* 263, 18574, 1988.

120. **Theofan, G. and Price, P. A.,** Bone Gla protein messenger ribonucleic acid is regulated by both 1,25-dihydroxyvitamin D$_3$ and 3′,5′-cyclic adenosine monophosphate in rat osteosarcoma cells, *Mol. Endocrinol.,* 3, 36, 1989.

121. **Stromstedt, P. E., Poellinger, L., Gustaffson, J. A., and Cralstedt-Duke, J.,** The glucocorticoid receptor binds to a sequence overlapping the TATA box of the human osteocalcin promoter: a potential mechanism for negative regulation, *Mol. and Cell Biol.,* 11, 3379, 1991.

122. **Nanes, M. S., Rubin, J., Titus, L., Hendy, G. N. and Catherwood, B. D.,** Interferon-γ inhibits 1,25-dihydroxyvitamin D$_3$-stimulated synthesis of bone Gla protein in rat osteosarcoma cells by a pretranslational mechanism, *Endocrinology,* 127, 588, 1990.

123. **Stein, G. S., Lian, J. B., Dworetzky, S. I., Owen, T. A., Bortell, R., Bidwell, J. P., and van Wijnen, A. J.,** Regulation of transcription-factor activity during growth and differentiation: involvement of the nuclear matrix in concentration and localization of promoter binding proteins, *J. Cell. Biochem.,* 47, 300, 1991.

124. **Bortell, R., Owen, T. A., van Wijnen, A. J., Bidwell, J. P., Gavazzo, P., Breen, E., DeLuca, H., Stein, G. S., and Lian, J. B.,** Vitamin D responsive protein/DNA interactions at multiple promoter regulatory elements contribute to the level of osteocalcin gene expression, *Proc. Natl. Acad. Sci. U.S.A.,* 1992.

125. **Terpening, C. M., Haussler, C. A., Jurutka, P. W., Galligan, M. A., Komm, B. S., and Haussler, M. R.,** The vitamin D-responsive element in the rat bone Gla protein gene is an imperfect direct repeat that cooperates with other *cis*-elements in 1,25-dihydroxyvitamin D$_3$-mediated transcriptional activation, *Mol. Endocrinol.,* 5, 373, 1991.

126. **Schüle, R., Rangarajan, P., Kliewer, S., Ransome, L. J., Bolado, J., Yang, N., Verma, I. M., and Evans, R. M.,** Functional antagonism between oncoprotein c-Jun and the glucocorticoid receptor, *Cell,* 62, 1217, 1990b.

127. **Owen, T. A., Bortell, R., Yocum, S. A., Smock, S. L., Zhang, M., Abate, C., Shalhoub, V., Aronin, N., Wright, K. L., van Wijnen, A. J., Stein, J. L., Curran, T., Lian, J. B., and Stein, G. S.,** Coordinate occupancy of AP-1 sites in the vitamin D responsive and CCAAT box elements by fos-jun in the osteocalcin gene: a model for phenotype suppression of transcription, *Proc. Natl. Acad. Sci. U.S.A.,* 87, 9990, 1990.

128. **Ozono, K., Liao, J., Kerner, S. A., Scott, R. A., and Pike, J. W.,** The vitamin D-responsive element in the human osteocalcin gene, *J. Biol. Chem.,* 265, 21881, 1990.

129. **Ross, T. K., Moss, V. E., Prahl, J. M., and DeLuca, H. F.,** A nuclear protein essential for binding of rat 1,25-dihydroxyvitamin D$_3$ receptor to its response elements, *Proc. Natl. Acad. Sci. U.S.A.,* in press.

130. **Brown, T. A. and DeLuca, H. F.,** Phosphorylation of the 1,25-dihydroxyvitamin D$_3$ receptor. A primary event in 1,25-dihydroxyvitamin D$_3$ action, *J. Biol. Chem.,* 265, 10025, 1990.

131. **Hsieh, J.-C., Jurutka, P. W., Galligan, M. A., Terpening, C. M., Haussler, C. A., Samuels, D. S., Shimizu, Y., Shimizu, N., and Haussler, M. R.,** Human vitamin D receptor is selectively phosphorylated by protein kinase C on serine 51, a residue crucial to its trans-activation function, *Proc. Natl. Acad. Sci. U.S.A.,* 88, 9315, 1991.

132. **Lian, J. B., Stein, G. S., Bortell, R., and Owen, T. A.,** Phenotype suppression: a postulated molecular mechanism for mediating the relationship of proliferation and differentiation by fos/jun interactions at AP-1 sites in steroid responsive promoter elements of tissue-specific genes, *J. Cell. Biochem.,* 45, 9, 1991.

133. **Candeliere, G. A., Prud'homme, J., and St-Arnaud, R.,** Differential stimulation of fos and jun family members by calcitriol in osteoblastic cells, *Mol. Endocrinol.,* 12, 1780, 1991.

134. **Diamond, M. I., Miner, J. N., Yoshinaga, S. K., and Yamamoto, K. R.,** Transcription factor interactions: selectors of positive or negative regulation from a single DNA element, *Science,* 249, 1266, 1990.

135. **Lian, J. B. and Stein, G. S.,** Transcriptional control of vitamin D regulated proteins, *J. Cell. Biochem.,* 49, 1992.

136. **Feldman, D. and Malloy, P. J.,** Hereditary 1,25-dihydroxyvitamin D resistant rickets: molecular basis and implications for the role of 1,25(OH)$_2$D$_3$ in normal physiology, *Mol. Cell. Endocrinol.,* 72, C57, 1990.

137. **Marx, S. J., Liberman, U. A., Eil, C., Gamblin, G. T., Degrange, D. A., and Balsan, S.,** Hereditary resistance to 1,25-dihydroxyvitamin D$_3$ suppresses parathyroid hormone secretion from bovine parathyroid cells in tissue culture, *Endocrinology,* 117, 2214, 1984.

138. **Liberman, U. and Marx, S. J.,** Vitamin D dependent rickets, in *Primer on Metabolic Bone Diseases and Disorders of Mineral Metabolism,* Fauvus, M. J., Ed., American Society of Bone and Mineral Research, Kelseyville, CA, 1990, 178.

139. **Glorieux, F. H.,** Rickets, the continuing challenge, *New Engl. J. Med.,* 325, 1875, 1991.

140. **Marks, S. J., Jr.,** Osteopetrosis-multiple pathways for the interruption of osteoclast functions, *Appl. Pathol.,* 5, 172, 1987.

141. **Shalhoub, V., Jackson, M. E., Lian, J. B., Stein, G. S., and Marks, S. C., Jr.,** Gene expression during skeletal development in three osteopetrotic rat mutations: evidence for osteoblast abnormalities, *J. Biol. Chem.,* 266, 9847, 1991.

142. **DeLuca, H. F.,** Newly discovered actions of 1,25-dihydroxyvitamin D_3 and osteoporosis: basis and treatment, *Proc. Soc. Exp. Biol. Med.,* 191, 1989.

143. **Avioli, L. V.,** Significance of osteoporosis: a growing international health care problem, *Calcif. Tissue Int.,* 49, S5, 1991.

144. **Eastell, R., Yergey, A. L., Vieira, N. E., Cedel, S. L., Kumar, R., and Riggs, B. L.,** Interrelationship among vitamin D metabolism, true calcium absorption, parathyroid function, and age in women: evidence of an age-related intestinal resistance to 1,25-dihydroxyvitamin A action, *J. Bone Min. Res.,* 6, 125, 1991.

145. **Franz, K. B.,** Abnormalities in parathyroid hormone secretion and 1,25-dihydroxyvitamin D_3 formation in women with osteoporosis, *New Engl. J. Med.,* 320, 1697, 1989.

146. **Silverberg, S. J., Shane, E., de la Cruz, L., Segre, G. V., Clemens, T. L., and Bilezikian, J. P.,** Abnormalities in parathyroid hormone secretion and 1,25-dihydroxyvitamin D_3 formation in women with osteoporosis, *New Engl. J. Med.,* 320, 277, 1989.

147. **Gallagher, J. C., Jerpbak, C. M., Lee, W. S. S., Johnson, K. A., DeLuca, H. F., and Riggs, B. L.,** 1,25-dihydroxyvitamin D_3: short- and long-term effects on bone and calcium metabolism in patients with postmenopausal osteoporosis, *Proc. Natl. Acad. Sci. U.S.A.,* 79, 3325, 1982.

148. **Geusens, P., Vanderschueren, D., Verstraeten, A., Dequeker, J., Devos, P., and Bouillon, R.,** Short-term course of $1,25(OH)_2D_3$ stimulates osteoblasts but not osteoclasts in osteoporosis and osteoarthritis, *Calcif. Tissue Int.,* 49, 168, 1991.

149. **Tsai, K. S., Heath, H., Kumar, R., and Riggs, B. L.,** Impaired vitamin D metabolism with aging in women. Possible role in pathogenesis of senile osteoporosis, *J. Clin. Invest.,* 73, 1668, 1984.

150. **Gallagher, J. C. and Riggs, B. L.,** Action of 1,25-dihydroxyvitamin D_3 on calcium balance and bone turnover and its effect on vertebral fracture rate, *Metabolism,* 39, 30, 1990.

151. **Aloia, J. F., Vaswani, A., Yeh, J. K., Ellis, K., Yasumura, S., and Cohn, S. H.,** Calcitriol in the treatment of postmenopausal osteoporosis, *Am. J. Med.,* 84, 401, 1988.

152. **Caniggia, A., Nuti, R., Galli, M., Lore, F., Turchetti, V., and Righi, G. A.,** Effect of a long-term treatment with 1,25-dihydroxyvitamin D_3 on osteocalcin in postmenopausal osteoporosis, *Calcif. Tissue Int.,* 38, 328, 1986.

153. **Gallagher, J. C., Riggs, B. L., Recker, R. R., and Goldgar, D.,** The effect of calcitriol on patients with postmenopausal osteoporosis with special reference to fracture frequency, *Proc. Soc. Exp. Biol. Med.,* 191, 287, 1989.

154. **Falch, J. A., Odegaard, O. R., Finnanger, A. M., and Matheson, I.,** Postmenopausal osteoporosis: no effect of three years treatment with 1,25-dihydroxycholecalciferol, *Acta Med. Scand.,* 221, 199, 1987.

155. **Ott, S. M. and Chestnut, C. H., III,** Calcitriol treatment is not effective in postmenopausal osteoporosis, *Ann. Int. Med.,* 110, 267, 1989.

156. **Manolagas, S. C., Hustmyer, F. G., and Yu, Z.-P.,** Immunomodulating properties of 1,25-dihydroxyvitamin D_3, *Kidney Int.,* 38, S9, 1990.

157. **Lemire, J. M. and Archer, D. C.,** 1,25-dihydroxyvitamin D_3 prevents the in vivo induction of murine experimental autoimmune encephalomyelitis, *J. Clin. Invest.,* 87, 1103, 1991.

158. **Kragballe, K., Gjertsen, B. T., De Hoop, D., Karlsmark, T., van de Kerkhof, P. C., Larko, O., Nieboer, C., Roed-Peterson, J., Strand, A., and Tikjb, G.,** Double-blind, right/left comparison of calcipotriol (MC903) and betamethasone valerate in treatment of psoriasis vulgaris, *Lancet,* 337, 193, 1991.

159. **Okano, M.,** 1 alpha,25-(OH)$_2$D$_3$ use on psoriasis and ichthyosis, *Int. J. Dermatol.*, 30, 62, 1991.

160. **Huckins, D., Felson, D. T., and Holick, M.,** Treatment of psoriatic arthritis with oral 1,25-dihydroxyvitamin D$_3$: a pilot study, *Arthritis Rheum.*, 33:1723-1727.

161. **Ostrem, V. K., Tanaka, Y., Prahl, J., DeLuca, H. F., and Ikekawa, N.,** 24- and 26-homo-1,25-dihydroxyvitamin D$_3$: preferential activity in inducing differentiation of human leukemic cells HL-60 in vitro, *Proc. Natl. Acad. Sci. U.S.A.*, 84, 2610, 1987.

162. **Paulson, S. K., Perlman, K., DeLuca, H. F., and Stern, P. H.,** 24- and 26-homo-1,25-dihydroxyvitamin D$_3$ analogs: potencies on in vitro bone resorption differ from those reported for cell differentiation, *J. Bone Min. Res.*, 5, 201, 1990.

163. **Zhou, J.-Y., Norman, A. W., Lubbert, M., Collins, E. D., Uskokovik, M. R., and Koeffler, H. P.,** Novel vitamin D analogs that modulate leukemic cell growth and differentiation with little effect on either intestinal calcium absorption or bone calcium mobilization, *Blood*, 74, 82, 1989.

164. **Binderup, L. and Bramm, E.,** Effects of a novel vitamin D analog MC903 on cell proliferation and differentiation in vitro and on calcium metabolism in vivo, *Biochem. Pharmacol.*, 37, 889, 1988.

165. **Bouillon, R., Allewaert, K., Tan, B. K., van Baelen, H., de Clercq, P., and Vandewalle, M.,** Biological activity of 1,25-dihydroxyvitamin D analogs: influence of relative affinities for DBP and vitamin D receptor, in *Vitamin D Gene Regulation Structure-Function Analysis and Clinical Application*, Norman, A. W., Bouillon, R., and Thomasset, M., Eds., Walter de Gruyter, Berlin, 1991, 131.

166. **Evans, D. B., Thavarajah, M., Uskokovic, M. R., and Kanis, J. A.,** Increased potency of 1,25-dihydroxyvitamin D$_3$ on human osteoblast-like cells following structural side-chain modification, *Bone*, 11, 439, 1990.

167. **Marie, P. J. Connes, D., Hott, M., and Miravet, L.,** Comparative effects of a novel vitamin D analogue MC-903 and 1,25-dihydroxyvitamin D$_3$ on alkaline phosphatase activity, osteocalcin and DNA synthesis by human osteoblastic cells in culture, *Bone*, 11, 171, 1990.

168. **Valaja, T., Mahonen, A., Pirskanen, A., and Maenpaa, P. H.,** Affinity of 22-oxa-1,25(OH)$_2$D$_3$ for 1,25-dihydroxyvitamin D receptor and its effects on the synthesis of osteocalcin in human osteosarcoma cells, *Biochem. Biophys. Res. Commun.*, 169, 629, 1990.

169. **Pernalete, N., Mori, T., Nishii, Y., Slatopolsky, E., and Brown, A. J.,** The activity of 22-oxacalcitriol in osteoblast-like (ROS 17/2.8) cells, *Endocrinology*, 129, 778, 1991.

170. **Tanaka, Y. and DeLuca, H. F.,** 26,26,26,27,27,27-hexafluoro-1,25-dihydroxyvitamin D$_3$: a highly potent, long-lasting analog of 1,25-dihydroxyvitamin D$_3$, *Arch. Biochem. Biophys.*, 229, 348, 1984.

171. **Kiriyama, T., Sumiaki, O., Ejima, E., Kurihara, N., Hakeda, Y., Ito, N., Izumi, M., Kumegawa, M., and Nagataki, S.,** Effect of a highly potent fluoro analog of 1,25-dihydroxyvitamin D$_3$ on human bone-derived cells, *Endocrinology*, 128, 81, 1991.

172. **Inaba, M., Okuno, S., Inoue, A., Nishizawa, Y., Morii, H., and DeLuca, H. F.,** DNA binding property of vitamin D$_3$ receptors associated with 26,26,26,27,27,27-hexafluoro-1,25(OH)$_2$D$_3$, *Arch. Biochem. Biophys.*, 268, 35, 1989.

173. **Heinrichs, A., Stein, J., Lian, J. B. and Stein, G. S.,** unpublished observations.

174. **Lian, J. B. and Stein, G. S.,** unpublished observations.

175. **Lian, J. B., Stein, G. S., and Uskokovic, M.,** unpublished observation.

176. **Tilyard, M. W., Spears, G. F. S., Thomson, J., and Dovey, S.,** Treatment of post-menopausal osteoporosis with calcitriol or calcium, *N. Engl. J. Med.*, 326, 357, 1992.

Chapter 19

RETINOIC ACID REGULATION OF GENE EXPRESSION IN ADIPOCYTES

Donald B. Jump, Gerald J. Lepar, and Ormond A. MacDougald

TABLE OF CONTENTS

I. INTRODUCTION

Retinol and retinoic acid have profound effects on a wide range of physiological and biochemical processes.[1-11] For example, the aldehyde derivative of retinol (11-*cis*-retinal) functions in vision.[5] Retinol is converted to retinoic acid (RA) in multiple tissues,[2-4] and RA has profound effects on pattern formation during early development, skeletal growth, cellular growth, and differentiation.[3,4,6-11] The diverse effects of RA are due to its regulation of the abundance and/or activity of specific proteins. As shown in Table 1, these proteins include hormones and growth factors,[12-14] cell surface and nuclear receptors,[15-17,19,22-26] and nuclear *trans*-acting factors,[18,20,21] specific enzymes,[18,27-33] plasma proteins,[38,39,42] and proteins involved in cellular architecture and other cellular functions.[34-36,40,43-47]

RA regulates cellular functions by binding to intracellular retinoic acid receptors (RR).[48-57] Two distinct families of RR have been identified, i.e., RAR and RXR, each with multiple isoforms. The RAR family consists of RARα and RARβ, which are found in many tissues, and RARγ, which occurs predominantly in skin and lung.[53] The RXR family consists of an α, β, and γ isoforms and RXRγ is found mainly in liver, lung, and muscle.[54] Even though there is low sequence homology between RARs and RXRs, each binds RA and DNA. Tissues like liver and white adipose tissue express multiple forms of the RRs.[48,49,55,56] The physiological significance of multiple receptor isoform expression in tissues has not been explained.

Both RARs and RXRs are structurally similar to steroid and thyroid hormone receptors (TR).[48-54] Both function as ligand-activated transcription factors in a manner similar to the regulation of gene expression described for steroid and thyroid hormones.[51] RRs contain Zn-finger protein sequence motifs which function in receptor binding to DNA, at retinoic acid response elements (RARE). The carboxyl terminal end of the molecule functions in ligand binding. In contrast to steroid receptors, RR do not interact with heat shock proteins, like HSP90,[57] proteins which prevent steroid receptor binding to DNA. These observations suggest that RR may function more like TRs. If analogous to TRs, RRs may be associated with the nuclear compartment in the absence of ligand. RA binding to nuclear receptors sets in motion a sequence of events that culminate in a change in transcription of the *cis*-linked gene. Such transcriptional changes lead to corresponding changes in mRNA coding for specific proteins and eventual changes in cell function.

The DNA targets for RR binding, i.e., RARE, have been the subject of intense investigation. Umesono et al.[58] first reported that the thyroid hormone response element (TRE) located upstream from the rat pituitary growth hormone (GH) gene was a target for RAR binding and action. The GH TRE used in these studies was palindromic and represented a modified version of the TRE identified in the GH gene. The notion that RAR might utilize TREs was reinforced by the finding that RA treatment of cultured pituitary cells

TABLE 1
Cellular Proteins Regulated by Retinoic Acid

Protein	Cell/tissue	Induced (I) or repressed (R)	Ref.
Hormones/Growth Factors			
Growth Hormone	Pituitary cells	I	12, 13
TGFβ2	Keratinocytes	I	14
Receptors/Nuclear Trans-Acting Factors			
1,25-(OH)$_2$D$_3$ Receptors	Rat osteosarcoma Ros 17/2	I	15
RARβ	HepG2 Hepatoma	I	16, 17
c-Fos	Rheumatoid synoviocytes	R	18
Progesterone receptors	T-47D Breast cancer	R	19
Zif268 transcription factor	Rat fetal calvaria	I	20
AP-2 Transcription factor	NT2 Teratocarcinoma	I	21
MSH Receptors	Cloudman S91 melanoma	I	22
Interleukin-6 receptors	Myeloma	R	23
Interleukin-2 receptors	Thymocyte	I	24
EGF receptors	Corneal endothelium	I	25
EGF receptors	Corneal epithelium	R	26
Enzymes			
Transglutaminase	Macrophages	I	27
Alkaline phosphatase	Rat osteosarcoma Ros 17/2.8	R	28
Collagenase	Rheumatoid synoviocytes	R	18
Phosphoenolpyruvate carboxykinase	H4IIE Hepatoma	I	29
G$_s$α	F9 Teratocarcinoma	I	30
Alcohol dehydrogenase	Hep3B hepatoma	I	31
t-Plasminogen activator	F9 Teratocarcinoma	I	32
Glycerophosphate Dehydrogenase	3T3-F442A Adipocytes	I	33
Other Proteins			
β$_1$ Laminin	F9 Teratocarcinoma	I	34–36
CRBP and CRABP	P16 Embryonal carcinoma	I	37
Complement factor H	L929 Fibroblast	I	38
Adipsin	3T3-F442A Adipocytes	R	39
Keratins, K5,K6, K14	Keratinocytes	R	40
Loricrin	Keratinocytes	R	41
Apolipoprotein AI	HepG2 hepatoma	I	42
Neurofilament (NF-M)	PC12 Neuroblastoma	I	43
GAP-43	PC12 Neuroblast	I	43
Collagen type X	Chick chondrocytes	I	44
S14 Protein	3T3-F442A Adipocytes	I	33
Collagen	F9 Teratocarcinoma	I	45
Fibronectin	F9 Teratocarcinoma	I	46
Vimentin	F9 Teratocarcinoma	I	47

elevated GH protein production, mRNA$_{GH}$ levels and GH gene transcription.[12,13] Subsequent reports showed that TR and RR formed heterodimers and bound to the GH TRE to induce GH gene transcription in an additive fashion.[13,62,63]

The finding that RA, as well as triiodothyronine (T$_3$), regulated the GH TRE suggested that other T$_3$-regulated genes might be targets for RA control or that RA-regulated genes might be affected by T$_3$. When similar studies were carried out using the myosin heavy chain TRE, the formation of heterodimers inhibited the T$_3$ activation of myosin heavy chain gene transcription.[62] Conversely, RA-regulated genes, like RARβ[16,17] and β1-laminin,[36] were not regulated by T$_3$. These results showed that structural features within response elements determine receptor binding.

The structure of response elements for TR, RAR, RXR, and vitamin D$_3$ receptors might be direct repeats rather than palindromic.[59-61] Each of the four receptors could recognize the same sequence, i.e., AGGTCA, when it was present in DNA as a direct repeat. However, receptor binding specificity was determined by the nucleotide spacing between the repeats. For example, RXR would bind to AGGTCA when one nucleotide separated the repeats,[60] while vitamin D$_3$ receptors, T$_3$ receptors, and RARs bound to DNA when the spacing between repeats was 3, 4, and 5 nucleotides, respectively.[59-60]

The basic components of the RA regulatory system are now being defined. However, recent studies in our laboratory suggest that there are additional cellular factors which contribute to the RA regulation of gene expression.[33] This review will show the evolution of thought that led us to investigate the role RA plays in adipocyte gene expression. We will begin with an examination of the deficient T$_3$ control of S14 gene expression in 3T3-F442A cells and progress to an analysis of RA action on S14 gene transcription in these cells. We will show that glucocorticoids and tissue-specific factors contribute significantly to the RA-mediated control of S14 gene transcription in cultured adipocytes.

II. REGULATION OF EXPRESSION OF THE S14 PROTEIN IN CULTURED ADIPOCYTES

The S14 protein (17 kDa; 4.9 pI) is expressed predominantly in tissues which synthesize triglycerides such as liver, lactating mammary gland, white and brown adipose tissues,[64-72] and cultured adipocytes.[73] In rat liver, transcription of the S14 gene is regulated by hormones,[67-70] dietary,[69,74-76] developmental,[77] and tissue-specific factors.[64,66,67] Hepatic S14 gene transcription was rapidly stimulated by T$_3$,[68] insulin,[70] dietary carbohydrate[69,70] and suppressed by dietary fat[74-76] and factors which elevated cellular levels of cAMP.[70] The overall pattern of control of S14 gene expression is typical of many proteins involved in lipid and carbohydrate metabolism.[78] For these reasons, the S14 gene can be used to study the tissue-specific, hormone, and nutrient control of gene transcription.

TABLE 2
T₃ Binding Capacity in Rat Liver,
3T3-F442A, and 3T3-L1 Adipocytes

Tissue/cells	T₃ Binding capacity[a] (#TR/nucleus)
Rat liver	3273 ± 299
3T3-F442A Adipocytes	420 ± 150
3T3-L1 Adipocytes	2773 ± 680

[a] T_3 binding capacity was measured in isolated rat liver or adipocyte nuclei as described.[73,83] Results are expressed as the number of T_3 receptors (TR)/ nucleus; mean ± SD. Number of determinations ranged from 3 to 5. Affinity of T_3 binding varied < twofold.

We were interested in developing a cell culture model to study both the hormonal and tissue-specific control of S14 gene transcription. The 3T3-F442A cell line was selected because of the purported role of the S14 protein in lipid metabolism.[67,71] When F442A cells differentiate from preadipocytes (fibroblast) to adipocytes,[79,80] there is a distinct change in morphology accompanied by: (1) an engorgement of triglyceride; (2) induction of enzymes involved in fatty acid and triglyceride synthesis; and (3) acquisition of responsiveness to both lipogenic and lipolytic hormones.[79-82] Thus, the 3T3-F442A cell line appeared to be well suited to study both the cellular and molecular mechanisms of fat cell differentiation and hormonal regulation of genes involved in lipid metabolism.

Our initial studies on S14 gene expression in cultured adipocytes revealed two surprising findings. First, mRNA$_{S14}$ was expressed in F442A adipocytes at vastly reduced levels (≤0.005 units) when compared to mouse liver or white adipose tissue (WAT) (≥0.3 units).[73] We would have expected mRNA$_{S14}$ to be much higher if the S14 protein played a critical role in triglyceride synthesis. Second, while glucocorticoids rapidly induced mRNA$_{S14}$ both *in vivo* and in F442A adipocytes, T_3 had no effect on S14 gene expression in F442A preadipocytes or adipocytes.[73] T_3 rapidly induced S14 gene expression in both mouse liver and WAT.[73]

Measurement of T_3 receptor binding capacity in isolated F442A adipocyte nuclei showed that TR were expressed at 420 ± 150 TR per nucleus (Table 2), representing an 85% reduction from the T_3 binding capacity when compared to rat liver. Since levels of TR expression in WAT are comparable to liver,[84] the low T_3 binding capacity indicated that F442A cells do not share the same phenotypic features as adipocytes in adult WAT.

Hausdorf et al.[85] and Moustaid and Sul[86] reported that S14 and fatty acid synthase (FAS) were induced by T_3 in 3T3-L1 adipocytes, respectively. 3T3-

Liver   WAT   F442A   L1

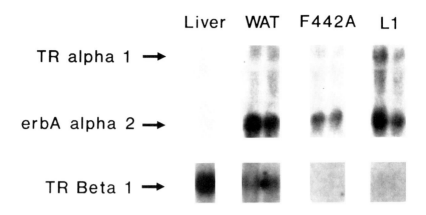

TR alpha 1 →

erbA alpha 2 →

TR Beta 1 →

FIGURE 1. T$_3$ receptor isoform analysis in mouse liver, white adipose tissue, 3T3-F442A, and 3T3-L1 adipocytes. Total RNA (25 μg) from mouse liver, epididymal fat (WAT), 3T3-F442A adipocytes (F442A), and 3T3-L1 adipocytes (L1) was electrophoretically separated on 1% agarose-2.2 *M* formaldehyde gels and transferred to nitrocellulose.[33] Northern blots were hybridized with cDNA inserts derived from rc-erbAα or rc-erbAβ plasmids (generously provided by H. Towle, Univ. of Minnesota). The rc-erbAα hybridized to two mRNAs: TRα1, 5 kb; c-erbA_2, 2.6 kb, while the rc-erbAβ probe hybridizes to 1 mRNA of 6.2 kb.[87] Duplicate samples were electrophoresed for WAT, F442A, and L1 adipocytes. Rat liver 18S and 28S ribosomal RNAs served as standards and were visualized by ethidium bromide staining.

L1 cells were also cloned from mouse embryonic fibroblast.[79] We have found that following a 72 h treatment of L1 adipocytes with T$_3$, mRNA$_{S14}$ was induced fivefold. Analysis of T$_3$ binding capacity in these cells showed that TR were expressed at 70% of the level observed in rat liver (Table 2). Thus, different adipocyte cell lines display different T$_3$ regulatory profiles.

Molecular cloning studies have shown that at least three ligand-binding isoforms of TR, i.e., c-*erb*Aα1 (TRα1), c-*erb*Aβ1 (TRβ1), and c-*erb*Aβ2 (TRβ2), are expressed in various rodent tissues.[87-92] While TRβ2 is pituitary specific, TRα1 and TRβ1 are expressed in several tissues.[87,88] In addition, three variants of these proteins (i.e., c-*erb*α2, c-*erb*α3, and rev-*erb*Aα) have been described which do not bind T$_3$.[88-93] Northern analysis was used to examine the TR isoforms expressed in mouse liver and WAT, and 3T3-F442A and 3T3-L1 adipocytes (Figure 1). A rat liver c-*erb*Aα cDNA hybridized to two mRNAs in liver, WAT, and both adipocyte cell lines. The 5-kb mRNA codes for TRα1, while the 2.5-kb mRNA codes for c-*erb*Aα2. The hybridization of rc-*erb*Aα to liver RNA is weak when compared to the hybridization signal in WAT, F442A, and L1. TRβ1 (6.2 kb) was expressed in both liver and WAT, but not in either adipocyte cell lines. The relative abundance of TRα1, c-*erb*Aα2, and TRβ1 in liver and L1 cells agrees favorably with previous reports.[87]

The differential T$_3$ responsiveness of F442A and L1 is due, at least in part, to differences in T$_3$ binding capacity. If these cells are "true" progenitors to adipocytes in WAT, then the absence of TRβ1 expression in F442A and

TABLE 3
Dexamethasone and Retinoic Acid Regulation of mRNA$_{S14}$ Levels in 3T3-F442A and 3T3-L1 Cells[a]

	3T3-F442A		3T3-L1	
Treatment	Units	Fold	Units	Fold
Vehicle	0.005 ± 0.001	1	0.004 ± 0.002	1
Retinoic acid	0.070 ± 0.013	14	0.160 ± 0.120	40
Dexamethasone	0.740 ± 0.013	148	0.240 ± 0.060	60
Dexamethasone + retinoic acid	1.800 ± 0.220	360	0.798 ± 0.079	200

[a] Adipocytes were exposed to 1 μM dexamethasone, retinoic acid, or a combination of treatments for 72 h.[33] Control cultures were exposed to vehicle (0.1% ethanol) for 72 h. RNA was extracted and analyzed for expression of mRNA$_{S14}$. Results are expressed as mean + SD. Number of determinations ranged from 6 to 12 for each treatment.

L1, and the low T_3 binding capacity in F442A cells, suggests that the T_3 regulatory system in adipocyte precursors might be under developmental control. The F442A and L1 cells might represent early to intermediate stages in development, and thus they may not be satisfactory models for T_3 action.

III. EFFECTS OF RA ON ADIPOCYTE GENE EXPRESSION

The studies on T_3 regulation of S14 gene expression in cultured adipocytes show that embryo-derived cells do not reflect the T_3 regulatory systems seen in WAT derived from adult animals. Based on the role of RA in development and differentiation[3,4,6-11] and the fact that RA regulates some T_3 responsive genes, we speculated that RA might regulate genes involved in lipid metabolism in these cells.

RA blocks the differentiation of both F442A and L1 preadipocytes (fibroblasts) to adipocytes,[94-97] indicating that RA was functional in these cells. We then determined whether RA regulated S14 gene expression in cultured F442A and L1 adipocytes. Expression of the S14 gene in F442A cells was dependent on the adipocyte phenotype and treatment of cells with glucocorticoids.[73] Thus, effects of RA on adipocyte S14 gene expression was determined in the presence and absence of the glucocorticoid agonist, dexamethasone (DEX) (Table 3). Treatment of F442A and L1 adipocytes with RA for 72 h induced mRNA$_{S14}$ 14- and 40-fold, respectively. DEX induced a 148- and 60-fold increase in mRNA$_{S14}$ abundance in F442A and L1 adipocytes, respectively. The combination of both treatments induced mRNA$_{S14}$ 360- and 200-fold in F442A and L1 adipocytes, respectively. While RA and DEX promoted S14 gene expression to different levels in the F442A and L1 cells, the induction pattern showed that both factors regulated S14 gene expression through independent pathways in each cell type.

Further investigation showed that RA and DEX interacted at the transcriptional level to regulate S14 gene expression in F442A adipocytes.[33] Treatment of F442A adipocytes with 0.1 nM DEX or 1 μM RA for 72 h induced a fivefold increase in both mRNA$_{S14}$ and S14 gene transcription. Following treatment of adipocytes with both DEX (0.1 nM) and RA (1 μM), S14 gene transcription and mRNA$_{S14}$ levels increased by 27- and 86-fold, respectively. When DEX was elevated to 1 μM in the presence of RA, S14 gene transcription and mRNA$_{S14}$ levels rose by 87- and 262-fold, respectively. Thus, RA and DEX synergistically regulate S14 gene expression. In the presence of 0.1 nM DEX, RA effects were dose-dependent (ED$_{50}$ = 5 nM) and rapid (50-fold within 4 h). This induction pattern indicates that the principal site for both RA and DEX action on S14 gene expression is at the level of S14 gene transcription. The synergistic regulation of S14 gene transcription suggests that RA may play an important modulatory role in DEX-mediated gene expression.

The specificity of RA action on adipocyte gene expression was determined by measuring the effects of RA on the expression of other genes. Changes in α glycerophosphate dehydrogenase (GPD), FAS, and β-actin mRNA levels and run-on activity were measured following a 72-h treatment of adipocytes with RA.[33] RA had no effect on FAS or β-actin gene expression, but suppressed GPD gene transcription and mRNA$_{GPD}$ levels by 75%.[73] In addition, others reported that RA suppressed adipsin gene expression through post-transcriptional regulatory mechanisms.[37] The lack of any significant effect of RA on β-actin or FAS indicated that RA does not induce generalized changes in adipocyte gene expression. More importantly, this limited analysis hints at a possible role for RA in modulating triglyceride synthesis in fat cells.

IV. RAR EXPRESSION IN CULTURED ADIPOCYTES

RA clearly regulates gene expression in both 3T3-F442A and 3T3-L1 adipocytes. In order to identify which RAR might mediate these changes in gene expression, α and β RAR cDNAs were used as hybridization probes for electrophoretically separated RNA. RARα was expressed as 2 mRNAs of 2.8 and 3.8 kb in both the L1 and F442A cells (Figure 2). No apparent difference in the level of RARα mRNA was detected between L1 or F442A cell lines or between the preadipocyte and adipocyte phenotypes. RARα was expressed at higher levels in cultured adipocytes than in liver.

In contrast to the RARα mRNA expression, analysis of RARβ mRNA levels show no detectable RNA in any of the L1 or F442A cells, but there were high levels of RARβ mRNA in liver. Zelent et al.[53] reported that RARβ was more abundant than RARα in liver. We have not determined whether other RAR isoforms, like RARγ[53] or RXR,[54] were expressed in F442A or L1 cells. However, Haq and Chytil[56] recently reported that RARα, RARβ, and RARγ were all expressed in rat WAT. We have also found that both

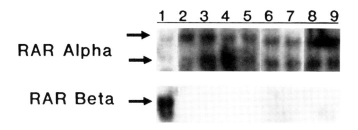

FIGURE 2. Retinoic acid receptor isoform analysis in mouse liver, 3T3-F442A, and 3T3-L1 cells. Total mRNA (25 μg) isolated from mouse liver (lane 1) and 3T3-F442A preadipocytes (lanes 2,3), 3T3-F442A adipocytes (lanes 4,5), 3T3-L1 preadipocytes (lanes 6,7), and 3T3-L1 adipocytes (lanes 8,9) was electrophoretically separated in 1% agarose-2.2 *M* formaldehyde gels.[33] Following electrophoresis RNA was transferred to nitrocellulose and hybridized with either ^{32}P-labeled inserts derived from pRARαO (RAR alpha) or pRARβ (RAR beta) plasmids. RARα hybridizes to two mRNA of 2.8 and 3.8 kb, while RARβ hybridizes to two mRNAs of 3.2 and 3.6 kb.[53] Rat liver 18S and 28S ribosomal RNAs served as standards and were visualized by ethidium bromide staining.

RARα and RARβ were expressed in rat white epididymal fat (not shown). If the mRNA levels coding for these various RAR isoforms reflect the level of active receptor in these cells, then RARα is the predominant RAR isoform expressed in both F442A and L1 cells. In addition, RARβ is upregulated by RA in some cell types.[16,17] We found no evidence for an effect of RA on RARβ expression in either preadipocytes or adipocytes (not shown).

V. IDENTIFICATION OF PROSPECTIVE RA AND GLUCOCORTICOID RESPONSIVE CIS-ACTING ELEMENTS

Both RA and DEX induced rapid and dose-dependent increases in adipocyte mRNA$_{S14}$ by activating S14 gene transcription.[33,73] RARα and glucocorticoid receptors (GR) might regulate S14 gene transcription through interaction with *cis*-linked response elements. In order to localize these elements, we transfected cells with plasmids containing 5' flanking regions isolated from the rat liver S14 gene. These regions were fused upstream from the chloramphenicol acetyltransferase (CAT) gene (i.e., S14-CAT) and used to measure changes in CAT activity induced by RA and DEX. Under these conditions, changes in CAT activity reflect changes in transcription of the CAT gene which is regulated by the S14 5' flanking sequences.

Adipocytes could not be transiently transfected, so stably transfected 3T3-F442A adipocyte cell lines were developed to examine S14-CAT gene expression. Fibroblasts were transfected with S14-CAT fusion genes in the presence of pSV2-Neo using the lipofection technique.[98] Stably transfected colonies were pooled and differentiated to adipocytes. Cells were treated with DEX or RA only after >80% of the cells acquired an adipocyte phenotype. Following hormone treatment, cell extracts were prepared and used to measure CAT activity.[99] Under these transfection conditions, endogenous receptors must be recruited to regulate integrated S14-CAT fusion genes.

Figure 3 illustrates a set of 5′ deletion mutations of the S14 promoter that were used to localize the boundaries of the DEX and RA response regions. Plasmids with 5′ end points at −4315, −2110, and −1588 bp were equally responsive to DEX (fourfold induction). However, plasmids with 5′ end points at −1075 and −290 bp were not responsive to DEX treatment. When an element extending from −1588 to −1075 bp was fused to the −290 bp S14 promoter, DEX control was conferred to the S14 promoter. Thus, the boundaries for the DEX regulatory unit controlling S14 gene transcription in adipocytes is located between −1588 and −1075 bp upstream from the 5′ end of the S14 gene.

Since the GH TRE also functioned as a RARE,[58,62,100] we examined the possibility that RA activated S14 gene transcription through the S14 TRE at −2.55 to 2.75 kb.[68,69,100] Using a similar strategy to locate RA-responsive regions, successive 5′ deletions from the S14 promoter showed that RA induction of CAT activity increased from 1.7- to 4-fold using plasmids with 5′ end points at −4315 and −1588 bp, respectively (Figure 3). The increased RA responsiveness with successive deletions may indicate the presence of elements located between −1588 and −4315 bp which modulate RA action. However, deletions with 5′ end points at −1075 and −290 showed no response to RA. Full RA activity was restored by ligating the −1588 to −1075 bp region to the −290 bp S14 promoter element. These results show that the boundaries for both the RA and glucocorticoid response units are within the same 513-bp fragment extending from −1588 to −1075 bp. These studies rule out the S14 TRE as a target for RA action.

VI. ROLE OF TISSUE-SPECIFIC FACTORS IN RA AND DEX CONTROL OF S14 GENE EXPRESSION

In preadipocytes, the S14 gene is quiescent and unresponsive to both DEX and RA treatment.[33,73] In contrast to genes like FAS and GPD,[79-82] S14 gene expression does not rise when cells differentiate from preadipocytes to adipocytes. However, once cells express the adipocyte phenotype, S14 gene transcription can be activated by both RA and DEX. Thus, S14 gene expression is dependent on both adipocyte-specific factors and hormone regulated factors.[33,73] This pattern of control can be due to: (1) tissue-specific regulation of RA and glucocorticoid regulatory networks; (2) regulation of key tissue-specific accessory proteins or transcription factors; or (3) a combination of these factors.

In order to define the molecular basis for the tissue-specific hormonal regulation of S14 gene transcription, we determined whether the glucocor-

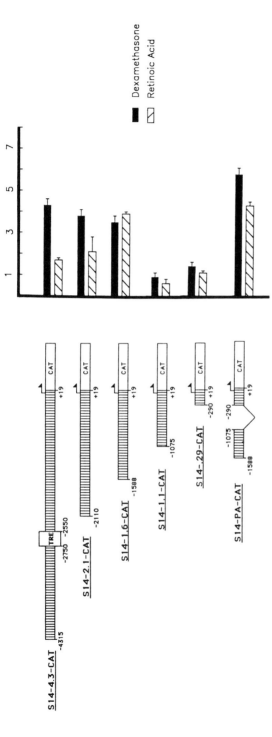

FIGURE 3. Analysis of retinoic acid and glucocorticoid induction of chloramphenicol acetyltransferase (CAT) expression in various deletion mutations of the S14 promoter. 3T3-F442A fibroblasts were stably transfected with plasmids containing various 5' deletions of the S14 promoter. The location of the TRE in pS14-4.3-CAT is between −2750 and −2550 bp. Stably transfected cells were differentiated to adipocytes and treated with 1 μ*M* RA or DEX for 72 h. CAT activity was assayed and results are expressed as fold induction, mean ± SD. *N* = 3 to 6.

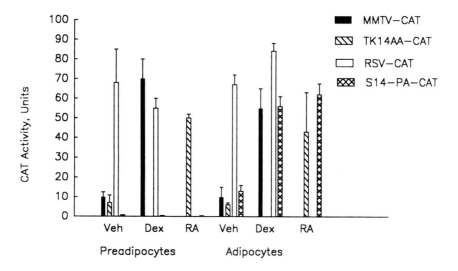

FIGURE 4. Tissue-specific expression of S14-CAT in 3T3-F442A cells. 3T3-F442A pre-adipocytes were stably transfected with either MAM-Neo-CAT (Clontech), TK14AA-CAT (generously provided by R. Koenig, University of Michigan), RSV-CAT (generously provided by S. Conrad, Michigan State Univ.), or S14-PA-CAT. Stably transfected preadipocytes and adipocytes were treated with vehicle, 1 μM DEX, or 1 μM RA for 72 h. Cells were harvested and assayed for CAT activity. Results are expressed as mean ± SE (N = 4) CAT units (pmoles of acetylated chloramphenicol formed per 100 μg protein per h).

ticoid regulatory network was operative in F442A preadipocytes. Accordingly, preadipocytes were transfected with pMAM-neo-CAT, a plasmid that contains the murine mammary tumor virus long terminal repeat (MMTV-LTR) ligated upstream from CAT. The MMTV-LTR contains glucocorticoid response elements (GRE) which bind ligand activated GRs and augment transcription from the *cis*-linked CAT gene.[101] Both preadipocytes and adipocytes were treated with vehicle or DEX (1 μM) for 72 h. The basal levels of expression, i.e., 10 units CAT, was not affected by adipocyte differentiation (Figure 4). DEX treatment increased CAT activity sevenfold in each phenotype indicating that GRs were functional in both preadipocytes and adipocytes.

In order to determine whether the RA regulatory network was operative in both phenotypes, F442A preadipocytes were stably transfected with the TK14AA-CAT plasmid. The TK14AA-CAT plasmid contains a modified version of the rat pituitary GH TRE ligated upstream from the thymidine kinase promoter and the CAT reporter gene.[102] The GH TRE binds RRs and functions as a RARE.[13,58,62,63] Basal expression of TK14AA-CAT (i.e., 8 units), was not affected by differentiation. However, RA treatment for 72 h induced CAT activity fivefold in both preadipocytes and adipocytes (Figure 4). Thus, the RA regulatory network is functional in both phenotypes.

As a final control study, RSV-CAT was stably transfected into F442A preadipocytes and CAT activity measured in both preadipocyte and adipocyte phenotypes in the presence and absence of DEX. Expression of CAT from RSV-CAT remained unaffected by differentiation status and hormone treatment (Figure 4). Thus, the effects of DEX and RA on CAT expression from the MAM-neo-CAT and TK14AA-CAT plasmids were due to transcriptional activation of the CAT gene and not to effects on enzyme stability or other generalized mechanisms.

In F442A adipocytes, the S14-PA-CAT plasmid is responsive to both RA and DEX (Figure 3). This plasmid contains a hormone responsive region (-1588 to -1075 bp) ligated upstream from the S14 promoter (-290 to $+19$ bp). Expression of S14-PA-CAT in preadipocytes was low (0.5 units) and remained unaffected by either DEX or RA treatment (Figure 4). Differentiation of F442A cells to adipocytes augmented the basal level of CAT activity tenfold and treatment with either RA or DEX enhanced CAT activity fourfold. Essentially the same results were obtained when L1 cells transfected with the various plasmids described above. The lack of hormonal control of S14 gene expression in F442A and L1 preadipocytes cannot be attributed to the absence of intact RA and glucocorticoid regulatory networks. Acquisition of hormone responsiveness of the S14 gene in both F442A and L1 cells requires the expression of adipocyte-specific factors that accompany differentiation of cells to adipocytes.

Analysis of the S14 promoter (-290 to $+19$ bp) shows that it functions as a weak promoter (i.e., <0.01 units) in both phenotypes and in either the absence or presence of hormones. While additional studies are still required, we do not view the S14 promoter as a primary target for tissue-specific control of S14 gene expression. The region extending from -1588 to -1075 bp is the most likely target for both hormonal and tissue-specific control. To examine this possibility, DNase I footprint analysis was used to determine whether changes in DNA-protein interaction occurred within this region during adipocyte differentiation.

Figure 5 illustrates the pattern of DNase I footprints in the -1588 to -1435 bp region. Two protected regions were detected at -1554 to -1530 bp (region 1) and -1477 and -1457 bp (region 2). Conversion of preadipocytes to adipocytes leads to greater protection of region 1 and less protection of region 2. This indicates that proteins which bind to these regions are under control during adipocyte differentiation. This analysis was expanded to other regions and is summarized in Figure 5. Four protected regions were detected within the 513 bp tissue-specific enhancer. Proteins binding to regions 1 and 4 are induced following differentiation of preadipocytes to adipocytes, while proteins interacting with region 2 are suppressed. Proteins interacting with region 3 remain unaffected by the differentiation status of the cells and are referred to as constitutive factors (CF1). These preliminary studies suggest that the adipocyte-specific hormonal control of S14 gene transcription is due to a change in nuclear proteins which interact with the S14 tissue-specific enhancer.

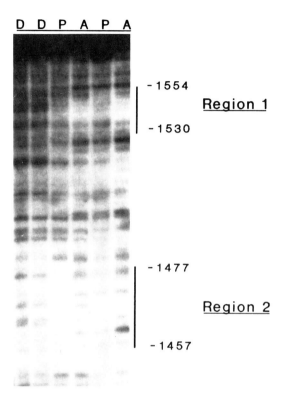

FIGURE 5.   DNase I footprint analysis of the S14 enhancer region. A DNA fragment extending from −1380 to −1588 bp relative to the S14 gene transcription start site was isolated. The noncoding strand was ^{32}P-labeled with T_4 polynucleotide kinase and used for DNase I footprint analysis.[103] In lanes labeled ''D'', the ^{32}P-labeled DNA was treated with dimethyl sulfate to generate a G-ladder. In lanes labeled ''P'' and ''A'', the ^{32}P-DNA was mixed with nuclear proteins (30 μg) extracted from L1 preadipocyte and L1 adipocyte nuclei, respectively. Samples were digested with DNase I; the ^{32}P-DNA was extracted and electrophoresed on a 8% acrylamide sequencing gel, dried, and exposed to X-ray film.

VII. MODEL FOR THE TISSUE-SPECIFIC, HORMONE, AND NUTRIENT CONTROLLED EXPRESSION OF THE S14 GENE IN CULTURED ADIPOCYTES

Figure 6 illustrates our current view of the arrangement of factors which regulate S14 gene transcription in cultured adipocytes. Both *in vivo* and *in vitro* studies show that transcription of the S14 gene is subject to complex, multifactorial control. Our studies have defined three regulatory regions located upstream from the 5′ end of the S14 gene. They include: (1) the promoter, extending from −290 to −1 bp which functions in the organization of the transcription initiation complex;[103] (2) the tissue-specific enhancer (TSE)

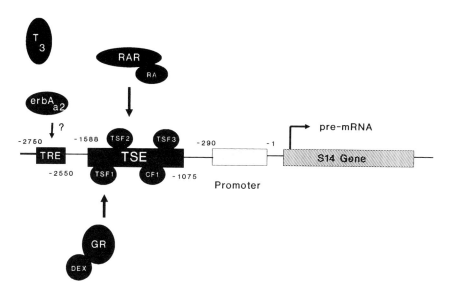

FIGURE 6. Organization of functional elements controlling S14 gene transcription in adipocytes. The diagram illustrates the S14 gene, transcription start site (horizontal arrow), the promoter region (-290 to -1 bp), the tissue specific enhancer (TSE) located between -1588 to -1075 bp, and the S14 thyroid hormone response element (TRE) between -2750 to -2550 bp. Ligand activated RR (RAR-RA) and glucocorticoid receptors (GR-DEX), T_3, and the c-*erb*Aα2 are illustrated. Tissue-specific factors (TSF1, TSF2, TSF3) and constitutive factor (GF1) are shown bound to the TSE.

extending from -1588 to -1075 bp, which also functions as a ligand-activated enhancer; and (3) the TRE extending from -2750 to -2550 bp which functions as a target for T_3 control in rat liver,[68,69,100] but not cultured adipocytes.

In the absence of enhancers, the S14 promoter is a weak promoter (<0.01 units CAT) that is not regulated by tissue-specific factors or hormones. The TSE functions both to increase the basal rate of S14 gene transcription in adipocytes and to confer responsiveness to both RA and DEX. Thus, the TSE appears to be a major functional unit controlling both tissue-specific and hormonal regulation of S14 gene transcription in adipocytes.

The TSE maps at the same location as a DNase I hypersensitive site (at -1525 to -1335 bp) identified as Hss-2 in rat liver.[67-70] DNase I hypersensitive sites typically mark the location of prospective *cis*-acting elements.[104] Chromatin structure studies showed that the organization of this region was regulated by tissue-specific factors[66] and by factors regulated by dietary carbohydrate.[69] This dietary carbohydrate control of S14 gene transcription requires insulin, but is inhibited by A-kinase regulated mechanisms.[70] Transfection analysis of cultured hepatocytes indicated that this region was a target for insulin-dependent, carbohydrate-regulated transcription factors.[105]

TABLE 4
Changes in DNA-Protein Interaction within the S14
Tissue-Specific Enhancer during Adipocyte
Differentiation

Region	Location[a]	Binding activity	Identification
1	−1554 to −1530	Increase	TSF1
2	−1477 to −1457	Decrease	TSF2
3	−1372 to −1352	No change	CF1
4	−1117 to −1094	Increase	TSF3

Note: TSF, Tissue-specific factors; CF1, Constitutive factor.

[a] Location of protected regions relative to the 5′ end of the S14 gene.

These diverse approaches indicate that in different cells, such as hepatocytes and adipocytes, the TSE region may be regulated by different regulatory networks. For example, we have found that the pattern of DNase I protection using liver and adipocyte nuclear extracts was not identical (not shown). Moreover, the TSE region functioned in the insulin/carbohydrate control of S14 gene expression in liver.[105] However, no such control has been detected in adipocytes (not shown). Alternatively, certain core factors which interact with the TSE may be required for this element to function in all tissues in which the S14 gene is expressed. S14 gene expression is regulated by glucocorticoids in liver, WAT, and F442A and L1 adipocytes.[73] Therefore, the glucocorticoid-regulated element(s) within the TSE may be a common functional target in all tissues where the S14 gene is expressed.

Perhaps the most interesting finding of this analysis was that neither DEX nor RA regulate S14 gene transcription in preadipocytes, despite the presence of functional RA and glucocorticoid regulatory networks. We attribute this to the differentiation-dependent expression of *trans*-acting factors which interact with the TSE. Tissue-specific factors which interact with regions 1, 2, and 4 and constitutive factors which interact with region 3 are likely to be involved in the tissue-specific control of S14 gene transcription in adipocytes. However, we do not know the identity of the proteins which interact with these regions or the precise role they play in regulating S14 gene transcription. Others have reported that AP-1 binding proteins, i.e., c-Fos/c-Jun[106,107] and C/EBP function in adipocyte-specific control of gene transcription.[108-110] We have no evidence for a direct interaction of these proteins with elements within the S14 TSE. Moreover, we have no evidence for direct interaction of RARα or GR with the TSE. DNA sequence analysis has not revealed canonical response elements for either RARE, RXRE, or GRE within the −1588 to −1075 bp region. It remains possible that neither RR or GR bind directly to the S14 DNA and that these regulatory networks control S14 gene transcription through the modulation of other transcription factors which interact with the

TSE. Clearly, additional studies will be required to decipher the molecular basis for the multifactorial control of S14 gene transcription in cultured adipocytes.

In contrast to rat liver, the S14 TRE does not function as a target for T_3-activated TR in F442A cells. Moreover, this region is not a target for RA control in F442A or L1 cells. However, in a tissue like liver or WAT, where both RR and TR receptors are present and the S14 gene is active, both T_3 and RA might regulate S14 gene transcription. This may be mediated by independent interaction of TR and RR with their corresponding response regions. Alternatively, TR and RR might form heterodimers and interact with one or both of the specific response elements. Under these conditions, T_3 and RA might antagonize or augment each other's action.

However, in F442A adipocytes, the low T_3 binding capacity would exclude T_3 as a potent inducer of S14 gene transcription. On the other hand, TRs are expressed at high levels in L1 adipocytes and these cells are responsive to T_3. Preliminary transfection analysis of L1 cells with S14-CAT fusion genes shows that the S14 TRE is not the target for T_3 action (not shown). This raises the possibility that high levels of c-*erb*Aα2 expression in both F442A and L1 cells might block TRα1 action at the S14 TRE. Alternatively, TRβ1 is not expressed in either F442A or L1 cell types, but is expressed in liver. The S14 TRE might be a target for TRβ1 and not TRα1 or c-*erb*Aα2. These issues remain unresolved and require further investigation.

What is clear from these studies is that the complex multifactorial control of S14 gene transcription can be explained, at least in part, by interaction of adipocyte specific *trans*-acting factors with defined *cis*-linked elements. However, these elements are not clustered near the 5' end of the S14 gene, but are dispersed over a 2.7 kb region. In fact, two spatially distinct ligand activated enhancers, i.e., TSE and TRE, have been defined. For these distal elements to function in the initiation of S14 gene transcription they must interact with the promoter. Regulating the interaction of these distal elements with promoter elements may represent another important step by which hormones, nutrients, and tissue-specific factors control the rate of initiation of S14 gene transcription.

VIII. SUMMARY: DOES RA FUNCTION IN INTERMEDIARY METABOLISM?

While RA action on cell growth and differentiation[34-36,54-60] has been well characterized, several recent reports show that RA regulates the level of mRNAs coding for proteins involved in intermediary metabolism. These proteins include phosphoenolpyruvate carboxykinase,[29] the S14 protein,[33] transglutaminase,[27] GPD,[33] alcohol dehydrogenase,[31] and $G_s\alpha$.[30] Although these findings are based on cell culture studies, they show that RA is a regulator factor in intermediary metabolism which must also function *in vivo*. The role of RA in intermediary metabolism is supported further by the presence of RA (as both 13-*cis*-RA and all-*trans*-RA) in plasma at 1 to 3 ng/ml or 3 to 10 nM[111,112] and by the wide tissue distribution of RR, particularly in lipogenic tissues such as liver and WAT. In addition to providing new insight into the interaction of the RA regulatory system with tissue-specific and glucocorticoid-regulated factors, the adipocyte cell culture model provides a means to investigate the metabolic consequences of RA action on adipocyte metabolism.

ACKNOWLEDGMENTS

The authors wish to thank Drs. Maija Zile and Henry Bayley for many helpful suggestions during the preparation of this manuscript. We also wish to thank Dr. P. Chambon, B. Spiegelman, H. Towle, L. Kedes, H-S. Sul, R. Koenig, and S. Conrad for the various cDNAs, expression vectors, and adipocyte cell lines used in these studies. This research was supported by grants from the National Institutes of Health (GM36851, DK43220), American Diabetes Association, The Upjohn Company, and Michigan State University.

REFERENCES

1. **Wolbach, S. B.**, Effects of vitamin A deficiency and hypervitaminosis A in animals, in *The Vitamins*, Vol. 1, Sebrell, W. B., Jr. and Harris, R. S., Eds., Academic Press, New York, 1954, 106.
2. **Ganguly, J., Rao, M. R. S., Murthy, S. K., and Sarada, K.,** Systemic mode of action of vitamin A, in *Vitamins and Hormones,* Vol. 38, Munson, P. L., Diczfalusy, E., Glover, J., and Olson, R. E., Eds., 1980, 1.
3. **Roels, O. A.,** Biochemical systems, in *The Vitamins*, 2nd ed., Vol. 1, Sebrell, W. H., Jr. and Harris, R. S., Eds., Academic Press, New York, 1967, 245.
4. **Wolf, G.,** Multiple functions of Vitamin A, *Physiol. Rev.,* 64, 873, 1984.
5. **Wald, G.,** The visual function of the Vitamins A, *Vitam. Horm.,* 18, 417, 1960.
6. **Napoli, J. L.,** The biogenesis of retinoic acid: a physiologically significant promoter of differentiation, in *Chemistry and Biology of Synthetic Retinoids,* Dawson, M. I. and Okamura, W. H., Eds., CRC Press, Boca Raton, FL, 229, 1990.
7. **Sani, B. B.,** Cellular retinoic acid binding proteins and action of retinoic acid, in *Chemistry and Biology of Synthetic Retinoids,* Dawson, M. I. and Okamura, W. H., Eds., CRC Press, Boca Raton, FL, 1990, 365.
8. **Wolf, G.,** Recent progress in vitamin A research: nuclear retinoic acid receptors and their interaction with gene elements, *J. Nutr. Biochem.,* 1, 284, 1991.
9. **Chytil, F.,** Retinoic acid: biochemistry and metabolism, *J. Am. Acad. Dermatol.,* 15, 741, 1986.
10. **Melton, D. A.,** Pattern formation during animal development, *Science,* 252, 234, 1991.
11. **Blomhoff, R., Green, M. H., Berg, T., and Norum, K. R.,** Transport and storage of vitamin A, *Science,* 250, 399, 1990.

12. **Morita, S., Fernadez-Mejia, C., and Melmed, S.,** Retinoic acid selectively stimulates growth hormone secretion and messenger ribonucleic acid levels in rat pituitary cells, *Endocrinology,* 124, 2052, 1989.

13. **Bedo, G., Santisteban, P., and Aranda, A.,** Retinoic acid regulates growth hormone gene expression, *Nature (London),* 339, 231, 1989.

14. **Glick, A. B., Flanders, K. C., Danielpour, D., Yuspa, S. H., and Sporn, M. B.,** Retinoic acid induces transforming growth factor β2 in cultured keratinocytes and mouse epidermis, *Cell Regul.,* 1, 87, 1989.

15. **Petkovich, P. M., Heersche, J. N. M., Tinker, D. O., and Jones, G.,** Retinoic acid stimulates 1,25-dihydroxyvitamin D_3 binding in rat osteosarcoma cells, *J. Biol. Chem.,* 259, 8274, 1984.

16. **Sucov, H. M., Murakami, K. K., and Evans, R. M.,** Characterization of an auto-regulated response element in the mouse RARβ gene, *Proc. Natl. Acad. Sci. U.S.A.,* 87, 5392, 1990.

17. **de The, H., Marchio, A., Tiollais, P., and Dejean, A.,** Differential expression and ligand regulation of the RAR α and β genes, *EMBO J.,* 8, 429, 1989.

18. **Lafyatis, R., Kim, S-J., Angel, P., Roberts, A. B., Sporn, M. B., Karin, M., and Wilder, R.,** Interleukin-1 stimulates and all-trans retinoic acid inhibits collagenase gene expression through its 5′ activator protein-1-binding site, *Mol. Endocrinol.,* 4, 973, 1990.

19. **Clarke, C. L., Roman, S. D., Graham, J., Koga, M., and Sutherland, R. L.,** Progesterone receptors regulated by retinoic acid in human breast cancer cell line T-47D, *J. Biol. Chem.,* 265, 12694, 1990.

20. **Suva, L. J., Ernst, M., and Rodan, G. A.,** Retinoic acid induces zif268 early gene expression in rat preosteoblastic cells, *Mol. Cell. Biol.,* 11, 2503, 1991.

21. **Luscher, B., Mitchell, P., Williams, T., and Tjian, R.,** Regulation of transcription factor AP-2 by the morphogen retinoic acid and second messengers, *Genes Dev.,* 3, 1507, 1989.

22. **Chakraborty, A. K., Orlo, S. J., and Pawelek, J. M.,** Stimulation of the receptor for melanocyte stimulating hormone by retinoic acid, *FEBS,* 276, 205, 1989.

23. **Sidel, N., Taga, T., Hirano, T., Kishimoto, T., and Saxon, A.,** Retinoic acid induced growth inhibition of a human myeloma cell line via down-regulation of IL-6 receptors, *J. Immunol.,* 146, 3809, 1991.

24. **Sidell, N. and Ramsdell, F. R.,** Retinoic acid upregulates interleukin-2 receptors on activated human thymocytes, *Cell. Immunol.,* 115, 299, 1988.

25. **Junquero, D., Modat, G., Coquelet, C., and Bonne, C.,** Retinoids enhance the number of EGF receptors in corneal endothelial cells, *Exp. Eye Res.,* 51, 49, 1990.

26. **Ubels, J. L., Iorfino, A., and O'Brien, W. J.,** Retinoic acid decreases the number of EGF receptors in corneal epithelium and Chang conjunctival cells, *Exp. Eye Res.,* 52, 763, 1991.

27. **Chiocca, E. A., Davies, P. J. A., and Stein, J. P.,** The molecular basis of retinoic acid action, *J. Biol. Chem.,* 263, 11584, 1988.

28. **Imai, Y., Rodan, S. B., and Rodan, G. A.,** Effects of retinoic acid on alkaline phosphatase mRNA, catecholamine receptors and G-proteins in ROS 17/2.8 cells, *Endocrinology,* 122, 456, 1988.

29. **Lucas, P. C., O'Brien, R. M., Mitchell, J. A., Davis, C. M., Imai, E., Forman, B. M., Samuels, H. H., and Granner, D. K.,** A retinoic acid response element is part of a pleiotropic domain in the PepCK gene, *Proc. Natl. Acad. Sci. U.S.A.,* 88, 2184, 1991.

30. **Chan, S. D. H., Strewler, G. J., and Nissensa, R. A.,** Transcription activation of Gsα gene expression by retinoic acid and parathyroid hormone-related protein in F9 terato-carcinoma cells, *J. Biol. Chem.,* 265, 20081, 1990.

31. **Duester, G., Shear, M. L., McBride, M. S., and Stewart, M. J.,** Retinoic acid response element in human alcohol dehydrogenase ADH3: Implication for regulation of retinoic acid synthesis, *Mol. Cell. Biol.,* 11, 1638, 1991.

32. **Darrow, L., Riches, R. J., Pecorino, L. T., and Strickland, S.,** Transcription factor SP1 is important for retinoic acid induced expression of tissue plasminogen activator gene during F9 teratocarcinoma cell differentiation, *Mol. Cell. Biol.,* 10, 5883, 1990.

33. **Lepar, G. J. and Jump, D. B.,** Retinoic acid and dexamethasone interact to regulate S14 gene transcription in 3T3-F442A adipocytes, *Mol. Cell. Endocrinology,* 77, 84, 65, 1992.

34. **Strickland, S., Smith, K. K., and Marotti, K. K.,** Hormonal induction of differentiation in teratocarcinoma stem cells: generation of parietal endoderm by retinoic acid and db-cAMP, *Cell,* 21, 347, 1980.

35. **Wang, S. Y. and Gudas, L. J.,** Isolation of cDNA clones specific for collagen IV and laminin from mouse teratocarcinoma, *Proc. Natl. Acad. Sci. U.S.A.,* 80, 5880, 1983.

36. **Vasios, G. W., Gold, J. D., Petkovich, M., Chambon, P., and Gudas, L. J.,** A retinoic acid response element is present in the 5' flanking region of the laminin β1 gene, *Proc. Natl. Acad. Sci. U.S.A.,* 86, 9099, 1989.

37. **Wei, L-N., Blaner, W. S., Goodman, D. S., and Nguyen-Huu, C. M.,** Regulation of cellular retinoid-binding proteins and their mRNAs during embryonal carcinoma cell differentiation induced by retinoic acid, *Mol. Endocrinol.,* 3, 454, 1989.

38. **Munoz-Canoves, P., Vik, D. P., and Tack, B. F.,** Mapping a retinoic acid responsive element in the promoter region of the complement factor H gene, *J. Biol. Chem.,* 265, 20065, 1990.

39. **Antras, J., Lasnier, F., and Pairault, J.,** Adipsin gene expression in 3T3-F442A adipocytes is post-transcriptionally down-regulated by retinoic acid, *J. Biol. Chem.,* 266, 1157, 1991.

40. **Stellmach, V., Leask, A., and Fuchs, E.,** Retinoid mediated transcriptional regulation of keratin genes in human epidermal and squamous carcinoma cells, *Proc. Natl. Acad. Sci. U.S.A.,* 88, 4582, 1991.

41. **Hohl, D., Lichti, U., Breitkreutz, D., Steinert, P. M., and Roop, D. R.,** Transcription of the human loricrin gene in vitro is induced by calcium and cell density and suppressed by retinoic acid, *J. Invest. Dermatol.,* 96, 414, 1991.

42. **Rottman, J. N., Widom, R. L., Nadal-Ginard, B., Mahdavi, V., and Karathanasis, S. K.,** A retinoic acid response element in the apolipoprotein A1 gene distinguishes between two different retinoic acid pathways, *Mol. Cell Biol.,* 11, 3814, 1991.

43. **Scheibe, R. J., Ginty, D. D., and Wagner, J. A.,** Retinoic acid stimulates differentiation of PC12 cell that are deficient in cAMP-dependent protein kinase A, *J. Cell Biol.,* 113, 1173, 1991.

44. **Pacifici, M., Golden, E. B., Iwamoto, M., and Adams, S. L.,** Retinoic acid treatment induces type X collagen gene expression in cultured chick chondrocytes, *Exp. Cell Res.,* 195, 38, 1991.

45. **Glover, A., Oshima, R. G., and Adamson, E. D.,** Epithelial layer formation in differentiation aggregates of F9 embryonal carcinoma cells, *J. Cell Biol.,* 96, 1690, 1983.

46. **Linder, S. K., Krondahl, U., Sennerstrom, R., and Ringerty, N. R.,** Retinoic acid induced differentiation of F9 embryonal carcinoma cells, *Exp. Cell Res.,* 132, 453, 1981.

47. **Lehlonen, E., Lehto, V. P., Badley, R. A., and Virtanen, I.,** Formation of vinculin plaques precedes other cytoskeletal changes during retinoic acid induced teratocarcinoma cells differentiation, *Exp. Cell Res.,* 144, 191, 1983.

48. **Petkovich, M., Brand, N. J., Krust, A., and Chambon, P.,** A human retinoic acid receptor which belongs to the family of nuclear receptors, *Nature (London),* 330, 444, 1987.

49. **Giguere, V., Ong, E. S., Segui, P., and Evans, R. M.,** Identification of a receptor for the morphogen retinoic acid, *Nature (London),* 330, 624, 1987.

50. **Brand, N., Petkovich, M., Krust, A., Chambon, P., de The, H., Marchio, A., Tiollais, P., and Dejean, A.,** Identification of a second human retinoic acid receptor, *Nature (London),* 332, 850, 1988.

51. **Evans, R. M.,** The steroid and thyroid hormone receptor superfamily, *Science,* 240, 889, 1988.
52. **Benbrook, D., Lenhardt, E., and Pfahl, M.,** A new retinoic acid receptor identified from a hepatocellular carcinoma, *Nature (London),* 333, 669, 1988.
53. **Zelent, A., Krust, A., Petkovich, M., Kastner, P., and Chambon, P.,** Cloning of murine α and β retinoic acid receptors and a novel γ receptor predominantly expressed in skin, *Nature (London),* 339, 714, 1989.
54. **Mangelsdorf, D. J., Ong, E. S., Dyck, J. A., and Evans, R. M.,** Nuclear receptor that identifies a novel retinoic acid response pathway, *Nature (London),* 345, 224, 1990.
55. **DeLuca, L. M.,** Retinoids and their receptors in differentiation, embryogenesis and differentiation, *FASEB J.,* 5, 2924, 1991.
56. **Haq, R. U. and Chytil, R.,** Expression of nuclear retinoic acid receptors in rat adipose tissue, *Biochem. Biophys. Res. Comm.,* 176, 1539, 1991.
57. **Dalman, F. C., Sturzenbecker, L. J., Levin, A. A., Lucas, D. A., Perdew, G. W., Petkovich, M., Chambon, P., Grippo, J. F., and Pratt, W. B.,** Retinoic acid receptor belongs to a subclass of nuclear receptors that do not form docking complexes with hsp90, *Biochemistry,* 30, 5605, 1991.
58. **Umesono, K., Giguere, V., Glass, C. K., Rosenfeld, M. G., and Evans, R. M.,** Retinoic acid and thyroid hormone induce gene expression through a common responsive element, *Nature (London),* 336, 262, 1988.
59. **Umesono, K., Murakami, K. K., Thompson, C. C., and Evans, R. M.,** Direct repeats as selective response elements for the thyroid hormone, retinoic acid and vitamin D_3 receptors, *Cell,* 65, 1255, 1991.
60. **Mangelsdorf, D. J., Umesono, K., Kliewer, S. A., Borgmeyer, U., Ong, E. S., and Evans, R. M.,** A direct repeat in the cellular retinol-binding protein type II gene confers differential regulation by RXR and RAR, *Cell,* 66, 555, 1991.
61. **Naar, A. M., Boutin, J.-M., Lipkin, S. M., Yu, V. C., Holloway, J. M., Glass, C. K., and Rosenfeld, M. G.,** The orientation and spacing of core DNA-binding motifs dictate selective transcriptional responses to three nuclear receptors, *Cell,* 65, 1267, 1991.
62. **Glass, C. K., Lipkin, S. M., Devary, O. V., and Rosenfeld, M. G.,** Positive and negative regulation of gene transcription by a retinoic acid-thyroid hormone receptor heterodimer, *Cell,* 59, 697, 1989.
63. **Forman, B. M., Yang, C.-R., Au, M., Casanova, J., Ghysdael, J., and Samuels, H. H.,** A domain containing leucine-zipper-like motifs mediate novel in vivo interaction between the thyroid hormone and retinoic acid receptors, *Mol. Endocrinol.,* 3, 1610, 1989.
64. **Jump, D. B., Narayan, P., Towle, H. C., and Oppenheimer, J. H.,** Rapid effects of triiodothyronine on hepatic gene expression, *J. Biol. Chem.,* 259, 2789, 1984.
65. **Jump, D. B. and Oppenheimer, J. H.,** High basal expression and 3,5, 3'-triiodothyronine regulation of mRNA$_{S14}$ in lipogenic tissues, *Endocrinology,* 117, 2259, 1985.
66. **Jump, D. B., Wong, N. C. W., and Oppenheimer, J. H.,** Chromatin structure and methylation state of a thyroid hormone responsive gene in rat liver, *J. Biol. Chem.,* 262, 778, 1985.
67. **Jump, D. B.,** Thyroid hormone regulation of rat liver S14 gene expression, in *Gene Regulation by Steroids Hormones IV,* Roy, A. K. and Clarke, J. H., Eds., Springer-Verlag, New York, 1989, 144.
68. **Jump, D. B.,** Rapid induction of rat liver S14 gene transcription by thyroid hormone, *J. Biol. Chem.,* 264, 4698, 1989.
69. **Jump, D. B., Bell, A., and Santiago, V.,** Thyroid hormone and dietary carbohydrate interact to regulate rat liver S14 gene transcription and chromatin structure, *J. Biol. Chem.,* 265, 3474, 1990.
70. **Jump, D. B., Bell, A., Lepar, G., and Hu, D.,** Insulin rapidly induces rat liver S14 gene transcription, *Mol. Endocrinol.,* 4, 1655, 1990.

71. **Oppenheimer, J. H., Schwartz, H. L., Mariash, C. N., Kinlaw, W. B., Wong, N. C. W., and Freake, H. C.,** Advances in our understanding of thyroid hormone action at the cellular level, *Endocrine Rev.,* 8, 288, 1987.

72. **Kinlaw, W. B., Ling, N. C., and Oppenheimer, J. H.,** Identification of rat S14 protein and comparison of its regulation with that of $mRNA_{S14}$ employing synthetic peptide antisera, *J. Biol. Chem.,* 264, 19779, 1989.

73. **Lepar, G. J. and Jump, D. B.,** Hormonal regulation of the S14 gene in 3T3-F442A cells, *Mol. Endocrinol.,* 3, 1207, 1989.

74. **Clarke, S. D., Armstrong, M. K., and Jump, D. B.,** Nutritional control of rat liver fatty acid synthase and S14 mRNA abundance, *J. Nutr.,* 120, 218, 1990.

75. **Clarke, S. D., Armstrong, M. K., and Jump, D. B.,** Dietary polyunsaturated fats uniquely suppress rat liver fatty acid synthase and S14 mRNA content, *J. Nutr.,* 120, 225, 1990.

76. **Blake, W. L. and Clarke, S. D.,** Suppression of rat hepatic fatty acid synthase and S14 gene transcription by dietary polyunsaturated fat, *J. Nutr.,* 120, 1727, 1990.

77. **Jump, D. B., Veit, A. M., Santiago, V. S., Lepar, G., and Herberholz, L.,** Transcriptional activation of the rat liver S14 gene during post-natal development, *J. Biol. Chem.,* 263, 7254, 1988.

78. **Goodridge, A. G.,** Dietary regulation of gene expression: enzymes involved in carbohydrate and lipid metabolism, *Annu. Rev. Nutr.,* 7, 157, 1987.

79. **Green, H. and Kehinde, O.,** Sublines of mouse 3T3 cells that accumulate lipid, *Cell,* 1, 113, 1974.

80. **Spiegelman, B. M., Frank, M., and Green, H.,** Molecular cloning of mRNA from 3T3 adipocytes, *J. Biol. Chem.,* 258, 10083, 1983.

81. **Bernlohr, D. A., Bolanowski, M. A., Kelly, T. J., Jr., and Lane, M. D.,** Evidence for an increase in transcription of specific mRNA during differentiation of 3T3-L1 pre-adipocytes, *J. Biol. Chem.,* 260, 5563, 1985.

82. **Wise, L. S., Sul, H. S., and Rubin, C. S.,** Coordinate regulation of the biosynthesis of ATP-citrate lyase and malic enzyme during adipocyte differentiation, *J. Biol. Chem.,* 259, 4827, 1984.

83. **Oppenheimer, J. H., Schwartz, H. L., and Surks, M. I.,** Tissue differences in the concentration of T_3 nuclear binding sites in the rat liver, kidney, pituitary, heart, brain, spleen and testes, *Endocrinology,* 95, 897, 1974.

84. **Buergi, U. and Abuehl, U.,** Nuclear triiodothyronine binding sites in rat adipocytes, *Metabolism,* 33, 326, 1984.

85. **Hausdorf, S., Clement, J., and Loos, U.,** Expression of S14 gene during differentiation of 3T3-L1 cells, *Horm. Metab. Res.,* 20, 723, 1988.

86. **Moustaid, N. and Sul, H. S.,** Regulation of expression of the fatty acid synthase gene in 3T3-L1 cells by differentiation and triiodothyronine, *J. Biol. Chem.,* 266, 18550, 1991.

87. **Murray, M. B., Zilz, N., McCreary, N. L., MacDonald, M. J., and Towle, H. C.,** Isolation and characterization of rat cDNA clones for two distinct thyroid hormone receptors, *J. Biol. Chem.,* 263, 12770, 1988.

88. **Hodin, R. A., Lazar, M. A., Wintman, B. I., Darling, D. S., Koenig, R. J., Larsen, P. R., Moore, D. D., and Chin, W. W.,** Identification of a thyroid hormone receptor that is pituitary specific, *Science,* 244, 76, 1989.

89. **Izumo, S. and Madhavi, V.,** Thyroid hormone receptor isoforms generated by alternative splicing differentially activate myosin HC gene transcription, *Nature (London),* 334, 539, 1988.

90. **Lazar, M. A., Hodin, R. A., Darling, D. S., and Chin, W. W.,** Identification of a rat c-erbA-related protein which binds deoxyribonucleic acid, but does not bind thyroid hormone, *Mol. Endocrinol.,* 2, 893, 1988.

91. **Mitsuhashi, T., Tennyson, G. E., and Nikodem, V. M.,** Alternative splicing generates messages encoding rat c-erbA proteins that do not bind thyroid hormone, *Proc. Natl. Acad. Sci. U.S.A.,* 85, 5804, 1988.

92. **Koenig, R. J., Lazar, M. A., Hodin, R. A., Brent, G. A., Larsen, P. R., Chin, W. W., and Moore, D. D.,** Inhibition of thyroid hormone action by a non-hormone binding c-erbA protein generated by alternative mRNA splicing, *Nature,* 337, 659, 1989.

93. **Cook, C. B. and Koenig, R. J.,** Expression of erbAα and β mRNAs in regions of adult rat brain, *Mol. Cell. Endocrinol.,* 70, 13, 1990.

94. **Kuri-Harcuch, W.,** Differentiation of 3T3-F442A cells into adipocytes is inhibited by retinoic acid, *Differentiation,* 23, 164, 1982.

95. **Castro-Munozledo, F., Marsch-Moreno, M., Beltran-Langarica, A., and Kuri-Harcuch, W.,** Commitment of adipocyte differentiation in 3T3 cells is inhibited by retinoic acid and the expression of lipogenic enzymes in modulated through cytoskeleton stabilization, *Differentiation,* 36, 211, 1987.

96. **Pairault, J. and Lasnier, F.,** Control of adipogenic differentiation of 3T3-F442A cells by retinoic acid, dexamethasone and insulin: a topographic analysis, *J. Cell Physiol.,* 132, 279, 1987.

97. **Stone, R. L. and Bernlohr, D. A.,** The molecular basis for inhibition of adipose conversion of murine 3T3-L1 cells by retinoic acid, *Differentiation,* 45, 119, 1990.

98. **Felgner, P. L., Gadek, T. R., Holm, M., Roman, R., Chan, H. W., Wenz, M., Northrop, J. P., Ringold, G. M., and Danielsen, M.,** Lipofection: a highly efficient, lipid-mediated DNA-transfection procedure, *Proc. Natl. Acad. Sci. U.S.A.,* 84, 7413, 1987.

99. **Gorman, C. M., Moffat, L. F., and Howard, B. H.,** Recombinant genomes which express chloramphenicol acetyltransferase in mammalian cells, *Mol. Cell Biol.,* 2, 1044, 1982.

100. **Zilz, N. D., Murray, M. B., and Towle, H. C.,** Identification of multiple thyroid hormone response elements located far upstream from the rat S14 promoter, *J. Biol. Chem.,* 265, 8136, 1990.

101. **Chandler, V. L., Maler, B. A., and Yamamoto, K. R.,** DNA sequences bound specifically by glucocorticoid receptor in vitro render a heterologous promoter hormone responsive in vivo, *Cell,* 33, 489, 1983.

102. **Brent, G., Larsen, P. R., Harney, J. W., Koenig, R. J., and Moore, D. D.,** Functional characterization of the rat growth hormone promoter elements required for induction by thyroid hormone with and without a co-tranfected β-type thyroid hormone receptor, *J. Biol. Chem.,* 264, 178, 1989.

103. **MacDougald, O. A. and Jump, D. B.,** Identification of functional cis-acting elements in the S14 promoter, *Biochem. J.,* 280, 761, 1991.

104. **Gross, D. S. and Garrard, W. T.,** Nuclease hypersensitive sites in chromatin, *Annu. Rev. Biochem.,* 57, 159, 1988.

105. **Jacoby, D. B., Zilz, N. D., and Towle, H. C.,** Sequences within the 5' flanking region of the S14 gene confer responsiveness to glucose in primary hepatocytes, *J. Biol. Chem.,* 264, 17623, 1989.

106. **Distel, R. J., Ro, H.-S., Rosen, B. S., Groves, D. L., and Spiegelman, B. M.,** Nucleoprotein complexes that regulate gene expression in adipocyte differentiation: direct participation of c-*fos*, *Cell,* 49, 835, 1987.

107. **Herrera, R. O., Robinson, H. S., Xanthopoulos, K. G., and Spiegelman, B. M.,** A direct role for c/EBP and AP-1 binding site in gene expression linked to adipocyte differentiation, *Mol. Cell Biol.,* 9, 5331, 1989.

108. **Christy, R. J., Yang, V. W., Ntambi, J. M., Geiman, D. E., Landschulz, W. H., Friedman, A. D., Nakabeppu, Y., Kelly, T. J., and Lane, M. D.,** Differentiation-induced gene expression in 3T3-L1 preadipocytes: CCAAT/enhancer binding protein interacts with and activates the promoters of adipocyte-specific genes, *Genes Dev.,* 3, 1323, 1989.

109. **Friedman, A. D., Landschulz, W. H., and McKnight, S. L.,** C/EBP activates the promoter of the serum albumin gene in cultured hepatoma cells, *Genes Dev.,* 3, 1314, 1989.
110. **McKnight, S. L., Lane, M. D., and Gluecksohn-Waelsch, S.,** Is CCAAT/enhancer binding protein a central regulatory of energy metabolism?, *Genes Dev.,* 3, 2021, 1989.
111. **Tang, G. and Russel, R. M.,** 13-cis-retinoic acid is an endogenous compound in human serum, *J. Lipids Res.,* 31, 175, 1990.
112. **Cullum, M. E. and Zile, M. H.,** Metabolism of all-trans-retinoic acid and all-trans-retinyl acetate. Demonstration of common physiological metabolites in rat small intestinal muscosa and circulation, *J. Biol. Chem.,* 260, 10590, 1985.

Chapter 20

THE ROLE OF VITAMIN K IN CLOTTING FACTOR BIOSYNTHESIS

J. W. Suttie

TABLE OF CONTENTS

0-8493-6961-4/93/$0.00 + $.50
© 1993 by CRC Press, Inc.

I. INTRODUCTION

Vitamin K was not discovered until the early 1930s when Dam[1] noted a hemorrhagic syndrome in chicks fed a lipid-free diet. This response could be cured by the addition of alfalfa meal to the diet or by the administration of a lipid extract of green plants. By 1939, a series of investigations led by Dam and his collaborators in Denmark, Almquist at Berkeley, and Doisy at St. Louis University had established that the form of the vitamin which is found in alfalfa, and now called vitamin K_1 or phylloquinone, was 2-methyl-3-phytyl-1,4-naphthoquinone. A large number of bacteria, including numerous faculative and obligatory anaerobes of the human lower bowel, synthesize a series of vitamers with unsaturated multiprenyl side chains at the 3-position. These were characterized soon after phylloquinone and were originally called vitamin K_2, but are now more correctly designated as menaquinones (Figure 1). This early history of the discovery and characterization of vitamin K has been adequately reviewed.[2-4]

The hemorrhagic condition that resulted from the dietary lack of vitamin K was shown by Dam et al.[5] to be associated with a decrease in plasma prothrombin activity, and, for some time, was thought to be due solely to a lowered concentration of prothrombin (factor II). Later, as they were discovered, it was shown that the synthesis of clotting factors VII, IX, and X, which are involved in the activation of prothrombin to thrombin, was also depressed in the deficient state.[4] These four plasma proteins were collectively called the "vitamin K-dependent clotting factors", and for years were thought to be the only proteins whose synthesis required the vitamin. It was not until the 1970s, after the biochemical role of the vitamin had been established, that two proteins now known to play an anticoagulant, rather than a procoagulant, role in hemostasis, protein C[6], and protein S[7] were discovered. The general role of the vitamin K-dependent procoagulants and anticoagulants in normal hemostasis is shown in Figure 2. Another vitamin K-dependent protein (protein Z) of unknown function is also found in plasma. A limited number of extrahepatic, vitamin K-dependent proteins, including osteocalcin or Bone Gla Protein (BGP) and Matrix Gla Protein (MGP), have been identified.[8] However, efforts to understand the enzymology of the vitamin K-dependent modification of proteins have concentrated on the activity expressed in hepatic tissue.

II. γ-CARBOXYGLUTAMIC ACID

Although vitamin K was known by the late 1930s to regulate the activity of a few specific proteins, the lack of a general understanding of the mechanism of protein biosynthesis prevented serious experimental approaches to the cellular and molecular mechanisms involved until the mid-1960s. By the early 1940s, the 4-hydroxy-coumarins were identified as indirect anticoagulants that functioned by antagonizing the action of vitamin K.[9] Studies

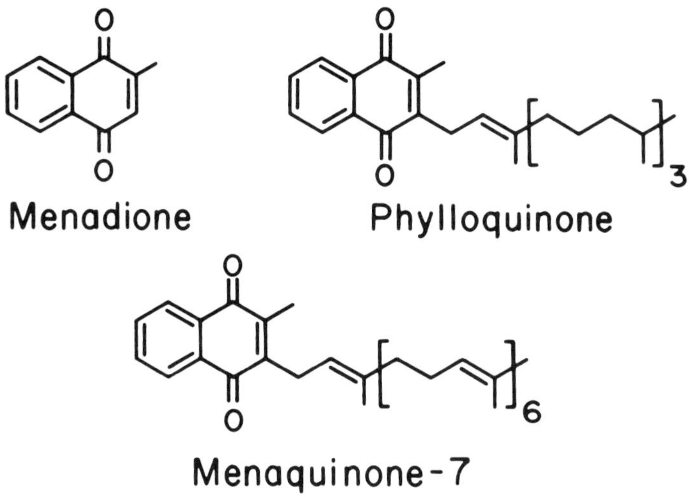

FIGURE 1. Biologically active forms of vitamin K. Phylloquinone, produced by plants, and menaquinone-7 (MK-7), an example of the long chain menaquinone produced by bacteria, are natural forms of the vitamin. Menadione is a synthetic compound which has biological activity after it is alkylated to MK-4 in the liver.

directed toward an understanding of the role of vitamin K concentrated on an understanding of the biosynthesis of prothrombin. These studies were aided by the ability to easily regulate the plasma activity of prothrombin by oral anticoagulant (usually warfarin) administration. Although an early hypothesis suggested that vitamin K regulated the expression of the prothrombin gene,[10] by the mid-1960s a number of animal model studies strongly suggested that vitamin K was involved in converting an intracellular inactive precursor of prothrombin to biologically active prothrombin.[11]

This hypothesis was strengthened by the observation of immunochemically similar, but biologically inactive, prothrombin molecules that increased in the plasma of anticoagulant-treated patients. Characterization of this "abnormal" or undercarboxylated prothrombin obtained from anticoagulant-treated bovine plasma revealed that its amino acid composition, amino and carboxyl terminal amino acids, and molecular weight appeared to be identical to those of prothrombin. This protein could be electrophoretically separated from normal prothrombin in the presence, but not the absence, of Ca^{2+}-containing buffers, and in contrast to normal vitamin K-dependent proteins, it was not adsorbed by insoluble barium salts. Calcium ions are required for normal prothrombin activation, and the abnormal prothrombin lacked the Ca^{2+} binding properties of normal prothrombin and did not demonstrate a Ca^{2+}-dependent association with phospholipid surfaces. This inability of the plasma-abnormal prothrombin to bind Ca^{2+} in solution suggested to investigators in

"Intrinsic Pathway"   "Extrinsic Pathway"

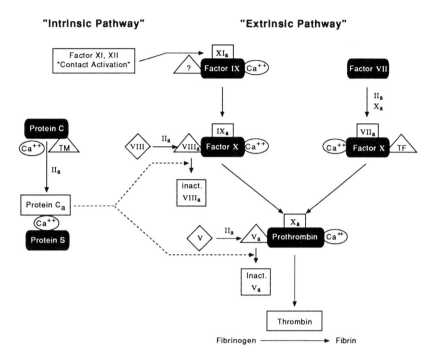

FIGURE 2. The role of vitamin K-dependent proteins in blood coagulation. The classical vitamin K-dependent clotting factors, II (prothrombin), VII, IX, and X, have procoagulant hemostatic roles and circulate as inactive zymogens which are converted to active serine proteases by specific peptide cleavage (e.g., X to Xa). Protein C and protein S are involved in the process by which the rapid generation of thrombin from prothrombin is brought under control by inactivation of accessory factors Va and VIIIa of the hemostatic system. The other clotting factors shown do not require vitamin K for their biosynthesis.

this field that the function of vitamin K was to attach a calcium-binding prosthetic group to a liver precursor of prothrombin.

Low molecular weight, calcium-binding peptides could be isolated from enzymatic digests of normal prothrombin, and these acidic peptides were subsequently shown by Stenflo[12] and Nelsestuen[13] to contain γ-carboxyglutamic acid (Gla), a previously unrecognized acidic amino acid. Further investigation of the chemistry of prothrombin revealed[14] that all 10 of what had once been thought to be glutamic acid residues in the first 42 residues of bovine prothrombin were modified in this fashion (Figure 3). These residues are required for the calcium-mediated interaction of the vitamin K-dependent proteins with a negatively charged phospholipid surface and are essential for the formation of the protein conformation that imparts this property. These studies, which led to the characterization of the vitamin K-dependent modification of prothrombin, have been reviewed.[15]

VITAMIN K-DEPENDENT PLASMA PROTEINS

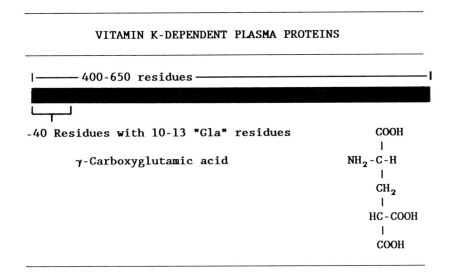

FIGURE 3. γ-Carboxyglutamic acid. The vitamin K-dependent proteins are typical glycoproteins that are unique in that they contain a number of γ-carboxyglutamyl (Gla) residues in a very homologous (see Table 1) amino terminal domain.

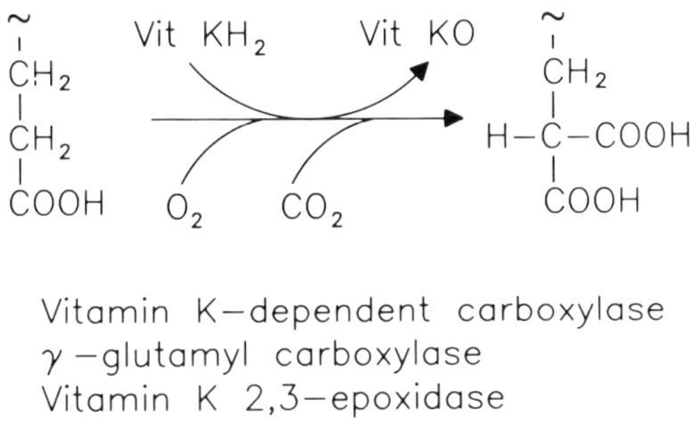

FIGURE 4. The liver microsomal vitamin K-dependent carboxylase. The enzyme utilizes the hydronaphthoquinone form of the vitamin (Vit KH_2) to drive the carboxylation of peptidyl bound glutamyl residues to γ-carboxyglutamyl residues. The coproduct of the reaction is vitamin K 2,3-epoxide (Vit KO).

A. THE VITAMIN K-DEPENDENT CARBOXYLASE

The enzyme responsible for the posttranslational conversion of glutamyl residues in newly synthesized proteins to γ-carboxyglutamyl residues in mature proteins (Figure 4) was first demonstrated[16] in a crude rat liver microsomal preparation. The activity is usually called the vitamin K-dependent carboxylase or γ-glutamyl carboxylase, and as seen in Figure 4, it could also be called a vitamin K-2,3-epoxidase. Early studies of the properties of the enzyme, which have been reviewed in detail,[17-21] demonstrated that the reaction required O_2, CO_2, and the reduced (hydronaphthoquinone) form of vitamin K. These early studies established that the only source of energy needed to drive this unique carboxylation reaction comes from the reoxidation of the reduced vitamin.

Most studies of the enzyme hve utilized crude microsomes solubilized in various detergents and short peptides containing Glu-Glu sequences as substrates. The enzyme is usually studied at 25°C or lower, and most studies have been carried out in very crude preparations. Reported measurements of kinetic contents have been variable but have been in the range of 75 μM for reduced vitamin K, 2 mM for $CO_2/HCO_3{}^-$, and 100 μM for O_2.[22] Most studies have utilized short peptides such as Phe-Leu-Glu-Glu-Leu or Boc-Glu-Glu-Leu-OMe as Glu site substrates. These have a K_m of a few mM in contrast to the low K_m values measured for the natural protein substrates of the enzyme.[22]

There have been few attempts to understand the mechanism of action of the enzyme by the use of steady-state kinetic measurements. It is difficult to saturate the enzyme with the nonvariable substrates, but Uotila[23] has studied the calf liver enzyme at a number of substrate concentrations. He concluded that the carboxylase has a sequential reaction mechanism which includes a quinternary complex of the enzyme with all of the four substrates. The data are consistent with an ordered steady-state addition of the substrates to the enzyme and suggest that CO_2 is either the first or last substrate to bind.

1. Mechanism of the Carboxylase Reaction

The vitamin K-dependent carboxylase appears to represent a unique biochemical mechanism that is not yet understood in any detail. Early studies considered two possibilities: that the vitamin was a cofactor utilized to abstract the hydrogen on the γ-position of the glutamyl residue to allow attack of CO_2 at this position, or that it was involved as a CO_2 carrier. Little evidence to support the latter hypothesis was ever obtained. Synthesis of the pentapeptide Phe-Leu-Glu-Glu-Leu, tritiated at the γ-carbon of each Glu residue,[24] was used to demonstrate that the enzyme catalyzed a vitamin KH_2- and O_2-dependent, but CO_2-independent release of tritium from this substrate. These data appeared to offer conclusive evidence that the role of the vitamin was not to activate or carry the CO_2 but to remove the γ-hydrogen of the Glu substrate, and to concurrently be converted to the epoxide. Epoxide formation is stimulated by the presence of a Glu site substrate of the carboxylase, and at saturating concentrations of CO_2 there is an apparent equivalent stoichiometry between epoxide formation and Gla formation.[25]

The enzyme will also catalyze a vitamin KH_2 and oxygen-dependent exchange of ^3H_2O into the γ-position of a Glu residue in the substrate Boc-Glu-Glu-Leu-OMe.[26,27] Exchange of ^3H with the γ-carbon hydrogen is decreased as the concentration of $HCO_3{}^-$ in the media is increased. Studies[28]

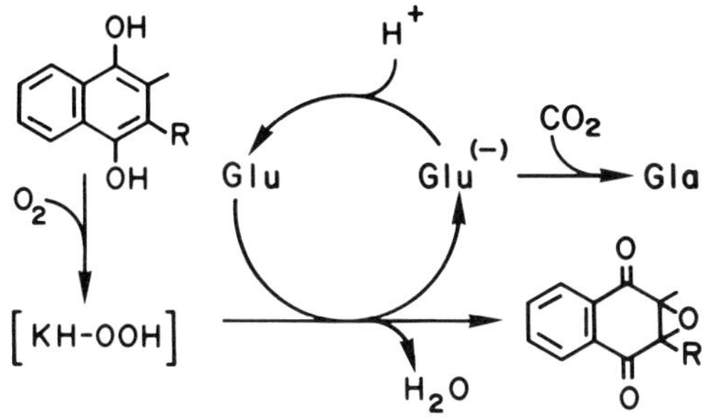

FIGURE 5. Proposed mechanism for the vitamin K-dependent carboxylase/epoxidase system. The available data strongly support the vitamin K-dependent formation of a carbanion at the γ-position of the Glu residue in a reaction coupled to epoxide formation. This event is followed by carboxylation in a step not involving the vitamin. Neither the chemical nature of the proposed oxygenated intermediate nor the mechanism by which hydrogen abstract is linked to epoxide formation can be determined from the available data. Recent data (see text) suggests that the key intermediate may be a 1,2-dioxetane which forms after the initial O_2 attack. In the absence of a Glu substrate, the enzyme will also carry out a low rate of epoxidation not linked to hydrogen abstraction.

utilizing γ-^{3}H-labeled Glu substrates have demonstrated a close association between epoxide formation, Gla formation, and γ-C-H bond cleavage. The efficiency of the carboxylation reaction, Gla/γ-C-H bond cleaved, was found to be independent of Glu substrate concentrations, and the data suggest that this ratio approaches unity at high CO_2 concentrations. These studies have also clearly demonstrated that the enzyme will carry out the half-reaction of epoxide formation at a diminished, but significant, rate in the absence of a Glu site substrate. The available data are consistent with the model shown in Figure 5, which indicates that the role of vitamin K is to abstract the χ-hydrogen to form a carbanion which can be attacked by CO_2. The available data do not offer conclusive proof that the abstraction event initially leaves a carbanion. It is possible that the initial event is a radical cleavage and that the carbon radical formed is subsequently reduced to a carbanion.

A chemical reaction, which results in the O_2-dependent intermolecular condensation of diethyl adipate to ethyl cyclopentanone-2-carboxylate and the formation of an α-keto epoxide from 2,4-dimethyl-1-naphthol has recently been proposed as an organic model for this enzymatic carboxylation.[29] The key step postulated is the formation of a dioxetane ring from a peroxy anion adduct which results from the initial attack of O_2. The dioxetane then decomposes to yield a strong base, the epoxy alkoxide, needed for proton abstraction. There is good support for the proposed pathway of the model reaction, and the available enzymological data are also consistent with this model. However, no direct evidence for the dioxetane intermediate is available for the enzymatic reaction.

A knowledge of the functional groups participating in the reaction would aid studies of mechanism, but these data will depend on the availability of significant amounts of pure enzyme. Progress in purifying the carboxylase activity was minimal for some time, but recent reports suggest that purification may have now been accomplished. A relatively simple procedure that yields a roughly 1000-fold purification of the enzyme has been described,[30] and it provides a convenient preparation for a number of studies of the enzyme. The initial report[31] that the enzyme had been purified to homogeneity has now been shown to represent the purification of the heat shock protein BiP, contaminated with the carboxylase. A second reported purification[32] of 7×10^3-fold has yielded a 95-kDa protein that has been cloned[33] and found not to be related in any significant degree to other known proteins. A third report[34] of a purification of 3.5×10^5-fold by a significantly different strategy has yielded a 98-kDa protein of apparent homogeneity. Which of these two proteins is the carboxylase enzyme will not be known until additional studies are completed.

2. Cell Biology of the Carboxylase

The γ-glutamyl carboxylase appears to be located on the luminal face of the rough endoplasmic reticulum and has been shown to utilize a high mannose form of prothrombin as a substrate.[22] The action of the enzyme can be shown to be definitely posttranslational under some conditions, but it is difficult to rule out the possibility that some cotranslational carboxylation may occur. Studies of the intracellular transport of prothrombin in the rat[35] have shown that prothrombin secretion occurs at a slower rate than that of albumin, and that the extent of carboxylation of prothrombin has some influence on secretion rate.

It has usually been assumed that the carboxylation of all vitamin K-dependent proteins proceeds in the same fashion, but this may not be the case. Wallin and Martin[36] have pointed out apparent differences between the processing of prothrombin and factor X by rat liver microsomes and have described some differences in the manner in which vitamin K-dependent proteins are processed by human and rat microsomes. The significance of these observations is not yet understood. An acquired vitamin K deficiency produced by anticoagulant administration results in the secretion into the plasma of large quantities of uncarboxylated and partially carboxylated forms of vitamin K-dependent clotting factors in both the human and bovine, but only low levels in the rat.[37] The specific alterations in the cell biology of protein secretion responsible for this interesting species difference have not been identified. However, the most likely explanation is that it is related to differences in the degradation rate of prothrombin precursors within the lumen of the endoplasmic reticulum of the two species.

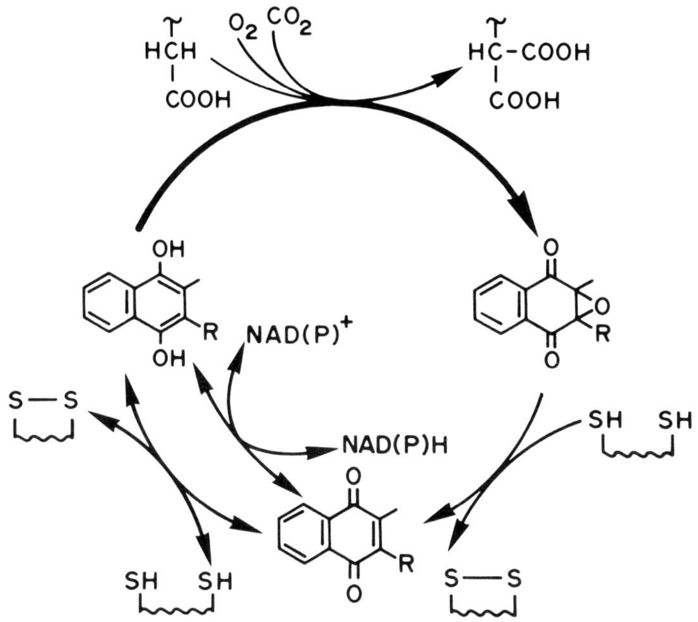

FIGURE 6. Vitamin K metabolism in liver. The vitamin K epoxide produced as a coproduct of the carboxylation reaction is reduced by dithiol-dependent epoxide and quinone reductases to the enzyme substrate form, the hydronaphthoquinone. These two reductive activities are the sites of action of the clinically useful 4-hydroxycoumarin anticoagulants, and it is possible that both activities are catalyzed by the same enzyme. The quinone form of the vitamin can also be reduced by pyridine nucleotide-dependent reductases that are much less coumarin sensitive.

3. Relationship of Carboxylation to Other Vitamin K-Metabolizing Enzymes

The reactions shown in Figure 6 summarize the known major metabolic transformation of vitamin K in rat liver microsomes and point out the relationship of the carboxylase to these other activities. At least in *in vitro* incubations, three forms of vitamin K, the quinone (vitamin K), the hydroquinone (vitamin KH_2), and the 2,3-epoxide (vitamin KO) can feed into a hepatic vitamin K cycle. Vitamin K can be reduced to vitamin KH_2 by a number of NAD(P)H-linked reductases including one that appears to be a microsomal-bound form of the extensively studied liver DT-diaphorase activity[38,39] and also by a dithiol-dependent reductase.[40] Vitamin KH_2 can serve as a substrate for the microsomal carboxylase/epoxidase activity that forms γ-carboxyglutamyl residues and converts the vitamin to vitamin KO. This epoxide is the substrate for another microsomal enzyme, the 2,3-epoxide reductase which reduces it to vitamin K. The epoxide reductase utilizes a sulfhydryl compound (dithiothreitol in *in vitro* incubations) as a reductant and

appears to be[41] the enzyme that is the site of the physiological action of the
4-hydroxycoumarins as anticoagulants. In crude *in vitro* systems, dithiothreitol will also reduce vitamin K to vitamin KH_2 in a 4-hydroxycoumarin-
sensitive reaction. It is likely that this reduction is catalyzed by the same
enzyme that reduces vitamin KO to vitamin K,[42] but final proof of this
relationship will depend on purification of the two activities. Recent evidence
suggests that the physiological reductant for the enzyme(s) that catalyzes the
reduction of vitamin K to vitamin KH_2 may be thioredoxin.[43,44]

III. STRUCTURE OF VITAMIN K-DEPENDENT PLASMA PROTEINS

The vitamin K-dependent plasma proteins are a family of structurally
related glycoproteins that represent less than 0.5% of the total plasma protein
pool. Their distinguishing feature is the presence of from 9 to 13 Gla residues
in an amino terminal domain. The majority of these proteins are zymogen
forms of a serine protease, and a number of them contain a second posttranslational modification in the form of β-hydroxyaspartate or β-hydrox-
yasparagine.

Early data, based on partial amino acid sequences, indicated that there
was a great deal of homology in the amino terminal, Gla-containing region
of the vitamin K-dependent proteins. The complete amino acid sequence of
prothrombin was established[45] by Magnusson in 1975, but because of the
small quantities of material available, determination of the primary sequence
of the other vitamin K-dependent proteins, particularly factor VII, represented
a formidable task in protein chemistry. By the late 1970s it was clear that the
primary sequence of these proteins could be more easily obtained by utilizing
the tools of molecular biology than through direct protein sequencing. Our
current understanding of the amino acid sequence of the vitamin K-dependent
proteins is, therefore, based on both direct peptide sequencing and, particularly
in the case of the human forms, an analysis of the cDNA structure.[46] The
data presented in Table 1 indicate the close structural homology of the amino
terminal Gla-containing region of some of the vitamin K-dependent proteins.
In addition to a homologous Gla domain, prothrombin (factor II), factor VII,
factor IX, factor X, and protein C share a serine protease domain containing
the reactive serine and neighboring histidine and aspartate residues first identified in pancreatic digestive enzymes such as trypsin, chymotrypsin, and
elastase. This serine protease homology is shared with the nonvitamin K-
dependent proteins of the plasma coagulation system, Kallikrein and factors
XI and XII, as well as plasmin and plasminogen activators of the fibrin lytic
systems.[46]

The availability of the complete sequence of the primary gene product of
all of the vitamin K-dependent proteins has provided information on sequence
homologies which were not previously apparent. Prothrombin contains two

"Kringle" domains. These looped sequences, named for their resemblance to the Danish pastry, are also found in the zymogen of the major enzyme of fibrinolysis, plasminogen, and its activators, urokinase, and tissue plasminogen activator, but not in the other vitamin K-dependent clotting factors. With the exception of prothrombin, the plasma vitamin K-dependent clotting factors also contain one or more epidermal growth factor (EGF) precursor-like domains. The second important posttranslational modification of the vitamin K-dependent protein, β-hydroxylation of aspartyl residues, is located in this region. Although it appears that this residue confers Ca^{2+}-binding properties to the vitamin K-dependent proteins that are independent of the Gla mediated interactions, the details of this role have not been determined.

The circulating forms of the vitamin K-dependent plasma proteins are, therefore, composed of a number of common domains.[46-49] A homologous region containing the Gla residues that gives these proteins their ability to form Ca^{2+}-dependent interactions with negatively charged phospholipids is present at the amino terminus. With the exception of protein S, these proteins circulate as zymogens and only in the case of prothrombin does their proteolytic activation release the Gla domain. In the case of prothrombin, the amino terminal Gla domain is followed by two Kringle domains, and the other proteins by two or more EGF domains. The remainder of these proteins consist of regions of unidentified function and the carboxy terminal serine protease domain. These relationships are illustrated in Figure 7.

The vitamin K-dependent proteins are relatively large, contain a significant amount of carbohydrate, and are readily degraded by protease action. Because of these factors, strongly defracting crystals of the intact proteins have been difficult to obtain, and progress in obtaining their tertiary structure by X-ray crystallography has moved forward slowly. Most of the effort has centered on prothrombin and has been directed toward obtaining the structure of the fragment 1 portion of bovine prothrombin. Early efforts yielded models of low resolution, and detailed locations of individual amino acids could not be determined. Park and Tulinsky[50] then obtained a 2.8 Å resolution structure of bovine fragment 1. This structure, and subsequent refinements, provided a clear picture of the Kringle region of fragment 1, but the first 35 residues of this peptide, which contain all of the Gla residues, were disordered. A second structure[51] eventually at 2.2 Å resolution[52] containing 7 Ca^{2+} atoms is now available. This solution has provided a clear model of the manner in which the charges of the Gla residues are ordered to provide two internal carboxylate surfaces of high negative charge density which hold the Ca^{2+} ions. A detailed explanation of the interactions between the remaining Gla residues, additional Ca^{2+} ions, and a negatively charged phospholipid surface is, however, still lacking. Molecular dynamics simulations based on the available crystallographic data have also provided insights into the conformational changes resulting from interactions with Ca^{2+} ions.[53]

TABLE 1
Amino Acid Sequence of the "Gla" Region of Selected Vitamin K-Dependent Plasma Proteins

Part 1 (positions +1 … +10 … +30)

	+1									+10						+30
F-II Human	Ala	Asn	Thr	—	Phe	Leu	Gla	Gla	Val	Arg	Lys	Gly	Asn	Leu	Gla	Arg
F-II Bovine	Ala	Asn	Lys	Gly	Phe	Leu	Gla	Gla	Val	Arg	Lys	Gly	Asn	Leu	Gla	Arg
F-II Chicken	Ala	Asn	Lys	Gly	Phe	Leu	Gla	Gla	Ile	Lys	Lys	Gly	Asn	Leu	Gla	Arg
F-II Rat	Ala	Asn	Ser	Gly	Phe	Leu	Gla	Gla	Met	Arg	Lys	Gly	Asn	Leu	Gla	A
F-X Human	Ala	Asn	Ser	—	Phe	Leu	Gla	Gla	Met	Lys	Lys	Gly	His	Leu	Gla	Arg
F-IX Human	Tyr	Asn	Ser	Gly	Lys	Leu	Gla	Gla	Phe	Val	Gln	Gly	Asn	Leu	Gla	Arg
F-VII Human	Ala	Asn	Ala	—	Phe	Leu	Gla	Gla	Leu	Arg	Pro	Gly	Ser	Leu	Gla	Arg
P-C Human	Ala	Asn	Ser	—	Phe	Leu	Gla	Gla	Leu	Arg	His	Ser	Ser	Leu	Gla	Arg
P-S Human	Ala	Asn	Ser	—	Leu	Leu	Gla	Gla	Thr	Lys	Gln	Gly	Ser	Leu	Gla	Arg

Part 2 (positions +20 … +40)

	+20											+40			
F-II Human	Gla	Cys	Val	Gla	Thr	Cys	Ser	Tyr	Gla	Gla	Ala	Phe	Gla	Ala	Leu
F-II Bovine	Gla	Cys	Leu	Gla	Pro	Cys	Ser	Arg	Gla	Gla	Ala	Phe	Gla	Ala	Leu
F-II Chicken	Gla	Cys	Leu	Gla	Thr	Cys	Asn	Tyr	Gla	Gla	Ala	Phe	Gla	Ala	Leu
F-II Rat	Gla	Cys	Val	Gla	Gln	Cys	Ser	Tyr	Gla	Gla	Ala	Phe	Gla	Ala	Leu
F-X Human	Gla	Cys	Met	Gla	Thr	Cys	Ser	Tyr	Gla	Gla	Ala	Arg	Gla	Val	Phe
F-IX Human	Gla	Cys	Met	Gla	Lys	Cys	Ser	Phe	Gla	Gla	Ala	Arg	Gla	Val	Phe
F-VII Human	Gla	Cys	Lys	Gla	Gln	Cys	Ser	Phe	Gla	Gla	Ala	Arg	Gla	Ile	Phe
P-C Human	Gla	Cys	Ile	Gla	Ile	Cys	Asp	Phe	Gla	Gla	Ala	Lys	Gla	Ile	Phe
P-S Human	Gla	Cys	Ile	Gla	Leu	Cys	Asn	Lys	Gla	Gla	Ala	Arg	Gla	Val	Phe

Part 3

F-II Human	Gla	Ser	Thr	Ala	Thr	Asp	Val	Phe	Trp	Ala	Lys	Tyr
F-II Bovine	Gla	Ser	Leu	Ser	Ala	Thr	Asp	Ala	Phe	Trp	Ala	Lys	Tyr
F-II Chicken	Gla	Ser	Thr	Val	Asp	Thr	Asp	Ala	Phe	Trp	—	—
F-II Rat	Gla	Ser	Pro	Gln	Asp	Thr	Asp	Val	Phe	Trp	Ala	Lys	Tyr
F-X Human	Gla	Asp	Ser	Lys	Asp	Thr	Asn	Gla	Phe	Trp	Asn	Lys	Tyr
F-IX Human	Gla	Asn	Thr	Arg	Gla	Thr	Gla	Gla	Phe	Trp	Lys	Gln	Tyr

F-VII	Human	Lys	Asp	Ala	Gla	Arg	Thr	Lys	Leu	Phe	Trp	Ile	Ser	Tyr
P-C	Human	Gln	Asn	Val	Asp	Asp	Thr	Leu	Ala	Phe	Trp	Ser	Lys	His
P-S	Human	Gla	Asn	Asp	Pro	Gla	Thr	Asp	Tyr	Phe	Tyr	Pro	Lys	Tyr

Note: References to the primary literature can be obtained in reviews[46-49] with the exception of the rat sequence[111] which is identical to that of the mouse.[112] Although the sequence of the Gla region of most of these proteins was determined by peptide sequences, complete structures for most were obtained only when cDNA structures were available.

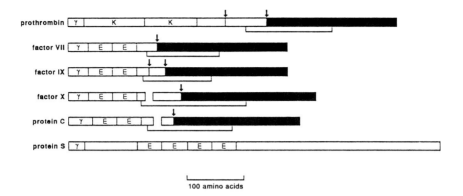

$$\underbrace{\hspace{6cm}}_{\text{100 amino acids}}$$

FIGURE 7. Structural domains of the vitamin K-dependent plasma clotting factors. The cleavage site(s) that give rise to the active proteases are indicated by arrows. Abbreviations used: γ, γ-carboxyglutamic acid-containing region; K, kringle; E, epidermal growth factor-like region. The serine protease catalytic region is shown as a solid bar. The structures are drawn to scale, and the 100-residue length is indicated. Adapted from Fung and MacGillivray.[49]

A. GENOMIC STRUCTURE OF VITAMIN K-DEPENDENT PROTEINS

The genes for the vitamin K-dependent proteins have been characterized by DNA sequence analysis, and the similarities of their structures have been reviewed.[46,49] The relationship between factors IX, X, and protein C are much closer than the relationship of this group to prothrombin. Although they all share strong similarities in the exon coding for the Gla and preproleader regions, they definitely diverge to two families downstream of the Gla exon. Not only does prothrombin contain three exons coding for the two Kringle regions rather than the EGF coding exons, but differences exist in the remainder of the gene. The serine protease (thrombin) region of prothrombin is coded for by five different exons which do not share homology in terms of intron sites with the factor IX family. The basis for the divergent evolutionary pathway of these closely related proteins is not yet defined, but a clear definition of the gene structure of all of the vitamin K-dependent proteins is nearly complete.[54]

B. CONGENITAL DEFICIENCIES OF VITAMIN K-DEPENDENT CLOTTING FACTORS

Classical hemophilia, a congenital factor VIII deficiency, is a sex-linked recessive characteristic that is the most common hereditary clotting defect in the human population. Severe hemophilia is found in about 1 per 25,000 of the population, and, when mildly affected patients are considered, the incidence of the disease might be as high as 1 in every 3000 to 4000 live births.[55] Factor VIII is not a vitamin K-dependent factor, but in many cases the bleeding tendency resulting from a factor VIII deficiency can be cured by increasing the concentrations of factor VII$_a$. The disease state has been known for many years to include a large number of variants, and these are rapidly being defined by modern methods of gene analysis.

Three of the classical vitamin K-dependent clotting factors, factors VII, IX, and X, were first discovered because of a congenital abnormality that resulted in a coagulation disorder. Christmas Disease, or hemophilia B, is a congenital deficiency of factor IX that is present in the human population at about 15% the incidence of hemophilia A. It is a sex-linked recessive bleeding disorder that is inherited in the same manner as hemophilia A. The disease encompasses a heterogeneous group of molecular disorders, and about half of the affected individuals have severely reduced circulating antigen levels while the remainder have circulating antigenically similar, but biologically less active forms of the protein.[56] Congenital deficiencies of prothrombin, factor X, and factor VII are much less common, are autosomally transmitted, and exist in a number of forms that may or may not be associated with circulating biologically inactive, but immunochemically related forms of the proteins.

As protein C and protein S function as natural anticoagulants rather than procoagulants, a congenital defect in these proteins might be expected to lead to a thrombotic rather than a hemorrhagic state. Griffin et al.[59] first described a patient with recurrent thrombophlebitis who was shown to have a decreased protein C concentration in plasma, and numerous cases of thrombin disorders associated with a heterozygous congenital protein C deficiency have been reported. It is now clear that structural variants or decreased concentrations of either protein C or protein S can result in thrombotic episodes. The extent of these congenital abnormalities in the population is not yet known, but it is likely that a significant portion of the population with recurrent thrombotic disorders of undiagnosed etiology may have an abnormality in one of these proteins.[58,59]

In addition to congenital defects of vitamin K-dependent proteins arising from a mutation in the gene directing the synthesis of a specific clotting factor, it is possible that an enzyme involved in the posttranslational vitamin K-dependent carboxylation of these proteins or in another biosynthetic processing step could be altered. Such an event might result in a multiple deficiency disease, and a limited number of cases of hemostatic disorders associated with altered concentrations or activities of all of the vitamin K-dependent proteins have been reported. One of the best studied[60] has been a female with severe functional abnormalities of all of the vitamin K-dependent coagulation factors. Immunologic and nonphysiologic measures indicated the presence of abnormal, dysfunctional proteins. This individual showed partial correction following administration of vitamin K and recurrence of clinical symptoms following discontinuation of therapy.

A mutation in a protein essential to intracellular protein processing, a defect in the carboxylase itself, or an interference with vitamin K metabolism are the most logical explanations for a functional abnormality of all of the vitamin K-dependent proteins. A child who maintains about 40% of the normal levels of prothrombin, factor VII, factor IX, and factor X on a daily intake of 10 mg of vitamin K has been described by Pauli et al.[61] This patient also exhibited a high circulating concentration of vitamin K epoxide, suggesting that the defect is an inability to metabolize the epoxide to vitamin K. This metabolic alteration is the same as that produced as the result of oral anti-coagulant therapy and would be consistent with a partial reversal of the congenital defect by high doses of vitamin K. Whether or not the same abnormality is present in other individuals with combined defects of vitamin K-dependent proteins, has not been established.

IV. THE PROPEPTIDE RECOGNITION SITE

The vitamin K-dependent proteins comprise approximately 1% of the total proteins synthesized by the liver and secreted into the plasma. This minor intracellular pool is, however, efficiently recognized, and specific Glu residues are selectively carboxylated by the vitamin K-dependent carboxylase located at the lumenal surface of the endoplasmic reticulum. Some indication of the basis for the selectivity of carboxylation was obtained when it was shown that the intracellular precursor form of prothrombin was larger and more basic than the secreted plasma form[62] and that isolated intracellular forms of prothrombin were much better substrates for the *in vitro* carboxylase than des-γ-carboxy plasma prothrombin.[63-65] These observations were consistent with structural information obtained through cDNA sequencing[46] which revealed that the primary gene product of all of the vitamin K-dependent plasma proteins contained a basic amino acid hydrophobic residue rich "propeptide" between the signal peptide region of the primary gene product and the amino terminal end of the mature plasma form of the protein. These observations suggest that this homologous amino terminal extension (Table 2) might play an important role in the recognition of these proteins by the microsomal carboxylase. This hypothesis was strengthened by the observation[66] that an amino terminal extension homologous to this region was also present in the precursor form of the vitamin K-dependent bone Gla protein.

A. CELLULAR EXPRESSION SYSTEMS AND PROPEPTIDE FUNCTION

The possibility that recombinant clotting factors, particularly factor VII, factor IX, and protein C, might have a role as hemostatic or antithrombotic drugs made them an attractive target for attempts to obtain biologically active proteins through the use of mammalian expression systems. Early studies demonstrated rF-IX production in hepatoma,[67,68] baby hamster kidney (BHK),[69]

TABLE 2
Amino Acid Sequences in the Propeptide Region of Human Vitamin K-Dependent Plasma Proteins

Protein			−20										−10
Prothrombin	Leu	Val	His	Ser	Gln	His	Val	Phe	Leu	Ala	Pro	Gln	Gln	Ala
Factor X	Leu	Leu	Leu	Gly	Glu	Ser	Leu	Phe	Ile	Arg	Arg	Glu	Gln	Ala
Factor IX	Leu	Ser	Ala	Glu	Cys	Thr	Val	Phe	Leu	Asp	His	Glu	Asn	Ala
Factor VII	Gln	Gly	Cys	Leu	Ala	Ala	Val	Phe	Val	Thr	Gln	Glu	Glu	Ala
Protein C	Pro	Ala	Pro	Leu	Asp	Ser	Val	Phe	Ser	Ser	Ser	Glu	Arg	Ala
Protein S	Leu	Pro	Val	Ser	Glu	Ala	Asn	Phe	Leu	Ser	Lys	Gln	Gln	Ala
									−1	+1				

Protein										
Prothrombin	Arg	Ser	Leu	Leu	Gln	Arg	Val	Arg	Arg	Ala
Factor X	Asn	Asn	Ile	Leu	Ala	Arg	Val	Thr	Arg	Ala
Factor IX	Asn	Lys	Ile	Leu	Asn	Arg	Pro	Lys	Arg	Tyr
Factor VII	His	Gly	Val	Leu	His	Arg	Arg	Arg	Arg	Ala
Protein C	His	Gln	Val	Leu	Arg	Ile	Arg	Lys	Arg	Ala
Protein S	Ser	Gln	Val	Leu	Val	Arg	Lys	Arg	Arg	Ala

Note: The amino terminal residue of the mature form of the plasma protein is designated as residue +1. The site of signal peptide cleavage for factor IX is between residues −19 and −18[81,113] and between −24 and −25 for protein C.[77] The position of cleavage has not been determined with certainty for all of the other factors.

or Chinese hamster ovary (CHO)70 cells, and rF-VII production in BHK cells.[71] The vitamin K-dependent proteins are complex and for full activity require γ-glutamyl carboxylation, glycosylation, disulfide bond formation, proteolytic processing, and often β-aspartyl hydroxylation. These problems and their relationship to the low levels of expression of biologically active recombinant clotting factors that were obtained in most of the early studies have been discussed.[72] Early attempts at expression of vitamin K-dependent proteins suggested that γ-glutamyl carboxylation often became limiting as cellular expression levels were increased. All of the vitamin K-dependent plasma proteins have now been expressed in cultured mammalian cell lines, and by selection of appropriate cell lines and expression vectors, increased quantities of biologically active recombinant clotting factors can now be produced. Of these efforts, the product closest to commercial use is probably rhF-VII which has been tested and found effective as a hemostatic agent in both animal models[73,74] and in limited clinical situations.[75,76]

The availability of these well-characterized cellular expression systems has been utilized to test the hypothesis that the homologous amino terminal propeptide of the vitamin K-dependent proteins is an important signal for γ-carboxylation. Both rPC[77] and rF-IX[78] secreted by cells transfected with constructs which had been modified so that the propeptide region was deleted were found to lack Gla residues. A second posttranslational modification of these proteins, β-aspartyl hydroxylation, was, however, not dependent on this region of the primary gene product.[79] Further studies[78,80] demonstrated that the well conserved -16 Phe, -10 Ala, and -6 Leu residues of the propeptide region (see Table 2) are of particular importance in directing efficient carboxylation. Residues -1 to -4 are presumably more important in directing the action of a propeptidase that is required for cleavage of the amino terminal extension prior to secretion. A natural F-IX variant with a mutation of the 4 Arg residue to a Gln has been shown[81] to secrete F-IX with the propeptide still attached. This variant lacks much of the normal biological activity, but does retain most of the Gla content of normal F-IX.[80] Further studies of the cellular expression of rF-II with specific mutations in the propeptide region[25] confirmed the importance of the -10 Ala and -16 Phe residues and demonstrated that the entire amino terminal region of the propeptide (residues -14 to -18) were important for efficient carboxylation. Although mutations in all residues was not investigated, it was demonstrated that mutations at the -14 Ala and -8 Val residues did not impair carboxylation. Much of the work in this area has recently been summarized.[83]

B. CARBOXYLASE SUBSTRATES AND PROPEPTIDE FUNCTION

The importance of specific regions or residues of the propeptide for carboxylase recognition has also been assessed through the synthesis of peptide substrates for the carboxylase. The enzyme is routinely assayed with a tri or penta peptide containing a Glu-Glu sequence such as Phe-Leu-Glu-Glu-Leu

(FLEEL) as a substrate.[22] These substrates have K_m values in the range of a few mM. Ulrich et al.[84] demonstrated that a 28-amino acid peptide consisting of residue -18 to $+10$ of human prothrombin and containing the $+6$ and $+7$ Glu sites was carboxylated with a K_m of 3.6 μM, about 500- to 1000-fold lower than FLEEL. Peptides comprising the first 10 residues of pro-thrombin ($+1$ to $+10$) of this region and a portion of the propeptide region (-10 to $+10$) were no better than FLEEL as a substrate. The importance of the -16 Phe residue of the propeptide region was demonstrated by the syn-thesis of the 28-residue peptide with Ala at the -16 position. Carboxylation of this substrate was significantly impaired when compared to the native Phe-containing peptide. An extension of these initial studies[85] demonstrated that, at least *in vitro*, the hF-II (prothrombin) propeptide was more efficient at directing carboxylation than the hF-IX propeptide, and that the carboxylation of peptides containing a Ser rather than an Arg at residues -1 or -4 is not drastically impaired. These are the residues altered in two characterized ge-netic variants of F-IX,[80] and the lack of involvement of these two residues in directing carboxylation has also been demonstrated through the use of substrates representing the -18 to $+41$ region of hF-IX.[86] This substrate, and variants of it, were produced as recombinant fusion proteins in *E. coli*. Price et al.[87] have identified a consensus sequence, Glu-X-X-X-Glu-X-Cys, within all known vitamin K-dependent proteins, and have suggested that it, as well as the propeptide region, serve as a recognition site for the substrate/carboxylase interaction. Comparison of the *in vitro* carboxylation of a peptide comprising residues -18 to $+36$ of prothrombin, which contains the con-sensus sequence, with the carboxylation of a -18 to $+10$ prothrombin hom-olog, which lacks this region, suggests[82] that this suggested consensus se-quence is not important in directing *in vitro* carboxylation.

C. DUAL ROLE OF THE PROPEPTIDE

Both the *in vitro* and cellular expression studies are consistent with the hypothesis that the carboxy terminal region of the propeptide contains a recognition site for binding to a propeptidase, and that residue -10 and the amino terminal region of the propeptide serves to recognize the carboxylase and to serve a "docking" function. Knobloch and Suttie[88] have shown that a 20-residue peptide containing the octadeca propeptide of human factor X strongly stimulates the activity of the rat liver vitamin K-dependent carbox-ylase when assayed with small peptide substrates such as FLEEL. This in-creased activity brought about by the noncovalently bound propeptide is mainly the result of a lowering of the K_m at the Glu binding site of the carboxylase.[89] These data suggest that this region of the vitamin K-dependent protein serves both a regulatory and a recognition function. When these studies were ex-tended,[90] it was demonstrated that the -10 Ala and -16 Phe that had been shown to be important in directing carboxylation in cellular expression systems and in covalently linked substrates were also important in influencing the interaction of enzyme and substrate at the Glu site. The ability of the free propeptide to stimulate the carboxylase is not limited to weak binding high K_m substrates. Des-γ-carboxy Bone Gla Protein (dBGP) is not a particularly good carboxylase substrate, but in the presence of free propeptide it is efficiently carboxylated[91] with a K_m of only 11 μM. These data indicating a role of the propeptide in modulating the Glu binding site of the enzyme do not suggest that the presumed docking function of the propeptide is not important. A second protein originally found in bone, MGP, does not have an amino terminal propeptide cleaved during protein processing. Rather, it has an internal sequence in the mature protein which is homologous to this region.[87] The K_m of dMGP as a substrate is only 0.2 μM, and its carboxylation is not stimulated, but is effectively inhibited by free propeptide,[91] presumably by displacing it from the enzyme. These data make it clear that the propeptide region of the few proteins that are substrates for the carboxylase serves a dual role, it anchors or docks the substrate with the enzyme, and also alters the affinity of the enzyme for the Glu residue which will be carboxylated.

Why both a docking and regulatory role for the propeptide should be required for normal carboxylation is not readily apparent. The manner by which the enzyme carboxylates its multisite substrates is not known, and it is possible that some Glu (potential Gla) sites may have a high affinity for the enzyme, or that it is an ordered series of events. The location of the Gla residues in partially carboxylated forms of prothrombin[92] suggests that the most amino terminal Gla site is most readily carboxylated but that there may also be high affinity sites elsewhere in the Glu region. The published studies of the *in vitro* carboxylation of numerous peptides with multiple Glu sites have not addressed the question of which of the possible sites are carboxylated. This question is not easily addressed in a substrate with ten potential Gla sites, but is more easily approached by studies of the carboxylation of dBGP which has only three carboxylated Glu residues. Preliminary data from this system[93] suggests that the most carboxy terminal Glu is preferentially carboxylated, but the influence of the propeptide region on the pattern of carboxylation has not yet been determined.

V. GENE EXPRESSION AND VITAMIN K-DEPENDENT PROTEINS

Both rat (H-35) and human (HepG2) hepatoma cell lines[94-96] have been shown to secrete prothrombin and other vitamin K-dependent proteins.[95,97] The general observation has been that warfarin treatment increases the intracellular pool of prothrombin and decreases secretion into the culture medium, while vitamin K administration increases secretion. It is not clear from the available data if these responses are related in any way to changes in the rate of expression of the prothrombin gene, or if they merely represent a response to the altered ability of the cells to effectively carry out the γ-glutamyl carboxylation step. The H-35 rat hepatoma synthesis systems have also been used[98] to demonstrate an increase of prothrombin synthesis upon the addition of the amino terminal, fragment-1 portion (residues 1 through 156) of pro-thrombin to the media. Indirect evidence suggested that this response was related to an effect on prothrombin gene expression, but no direct evidence was presented, and the observation has not been repeated.

Plasma concentrations of vitamin K-dependent clotting factors are nor-mally maintained in a narrow range, and there is some evidence for a humoral factor that is involved in this control. Karpatkin and his co-workers[90,100] have demonstrated that the intravenous injection of a small amount of plasma obtained from an anticoagulant-treated rabbit with a low level of vitamin K-dependent clotting factors into a normal rabbit will result in an increase in the activities of these factors in the recipient rabbit. The humoral factor responsible for this response has been called a coagulopoietin(s). This response has also been observed in the rat,[101] and there are indications[101-103] that this treatment increases biological activity of prothrombin more than prothrombin antigen. It has been demonstrated[104] that this factor appears early during the development of a deficiency state and that the activity is variable between different animals. The chemical nature of the proposed coagulopoietin is not known, and the current data do not clearly differentiate an effect on expression of the vitamin K-dependent proteins, or of components of the carboxylase system.

The concentration of undercarboxylated prothrombin in normal human plasma is very low and increases substantially with oral anticoagulation or vitamin K deficiency, and slightly in many patients with liver disease.[105] It has been shown[106,107] that this protein is increased more dramatically in cases of primary hepatocellular carcinoma than in other liver disorders, and its presence can be used to screen for this disease. This increase is independent of vitamin K status, and the molecular basis of the observation has been studied in a rat model.[108] A series of Morris hepatoma tumors were trans-planted into recipient Buffalo strain rats. Transplantation of some but not all tumor lines resulted in increased levels of circulating undercarboxylated pro-thrombin, and the tumor mass had low levels of carboxylase. The tumor mass of those lines not secreting this marker had much higher levels of the car-boxylase. The data suggest that the secretion of abnormal prothrombin by hepatocellular tumors is the result of normal expression of the prothrombin gene and a failure of the tumor to fully express the carboxylase gene.

These examples indicate that neither the role of vitamin K, if any, in regulating gene expression, nor the role of other factors, which influence gene expression on vitamin K metabolism, is yet understood. Progress in this area should, however, be rapid. The vitamin K-dependent proteins themselves have been cloned, and mRNA probes have been successfully used to follow the regulation of prothrombin synthesis during fetal and neonatal develop-ment.[109,110] The currently confused status of the possible regulation of clotting factor genes by vitamin K status should therefore soon be clarified. The apparent purification and cloning of the carboxylase will also make available the tools with which to probe the factors which regulate the expression of this unique and important enzyme.

REFERENCES

1. **Dam, H.,** Hemorrhages in chicks reared on artificial diets: a new deficiency disease, *Nature (London),* 133, 909, 1934.
2. **Suttie, J. W.,** Vitamin K, in *Handbook of Vitamins,* 2nd ed., Machlin, L. J., Ed., Marcel Dekker, New York, 1991.
3. **Bentley, R. and Meganathan, R.,** Biosynthesis of vitamin K (menaquinone) in bacteria, *Microbiol. Rev.,* 46, 241, 1982.
4. **Suttie, J. W.,** Vitamin K, in *The Fat-Soluble Vitamins,* Diplock, A. T., Ed., William Heinemann, London, 1985, 225.
5. **Dam, H., Schonheyder, F., and Tage-Hansen, E.,** Studies on the mode of action of vitamin K, *Biochem. J.,* 30, 1075, 1936.
6. **Stenflo, J.,** A new vitamin K-dependent protein. Purification from bovine plasma and preliminary characterization, *J. Biol. Chem.,* 251, 355, 1976.
7. **Di Scipio, R. G., Hermodson, M. A., Yates, S. G., and Davie, E. W.,** A comparison of human prothrombin, factor IX (Christmas factor), factor X (Stuart factor), and protein S, *Biochemistry,* 16, 698, 1977.
8. **Price, P. A.,** Role of vitamin K-dependent proteins in bone metabolism, *Annu. Rev. Nutr.,* 8, 565, 1988.
9. **Link, K. P.,** The discovery of dicumarol and its sequels, *Circulation,* 19, 97, 1959.
10. **Olson, R. E.,** Vitamin K induced prothrombin formation: antagonism by actinomycin D, *Science,* 145, 926, 1964.
11. **Suttie, J. W.,** Metabolism and properties of a liver precursor to prothrombin, *Vitam. Horm.,* 32, 463, 1974.
12. **Stenflo, J., Fernlund, P., Egan, W., and Roepstorff, P.,** Vitamin K dependent modifications of glutamic acid residues in prothrombin, *Proc. Natl. Acad. Sci. U.S.A.,* 71, 2730, 1974.
13. **Nelsestuen, G. L., Zytkovicz, T. H., and Howard, J. B.,** The mode of action of vitamin K. Identification of γ-carboxyglutamic acid as a component of prothrombin. *J. Biol. Chem.,* 249, 6347, 1974.
14. **Magnusson, S., Sottrup-Jensen, L., Petersen, T. E., Morris, H. R., and Dell, A.,** Primary structure of the vitamin K-dependent part of prothrombin, *FEBS Lett.,* 44, 189, 1974.
15. **Stenflo, J. and Suttie, J. W.,** Vitamin K-dependent formation of γ-carboxyglutamic acid, *Annu. Rev. Biochem.,* 46, 157, 1977.
16. **Esmon, C. T., Sadowski, J. A., and Suttie, J. W.,** A new carboxylation reaction. The vitamin K-dependent incorporation of $H^{14}CO_3^-$ into prothrombin, *J. Biol. Chem.,* 250, 4744, 1975.
17. **Suttie, J. W.,** Mechanism of action of vitamin K: synthesis of Y-carboxyglutamic acid, *CRC Crit. Rev. Biochem.,* 8, 191, 1980.
18. **Johnson, B. C.,** Post-translational carboxylation of preprothrombin, *Mol. Cell Biochem.,* 38, 77, 1981.

19. **Vermeer, C. and de Boer-van den Berg, M. A. G.,** Vitamin K-dependent carboxylase, *Haematologia,* 18, 71, 1985.
20. **Olson, R. E.,** The function and metabolism of vitamin K, *Annu. Rev. Nutr.,* 4, 281, 1984.
21. **Friedman, P. A. and Przysiecki, C. T.,** Vitamin K-dependent carboxylation, *Int. J. Biochem.,* 19, 1, 1987.
22. **Suttie, J. W.,** Vitamin K-dependent carboxylation of glutamyl residues in proteins, *BioFactors,* 1, 55, 1988.
23. **Uotila, L.,** Vitamin K-dependent carboxylase from calf liver: studies on the steady-state kinetic mechanism, *Arch. Biochem. Biophys.,* 264, 135, 1988.
24. **Friedman, P. A., Shia, M. A., Gallop, P. M., and Griep, A. E.,** Vitamin K-dependent γ-carbon-hydrogen bond cleavage and the non-mandatory concurrent carboxylation of peptide bound glutamic acid residues, *Proc. Natl. Acad. Sci. U.S.A.,* 76, 3126, 1979.
25. **Larson, A. E., Friedman, P. A., and Suttie, J. W.,** Vitamin K-dependent carboxylase: stoichiometry of carboxylation and vitamin K 2,3-epoxide formation, *J. Biol. Chem.,* 256, 11032, 1981.
26. **McTigue, J. J. and Suttie, J. W.,** Vitamin K-dependent carboxylase: demonstration of a vitamin K- and O_2-dependent exchange of ^3H from ^3H_2O into glutamic acid residues, *J. Biol. Chem.,* 258, 12129, 1983.
27. **Anton, D. L. and Friedman, P. A.,** Fate of the activated γ-carbon-hydrogen bond in the uncoupled vitamin K-dependent γ-glutamyl carboxylation reaction, *J. Biol. Chem.,* 258, 14084, 1983.
28. **Wood, G. M. and Suttie, J. W.,** Vitamin K-dependent carboxylase. Stoichiometry of vitamin K epoxide formation, γ-carboxyglutamyl formation, and γ-glutamyl-3H cleavage, *J. Biol. Chem.,* 263, 3234, 1988.
29. **Dowd, P., Ham, S. W., and Geib, S. J.,** Mechanism of action of vitamin K, *J. Am. Chem. Soc.,* 113, 7734, 1991.
30. **Harbeck, M. C., Cheung, A. Y., and Suttie, J. W.,** Vitamin K-dependent carboxylase: partial purification of the enzyme by antibody affinity techniques, *Thrombosis Res.,* 56, 317, 1989.
31. **Hubbard, B. R., Ulrich, M. M. W., Jacobs, M., Vermeer, C., Walsh, C., Furie, B., and Furie, B. C.,** Vitamin K-dependent carboxylase: affinity purification from bovine liver by using a synthetic propeptide containing the γ-carboxylation recognition site, *Proc. Natl. Acad. Sci. U.S.A.,* 86, 6893, 1989.
32. **Wu, S-M., Morris, D. P., and Stafford, D. W.,** Identification and purification to near homogeneity of the vitamin K-dependent carboxylase, *Proc. Natl. Acad. Sci. U.S.A.,* 88, 2236, 1991.
33. **Wu, S-M., Cheung, W-F., Frazier, D., and Stafford, D.,** Cloning and expression of the cDNA for human γ-glutamyl carboxylase, *Science,* 254, 1634, 1991.
34. **Berkner, K. L., Harbeck, M., Lingenfelter, S., Bailey, C., Sanders-Hinck, C. M., and Suttie, J. W.,** Purification and identification of the bovine liver γ-carboxylase, *Proc. Natl. Acad. Sci. U.S.A.,* 89, 6242, 1992.
35. **Kvalvaag, A. H., Tollersrud, O. K., and Helgeland, L.,** A study on the intracellular transport of prothrombin, albumin and transferrin in rat, *Biochim. Biophys. Acta,* 937, 319, 1988.
36. **Wallin, R. and Martin, L. F.,** Early processing of prothrombin and factor X by the vitamin K-dependent carboxylase, *J. Biol. Chem.,* 263, 9994, 1988.
37. **Shah, D. V., Swanson, J. C., and Suttie, J. W.,** Abnormal prothrombin in the vitamin K-deficient rat, *Thrombosis Res.,* 35, 451, 1984.
38. **Wallin, R., Gebhardt, O., and Prydz, H.,** NAD(P)H dehydrogenase and its role in the vitamin K (2-methyl-3-phytyl-1,4-naphthoquinone)-dependent carboxylation reaction, *Biochem. J.,* 169, 95, 1978.

39. **Fasco, M. J. and Principe, L. M.,** Vitamin K_1 hydroquinone formation catalyzed by DT-diaphorase, *Biochem. Biophys. Res. Commun.,* 104, 187, 1982.

40. **Fasco, M. J. and Principe, L. M.,** Vitamin K_1 hydroquinone formation catalyzed by a microsomal reductase system, *Biochem. Biophys. Res. Commun.,* 97, 1487, 1980.

41. **Suttie, J. W.,** Vitamin K antagonists, in *Hemostasis and Thrombosis: Basic Principles and Clinical Practice,* 3rd ed., Colman, R. W. and Hirsh, J., Eds., J. B. Lippincott, Philadelphia, in press.

42. **Gardill, S. L. and Suttie, J. W.,** Vitamin K epoxide and quinone reductase activities: evidence for reduction by a common enzyme, *Biochem. Pharmacol.,* 40, 1055, 1990.

43. **van Haarlem, L. J. M., Soute, B. A. M., and Vermeer, C.,** Vitamin K-dependent carboxylase. Possible role for thioredoxin in the reduction of vitamin K metabolites in liver, *FEBS Lett.,* 222, 353, 1987.

44. **Silverman, R. B. and Nandi, D. L.,** Reduced thioredoxin: a possible physiological cofactor for vitamin K epoxide reductase. Further support for an active site disulfide, *Biochem. Biophys. Res. Commun.,* 155, 1248, 1988.

45. **Magnusson, S., Petersen, T. E., Sottrup-Jensen, L., and Claeys, H.,** Complete primary structure of prothrombin: Isolation, structure and reactivity of ten carboxylated glutamic acid residues and regulation of prothrombin activation by thrombin, in *Proteases and Biological Control,* Vol. 2, Reich, E., Rifkin, D. B., and Shaw, E., Eds., Cold Spring Harbor Conf. on Cell Proliferation, 1975, 123.

46. **Davie, E. W.,** The blood coagulation factors: their cDNAs, genes, and expression, in *Hemostasis and Thrombosis: Basic Principles and Clinical Practice,* 2nd ed., Colman, R. W., Hirsh, J., Mardar, V. J., and Salzman, E. W., Eds., Lippincott, Philadelphia, 1987, 242.

47. **Walz, D. A., Hewett-Emmett, D., and Guillin, M-C.,** Amino acid sequences and molecular homology of the vitamin K-dependent clotting factors, in *Prothrombin and Other Vitamin K Proteins,* vol. I, Seegers, W. H. and Walz, D. A., Eds., CRC Press, Boca Raton, FL, 1986, 125.

48. **Furie, B. and Furie, B. C.,** The molecular basis of blood coagulation, *Cell,* 53, 505, 1988.

49. **Fung, M. R. and MacGillivray, R. T. A.,** Organization of the genes coding for the vitamin K-dependent clotting factors, in *Current Advances in Vitamin K Research,* Suttie, J. W., Ed., Elsevier, New York, 1988, 143.

50. **Park, C. H. and Tulinsky, A.,** Three dimensional structure of the Kringle sequence: structure of prothrombin fragment 1, *Biochemistry,* 25, 3977, 1986.

51. **Soriano-Garcia, M., Park, C. H., Tulinsky, A., Ravichandran, K. G., and Skrzypczak-Jankun, E.,** Structure of Ca^{2+} prothrombin fragment 1 including the conformation of the Gla domain, *Biochemistry,* 28, 6805, 1989.

52. **Soriano-Garcia, M., Padmanabhan, K., deVos, A. M., and Tulinsky, A.,** The Ca^{2+} ion and membrane binding structure of the Gla-domain of Ca-prothrombin fragment 1, *Biochemistry,* in press.

53. **Charifson, P. S., Darden, T., Tulinsky, A., Hughey, J. L., Hiskey, R. G., and Pedersen, L. G.,** Solution conformations of the γ-carboxyglutamic acid domain of bovine prothrombin fragment 1, residues 1-65, *Proc. Natl. Acad. Sci. U.S.A.,* 88, 424, 1991.

54. **Patthy, L.,** Evolutionary assembly of blood coagulation proteins, *Sem. Thrombos. Haemostas.,* 16, 245, 1990.

55. **Rizza, C. R.,** The clinical features of clotting factor deficiencies, in *Human Blood Coagulation, Haemostasis and Thrombosis,* Biggs, R., Ed., Blackwell Scientific, Oxford, 1972, 210.

56. **Shapiro, S. S. and Martinez, J.,** Molecular variations of coagulation proteins, in *Hemostasis and Thrombosis,* Bowie, E. J. W. and Sharp, A. A., Eds., Butterworths, London, 1985, 197.

57. **Griffin, J. H., Evatt, B., Zimmerman, T. S., Kleiss, A. J., and Wideman, C.,** Deficiency of protein C in congenital thrombotic disease, *J. Clin. Invest.,* 68, 1370, 1981.

58. **Clouse, L. H. and Comp, P. C.,** The regulation of hemostasis: the protein C system, *New Engl. J. Med.,* 314, 1298, 1986.

59. **Esmon, C. T.,** The regulation of natural anticoagulant pathways, *Science,* 235, 1348, 1987.

60. **Chung, K-S, Bezeaud, A., Goldsmith, J. C., McMillan, C. W., Menache, D., and Roberts, H. R.,** Congenital deficiency of blood clotting factors II, VII, IX, and X, *Blood,* 53, 776, 1979.

61. **Pauli, R. M., Lian, J. B., Mosher, D. F., and Suttie, J. W.,** Association of congenital deficiency of multiple vitamin K-dependent coagulation factors and the phenotype of the warfarin embryopathy: clues to the mechanism of teratogenicity of coumarin derivatives, *Am. J. Hum. Genet.,* 41, 566, 1987.

62. **Swanson, J. C. and Suttie, J. W.,** Prothrombin biosynthesis: characterization of processing events in rat liver microsomes, *Biochemistry,* 24, 3890, 1985.

63. **Soute, B. A. M., Vermeer, C., De Metz, M., Hemker, H. C., and Lijnen, H. R.,** In vitro prothrombin synthesis from a purified precursor protein. III. Preparation of an acid-soluble substrate for vitamin K-dependent carboxylase by limited proteolysis of bovine descarboxyprothrombin, *Biochim. Biophys. Acta,* 676, 101, 1981.

64. **Shah, D. V., Swanson, J. C., and Suttie, J. W.,** Vitamin K-dependent carboxylase: effect of detergent concentrations, vitamin K status, and added protein precursors on activity, *Arch. Biochem. Biophys.,* 222, 216, 1983.

65. **Evans, M. R., Sung, M. W. P., and Esnouf, M. P.,** The early stages in the biosynthesis of prothrombin, *Biochem. Soc. Trans.,* 12, 1051, 1984.

66. **Pan, L. C. and Price, P. A.,** The propeptide of rat bone γ-carboxyglutamic acid protein shares homology with other vitamin K-dependent protein precursors, *Proc. Natl. Acad. Sci. U.S.A.,* 82, 6109, 1985.

67. **de la Salle, H., Altenburger, W., Elkaim, R., Dott, K., Dieterle, A., Drillien, R., Cazenave, J-P., Tolstoshev, P., and Lecocq, J-P.,** Active γ-carboxylated human factor IX expressed using recombinant DNA techniques, *Nature (London),* 316, 268, 1985.

68. **Anson, D. S., Austen, D. E. G., and Brownlee, G. G.,** Expression of active human clotting factor IX from recombinant DNA clones in mammalian cells, *Nature (London),* 315, 683, 1985.

69. **Busby, S., Kumar, A., Joseph, M., Halfpap, L., Insley, M., Berkner, K., Kurachi, K., and Woodbury, R.,** Expression of active human factor IX in transfected cells, *Nature (London),* 316, 271, 1985.

70. **Kaufman, R. J., Wasley, L. C., Furie, B. C., Furie, B., and Shoemaker, C. B.,** Expression, purification, and characterization of recombinant γ-carboxylated factor IX synthesized in Chinese hamster ovary cells, *J. Biol. Chem.,* 261, 9622, 1986.

71. **Thim, L., Bjoern, S., Christensen, M., Nicolaisen, E. M., Lund-Hansen, T., Pedersen, A. W., and Hedner, U.,** Amino acid sequence and posttranslational modifications of human factor VIIa from plasma and transfected baby hamster kidney cells, *Biochemistry,* 27, 7785, 1988.

72. **Yan, S. C. B., Grinnell, B. W., and Wold, F.,** Post-translational modifications of proteins: some problems left to solve, *Trends Biochem. Sci.,* 14, 264, 1989.

73. **Brinkhous, K. M., Hedner, U., Garris, J. B., Diness, V., and Read, M. S.,** Effect of recombinant factor VIIa on the hemostatic defect in dogs with hemophilia A, hemophilia B, and von Willebrand Disease, *Proc. Natl. Acad. Sci. U.S.A.,* 86, 1382, 1989.

74. **Diness, V., Lund-Hansen, T., and Hedner, U.,** Effect of recombinant human FVIIA on warfarin-induced bleeding in rats, *Thrombosis Res.,* 59, 921, 1990.

75. **Hedner, U., Glazer, S., Pingel, K., Alberts, K. A., Blomback, M., Schulman, S. and Johnsson, H.,** Successful use of recombinant factor VIIa in patient with severe haemophilia A during synovectomy, *Lancet,* 2, 1193, 1988.

76. **Macik, B. G., Hohneker, J., Roberts, H. R., and Griffin, A. M.,** Use of recombinant activated factor VII for treatment of a retropharyngeal hemorrhage in a hemophilic patient with a high titer inhibitor, *Am. J. Hematol.,* 32, 232, 1989.

77. **Foster, D. C., Rudinski, M. S., Schach, B. G., Berkner, K. L., Kumar, A. A., Hagen, F. S., Sprecher, C., Insley, M., and Davie, E. W.,** Propeptide of human protein C is necessary for gamma-carboxylation, *Biochemistry,* 26, 7003, 1987.

78. **Jorgensen, M. J., Cantor, A. B., Furie, B. C., Brown, C. L., Shoemaker, C. B., and Furie, B.,** Recognition site directing vitamin K-dependent γ-carboxylation residues on the propeptide of factor IX, *Cell,* 48, 185, 1987.

79. **Rabiet, M-J., Jorgensen, M. J., Furie, B., and Furie, B. C.,** Effect of propeptide mutations on post-translational processing of factor IX. Evidence that β-hydroxylation and γ-carboxylation are independent events, *J. Biol. Chem.,* 262, 14895, 1987.

80. **Handford, P. A., Winship, P. R., and Brownlee, G. G.,** Protein engineering of the propeptide of human factor IX, *Protein Eng.,* 4, 319, 1991.

81. **Bentley, A. K., Rees, D. J. G., Rizza, C., and Brownlee, G. G.,** Defective propeptide processing of blood clotting factor IX caused by mutation of arginine to glutamine at position −4, *Cell,* 45, 343, 1986.

82. **Huber, P., Schmitz, T., Griffin, J., Jacobs, M., Walsh, C., Furie, B., and Furie, B. C.,** Identification of amino acids in the γ-carboxylation recognition site on the propeptide of prothrombin, *J. Biol. Chem.,* 265, 12467, 1990.

83. **Furie, B. and Furie, B. C.,** Molecular basis of vitamin K-dependent γ-carboxylation, *Blood,* 75, 1753, 1990.

84. **Ulrich, M. M. W., Furie, B., Jacobs, M. R., Vermeer, C., and Furie, B. C.,** Vitamin K-dependent carboxylation. A synthetic peptide based upon the γ-carboxylation recognition site sequence of the prothrombin propeptide is an active substrate for the carboxylase in vitro, *J. Biol. Chem.,* 263, 9697, 1988.

85. **Hubbard, B. R., Jacobs, M., Ulrich, M. M. W., Walsh, C., Furie, B., and Furie, B. C.,** Vitamin K-dependent carboxylation. In vitro modification of synthetic peptides containing the γ-carboxylation recognition site, *J. Biol. Chem.,* 264, 14145, 1989.

86. **Wu, S-M., Soute, B. A. M., Vermeer, C., and Stafford, D. W.,** *In vitro* γ-carboxylation of a 59-residue recombinant peptide including the propeptide and the γ-carboxyglutamic acid domain of coagulation factor IX. Effect of mutations near the propeptide cleavage site, *J. Biol. Chem.,* 265, 13124, 1990.

87. **Price, P. A., Fraser, J. D., and Metz-Virca, G.,** Molecular cloning of matrix Gla protein: implications for substrate recognition by the vitamin K-dependent γ-carboxylase, *Proc. Natl. Acad. Sci. U.S.A.,* 84, 8335, 1987.

88. **Knobloch, J. E. and Suttie, J. W.,** Vitamin K-dependent carboxylase. Control of enzyme activity by the ''propeptide'' region of factor X, *J. Biol. Chem.,* 262, 15334, 1987.

89. **Cheung, A., Engelke, J. A., Sanders, C., and Suttie, J. W.,** Vitamin K-dependent carboxylase: influence of the ''propeptide'' region on enzyme activity, *Arch. Biochem. Biophys.,* 274, 574, 1989.

90. **Cheung, A., Suttie, J. W., and Bernatowicz, M.,** Vitamin K-dependent carboxylase: structural requirements for propeptide activation, *Biochim. Biophys. Acta,* 1039, 90, 1990.

91. **Engelke, J. A., Hale, J. E., Suttie, J. W., and Price, P. A.,** Vitamin K-dependent carboxylase: utilization of decarboxylated bone Gla protein and matrix Gla protein as substrates, *Biochim. Biophys. Acta,* 1078, 31, 1991.

92. **Liska, D. J. and Suttie, J. W.,** Location of γ-carboxyglutamyl residues in partially carboxylated prothrombin preparations, *Biochemistry,* 27, 8636, 1988.

93. **Benton, M. E., Price, P. A., and Suttie, J. W.,** The processivity of the vitamin K-dependent carboxylase, *FASEB J.,* abstr., in press.

94. **Munns, T. W., Johnston, M. F. M., Liszewski, M. K., and Olson, R. E.,** Vitamin K-dependent synthesis and modification of precursor prothrombin in cultured H-35 hepatoma cells, *Proc. Natl. Acad. Sci. U.S.A.,* 73, 2803, 1976.

95. **Fair, D. S. and Bahnak, B. R.,** Human hepatoma cells secrete single chain factor X, prothrombin, and antithrombin III, *Blood,* 64, 194, 1984.

96. **Karpatkin, S., Finlay, T. H., Ballesteros, A. L., and Karpatkin, M.,** Effect of warfarin on prothrombin synthesis and secretion in human HEP G2 cells, *Blood,* 70, 773, 1987.

97. **Fair, D. S. and Marlar, R. A.,** Biosynthesis and secretion of factor VII, protein C, protein S, and the protein C inhibitor from a human hepatoma cell line, *Blood,* 67, 64, 1986.

98. **Graves, C. B., Munns, T. W., Carlisle, T. L., Grant, G. A., and Strauss, A. W.,** Induction of prothrombin synthesis by prothrombin fragments, *Proc. Natl. Acad. Sci. U.S.A.,* 78, 4772, 1981.

99. **Friedman, E. W., Karpatkin, M., and Karpatkin, S.,** Evidence suggesting the regulation of coagulation factor levels in rabbits by a transferable plasma agent, *Blood,* 48, 949, 1976.

100. **Karpatkin, M. and Karpatkin, S.,** Humoral factor that specifically regulates prothrombin (factor II) levels (coagulopoietin-II), *Proc. Natl. Acad. Sci. U.S.A.,* 76, 491, 1979.

101. **Shah, D. V., Nyari, L. J., Swanson, J. C., and Suttie, J. W.,** Effect of a humoral factor on plasma prothrombin and vitamin K-dependent liver carboxylase levels in the rat, *Thrombosis Res.,* 19, 111, 1980.

102. **Karpatkin, S., Chang, R-J., Pierce, W., and Karpatkin, M.,** Effect of coumadin-induced coagulopoietin plasma on vitamin K-dependent carboxylation of liver microsomes, *Br. J. Haematol.,* 55, 673, 1983.

103. **Owens, M. R. and Cimino, C. D.,** A plasma factor enhances activity of vitamin K-dependent coagulation proteins, *Thromb. Haemostas.,* 50, 749, 1983.

104. **Shah, D. V. and Suttie, J. W.,** Coagulopoietin activity in rat plasma: influence of degree of hypoprothrombinemia and reproducibility of the response, *Thromb. Res.,* 39, 43, 1985.

105. **Blanchard, R. A., Furie, B. C., Jorgensen, M., Kruger, S. F., and Furie, B.,** Acquired vitamin K-dependent carboxylation deficiency in liver disease, *New Engl. J. Med.,* 305, 242, 1981.

106. **Liebman, H. A., Furie, B. C., Tong, M. J., Blanchard, R. A., Lo, K-J., Lee, S-D., Coleman, M. S., and Furie, B.,** Des-γ-carboxy (abnormal) prothrombin as a serum marker of primary hepatocellular carcinoma, *New Engl. J. Med.,* 310, 1427, 1984.

107. **Naraki, T. and Watanabe, K.,** Quantitative measurement of PIVKA-II by the enzyme immunoassay diagnostic kit Eitest Mono P-II, in *Vitamin K-Dependent Proteins and Their Metabolic Roles,* Elsevier, New York, 1990, 83.

108. **Shah, D. V., Engelke, J. A., and Suttie, J. W.,** Abnormal prothrombin in the plasma of rats carrying hepatic tumors, *Blood,* 69, 850, 1987.

109. **Kisker, C. T., Perlman, S., Bohlken, D., and Wicklund, B.,** Measurement of prothrombin mRNA during gestation and early neonatal development, *J. Lab. Clin. Med.,* 112, 407, 1988.

110. **Jamison, C. S. and Degen, S. J. F.,** Prenatal and postnatal expression of mRNA coding for rat prothrombin, *Biochim. Biophys. Acta,* 1088, 208, 1991.

111. **Dihanich, M. and Monard, D.,** cDNA sequence of rat prothrombin, *Nucleic Acids Res.,* 18, 4251, 1990.

112. **Degen, S. J. F., Shaefer, L. A., Jamison, C. S., Grant, S. G., FitzGibbon, J. J., Pai, J-A., Chapman, V. M., and Elliott, R. W.,** Characterization of the cDNA coding for mouse prothrombin and localization of the gene on mouse chromosome 2, *DNA Cell Biol.,* 9, 487, 1990.

113. **Diuguid, D. L., Rabiet, M. J., Furie, B. C., Liebman, H. A., and Furie, B.,** Molecular basis of hemophilia B: a defective enzyme due to an unprocessed propeptide is caused by a point mutation in the factor IX precursor, *Proc. Natl. Acad. Sci. U.S.A.,* 83, 5803, 1986.

Chapter 21

ROLE OF ASCORBATE IN REGULATING THE COLLAGEN PATHWAY

Richard I. Schwarz

TABLE OF CONTENTS

0-8493-6961-4/93/$0.00 + $.50

I. INTRODUCTION

Ascorbate is a small molecule with unusual biological activities. It has long been recognized as a reducing agent and for its ability to cure scurvy, yielding its alternate name, Vitamin C.[18,22] Even though these are important attributes, they have overshadowed its other properties. For instance, with the advent of cell culture, it became clear that while humans required ascorbate for survival, cultures of human cells proliferate well in its absence. Consequently, ascorbate is the only required vitamin needed solely for the maintenance of differentiated functions that are necessary at the tissue level. One such function requiring ascorbate is collagen biosynthesis, and it is this interaction of ascorbate with the collagen pathway that is the primary subject of this chapter. The focus will be to resolve where and when changes take place in the collagen pathway after the addition of ascorbate that lead to an increase in collagen biosynthesis:

type-specific procollagen transcription ↔ procollagen mRNA pool size ↔ translation rates ↔ internal procollagen pool size ↔ secretion rates ↔ extracellular cleavage of propeptides ↔ collagen fiber formation and extracellular modifications

But before focusing in more detail on the collagen pathway, one needs a better overview of how ascorbate itself is regulated and how other functions are affected by this molecule. While it is generally realized that humans, other primates, and guinea pigs are incapable of making ascorbate because of a mutation in a required enzyme (L-gulonolactone oxidase),[18] it is less well known that most cells, even in a competent organism, lack the capacity to make ascorbate as well. Instead, the vitamin is made in either the liver or the kidney and transported to other tissues via the blood.[22] One consequence is that the ascorbate concentration in each tissue varies depending on the ability to actively transport ascorbate. In this regard ascorbate acts more like a hormone than a vitamin. So it is not difficult to see why in a competent organism, where most cells do not produce ascorbate, that a mutation that deletes the synthetic capacity would have little effect, as long as the diet had sufficient vitamin C. The reason behind this complex regulation is not known. However, it may relate to the postulated role of ascorbate as a signaling molecule between cell types.[28,36,80]

In addition to its hormone and signaling potential, one finds that ascorbate has many critical biological activities in different tissues in the body. Ascorbate acts as a hydrophilic protector of biological molecules from oxidative damage. Lipids in the blood are shielded from oxidation by the presence of ascorbate.[21,42] It also acts as a biological reducing agent with the ability to reduce ferric to ferrous and cupric to cuprous. Other reducing agents like NADH and glutathione are ineffective. This accounts for its role as a cofactor in many oxygen-utilizing reactions that are catalyzed by enzymes containing these ions: dopamine hydroxylation,[15] carnitine hydroxylation,[62] alpha-amidation of neuroendocrine peptides,[17] and proline hydroxylation.[57,58] The latter has linked ascorbate to the collagen pathway because in collagen half the proline residues are hydroxylated. Since the presence of hydroxyproline in the collagen molecule stabilizes the normal triple helical conformation, the action of ascorbate in altering the collagen pathway would appear to be straightforward. As will be discussed below, however, this is not the case. The collagen pathway is under complex regulation including feedback control. Add to this that collagen production is regulated by other parameters, and the resolution of specific interactions can be difficult unless the experimental conditions are well understood.

To simplify the discussion, this review will concentrate on the direct interaction of ascorbate with type I collagen expression in fibroblasts. This is the collagen type, the cell type, and the mode of action of ascorbate that has been studied in enough detail to yield a coherent picture. This information will then be used as the background for the less developed examples of how ascorbate can affect the expression of other collagen types. This review will also touch on ascorbate's role in altering collagen expression indirectly. In this case, a noncollagen pathway is affected by ascorbate and its product goes on to trigger collagen production.

What will become clear is that ascorbate is a simple molecule, but defining its role as the cure for scurvy has overlooked the potential of even a simple molecule to alter the differentiated state of a cell. As details of the cell biology of ascorbate action have accumulated, it has forced a new perspective on how the cell regulates the synthesis of critical protein-like collagen.

II. HISTORICAL (IMPORTANCE OF CELL CULTURE CONDITIONS AND DEVELOPMENTAL PROGRAMMING)

Ascorbate has been known to play a role in collagen regulation because of its association with the nutritional disease scurvy. Poor wound healing, loose teeth, and other symptoms of the disease can be traced back to an impairment in collagen expression. In the late 1920s Szent-Gorgyi recognized that ascorbate could be vitamin C,[74,75] and in the early 1960s Green and Goldberg showed that cells in culture produced more collagen in the presence of ascorbate (actually, their assays could only show an increase in the number of hydroxyproline residues in collagen; a point discussed more extensively below).[25,26] In 1965, Peterkofsky and Udenfriend found that ascorbate was an essential cofactor in the hydroxylation reaction.[58] This fact finally linked ascorbate to a step in the collagen pathway. Several groups then went on to show that proline hydroxylation stabilizes the triple helical conformation of the collagen molecule (through increased hydrogen bonding) which gave further importance to this step.[4,16,27]

Attempts to further clarify how ascorbate interacted with the collagen pathway focused on cells in culture. But as more detailed questions were asked, more inconsistencies were found in the ability of ascorbate to induce collagen production. Part of the problem could be traced to the collagen assay. In early studies collagen production was measured by the ratio of hydroxy-proline to proline. This could be used as an indicator because in fibroblasts hydroxyproline is found almost exclusively in collagen. However, this assay can be misleading because a cell after addition of ascorbate could change the level of hydroxylation while still producing the same number of collagen molecules. Peterkofsky and Diegelmann in 1971 resolved this problem by developing an assay based on the use of a bacterial collagenase to degrade collagen which is independent of the level of hydroxylation.[56] Even with this improved assay, cell cultures continued to give inconsistent results: some cell types increased their collagen production, some showed no effect, and one cell type was even reported to lower its collagen production.[52,54,65,68]

At this point, if anything was clear, it was that the interaction of ascorbate with the cell was complex. Part of the problem was that other cell culture parameters could affect collagen expression and negate the positive role of ascorbate. This was most clearly shown by Peterkofsky when she took calvaria fibroblasts from the chick embryo and put them in culture.[55] In the embryo these cells produced 60% of their total protein production as procollagen, but in culture, even in the presence of ascorbate, procollagen synthesis dropped to 3%. This experiment raised the question of what could be learned about the mode of action of a positive modulator in the presence of stronger and unknown negative regulators. Another cell culture phenomenon that affected the results was that as cells were passaged for many generations, collagen production decreased. The cell strains first analyzed by Green and Goldberg in the 1960s made ~15% collagen;[26] cell lines, such as 3T3 cells, derived from these strains made 8%.[26] And 10 years later, these lines were only making a couple of percent collagen.[54] So again, even under the same cell culture conditions, cells can lower their collagen production in a way that is not reversible by ascorbate.

In addition to cell culture artifacts, the developmental programming of the cell can alter the responsiveness of a cell to ascorbate. For type I collagen, the major collagen-producing fibroblasts are in tissues that are the structural organizers of the body: bone, tendon, skin, etc. The size of these tissues has to be closely regulated to the overall size of the whole organism.[71] Therefore in early development, when growth is rapid, these tissues grow rapidly; and consequently in the adult, when growth of the whole organism has ceased, so does the growth of these tissues. One would predict that net collagen production would follow a similar pattern being very high in the embryo and falling to a low maintenance level in the adult (Figure 1). Here again, the developmental programming of a cell to lower collagen production restricts the stimulatory effects that one can expect due to ascorbate addition.

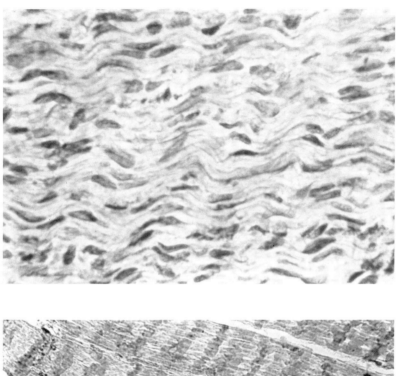

FIGURE 1. Stained frozen sections of chick leg tendons from 16-d embryos (top) and from a 4-month old chicken (bottom: 2.5× lower magnification). Two critical changes occur over this relatively short-time span as the organism goes from a stage of rapid growth to attainment of full size. First, the tissue goes from being hypercellular to being hypocellular. Second, the level of collagen synthesis drops (see Reference 71). The consequence is that the capacity of the tissue to produce collagen is dramatically reduced.

In hindsight, one could predict that the strongest ascorbate effect would be seen in embryonic cells in a cell culture environment permissive for maintaining differentiated function. Indeed, one finds that primary avian tendon (PAT) cells from 16-d chick embryos grown in low serum medium increase their procollagen production from ~10 to ~50% of total protein synthesis on addition of ascorbate.[68] Other cells producing type I procollagen, where a relatively high procollagen rate has been maintained in culture, also show a strong ascorbate effect on collagen production. What has made this so interesting is that there is only evidence for the direct action of ascorbate on the posttranslational modification step of prolyl hydroxylation. The implication is that the general increase in the collagen pathway occurs because most steps are tightly coupled allowing feedback regulation.[66] The strong evidence for this type of control is presented below.

III. ASCORBATE INTERACTION WITH THE COLLAGEN PATHWAY—SECRETION

For 25 years, researchers have been adding ascorbate to cell cultures and analyzing its effects on the collagen pathway. While many changes have been observed, the most consistent effect has been that ascorbate causes cells to maximally hydroxylate their proline residues—between 40 and 50% of the prolines are hydroxylated in the collagen molecule. It had been suggested in the 1950s based on the stability of the collagen helix in different species[16] and then shown in the 1970s based on resistance to protease digestion[4,27] that hydroxylation of proline residues stabilizes the triple helical conformation,[19] a hallmark of the collagen molecule. It is now known that the enzyme, prolyl hydroxylase, prefers the nonhelical form and the reaction is therefore self-limiting as increasing hydroxylation triggers the change to a helical form. The question can then be asked whether this conformation change has any additional effects on the collagen pathway.[2,40]

From early studies using electron microscopy to visualize the changes occurring during wound healing in the scorbutic guinea pig, there were indications that collagen secretion was retarded.[63] This was deduced from swelling inside cells of what was presumed to be unsecreted collagen. Further evidence came from experiments where the hydroxylation reaction was inhibited by using the ferrous ion chelator, α,α'-dipyridyl. Ferrous ion is a necessary cofactor in the hydroxylation reaction; so, removing iron from the reaction would be equivalent to depriving the cells of ascorbate (the role of ascorbate in the hydroxylation reaction is to keep the iron in the reduced state). In these experiments, α,α'-dipyridyl depressed collagen secretion.[40]

In cell culture with 3T3 and chick embryo fibroblasts,[55] Peterkofsky could show that the presence of ascorbate stimulated the secretion of procollagen. In this case, newly synthesized procollagen was tracked by using a short pulse of radioactive proline. The fraction of labeled procollagen in the medium was found to be greater when cells were treated with ascorbate.

Secretion, however, is a complicated parameter that can be easily mis-interpreted. Secretion, unlike the hydroxylation reaction, does not perma-nently alter the molecule, only its location. Therefore, it is the rate of the reaction that is critical. The slow step in procollagen secretion follows first order kinetics (see below). Therefore, the rate is dependent on a rate constant, which changes depending on the presence or absence of ascorbate, and on the procollagen pool size within the cell. The pool size, itself, is dependent on translation, secretion, and internal degradation rates. One consequence is that a cell cannot secrete faster than it can translate except for short transition periods. In the case of 3T3 cells above, where collagen translation rates do not change with ascorbate, the flux of radioactive procollagen to the outside depends on when ascorbate was added and the length of the labeling period used to measure secreted procollagen.[2,55]

This is probably best demonstrated with PAT cells where there are more extensive experimental data. When ascorbate was given to these cells, there was no change in translation rates for several hours,[65] but in an hour or less, the secretion rates increased dramatically (Figure 2).[69] This increase in the secretion rate, but not in the translation rate, caused the internal procollagen pool to fall. As the internal procollagen pool dropped, this reduced the overall rate of secretion so that it again matched the translation rate. With these changes taking place one can see that the length of the radiolabeling period can make a difference in the fraction of the label that is seen inside vs. outside the cell. Except for the short transition period after addition of ascorbate, the flux of procollagen molecules to the outside remains unchanged until the cell can increase its rate of procollagen translation. This dependency on both a secretion rate constant and the procollagen pool size in determining the flux out of the cell has led to confusion in the literature about whether ascorbate does affect the procollagen secretion process.

A significant improvement in the methodology of measuring procollagen secretion rates was made by Kao et al.[33] by adapting a pulse-chase protocol to this problem. Because during the chase the pool of radioactive molecules does not increase, one can focus directly on the rate of loss of radioactive procollagen from the cell. Their data, using freshly isolated tendon cells in short-term suspension culture, showed that procollagen was secreted from the cell with first order kinetics. What was surprising about their experiments was that procollagen secretion is described by two independent rate constants: a fast rate where the half-life of a procollagen molecule in the internal pro-collagen pool is ~15 min (about 60% of the molecules used this route), and a slow rate where the half-life is ~115 min. A variety of interpretations are possible to explain this data, but a simple one will suffice for this discussion. There is a common pool for newly synthesized procollagen inside the en-doplasmic reticulum (ER). From this pool, procollagen molecules are selected for secretion by either a fast or slow route. What made this finding of two routes even more fascinating was the observation that when proline hydroxy-lation was inhibited using α,α'-dipyridyl, the consequence was to completely

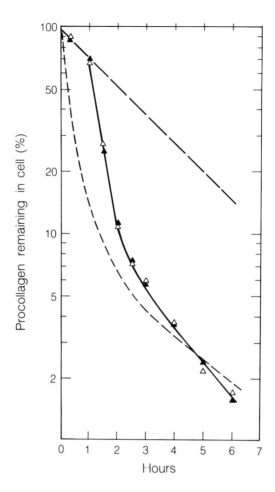

FIGURE 2.   Secretion of ³H-labeled procollagen from PAT cells during the chase period when ascorbate was added only at the time of the chase (the open and closed triangles are independent determinations). The short dashed line shows the secretion profile of cells always in the presence of ascorbate and the long dashed line shows the secretion profile of cells in the absence of ascorbate (for more details of these controls see Reference 69). After ~1 h, the cells are able to hydroxylate the already synthesized procollagen and secrete it at the fast rate.

shift procollagen secretion to the slow route.[34] The implication of this result is that a step in the fast route, possibly a receptor, requires the triple helical conformation of the procollagen molecule in order to proceed.

These pulse-chase experiments raised the question of what the positive effect of ascorbate addition would be on the process of procollagen secretion and whether it would be directly opposite to those obtained using the inhibitor, α,α′-dipyridyl. In the experiments above, freshly isolated chick embryo tendon cells were used, and one cannot ask the question directly. In the chick embryo, ascorbate is made in the kidney and liver, and freshly isolated tendon cells are, therefore, already maximally stimulated with ascorbate. One way to circumvent this problem is to grow the cells in culture without ascorbate for more than 36 h.[65] When pulse-chase experiments were performed using tendon cells in culture, the data obtained by ascorbate addition to scorbutic cultures was completely compatible with the inhibitor studies.[69] With ascorbate, the cells show a biphasic secretion profile with half-lives of ~20 min and ~120 min. In confluent cultures almost all the procollagen (~90%) leaves the cell by the fast route. This may mean that in freshly isolated cells, assayed in suspension culture, that the round shape caused by being unattached to a substratum may partially inhibit the fast route. In any case, scorbutic tendon cells secrete all of their procollagen by the slow route. Because ascorbate is actively and rapidly transported into the cell, one can gauge how quickly secretion rates can change by addition of ascorbate only at the time of the chase (Figure 2). In an hour or less the cells stopped secreting at the slow rate and started secreting the already synthesized procollagen molecules at the high rate. Because the prolyl hydroxylase enzyme is located in the ER,[13,14] this meant that the accumulated pool of procollagen must also be in the ER and that the rate-controlling step in the secretion process is leaving the ER. These experiments also showed that the underhydroxylated procollagen molecule shares the same pool and is capable of switching to the fast route if it becomes hydroxylated and triple helical.

The reason why a triple helical molecule should be secreted six times faster than its nonhelical counterpart is not understood but is critical to a major action of ascorbate on the collagen pathway. The mechanism of procollagen secretion has been debated for 25 years; yet it remains unclear whether procollagen is secreted through the Golgi or more directly out of the cell. Early experiments by Ross et al. argued that the kinetics of labeling during a pulse-chase experiment did not support transport through the Golgi.[63] Marchi and LeBlond tracked procollagen and other proteins using ³H-proline labeling in fibroblast cultures, and they presented evidence that the Golgi was involved.[39] Antibody studies have localized procollagen in the Golgi. Olsen et al. showed by antibody staining that procollagen molecules could be seen in the Golgi in tendon cells under conditions that sufficiently disrupted the membranes and allowed antibody penetration.[43-45,53] On the other hand, Pierce et al.[59] using antibodies could not find type IV procollagen in the Golgi of parietal yolk sack cells. Since EM immunostaining is subject to many potential artifacts depending on the cell system and the antibody, definitive conclusions are hard to draw using this technique alone. Moreover, the kinetic data presented above would argue that procollagen can be secreted in more than one way, making the interpretation of this earlier data even more difficult. In addition, many other questions can be raised. Why is the Golgi apparatus

TABLE 1
Temperature Sensitivity of Secretion from
PAT Cells

°C	Half-life, −asc procollagen	Half-life, +asc procollagen	Half-life noncollagen
41	120[a]	20[b]	70
39	120[a]	20[b]	75
32	120[a]	20[b]	140
29	*	36	*
25	*	*	*

Note: * Not secreted with first order kinetics.

[a,b] Average value presented; differences between temperatures were not statistically significant.

so small in tendon cells when half their total protein production is secreted?[69] Why are secretory vesicles in tendon cells difficult to see by electron microscopy? Why does secreted procollagen remain sensitive to endoglycosidase H,[11] an indication that sugar rearrangements that should have occurred in the Golgi have not taken place in the procollagen molecule? Why is the fast rate of procollagen secretion so insensitive to temperature (Table 1), something one would not expect if secretion were occurring by an energy intensive process such as vesicle transport? Until more is understood about the process of procollagen secretion, it is difficult to give a physical meaning to the sixfold differences in secretion rates that appear to be critically dependent on whether the procollagen molecule is triple helical.

IV. ASCORBATE INTERACTION WITH THE COLLAGEN PATHWAY—TRANSLATION

The addition of ascorbate to fibroblast cultures has been reported to increase, decrease, and not alter the amount of procollagen synthesized.[52,54,65,68] Needless to say, such diversity of action has not led to a consensus for the role of ascorbate in regulating this step. For reasons that will become clear, only cells devoting a significant portion of their total protein synthesis to procollagen are capable of a dramatic response. Over the last 15 years as cell culture techniques have improved, ascorbate has been shown to be a strong inducer in chick tendon[64-68] and bone cells,[1,47] and human skin fibroblast.[8,41,60,76] When these cell types were observed more closely, at least part, and sometimes all, of the increase in procollagen production could be attributed to an increase in procollagen mRNA translation rates.

One of the first indications for actual translational regulation came from the inhibitor studies using α,α′-dipyridyl as described above. One would expect that when secretion rates were reduced by addition of the inhibitor that the pool of procollagen inside the cell would build up dramatically. Instead, Kao et al.[34] observed that the internal procollagen pool remained constant. They concluded that the inhibitor decreased procollagen translation rates (procollagen degradation[5] could have played a role but was not discussed). This possibility for translational regulation could be tested directly using this same cell type in primary culture (PAT cells).[64] In this case α,α'-dipyridyl was given to cells fully induced with ascorbate, making half their total protein production as procollagen, and to scorbutic cells, as a control. The data was plotted as the ratio between counts incorporated into procollagen and those into noncollagen protein (Figure 3). This controlled for the non-specific decrease in protein synthesis ($\sim$40%)[34,64] resulting from a toxic effect of chelation of ferrous ion by the inhibitor. As might be predicted, scorbutic cells showed no specific change in procollagen production, confirming that a cell inhibited at the proline hydroxylation step by a lack of ascorbate cannot be further inhibited by a lack of ferrous ion. In contrast, fully induced cells showed a specific and rapid reduction in procollagen translation rates. By 4 h the level had dropped by 2/3 and then leveled off with increasing time. In the short time period of this experiment, there was no significant change in procollagen mRNA levels.

The question can be raised that if there is a feedback linkage between translation and secretion rates why is the drop in collagen production on addition of α,α'-dipyridyl only 2/3 of the scorbutic rate. The answer is that the linkage between these steps is better than it first appears. When one takes into account that the inhibitor nonspecifically inhibits 1/3 of all protein synthesis, then the procollagen translation rates need only drop by 2/3 to bring them into synchrony with the ability of the cell to secrete what is made.

Translation regulation is also observed when one induces procollagen production with ascorbate (Figure 4). In this case, the efficiency of translating procollagen mRNA increases by $\sim$2-fold and this continues until the cells reach a fully induced state where upon it returns to the original rate.[64] In the context of the kinetics of the induction process, secretion rates are rapidly altered by ascorbate ($<$1 h), but full induction of translation rates requires a much longer time ($\sim$48 h) in large part because of the slow increase in procollagen mRNA levels (starting at 12 h and requiring $\sim$72 h to complete).[64] Until translation rates can equal secretion rates the pool of procollagen within the ER is depressed, and a consequence is that the cell raises its translational efficiency.

The kinetics of the induction process are incompatible with changes in degradation rates playing a significant role.[5] One might assume that scorbutic cells producing nonhelical collagen would have more degradation. This follows from the fact that nonhelical collagen (gelatin) is susceptible to most proteases, but the triple helical collagen molecule requires specific proteases, called collagenases, to denote their unusual ability to attack the triple helical

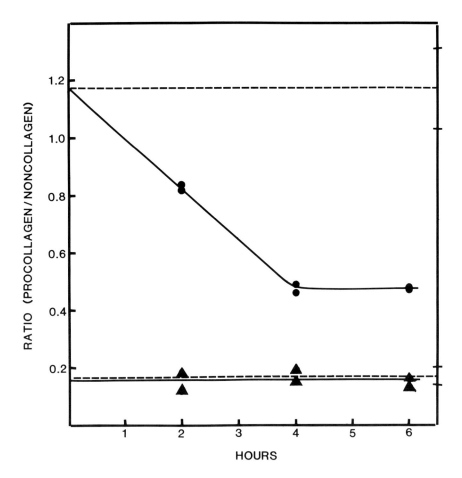

FIGURE 3. Effect of adding α,α'-dipyridyl (0.3 mM) to ascorbate-induced (●) and uninduced (▲) PAT cells.[64] The rates of procollagen synthesis have been normalized to that of noncollagen protein synthesis to correct for the nonspecific inhibition of protein synthesis (~40% in 6 h). Each point in the graph reflects a 2-h pulse ending at the time indicated. The fully induced and uninduced controls are shown as dashed lines $\pm$ SD on the right ordinate.

conformation. As logical as this hypothesis appears to be, the relevant data do not support it. Ascorbate acts quickly to fully hydroxylate procollagen and thereby stabilize the triple helical conformation. This change, if the model were correct, would reflect itself in the ascorbate induction process as an early jump in the levels of net procollagen translation. No such jump is observed, and, instead , net translation rates are not significantly increased in the first few hours after addition of ascorbate.[65]

Pinnell et al.,[60] using human skin fibroblast, have also observed strong evidence for translational control without evidence for a change in degradation

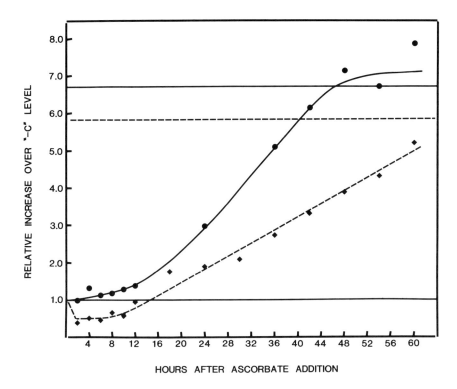

FIGURE 4. Kinetics of the relative increase in procollagen synthesis (ratio of procollagen counts incorporated to noncollagen counts; solid line) and the accumulation of procollagen mRNA after addition of ascorbate to PAT cells (dashed line).[64] Horizontal lines delineate the average maximum values observed (6.8 ± 0.5 [SD] for the protein; 5.9 ± 0.4 for the RNA). During the transition to the induced state, about twice as much procollagen is synthesized for a given amount of procollagen mRNA. The initial drop in procollagen mRNA levels is only occasionally observed and is probably an artifact of the extraction procedure for cytoplasmic RNA using Triton X-100. This weak detergent only extracts a part of the procollagen mRNA that is membrane-bound. Ascorbate addition appears to strengthen the interaction between the procollagen mRNA and the membrane-bound ribosome reducing its extractability. Consistent with this view is the observation that when the detergent SDS is used to extract all the RNA no drop is observed.

rates. On addition of ascorbate, these cells increased their levels of procollagen synthesis by fourfold, but procollagen mRNA levels only increased twofold. Using lysine hydroxylation as a marker of collagen, which was insensitive to the presence of ascorbate, they could show directly that procollagen degradation was not a factor in the translational induction by ascorbate.

Again, using human skin fibroblasts, Chan et al.[8] observed similar changes where collagen synthesis increased six- to sevenfold with ascorbate but procollagen mRNA only increased two- to threefold. Interestingly, after addition of ascorbate, full induction of procollagen synthesis occurred in 10 h and this period was too short for an increase in procollagen mRNA levels to be observed. While this appeared to be much more rapid than the induction in PAT cells, it is not inconsistent with that data. The reason is that procollagen mRNA levels in the human fibroblasts never dropped more than approximately twofold. Since translational efficiency can increase approximately twofold after ascorbate addition, there is no need for mRNA levels to increase in order to obtain full induction. It is the slow increase in mRNA levels that retards the induction process in PAT cells.

Owen et al.[47] have shown that for rat osteoblasts there is a strong ascorbate induction of collagen synthesis from ~2.5 to ~30% after ~36 h. Even though detailed kinetics were not done and no comparisons were made with procollagen mRNA levels, this study again shows that ascorbate can have a dramatic effect on the collagen pathway and the amount of procollagen synthesized.

One can ask why the earlier results with ascorbate were so variable when the more recent studies show a common response to ascorbate. One important difference is that earlier studies used fibroblastic cell lines making low levels of collagen. As discussed before, this could be reflecting cell culture conditions that are dominantly negative for collagen expression.[77] Or it could be reflecting the fact that long-term cultures lower their collagen production by a mechanism that is not reversible by ascorbate addition. In addition, a cell making 1% procollagen no longer taxes the secretory machinery of the cell. At that level of synthesis, changes in the secretion rate may still alter the pool size, but the initial pool would be so small that these changes may not feed back to alter translation rates. Therefore, even though a cell making low levels of collagen could respond to ascorbate, its lack of increase in collagen production could be due to multiple factors.

Finally, a discussion of translational control and collagen regulation would not be complete without reference to some of the unusual features of the translational machinery of the cell. As might be expected, embryonic tendon cells have a high proportion of membrane-bound ribosomes to support the high level of procollagen translation rates.[31,66] This only drops slowly when cells are grown without ascorbate in contrast to the more rapid drop in procollagen synthesis. This leads to a large disparity between the relative number of membrane-bound ribosomes and the level of secreted proteins. A finding that promotes the concept that the number of membrane-bound ribosomes is a characteristic trait of the differentiated state at the time when the cells were isolated. Similarly, the pool of proline and glycine tRNA is elevated in PAT cells and this too is unaffected by the absence of ascorbate but does drop as cells age developmentally.[46] Also, there is evidence to support the model that there are specific sites on membrane-bound ribosomes for each procollagen chain helping to organize the interaction between collagen chains that results in a type I collagen molecule with two α_1 and one α_2 chains.[78,79] Needless to say, much more needs to be known about procollagen translation and its regulation before these observations can be understood in molecular terms.

V. ASCORBATE INTERACTION WITH THE COLLAGEN PATHWAY—mRNA LEVELS

Despite the common assumption that mRNA levels are the rate-controlling step in a protein induction process, its position at the end of this discussion reflects the specific kinetics of the ascorbate induction process. In this case, the last pool to increase after addition of ascorbate is procollagen mRNA levels.[8,64] After ascorbate addition to PAT cells, a sixfold increase in secretion rates is followed by a sixfold increase in translation rates which is followed and overlapped by a sixfold increase in procollagen mRNA levels. As seen in Figure 4 procollagen mRNA levels in PAT cells begin to increase only after 12 h and linearly increase over the next 72 h.[64] As was discussed earlier in human fibroblasts (Chen et al.)[8] where there is only a twofold change in mRNA levels between induced and uninduced states, then this can be compensated for by translational efficiency. The consequence is that the induction process appears to be more rapid. With PAT cells where there is a sixfold change, mRNA levels have to increase in order to achieve full induction. The limitations of having only a single copy gene restricts the amount of transcription and accounts for the slow increase in the procollagen mRNA pool. Transcription rates are increased threefold in fully induced PAT cells as compared to those cells maintained without ascorbate.[38] The pool increases ~6-fold because the half-life of the procollagen mRNA also increases 2-fold from approximately 11 to 20 h.[38]

This raises the important question of how the procollagen mRNA pool is controlled. One must consider that there are in reality two pools—one is a free (a relative term) pool and the other is membrane-bound as part of the translation complex. If one assumes the latter has a long half-life (>20 h) and the former a short half-life (<11 h), then the increased stability of the mRNA after ascorbate addition may only be reflecting the increased translational activity of the procollagen mRNA. Conceptually, this ties mRNA half-life to increases in translational activity and would be a direct link between ascorbate's participation in the hydroxylation step and part of the increase in procollagen mRNA levels. However, this does not resolve how ascorbate increases transcriptional activity of the procollagen genes (α_1 and α_2).

Classically, one would postulate an interaction between ascorbate and a factor that would bind to the promoter region of the collagen gene and stimulate transcription.[61] Alternatively, one can use the lessons learned from other steps in the pathway and postulate a feedback mechanism revolving around the free pool of procollagen mRNA. In the latter case, as the free pool of procollagen mRNA gets depleted by participation in the translational complex, this could trigger more transcription of procollagen mRNA. Even though there is no direct evidence for this feedback mechanism, it has two indirect lines of evidence supporting it. First is the 12-h lag before procollagen mRNA levels start to increase.[64] In the feedback model, the lag is predicted because the other steps would have to be induced before the transcriptional activity could increase. In the case of a transcription factor model, a fairly rapid increase in transcriptional activity would be expected. Ascorbate is rapidly taken up by the cell;[6,69] so, there is no simple explanation for the delay in mRNA induction using this model. Second is the fact that isoascorbate, an isomer whose uptake by the cell is strongly selected against, once inside the cell is as active as ascorbate at inducing procollagen production.[35] These data support the idea that the proline hydroxylation step can not discriminate between the isomers. Since other reducing agents can participate in the reaction *in vitro* it is questionable whether a specific recognition site exists. Since the isomers have the same reducing potential, they are interchangeable at this step. However, one can question whether a transcription factor that specifically bound ascorbate would also bind its isomer with equal efficiency. Again, these questions only make it clear that more needs to be known about the controls regulating procollagen mRNA levels and how these controls are specifically changed during the ascorbate induction process.

VI. ASCORBATE INTERACTION WITH THE COLLAGEN PATHWAY—OTHER COLLAGEN TYPES

The role of ascorbate in the induction of type I collagen production is only now becoming clear and this is the most studied system. The effect of ascorbate addition on other collagen types shows both similarities and differences to the type I example. Some of the differences can be explained by the uniqueness of that particular collagen molecule, other differences may reflect deviations between the specific collagen pathways, and some differences may only reflect an inability to optimize the cell culture conditions for that cell type. With few studies to refer to, the chances for misinterpretation increase. Nevertheless, an analysis of the effect of ascorbate on another collagen type gives an opportunity to categorize the changes in terms of their generality for all collagens from those changes that are more type specific. The information not only reveals differences in the ascorbate induction process but also uniquenesses in the regulation of specific collagen types.

The one change after ascorbate addition that appears to be universal across collagen types is the induction of full hydroxylation of proline residues.[32] The question of whether this change goes on to stimulate procollagen secretion rates is not easily answered for all collagen types due to a lack of data. For instance, types II and X secrete at a higher rate when their chains are fully hydroxylated,[50] but type IV appears to be unaffected.[32] The change in secretion rates with ascorbate addition for types II and X is around sixfold.[50] However, the fast rate has a half-life for secretion of ~120 min which is comparable to the slow rate for type I.[32] So the mechanism for secretion appears to be different for each collagen type. Type IV collagen behaves uniquely but is consistent with its unusual characteristics. Secretion rates are postulated to change for type I collagen because the more hydroxylated chains stabilize the triple helical conformation. With type IV collagen, even fully hydroxylated molecules exist as single chains and may be secreted as single chains.[12,50] Therefore, using the model that the increase in secretion rates is a result of the conformation change of the procollagen molecule, the observation that the secretion of typed IV collagen is not significantly changed by ascorbate is consistent with that model (although more conclusive data is needed to prove the point).

The effect of ascorbate on procollagen translation rates is even more difficult to establish with other cell types. For instance, indirect evidence has shown that ascorbate causes a remodeling of the ER in chondrocytes making collagen types II and X.[51] In type X more procollagen is made after ascorbate, but the kinetics have not been studied in enough detail to show a linkage between this step and secretion rates.[37] With type II the analysis is quite complex. Ascorbate has not been shown to increase synthesis and in some cases there is a decrease.[37] Part of the problem is a consequence of chondrocytes being able to produce both type I and type II collagen.[20,24] The presence of ascorbate, especially for cells grown on plastic, can encourage the transition to a more "fibroblastic" phenotype-reducing type II production.[3] It remains to be seen if a chondrocyte, in suspension culture or in other conditions that stabilize the chondrocyte phenotype, will respond to ascorbate by increasing their type II procollagen translation rates.

The effect of ascorbate on mRNA levels remains to be understood in more detail. In the case of chondrocytes, type X mRNA increases 14-fold, 7 days after ascorbate addition. In the same cells, type II mRNA drops 50% but this occurs independently of ascorbate.[37]

What conclusions can be drawn about the role of ascorbate in regulating other collagen types? Even though there are strong analogies in some situations to the type I example, more studies will have to be done especially with cell types in cell culture conditions permissive for high procollagen expression. As was seen with secretion and type IV collagen, the differences as well as the similarities to the type I situation will be informative about how specific cell types regulate each collagen type.

VII. ASCORBATE INDUCTION OF THE COLLAGEN PATHWAY—OVERVIEW

Ascorbate can induce a major increase in type I collagen production. Conversely, deprivation of ascorbate in humans causes scurvy with a major symptom being an inability to produce type I collagen. It seems strange that a major protein would be regulated by a molecule whose primary function is that of a reducing agent. Moreover, why should an animal such as a chicken that can make its own ascorbate, and presumably never be in a deprived state, still show such a complex inductive response when this artificial situation occurs in culture?

There are no definitive answers to these questions. Nevertheless, a partial explanation comes from the tight interaction between steps in the collagen pathway allowing feedback regulation,[60,66] and from the fact that collagen is the major structural protein of the body, and hence will be under intense regulation. Having a pathway which is tightly coupled allows a tremendous variety in regulatory controls and an ability to match small variations outside the cell to changes inside the cell. In contrast, regulation at the transcriptional level is much cruder when constrained by having only a single copy gene and the requirement for high procollagen mRNA translation rates.[38] It requires long times (3 d)[64] to fully induce mRNA levels (low transcription rates) or to reduce mRNA levels (long half-life of the mRNA, ~1 d)[38] in PAT cells. During development of the chick, changes in mRNA levels that required days to become effective would overlap major alterations in tissue structure. Therefore, in order to adequately regulate collagen production in this time frame requires that collagen production be regulated at additional steps. The regulation by ascorbate is an example of ways collagen regulation can occur at other points in the pathway to fine-tune the process.

The hydroxylation step may have become an important step in regulating the collagen pathway not because of the interaction of ascorbate, but because cell density exerts its control at the same point.[29,30,70] Cultured PAT cells will only make a high percentage of procollagen at a high cell density.[64,68] One can show that one subunit of the prolyl hydroxylase enzyme is regulated by cell density.[30] Moreover, in a tissue, such as a tendon that goes through large variations in cell density over the course of development (Figure 1), regulation at the proline hydroxylation step could be an important part of the developmental control of collagen production. Because ascorbate interacts at the same point, it may only be accidental that the collagen pathway is so dramatically affected by this molecule.

Nevertheless, because the potential exists for ascorbate to regulate the pathway and because ascorbate levels are tightly controlled in the blood stream, ascorbate could act as a messenger to induce collagen production in some cells. There is some support for this[28,36,80] and as more is understood about development regulation, this possibility may become more significant.

VIII. INDIRECT ROLES OF ASCORBATE IN REGULATING COLLAGEN PRODUCTION

Because ascorbate affects many pathways and because the collagen pathway can be altered at many levels, it is not surprising that ascorbate can affect collagen production indirectly. In these cases, the primary action of ascorbate is not on the collagen pathway. As a consequence of that change, however, there is an alteration in collagen synthesis. There are probably many such interactions, but two well-documented cases will be discussed below.

The first involves the complete removal of ascorbate from the diet of the guinea pig. The consequence is that the guinea pig stops eating after two weeks and literally starves.[9] This unusual behavior may result from an imbalance caused by a lack of an essential vitamin. In any case, collagen production drops rapidly and can be shown to be equally affected by food deprivation in the presence of ascorbate.[9] This is an unusual case. Lack of appetite in the guinea pig is associated only with complete removal of the vitamin from the diet and a comparable effect is not observed in humans. Nevertheless, it is a good example of how in the whole organism the ascorbate effect can become quite complex. Peterkofsky and her co-workers have analyzed this problem in detail and they have found that the lack of vitamin C causes a change in the blood.[49] The serum from fasted animals can be shown to reduce the collagen production in cells in culture. The active factor appears to bind to and inhibit the activity of insulin-like growth factor I in the fasted serum.[4,8] How this factor alters the collagen pathway is still under investigation.

The second case involves the oxidation of lipids to aldehydes.[7,10,23] Chojkier et al.[7,10] have shown that aldehydes can induce collagen production in certain cells in culture. The human fibroblasts used were making about 3% collagen, and this increased approximately twofold by either ascorbate or aldehyde addition. In this case, the role of ascorbate is difficult to interpret since in the highly oxygenated environment of cell culture, ascorbate breaks down and can promote the oxidation of the lipids to aldehydes. This mechanism of collagen stimulation may be important in wound repair and in the fibrosis due to injury seen in the lung and other tissues. It is probably not a major route of ascorbate stimulation of collagen synthesis. Ascorbate is thought to prevent lipid oxidation under normal circumstances.[21] In addition, ascorbate and isoascorbate would be expected to have equal activities in stimulating collagen if oxidation was the critical factor since the isomers are oxidized outside the cell at about the same rate.[35] In fact, isoascorbate is about three- to fivefold less effective than ascorbate at stimulating PAT cells because of its inability to be taken up effectively by the cell.[35] Nevertheless, one has to keep in mind when dealing with multifaceted molecules and multiregulated pathways, alternative mechanisms have to be expected and can play important roles in specific cases.

IX. CONCLUSIONS

Ascorbate acts as a specific inducer of collagen production in many cells producing type I collagen. If it had worked directly on transcriptional regulation of the collagen gene, it would have been a mainstream research topic, but it would have revealed little new about the regulation of the collagen pathway except perhaps a new promotor binding site. The fact that a molecule with little more than a good reducing potential can act to specifically stimulate procollagen synthesis requires a new look at protein regulation. The critical concept that this molecule keeps teaching us about the collagen pathway (type I) is that it is tightly coupled through feedback regulation. Therefore, ascorbate levels (or cell density) by controlling the activity of the proline hydroxylation step can control the flow through the whole pathway. Hydroxylation levels control whether the collagen molecule is triple helical. This conformation changes the secretion rates by sixfold. This in turn can alter the pool inside the ER; the pool size alters translational efficiency; and translational efficiency alters the half-life of the procollagen mRNA.

It is also clear that not all collagen types respond to ascorbate in the same way. But this may be an advantage in that the differences may reveal even more about how cells can control tissue-specific functions. The action of ascorbate on other collagen types promises to be an area that will yield important new information about specific points of control within the collagen pathway.

At the same time one has to be aware that in the whole organism ascorbate can affect many biological pathways in nonmesenchymal tissues, and these changes can feedback to alter collagen production. Likewise, the ability of the molecule to form free radicals in an oxygenated environment can modulate the collagen pathway indirectly. In the end, as we better understand the complexities of ascorbate action on the cell, this will all add to our understanding of how the differentiated state can be controlled.

After 70 years since the discovery of ascorbate, we can finally give a molecular model for why scurvy is a collagen disease. But to its credit, this specific inducer of collagen synthesis has been a key to unlocking the complex controls that regulate the differentiated state of the cell.

REFERENCES

1. **Aronow, M. A., Gerstenfeld, L. C., Owen, T. A., Tassinari, M. S., Stein, G. S., and Lian, J. B.,** Factors that promote progressive development of the osteoblast phenotype in cultured fetal rate calvaria cells, *J. Cell. Physiol.,* 143, 213, 1990.
2. **Bates, C. J., Bailey, A. J., Prynne, C. J., and Levine, C. I.,** The effect of ascorbic acid on the synthesis of collagen precursor by 3T6 mouse fibroblasts in culture, *Biochim. Biophys. Acta,* 278, 372, 1972.
3. **Benya, P. D. and Shaffer, J. D.,** Dedifferentiated chondrocytes reexpress the differentiated collagen phenotype when cultured in agarose gels, *Cell,* 30, 215, 1982.
4. **Berg, R. A. and Prockop, D. J.,** The thermal transition of a nonhydroxylated form of collagen. Evidence for a role for hydroxyproline in stabilizing the triple-helix of collagen, *Biochem. Biophys. Res. Commun.,* 52, 115, 1973.
5. **Bienkowski, R. S., Curran, S. F., and Berg, R. A.,** Kinetics of intracellular degradation of newly synthesized collagen, *Biochemistry,* 25, 2455, 1986.
6. **Blanck, T. J. J. and Peterkofsky, B.,** The stimulation of collagen secretion by ascorbate as a result of increased proline hydroxylation in chick embryo fibroblasts, *Arch. Biochem. Biophys.,* 171, 259, 1975.
7. **Brenner, D. A. and Chojkier, M.,** Acetaldehyde increases collagen gene transcription in cultured human fibroblasts, *J. Biol. Chem.,* 262, 17690, 1987.

8. **Chan, D. Lamande, S. R., Cole, W. G., and Bateman, J. F.,** Regulation of procollagen synthesis and processing during ascorbate-induced extracellular matrix accumulation in vitro, *Biochem. J.,* 269, 175, 1990.

9. **Chojkier, M., Spanheimer, R., and Peterkofsky, B.,** Specifically decreased collagen biosynthesis in scurvy dissociated from an effect on proline hydroxylation and correlated with body weight loss, *J. Clin. Invest.,* 72, 826, 1983.

10. **Chojkier, M., Houglum, K., Solis-Herruzo, J., and Brenner, D. A.,** Stimulation of collagen gene expression by ascorbic acid in cultured human fibroblasts: a role for lipid peroxidation, *J. Biol. Chem.,* 264, 16957, 1989.

11. **Clark, C. C.,** The distribution and initial characterization of oligosaccharide unit on the COOH-terminal propeptide extensions of the pro- 1 and pro- 2 chains of type I procollagen, *J. Biol. Chem.,* 254, 10798, 1979.

12. **Crouch, E. and Bornstein, P.,** Characterization of a type IV procollagen synthesized by human amniotic fluid cells in culture, *J. Biol. Chem.,* 254, 4197, 1979.

13. **Cutroneo, K. R., Guzman, N. A., and Sharawy, M. M.,** Evidence for a subcellular vesicular site of collagen prolyl hydroxylation, *J. Biol. Chem.,* 249, 5989, 1974.

14. **Diegelmann, R. F., Bernstein, L., and Peterkofsky, B.,** Cell-free collagen synthesis on membrane-bound polysomes of chick embryo connective tissue and the localization of prolyl hydroxylation on the polysome-membrane complex, *J. Biol. Chem.,* 248, 6514, 1973.

15. **Diliberto, E. J., Jr., Daniels, A. J., and Viveros, O. H.,** Multicompartmental secretion of ascorbic acid and its dual role in dopamine-γ-hydroxylation, *Am. J. Clin. Nutr.,* 54, 1163S, 1991.

16. **Doty, P. and Nishihara, T.,** The molecular properties and thermal stability of soluble collagen, in *Recent Advances in Gelatin and Glue Research,* Stainsberg, G., Ed., Pergamon Press, London, 1957, 92.

17. **Eipper, B. A. and Mains, R. E.,** The role of ascorbate in the biosynthesis of neuroendocrine peptides, *Am. J. Clin. Nutr.,* 54, 1153S, 1991.

18. **England, S. and Seifter, S.,** The biochemical functions of ascorbic acid, *Annu. Rev. Nutr.,* 6, 365, 1986.

19. **Fessler, L. I. and Fessler, J. H.,** Protein assembly of procollagen and effects of hydroxylation, *J. Biol. Chem.,* 249, 7637, 1974.

20. **Finer, M. H., Gerstenfeld, L. C., Young, D., Doty, P., and Boedtker, H.,** Collagen expression in embryonic chicken chondrocytes treated with phorbol myristate acetate, *Mol. Cell. Biol.,* 5, 1415, 1985.

21. **Frei, B.,** Ascorbic acid protects lipids in human plasma and low-density lipoprotein against oxidative damage, *Am. J. Clin. Nutr.,* 54, 1113S, 1991.

22. **Friedrich, W.,** in *Vitamins,* Walter de Gruyter, Berlin, 1988, 929.

23. **Geesin, J. C., Gordon, J. S., and Berg, R. A.,** Retinoids affect collagen synthesis through inhibition of ascorbate-induced lipid peroxidation in cultured human dermal fibroblasts, *Arch. Biochem. Biophys.,* 278, 350, 1990.

24. **Gerstenfeld, L. C., Finer, M. H., and Boedtker, H.,** Quantitative analysis of collagen expression in embryonic chick chondrocytes having different developmental fates, *J. Biol. Chem.,* 264, 5112, 1989.

25. **Green, H. and Goldberg, B.,** Collagen synthesis by human fibroblast strains, *Proc. Soc. Exp. Biol. Med.,* 117, 258, 1964.

26. **Green, H., Goldberg, B., and Todaro, G.,** Differentiated cell types and the regulation of collagen synthesis, *Nature (London),* 212, 631, 1966.

27. **Jimenez, S. A. and Yankowski, R.,** Role of molecular conformation on secretion of chick tendon procollagen, *J. Biol. Chem.,* 253, 1420, 1978.

28. **Kalcheim, C. and Leviel, V.,** Stimulation of collagen production in vitro by ascorbic acid released from explants of migrating avian neural crest, *Cell Diff.,* 22, 107, 1988.

29. **Kao, W.-Y.W., Kao, W.-C. C., and Schwarz, R. I.,** Prolyl hydroxylase production can be uncoupled from the regulation of procollagen synthesis, *Exp. Cell Res.,* 157, 265, 1985.

30. **Kao, W. W-Y. and Berg, R. A.,** Cell density-dependent increase in prolyl hydroxylase activity in cultured L-929 cells requires de novo protein synthesis, *Biochim. Biophys. Acta,* 586, 528, 1979.

31. **Kao, W. W-Y., Flaks, J. G., and Prockop, D. J.,** Primary and secondary effects of ascorbate on procollagen synthesis and protein synthesis by primary cultures of tendon fibroblasts, *Arch. Biochem. Biophys.,* 173, 638, 1976.

32. **Kao, W. W-Y., Mai, S. H., Chou, K-L., and Ebert, J.,** Mechanism for the regulation of post-translational modifications of procollagens synthesized by matrix-free cells from chick embryos, *J. Biol. Chem.,* 258, 7779, 1983.

33. **Kao, W. W-Y., Berg, R. A., and Prockop, D. J.,** Kinetics for the secretion of pro-collagen by freshly isolated tendon cells, *J. Biol. Chem.,* 252, 8391, 1977.

34. **Kao, W. W-Y., Prockop, D. J., and Berg, R. A.,** Kinetics for the secretion of nonhelical procollagen by freshly isolated tendon cells, *J. Biol. Chem.,* 254, 2234, 1979.

35. **Kipp, D. and Schwarz, R. I.,** Effectiveness of isoascorbate versus ascorbate as an inducer of collagen synthesis in primary avian tendon cells, *J. Nutr.,* 120, 185, 1990.

36. **Knaack, D., Podleski, T. R., and Salpeter, M. M.,** Ascorbic acid and acetylcholine receptor expression, *Ann. N.Y. Acad. Sci.,* 498, 77, 1987.

37. **Leboy, P. S., Vaias, L., Uschmann, B., Golub, E., and Adams, S. L.,** Ascorbic acid induces alkaline phosphatase, type X collagen, and calcium deposition in cultured chick chondrocytes, *J. Biol. Chem.,* 264, 17281, 1989.

38. **Lyons, B. L. and Schwarz, R. I.,** Ascorbate stimulation of PAT cells causes an increase in transcription rates and a decrease in degradation rates of procollagen mRNA, *Nucleic Acids Res.,* 12, 2569, 1984.

39. **Marchi, F. and Leblond, C. P.,** Radioautographic characterization of successive com-partments along the rough endoplasmic reticulum-golgi pathway of collagen precursors in foot pad fibroblasts of [³H]proline-injected rats, *J. Cell Biol.,* 98, 1705, 1984.

40. **Margolis, R. L. and Lukens, L. N.,** The role of hydroxylation in the secretion of collagen by mouse fibroblasts in culture, *Arch. Biochem. Biophys.,* 147, 612, 1971.

41. **Murad, S., Grove, D., Lindberg, K. A., Reynolds, G., Sivarajah, A., and Pinnell, S. R.,** Regulation of collagen synthesis by ascorbic acid, *Proc. Natl. Acad. Sci. U.S.A.,* 78, 2879, 1981.

42. **Niki, E.,** Action of ascorbic acid as a scavenger of active and stable oxygen radicals, *Am. J. Clin. Nutr.,* 54, 1119S, 1991.

43. **Nist, C., von der Mark, K., Hay, E. D., Olsen, B. R., Bornstein, P., Ross, R., and Dehm, P.,** Localization of procollagen in chick corneal and tendon fibroblasts with ferritin-conjugated antibodies, *J. Cell Biol.,* 65, 75, 1975.

44. **Olsen, B. R., Berg, R. A., Kishida, Y., and Prockop, D. J.,** Collagen synthesis: localization of prolyl hydroxylase in tendon cells detected with ferritin-labeled antibodies, *Science,* 182, 825, 1973.

45. **Olsen, B. R., Berg, R. A., Kishida, Y., and Prockop, D. A.,** Further characterization of embryonic fibroblasts and the use of immunoferritin techniques to study collagen biosynthesis, *J. Cell Biol.,* 64, 340, 1975.

46. **Ouenzar, B., Weill, D., Agoutin, B., Keith, G., and Heyman, T.,** Relative content of isoaccepting tRNAs for glycine and proline in avian tendon cells with different rates of procollagen synthesis, *Biochim. Biophys. Acta,* 950, 429, 1988.

47. **Owen, T. A., Aronow, M., Shalhoub, V., Barone, L. M., Wilming, L., Tassinari, M. S., Kennedy, M. B., Pockwinse, S., Lian, J. B., and Stein, G. S.,** Progressive development of the rate osteoblast phenotype in vitro: reciprocal relationships in expres-sion of genes associated with osteoblast proliferation and differentiation during formation of the bone extracellular matrix, *J. Cell. Physiol.,* 143, 420, 1990.

48. **Oyamada, I., Palka, J., Schalk, E. M., Takeda, K., and Peterkofsky, B.,** Scorbutic and fasted guinea pig sera contain an insulin-like growth factor I-reversible inhibitor of proteoglycan and collagen synthesis in chick embryo chondrocytes and adult human skin fibroblasts, *Arch. Biochem. Biophys.,* 276, 85, 1990.

49. **Oyamada, I., Bird, T. A., and Peterkofsky, B.,** Decreased extracellular matrix production in scurvy involves a humoral factor other than ascorbate, *Biochem. Biophys. Res. Commun.,* 152, 1490, 1988.

50. **Pacifici, M.,** Independent secretion of proteoglycans and collagens in chick chondrocyte cultures during acute ascorbic acid treatment, *Biochem. J.,* 272, 193, 1990.

51. **Pacifici, M. and Iozzo, R. V.,** Remodeling of the rough endoplasmic reticulum during stimulation of procollagens secretion by ascorbic acid in cultured chondrocytes, *J. Biol. Chem.,* 263, 2483, 1988.

52. **Paz, M. A. and Gallop, P. M.,** Collagen synthesized and modified by aging fibroblasts in culture, *In Vitro,* 11, 302, 1975.

53. **Pesciotta, D. M. and Olsen, B. R.,** The cell biology of collagen secretion, in *Immunochemistry of the Extracellular Matrix,* Vol. II, Furthmayr, H., Ed., CRC Press, Boca Raton, FL, 1982, 1.

54. **Peterkofsky, B.,** The effect of ascorbic acid on collagen polypeptide synthesis and proline hydroxylation during the growth of cultured fibroblasts, *Arch. Biochem. Biophys.,* 152, 318, 1972.

55. **Peterkofsky, B.,** Regulation of collagen secretion by ascorbic acid in 3T3 and chick embryo fibroblasts, *Biochem. Biophys. Res. Commun.,* 49, 1343, 1972.

56. **Peterkofsky, B. and Diegelmann, R.,** Use of mixture of protease-free collagenase for specific assay of radioactive collage in the presence of other proteins, *Biochemistry,* 10, 988, 1971.

57. **Peterkofsky, B. and Udenfriend, S.,** Conversion of proline to collagen hydroxyproline in a cell-free system from chick embryo, *J. Biol. Chem.,* 238, 3966, 1963.

58. **Peterkofsky, B. and Udenfriend, S.,** Enzymatic hydroxylation of proline in microsomal polypeptide leading to formation of collagen, *Proc. Natl. Acad. Sci. U.S.A.,* 53, 335, 1965.

59. **Pierce, G. B., Jones, A., Orfanakis, N. G., Nakane, P. K., and Lustig, L.,** Biosynthesis of basement membrane by parietal yolk sac cells, *Differentiation,* 23, 60, 1982.

60. **Pinnell, S. R., Murad, S., and Darr, D.,** Induction of collagen synthesis by ascorbic acid, *Arch. Dermatol.,* 123, 1684, 1987.

61. **Quinones, S. R., Neblock, D. S., and Berg, R. A.,** Regulation of collagen production and collagen mRNA amounts in fibroblast in response to culture conditions, *Biochem. J.,* 239, 179, 1986.

62. **Rebouche, C. J.,** Ascorbic acid and carnitine biosynthesis, *Am. J. Clin. Nutr.,* 54, 1147S, 1991.

63. **Ross, R. and Benditt, E. P.,** Wound healing and collagen formation, *J. Cell Biol.,* 27, 83, 1965.

64. **Rowe, L. B. and Schwarz, R. I.,** Role of procollagen mRNA levels in controlling the rate of procollagen synthesis, *Mol. Cell. Biol.,* 3, 241, 1983.

65. **Schwarz, R. I., Mandell, R. B., and Bissell, M. J.,** Ascorbate induction of collagen synthesis as a means for elucidating a mechanism for quantitative control of tissue-specific function, *Mol. Cell. Biol.,* 1, 843, 1981.

66. **Schwarz, R. I., Kleinman, P., and Owens, N.,** Ascorbate can act as an inducer of the collagen pathway because most steps are tightly coupled, *N. Y. Acad. Sci.,* 498, 172, 1987.

67. **Schwarz, R. I., Colarusso, L., and Doty, P.,** Maintenance of differentiation in primary avian tendon cells, *Exp. Cell Res.,* 102, 63, 1976.

68. **Schwarz, R. I. and Bissell, M. J.,** Dependence of the differentiated state on the cellular environment: modulation of collagen synthesis in tendon cells, *Proc. Natl. Acad. Sci. U.S.A.,* 74, 4453, 1976.

69. **Schwarz, R. I.,** Procollagen secretion meets the minimum requirements for the rate-controlling step in the ascorbate induction of procollagen synthesis, *J. Biol. Chem.,* 260, 3045, 1985.

70. **Schwarz, R. I.,** Cell-to-cell signaling in the regulation of procollagen expression in primary avian tendon cells, *In Vitro Cell. Dev. Biol.,* 27a, 698, 1991.

71. **Schwarz, R. I., Farson, D. A., and Bissell, M. J.,** Requirements for maintaining the embryonic state of avian tendon cells in culture, *In Vitro,* 15, 941, 1979.

72. **Spanheimer, R. G., Bird, T. A., and Peterkofsky, B.,** Regulation of collagen synthesis and mRNA levels in articular cartilage of scorbutic guinea pigs, *Arch. Biochem. Biophys.,* 246, 33, 1986.

73. **Spanheimer, R. C. and Peterkofsky, B.,** A specific decrease in collagen synthesis in acutely fasted, vitamin C-supplemented, guinea pigs, *J. Biol. Chem.,* 260, 3955, 1985.

74. **Szent-Gyorgyi, A.,** Observations on the function of peroxidase systems and the chemistry of the adrenal cortex. Description of a new carbohydrate derivative, *Biochem. J.,* 22, 1387, 1928.

75. **Szent-Gyorgyi, A. and Haworth, W. N.,** 'Hexuronic acid' (ascorbic acid) as the antiscorbutic factor, *Nature,* 131, 24, 1933.

76. **Tajima, S. and Pinnell, S. R.,** Regulation of collagen synthesis by ascorbic acid. Ascorbic acid increases type I procollagen mRNA, *Biochem. Biophys. Res. Comm.,* 106, 632, 1982.

77. **Valmassoi, J. and Schwarz, R. I.,** High serum levels interfere with the normal differentiated state of avian tendon cells by altering translational regulation, *Exp. Cell Res.,* 176, 268, 1988.

78. **Veis, A. and Kirk, T. Z.,** The coordinate synthesis and cotranslational assembly of type I procollagen, *J. Biol. Chem.,* 264, 3884, 1989.

79. **Veis, A., Leibovich, S. J., Evans, J., and Kirk, T. Z.,** Supramolecular assemblies of mRNA direct the coordinated synthesis of type I procollagen chains, *Proc. Natl. Acad. Sci. U.S.A.,* 82, 3693, 1985.

80. **Vogel, Z., Daniels, M. P., Chen, T., Xi, Z-Y., Bachar, E., Ben-David, L., Rosenberg, N., Krause, M., Duksin, D., and Kalcheim, C.,** Ascorbate-like factor from embryonic brain. Role in collagen formation, basement membrane deposition, and acetylcholine receptor aggregation by muscle cells, *Ann. N. Y. Acad. Sci.,* 498, 13, 1987.

Chapter 22

DIETARY REGULATION OF METALLOTHIONEIN EXPRESSION

Neil F. Shay and Robert J. Cousins

TABLE OF CONTENTS

0-8493-6961-4/93/$0.00 + $.50

I. INTRODUCTION

The divalent cation zinc is a required component of plant and animal cells. The full spectrum of biological roles for this nutrient is only now becoming appreciated. Zinc has been demonstrated to be an integral part of at least 60 to 80 different enzymes.[1] It functions within enzymes providing either a structural or catalytic role. Carbonic anhydrase, for example, has a zinc ion coordinated within the active site of the enzyme, which has the ability to catalyze over 36 million hydrations of carbon dioxide per second.[2] Zinc "fingers" are polypeptide sequences within proteins containing cysteine and histidine residues which allow for coordination of a zinc ion.[3] These structures have been shown to interact with certain DNA sequences and function as part of the DNA binding domain of transcription factors including TFIIIA of *Xenopus*,[4] segmentation genes Kruppel[5] and Hunchback[6] of *Drosophila*, and the human gene ZFY.[7] It has been proposed recently that the availability of zinc for these fingers may be one way that the DNA-binding affinity of these transcription factors, such as Sp1, may be regulated, thereby regulating transcription itself.[8] The zinc binding properties of the LIM motif,[9] which is present in putative transcription factors,[10] is an example of another potential zinc function at the gene level.

Decreased availability of zinc for cells can lead to dysfunction of one or more of these zinc-requiring systems, which then leads to evidence of zinc deprivation. Laboratory animals show the characteristic features of zinc deficiency when supplied with a defined diet reduced in zinc content (e.g., 1 mg Zn per kg diet). Alopecia, dermatitis, and loss of appetite are typically seen as animals become zinc deficient.[11] The murine "lethal milk" (lm) syndrome is a genetic model for zinc deficiency.[12] Lethal milk dams are unable to provide adequate amounts of zinc via milk for their offspring, despite having adequate levels of zinc themselves. Without intervention, 1m pups die. With zinc supplementation, the pups grow in a normal fashion. This zinc-responsive genetic lesion is most likely to reside in the mammary gland. In humans, impaired growth and development, skin lesions, and altered immune function are the most common signs of zinc deficiency.[11] The autosomal recessive disease acrodermatitis enteropathica (AE) is the result of an inability to efficiently absorb zinc from the diet.[13] AE is characterized by alopecia and orificial skin lesions along with other signs of zinc deficiency. The condition is successfully treated with an oral zinc supplement, apparently allowing absorption of sufficient zinc to overcome an impaired intestinal uptake. A similar condition is bovine Adema disease, which is characterized by decreased zinc absorption and immune dysfunction.[14] Collectively, these examples of congenital abnormalities are given to demonstrate that a variety of genes have effects on the metabolic handling and/or utilization of zinc.

The focus of this chapter addresses the effects of dietary zinc on gene expression. The genetic defects described above are examples of altered zinc metabolism and/or alteration of a zinc-requiring cellular function. Although yet undetermined, this presumably involves altered protein structure or levels. In contrast, there is substantial data on the regulation of metallothionein by dietary zinc under practical conditions to suggest zinc has regulatory control over some genes at the transcriptional level. There is recent evidence that the response elements are not unique to this gene. Therefore, the mechanism of transcriptional regulation of metallothionein expression by this micronutrient may, in fact, parallel that for other zinc-responsive genes.

II. METALLOTHIONEIN

Metallothionein (MT) typically exists simultaneously as two distinct yet very similar proteins, termed MT-1 and MT-2. In some species, for example, humans, there may be several subclasses of different MT isoforms.[15] MTs are hydrophilic, low molecular weight proteins (6 to 7 kDa) with a high content (23 to 33 mol%) of cysteine residues. The high cysteine content allows for the binding of heavy metals (usually 7 to 12 atoms/mole) via clusters of thiolate bonds. The hydrophilic nature of the MT protein along with the absence of disulfide bonds, despite the high cysteine content, makes MT a heat stable protein.

A. ROLES OF METALLOTHIONEIN

Since the discovery of metallothionein, the function of this protein has been debated and a number of potential roles have been postulated and tested. The ability to serve as a cellular detoxification system has been well documented.[15] When a cell or organism is challenged with increased amounts of a potentially damaging heavy metal, such as cadmium, metallothionein synthesis is increased. The increased MT may then bind the cadmium, thus reducing the toxic effect of the heavy metal. What is unclear is if this mechanism evolved as a protective mechanism against heavy metals or if it conveniently serves this role in addition to another role, such as the regulation of free zinc levels within cells or protection against free-radical damage.[16] Zinc, a micronutrient required in the diet,[11] also activates the increase in MT synthesis. Increases in MT protein allows for binding of excess zinc, thereby serving as a homeostatic regulatory system for zinc.[16] It remains a largely unanswered question whether the regulation of the required metal zinc, or the toxic metals, such as mercury or cadmium, has been the driving force during the evolution of the metallothionein regulatory system. Antiquity of the metallothionein gene suggests the protein is serving a required, essential role.[15] Others have proposed roles for MT serving as an intracellular free radical scavenger[17] or part of a system protecting the cells from insults such as ionizing radiation or alkylating agents.[18-20]

B. TRANSCRIPTIONAL REGULATION OF METALLOTHIONEIN

The metallothionein cDNA and gene were identified by Palmiter and co-workers.[21] As the protein was already known to be regulated by zinc, the identification of metallothionein promoters allowed for the regulatory mechanism of metal induction to be elucidated. By combining the results of DNA sequence information with the results of deletion mapping studies, unique, short DNA sequences were found that mediated the metal response. These metal response elements, or MREs, were identified in the 5′ regulatory region just upstream from the transcription start site. They were identified in transgenic mice using MT gene constructs containing deletions within the 5′ region with thymidine kinase as the reporter gene, and identifying when a construct had either maintained or lost the characteristic metal response upon addition of zinc or cadmium.[22,23] It was apparent that these MREs were sites for *trans*-acting transcription factors to bind and enhance the basal rate of transcription of the MT genes. The various MT isoforms, and even the individual MREs within one MT gene, have been shown to have a varied response to induction by different metals.[24]

Putative protein factors that bind to the MREs have been identified by a variety of groups using DNA band shift analyses, but it is presently unclear if there are several different MRE binding proteins or if some of these groups have perhaps identified proteolytic fragments of a larger MRE binding protein. Published reports identifying MRE binding proteins as 108,[25] 74,[26] or 39 kDa[27] emphasize the potential heterogeneity of the results. The DNA sequence of a MRE binding protein has not been cloned, nor has an antibody to such a protein been produced. Study of the purified MRE binding proteins will be fundamental to the understanding of how minerals regulate gene expression in mammalian systems.

C. TURNOVER OF METALLOTHIONEIN

The turnover of MT has not been studied as well as the induction process. However, evidence has shown that the type of metal ions bound to MT has a direct influence on the stability of the protein to proteolytic activity. In general, the cellular content of MT protein is a composite of dietary and hormonal inputs vs. metal release and other factors that influence degradation. Readers are referred to more complete discussions of MT turnover than space permits in this chapter.[15,16]

III. DIETARY REGULATION OF METALLOTHIONEIN EXPRESSION

From the discovery of metallothionein as a renal cadmium-associated protein, the relationship between metals and MT has been a primary focus of research on this protein. Along with oral doses or injections of metals into test animals, it has been found that physiologically relevant levels of metals (e.g., zinc) in the diet regulate levels of metallothionein in the kidney, liver, intestine, and other tissues.[15,16] Increases in the abundance of MT is much more striking in kidney, liver, and intestine, most likely due to changes in the metabolic compartment that regulates transcriptional control by metals. Early experiments done to quantitate changes in metallothionein levels by dietary zinc used ^{35}S and/or [65]Zn labeling of cellular protein followed by chromatography of cellular extracts. The increase in MT synthesis has since been shown to be the result of increased MT gene transcription, with maximal levels of MT mRNA abundance occurring a few hours before the increased levels of protein are detected.[28,29]

A. METALLOTHIONEIN LEVELS AND ZINC UPTAKE

Early work tested the alteration in metallothionein as a result of acute changes in dietary zinc concentrations.[31,32] A 1-d zinc-deficient diet (<1 mg zinc per kg) fed to rats resulted in a reduction of the metallothionein protein levels from normal to nearly undetectable levels. Repletion of zinc with a diet containing 150 mg zinc per kg for one additional day boosted metallothionein levels approximately 10-fold over the initial values. Subsequently, another 1 d of zinc depletion dropped MT levels to near trace levels. Also, when rats were deprived of zinc for 7 d and subsequently repleted with 25, 75, or 150 mg zinc per kg diet, both hepatic and intestinal MT increased in proportion to the dietary level. The interpretations made from these studies were (1) MT is rapidly synthesized to bind excess zinc, (2) MT serves as a coordinating ligand to introduce new zinc into the intracellular pool, or (3) MT serves in both functions. In another study, diets containing 1 to 2 g zinc per kg diet fed to rats for up to 8 weeks was necessary to see dramatic increases in liver MT levels. These superinduced MT levels could then be dropped to trace amounts after only 3 d of feeding on a zinc-deficient diet.[32] Actinomycin D would block changes in metallothionein caused by increasing the dietary zinc supply.[30,31]

The temporal relationship between MT mRNA and protein induction with realistic dietary levels of zinc was more closely examined by McCormick et al.[28] Rats were depleted of zinc for 4 d, then, after an overnight fast, were fed by gavage a diet containing 1 ($-Zn$) or 125 ($+Zn$) mg zinc per kg. In response to the $+Zn$ feeding, serum zinc values increased dramatically, peaking at 3 h. They dropped slowly thereafter, not returning to normal levels for 15 h. During this period, the liver cytosol zinc content increased steadily, a portion being associated with increases in both MT-1 and MT-2. MT abundance was threefold higher in the $+Zn$ rats as compared to the $-Zn$ controls. The increase in MT synthesis, as determined by presence of the MT mRNA in polyribosomes, preceded the accumulation of zinc in the liver by a few hours. The rate of MT synthesis increased to a maximum of 4.5-fold over the initial rate 10 h after administration of the $+Zn$ diet. This pattern of events for MT synthesis and zinc accumulation was also seen in the intestine, where MT synthesis was elevated 4-fold in rats between 6 and 9 h after feeding the $+$Zn diet.[33] This coincided with maximal levels of translatable intestinal MT mRNA.

Food restriction in rats has also been shown to increase liver metallothionein levels.[33,34] These were increased up to tenfold in rats fed zinc-adequate or zinc-deficient diets prior to fasting. It has not been demonstrated, however, if this response is due to the liberation of zinc into the systemic circulation from other body stores of zinc, such as muscle, redistribution of liver zinc, a hormonal response due to the fasting, or to a combination of these factors.

Immunological methods have received limited use to measure metallothionein levels in response to dietary treatments. A notable example is the RIA data generated by Bremner and co-workers.[35] Using polyclonal ovine antisera against rat MT-1, they demonstrated that dietary zinc depletion ($<$1 mg zinc per kg) reduced plasma and liver MT-1 levels. Food restriction also elevated liver MT-1 levels. Plasma MT-1 is probably a reflection of hepatic synthesis in response to upregulation of expression. Recently, the introduction of an ELISA method for quantitating human MT protein levels by Grider et al.[36] has given nutritionists a tool that allows the measurement of MT levels in human subjects. The results show that dietary zinc restriction to $<$0.4 mg/day reduces by 59% erythrocyte MT levels in male subjects.[37] Similarly, zinc supplementation increases erythrocyte MT by more than sixfold. These data are assumed to reflect increased MT gene expression in progenitor cells in marrow in response to dietary zinc intake and subsequent release of the MT-containing reticulocytes into the circulation. Erythrocyte MT levels remain elevated as long as supplementation continues, then fall with a half-life of about 10 d. Since this is shorter than the life span of human erythrocytes (ca. 120 d), proteolysis or another event rendering the protein nonantigenic must occur. Preliminary data with rats fed levels of dietary zinc between (1 vs. 180 mg zinc per kg) support the regulation of MT expression in marrow cells.[38]

There is a large volume of data on the inducibility of the metallothionein gene. Only a few papers, which chronicle the evolution of methodology applied to dietary regulation, have been cited here. A number of reviews can be examined for a more complete presentation of the topic.[15,16,24]

B. METALLOTHIONEIN mRNA STUDIES

Blalock et al.[29] reported on a comprehensive set of experiments in rats fed a diet with either deficient, adequate, or supplemented levels of both copper and zinc. Rats were fed one of the nine different combinations of copper (1, 6, or 36 mg/kg) and zinc (5, 30, or 180 mg/kg) using a factorial design. Rats were fed the diet for 14 d, and RNA was purified from liver, kidney, small intestine, and the brain. Oligonucleotide probes were synthesized corresponding to the 5′ region of both rat MT-1 and -2 mRNA sequences. After labeling the oligonucleotides with ^{32}P-ATP, they were used as probes

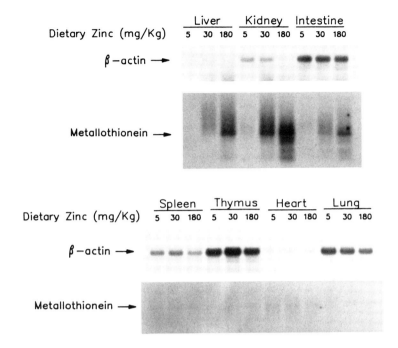

FIGURE 1. Northern blot analysis of metallothionein mRNA (bottom bands) and β-actin mRNA (top bands). Total RNA was extracted from liver, kidney, intestine, thymus, spleen, lung, and heart of rats that were fed diets supplying 5, 30, or 180 mg zinc per kg (as indicated) for 2 weeks. Each dietary group was composed of five rats. Each lane had 7.4 μg of RNA. Hybridization was to a mixed 60-base probe for metallothionein-1 and -2 mRNA plus a 60-base probe for β-actin mRNA. (From Cousins, R. J. and Lee-Ambrose, L. M., *J. Nutr.*, 122, 56, 1992. With permission.)

for MT mRNA in the RNA samples affixed to nitrocellulose filters. After hybridization with a 21-mer and washing, the filters were used for autoradiography. Then ^{32}P in each spot was quantitated using liquid scintillation counting. It was found that the kidney, liver, and intestine all demonstrated increases in MT mRNA abundance due to increased zinc in the diet, with no effect seen in brain MT levels. The abundance of MT mRNA as determined by autoradiography was highest in kidney and lower in the other tissues. However, estimations of numbers of MT mRNA molecules per cell due to differing amounts of zinc intake yielded relatively similar differences in these tissues. Data from both methods of presentation show that copper at these realistic dietary intakes does not affect MT expression in rats.

More recent data using 60-mer individual probes for MT-1 and MT-2 mRNAs have shown the same general response to dietary zinc at the deficient, adequate and supplemental intake level.[39] As shown in Figure 1, the decreasing order of abundance is kidney > liver > intestine > heart > spleen. These

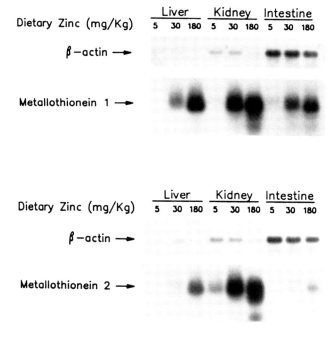

FIGURE 2. Northern blot analysis of metallothionein-1 mRNA, metallothionein-2 mRNA, and β-actin. Top panels: Metallothionein-1 mRNA (bottom bands) and β-actin (top bands). Bottom panels: Metallothionein-2 mRNA (bottom bands) and β-actin (top bands). Total RNA was extracted from liver, kidney, and intestine of rats that were fed diets supplying 5, 30, or 180 mg zinc per kg (as indicated) for 2 weeks. Each dietary group was composed of five rats. Each lane had 7.4 μg of RNA. Hybridization was to individual 60-base probes for metallothionein-1 or -2 mRNA plus a 60-base probe for β-actin mRNA. (From Cousins, R. J. and Lee-Ambrose, L. M., *J. Nutr.*, 122, 56, 1992. With permission.)

probes do not reveal differences in expression in thymus or lung. Of interest is that it appears that MT-1 mRNA is the more abundant message (Figure 2). The significance of that difference is unclear at this time.

Huber and Cousins[40] demonstrated that maternal dietary zinc deficiency can influence fetal gene expression. Pregnant rats were fed purified diets containing 1, 5, 30, or 180 mg zinc per kg representing deficient, marginal, adequate, and supplemented levels of zinc, respectively. At the 20th d of gestation, metallothionein mRNA and protein levels were measured in the maternal liver, kidney, and placenta as well as the fetal liver. Fetal liver MT mRNA and protein levels were found to be depressed when dams were fed the deficient and marginal zinc diets (Figure 3). Maternal levels of MT mRNA and protein were likewise regulated by zinc, with the lowest amounts found in the rats fed 1 and 5 mg zinc per kg diets and highest in rats fed the 180 mg zinc per kg diet. The significance of this study is apparent considering

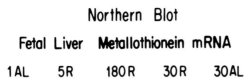

Northern Blot

Fetal Liver Metallothionein mRNA

1AL   5R   180R   30R   30AL

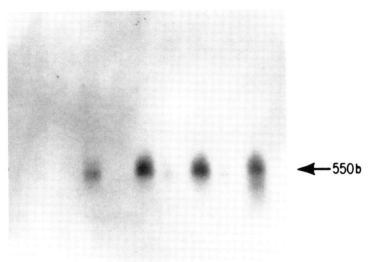

←—550b

FIGURE 3. Northern blot analysis of fetal liver metallothionein mRNA. Pregnant rats were fed dietary zinc at 1, 5, 30, or 180 mg/kg diet from days 13 through 20 of gestation. Intake was either restricted (R) to that of the 1 mg zinc per kg group or *ad libitum* (AL). Total RNA from fetal liver was fractionated in 1.1% agarose gels. Hybridization of RNA transferred by Northern blot was to a ^{32}P-labeled 60-base probe oligonucleotide DNA probe for metallothioneins 1 and 2. The 550b metallothionein mRNA band is indicated. (From Huber, K. L. and Cousins, R. J., *J. Nutr.*, 118, 1570, 1988. With permission.)

the impairment of fetal development due to zinc deficiency.[41] The metallothionein levels in fetal liver were many fold higher than levels found in adult male and female rats, correlating with previous reports from other laboratories.[15] This increased level of MT may be serving to provide adequate amounts of zinc to zinc-requiring metalloproteins and cellular sites necessary in the developmental program of the fetus. Alternatively, the MT may be serving as a "sink" for zinc and other metals to supply the neonates with adequate mineral levels for an initial period after birth.

C. OTHER STUDIES

Although metallothionein has been consistently considered a cytosolic protein, immunocytochemical localization[42] has confirmed the presence of metallothionein in the nucleus. Whether this is real or an artifactual contamination has not been definitively addressed. However, with the identification recently of zinc-finger-containing transcription factors, the possibility of metallothionein playing a role in nuclear regulation of zinc may exist. In fact, Zeng et al.[8] have speculated that metallothionein may play a role in the regulation of zinc available for these zinc-containing nuclear metalloproteins, perhaps adding another level of control to the process of gene regulation. Specifically, they have shown that apo-metallothionein can remove zinc from the transcription factor Sp1, decreasing the DNA-binding activity of the protein. Since the binding constant of MT for metals is considerably greater than other proteins, they propose metal exchange from Sp1 to apo MT occurs. Presumably, the nuclear level of MT protein could regulate the availability of Zn for Sp1 and influence regulation of specific genes.

Cousins and Lee-Ambrose[39] have investigated the relationship between the dietary regulation of metallothionein and the uptake of zinc into the nucleus in a variety of rat tissues to determine the changes in nuclear zinc brought about by dietary change. Growing male rats were fed 5, 30, or 180 mg zinc per kg for a 2-week period. Nuclei purified by sucrose gradient methods from the liver, kidney, and spleen had approximately the same total zinc content as measured by atomic absorption. Subsequently, a constant dietary zinc intake of either 5, 30, or 180 mg zinc per kg was provided to rats for 2 d, followed by gavage of liquified diet containing those same levels of zinc. By tagging the gavaged zinc with ^{65}Zn, it was shown that substantial amounts of dietary zinc entered cell nuclei shortly after the oral feeding. Two hours after this oral dose, nuclei were isolated from the tissues. Intestine and liver accumulated substantial amounts of zinc and directly reflected the dietary zinc level as shown in Table 1. For example, nuclei from kidney accumulated up to 6.2% of the zinc in the oral dose during the 2-h period. Additionally, chromatography of extracts from liver nuclei showed that ^{65}Zn introduced into the portal supply bound to specific fractions of nuclear proteins, including one specific fraction that could contain the zinc-containing metal responsive element binding protein (MRE-BP) that stimulates transcription of the metallothionein genes. Clearly, this study shows the rapidity of the association of dietary zinc with nuclear proteins, consistent with the known temporal regulation of the metallothionein gene by zinc administration. Westin and Schaffner[43] showed that a MREd binding protein from nuclear extracts of HeLa cells binds DNA in a zinc-dependent fashion. Andersen et al.[27] and Searle[23] have made similar observations. This supports the hypothesis that an influx of zinc from the diet rapidly reaches the nucleus and influences transcription. The interrelationships among dietary zinc, MREs, and MRE-BP are presented in Figure 4.

At the interface between contemporary molecular biology and traditional dietary studies is the elegant transgenic mouse work done by Palmiter and others.[22,23] Chimeric fusion genes were created using recombinant techniques.

TABLE 1
Nuclear Zinc Derived from
Dietary Zinc[a]

Dietary Zn intake (mg Zn/kg)	(% of total nuclear zinc derived from dietary zinc		
	Liver	Kidney	Spleen
5	0.1	0.3	0.1
30	0.8	0.9	0.3
180	4.8	6.2	1.6

[a] Rats were fed 30 mg zinc per kg of diet and total zinc concentration of nuclei purified by sucrose gradient methods was measured. Other rats were fed a purified diet with 5, 30, or 180 mg zinc per kg containing ^{65}Zn 2 h prior to purification of nuclei. Dietary zinc uptake was calculated from the specific activity of the labeled diet.

Derived from Cousins, R. J. and Lee-Ambrose, L. M., *J. Nutr.*, 122, 56, 1992. With permission.

These genes had the mouse MT-1 promoter fused to the coding sequences of a variety of genes including rat and human growth hormones. The gene constructs were injected into the pronucleus of fertilized mouse ova. The microinjected ova were then reimplanted into the oviducts of surrogate mothers. Upon screening the offspring, it was found that approximately 25% were expressing the transgenes. The level of growth hormone assayed in the transgenic animals was found to be regulated by zinc levels in coordination with the native MT genes. This technology has many applications where dietary zinc could provide the signal for transcriptional activation of transgenes. Zinc is relatively nontoxic, and we have shown that the native MREs provide excellent regulation by this micronutrient at the physiological range of dietary intakes.

Of equal importance to zinc regulation of transgenes is the effect dietary zinc could potentially have in metallothionein-related drug resistance. There is convincing evidence that resistance to many alkylating agents (e.g., cisplatin) used in chemotherapy is directly related to MT expression.[18-20] If methods could be developed to block the role of dietary zinc in this activation, drug resistance may be decreased.

IV. OTHER GENES REGULATED BY DIETARY ZINC

Recently, Yiangou et al.[44] determined that several acute-phase protein genes are induced by heavy metals. The mRNAs for C-reactive protein and alpha-1-acid glycoprotein were found to be induced severalfold in the livers

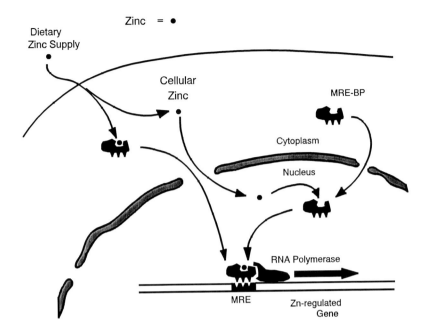

FIGURE 4. Zinc-regulated gene expression. Diagram illustrates zinc-dependent gene transcription. Zinc (●) supplied directly by the diet and indirectly from the endogenous supply enters cells and is most likely found bound to metallothionein or other zinc-binding proteins. Metallothionein may buffer or regulate the available free zinc supply. The zinc may bind in the cytoplasm to the metal responsive element-binding protein (MRE-BP) or may be carried into the nucleus by metallothionein or other proteins before interacting with the MRE-BP. Once the MRE-BP has bound one or more zinc ions, it presumably has a higher affinity for the MRE sequences of specific promoters and enhances transcription of the Zn-regulated gene by RNA polymerase. The MRE-BP is shown here directly interacting with RNA polymerase—evidence for this interaction has not yet been conclusively demonstrated.

of Balb/c mice treated with a variety of heavy metals given by i.p. injection. Mercury elicited the strongest response, followed in order of response by Cd > Pb > Cu > Ni > Zn > Mg. Other acute-phase reactants, such as alpha-1-antitrypsin and alpha-1-antichymotrypsin, were unaffected by the metal treatment. Additionally, albumin, which is downregulated during acute-phase, was not repressed by the metal treatment, which indicates that the induction of these mRNAs was not due to a metal-mediated inflammatory response. Adrenalectomized animals were also challenged with metals, and the response was still present, indicating that the induction by metals was not mediated by the glucocorticoid induction pathway. They also noted that some DNA sequence homologies exist between the MREs of the MT genes and sequences in the promoter region of the alpha-1-acid glycoprotein and C-reactive protein genes. Metal induction potential of these acute phase genes were tested in a transiently tranfected cell system. Approximately 600 bp of the alpha-1-acid glycoprotein promoter sequence was linked to the bacterial chloramphenicol (CAT) gene in a pP5-CAT transfection vector. This plasmid was then used to transiently transfect human HepG2 cells growing in culture. Addition of mercury to the culture medium (1 μM to 1 nM) was sufficient to stimulate CAT activity three- to fourfold in the HepG2 cells. This cell system, free of the extraneous influences of hormones that would be produced at other sites *in vivo*, strongly suggests that these genes are being directly regulated by the presence of metals, and the site of interaction is probably the MRE-like sequences found in the promoter region of these genes. Certainly, analysis of deletion mutants by transfection and CAT assay will determine which sites within the promoter are responsible for the metal induction. Dietary regulation of these genes has not been examined, but would appear to be a promising area to study.

Shay and Cousins[45] have conducted a "plus-minus" or differential hybridization experiment with the express intent of finding mRNAs in the intestine and skin that are regulated by zinc levels in the diet. Rats were fed a purified diet for 14 d containing either a deficient (1 mg zinc per kg) or adequate (30 mg zinc per kg) level of zinc, after which mRNA was purified from the intestine and skin. cDNA plasmid libraries were made and screened with ^{32}P-labeled cDNA probes derived from the zinc-deficient and -adequate pools of mRNA. We have identified several mRNAs that are regulated by the zinc levels in the diet and are currently characterizing these mRNAs by measuring their relative abundance under a variety of dietary conditions, determining their tissue specificities, and sequencing the respective cDNAs to determine the identity of these mRNAs or other homologous sequences identified by DNA sequence data base search.

Finally, Hempe and Cousins[46] discovered a low molecular weight zinc-binding protein present in rat intestinal mucosa cells that could be a candidate protein for an intracellular carrier of zinc. The protein could potentially be functioning as a transporter of zinc from the luminal brush border to the circulatory system.[47] The protein was identified using [65]Zn ligand blotting of SDS-PAGE/size-fractionated protein as well as dot blotting of cytosolic protein fractions size-fractionated by Superose-HPLC. The protein was blotted onto a PVDF membrane following SDS-PAGE, excised, and used for protein microsequencing. Analysis of the N terminal amino acid sequence allowed the known protein data bases to be searched, and it was found that this protein was the "Cysteine-Rich Intestinal Protein" (CRIP) identified as a cDNA sequence by Birkenmeier and Gordon[48] in 1986. They screened a rat intestine cDNA library to identify cDNA clones induced during weaning. The CRIP cDNA was isolated, determined that it was nearly tissue-specific to the intestine, and that the mRNA was induced from low basal levels during suckling to a higher, moderately abundant concentration after weaning.

Birkenmeier and Gordon[48] noted some homology between CRIP and certain bacterial ferredoxins, whereas the similarity between CRIP and the histidine and cysteine residues of the ''LIM'' motif was recognized by Freyd et al.[10] The latter is found in some transcription factors and has some features in common with the consensus sequence of zinc-finger proteins. Certainly, the structural features of this protein coupled with the nearly absolute localization to the intestine and induction at weaning give us strong circumstantial evidence that this zinc-binding protein is crucially important to nutrient absorption. It will be both interesting and a challenge to design experiments that define the nutritional regulation of CRIP.

V. SUMMARY

Metallothionein has proved to be enigmatic to those who have chosen to study it. The mechanism of MT gene regulation has been clearly defined. It is transcriptionally activated not only by numerous hormones and other cellular factors, particularly in a tissue-specific mode by various cytokines, but uniquely by dietary zinc. Data suggest that dietary copper does not have a major effect on MT gene expression at practical intake levels. This metal regulation is conferred by MRE sequences in the promoter region that apparently bind the putative MRE-BP that ''senses'' the metal status within the cell and elevates transcription of MT genes when bound to the MRE sequences. Several groups have identified candidate MRE-binding proteins. It is not clear at the present time if there is a single or multiple species of this protein. Metallothionein gene regulation has been so well understood that it has served not only as a model for nutritionists interested in gene regulation but for all biologists interested in gene expression. The metal regulation of the MT promoter has been used in a variety of chimeric constructs allowing genes of interest to be put under the control of zinc concentrations in transfected cell systems. These systems have been so widely accepted that transfection vectors are now commercially available using the MT promoter to control inserted cDNA coding sequences.

In contrast to the mechanism of expression, the function of metallothionein has proved to be a frustration. Clearly, the protein is involved in toxicity protection,[21] and recent genetic studies done in yeast suggest that the major function of metallothionein may be as a free radical scavenger,[20] a function that has been investigated for a number of years by various groups. The metal regulation of MT also confers on it a role in metal homeostasis *in vivo*, and we have tried to present here some results from investigations involving dietary regulation of metallothionein by zinc and the implications of that regulation on cellular zinc availability, in particular.

Recently, several genes have been identified that are either regulated by metals or are serving important functions in the absorption or homeostasis of minerals. With the current pace of advances in molecular cell biology, it is hoped that the roles of these genes in nutrition will be rapidly determined.

REFERENCES

1. **Vallee, B. L. and Galdes, A.,** The metallobiochemistry of zinc enzmes, *Adv. Enzymol.,* 57, 283, 1984.
2. **Lehninger, A. L.,** *Principles of Biochemistry,* Worth Publishers, New York, 1982, 220.
3. **Miller, J., McLachlan, A. D., and Klug, A.,** Repetitive zinc-binding domains in the protein transcription factor IIIA from Xenopus oocytes, *EMBO J.,* 4, 1609, 1985.
4. **Engelke, D. R., Ng, S.-Y., Shastry, B. S., and Roeder, R. G.,** Specific interaction of a purified transcription factor with an internal control region of 5S RNA genes, *Cell,* 19, 717, 1980.
5. **Rosenberg, U. B., Schroder, C., Preiss, A., Kienlen, A., Cote, S., Riede, I., and Jackle, H.,** Structural homology of the product of the Drosophila Kruppel gene with Xenopus transcription factor IIIA, *Nature,* 319, 336, 1986.
6. **Tautz, D., Lehman, R., Schnurch, H., Schuh, R., Seifert, E., Kienlin, A., Jones, K., and Jackle, H.,** Finger protein of novel structure encoded by hunchback, a second member of the gap class of Drosophila segmentation genes, *Nature (London),* 327, 383, 1987.
7. **Page, D. C., Mosher, R., Simpson, E. M., Fisher, E. M. C., Mardon, G., Pollack, J., McGillivray, B., de la Chapelle, A., and Brown, L. G.,** The sex-determining region of the human Y chromosome encodes a finger protein, *Cell,* 51, 1091, 1987.
8. **Zeng, J., Heuchel, R., Schaffner, W., and Kagi, J. H. R.,** Thionein (apometallothionein) can modulate DNA binding and transcription activation by zinc finger containing factor Sp1, *FEBS Lett.,* 279, 310, 1991.
9. **Li, P. M., Reichert, J., Freyd, G., Horvitz, H. R., and Walsh, C. T.,** The LIM region of a presumptive Caenorhabditis elegans transcription factor is an iron-sulfur and zinc-containing metallodomain, *Proc. Natl. Acad. Sci. U.S.A.,* 88, 9210, 1991.
10. **Freyd, G., Kim, S. K., and Horvitz, H. R.,** Novel cysteine-rich motif and homeodomain in the product of the Caenorhabditis elegans cell lineage gene lin-II, *Nature (London),* 344, 876, 1990.
11. **Hambidge, K. M., Casey, C. E., and Krebs, N. F.,** Zinc, in *Trace Elements in Human and Animal Nutrition II,* Mertz, W., Ed., Academic Press, Orlando, 1986, 1.
12. **Piletz, J. E. and Ganschow, R. E.,** Zinc deficiency in murine milk underlies expression of the lethal-milk (lm) mutation, *Science,* 199, 181, 1978.
13. **Lombeck, I., Schnippering, H. G., Ritzl, F., Feinendegen, L. H., and Bremner, H. J.,** Absorption of zinc in acrodermatitis enteropathica, *Lancet,* 1, 855, 1975.
14. **Flagstad, T.,** Zinc absorption in cattle with a dietary picolinic acid supplement, *J. Nutr.,* 111, 1996, 1981.
15. **Kagi, J. H. R. and Kojima, Y.,** Chemistry and biochemistry of metallothionein, in *Metallothionein II,* Experientia Supplementum, Vol. 52, Kagi, J. H. R. and Kojima, Y., Eds., Birkhauser Verlag, Basel, 1987, 25.
16. **Dunn, M. A., Blalock, T. L., and Cousins, R. J.,** Metallothionein, *Proc. Soc. Exp. Biol. Med.,* 185, 107, 1987.
17. **Thornalley, P. J. and Vasak, M.,** Possible role for metallothionein in protection against radiation-induced oxidative stress. Kinetic and mechanism of its reaction with superoxide and hydroxyl radicals, *Biochim. Biophys. Acta,* 827, 36, 1985.
18. **Lohrer, H. and Robson, T.,** Overexpression of metallothionein in CHO cells and its effect on cell killing by ionizing radiation and alkylating agents, *Carcinogenesis,* 10, 2279, 1989.
19. **Kelley, S. L., Basu, A., Teicher, B. A., Hacker, M. P., Hamer, D. H., and Lazo, J. S.,** Overexpression of metallothionein confers resistance to anticancer drugs, *Science,* 241, 1813, 1988.

20. **Kaina, B., Lohrer, H., Karin, M., and Herrlich, P.,** Overexpressed human metal-lothionein IIA gene protects Chinese hamster ovary cells from killing by alkylating agents, *Proc. Natl. Acad. Sci. U.S. A.,* 87, 2710, 1990.

21. **Durnam, D. M. and Palmiter, R.,** Transcriptional regulation of the mouse metallothi-onein-I gene by heavy metals, *J. Biol. Chem.,* 256, 6712, 1981.

22. **Stuart, G. W., Searle, P. F., Cen, H. Y., Brinster, R. L. and Palmiter, R. D.,** A 12-base-pair DNA motif that is repeated several times in metallothionein gene promoters confers metal regulation to a heterologous gene, *Proc. Natl. Acad. Sci. U.S.A.,* 81, 7318, 1984.

23. **Searle, P. F.,** Zinc dependent binding of a liver nuclear factor to metal response element MRE-a of the mouse MT-1 gene and variant sequences. *Nucleic Acids Res.,* 18, 4683, 1990.

24. **Hamer, D.,** Metallothionein, *Annu. Rev. Biochem.,* 55, 913, 1986.

25. **Seguin, C. and Prevost, J.,** Detection of a nuclear protein that interacts with a metal regulatory element of the mouse MT-1 gene, *Nucleic Acids Res.,* 16, 10547, 1988.

26. **Imbert, J., Zafarullah, M., Culotta, V. C., Gedamu, L., and Hamer, D.,** Transcrip-tion factor MBF-I interacts with metal regulatory elements of higher eucaryotic MT genes, *Mol. Cell. Biol.,* 9, 5315, 1989.

27. **Andersen, R. D., Taplitz, S. J., Oberbauer, A. M., Calame, K. L., and Herschman, R.,** Metal-dependent binding of a nuclear factor to the rat MT-1 promoter, *Nucleic Acids Res.,* 18, 6049, 1990.

28. **McCormick, C. C., Menard, M. P., and Cousins, R. J.,** Induction of hepatic MT by feeding zinc to rats of depleted zinc status, *Am. J. Physiol.,* 240, E414, 1981.

29. **Blalock, T. L., Dunn, M. A., and Cousins, R. J.,** Metallothionein gene expression in rats: tissue-specific regulation by dietary copper and zinc, *J. Nutr.,* 118, 222, 1988.

30. **Richards, M. P. and Cousins, R. J.,** Metallothionein and its relationship to the me-tabolism of dietary zinc in rats, *J. Nutr.,* 106, 1591, 1976.

31. **Richards, M. P. and Cousins, R. J.,** Zinc-binding protein: relationship to short term changes in zinc metabolism, *Proc. Soc. Exp. Biol. Med.,* 153, 52, 1976.

32. **Whanger, P. D. and Weswig, P. H.,** Effect of some copper antagonists on induction of ceruloplasmin in the rat, *J. Nutr.,* 100, 341, 1970.

33. **Menard, M. P., McCormick, C. C., and Cousins, R. J.,** Regulation of intestinal metallothionein biosynthesis in rats by dietary zinc, *J. Nutr.,* 111, 1353, 1981.

34. **Bremner, I. and Davies, N. T.,** The induction of metallothionein in rat liver injection and restriction of food intake, *Biochem. J.,* 149, 733, 1975.

35. **Sato, M., Mehra, R. K., and Bremner, I.,** Measurement of plasma metallothionein-1 in the assessment of the zinc status of zinc-deficient and stressed rats, *J. Nutr.,* 114, 1683, 1984.

36. **Grider, A., Kao, K. J., Klein, P. A., and Cousins, R. J.,** Enzyme-linked immuno-sorbent assay for human metallothionein: correlation of induction with infection, *J. Lab. Clin. Med.,* 113, 221, 1989.

37. **Grider, A., Bailey, L. B., and Cousins, R. J.,** Erythrocyte metallothionein as an index of zinc status in humans, *Proc. Natl. Acad. Sci. U.S.A.,* 87, 1259, 1990.

38. **Huber, K. L. and Cousins, R. J.,** K562 cells respond to zinc and alter metallothionein gene expression during differentiation, *FASEB J.,* 6, 873A, 1992.

39. **Cousins, R. J. and Lee-Ambrose, L. M.,** Nuclear zinc uptake and interactions and metallothionein gene expression are influenced by dietary zinc in rats, *J. Nutr.,* 122, 56, 1992.

40. **Huber, K. L. and Cousins, R. J.,** Maternal zinc deprivation and interleukin-1 influence metallothionein gene expression and zinc metabolism of rats, *J. Nutr.,* 118, 1570, 1988.

41. **Hurley, L. S.,** Teratogenic aspects of manganese, zinc and copper nutrition, *Physiol. Rev.,* 61, 249, 1981.

42. **Cherian, M. G., Templeton, D. M., Gallant, K. R., and Banerjee, D.,** Biosynthesis and metabolism of metallothionein in rat during perinatal development, in *Metallothionein II,* Experientia Supplementum, Vol. 52, Kagi, J. H. R. and Kojima, Y., Eds., Birkhauser Verlag, Basel, 1987, 499.

43. **Westin, G. and Schaffner, W.,** A zinc-responsive factor interacts with a metal-regulated enhancer element (MRE) of the mouse metallothionein-1 gene, *EMBO J.,* 7, 3763, 1988.

44. **Yiangou, M., Ge, X., Carter, K. C., and Papaconstantinou, J.,** Induction of several acute-phase protein genes by heavy metals: a new class of metal-responsive genes, *Biochemistry,* 30, 3798, 1991.

45. **Shay, N. F. and Cousins, R. J.,** Identification and cloning of zinc-regulated mRNAs from rat intestine, *FASEB J.,* 6, A483, 1992.

46. **Hempe, J. M. and Cousins, R. J.,** Cysteine-rich intestinal protein binds zinc during transmucosal zinc transport, *Proc. Natl. Acad. Sci. U.S.A.,* 88, 9671, 1991.

47. **Hempe, J. M. and Cousins, R. J.,** Cysteine-rich intestinal protein and intestinal metallothionein: an inverse relationship as a conceptual model for zinc absorption in the rat, *J. Nutr.,* 122, 89, 1992.

48. **Birkenmeier, E. H. and Gordon, J. I.,** Developmental regulation of a gene that encodes a cysteine-rich intestinal protein and maps near the murine immunoglobulin heavy chain locus, *Proc. Natl. Acad. Sci. U.S.A.,* 83, 2516, 1986.

Chapter 23

IRON-DEPENDENT REGULATION OF FERRITIN SYNTHESIS

H. N. Munro, Z. Kikinis, and R. S. Eisenstein

TABLE OF CONTENTS

0-8493-6961-4/93/$0.00 + $.50

I. OVERVIEW

Since the middle of the last century, it has been known that tissues contain iron-positive granules called hemosiderin.[1] In 1937, Laufberger isolated and crystallized another iron-rich compound from horse spleen and named it ferritin to reflect its high iron content.[2] It has since been shown that ferritin is a ubiquitous constituent of animal, plant, and microorganismal cells. Ferritin consists of a protein shell made from subunits, usually 24 in number, surrounding a cavity in which iron can be stored. Harrison et al.[3] provide an excellent account of the physical features of ferritin and its function in iron storage. In 1974 to 1975, two groups[4,5] recognized heterogeneity in the subunits of the protein shell and eventually these were classified either as light (L) or heavy (H) based on their slightly different molecular weights.

Drysdale and Munro[6] had shown in 1966 that iron induces ferritin synthesis in the liver of the rat by a mechanism that is not quenched by actinomycin D and therefore represents a posttranscriptional event. In 1976, Zahringer et al.[7] confirmed and extended this observation by showing that in rats not receiving extra iron a large number of ferritin mRNAs lie dormant in their liver cell sap. When the rats were given iron supplements, these reserve mRNAs became attached to ribosomes in the liver and started to translate ferritin mRNA actively. They proposed that, at low levels of intracellular iron, a specific protein binds to the anterior end (5' untranslated region, 5'UTR) of ferritin mRNA, thus blocking initiation of ferritin protein synthesis. When more iron enters the cell, the regulatory protein loosens so that ferritin protein synthesis can proceed to meet the need for more ferritin molecules to trap the excess iron. These and other studies of ferritin up to 1978 are reviewed elsewhere.[8] The present chapter deals in detail with research performed during the past decade which has considerably increased our understanding of the mechanisms involved in the translational regulation of ferritin mRNA availability in order to achieve iron homeostasis.

II. FERRITIN SUBUNITS: AMINO ACID SEQUENCES, MESSENGER RNAs, GENES

Using sequenator analysis or more commonly deducing amino acids from nucleotide sequences involving cDNA and occasionally ferritin genes, a number of investigators have reported the amino acid sequence of the L and H subunits of a series of mammals and the chicken as summarized elsewhere,[9] to which can be added the more recently analyzed L subunit chain of the mouse[10] and also the three chains of the bullfrog[11] and the H chain of the snail.[12] If we assign the human L subunit a value of 100, the L chain of human liver and spleen are essentially identical (98% and 97%, respectively), whereas the L chain of horse spleen shows only 88% identity, rabbit liver 85%, rat liver 84%, and mouse liver 82% similarity. In contrast, taking human

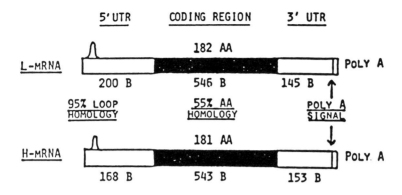

RAT FERRITIN L-GENE

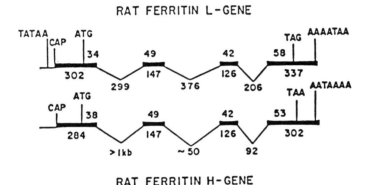

RAT FERRITIN H-GENE

FIGURE 1. Upper diagram: Comparison of the number of nucleotide bases (B) and amino acids (AA) of the coding regions of rat ferritin L and H subunit mRNAs. Also shown are the features of the 5' and 3' untranslated regions with loop and stem structure at the same position in the 5' regions of both mRNAs. Lower diagram: Comparison of the genomic structures of the expressed rat ferritin L and H subunit genes. The exons are indicated by thick bars and the introns by thin lines. The location of the TATA box, initiating methionine codon (ATG), termination codons (TAG,TAA), and polyA polymerase (AATAA) are indicated. (From Munro, H. N. et al., *Ann. N.Y. Acad. Sci.*, 526, 117, 1988. With permission.)

liver H subunit as 100, rat liver shows 95% identity of sequence, mouse 94%, and chicken 93%, thus suggesting more conservation.

The major features of the mRNAs of rat ferritin L and H subunits are shown in Figure 1. Despite closeness in size, the reading frames for rat L and H are only 55% homologous; a similar divergence is seen when human L and H subunits are compared, namely 58%. However, when the Chou and Fasman rules[13] are applied, both subunits show five alpha-helices, thus supporting a common ancestral gene. This suggests that L and H subunits may interchange in the construction of the protein shell and thus give rise to numerous isoferritins.

Figure 1 also shows that L and H subunit mRNAs have long 5' UTRs which, for the most part, show little homology except for a sequence of 28 nucleotides close to the cap region. This region is highly conserved between rat H and L and also between species as widely divergent as man and the bullfrog.[14] On the basis of the reported sequences of this region for the L chain of the human, rabbit, rat, and mouse, the consensus becomes:

UCUUGCUUCAA<u>CAGUGU</u>UUGGACGGAAC   L chain
UC<u>C</u>UGCUUCAA<u>CAGUGC</u>UUGGACGGAAC   H chain

The H ferritin of human, rat, mouse, and chicken are similar except for substitution of C for U in two places in all four H sequences. On computer analysis, the conserved sequence is able to form a double-stranded stem supporting a loop (underlined) of six nucleotides. This sequence will be further discussed under the section on regulation of ferritin mRNAs by iron.

Regarding the L and H ferritin genes of the rat, Southern blots show more than 20 copies of the L gene[15] and 5 of the H gene.[16] Figure 1 displays the exon and intron patterns of the single expressed L and H genes of the rat, the others being considered as pseudogenes. The latter were evident when a selection of L genes of the rat were sequenced and were found to contain no introns, but to have polyadenylic acid tails and direct repeats in the adjacent DNA, all of which is compatible with the random insertion of processed pseudogenes into DNA by a retrovirus. Figure 1 shows that 3 introns (thin lines) separate the 4 exons (thick lines) which encompass the 5 alpha-helices of the L and H subunits, the small fifth alpha helix being present with the fourth alpha helix on the last exon. Note also that the first and the last exons also carry the 5' and 3' UTRs, respectively, of the mRNA. Irrespective of whether the genes are from L or H subunits, or are from man, rat, or chicken, the middle two exons have identical lengths of 147 and 126 nucleotides, respectively, representing 49 and 42 amino acids in the completed protein. This is compelling evidence that the L and H genes share a common ancestry.

III. REGULATION OF FERRITIN SYNTHESIS BY IRON

The levels of control of ferritin gene expression are illustrated in Figure 2. This involves regulation of transcription and also of translation in response to iron and possibly to other factors.

A. TRANSLATION
1. The Ferritin Model: The Iron-Responsive Element
As described in the introduction, injection of iron salts into the peritoneal cavity of the rat is quickly followed by the synthesis of extra ferritin subunits by a mechanism that is not quenched by prior injection of actinomycin D.[8]

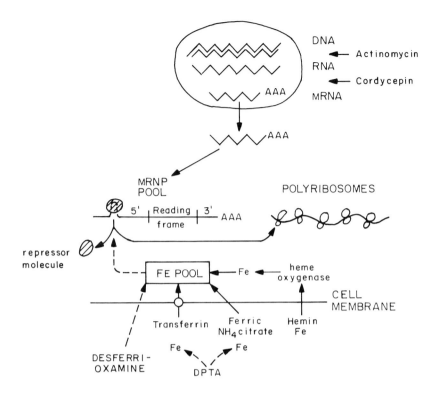

FIGURE 2. Factors involved in regulation of ferritin subunit mRNAs. These include the supply of mRNA from the nucleus, segregation of inactive ferritin mRNA, and its release for polyribosome function when excess iron enters the cell. This is determined by the effect of the entering iron on the free iron pool which can be restricted by chelation with desferioxamine which enters the cell, but not with DPTA, which cannot enter. Note that iron can come from transferrin, inorganic salts, or hemin, the latter through the action of heme oxygenase. (From Munro, H. N. et al., *Ann. N.Y. Acad. Sci.,* 526, 117, 1988. With permission.)

This mechanism is therefore unlikely to involve transcription but rather translation. The interpretation was strengthened when a large stock of dormant mRNAs of ferritin H and L subunits were identified in the cell sap of rats untreated with iron.[7] The stored mRNA represents a reserve needed for rapid availability of ferritin when the cell receives excessive amounts of iron. When more sophisticated techniques became available, the dormant and the active ferritin mRNA could be separated into free ribonucleoprotein (mRNP) particles and polysome-attached mRNA. Density gradients were used to separate ferritin mRNAs at different levels on the gradient ranging from RNPs particles at the top to large polyribosomes at the bottom. The ferritin mRNA was then quantitated with Northern blots. Figure 3 illustrates[17] such a gradient for separating L-mRNA from rat liver cell sap according to whether the animal had received no extra iron (A) or had been injected with iron (B). As measured

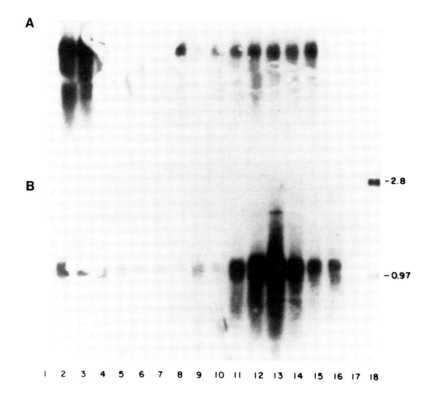

FIGURE 3. Cytoplasmic distribution of the rat ferritin L subunit mRNA in control (above) and iron-treated rats (below). Rats were injected intraperitoneally with ferric ammonium citrate or saline (controls) and killed 3.5 h later. Sucrose gradients of liver homogenates were run. The gradients were divided into 18 fractions from the top ("free" mRNP) through monosomes to the bottom (large polysomes). Each fraction was extracted from RNA, which was then run on the Northern blot gel to show the quantity of L subunit RNA in each fraction using hybridisation with labeled L subunit cDNA. The gel was then autoradiographed. (From Aziz, N. and Munro, H. N., *Nucleic Acid Res.*, 14, 915, 1986. With permission.)

by Northern blotting of different fractions of the gradient, it will be seen that, in the case of the uninjected rats, most of the L-mRNA is in the free mRNP fraction where it lies dormant. On the other hand, rats killed 3 h after injection of iron show extensive translocation of L-mRNA to the polyribosome fractions. This procedure was then repeated for ferritin H-mRNA which showed a similar response to iron administration. Thus, the basic picture presented by Zahringer et al.[7] is substantially correct.

It can be accepted that cells accumulate large stocks of ferritin L and H-mRNAs in order to respond rapidly to increases in intracellular iron levels which would otherwise cause peroxidation of lipids and DNA and result in other disruptive actions. For example, Dougherty et al.[18] describe experiments in which rats received a single peritoneal dose of iron salts and responded by exhaling ethane made from peroxidation of membrane-associated omega-3-fatty acids. The resulting generation of free radicals could well be a factor in aging.[19]

The mechanism linking intracellular iron level to synthesis of ferritin can now be considered. The occurrence of a highly conserved sequence near the cap region of the 5′ UTR (Figure 1) was examined by computer for the presence of a stem-loop structure which was indeed found by us in the first 79 nucleotides of the 5′ untranslated region of L ferritin.[20] This finding has been confirmed by several investigators including ourselves, using newer computer programs with which details emerge regarding the stem-loop structure. This motif, a candidate for being involved in the response to iron, is usually referred to as the iron regulatory element or IRE. In order to function as a regulator, it has to act as the *cis* element to which a specific protein acts as the *trans* element.

First, it was necessary to confirm that this consensus sequence is involved in responding to the iron in the cell.[21] Accordingly a chimera was made in which the 5′ and 3′ UTRs from rat liver L-mRNA are joined to the reading frame of the bacterial protein CAT (chloramphenicol acetyl transferase). Using retrovirus-mediated gene transfer, this construct and modifications of it were inserted into rat hepatoma cells in culture. The inserted sequences included the intact 5′ UTRf (5CAT3), then a sequence with deletion of the first 67 nucleotides of the 5′ UTR (5sCAT3) and finally another with removal of the 3′ UTR (5CAT). Cells with the complete insert (5CAT3) showed a response to extra iron in the medium by synthesizing more CAT (Figure 4). Deletion of the first 67 bases (5sCAT3) resulted in the disappearance of CAT responsiveness to iron, showing that the regulatory sequence occurs within these 67 nucleotides. The removal of the 3′ UTR (5CAT) did not affect responsiveness to iron. On the other hand, Dickey et al.[22] studied the loss of the 3′-region from the ferritin mRNA of the tadpole and concluded that it played a part in the response of ferritin synthesis to iron. Using human H-mRNA with CAT and also growth hormone as reported proteins, Caughman et al.[23] reached conclusions similar to ours. In a more recent paper,[24] they varied the position of the stem-loop structure along the 5′ UTR and concluded that its normal location near the cap was essential to its regulatory function, possibly because it interfered in that position with initiation of protein synthesis.

The *trans* element interacting with the stem-loop structure is a protein present in the cell sap. Using rat L-mRNA, Leibold and Munro[25] prepared a ^{32}P-labeled transcript of the initial 65 nucleotides of the 5′ UTR which included the conserved 28 nucleotides. The transcripts were added to cell sap harvested from rat liver and rat heart and from hepatoma cells in culture and were allowed to react. The products were then separated on polyacrylamide gels. Figure 5 illustrates an experiment in which cell sap was extracted from rat hepatoma cells either before or after iron treatment and was then reacted with [^{32}P]-5′LRNA containing the 28-nucleotide sequence. As compared to the migration of RNA alone on the gel, the binding of RNA to the protein retarded

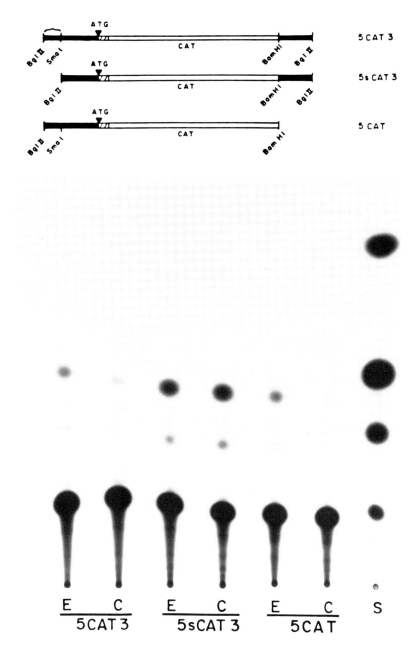

FIGURE 4.   Inducibility of the bacterial enzyme chloramphenicol acetyl transferase (CAT) by iron using the intact ferritin-CAT chimera (5CAT3) and the chimera with deletion of the first 67 nucleotides of the 5′ UTR (5sCAT3) or deletion of the 3′ UTR (5CAT). These were permanently transfected into FTO-2B cells, and iron in the form of hemin-treated (E) or control cells (C) were incubated for 14 h followed by CAT assay. (From Aziz, N. and Munro, H. N., *Proc. Natl Acad. Sci. U.S.A.*, 84, 8478, 1987. With permission.)

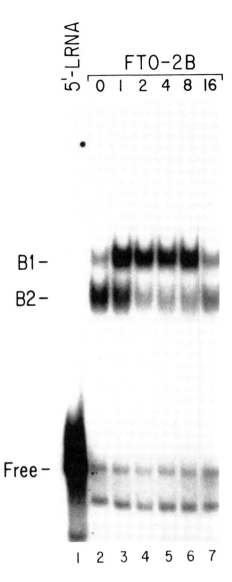

FIGURE 5. RNA-protein complex formation with S100 extracts from hepatoma cells (FTO-2B) treated with ferric ammonium citrate for 0, 1, 2, 4, 8, and 16 h. Positions of complexes B1 and B2 and free RNA are indicated. (From Leibold, E. A. and Munro, H. N., *Proc. Natl. Acad. Sci. U.S.A.,* 85, 2171, 1988. With permission.)

the migration of the labeled 5′ UTR-protein complex which appeared as two bands, B-1 and B-2. When the cells were incubated with extra iron, there was extensive loss of B-2 binding and a gain in B-1 binding. When the fractions B-1 and B-2 were digested with RNAse T1, about 40 nucleotides were protected from digestion including the 28-nucleotide section. Cross-linking using UV irradiation showed that proteins B-1 and B-2 were both about 90 kDa. in molecular weight. Rouault et al.,[26] using human H-mRNA, also showed binding to human cell sap protein. In an extensive survey of the binding proteins in 12 different species, Rothenberger et al.[27] showed it to be present in cytoplasmic extracts from human, mouse, chicken, frog, fish, and fly, but was absent from bacterial, yeast, and plant sources. The proteins found in animal sources varied from 95 to 100 kDa. Finally, like the rat,[25] the mouse also displayed two binding proteins, thus confirming another deviation confined to these rodents, namely an insertion of eight amino acids into the L subunit of the rat[9] and the mouse[10] but not found in other species.[9]

The next step was to purify the binding protein in order to characterize and identify its structure. Rouault et al.[26,29,30] describe a procedure in which the regulatory RNA stem-loop structure (IRE) is used to trap the human binding protein (IRE-BP). In order to do so, the segment of RNA containing the IRE is reacted with biotin which makes the product capable of attaching to avidin bound to an agarose column. After two cycles, an almost pure product was obtained which migrated on an SDS/PAGE gel as a single band with a molecular weight of 90 kDa. A peptide sequence from this purified protein was used to prepare oligonucleotides to screen a human cDNA library. A clone was identified which encoded a protein of 87 kDa.[29] We[31] also employed biotinylation affinity chromatography to isolate the rat B-1 binding protein, the yield being considerably increased by including six head-to-tail IRE repeats. This protein had a molecular weight of 95 kDa and its N-terminal amino acid sequence was close to that of the human.[29,30] Walden et al.[28] employed a biochemical method using nucleotide binding dyes to isolate a single polypeptide chain with a mol wt of about 90 kDa from rabbit liver.

Rouault et al.[29] noted in the amino acid sequence of their protein that it included cysteine residues in motifs suggestive of those found in iron-sulfur proteins. In particular, the mitochondrial enzyme aconitase shows 30% homology in its amino acid sequence compared with human binding protein.[32] This enzyme is active when all four Fe-S are present but can be inactivated by loss of one iron atom. They therefore theorized that the gain or loss of this iron atom may be the mechanism by which the IRE-BP senses iron levels in the cell. The degree of homology of IRE-BP and aconitase is also emphasized by Hentze and Argos[34] who remark that another protein, isopropylmalate isomerase, shows 23% homology with the binding protein sequence. It would be important to sequence the cytosolic form of aconitase since this is the cell compartment in which the binding protein is located. The activity of the binding protein may also vary with sulfhydryl levels in the cytosol[34] and indeed Barton et al.[35] present a mechanism in response to iron level involving multiple pools of IRE-BP, their characteristics, and kinetics including the important role of sulfhydryl groups (Figure 6).

As a step towards understanding how iron regulates the binding protein, studies have focused on hemin and on chelatable cytosolic iron as likely candidates. Figure 2 shows that chelatable free iron can be replenished from

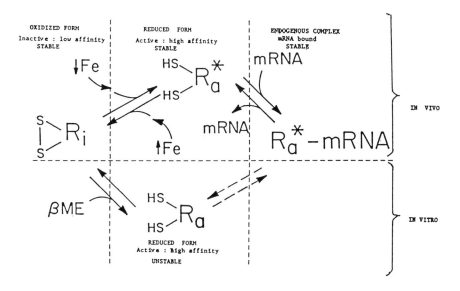

FIGURE 6. Scheme for interconversions of the pools of IRE-binding protein (IRE-BP) in response to cytosolic levels of iron. When the iron content of the cytosol is high, IRE-BP occurs as an inactive form (R_i) with low affinity for the IRE motif, thus allowing ferritin subunit synthesis to proceed. The sulfhydryl groups of R_i are in the oxidized state. In response to a reduction in cytosolic iron level, the R_i can be converted *in vivo* to high affinity active protein (R_a^*) with reduced sulfhydryl groups. Since this form binds tightly to the 28-nucleotide IRE, it prevents translation of the ferritin mRNAs and is thus converted to the third pool, the endogenous complex made up of R_a^* and mRNA. Note that by adding thiol reductants the first pool (R_i) can also be converted *in vitro* to an active form (R_a) which however is unstable. (Diagram kindly provided by Dr. Hugh A. Barton.)

iron-loaded transferrin, from inorganic iron, from ferric pyridoxalisonicotinoyl hydrazone (FEPIH, not illustrated in the figure), and from hemin. Using *in vitro* studies in which hemin was incubated with purified IRE-BP, Lin et al.[36] were able to prevent the inhibitory action of binding protein on translation of ferritin mRNA. This action of hemin was not affected by chelators of iron, indicating that release of free iron was not involved. In a later publication, they[37] were able to show direct binding of hemin to a specified site on the binding protein surface. Haile et al.[38] confirm that hemin added *in vitro* inhibits the binding of the protein IRE-BP to the IRE motif in ferritin 5' UTR. However, they show that hemin *in vitro* also binds nonspecifically to other RNA and DNA binding proteins and to a variety of nucleases. Contrary to the *in vitro* studies of Lin et al.,[36,37] Rogers and Munro[39] showed that, using rat hepatoma cells in culture, addition of hemin to the medium increased translation of ferritin mRNA, but that this action of hemin could be prevented by adding the chelator desferal to the medium, thus suggesting that hemin has to yield free iron in order to stimulate ferritin synthesis.

TABLE 1
Contrasting Effects of ALA on the Iron-Dependent Induction of Ferritin and of Heme Oxygenase Synthesis in Rat Fibroblasts

FEPIH 13 µM	ALA µM	SA 1 mM	H-F Synthesis cpm	HO Synthesis cpm
−	—	—	140	170
+	—	—	690	290
+	100	—	270	560
+	500	—	110	590
+	1000	−	130	1010
+	—	+	520	460
+	100	+	530	480
+	500	+	400	480
+	1000	+	480	450

Note: ALA, 5-amino levulinic acid; SA, the inhibitor succinyl acetone; HF, H ferritin subunit; HO, heme oxygenase.

This thesis was explored in depth by Eisenstein et al.[40] who used rat fibroblasts and rat hepatoma cells in culture to determine the effects of hemin, other metalloporphyrins and nonheme iron (FEPIH) on the synthesis of ferritin and on heme oxygenase, the latter being the enzyme that releases iron from hemin. It was assumed that the level of activity of each enzyme reflects the level of substrate, namely heme for heme oxygenase and chelatable iron for ferritin. (This has been confirmed by later experiments [unpublished] in which ^{55}Fe was used to allow identification of the amounts of iron in different cell components.) As shown in the published paper,[40] addition of hemin to the medium induced both proteins, the oxygenase peaking first and releasing iron for the subsequent increase in ferritin synthesis. Tin-mesoporphyrin, an inhibitor of heme breakdown by the oxygenase, delayed the induction of ferritin. On the other hand, desferal, the chelator of free iron, inhibited ferritin synthesis without affecting syntheses of heme oxygenase, thus suggesting that free chelatable iron is the dominant factor in ferritin synthesis. This was confirmed with free iron from added FEPIH, which selectively stimulated ferritin synthesis but not heme oxygenase. Finally, addition to the cultured cells of aminolevulinic acid (ALA), a regulator of the rate of heme synthesis, increased the demands for iron by the heme synthesis pathway and thus reduced the amount available for ferritin synthesis. These responses can be prevented by addition to the medium of succinyl acetone (SA), an inhibitor of ALA utilization (Table 1). Altogether these experiments show that ferritin synthesis is regulated by chelatable free iron and not by hemin, making the free cytosolic iron the preferred regulator of the IRE-BP.

Finally, optimum structure of the IRE for interaction with the IRE-BP will be explored. Using the most recent Zuker[41] computer-generated program

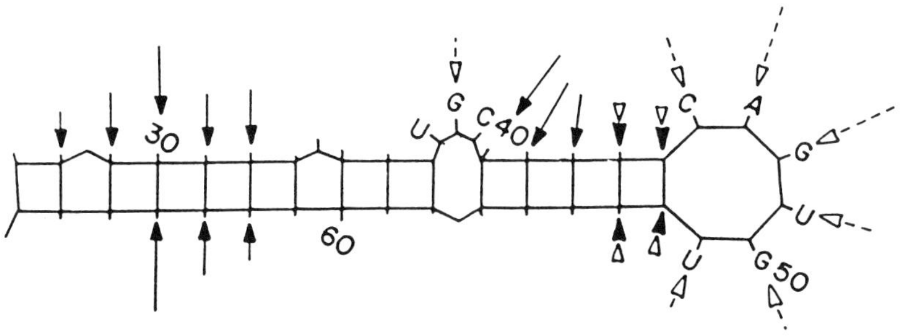

FIGURE 7. Suboptimal folding analysis of the nucleotides 26 through 68 of ferritin L-mRNA using the Zucker program.[41] Single-strand cleavage sites (T1 and S1) are denoted by the broken arrows, double-strand cleavage sites (V1) by the solid arrows. The intensity of cleavage is indicated by the length of the arrow. (From Bettany, A. J. E. et al., in press. With permission.)

for predicting suboptional folding of RNA (Figure 7), we[42] have analyzed the stem-loop structure of the IRE of the rat L subunit and have compared it with the cleavage sites for single-stranded areas of the IRE using enzymes T-1 and S-1, represented in the Figure 7 by dotted line arrows on the six-membered loop and also the major bulge of three unpaired nucleotides (38 to 41) on one side of the stem. Similarly, scission of double-stranded areas was accomplished by using enzyme V-1 and is represented by solid arrows. Wang et al.[43] explored the structure of the bullfrog H subunit using the chemical nuclease 1,10-phenanthroline-copper which cleaves single-stranded and bulged areas. This supplemented the evidence from their application of T-1 while V-1 was applied for identification of the double-stranded areas. In both of these studies the predicted stem-loop structure has been confirmed.

In addition to studies of secondary structure described above, there have been several reports of nucleotide replacement or deletion and the effect of this manoever on IRE function. One example comes from a comparison of the IRE of the L and the H subunits of mammals. As shown earlier,[14] all H subunit IREs of mammals so far examined have a cytidylic acid replacing the uridylic acid of the L subunit IREs in two places (one in the sixth position of the CAGUGU loop) without apparent loss of function. Barton et al.[35] replaced two nucleotides in the loop without greatly altering the affinity of the IRE to bind the specific protein as determined by *in vitro* competition and saturation assays. Similarly, exchange of adenylic acid for cytidylic acid in the unpaired bulge also left protein binding unaffected. In contrast, Hentze et al.[45] *deleted* a cytidylic acid from the loop and from the stem of the human H-IRE. The result with each deletion was loss of response to iron, which may be attributed to the alteration in secondary structure of the IRE rather than to loss of a specific nucleotide. An even more extensive loss of binding occurred when one leg of the stem of rat L-mRNA was replaced by other nucleotides that did not pair with the other leg of the stem, but was restored by inserting a new leg of nucleotides compatible with the other new leg.[42] Thus one stem can be replaced with another of a different nucleotide composition and can function so long as the nucleotides in both legs are able to pair. This leads to the conclusion that the secondary conformation of the IRE stem is more important than the primary sequence of its nucleotides so long as they form a structure of paired nucleotides of adequate length.[35] This is also the conclusion drawn by Leibold et al.[46] who applied competition assays to obtain the 50% binding affinity between ferritin native IRE and mutant IRE. These conclusions require testing *in vivo*.

2. Other Systems Involving the Iron-Responsive Element

The IRE motif performs a different function in regulating the stability of the transferrin receptor (TfR) in response to changes in the intracellular iron level.[47,48] In the 3' UTR of the receptor, five stem-loop structures resembling the single IRE in the 5' UTR of ferritin mRNA have been identified.[49,50] Three of these provide the function of regulating TfR mRNA breakdown in response to changes in intracellular iron levels.[50] When cells are treated with the iron-chelator desferal, the TfR mRNA becomes more stable and thus favors the entry of more iron into the cell, whereas when iron uptake into the cell increases, the TfR mRNA becomes unstable and diminishes in amount so that entry of more iron is discouraged. These changes in stability of the mRNA of TfR are due to binding of a specific protein which recognizes the 3' IREs and either increases (stabilizes) or relaxes (destabilizes) its binding according to the iron level. Both the IREs[49] and the protein[51] associated with the 3' UTR are able to function if transferred to the 5' UTR where they then regulate ferritin subunit translation in spite of having quite different nucleotide pairs in their stems when compared with the stem of the 5' IRE regulating ferritin subunit synthesis. Note that the same binding protein coordinates both the synthesis of ferritin and the amount of transferrin receptor in relation to changes in intracellular iron status.

An iron-regulated protein restricted to erythroid cells is the erythroid 5-aminolevulinic acid synthase (eALAS), the rate-limiting enzyme in protoporphyrin synthesis. In addition to this erythroid-specific form, there is also a ubiquitous ALAS which is activated by porphyrinogenic drugs and is regulated by hemin. The mRNA of the erythroid ALAS displays an IRE motif in its 5' UTR,[52,53] whereas the housekeeping form of this enzyme does not have this motif. In order to test the importance of individual nucleotides in the loop CAGUGC in the eALAS mRNA, Cox et al.[52] replaced it with GGAUCC and observed that the mutant IRE no longer competed with H ferritin IRE. However, positions 1 and 6 of their mutant loop were occupied by nucleotides G and C opposing one another at the top of the stem to which the G-C pair created by the mutation contributed another nucleotide pair while reducing the loop to four nucleotides, a possible cause of the loss of function of the mutant. In the course of their exploration of the IRE in eALAS,

Dandekar et al.[53] performed a computer-assisted identification of putative IREs which turned up in large numbers from which these investigators offer potentially interesting examples. One such is aconitase, a member of the Fe-S family thought to include the IRE-binding protein.[32] Another example recorded by Tanzi and Hyman[54] identifies a stem-loop structure in the gene for the amyloid protein precursor in Alzheimer's disease. This motif occurs at the point where the much smaller amyloid protein is released from the precursor and thus may regulate either the amount of the precursor protein or the amount of amyloid released.

B. TRANSCRIPTION
1. Response of H and L Genes to Iron

The regulatory stem-loop motif in the 5' UTR of both the L- and H-mRNAs of ferritin implies that activation of translation of these two subunits will create new ferritin molecules with H and L subunits in proportion to the dormant H- and L-mRNAs. However, in an experiment in which new ferritin subunits in rat liver were labeled with ^{14}C-leucine, we[9,55] obtained a ratio of new L to H subunits which increased substantially after an intraperitoneal dose of iron dextran, implying either preferential synthesis of L-mRNA or decreased rate of breakdown of L-mRNA. Using HeLa cells in culture, Cairo et al.[56] found that transcription of L-mRNA increased after adding more iron to the medium, but for technical reasons were unable to monitor the response of H-mRNA. Using transcription run-off assays for measuring L and H mRNA synthesis, we[57] were able to show a preferential increase in the rate of transcription of L compared with H-mRNA by rat liver nuclei. The pattern of increased abundance of labeled L-mRNA in the nuclei then appeared in the ribonucleoprotein pool (Figure 2) and later in the polyribosomal fraction of the cell. This resulted in production of isoferritins richer in the L subunit, and therefore better able to store iron.[55] The different effects of iron on transcription and on translation thus allow flexibility in the creation of ferritin spheres with various L/H ratios.

2. Response of H and L Genes to Other Agents

Factors other than iron can alter the expression of ferritin genes, usually differentially favoring the H subunit. In 1977, Konijn and Hershko[58] produced an acute phase reaction by injecting turpentine into rats. In addition to established acute phase liver proteins, they observed an increase in liver ferritin 5 to 10 h after injection. Campbell et al.[59] attributed this to a translational response to inflammation. Subsequent work has identified interleukin-1 (IL-1) as the agent released by macrophages in the injured tissues that causes the acute phase response.[60] Rogers et al.[61] added IL-1 to human hepatoma cells in culture and within 2 h obtained increases in the amounts of both H and L ferritin subunits without changing the amount of L-mRNA (H-mRNA was not reported). Consequently, the response to IL-1 must have involved the transfer of dormant mRNP to active polyribosomes as discussed earlier under translation. Cell iron participated in the response to IL-1 only when it was reduced by desferal to a level that prevented translocation of the mRNA to an active status. In studies on muscle cells in culture, Wei et al.[62] compared the action of IL-1 and tumor necrosis factor (TNF) on the abundance of H-mRNA which increased in amount following addition of each cytokine. Since IL-1 and TNF bind to different cell receptors, it is not surprising that their actions were additive. In a subsequent paper,[63] the same group found that actinomycin inhibited the action of TNF on the abundance of H-mRNA, implying an effect of TNF on transcription. In addition they[64] also analyzed the regulation of transferrin receptors in response to TNF and to IL-1. Using human fibroblasts exposed to TNF and to IL-1, they observed increases in TfR mRNA and protein after a lag of 12 to 24 h. In the same experiment, however, ferritin H-mRNA increased within 4 h and the H-protein peaked at 8 h, after which synthesis of both H and L proteins declined rapidly. Consequently, the late increase in TfR levels cannot be credited with providing more iron for the cell within the first 4 to 8 h. It can be concluded that TNF and IL-1 increase H subunit levels through transcription provided that ferritin mRNA translation is not obstructed by too low an intracellular iron level.

Finally, changes in H and L subunit abundance have been linked to cell differentiation. Chou et al.[65] used a human promyelocyte cell line that differentiated *in vitro* when incubated with TPA to develop into macrophages or DMSO with amiloride to yield neutrophils. In each case there was an increase in both H- and L-mRNA, the increase in the H-mRNA being severalfold greater. Testa et al.[66] also described a cell proliferation model of ferritin involving human lymphocytes stimulated by phytohemagglutinin (PHA) and also a cell differentiation model of monocytes maturing to macrophages. In each case, they evaluated levels of TfR expression, ferritin (protein and mRNA), and IRE-BP activity. It is of interest that the cell proliferation model (lymphocytes) in an iron-supplemented medium responded to PHA with an increase limited to H-mRNA, whereas the differentiation model (monocytes) showed similar large changes in H and L ferritin mRNA when iron was added to the medium. Liau et al.[67] used the vascular smooth vessel cell to show that growth-arrest results in increases in the ferritin H chain which cannot be attributed to a rise in the intracellular iron level. They present evidence suggesting cAMP as a likely factor acting at the posttranscriptional level to increase the H chain selectively.

IV. CONCLUSION

The translational control of ferritin subunit synthesis and the stability of transferrin receptors provide a well-authenticated linking of factors regulating intracellular iron content. This relationship emphasizes ferritin shells with L subunit abundance as favoring iron storage which can be provided by increasing L-mRNA transcription preferentially. However several publications have shown that ferritin can preferentially increase mRNA after inflammation. Two cautions regarding such work must be voiced. First, if there is insufficient free iron in the target cell, transcribed H- or L-mRNA will not undergo translation unless there is iron supply that allows the IRE-BP to dissociate from the IRE, a limitation recognized by one group.[62] Second, it is not often remembered that mRNA and protein can also be regulated by varying the rate of their breakdown, which therefore should be measured. Third, a number of authors has examined the importance of specific nucleotides in the conserved 28 nucleotides of the IRE, using gel retardation, etc., as the criteria *in vitro*. Based on such tests it has been concluded that nucleotides are mostly replaceable in the stem and perhaps some in the loop of the IRE. It has generally been found that this is true for the stem, but the six-membered loop should be tested by *in vivo* criteria. The perfect retention of the sequence CACUGN, stretching from snail[12] to man,[9] is impressive, whereas the stems of snail and of man are only 40% concordant. This makes imperative *in vivo* studies of loop structure using nucleotide replacement.

Finally, Dickey et al.[11] have recognized the cellular function may determine the proportion of H and L subunits in the ferritin shell depending on whether the cells require a housekeeping ferritin or a cell-specific ferritin. This projects us into the realm of different H and L subunit abundances in different tissues. For example heart ferritin has a larger size of ferritin shell in which the H subunit predominates.[55] *Ex Afrika semper aliquid novi* (Pliny the Elder).

REFERENCES

1. **Wixom, R. L., Prutkin, L., and Munro, H. N.,** Hemosiderin: nature, formation and significance, in *International Review Experimental Pathology,* Richter, G. W. and Epstein, M. A., Eds., Academic Press, New York, 1980, 22, 193.
2. **Laufberger, M. V.,** Sur la cristallisation de la ferritine, *Bull. Soc. Chim. Biol.,* 19, 1575, 1937.
3. **Harrison, P. M., Andrews, S. C., Artymiuk, P. J., Ford, G. C., Lawson, D. M., Smith, J. M. A., Treffry, A., and White, J. L.,** Ferritin, in *Iron Transport and Storage,* Ponka, P., Schulman, H. M., and Woodworth, R. C., Eds., CRC Press, Boca Raton, FL, 1990, 81.
4. **Linder, M. C., Moor, J. B., and Munro, H. N.,** Subunit heterogeneity in rat liver apoferritin: differential responses of the subunits to iron administration, *J. Biol. Chem.,* 249, 7707, 1974.
5. **Adelman, T. G., Arosio, P., and Drysdale, J. W.,** Multiple subunits in human ferritin: evidence for hybrid molecules, *Biochem. Biophys. Res. Commun.,* 63, 1056, 1975.
6. **Drysdale, J. W. and Munro, H. M.,** Regulation of synthesis and turnover of ferritin in rat liver, *J. Biol. Chem.,* 241, 3630, 1966.

7. **Zahringer, J., Baliga, B. S., and Munro, H. N.,** Novel mechanism for translational control in regulation of ferritin synthesis by iron, *Proc. Natl. Acad. Sci. U.S.A.,* 73, 857, 1976.

8. **Munro, H. N. and Linder, M. C.,** Ferritin: structure, biosynthesis and role in iron metabolism, *Physiol. Rev.,* 58, 317, 1978.

9. **Munro, H. N., Leibold, E. A., Aziz, N., Murray, M. T., White, K., and Rogers, J.,** Ferritin gene structure and expression, in *Iron Transport and Storage,* Ponka, P., Schulman, H. M., and Woodworth, R. C., Eds., CRC Press, Boca Raton, FL, 1990, 134.

10. **Beaumont, C., Dugast, I., Renaudie, F., Souroujun, M., and Grandchamp, B.,** Transcriptional regulation of ferritin H and L subunits in adult erythroid and liver cells from the mouse, *J. Biol. Chem.,* 264, 7498, 1989.

11. **Dickey, L. F., Sreedharan, S., Theil, E. C., Didsbury, J. R., Wang, Y-H., and Kaufman, R. E.,** Differences in the regulation of messenger RNA for housekeeping and specialized cell ferritin: a comparison of three different ferritin complementary DNAs, the corresponding subunits and identification of the first processed pseudogenes in amphibia, *J. Biol. Chem.,* 262, 7901, 1987.

12. **Von Darl, M. and Bottke, W.,** Vollstandige Basensequenz des langsten Somaferritin-clons: Aminosauresequenz des Somaferritins. EMBL-accession Nr X 56778, 1991.

13. **Chou, P. Y. and Fasman, G. D.,** Empirical predictions of protein conformation, *Annu. Rev. Biochem.,* 47, 251, 1978.

14. **Eisenstein, R. S., Bettany, A. J. E., and Munro, H. N.,** Regulation of ferritin gene expression, in *Advances in Inorganic Biochemistry,* Eichhorn, G. L. and Marzilli, L. G., Eds., Elsevier, New York, 1990, 8, 91.

15. **Leibold, E. A. and Munro, H. N.,** Characterization and evolution of the expressed rat ferritin light subunit gene and its pseudogene family, *J. Biol. Chem.,* 262, 7335, 1987.

16. **Murray, M. T., White, K., and Munro, H. N.,** Conservation of ferritin heavy subunit gene structure: implications for the regulation of ferritin gene expression, *Proc. Natl. Acad. Sci. U.S.A.,* 84, 7438, 1987.

17. **Aziz, N. and Munro, H. N.,** Both subunits of rat liver ferritin are regulated at a translational level by iron induction, *Nucleic Acids Res.,* 14, 915, 1986.

18. **Dougherty, J. J., Croft, W. A., and Hoekstra, W. G.,** Effects of ferrous chloride and iron dextran on lipid peroxidation in vivo in vitamin E and selenium adequate and deficient rats, *J. Nutr.,* 111, 1784, 1981.

19. **Floyd, R. A.,** Role of oxygen free radicals in carcinogenesis and brain ischemia, *FASEB J.,* 4, 258, 1990.

20. **Munro, H. N., Leibold, E. A., Vass, J. K., Aziz, N., Rogers, J., Murray, M., and White, K.,** Ferritin gene structure and expression, in *Proteins of Iron Storage and Transport,* Spik, G., Montreuil, J., Crichton, R. R., and Mazurier, J., Eds., Elsevier, Amsterdam, 1985, 331.

21. **Aziz, N. and Munro, H. N.,** Iron regulates ferritin mRNA translation through a segment of its 5′ untranslated region, *Proc. Natl. Acad. Sci. U.S.A.,* 84, 8478, 1987.

22. **Dickey, L. F., Wang, Y-H, Schull, G. E., Wortman, I. A., and Theil, E. C.,** The importance of the 3′ untranslated region in the translational control of ferritin mRNA, *J. Biol. Chem.,* 263, 3071, 1988.

23. **Caughman, S. W., Hentze, M. W., Rouault, T. A., Harford, J. B., and Klausner, R. D.,** The iron-responsive element is the single element responsible for iron-dependent translational regulation of ferritin biosynthesis: evidence for function as the binding site for a translational repressor, *J. Biol. Chem.,* 263, 19048, 1988.

24. **Goessen, B., Caughman, S. W., Harford, J. B., Klausner, R. D., and Hentze, M. W.,** Translational repression by a complex between the iron-responsive element of ferritin mRNA and its specific cytoplasmic binding protein is position-dependent in vivo, *EMBO J.,* 9, 4127, 1990.

25. **Leibold, E. A. and Munro, H. N.,** Cytoplasmic protein binds in vitro to a highly conserved sequence of the 5′ untranslated region of ferritin heavy and light subunit mRNAs, *Proc. Natl. Acad. Sci. U.S.A.,* 85, 2171, 1988.

26. **Rouault, T. A., Hentze, M. W., Caughman, S. W., Harford, J. B., and Klausner, R. D.,** Binding of a cytosolic protein to the iron responsive element of human ferritin messenger RNA, *Science,* 241, 1207, 1988.

27. **Rothenberger, S., Mullner, E. W., and Kuhn, L. C.,** The mRNA-binding protein which controls ferritin and transferrin receptor expression is conserved during evolution, *Nucleic Acids Res.,* 18, 1175, 1990.

28. **Walden, W. E., Patino, M. M., and Gaffield, L.,** Purification of a specific repressor of ferritin mRNA translation from rabbit liver, *J. Biol. Chem.,* 264, 13765, 1989.

29. **Rouault, T. A., Hentze, M. W., Haile, D. J., Harford, J. B., and Klausner, R. D.,** The iron-responsive element binding protein: a method for the affinity purification of a regulatory RNA-binding protein, *Proc. Natl. Acad. Sci. U.S.A.,* 86, 5768, 1989.

30. **Rouault, T. A., Tang, C. K., Kaptain, S., Burgess, W. H., Haile, D. J., Samaniego, F., McBride, O. W., Harford, J. B., and Klausner, R. D.,** Cloning of the cDNA encoding an RNA-regulatory protein: the human iron-response element binding protein, *Proc. Natl. Acad. Sci. U.S.A.,* 87, 7958, 1990.

31. **Bomford, A. B., Barton, H. A., Pettingell, W. H., Munro, H. N., and Eisenstein, R. S.,** The iron-responsive element-binding protein of the rat: biochemical and functional studies of the purified protein. (Submitted).

32. **Rouault, T. A., Stout, C. D., Kaptain, S., Harford, J. B., and Klausman, R. D.,** Structural relationship between an iron-regulated RNA-binding protein (IRE-BP) and aconitase: functional implications, *Cell,* 64, 881, 1991.

33. **Kaptain, S., Downey, W. E., Tang, C., Philpott, C., Haile, D., Orloff, D. G., Harford, J. B., Rouault, T. A., and Klausner, R. D.,** A regulated RNA binding protein also possesses aconitase activity, *Proc. Natl. Acad. Sci. U.S.A.,* 88, 10109, 1991.

34. **Hentze, M. W., Rouault, T. A., Harford, J. B., and Klausner, R. D.,** Oxidation-reduction and the molecular mechanism of a regulatory RNA-protein interaction, *Science,* 244, 357, 1989.

35. **Barton, H. A., Eisenstein, R. S., Bomford, A., and Munro, H. N.,** Determinants of the interaction between the iron-responsive element-binding protein and its binding site in rat L-ferritin mRNA, *J. Biol. Chem.,* 265, 7000, 1990.

36. **Lin, J. J., Daniels-McQueen, S., Patino, M. M., Gaffield, L., Walden, W. E., and Thach, R. E.,** Derepression of ferritin messenger RNA translation by hemin in vitro, *Science,* 247, 74, 1990.

37. **Lin, J. J., Patino, M. M., Garfield, L., Walden, W. E., Smith, A., and Thach, R. E.,** Cross-linking of hemin to a specific site on the 90-kd repressor protein, *Proc. Natl. Acad. Sci. U.S.A.,* 88, 6068, 1991.

38. **Haile, D. J., Rouault, T. A., Harford, J. B., and Klausner, R. D.,** The inhibition of the iron responsive element RNA-protein interaction by heme does not mimic in vivo iron regulation, *J. Biol. Chem.,* 265, 12786, 1990.

39. **Rogers, J. and Munro, H. N.,** Translation of light and heavy subunit mRNAs is regulated by intracellular chelatable iron levels in the rat hepatoma cells, *Proc. Natl. Acad. Sci. U.S.A.,* 84, 2277, 1987.

40. **Eisenstein, R. S., Garcia-Mayol, D., Pettingell, W., and Munro, H. N.,** Regulation of ferritin and heme oxygenase synthesis in rat fibroblasts by different forms of iron, *Proc. Natl. Acad. Sci. U.S.A.,* 88, 688, 1991.

41. **Jaeger, J. A., Turner, D. H. and Zuker, M.,** Predicting optimal and suboptimal secondary structure for RNA, *Methods Enzymol.,* 183, 281, 1990.

42. **Bettany, A. J. E., Eisenstein, R. S., and Munro, H. N.,** Role of RNA secondary structure in the interaction between the iron regulatory element of ferritin mRNA and its binding protein, in press.

43. **Wang, Y-H, Sczekan, S. R., and Theil, E. C.,** Structure of the 5' untranslated regulatory region of ferritin mRNA studied in solution, *Nucleic Acids Res.,* 18, 4463, 1990.

44. **Harrell, C. M., McKenzie, A. R., Patino, M. M., Walden, W. E., and Theil, E. C.,** Ferritin mRNA: interactions of iron regulatory element with translational regulator protein P-90 and the effect on base-paired flanking regions, *Proc. Natl. Acad. Sci. U.S.A.,* 88, 4166, 1991.

45. **Hentze, M. W., Caughman, S. W., Casey, J. L., Koeller, D. M., Rouault, T. A., Harford, J. B., and Klausner, R. D.,** A model for the structure and functions of iron-responsive elements, *Gene,* 72, 201, 1988.

46. **Leibold, E. A., Laudano, A., and Yang, Y.,** Structural requirements of iron responsive elements for binding of the protein involved in both transferrin receptor and ferritin mRNA post-transcriptional regulation, *Nucleic Acids Res.,* 18, 1819, 1990.

47. **Mullner, E. W. and Kuhn, L. C.,** A stem-loop in the 3' untranslated region mediates iron-dependent regulation of transferrin receptor mRNA stability in the cytoplasm, *Cell,* 53, 815, 1988.

48. **Mullner, E. W., Neupert, B., and Kuhn, L. C.,** A specific mRNA binding factor regulates the iron-dependent stability of cytoplasmic transferrin receptor mRNA, *Cell,* 58, 373, 1989.

49. **Casey, J. L., Hentze, M. W., Koeller, D. M., Caughman, S. W., Rouault, T. A., Klausner, R. D., and Harford, J. B.,** Iron-responsive elements: regulatory RNA sequences that control mRNA levels and translation, *Science,* 240, 924, 1988.

50. **Casey, J. L., Koeller, D. M., Ramin, V. C., Klausner, R. D., and Harford, J. B.,** Iron regulation of transferrin receptor mRNA levels requires iron-responsive elements and a rapid turnover determinant in the 3' untranslated region of the mRNA, *EMBO J.,* 8, 3693, 1989.

51. **Koeller, D. M., Casey, J. L., Hentze, M. W., Gerhardt, E. M., Chan, L-N. L., Klausner, R. D., and Harford, J. B.,** A cytosolic protein binds to structural elements within the iron regulatory region of the transferrin receptor mRNA, *Proc. Natl. Acad. Sci. U.S.A.,* 86, 3574, 1989.

52. **Cox, T. C., Bawden, M. J., Martin, A., and May, B. K.,** Human erythroid 5-aminolevulinate synthase: promoter analysis and identification of an iron-responsive element in the mRNA, *EMBO J.,* 10, 1891, 1991.

53. **Dandekar, T., Stripecke, R., Gray, N. K., Goosen, B., Constable, A., Johansson, H., and Hentze, M.,** Identification of a novel iron-responsive element in murine and human 5-erythroid amino levulinic acid synthase mRNA, *EMBO J.,* 10, 1903, 1991.

54. **Tanzi, R. E. and Hyman, B. T.,** Alzheimer's mutation, *Nature (London),* 350, 564, 1991.

55. **Bomford, A., Conlon-Hollingshead, C., and Munro, H.,** Adaptive responses of rat tissue isoferritins to iron administration, *J. Biol. Chem.,* 256, 948, 1981.

56. **Cairo, G., Bardella, L., Schiaffonati, L., Arosio, P., Levi, S., and Barnelli-Zazzera, A.,** Multiple mechanisms of iron-induced ferritin synthesis in HeLa cells, *Biochem. Biophys. Res. Commun.,* 133, 314, 1985.

57. **White, K. and Munro, H. N.,** Induction of ferritin subunit synthesis by iron is regulated at both the transcriptional and translational levels, *J. Biol. Chem.,* 263, 8938, 1988.

58. **Konijn, A. M. and Hershko, C.,** Ferritin synthesis in inflammation. I. Pathogenesis of impaired iron release, *Br. J. Haematol.,* 37, 7, 1977.

59. **Campbell, C. H., Solgonick, R., and Linder, M. C.,** Translational regulation of ferritin synthesis in rat spleen: effects of iron and inflammation, *Biochem. Biophys. Res. Commun.,* 160, 453, 1989.

60. **Dinarello, C. A.,** Biology of interleukin-1, *FASEB J.,* 2, 108, 1988.

61. **Rogers, J. T., Bridges, K. R., Durmowicz, G. P., Glass, J., Auron, P. E., and Munro, H. N.,** Translational control during the acute phase response; ferritin synthesis in response to interleukin-1, *J. Biol. Chem.,* 265, 14572, 1990.

62. **Wei, Y., Miller, S. C., Tsuji, Y., Torti, S. V., and Torti, F. M.,** Interleukin 1 induces ferritin heavy chain in human muscle cells, *Biochem. Biophys. Res. Commun.,* 169, 289, 1990.

63. **Miller, L. L., Miller, S. C., Torti, S. V., Tsuji, Y., and Torti, F. M.,** Iron-independent induction of ferritin H chain by tumor necrosis factor, *Proc. Natl. Acad. Sci. U.S.A.,* 88, 4946, 1991.

64. **Tsuji, Y., Miller, L. L., Miller, S. C., Torti, S. V., and Torti, F. M.,** Tumor necrosis factor-alpha and interleukin 1-alpha regulate transferrin receptor in human diploid fibro-blasts; relationship to the induction of ferritin heavy chain, *J. Biol. Chem.,* 266, 7257, 1991.

65. **Chou, C-C., Gatti, R. A., Fuller, M. L., Concannon, P., Wong, A., Chada, S., Davis, R. C., and Salser, W. A.,** Structure and expression of ferritin genes in a human promyelocyte cell line that differentiates *in vitro, Mol. Cell. Biol.,* 6, 566, 1986.

66. **Testa, U., Kuehn, L., Petrini, M., Quaranta, M. T., Pelosi, E., and Peschle, C.,** Differential regulation of iron regulatory element-binding protein(s) in cell extracts of activated lymphocytes versus monocytes-macrophages, *J. Biol. Chem.,* 266, 13925, 1991.

67. **Liau, G., Chan, L. M., and Feng, P.,** Increased ferritin gene expression is both promoted by cAMP and a marker of growth arrest in rabbit vascular smooth muscle cells, *J. Biol. Chem.,* 266, 18819, 1991.

Chapter 24

VITAMIN B$_6$ MODULATION OF STEROID-INDUCED GENE EXPRESSION

Douglas B. Tully, Victoria E. Allgood, and John A. Cidlowski

TABLE OF CONTENTS

I. INTRODUCTION

The steroid hormones, glucocorticoid, estrogen, progesterone, androgen, and aldosterone, regulate many physiological processes in various tissues and cell types throughout the body. These hormones are intimately involved in numerous fundamental biological processes including growth, development, reproduction, regulation of metabolic state, and maintenance of homeostasis.[1-3] Steroid hormones elicit their biological effects at the single cell level by regulating the expression of distinct target genes. Each type of steroid hormone binds specifically to a unique intracellular receptor protein, and this binding initiates transformation of the receptor into a form which is capable of selective interaction with DNA regulatory sequences which are typically contained within or near the promoter regions of hormone-responsive genes. Interaction of the steroid-receptor-complex with these DNA regulatory sites called hormone response elements results in modulation of target gene expression, and ultimately leads to alterations in the protein composition and phenotype of steroid-responsive cells.

Because steroid hormone receptors are critically involved in so many physiological and biological processes, there has been considerable interest in determining both the mechanisms through which steroid hormone action is mediated and related regulatory mechanisms which serve to control the degree of cellular responsiveness to steroid hormones. Numerous factors have been shown to influence the capacity of cells to respond to steroid hormone stimulation including the number of intracellular receptors,[4,5] the state of cellular differentiation,[6-9] the stage of the cell cycle,[10-12] as well as nutritional components such as butyric acid,[8,9] and vitamin B_6.[13-15] Such interactions between nutritional factors, such as vitamin B_6, and the endocrine system will serve as the focus for this chapter. For example, early evidence from whole animal studies suggests that glucocorticoid induction of tyrosine aminotransferase is decreased in response to the administration of pharmacological doses of vitamin B_6 and elevated under conditions of vitamin deficiency.[16] However, these studies measured enzymatic activity, and since tyrosine aminotransferase is itself a pyridoxal phosphate dependent enzyme,[17] these results cannot distinguish between an effect of the vitamin on glucocorticoid receptor function or an effect of the vitamin on tyrosine aminotransferase enzyme activity. More recent evidence demonstrates that vitamin B_6 alters the steroid responsiveness of certain cell types by modulating the ability of steroid hormone receptors to activate target gene expression.[18] These recent observations coupled with *in vitro* studies showing that pyridoxal phosphate affects various biophysical properties of steroid receptors serve as the basis for the idea that intracellular vitamin B_6 is a physiological regulator of steroid hormone action. In this chapter, we will review the data on which this hypothesis is founded and discuss possible mechanisms through which this essential vitamin may influence steroid hormone action.

II. THE STEROID HORMONE RECEPTORS AND THEIR MECHANISM OF ACTION

The steroid hormone receptors belong to a superfamily of ligand-activated transcription factors which exert their biological effects through regulation of target gene expression.[19,20] Despite the diverse array of physiological events mediated by different steroid hormones, receptors for the glucocorticoid, estrogen, progesterone, and androgen hormones exhibit many structural and functional similarities.[3] In early studies which employed limited proteolytic digestion of steroid receptor preparations, it was demonstrated that the steroid hormone receptors all share a common domain structure.[21-23] Within the past 6 years, cDNAs encoding the glucocorticoid, estrogen, progesterone, and androgen hormone receptors from many different species have been cloned. DNA sequence analysis has demonstrated that considerable DNA sequence homology exists among the different members of this family of ligand-dependent transcription factors. In addition, DNA and amino acid sequence analyses have confirmed earlier suggestions that the steroid receptors share a common modular organization.[3,24-27] Structure-function studies and mutational analyses have subsequently verified that all of the steroid hormone receptors are composed of three specific domains. The NH_2-terminal region of the receptor molecules is the most variable in size and character. This domain of the receptor proteins contains antigenic determinants which permit antibody discrimination among the different classes of steroid hormone receptor,[28-30] and, not surprisingly, displays the lowest degree of DNA and amino acid sequence homology.[3] Despite this variability, however, the NH_2-terminal domain of the receptors appears to be required for maximal receptor activity and may mediate some aspect of the modulatory effects receptors have on transcriptional activation of target genes.[31,32] A region of somewhat greater sequence homology exists within the COOH-terminal portion of the receptor molecules. This region of the receptors has been shown to be extremely important for steroid hormone action, because it is this region of the proteins which is responsible for high affinity, specific ligand binding.[3,33] Within this ligand binding domain there exist two short segments of 42 and 22 amino acids which are hydrophilic in nature and conserved to some extent among all members of the receptor superfamily. The fact that these sequence segments exhibit a relatively high degree of sequence homology among the different steroid receptors suggests that they are of some functional significance, and it has been proposed that these regions may be involved directly in ligand binding, dimerization of receptor molecules, or interaction of receptor with other proteins.[33,34] However, establishing a definitive role for these short sequences within the ligand binding domains has proven to be difficult, since deletions, insertions, or mutations within the steroid binding domain have generally interfered with steroid binding activity. The region of greatest sequence homology, however, exists between the variable NH_2-terminal and the ligand binding domains of the receptors.[3,35] This highly conserved region consists of 66 amino acids which permit the receptor molecules to recognize and bind to specific DNA regulatory sequences, the steroid response elements, located within or near to the promoter regions of steroid responsive genes. Contained within this DNA binding domain are nine highly conserved cysteine residues, eight of which are thought to form two zinc fingers, with each finger coordinately binding one zinc atom.[36] It is believed that one zinc finger determines target gene specificity by distinguishing between different steroid response elements, while the other zinc finger contacts the sugar-phosphate backbone of the DNA helix to stabilize binding of the receptor to the steroid response element.[37-40]

It is remarkable that despite such a high degree of conservation of amino acid sequence and protein structure, different steroid receptors remain capable of distinguishing between different hormonally regulated genes, activating transcription only of the appropriate target genes in the appropriate cell or tissue type. It is possible that other regions of the receptors may participate in determining target gene specificity. Alternatively, cell- or tissue-specific factors may participate in regulating receptor function, perhaps through direct interaction with the receptor molecules. Although only a speculative notion, the NH_2-terminal region, which exhibits the most variability among different receptors, may be responsible for promoting such receptor-specific functions. This could explain variations in the cell- or tissue-specific expression of many hormonally regulated genes. For example, expression of the tyrosine aminotransferase gene is selectively induced in parenchymal cells of liver, but not in other tissues, following glucocorticoid treatment,[41] One could speculate that a liver-specific factor may interact with the NH_2-terminal portion of the glucocorticoid receptor to allow glucocorticoid-specific induction of the tyrosine aminotransferase gene in the liver, while absence of an analogous factor in other cell types would inhibit glucocorticoid induction of tyrosine aminotransferase gene expression in nontarget tissues. Alternative explanations may also apply, however, and in the case of the tyrosine aminotransferase gene it has been reported that tissue specific methylation of CpG dinucleotides in certain regions of DNA may inhibit expression of tyrosine aminotransferase gene in nonexpressing cells.[42] Theories such as these are currently being investigated in an effort to determine how specific gene expression is mediated in the presence of similar steroid receptors.

Nonetheless, given the extensive sequence and structural homology that exists among the different steroid hormone receptors, it is not surprising to learn that the receptors all act through essentially the same pathway to exert their diverse biological functions. Each of the steroid hormone receptors is believed to occur in an oligomeric complex containing two molecules of heat shock protein 90 and perhaps a smaller heat shock protein or other protein in addition to receptor.[43,44] While the receptor is associated with this oligomeric complex, it is maintained in an unactivated state, which is operationally defined as being a non-DNA-binding state. When the receptor protein binds its specific steroid ligand, the receptor undergoes a poorly understood process termed transformation or activation.[45-47] Although the precise nature of the changes that occur in steroid receptor proteins following ligand binding and activation have not been fully elucidated, it is likely that transformation involves dissociation of the receptor from its oligomeric complex with heat shock proteins and may also involve a conformational change in the receptor protein itself. The primary point that is generally accepted is that the activated, liganded receptor has acquired high affinity for the specific DNA regulatory sites that typically reside within or near the promoter regions of hormone responsive genes. Binding of activated steroid receptors to these DNA regulatory sites results in alteration of the level of expression of specific target genes.[48-51] In most cases, expression of target genes is specifically induced in response to hormone treatment, although it is clear that expression of some genes is physiologically suppressed following hormone exposure.[52,53] It is through this regulation of gene expression that steroid hormones exert their primary biological effects.

Efforts directed toward understanding the regulation of steroid hormone action have revealed that several additional mechanisms contribute to the overall degree of hormone response in many cell types. Alterations in the ligand binding capacity of the receptor,[35] the number of functional receptors in a particular tissue or cell type,[4,5] the ability of the receptor to undergo translocation to or exit from the nucleus,[54] and the affinity of the receptor for binding to DNA[55] have all been shown to affect the level of gene expression which occurs following exposure of cells or tissues to steroid hormones. Additional factors including the state of differentiation or development of the cell or tissue,[6,7] the stage of cell cycle,[10] the concentration of serum components,[56] intracellular second messenger levels,[56,57] and the concentrations of nutritional components such as butyric acid[8,9] and vitamin B_6[13-15] have also been shown to influence the degree of hormone responsiveness of selected tissue types.

III. CLASSICAL ACTIONS OF VITAMIN B_6

Vitamin B_6 is a water soluble vitamin essential for normal growth, development, and metabolism.[14,58,59] The physiologically active form of the vitamin pyridoxal 5'-phosphate (pyridoxal phosphate) is synthesized in liver from the dietary precursors pyridoxine, pyridoxal, and pyridoxamine, which are biologically inactive.[60-62] Active pyridoxal phosphate is released into circulation in association with albumin, which is thought to protect the vitamin from degradation while in circulation.[63-65] However, since the physiologically active, phosphorylated form of the vitamin is unable to cross cell membranes, target tissues contain membrane-associated alkaline phosphatases which convert the pyridoxal phosphate to pyridoxal. Pyridoxal readily passes through the cell membrane, after which active pyridoxal phosphate is regenerated by pyridoxal kinase enzymes resident within target cells.[66]

Classically, pyridoxal phosphate is known for its role as an essential cofactor for a diverse array of enzymes including aminotransferase (transaminase), racemase, aldolase, dehydratase, cystathionase, and decarboxylase enzymes involved in numerous aspects of intermediary metabolism.[67-70] Pyridoxal phosphate-dependent, amino acid-specific aminotransferase and decarboxylase enzymes carry out many of the reactions of amino acid metabolism[67-70] and neurotransmitter production.[71-74] The vitamin also plays an important role as a cofactor for enzymes involved in lipid metabolism,[75] gluconeogenesis,[75] and immune function.[76,77] One of the most extensively characterized pyridoxal phosphate-dependent enzymes is aspartate aminotransferase, for which both cytoplasmic and mitochondrial isozymes have been purified and crystallized.[78-79] X-ray crystallographic and other studies with these isozymes have demonstrated that pyridoxal phosphate physically associates with certain lysine amino acids of the enzymes.[78-81] Specifically, the C4-carboxaldehyde group of pyridoxal phosphate interacts through Schiff's base formation with the epsilon amino group of a specific lysine residue at the active site of the enzyme. With the cofactor covalently attached to the active site lysine, the pyridine ring of pyridoxal phosphate serves as an electron sink providing a low-energy carbanionic path to catalyze reactions at the alpha-, beta-, or gamma-carbons of the specific amino acid substrate.[67-70] While the role pyridoxal phosphate plays as an enzyme cofactor is clearly its most well characterized function, we have recently reported an additional role for this important vitamin in modulating hormone-dependent gene expression.[14,18] These new studies suggest that vitamin B_6 may also function as an important physiological mediator of endocrine function.

IV. EFFECTS OF THE VITAMIN ON PHYSICAL PROPERTIES OF STEROID RECEPTORS

Consideration of a potential physiologic interplay between vitamin B_6 and steroid receptors is actually a relatively new field of investigation which has its roots in classical biochemistry rather than traditional physiology or nutrition. In much of this early work, pyridoxal phosphate was used as a biochemical reagent rather than in physiologic quantities as a vitamin. Nonetheless, these early experiments using pyridoxal phosphate as a chemical reagent have contributed much to our understanding of the structure and function of steroid hormone receptors and have laid a foundation for current investigations of possible physiologic interactions between this important vitamin and the endocrine system. In 1978, Nishigori et al. first demonstrated that binding of partially purified progesterone receptor to ATP-Sepharose could be inhibited by treatment of the receptor preparation with pyridoxal phosphate.[82] Modak had previously reported that relatively low concentrations of pyridoxal 5′-phosphate on the order of 0.5 to 1.0 mM acted competitively with nucleoside triphosphates to inhibit the activity of a variety of DNA polymerases,[83] and it is probably this report that first prompted Nishigori and Toft to examine the effects of pyridoxal phosphate on progesterone receptor.

Subsequently, Cake et al. reported that pyridoxal phosphate could inhibit the *in vitro* association of partially purified rat liver glucocorticoid receptor with DNA cellulose, phosphocellulose, and isolated liver nuclei.[84] However, these studies did not distinguish between two possible effects of the vitamin: inhibition of the receptor activation process or inhibition of receptor association with these synthetic matrices. Early work from our laboratory addressed these concerns using intact thymic lymphocytes, which contain specific, high affinity, saturable glucocorticoid receptors.[85] Exposure of thymocytes to the synthetic glucocorticoid hormone dexamethasone results in activation and nuclear localization of glucocorticoid receptors. However, glucocorticoid receptors localized to the nucleus under physiological conditions of hormone treatment were efficiently extracted from nuclei by treatment with 5 mM pyridoxal phosphate. Only the physiologically active form of the vitamin, pyridoxal 5′-phosphate, was effective; neither precursors (pyridoxine, pyridoxamine phosphate), metabolites (pyridoxal), nor nonphysiological analogs (5-deoxypyridoxal) were capable of extracting nuclear receptors. These data suggested that pyridoxal phosphate does not affect the process of receptor activation, but instead acts to promote release of receptors previously contained within or associated with the nucleus. This study also demonstrated that the C4-carboxaldehyde group of pyridoxal phosphate is required for efficient extraction of nuclear receptors.[85] As described above, it is through this functional moiety that pyridoxal phosphate interacts with the enzymes for which it serves as cofactor. Reduction of pyridoxal phosphate with either hydroxylamine or semicarbazide produces derivatives of pyridoxal phosphate which are incapable of extracting nuclear receptors, as shown in Figure 1. These data were the first to demonstrate a direct correlation between the ability of pyridoxal phosphate to interact with proteins, via Schiff's base formation, and its ability to extract glucocorticoid receptors previously localized to the nucleus. In subsequent studies, *in vitro* exposure to pyridoxal phosphate has been shown to inhibit the association of progesterone,[86] estrogen,[87] and androgen[88] receptors with DNA and isolated nuclei and to inhibit both DNA- and steroid-binding by glucocorticoid receptor.[89] Further, whole animals with elevated levels of vitamin B$_6$ exhibit decreased levels of uterine nuclear estrogen receptors,[90] while vitamin B$_6$ deficiency has been shown to result in enhanced nuclear localization of both uterine estrogen receptor[91,92] and prostatic androgen receptor.[93] Thus, with the caveat that the high concentrations of vitamin B$_6$ used in these experiments really do not prove a direct effect of the vitamin on receptor, pyridoxal phosphate treatment appears to modulate at least two functions common to all the steroid receptors: DNA binding and nuclear localization.

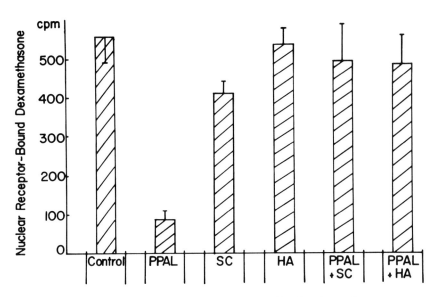

FIGURE 1.   Inhibition of pyridoxal phosphate extraction of nuclear dexamethasone receptors by hydroxylamine and semicarbazide. Nuclei from 20 μl of cells previously exposed to [³H] dexamethasone ± unlabeled dexamethasone were extracted with 200 μl of the following solutions for 1 h at 3°C. Controls represent initial binding values and were not extracted. All solutions were 1.5 m*M* MgCl₂ (pH 7.0 ± 0.05). PPAL, 5 m*M* pyridoxal phosphate; SC, 10 m*M* semicarbazide; HA, 10 m*M* hydroxylamine; PPAL + SC, 5 m*M* pyridoxal phosphate and 7.5 m*M* semicarbazide (pH 7.4); PPAL + HA, 5 m*M* pyridoxal phosphate and 7.5 m*M* hydroxylamine. Each value represents the mean ± SD from four separate determinations. (From Cidlowski, J. A. and Thanassi, J. W., *Biochem. Biophys. Res. Commun.*, 82, 1140, 1978. With permission.)

The observation that pyridoxal phosphate affects nuclear binding of glucocorticoid receptors prompted further examination into the mechanism through which pyridoxal phosphate acts on steroid hormone receptors. Thus, we examined the effect of pyridoxal phosphate treatment on the conformation or subunit composition of the glucocorticoid receptor isolated from HeLa S₃ cells.[94] These cells are of human origin and contain high affinity, saturable glucocorticoid receptors. The unactivated cytosolic receptor prepared from these cells was known from previous studies to migrate on sucrose density gradients with a sedimentation coefficient of 8S, while the activated or nuclear form of receptor sediments as a 4S complex. Treatment of the unactivated, cytosolic (8S) form of receptor with pyridoxal phosphate, followed by reduction with sodium borohydride, produces a receptor form with a sedimentation profile similar to the activated or nuclear receptor, migrating with a sedimentation coefficient of 4S. This effect on the sucrose density gradient sedimentation profile of human glucocorticoid receptor was specific for the physiologically active form of the vitamin and also required the C4-carboxaldehyde group; vitamin analogs lacking the C4-carboxaldehyde group, pyridoxine and pyridoxamine phosphate, did not alter the sedimentation coefficient of the receptor, supporting the idea that pyridoxal phosphate may interact directly with the glucocorticoid receptor through Schiff's base formation, to alter its sedimentation profile. Other lysine-reactive compounds, including dinitrofluorobenzene, picryl sulfonic acid, succinate, or p-nitrophenol acetate, had no effect on receptor sedimentation profiles.[94] Thus, the glucocorticoid receptor exhibits an alteration in conformation or subunit composition as a result of exposure specifically to the physiologically active form of vitamin B_6, pyridoxal phosphate. The requirement for the carboxaldehyde moiety of pyridoxal phosphate further supports the notion of a direct interaction between the receptor and the vitamin. Pyridoxal phosphate treatment has also been shown to have similar effects on other steroid receptors. Both progesterone[86] and estrogen receptors[87] exhibit a reduction in sedimentation coefficient on sucrose density gradients following treatment with pyridoxal phosphate. These observations suggest that pyridoxal phosphate may act through a single, common mechanism to influence multiple members of the steroid hormone receptor family of proteins. However, it still remained unclear whether the vitamin was acting directly on the receptor or on another protein in the oligomeric complex.

V. EVIDENCE FOR A DIRECT INTERACTION BETWEEN PYRIDOXAL PHOSPHATE AND THE GLUCOCORTICOID RECEPTOR

Subsequent work from our group was directed towards determining if, in fact, a direct interaction between pyridoxal phosphate and the glucocorticoid receptor exists. Initial experiments addressing this question examined the effect of pyridoxal phosphate treatment on site-specific, limited proteolytic digestion of the rat thymic glucocorticoid receptor.[23] As shown in Figure 2, the unactivated receptor elutes from S-200 Sephacryl as a 56 to 60 Å form, while treatment of the receptor extract with exogenous trypsin generates a much smaller 18 to 20 Å form of receptor. Treatment of unactivated receptor preparations with pyridoxal phosphate generates an intermediately sized receptor, which elutes as a 42 Å form. This shift in size of conformation or subunit composition is consistent with the observed shift in sucrose density gradient sedimentation following treatment of receptor preparations with pyridoxal phosphate. Surprisingly, however, pretreatment of unactivated glucocorticoid receptor preparations with pyridoxal phosphate prior to treatment with trypsin prevented the generation of the smaller 18 to 20 Å form of receptor. This protection from trypsinolysis was specific for the active form of the vitamin, since neither pyridoxine nor pyridoxal were capable of preventing cleavage with trypsin. These observations suggest that pyridoxal phosphate may bind directly to the glucocorticoid receptor at or near a site which is susceptible to cleavage with trypsin, thereby preventing proteolytic digestion.

The idea of a direct interaction of pyridoxal phosphate with the glucocorticoid receptor was further examined using an antibody which specifically

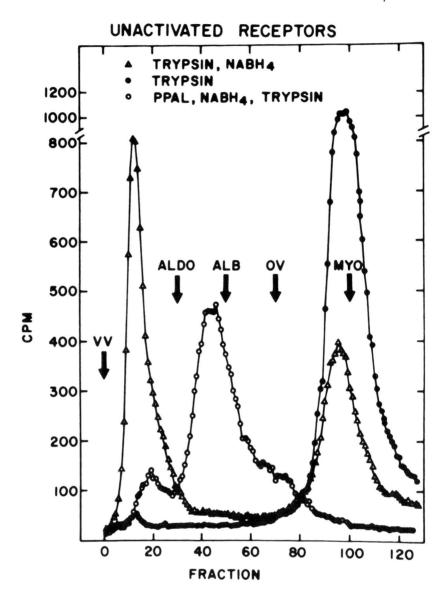

FIGURE 2. The influence of trypsin on the metabolism of rat thymocyte glucocorticoid receptor. Cytosols from thymocytes incubated with [³H] dexamethasone were treated with Worthington trypsin 10 μg/ml for 30 min at 4°C. The trypsin reaction was stopped by the addition of 1 mg of soybean trypsin inhibitor (Sigma), and samples were treated with dextran-coated charcoal as described under Methods prior to chromatography on Sephacryl. The closed circles were from untreated (unactivated) cytosol, and the open triangles were from samples to which a small amount of solid sodium borohydride was added prior to trypsin. The open circles were derived from samples of cytosol treated with 5 m*M* pyridoxal phosphate for 3 h at 0 to 4°C, reduced by the addition of solid sodium borohydride, and then trypsin treated. The patterns are representative of three similar experiments. (From Cidlowski, J. A., *Biochemistry,* 19, 6162, 1980. With permission.)

recognizes pyridoxal phosphate.[95] In early experiments with this antibody, the sucrose density gradient sedimentation profile of the pyridoxal phosphate-treated glucocorticoid receptor incubated with antibody was altered, such that the receptor sedimented as a much larger complex. In contrast, incubation with the antibody failed to alter the sedimentation profile of glucocorticoid receptors from extracts that were not treated with pyridoxal phosphate. These results provided further evidence that pyridoxal phosphate may interact directly with the glucocorticoid receptor.

Subsequent two-dimensional gel analysis of glucocorticoid receptor extracts treated with pyridoxal phosphate demonstrated that the same protein present in HeLa cells which binds radiolabeled glucocorticoid hormone is also recognized specifically by the antibody against pyridoxal phosphate. These observations made using the anti-pyridoxal phosphate antibody, coupled with the data on pyridoxal phosphate protection of receptor against proteolysis by trypsin, provide the most direct evidence to date that pyridoxal phosphate can interact directly with the glucocorticoid receptor. It follows, then, that the effects of pyridoxal phosphate on many activities and physical properties of steroid hormone receptors (DNA binding, nuclear localization, and conformation or subunit composition) may be mediated through a direct interaction of the vitamin with each of the individual steroid receptors. These observations also suggest that the vitamin effects are mediated through the receptor molecules themselves and not through other receptor-associated factors.

Based on these findings, we decided to examine glucocorticoid receptors treated *in vitro* with or without pyridoxal phosphate by isoelectric focusing. If pyridoxal phosphate interacts directly with the glucocorticoid receptor, one might expect that the isoelectric point of the receptor would shift due to elimination of the positive charge on the side chain of the bound lysine and the concurrent addition of the negatively charged phosphate moiety of pyridoxal phosphate. The results obtained from isoelectric focusing of pyridoxal phosphate-treated receptor extracts indicated that the isoelectric point of the receptor was shifted to more acidic forms as a result of *in vitro* treatment with pyridoxal phosphate.[15] This result is consistent with the hypothesis that pyridoxal phosphate interacts directly with a lysine residue of the receptor, and so alters its charge as anticipated, thus rendering the receptor protein more acidic. This pyridoxal phosphate-induced alteration in the isoelectric point of the receptor may have important biological ramifications. A more acidic form of the receptor would probably tend to interact less favorably with other acidic molecules, such as DNA. Thus, the specific alteration in surface charge that would result upon pyridoxal phosphate binding may serve as the mechanism by which DNA binding and nuclear localization of the glucocorticoid receptor is inhibited by treatment with pyridoxal phosphate. A caveat in this series of experiments is that these results are still based on *in vitro* treatment of receptor extracts with the vitamin and may not reflect how pyridoxal phosphate would react *in vivo*. To ascertain whether the isoelectric point of glucocorticoid receptor would be altered by pyridoxal phosphate *in vivo*, glucocorticoid receptor extracts were prepared from HeLa cells cultured under conditions of elevated intracellular concentrations of pyridoxal phosphate and subjected to isoelectric focusing. Results of this experiment indicated that the *in vivo* treated glucocorticoid receptor also focuses as a more acidic form.

This data was thus the first to demonstrate that the glucocorticoid receptor may be influenced not only by *in vitro* treatment with pyridoxal phosphate, but also by *in vivo* exposure. Although this observed effect of the vitamin on the isoelectric point or surface charge of the glucocorticoid receptor measured a physical property of the receptor protein and did not directly examine physiological functions of the receptor, it stimulated the provocative idea that pyridoxal phosphate may be able to act *in vivo* to affect physiological functions of the glucocorticoid receptor including DNA binding and modulation of target gene expression.

VI. *IN VIVO* EVIDENCE FOR VITAMIN B₆ EFFECTS ON STEROID HORMONE RECEPTOR-MEDIATED GENE EXPRESSION

To ascertain whether alterations in the *in vivo* concentrations of pyridoxal phosphate could, in fact, affect the ability of the glucocorticoid receptor to activate expression of a glucocorticoid-regulated target gene, we introduced into HeLa cells a glucocorticoid responsive reporter plasmid, pGMCS. This plasmid contains the entire coding sequence for the bacterial enzyme chloramphenicol acetyltransferase (CAT). Expression of the CAT gene in this plasmid is controlled by a promoter obtained from the long terminal repeat of the mouse mammary tumor virus which is known to confer hormone responsiveness upon heterologous genes.[96] In preliminary studies transfecting this plasmid into HeLa cells, we determined that induction of CAT activity occurs in direct response to glucocorticoid hormone treatment, is mediated by the glucocorticoid receptor, and is not influenced by other steroid hormones.[19] In parallel studies, we determined conditions under which the intracellular concentration of pyridoxal phosphate could be altered.[19] We found that supplementation of the cell culture medium with the pyridoxal phosphate synthesis precursor pyridoxine produced an increase in intracellular pyridoxal phosphate concentration, while culture in medium supplemented with a pyridoxal phosphate synthesis inhibitor, 4-deoxypyridoxine, resulted in a slight intracellular pyridoxal phosphate deficiency. When HeLa cells transfected with the glucocorticoid-responsive plasmid pGMCS were grown in medium supplemented with pyridoxine, the level of glucocorticoid-induced CAT activity was suppressed, as shown in the left panel of Figure 3; in contrast, cells cultured under conditions of mild vitamin deficiency exhibited an enhanced level of glucocorticoid-induced CAT activity. These effects were

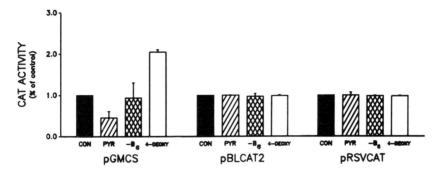

FIGURE 3. Effect of alterations in intracellular pyridoxal phosphate concentration on CAT activity derived from glucocorticoid-responsive and nonresponsive plasmids in transfected HeLa cells. Following transfection with either the glucocorticoid-responsive plasmid pGMCS or a glucocorticoid-insensitive plasmid, pBLCAT2 or pRSVCAT, HeLa cells were cultured in unaltered medium (CON), pyridoxine-deficient medium (-B6), or medium supplemented with either 1 mM pyridoxine (PYR) or 5 mM 4-deoxypyridoxine (4-DEOXY) for 48 h. Cells were then stimulated with 100 nM dexamethasone for 8 h and assayed for CAT activity. The values shown are representative of the mean ± SE from two independent transfections, with CAT activity from cells grown in unaltered medium assigned a value of 1.0, and CAT activity from cells grown in other media expressed as a percentage of this value. The first 4 lanes are from cells transfected with pGMCS, lanes 5 through 8 from cells transfected with pBLCAT2, and lanes 9 through 12 from cells transfected with pRSVCAT. (From Allgood, V. E., Powell-Oliver, F. E., and Cidlowski, J. A., *J. Biol. Chem.*, 265, 12424, 1990. With permission.)

observed under vitamin concentrations in which neither the cellular concentration of glucocorticoid receptor nor its capacity to bind steroid were affected, suggesting that changes in the amount of CAT activity were probably indicative of alterations in the ability of the glucocorticoid receptor to induce CAT gene expression. However, it was not clear if the effects of altered pyridoxal phosphate concentration represented specific effects on glucocorticoid receptor-dependent gene expression or perhaps reflected nonspecific actions on other transcription factors involved in the induction of CAT gene expression or CAT enzymatic activity. To distinguish between these possibilities, we examined the effect of variations in intracellular pyridoxal phosphate concentration on the level of CAT activity derived from other transfected CAT reporter plasmids. For this analysis, we measured CAT enzymatic activity derived from CAT reporter plasmids regulated by constitutive, hormone-independent promoters.[19] Data from these experiments, shown in the middle and right panels of Figure 3, demonstrated that the same pyridoxal phosphate concentrations which influence the level of glucocorticoid-induced gene expression from the pGMCS reporter plasmid have no effect on the level of CAT activity derived from reporter plasmids containing constitutive, hormone-independent promoters. Thus, intracellular pyridoxal phosphate does not act indiscriminately to influence gene transcription in a nonspecific manner, but instead acts specifically to modulate the level of glucocorticoid receptor-mediated gene expression.

Based on the sequence homology and structural similarities between the glucocorticoid and other steroid hormone receptors, as well as the similarity in *in vitro* effects on other steroid hormone receptors resulting from pyridoxal phosphate treatment, we examined the possibility that alterations in intracellular pyridoxal phosphate concentration might also affect the physiological capacity of the progesterone, androgen, and estrogen receptors to activate target gene expression.

We first investigated the effect of alterations in pyridoxal phosphate concentration on gene expression mediated by the androgen receptor endogenously expressed in E8.2 cells, a derivative of the L929 cell line, and progesterone receptor endogenously expressed in T47D cells. Extensive DNA binding and transcriptional activation studies have demonstrated that the glucocorticoid, androgen, and progesterone receptors can all act through the same hormone response element present in the mouse mammary tumor virus promoter to activate gene expression.[96-98] Therefore, we transfected these cells with the same hormone-inducible CAT reporter construct described above, pGMCS, and examined the effect of altered pyridoxal phosphate concentrations on the level of androgen- and progesterone-induced gene expression. Results from these experiments demonstrated that gene expression mediated through the androgen receptor in E8.2 cells and progesterone receptor-mediated gene expression in T47D cells were both affected in a manner analogous to that which we had observed with the glucocorticoid receptor.[99] That is, the level of androgen- or progesterone-induced CAT gene expression was suppressed when cells were cultured under conditions which elevate intracellular pyridoxal phosphate concentration, and hormone-induced gene expression was enhanced under conditions which produce pyridoxal phosphate deficiency. These observations are presented in Figure 4. Thus, the effects of pyridoxal phosphate concentration on steroid receptor-mediated transcriptional activation were not restricted to the glucocorticoid receptor, but extended to other members of the steroid hormone receptor family, including at least the androgen and progesterone receptors.

Pyridoxal phosphate treatment had also been shown to exert some effect on physical properties of the estrogen receptor,[87] and we examined the effect of pyridoxal phosphate concentration on the level of estrogen receptor-mediated gene expression. For these studies, we cotransfected HeLa cells with an expression vector directing production of the estrogen receptor along with an estrogen receptor-responsive CAT reporter plasmid. Results from these studies confirmed our hypothesis that alterations in intracellular pyridoxal phosphate concentration would similarly affect the level of estrogen-induced CAT gene expression. The level of estrogen receptor-mediated gene expression was reduced under conditions of elevated pyridoxal phosphate concentration and enhanced under the opposite condition of pyridoxal phosphate deficiency,[99] as shown in Figure 5. These observations demonstrate that pyridoxal phosphate modulation of steroid receptor-mediated gene expression is not restricted to the glucocorticoid receptor-regulated plasmid which contains

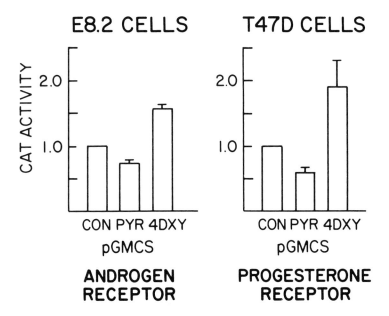

FIGURE 4. Vitamin B6 influences transcriptional activation of the androgen and progesterone receptors. Left: 16 h after transfection with pGMCS, E8.2 cells were cultured for 48 h in either control medium (CON), medium supplemented with 2 mM pyridoxine (PYR), or medium supplemented with 3 mM 4-deoxypyridoxine (4DXY). Cells were then treated with 10 nM R1881 or vehicle for 12 h prior to CAT activity determination. Right: 16 h after transfection with pGMCS, T47D cells were cultured for 48 h in either control medium, medium supplemented with 3 mM pyridoxine, or medium supplemented with 2 mM 4-deoxypyridoxine prior to 24-h treatment with 1 nM progesterone or vehicle and CAT activity determination. The data shown are representative of at least three individual experiments. (From Allgood, V. E. and Cidlowski, J. A., *J. Biol. Chem.*, 267, 3819, 1992. With permission.)

sequences derived from the mouse mammary tumor virus promoter, and that variation of intracellular pyridoxal phosphate concentration also affects transcription mediated through other hormonally regulated promoters.

VII. SUMMARY

These studies on transcriptional activation by the glucocorticoid, androgen, progesterone, and estrogen receptors provide the most direct evidence to date that pyridoxal phosphate, the active form of vitamin B$_6$, acts not only to affect some physical properties of steroid hormone receptors, but also influences the physiological function of these molecules. The demonstration that alterations in intracellular pyridoxal phosphate concentration have dramatic effects on hormonally regulated target gene expression suggests that vitamin B$_6$ may act as a physiological regulator of steroid hormone action.

Despite the efforts of many laboratories, surprisingly little is known about how steroid hormone action may be modulated in different tissues and cell types. Although the specific mechanisms through which receptor-mediated

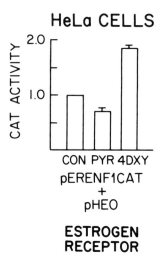

FIGURE 5. Estrogen receptor-mediated gene expression is modulated by alterations in intracellular pyridoxal phosphate concentration. HeLa cells were transfected with the estrogen receptor expression vector pHEO and the estrogen-regulated CAT reporter pERENF1CAT. Cells were exposed to control medium (CON) 16 h after transfection medium, medium supplemented with 1 mM pyridoxine (PYR), or medium supplemented with 5 mM 4-deoxypyridoxine (4DXY) prior to 24-h treatment with 5 nM 17 beta-estradiol or vehicle and CAT activity determination. The data shown reflect results from at least three individual transfection experiments. (From Allgood, V. E. and Cidlowski, J. A., *J. Biol. Chem.*, 267, 3819, 1992. With permission.)

gene expression is induced have been much more clearly identified, much less is known about the mechanisms cells may employ to regulate the magnitude of their response to steroid hormones. The observations which have been described here are consistent with the hypothesis that pyridoxal phosphate interacts with the steroid hormone receptors, and, in so doing, alters both physical properties of the receptors and their transcriptional activation capacity. Thus by regulating the intracellular concentration of pyridoxal phosphate, differentiated cell types may be able to modify the magnitude of their response to steroid hormone stimulation. Identification of a site on the steroid receptors with which pyridoxal phosphate interacts will facilitate proof of this hypothesis. However, it is clear that, regardless of mechanism, vitamin B$_6$ acts to modulate the capacity of many different steroid hormone receptors to induce target gene expression. Thus, vitamin B$_6$ has an important role in the regulation of steroid hormone action.

REFERENCES

1. **Beato, M.,** Transcriptional control by nuclear receptors, *FASEB J.,* 5, 2044, 1991.
2. **Munck, A., Mendel, D. B., Smith, L. I., and Orti, E.,** Glucocorticoid receptors and actions, *Am. Rev. Respir. Dis.,* 141, S2, 1990.
3. **Evans, R. M.,** The steroid and thyroid hormone receptor superfamily, *Science,* 240, 889, 1988.
4. **Vanderbilt, J. N., Miesfeld, R., Maler, B. A., and Yamamoto, K. R.,** Intracellular receptor concentration limits glucocorticoid-dependent enhancer activity, *Mol. Endocrinol.,* 1, 68, 1987.
5. **Gehring, U., Mugele, K., and Ulrich,** Cellular receptor levels and glucocorticoid responsiveness of lymphoma cells, *J. Mol. Cell. Endocrinol.,* 36, 107, 1984.
6. **Ballard, P. L.,** Glucocorticoids and differentiation, in *Glucocorticoid Hormone Action, Monographs in Endocrinology,* Vol. 12, Baxter, J. D. and Rousseau, G. G., Eds., Springer-Verlag, Berlin, 1979, 493.
7. **Kalinyak, J. E., Griffin, C. A., Hamilton, R. W., Bradshaw, J. G., Perlman, A. J., and Hoffman, A. R.,** Developmental and hormonal regulation of glucocorticoid receptor messenger RNA in the rat, *J. Clin. Invest.,* 84, 1843, 1989.
8. **Littlefield, B. A., Cidlowski, N. B., and Cidlowski, J. A.,** Modulation of glucocorticoid effects and steroid receptor binding in butyrate-treated HeLa S_3 cells, *Arch. Biochem. Biophys.,* 201, 174, 1980.
9. **Littlefield, B. A. and Cidlowski, J. A.,** Increased steroid responsiveness during sodium butyrate-induced ''differentiation'' of HeLa S_3 cells, *Endocrinology,* 114, 566, 1984.
10. **Cidlowski, J. A. and Michaels, G. A.,** Alteration in glucocorticoid binding site number during the cell cycle in HeLa cells, *Nature (London),* 266, 643, 1977.
11. **Tomkins, G. M., Gelehrter, T., Granner, D., Martin, D., Samuels, H. H., and Thompson, E. B.,** Control of gene expression in higher organisms, *Science,* 166, 1474, 1969.
12. **Griffin, M. J. and Ber, R.,** Cell cycle events in the hydrocortisone regulation of alkaline phosphatase in HeLa S_3 cells, *J. Cell Biol.,* 40, 297, 1969.
13. **Allgood, V. E. and Cidlowski, J. A.,** Novel role for vitamin B_6 in steroid hormone action: a link between nutrition and the endocrine system, *J. Nutr. Biochem.,* 2, 641, 1991.
14. **Allgood, V. E., Powell-Oliver, F. E., and Cidlowski, J. A.,** Influence of vitamin B_6 on the structure and function of the glucocorticoid receptor, *Ann. N.Y. Acad. Sci.,* 585, 452, 1990.
15. **Compton, M. M. and Cidlowski, J. A.,** Vitamin B_6 and glucocorticoid action, *Endocrine Rev.,* 7, 140, 1986.
16. **DiSorbo, D. M. and Litwack, G.,** Changes in the intracellular levels of pyridoxal 5'-phosphate affect the induction of tyrosine aminotransferase by glucocorticoids, *Biochem. Biophys. Res. Commun.,* 99, 1203, 1981.
17. **Valeriote, F. A., Auricchio, F., Tomkins, G. M., and Riley, D.,** Purification and properties of rat tyrosine aminotransferase, *J. Biol. Chem.,* 244, 3618, 1969.
18. **Allgood, V. E., Powell-Oliver, F. E., and Cidlowski, J. A.,** Vitamin B_6 influences glucocorticoid receptor-mediated gene expression, *J. Biol. Chem.,* 265, 12424, 1990.
19. **Beato, M.,** Gene regulation by steroid hormones, *Cell,* 56, 355, 1989.
20. **O'Malley, B. W.,** The steroid receptor superfamily: more excitement predicted for the future, *Mol. Endocrinol.,* 4, 363, 1990.
21. **Reichman, M. E., Foster, C. M., Eisen, L. P., Torain, B. F., and Simons, S. S.,** Limited proteolysis of covalently labeled glucocorticoid receptors as a probe of receptor structure, *Biochemistry,* 23, 5376, 1984.
22. **Sherman, M. R., Moran, M. C., Tuazon, F. B., and Stevens, Y.-E.,** Structure, dissociation, and proteolysis of mammalian steroid receptors, *J. Biol. Chem.,* 258, 10366, 1983.

23. **Cidlowski, J. A.,** Pyridoxal phosphate induced alterations in glucocorticoid receptor metabolism by proteases, *Biochemistry,* 19, 6162, 1980.

24. **Bellingham, D. L. and Cidlowski, J. A.,** The glucocorticoid receptor: functional implications of protein and gene structure, in *Steroid Receptors and Disease,* Sheridan, P. J., Blum, K., and Trachtenberg, M. C., Eds., Marcel-Dekker, New York, 1988, 97.

25. **Gronemeyer, H., Green, S., Kumar, V., Jeltsch, J.-M. and Chambon, P.,** Structure and function of the estrogen receptor and other members of the nuclear receptor family, in *Steroid Receptors and Disease,* Sheridan, P. J., Blum, K., and Trachtenberg, M. C., Eds., Marcel-Dekker, New York, 1988, 153.

26. **Beato, M.,** Gene regulation by steroid hormones, *Cell,* 56, 335, 1989.

27. **Krust, A., Green, S., Argos, P., Kumar, V., Walter, P., Bornert, J. M., and Chambon, P.,** The chicken oestrogen receptor sequence: homology with v-erba and the human oestrogen and glucocorticoid receptors, *EMBO J.,* 5, 891, 1986.

28. **Carlstedt-Duke, J., Stromstedt, P.-E., Wrange, O., Bergman, T., Gustafsson, J.-A., and Jornvall, H.,** Domain structure of the glucocorticoid receptor protein, *Proc. Natl. Acad. Sci. U.S.A.,* 84, 4437, 1987.

29. **Rusconi, S. and Yamamoto, K. R.,** Functional dissection of the hormone and DNA binding activities of the glucocorticoid receptor, *EMBO J.,* 6, 1309, 1987.

30. **Carlstedt-Duke, J., Okret, S., Wrange, O., and Gustafsson, J.-A.,** Immunochemical analysis of the glucocorticoid receptor: identification of a third domain separate from the steroid-binding and DNA-binding domains, *Proc. Natl. Acad. Sci. U.S.A.,* 79, 4260, 1982.

31. **Tora, L., White, J., Brou, C., Tasset, D., Webster, N., Scheer, E., and Chambon, P.,** The human estrogen receptor has two independent nonacidic transcriptional activation functions, *Cell,* 59, 477, 1989.

32. **Hollenberg, S. M. and Evans, R. M.,** Multiple and cooperative trans-activation domains of the human glucocorticoid receptor, *Cell,* 55, 899, 1988.

33. **Carson-Jurica, M. A., Schrader, W. T., and O'Malley, B. W.,** Steroid receptor family: structure and functions, *Endocrine Rev.,* 11, 201, 1990.

34. **Forman, B. M. and Samuels, H. H.,** Interactions among a subfamily of nuclear hormone receptors: the regulatory zipper model, *Mol. Endocrinol.,* 4, 1293, 1990.

35. **Green, S. and Chambon, P.,** Nuclear receptors enhance our understanding of transcription regulation, *Trends Genet.,* 4, 309, 1988.

36. **Freedman, L. P., Luisi, B. F., Korszun, Z. R., Basavappa, R., Sigler, P., and Yamamoto, K. R.,** The function and structure of the metal coordination sites within the glucocorticoid receptor DNA binding domain, *Nature (London),* 334, 543, 1988.

37. **Umesono, K. and Evans, R. M.,** Determinants of target gene specificity for steroid/thyroid hormone receptors, *Cell,* 57, 1139, 1989.

38. **Danielsen, M., Hinck, L., and Ringold, G. M.,** Two amino acids within the knuckle of the first zinc finger specify DNA response element activation by the glucocorticoid receptor, *Cell,* 57, 1131, 1989.

39. **Mader, S., Kumar, V., Verneuil, H., and Chambon, P.,** Three amino acids of the oestrogen receptor are essential to its ability to distinguish an oestrogen from a glucocorticoid responsive element, *Nature (London),* 338, 271, 1989.

40. **Green, S., Kumar, V., Theulaz, I., Wahli, W., and Chambon, P.,** The N-terminal DNA-binding 'zinc finger' of the oestrogen and glucocorticoid receptors determines target gene specificity, *EMBO J.,* 7, 3037, 1988.

41. **Schmid, E., Schmid, W., Jantzen, M., Mayer, D., Jastorff, B., and Schutz, G.,** Transcription activation of the tyrosine aminotransferase gene by glucocorticoids and cAMP in primary hepatocytes, *Eur. J. Biochem.,* 165, 499, 1987.

42. **Becker, P. B., Ruppert, S., and Schutz, G.,** Genomic footprinting reveals cell type-specific DNA binding of ubiquitous factors, *Cell,* 51, 435, 1987.

43. **Mendel, D. B. and Orti, E.,** Isoform composition and stoichiometry of the ~90-kDa heat shock protein associated with glucocorticoid receptors, *J. Biol. Chem.,* 263, 6695, 1988.

44. **Denis, M., Wikstrom, A.-C., Gustafsson, J.-A.,** the molybdate-stabilized nonactivated glucocorticoid receptor contains a dimer of Mr 90,000 non-hormone-binding protein, *J. Biol. Chem.,* 262, 11803, 1987.

45. **Currie, R. A. and Cidlowski, J. A.,** Physicochemical properties of the cytoplasmic glucocorticoid receptor complex in HeLa S_3 cells, *J. Steroid Biochem.,* 16, 419, 1982.

46. **Holbrook, N. J., Bodwell, J. E., Jeffries, M., and Munck, A.,** Characterization of nonactivated and activated glucocorticoid receptor complexes from intact rat thymus cells, *J. Biol. Chem.,* 258, 6477, 1983.

47. **Vedeckis, W. V.,** Subunit dissociation as a possible mechanism of glucocorticoid receptor activation, *Biochemistry,* 22, 1983, 1983.

48. **Burnstein, K. and Cidlowski, J. A.,** Regulation of gene expression by glucocorticoids, *Annu. Rev. Physiol.,* 51, 683, 1989.

49. **Yamamoto, K. R. and Alberts, B. M.,** Steroid receptors: elements for modulation of eukaryotic transcription, *Annu. Rev. Biochem.,* 45, 721, 1975.

50. **Rousseau, G. G.,** Control of gene expression by glucocorticoid hormones, *Biochem. J.,* 224, 1, 1984.

51. **Ringold, G. M.,** Steroid hormone regulation of gene expression, *Annu. Rev. Pharm. Toxicol.,* 25, 529, 1985.

52. **Drouin, J., Sun, Y. L., and Nemer, M.,** Glucocorticoid repression of proopiomelanocortin gene transcription, *J. Steroid Biochem.,* 34, 63, 1989.

53. **Mordacq, J. C. and Linzer, D. I. H.,** Co-localization of elements required for phorbol ester stimulation and glucocorticoid repression of proliferin gene expression, *Genes Develop.,* 3, 760, 1989.

54. **Picard, D. and Yamamoto, K. R.,** Two signals mediate hormone-dependent nuclear localization of the glucocorticoid receptor, *EMBO J.,* 6, 3333, 1987.

55. **Nordeen, S. K., Suh, B. J., Kuhnel, B., and Hutchison, C. A., III,** Structural determinants of a glucocorticoid receptor recognition element, *Mol. Endocrinol.,* 4, 1866, 1990.

56. **Aronica, S. M. and Katzenellenbogen, B. S.,** Progesterone receptor regulation in uterine cells: stimulation by estrogen, cyclic adenosine 3′,5′-monophosphate, and insulin-like growth factor I and suppression by antiestrogens and protein kinase inhibitors, *Endocrinology,* 128, 2045, 1991.

57. **Groul, D. J., Harrigan, M. T., and Bourgeois, S.,** Modulation of glucocorticoid-induced responses by cyclic AMP in lymphoid cell lines, in *Gene Regulation by Steroid Hormones,* Vol. 4, Roy, A. K. and Clark, J. H., Eds., Springer-Verlag, Berlin, 1989, 41.

58. **Tryfiates, G. P.,** *Vitamin B_6 Metabolism and Role in Growth,* Food and Nutrition Press, Westport, CT, 1980, 1.

59. **Merrill, A. H. and Henderson, J. M.,** Diseases associated with deficiencies in vitamin B_6 metabolism or utilization, *Annu. Rev. Nutr.,* 7, 137, 1987.

60. **Merrill, A. H., Henderson, J. M., Wang, E., McDonald, B. W., and Millikan, W. J.,** Metabolism of vitamin B_6 by human liver, *J. Nutr.,* 114, 1664, 1964.

61. **McCormick, D. B.,** Two interconnected B vitamins: riboflavin and pyridoxine, *Physiol. Rev.,* 69, 1170, 1989.

62. **Wada, H. and Snell, E. E.,** The enzymatic oxidation of pyridoxine and pyridoxamine phosphates, *J. Biol. Chem.,* 236, 2089, 1961.

63. **Dempsey, W. B. and Christensen, H. N.,** The specific binding of pyridoxal 5-phosphate to bovine plasma albumin, *J. Biol. Chem.,* 237, 1113, 1962.

64. **Anderson, B. B., Newmark, P. A., and Rawlins, M.,** Plasma binding of vitamin B_6 compounds, *Nature (London),* 250, 502, 1974.

65. **Lumeng, L. and Li, T. K.,** Mammalian vitamin B_6 metabolism: regulatory role of protein binding and the hydrolysis of pyridoxal 5′-phosphate in storage and transport, in *Vitamin B_6 Metabolism and Role in Growth,* Tryfiates, G. P., Ed., Food and Nutrition Press, Westport, CT, 1980, 27.

66. **Brin, M.,** Vitamin B_6: Chemistry, absorption, metabolism, catabolism, and toxicity, in *Human Vitamin B_6 Requirements,* National Academy of Sciences, Washington, D.C., 1976.

67. **Leklem, J. E.,** Vitamin B_6, in *Handbook of Vitamins,* 2nd ed., Machlin, L. J., Ed., Marcel Dekker, New York, 1991, chap. 9.

68. **Hayashi, H., Wada, H., Yoshimura, T., Esaki, N., and Soda, K.,** Recent topics in pyridoxal 5′-phosphate enzyme studies, *Annu. Rev. Biochem.,* 59, 87, 1990.

69. **Sauberlich, H. E.,** Biochemical systems and biochemical detection of deficiency, in *The Vitamins,* Sebrell, W. H. and Harris, R. S., Eds., Academic Press, New York, 1968, 31.

70. **Walsh, C.,** Enzymatic reactions requiring pyridoxal phosphate, in *Enzymatic Reaction Mechanisms,* W. H. Freeman, San Francisco, 1979, chap. 24.

71. **Buffoni, F.,** Histamine and related amine oxidases, *Pharmacol. Rev.,* 18, 1163, 1966.

72. **Kahlson, G. and Rosengren, E.,** Histamine, *Annu. Rev. Pharmacol.,* 5, 305, 1965.

73. **Lovenberg, W., Weissbach, H., and Udenfriend, S.,** Aromatic L-amino acid decarboxylase, *J. Biol. Chem.,* 237, 89, 1962.

74. **Roberts, E. and Frankel, S.,** Gamma-amino butyric acid in brain: its formation from glutamic acid, *J. Biol. Chem.,* 187, 55, 1950.

75. **Leklem, J. E.,** Vitamin B_6 metabolism and function in humans, in *Clinical and Physiological Applications of Vitamin B_6* , Leklem, J. E. and Reynolds, R. D., Eds., Alan R. Liss, New york, 1988, 3.

76. **Axelrod, A. E.,** Immune processes in vitamin deficiency states, *Am. J. Clin. Nutr.,* 24, 265, 1971.

77. **Chandra, R. K. and Purl, S.,** Vitamin B6 modulation of immune responses and infection, in *Vitamin B6: its role in health and disease,* Reynolds, R. D. and Leklem, J. E., Eds., Alan R. Liss, New York, 1984, 163.

78. **Harutyunyan, E. G., Malashkevich, V.N., Kochkina, V.M., and Torchinsky, Y. M.,** Conformational changes in chicken heart cytosolic aspartate aminotransferase as revealed by X-ray crystallography, in *Chemical and Biological Aspects of Vitamin B_6,* Evangelopoulos, A. E., Ed., Alan R. Liss, New York, 1984, 205.

79. **Jansonius, J. N., Eichele, G., Ford, G. C., Kirsch, J. F., Picot, D., Thaller, C., and Vincent, M. G.,** Crystallographic studies on the mechanism of action of mitochondrial aspartate aminotransferase, in *Chemical and Biological Aspects of Vitamin B_6,* Evangelopoulos, A. E., Ed., Alan R. Liss, New York, 1984, 195.

80. **Gehring, H., Sandmeier, E., Kirsten, H., Carrassi, R., Christen, P., and Jansonius, J. N.,** Aspartate aminotransferase: studies on conformational adaptations and ^{3}H-transfer during transamination, in *Chemical and Biological Aspects of Vitamin B_6 Catalysis,* Evangelopoulos, A. E., Ed., Alan R. Liss, New York, 1984, 223.

81. **Taylor, J. E., Metzler, D. E., and Arnone, A.,** Modeling inhibitors in the active site of aspartate aminotransferase, *Ann. N.Y. Acad. Sci.,* 585, 58, 1990.

82. **Nishigori, H., Moudgil, V. K., and Toft, D.,** Inactivation of avian progesterone receptor binding to ATP-Sepharose by pyridoxal 5′-phosphate, *Biochem. Biophys. Res. Commun.,* 80, 112, 1978.

83. **Modak, M. J.,** Observations on the pyridoxal 5′-phosphate inhibition of DNA polymerases, *Biochemistry,* 15, 3620, 1976.

84. **Cake, M. H., DiSorbo, D. M., and Litwack, G.,** Effect of pyridoxal phosphate on the DNA binding site of activated hepatic glucocorticoid receptor, *J. Biol. Chem.,* 253, 4886, 1978.

85. **Cidlowski, J. A. and Thanassi, J. W.,** Extraction of nuclear glucocorticoid receptor complexes with pyridoxal phosphate, *Biochem. Biophys. Res. Commun.,* 82, 1140, 1978.

86. **Nishigori, H. and Toft, D.,** Chemical modification of the avian progesterone receptor by pyridoxal 5′-phosphate, *J. Biol. Chem.,* 254, 9155, 1979.
87. **Muldoon, T. G. and Cidlowski, J. A.,** Specific modification of rat uterine estrogen receptor by pyridoxal 5′-phosphate, *J. Biol. Chem.,* 255, 3100, 1980.
88. **Hiipakka, R. A. and Liao, S.,** Effect of pyridoxal phosphate on the androgen receptor from rat prostate: inhibition of receptor aggregation and receptor binding to nuclei and to DNA-cellulose, *J. Steroid Biochem.,* 13, 841, 1980.
89. **Silva, C. M., Tully, D. B., Petch, L. A., Jewell, C. M., and Cidlowski, J. A.,** Application of a protein-blotting procedure to the study of human glucocorticoid receptor interactions with DNA, *Proc. Natl. Acad. Sci. U.S.A.,* 84, 1744, 1987.
90. **Holley, J., Bender, D. A., Coulson, W. F., and Symes, E. K.,** Effects of vitamin B_6 nutritional status on the uptake of [^{3}H]-oestradiol into the uterus, liver, and hypothalamus of the rat, *J. Steroid Biochem.,* 18, 161, 1983.
91. **Bowden, J. F., Bender, D. A., Coulson, W. F., and Symes, E. K.,** Increased uterine uptake and nuclear retention of [^{3}H] oestradiol through the oestrous cycle and enhanced end-organ sensitivity to oestrogen stimulation in vitamin B_6 deficient rats, *J. Steroid Biochem.,* 25, 359, 1986.
92. **Bunce, G. E. and Vessal, M.,** Effect of zinc and/or pyridoxine deficiency upon oestrogen retention and oestrogen receptor distribution in the rat uterus, *J. Steroid Biochem.,* 26, 303, 1987.
93. **Symes, E. K., Bender, D. A., Bowden, J. F., and Coulson, W. F.,** Increased target tissue uptake of, and sensitivity to, testosterone in the vitamin B_6 deficient rat, *J. Steroid Biochem.,* 20, 1089, 1984.
94. **O'Brien, J. M. and Cidlowski, J. A.,** Interaction of pyridoxal phosphate with glucocorticoid receptors from HeLa S_3 cells, *J. Steroid Biochem.,* 14, 9, 1981.
95. **Viceps-Madore, D., Cidlowski, J. A., Kittler, J. M., and Thanassi, J. W.,** Preparation, characterization, and use of monoclonal antibodies to vitamin B_6, *J. Biol. Chem.,* 258, 2689, 1983.
96. **Chandler, V. L., Maler, B. A., and Yamamoto, K. R.,** DNA sequences bound specifically by glucocorticoid receptor *in vitro* render a heterologous promoter hormone responsive *in vivo, Cell,* 33, 489, 1983.
97. **Cato, A. C. B., Skroch, P., Weinmann, J., Butkeraitis, P., and Ponta, H.,** DNA sequences outside the receptor-binding sites differently modulate the responsiveness of the mouse mammary tumour virus promoter to various steroid hormones, *EMBO J.,* 7, 1403, 1988.
98. **Ham, J., Thomson, A., Needham, M., Webb, P., and Parker, M.,** Characterization of response elements for androgens, glucocorticoids and progestins in mouse mammary tumour virus, *Nucleic Acids Res.,* 16, 5263, 1988.
99. **Allgood, V. E. and Cidlowski, J. A.,** Vitamin B6 modulates transcriptional activation by multiple members of the steroid hormone receptor superfamily, *J. Biol. Chem.,* in press.

INDEX